Genetic Analysis

Genetic Analysis

Genes, Genomes, and Networks in Eukaryotes

Philip Meneely

Haverford College

Third Edition

OXFORD
UNIVERSITY PRESS

Great Clarendon Street, Oxford, OX2 6DP,
United Kingdom

Oxford University Press is a department of the University of Oxford.
It furthers the University's objective of excellence in research, scholarship,
and education by publishing worldwide. Oxford is a registered trade mark of
Oxford University Press in the UK and in certain other countries

First Edition 2009
Second Edition 2014

Impression: 1

Published in the United States of America by Oxford University Press
198 Madison Avenue, New York, NY 10016, United States of America

British Library Cataloguing in Publication Data
Data available

Library of Congress Control Number: 2019945060

ISBN 978–0–19–880990–6

Printed and bound in the UK
by TJ International Ltd

DETAILED CONTENTS

BRIEF CONTENTS

PREFACE TO THE THIRD EDITION

As with the previous editions, the goal of this book is to describe the logic of genetic analysis as it applies in an era when genome sequences have become a standard research tool. The advances in genetic analysis over the three editions of this book are remarkable. The first edition appeared when sequencing genomes was becoming widespread; the second edition included much more information gained from the annotations of these genomes. With the third edition, our ability to make targeted changes—that is, to edit the genome—has emerged as the newest powerful tool in genetic analysis. But everything we learn from these new methods should be interpreted in light of what was known previously from work in classical genetics and molecular biology. As recounted in the book, the mantra of genetic analysis for decades has been: "Find a mutant!" In the current age, genome editing has modified this to "Make the mutant that you want!" But the same analytical principles apply no matter what methods have been used to connect mutant phenotypes with molecular changes. Geneticists worked successfully and intelligently for many years on many different biological questions before it became possible to find the DNA sequence of even a single gene, let alone an entire genome, and certainly before it was possible to make targeted edits to the genome. The new information and experimental methods do not replace this previous work, but rather add a rich dimension to it. That richness is what I hope to have captured in this book.

Course suitability

I have taught a course entitled "Advanced Genetic Analysis" for undergraduates at Haverford College for almost twenty-five years. These students have taken an introductory course in basic genetics and molecular biology and have a general familiarity with biochemistry and cell biology as well. The book is based on that course, so it is written with that level of preparation in mind. Although the structure is designed for one semester, this book covers far more topics than I include in my course. Even with that, I am keenly aware of the many topics that I have had to omit to keep the book to a manageable length and within my own areas of knowledge.

I have retained a focus on five model organisms, *Saccharomyces cerevisiae*, *Arabidopsis thaliana*, *Caenorhabditis elegans*, *Drosophila melanogaster*, and *Mus musculus*, with occasional references or examples drawn from other well-studied model organisms or from humans; in this edition, Chapter 5, about identifying the DNA sequence affecting a variant phenotype, is now largely focused on human genetics. For those who work on a different model organism and feel slighted that I have not included, for example, fission yeast or zebrafish, I sincerely apologize for not including more about these organisms. I tend to be syncretic, so I would like to have the ability and the space to talk about many more model organisms than I do. In fact my mid-term exam for the past several years has consisted in provide students with a recent paper about the genome of an organism, such as an octopus, a sea horse, a gar, a sponge, an acorn worm, or some other organism that has not been widely used for genetic analysis and that we have not discussed in class. I would then ask them to apply what they have learned from our analysis of model organisms to these completely different species. Since, in Charles Darwin's famous words, "[t]here is grandeur in this view of life," genome analysis is a type of art appreciation for biologists.

Organization and changes in the third edition

While the overall approach to the topics is similar to that of the previous editions, the organization has been changed substantially. Some topics that were included in the second edition, in particular about natural variation, are no longer included; many of these topics were moved, with revision, into the introductory genetics book that I have co-authored. At the suggestion of reviewers, some topics useful for thinking about mutant phenotypes were reintroduced; some of this was included in the first edition but not in the second, so these are now re-examined with a different focus, made possible in the age of genome editing. All of the chapters and every figure have been reviewed and revised, even if these revisions may not always be apparent. There are now twelve chapters in three units.

The first unit, "Genes, Genomes, and Organisms," retains a historical overview of genetic principles in Chapter 1. Much of this information will have been covered more thoroughly in an introductory genetics course, although probably with a different emphasis or synthesis. Chapter 2 is a substantially new chapter on chromosomes and epigenetics, a rapidly expanding topic in current genetic analysis. The concept, and even the examples, of epigenetic changes are not themselves new, but our knowledge of such examples and the connections to chromosome structure are completely new. Only some of this material was included in the second edition, as part of Chapter 3 on genome annotation. Genome annotation has now advanced to the point that we can focus on the underlying principles of epigenetic change rather than on the experimental methods used to study such effects. Chapter 3 again introduces the five model organisms that are very widely used in genetic analysis and that provide nearly all the examples used elsewhere in the book.

The second unit, "Mutants and Phenotypes," comprises roughly half of the book overall, some of the topics being brought forward and revised from the previous editions and others being new to this edition. Chapter 4, again, combines information on genetic screens and discusses the principles of mutant analysis. Chapter 5 describes how the variant or mutant phenotypes are connected to the underlying DNA sequence, both experimentally and conceptually, human genetic analysis being the most prominent here. For model organisms with sequenced genomes, "cloning a gene" now typically involves identifying the sequence among the annotated genome, so the emphasis has changed from previous editions. Chapter 6 is a discussion of how phenotypes and genotypes are connected to one another. This is a completely new version of some topics that were included in the first edition, but not in the second edition. Geneticists working in the pre-molecular age and who probably never envisioned the genomic age thought carefully about the mutant phenotypes they observed and how such phenotypes could be studied and interpreted. This chapter attempts to make some of that so-called "classical genetics" accessible and applicable to the current student of genetics. Chapters 7 and 8 are about transgenic organisms and genome editing. The long-standing examples of genome editing in yeast and in mice are described in Chapter 7. These approaches set the stage for a new chapter on genome editing as a widespread tool for many organisms, that is, on CRISPR. The specific technology associated with genome editing will certainly change—Chapter 8 was updated multiple times as the book was being written and likely could be updated again—so this chapter attempts to put genome editing into the broader context of genetic analysis as it has been done for the past century. Chapter 9 is an updated chapter on genome-wide mutant screens, concluding this unit on the generation and analysis of phenotypes that arise from changes in single genes.

The third unit, "Interactions, Pathways, and Networks," focuses on gene interactions, that is, on the interpretation of the phenotypes of double and multiple mutants rather than on single mutant strains. Chapter 10, on suppressors and enhancers, and Chapter 11, on epistatic pathways, are revisions of the corresponding chapters from the previous editions. Chapter 12 is a much updated perspective on gene interactions networks studied on genome-wide scales. I regard this as one of the most intellectually provocative topics for modern geneticists, as we continue to attempt to connect molecular changes with phenotypes and to think about how the genotype of an individual, for all of its genes rather than for only a few well-defined genes, gives rise to the phenotypes we observe.

Pedagogical features

The book is designed to be accessible to different levels of expertise, on the basis of my experience of teaching these topics to undergraduates. For example, I have tried to maintain a tone that is conversational and familiar without being overly casual. For those who want more detail or want to find the source of some of the information, each chapter includes some references and sources I have relied on. Many of these references are embedded in literature links within the text, and additional readings are supplied at the end of each chapter.

To improve its use by students, the book also includes a glossary and an index. In the text, **glossary** terms are emboldened.

Nearly every chapter includes at least one "Tool Box," and a few chapters have two such boxes. These boxes discuss an underlying technique or an experimental strategy that is relevant to the chapter, for instance using zebrafish as a model organism to study seahorses (Chapter 2) or the role of balancer chromosomes (Chapter 4). Chapter 8, on genome editing, also includes a Human Angle Box (a device employed in Meneely et al., *Genetics: Genes, Genomes, and Evolution*) to consider a few of the ethical issues that arise with genome editing in humans.

While experimental examples are used throughout the book, most chapters also have an associated "Case Study." Each of these looks at one particular example in detail, using the original papers and data as much as possible, to show how the strategies described in the chapter have been used in specific cases. These have been substantially revised from the previous editions, with an emphasis on their use as a pedagogical tool. Most significantly for pedagogical purposes, each case study now includes ten or more study questions in a section entitled "Thinking through the Experiment," which is designed to encourage the reader to engage more thoughtfully and analytically with the experiments being described. Some of these questions are relatively straightforward, while others ask about other interpretations, alternative ways in which the experiments might have been done, and so on. These are the types of questions that I usually ask the students in class as we work through the experiments. The case study examples were chosen not because they are familiar or groundbreaking (although some of them were) but because they illustrate how geneticists used the strategies the book describes to address a specific biological question. What better way to describe a strategy that to analyse an example in which it actually worked?

Genetics is a rapidly changing field, of course. The online resources will provide updates and corrections, as well as additional information and activities for each chapter to enhance student learning. As in the first two editions, the online resources include one or more guided journal clubs for each chapter. In these journal clubs, a student is guided through an original research paper, frequently one associated with the case study, via a series of questions with answers. I often use such journal clubs in my courses to help students learn to read the original papers and analyse the original data, so I encourage instructors to use these as well. Practice problems for each chapter are also included; these range from testing definitions and fundamental concepts to more advanced questions. These study questions focus primarily on the main narrative, so they are designed to be used together with the chapter to provide depth of coverage, while the case studies and journal clubs are designed to be used together to provide greater depth for interested or more advanced students.

Online resources

This book has been written with web usage in mind, so the reader is urged to consult the website associated with the book (www.oup.com/he/meneely3e) for updates and additional resources as they are developed. Key resources on this site include the following.

For registered adopters of the book

- figures from the book in electronic format, ready to download;
- journal club: suggested papers and discussion questions linked to topics covered in the book.

For students

- topical updates: key updates on topics or tools presented in the book, to keep you up to date with the latest developments in the field;
- additional case studies and text boxes that complement and add to those found in the book;
- practice problems designed to test your knowledge of the concepts presented and to help you master them.

ACKNOWLEDGMENTS

I have benefited greatly from the diligent work of Lucy Wells, Jonathan Crowe, and the editorial team at OUP. As I said and wrote previously, I never imagined writing a textbook, let alone two of them with multiple editions. Thank you for your confidence in me and for your patience.

For nearly twenty-five years, I have had the great privilege to teach both introductory and advanced genetics to some of the brightest and most industrious undergraduates that any professor could want. This book reflects their contributions and questions throughout. These students helped me learn how to think about biological questions more carefully and how genetic approaches will always be relevant to experimental biology. I became a much better communicator and research scientist thanks to these students, and I cannot imagine a job that is more enjoyable than the one I have had. Thank you.

Foremost among those to whom I wish to thank are the members of my family, for their support of this book and their patience with my love of genetics. They realize, and occasionally even appreciate, that any dinner table conversation or walk around the park might turn into a discourse on some principles of genetics, including some that they found interesting. My wife Deb encouraged me to write and accepted that most evenings and weekends I would be working on this book or on another project involving genetics. My children have married and my family has grown to the next generation since the first edition appeared. Nothing else in my experience with genetics is as enriching as seeing these next generations.

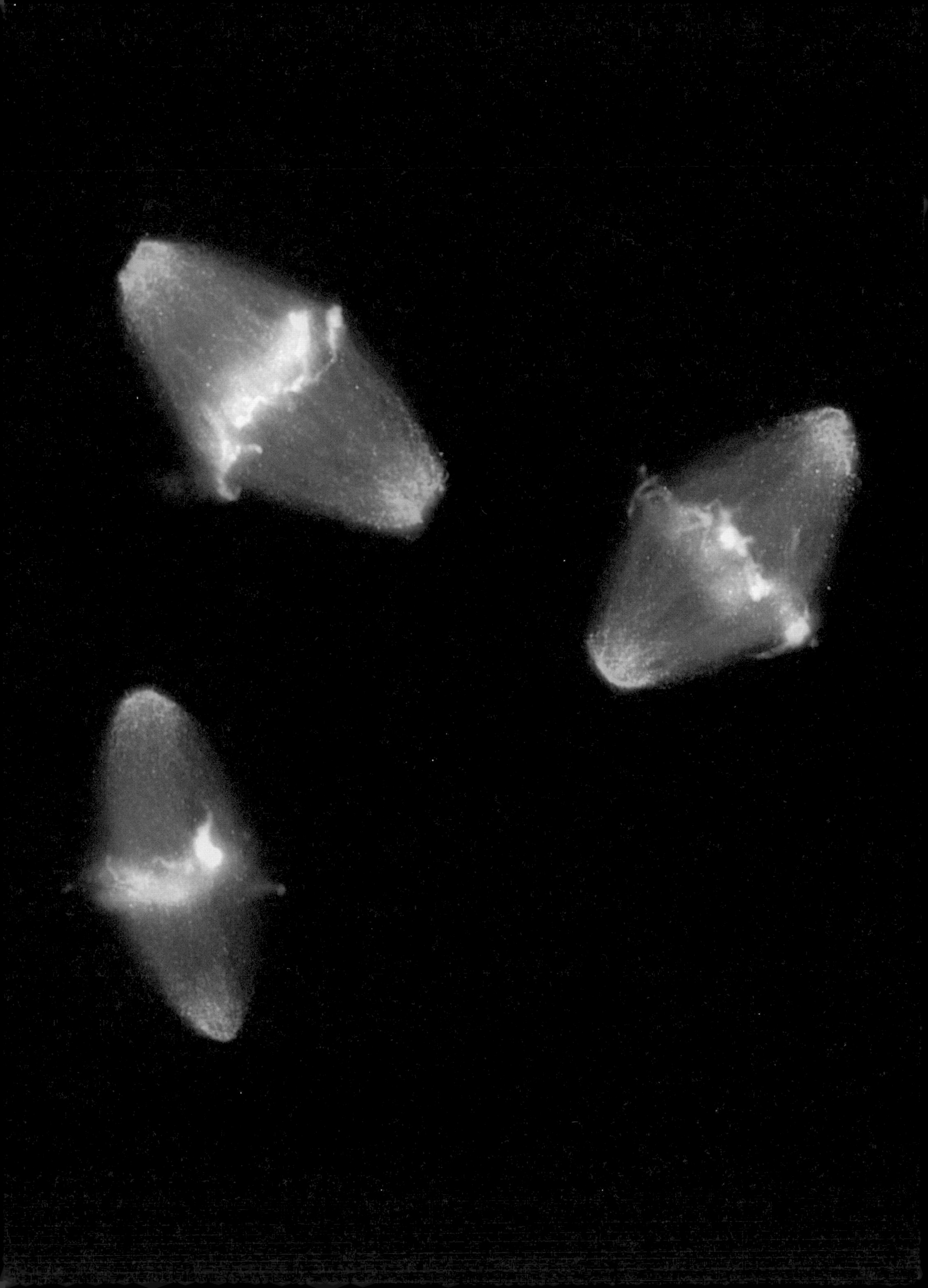

Unit I
Genes, Genomes, and Organisms

1

CHAPTER 1

The basis for genetic analysis

TOPIC SUMMARY

The fundamental findings from introductory genetics include the elucidation of properties of genes such as inheritance, allelism, linkage, and mutation. These properties are extended to gene functions through the actions of RNA, and especially of proteins, which have formed the foundation of modern genetics for the past sixty years. However, the analysis of genomes in recent years has revealed that most of the genome is transcribed and that many genes encode RNA as their final product rather than polypeptides. The functions and biological roles of most of these newly discovered genes are not yet known. Although we no longer have a simple definition that fits most genes, neither did previous generations of geneticists. Nonetheless, we can continue to use genetic analysis based on mutant phenotypes as an entry point for studying complicated biological processes.

INTRODUCTION

Genetics has a long and rich history. Parents have always been interested in knowing how their children will be like them and different from them. Our earliest forebears domesticated plants and animals and favorable traits among the offspring were selected for. References to hereditary traits are found in the ancient literature of nearly every culture. Modern geneticists, eager to take the next steps forward, stand at the leading end of a long line of previous investigators, some known to us but most of them unknown.

While observing inheritance is an old science, recognizing the patterns of inheritance and using genetics as a means of experimental analysis only began in the twentieth century. The outline of the story is well known. During the 1850s and1860s, in the town of Brno, in what is now the Czech Republic, Gregor Mendel, an Augustinian monk with training in botany from the University of Vienna, grew vegetables in the abbey garden plot. He engaged in a lengthy series of plant breeding experiments with peas and meticulously recorded his results. Mendel presented his findings to the local natural history society and sent a paper with his results and their interpretation to the great botanist Naegeli, who failed to realize their significance. Published in the proceedings of the local natural history society, Mendel's paper lay in obscurity for thirty-five years before being simultaneously and independently rediscovered and referenced in 1900 by de Vries, Correns, and Tschermak, who recognized its importance. Mendel himself went on to work on honeybees, meteorology, and the administration of the abbey and produced no more papers on peas or heredity. He died in 1884, years before his work would be rediscovered and connected to the larger scientific world that had come to encompass cytology, embryology, and evolution.

The British naturalist William Bateson was introduced to Mendel's work by deVries and became one of Mendel's earliest and most effective champions. Bateson was the first to use the word "genetics" to describe the new science of heredity; his friendship with Reginald Punnett led to a commonly used method to predict the outcomes of crosses; and he published an early textbook of genetics, *Mendel's Principles of Heredity*, in 1909. Outstanding biologists on both sides of the Atlantic Ocean quickly began working on genetics. At first, the emphasis was on the laws of genetics themselves and on showing that they applied to many different traits in all organisms, including humans. Soon afterwards, genetics became a set of methods that could be used to analyse other experimental matters, for example sex determination, color patterns, behavior, fecundity, human diseases, and many others. Since a defining property of living things is the ability to reproduce, genetics illuminates every other area of biology and lies at the center of many other biological enquiries.

The fundamental rules by which genes were inherited were laid out by Mendel, but the analytical challenge was to explain the inheritance and behavior of *phenotypes* in terms of the inheritance and behavior of *genes*, or the genotype. This remains among the central challenges of genetics. But because the inheritance behavior of genes was known, the principles by which genes are expressed to produce phenotypes could also be known, at least in part. With the combination of the predictable inheritance pattern and the insights into gene expression and phenotypes, genetics could be used to analyse other biological questions. **Genetic analysis**—that is, the use of the principles of genetics to investigate other biological questions—was present almost from the beginning of the history of the gene.

In this respect, nothing has changed for genetic analysis in the twenty-first century. Those of us who rely on genetics as an analytical approach to explain some complicated phenotype believe that we are right in our interpretation. However, we also expect to see some of our favorite ideas elaborated upon, or even overturned by new findings; in fact we may look forward to it, because it means that we have encountered something novel. Like those who worked before us, we have confidence that we are correct yet incomplete in our understanding of the genetic basis of a biological process.

1.1 The logic of genetic analysis: An historical overview

In order to discuss genetic analysis, it may help to define one of its most basic terms. What is a gene? The question is deceptively simple. Even those of us who have studied genes for most of our adult lives would struggle to present a full and comprehensive definition of what we study. In fact, as more genomes are analysed, some geneticists have suggested that a single definition that fits every gene may not be possible. We usually fall back on a definition that includes what a gene does—that is, its phenotype or the process that it affects—rather than attempting to cover all elements of its nature. Our description of a gene may include a mutant phenotype, an expression pattern, a DNA sequence, an inheritance pattern, the frequencies of an allele in a population, the role that is plays in an information network, and more. The definition that is being used often depends on the property of the gene that is being investigated. Even more intriguing is that the definition being offered now might have to be modified, as more genomes are analysed and more genes are discovered.

So what is a gene? Most biologists have an intuitive sense of the term "gene," even if they can't articulate a definition that encompasses all known genes. Let's approach this question from the perspectives of some of those who have defined what we know about genes so far.

As we take this historical tour of the fundamental principles of genetics, we will consider a highly selective summary of the findings and interpretations. We assume that much of this material will be familiar but that our perspective could be new. Thus we will emphasize the key findings without much additional explanation or background.

1.2 Genes are the units of inheritance, or Mendel was right—up to a point

Mendel was right. Genes—a term coined by Wilhelm Johannsen nearly forty years after Mendel's experiments—are inherited as individual particles or as cellular elements, and a diploid organism has two copies of each gene. Mendel recognized that these individual-particle genes can have different forms, which we now call **alleles**. He recognized from his experiments that the genes affecting different phenotypes—such as tall versus dwarf plants, or round versus wrinkled seeds—are inherited separately and independently of each other. In addition, he found in his experiments that different genes acted independently of each other in affecting the phenotype. That is, the four phenotypic combinations of tall round, tall wrinkled, dwarf round, and dwarf wrinkled were found in predictable frequencies because the height of the plant did not affect the shape of its peas. Mendel identified two of the primary components of a gene, its inheritance and its activity, and his work showed that these can be studied separately of each other.

We now can recognize that the inheritance patterns of phenotypes that Mendel observed are the outcome of the behavior of chromosomes in meiosis. This is diagrammed in Figure 1.1. As shown in Figure 1.1 Part A, a diploid eukaryote has two homologs of each chromosome—each homolog with its own allele—that separate from each other in the first meiotic division. The sister chromatids separate in the second meiotic division and four haploid products are produced, each with one of the sister chromatids. This is known as Mendel's rule of segregation: the alleles *segregate* from each other in meiosis I.

In addition, the assortment of each pair of homologous chromosomes is independent of the assortment of any other pair of homologous chromosomes, as shown in Figure 1.1 Part B. In cytological terms, the orientation of one pair of chromosomes on the meiotic spindle is independent of the orientation of another pair of chromosomes. Thus, genes on different chromosomes—as all of Mendel's genes appear to be—are inherited independently of each other. This is Mendel's rule of independent assortment.

Of course, on these two points of independent inheritance and independent activity of different genes, Mendel was not entirely correct. We now know that each chromosome has many genes linked together whose inheritance is not independent of each other. Mendel did not encounter linkage for the genes he studied or, if he did encounter a pair of linked genes, he did not use them in reporting his results. In retrospect, working with unlinked genes was a fortunate circumstance. The simplicity of Mendel's results was crucial to the rapid spread of genetics upon their rediscovery. More complicated results that included linkage might well have been more difficult to accept.

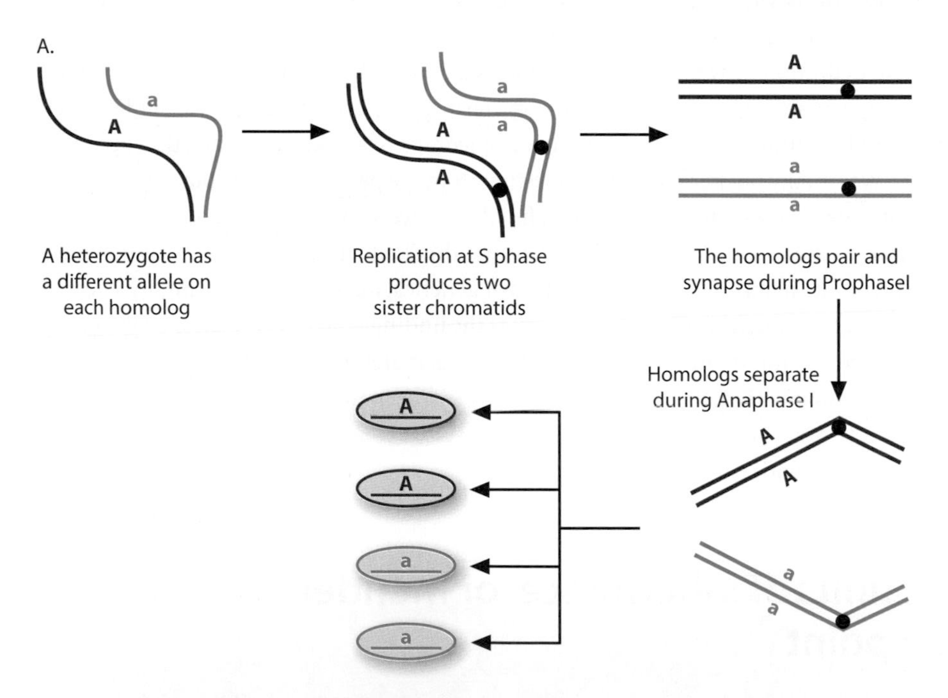

Figure 1.1 Part A. Mendel's rule of segregation. A pair of homologous chromosomes is shown with different alleles of a gene. Following replication at pre-meiotic S phase, each chromosome has two sister chromatids. The homologous chromosomes pair at prophase I of meiosis and separate at anaphase I. During meiosis I, the sister chromatids cohere with each other but the homologous chromosomes separate. The sister chromatids separate during meiosis II, resulting in four haploid products, each with one sister chromatid.

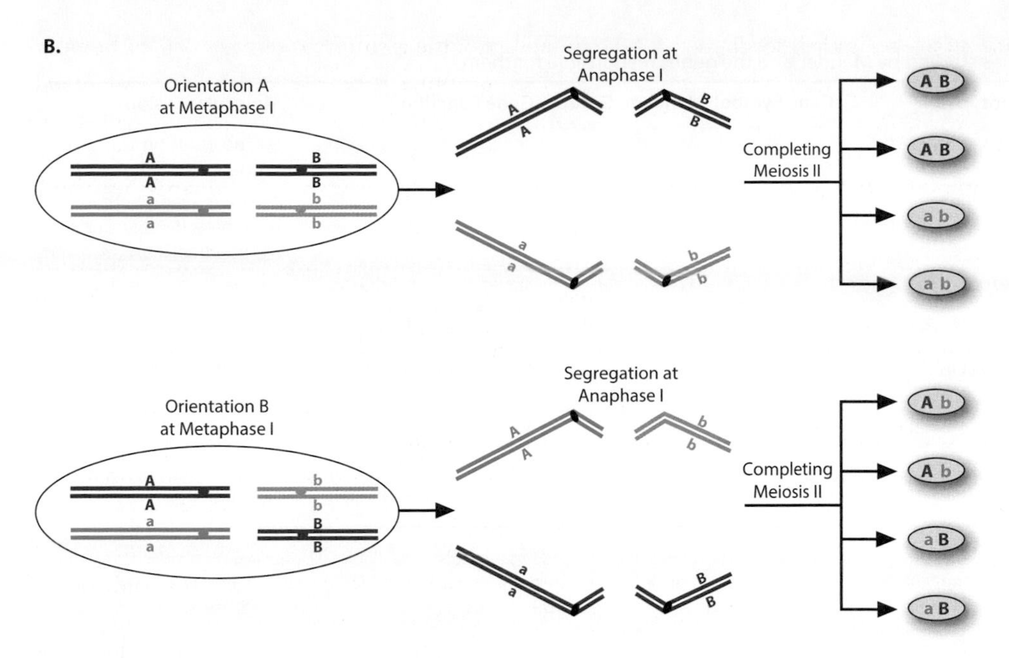

Figure 1.1 Part B. Mendel's rule of independent assortment. Two different pairs of homologous chromosomes are shown, one in shades of blue with two alleles of the *A* gene and the other in red and brown with two alleles of the *B* gene. When the homologous chromosomes pair at meiosis I, either of the two orientations of the non-homologous chromosomes can occur. With the chromosomes in orientation A, the *A* and *B* alleles segregate together at meiosis I. With the chromosomes in orientation B, the *A* and *b* alleles segregate together at meiosis I. Because the two orientations occur with the same probability, four types of haploid products result, *AB*, *ab*, *Ab*, and *aB*, with equal frequencies.

We don't know for sure what genes Mendel was actually working with, but this has been re-examined with our current knowledge of the genetic map of peas.

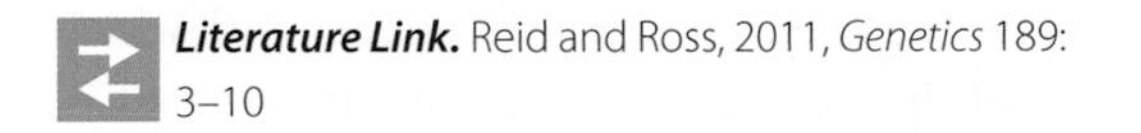
Literature Link. Reid and Ross, 2011, *Genetics* 189: 3–10

In Table 1.1 we have listed the genes that Mendel probably used for six of his seven traits. Of those six genes, two are on the same chromosome but do not show linkage in most crosses; in fact Mendel does not report data for those two genes among his dihybrid crosses. The other four genes lie on separate chromosomes. Mendel did not encounter linkage because, fortuitously, the genes he studied are unlinked.

Likewise, phenotypes that arise from the interactions between genes were not part of Mendel's results. As with the linkage results, we know that his choice of phenotypes was similarly fortuitous or prescient. If Mendel had not recognized that his phenotypes were two discrete alternatives for a gene, his results would certainly have been lost. Mendel did not work with complex phenotypes that arise from the action of multiple different genes and from the interaction between genotypes and the environment, even though complex phenotypes are more familiar than single gene traits to most people observing inheritance. Peas certainly have complex phenotypes—the number of pods on a plant is an obvious one—and Mendel may well have seen these differences. If so, he did not attribute them to genetic differences. Mendel read Darwin—the abbey's copy of *The Origin of Species* with Mendel's handwritten notes in the margin is on display in the Mendel Museum in Brno—but the evolutionary synthesis explaining complex phenotypes in Mendelian terms lay many decades in the future.

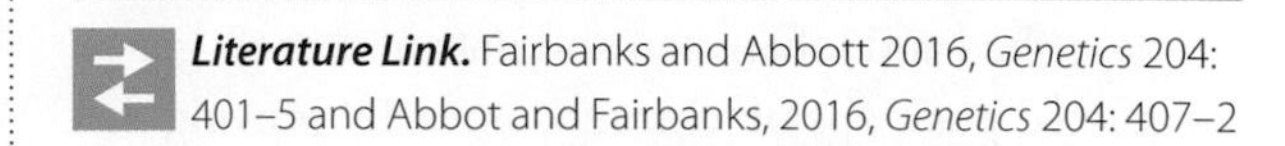
Literature Link. Fairbanks and Abbott 2016, *Genetics* 204: 401–5 and Abbot and Fairbanks, 2016, *Genetics* 204: 407–2

Mendel was right. But the areas that he left incomplete were the subjects of experiments for another set of great geneticists.

Table 1.1 The phenotypes studied by Mendel, and the genes responsible for them

Mendel's trait	Phenotypes	Gene Symbol	Linkage Group	Gene Function	Molecular Lesion
Seed shape	Round vs wrinkled	R	V	Starch branching enzyme 1	800 bp insertion of Ac/Ds sequence
Stem length	Tall vs dwarf	LE	III	GA-3 oxidase 1	Ala-thr missense mutation near the active site, due to a G to A transition
Cotyledon color	Yellow vs green	I	I	Stay-green gene, due to reduced chlorophyll break-down in dominant allele	6 bp in-frame insertion
Seed coat and flower color	Purple vs white	A	II	bHLH transcription factor	Splice donor site for intron 6
Flower position	Axial vs terminal	FA	IV	Meristem function	Not cloned
Pod color	Green vs yellow	GP	V	Chloroplast structure	Not cloned
Pod form	Inflated vs constricted	V? or P?	III or VI	Sclerenchyma formation	Not cloned

Adapted from Reid and Ross 2011, *Genetics* 189: 3–10. These are varieties of peas that fit Mendel's descriptions and that were being grown in Eastern Europe during the middle of the nineteenth century. Inflated vs. constricted pea pod shape has been used as a horticultural trait for centuries. Either of two genes is a good candidate for being the variant that Mendel used. The V gene is 12.6 map units from the LE gene, so linkage would have been expected; however, Mendel did not use or did not report data on this dihybrid combination.

1.3 Genes are found on chromosomes, or Morgan and the Fly Room were right—mostly

Thomas Hunt Morgan, his wife Lillian, his students Calvin Bridges, Alfred Sturtevant, H. J. Muller, and others comprised the denizens of the different "fly rooms" at Columbia University in New York and then at California Institute of Technology in Pasadena, where they used *Drosophila melanogaster* (rather than peas) to expand on what Mendel had learned about the inheritance and activity of genes. The core group was a colorful cast of characters with distinct personalities, each of them brilliant.

Literature Link. Ganetzy and Hawley, 2016, *Genetics* 202: 15–23

Morgan and the other geneticists in the Fly Room were right. As noted above, they recognized that Mendelian inheritance describes the behavior of chromosomes at meiosis. The cellular elements whose inheritance Mendel described and Johannsen and Batson named "genes" lie at fixed positions on chromosomes. By proving that genes lie on chromosomes, Morgan and his students were introduced to many other properties of genes, a brief list of which is given below and summarized in Figures 1.2 and 1.3.

- Genes lie at positions on chromosomes, and genes on the same chromosome will tend to be inherited together (Figure 1.2 Part A).
- Recombination or crossing over between the chromosomes of an homologous pair means that alleles of genes on the same chromosome are not *always* inherited together. In fact, the frequency with which alleles of different genes are co-inherited—that is, linked in their inheritance pattern—can be used to infer the relative positions of the genes on the chromosome. A map of the position of the genes on a chromosome can be constructed using the frequency of crossing over between two genes (Figure 1.2 Part B).
- Although a diploid individual has only two alleles for a gene, as Mendel observed, a population of individuals can display more than two alleles for a gene—in fact a gene can have a nearly limitless number of different alleles; some examples are summarized in Figure 1.3. The same gene can have mutant alleles that are dominant-to-wild type (another concept from *Drosophila*) and others that are recessive-to-wild type. Most mutant alleles are recessive-to-wild type

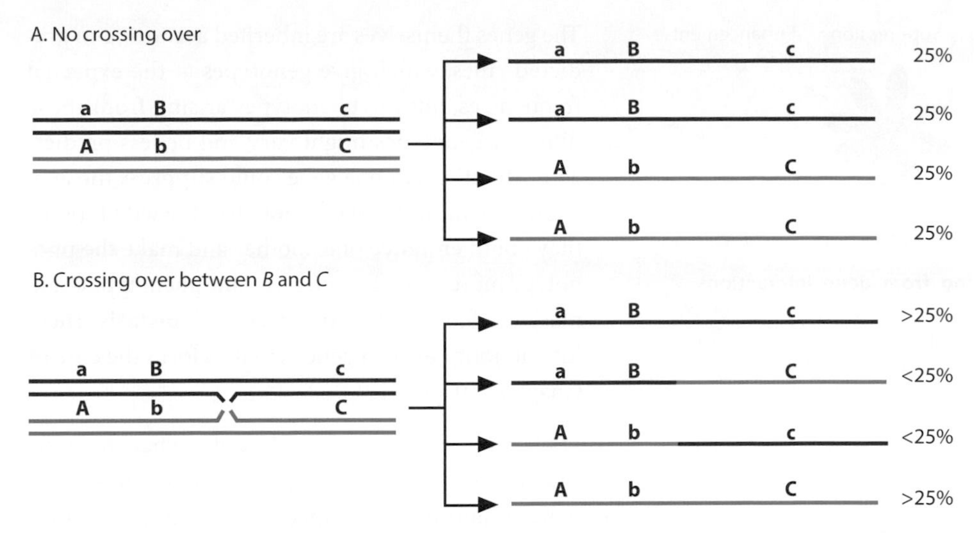

Figure 1.2 Linkage and crossing over. A pair of homologous chromosomes is shown from an individual that is heterozygous for the genes *A*, *B*, and *C*. As shown in Part A, if no recombination occurs between any pair of the genes, the alleles will be inherited together, half of the haploid gametes having the genotype of each parent. If, as shown in Part B, recombination occurs in the interval between genes *B* and *C*, some of the gametes will have the parental arrangements *a B c* and *A b C*, whereas others will have the recombinant arrangements *a B C* and *A b c*. The percentage of gametes that have the recombinant arrangement defines the genetic map distance between genes *B* and *C*. In normal meiosis, all homologous pairs of chromosomes have at least one crossover event; a genetic map is based on the probability that crossing over occurred in a particular interval.

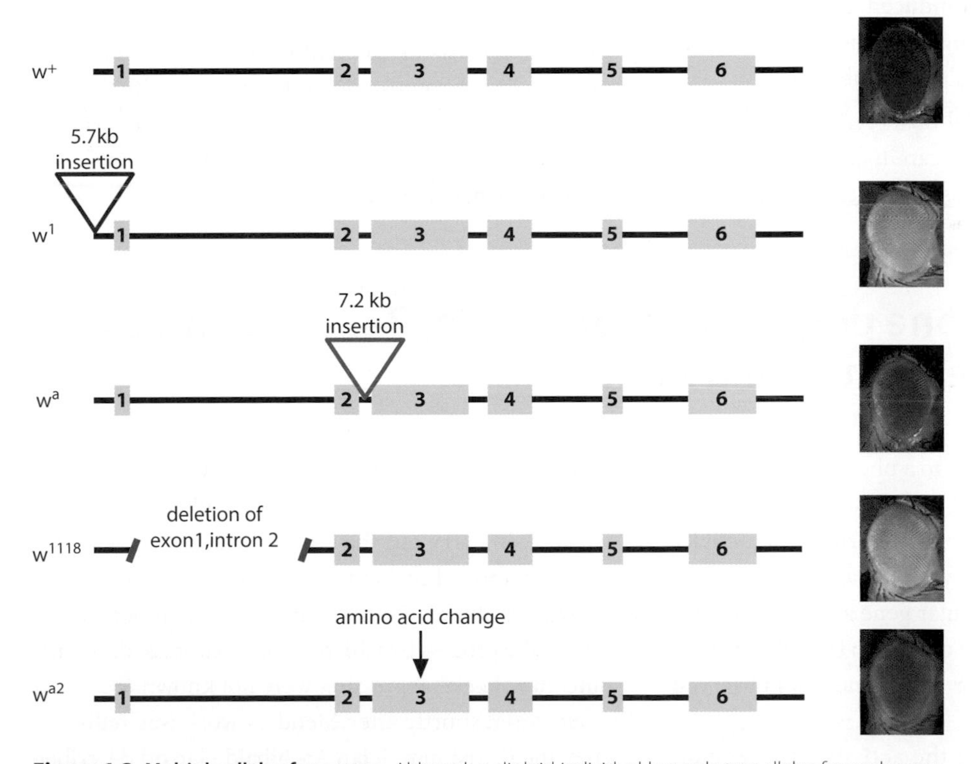

Figure 1.3 Multiple alleles for a gene. Although a diploid individual has only two alleles for each gene, many possible alleles can be found, since any part of the DNA sequence can be changed or disrupted. Some alleles of the *white* gene in *Drosophila* are shown. Exons are numbered light blue boxes, separated by the introns. The wild-type gene has six exons, drawn approximately to scale. The molecular lesions and phenotypes of some of the many *white* mutant alleles are shown, although the precise amino acid substitution found in w^{1118} is not recorded in Flybase. Images of the eyes in wild-type flies and in mutant flies are taken from Flybase.

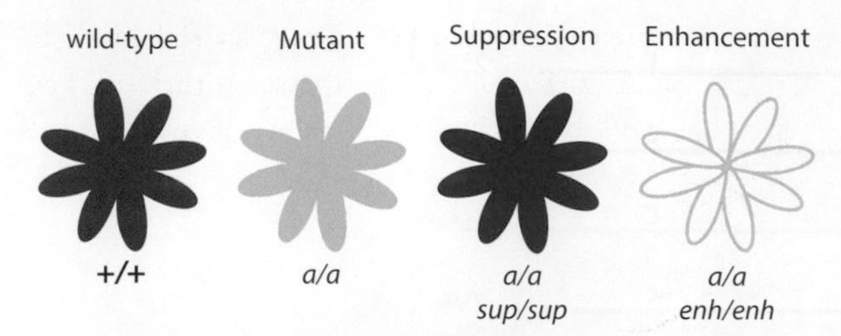

Figure 1.4 Phenotypes arising from gene interactions. A hypothetical flower is shown whose wild-type color is purple. A mutation in the *a* gene results in a lighter purple color. An interacting mutation in a second gene can suppress the mutant phenotype, resulting in a double mutant (*a/a; sup/sup*) more similar to the wild type in color, or can enhance the mutant phenotype, resulting in a double mutant (*a/a; enh/enh*) with a more mutant flower, in this case pale purple.

and arise from a loss or reduction in gene activity. The residual gene activity of mutant alleles can be inferred and different mutant alleles can be placed in a series, as explained more fully in Chapter 4.

- The alleles of a gene are not themselves permanent. Genes can be mutated to new and different allelic forms. Mutations occur spontaneously, by natural processes, but can also be induced through radiation and chemicals. These newly arisen mutations are inherited just like other alleles of the gene.
- Contrary to what Mendel observed, the activity or the phenotype of one gene can in fact affect the phenotype of another gene, as illustrated in Figure 1.4. The genes themselves are inherited according to predicted rules, which give genotypes at the expected frequencies, but the phenotypes arising from those different genotypes might vary and be less predictable. Mutations in one gene could **suppress** those in another gene and make it more like the wild type, or they could **enhance** one another and make the phenotype more severely mutant, or one phenotype could mask the other one in the process of **epistasis**. These interactions between gene activities form the core of Chapters 10 and 11.

This brief list only begins to describe what the early *Drosophila* geneticists learned. A few of these findings have had to be modified somewhat; most notably, transposable elements are components of every genome and do not sit at constant locations on the chromosome. However, many of the classical genetics concepts in this book—that is, those aspects of genetic analysis that can be performed using mutant alleles—have their origins in the work on Morgan and his students in the Fly Room.

But their work was still incomplete. A gene can be mutated; it exists in different forms; it lies in a defined position on a chromosome; and it interacts with the phenotype of another gene. But how does the gene carry out its activities? There was room for experimentation by many more geneticists.

1.4 One gene, one polypeptide, or Garrod, Beadle, Ephrussi, and Tatum were right—for a while

If we say that a gene gives rise to a phenotype—or, more precisely, that an allele of the gene results in noticeable variation in a phenotype—we are describing the activity of the gene. Mendel worked with tall and dwarf plants, so the activity of that particular gene affects the height of the plant. But how exactly does it do this? What is the biochemical explanation for the phenotypes observed from different alleles?

The concept of inferring the activity of a gene on the basis of its variant phenotypes is discussed in more detail in Chapter 6. For our purposes now, we can simply say that genes carry out cellular functions and that the activity of a gene is largely distinct from its pattern of inheritance. These cellular functions arise because the gene is expressed into another molecule or molecules, which then carry out the biochemical activities. While this was implied in Mendel's inheritance patterns and in fly phenotypes, the macromolecular composition of a gene, the process by which a gene is expressed, and the outcome of gene expression were not known.

However, shortly after Mendel's work was rediscovered, the British physician Archibald Garrod described "in-born errors of metabolism" in infants, including the biochemical bases for the rare genetic diseases phenylketonuria (PKU) and alkaptoneuria. With the help of Bateson, Garrod recognized that these were recessive traits in humans and postulated that affected children were missing some key metabolic component. Thirty years later, George Beadle and his co-workers Boris Ephrussi and Edward Tatum used pigmentation mutants in the *Drosophila* eye and metabolic mutants in the fungus *Neurospora crassa* to examine the biochemical basis for the metabolic defects that resulted in different phenotypes. This work led to the hypothesis that genes encode enzymes—which is summarized in the phrase "one gene, one enzyme." This work is illustrated in Figure 1.5.

Although the general conclusion is right and still provides a helpful starting point for thinking about genes and phenotypes, the "one gene, one enzyme" hypothesis has been much modified in subsequent research. The first modification is that the term "enzyme" has been replaced by "polypeptide." The different alleles observed by classical genetics were different forms of the same polypeptide, with changes that altered or eliminated its function by changing or truncating the amino acid sequence. The change to "one gene, one polypeptide" was made soon after the Beadle and Tatum original findings and stood up well for many years.

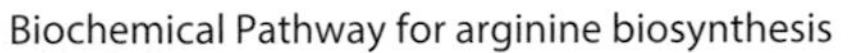

In a lecture delivered in 1957, Francis Crick laid out the central dogma about gene expression that still lies at the heart of how most of us think about genetics and gene activity, as summarized in Figure 1.6.

Literature Link. Crick, F. 1958, *Symposia of the Society for Experimental Biology* 12: 138–63

The properties of genes are explained by the biochemical properties of DNA. A gene has its functional information encoded in the nucleotide base sequence of its DNA. This DNA sequence is transcribed into an RNA molecule whose nucleotide base sequence is complementary to one of the two strands of DNA and identical to the other DNA strand (allowing for the replacement of thymidine in DNA with uracil in RNA, and for the use of ribonucleotides rather than deoxyribonucleotides). The RNA sequence is then translated into a sequence of amino acids, which comprise the polypeptide. There is colinearity between DNA sequences, RNA sequences, and amino acid sequences; the correspondence provided by base pairing (for nucleic acids) and the genetic code (for translation into an amino acid sequence) is such that, if we know the

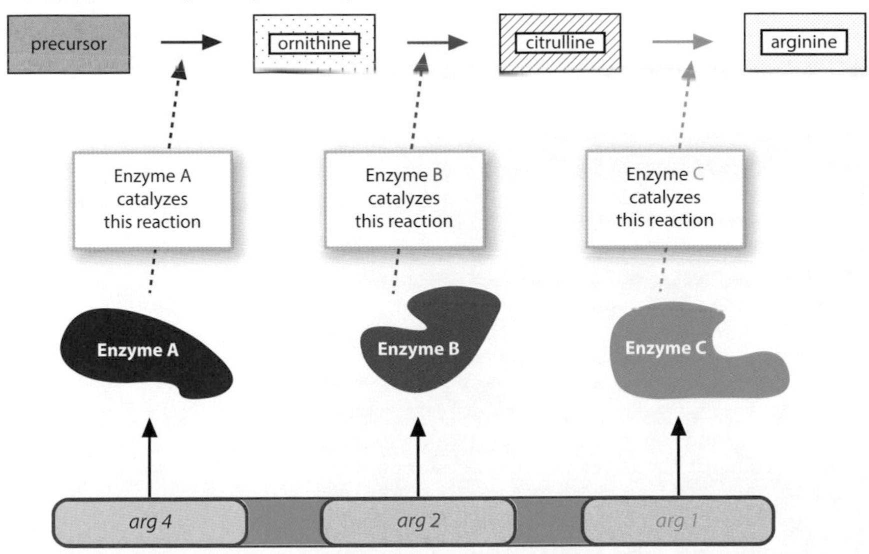

Figure 1.5 One gene, one protein. A summary of Beadle and Tatum's experiment in *Neurospora*. *Neurospora* synthesizes the amino acid arginine from a precursor molecule in three steps, with the enzymes responsible for catalysing each step shown in different colors. This is shown as the biochemical pathway at the top of the figure. Beadle and Tatum identified mutant strains of *Neurospora* that could not synthesize arginine, and used these to identify three genes, *arg4*, *arg2*, and *arg1*, shown at the bottom. Each encodes one of the enzymes that catalyses a step in this biosynthetic pathway. From Meneely et al., *Genetics: Genes, genomes and evolution* (Oxford University Press, 2017), adapted from Figure 15.3 there.

sequence of subunits in one of these macromolecules, we can infer the sequence of subunits in the other two macromolecules. This is a powerful explanation for key components of gene expression and for the definition of a gene.

But now even this generalization has to be treated with caution, and our definition of a gene is challenged once again, as the exceptions to this principle proliferate. One important exception began with the recognition of RNA splicing about twenty years after the central dogma was proposed. When RNA splicing itself was first described—that is, that the initial transcript from a gene in a multicellular organism is edited to the much smaller mRNA before translation—questions about the colinearity of genes and polypeptides were re-examined. The result was to reaffirm the colinear order between the DNA sequence and the amino acid sequence, but with intervening sequences dividing that order, as shown in the example in Figure 1.7.

The coding sequence of the gene is interrupted by sequences that are transcribed, then removed from the transcript. These intervening sequences—the

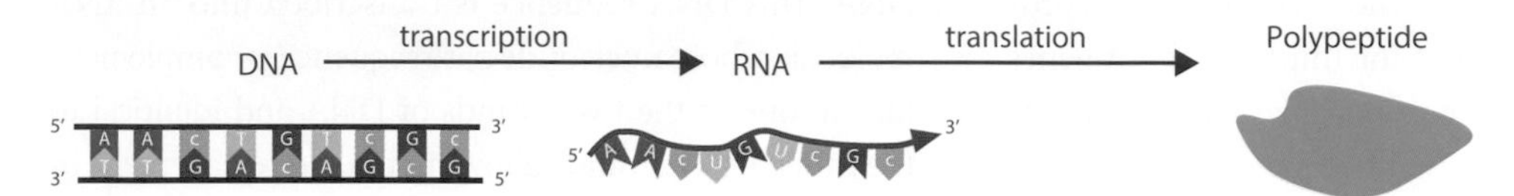

Figure 1.6 The central dogma of molecular biology. The arrows indicate the direction of information flow. DNA provides the template to make RNA via transcription, while RNA provides the template to make a polypeptide via translation. From Meneely et al., *Genetics: Genes, genomes and evolution* (Oxford University Press, 2017), adapted from Figure 2.7 there.

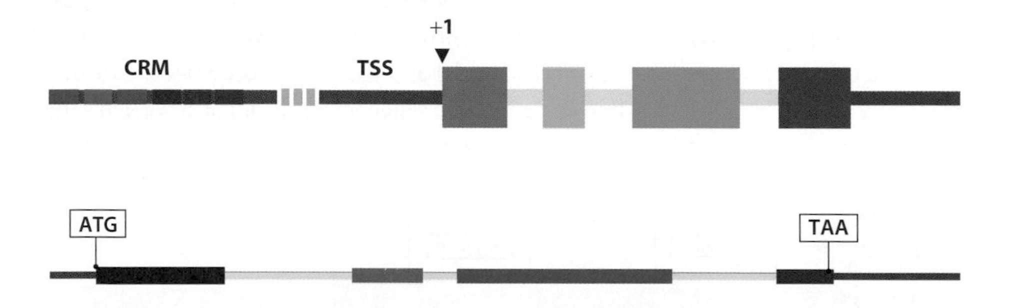


```
aaaaccgacgagaaaacaacATGAAATCTATTATCATTTTATCAGTTTTCTTAGTTTCTTCCGCGTT
GGCTGCCAGTTCTGATAAGGATACTGAATCAAAAGTGGCCAGATATTTGAAATCCTGTGGACTTATC
ACTTCATTGAAGACTACTAAATGTCTTAAGgtgggttttataaaattgttcaagcaggcttcaattg
tcatgattgctcatgcagttcttattttggtggaatttcaaacatggagatttatcacagtagtcct
tttttgttgtacttgctgcatttcgaactatttccgtcattttaataaaaaaacacttttagAGAA
CCGATGAAGTAGATAAGGAGCTCAAAGCTCTCGTAAAGAGCAAAGACAAGGATCTCTCAAAAGTCAA
GGCTTCGTCCAAAGACACCATGgtgggtaacatcagtgatattaaaaacttctcaaatgtgtttcag
GAATGCATTAAAGACCTGAAATCTAAACCACTCGAAGAACATCACAAGGAGACTCTCGACTTCTGTG
GTCGTCCTCTTTTCCAGGACGAGTTCAAGGATTGCAGCAAGAAATTGCAGGCATTGAAGAAAAGTGA
TAAGGATGCTGCAAAGTGTTTGGAAGATTTTAAATCAGAGACAAAGAAGTCGACAAAGGACACTTGC
ACATTTTTGAAGGGCTCGAAAGATTGTATCAAGGCTCATATTAAGAAAGAGTGCGGGGATGATAAGT
TGCAGGGATGGCAGAAGgtaagcggatgatgtcttagaaatcggtttaggcttcggcgcaggcctta
aacttagatgcttaggcttaggcttagagagcctcaaactttcattttcatttccaccaacttttct
aaatttctatttttcagTTTGCTCAAGACTCGTTCAAAGAGAACGAATGCGAGAAGGCAATTGGAGC
AAAATGGAACTAActtcattaaatgttcttcgatttcttgttccttgttacagtaaccctgcaaaac
tgcaataattaatttaaaagtctttaaaattccggttgagtacccattcctccctcctacattcttc
atttcttcaaagggtttctctccaggtaccaataaaattttaacgcggtca
```

Figure 1.7 The colinearity of genes, transcript, and polypeptides. A gene from *C. elegans* referred to as D1086.3 is shown as an example; the function of this gene is not known. The top diagram has the exons in different colors and the introns in orange, with the ATG start codon and the TAA stop codon noted. The transcript in the middle has the exon sequences in uppercase letters highlighted in the same colors as at the exon boxes at the top, with the introns in lowercase letters highlighted in orange. Sequences highlighted in gray are the untranslated regions at the 5′ and 3′ ends of the transcript. The amino acid sequence at the bottom illustrates which parts of the polypeptide are encoded by each exon. From Meneely et al., *Genetics: Genes, genomes and evolution* (Oxford University Press, 2017), adapted from Figure 2.20 there.

introns—could be much longer than the translated or expressed sequences, the exons. In fact, the number and length of introns has greatly increased with evolutionarily complexity. In the yeast *Saccharomyces cerevisiae*, only about 5 percent of the genes have an intron, and those genes with introns typically have only one intron. In humans, genes without introns are a distinct minority, and most genes have many introns that are longer in total sequence than sum of the exons. However, even with RNA splicing, the principle of "one gene, one polypeptide" still held. The exceptions to the rule had not yet been found.

Alternative splicing results in "one gene, many polypeptides"

The mere presence of introns does not violate the "one gene, one polypeptide" principle; but alternative patterns of splicing together the exons definitely do make a violation. Some types of alternative splicing are shown in Figure 1.8. Alternative splicing of transcripts is the rule rather than the exception, at least for genes in most plants and animals. It is hard to know the exact percentage of genes that are alternatively spliced, since new splice variants might still be found even for well-studied genes and genomes; but different splice variants have been identified for more than three fourths of mammalian genes. In addition, not only are most genes alternatively spliced, but the number of different transcripts from the same gene is fairly large. Data for the human genome find a median of 6.3 alternatively spliced transcripts for human genes, so we can no longer say that one gene makes only one transcript.

Alternative splicing often occurs in the non-translated parts of the transcript, so those splice variants may differ in RNA stability or expression, but the polypeptide products they encode will have the same amino acid sequences; one gene still makes one polypeptide, despite the splice variants in the RNA. However, alternative splicing is also common for the protein-coding portions of the transcript. For human genes, the median number of splice variants affecting the exons is 3.9, although this number, too, is likely to represent an underestimate. Similar numbers are found in other multicellular organisms, but the range in the number of splice variants per gene is probably more significant than the median number. In humans there are hundreds of genes with more than twenty-five splice variants, and at least a thousand genes with ten or more splice variants. *Drosophila* has about fifty genes with more than sixty different splice variants. Since genes are

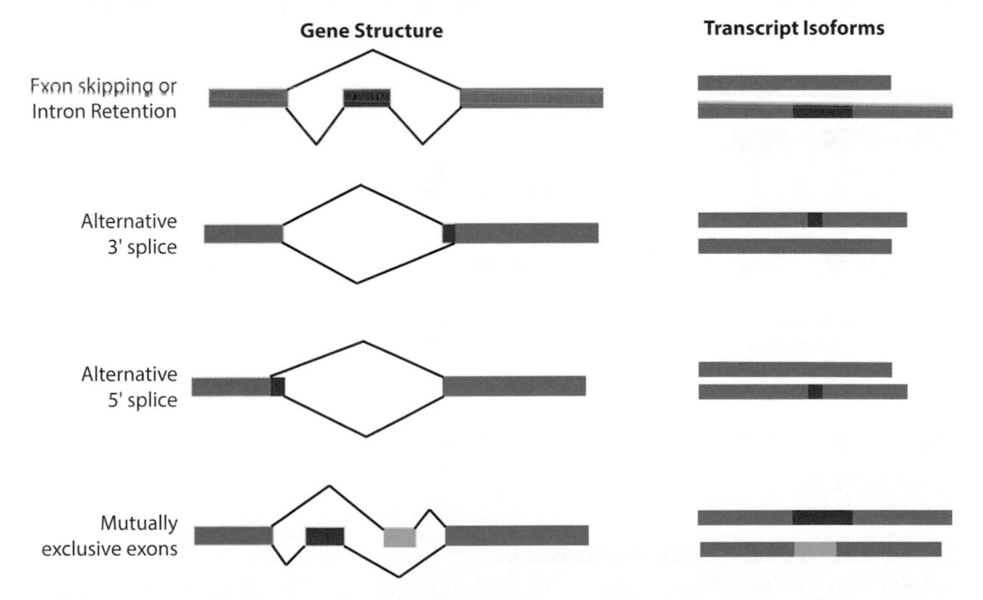

Figure 1.8 Alternative splicing of coding regions. Four different types of alternative splicing are shown, each of which changes the amino acid sequence of the proteins. The common exons are shown in green, the alternative exons in purple or light blue. Splice patterns are shown by the lines connecting exons, and the transcript isoforms show the outcome of the alternative splicing. With exon skipping or intron retention, one transcript isoform includes an exon that the other one lacks; exon skipping and intron retention have similar effects on the transcripts, but the mechanisms are different. Alternative 3′ or 5′ splicing produces isoforms in which one transcript has additional bases at either the 3′ or the 5′ side of the exon. With mutually exclusive exons, some transcript isoforms include one exon (in purple) but not the other (in light blue), while others have the opposite splice pattern; no transcript includes both the purple and the light blue exon.

spliced in many different ways, these genes can encode many related but different polypeptides. Exons in alternatively spliced transcripts may differ by a few codons at the upstream donor or downstream acceptor sites, resulting in a few amino acid differences in the polypeptide products at these locations. More profoundly, exons may be present in some splice variants and absent in others, or introns can be retained in some transcripts and not in others. (While the outcomes of these two alternatives are similar, in that polypeptides are made that either include or exclude some amino acids, the processes of exon skipping and intron retention are mechanistically distinct.)

So the principle that one gene encodes the information for one polypeptide has broken down. A more accurate but less memorable version for humans would be "one gene, probably about four or five different polypeptides and possibly even more transcripts."

The difference in splice variants has physiological significance for many fundamental biological processes. One well-studied example is sex determination in *Drosophila melanogaster*, as shown in Figure 1.9. Males and females are quite different in morphology and behavior, and the differences are easily spotted—students who never seen a fruit fly before can learn to distinguish the sexes reliably in a few minutes. At the level of the fly, the differences are extensive—genitalia, germline, mating behavior, cuticle pigmentation, and so on. At the level of the gene, these biological differences between males and females arise from the use of a splice acceptor site in the gene *doublesex* (*dsx*). The gene is transcribed in both sexes, at about the same level; however, the transcripts are spliced differently in 1X and 2X embryos (which will become males and females). The first three exons are common to both 1X and 2X embryos, but an alternative splice results in a transcript in 1X embryos with two exons that are absent in 2X embryos, and in a 2X version that has an exon that is absent in 1X embryos. The amino acid sequences of the C-terminal portion of the polypeptides in males and females are thus different, and the proteins encoded by the same gene have different functions.

Alternative splicing is so common that it has important implications for genetics, genomics, and evolution. Sex determination in *Drosophila* presents an example in which different splice variants result in biologically significant differences in protein function. Many more examples of functionally significant alternative splices could be described, and nearly every biological process in multicellular organisms includes several genes that are alternatively spliced. From the point of view of genome annotation, alternative splicing is usually detected by

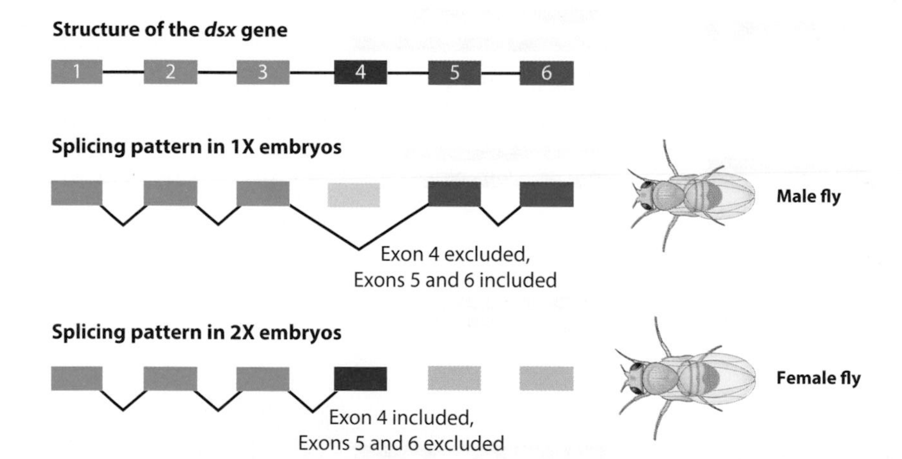

Figure 1.9 Alternative splicing and sex determination in *Drosophila melanogaster*. The *dsx* gene in *Drosophila* has six exons, as shown. Mutually exclusive alternative splicing results in different polypeptides being made in 1X (male) and 2X (female) embryos, which gives rise to the sexual dimorphism in adult flies. The first three exons, shown in green, are not sex-specific. In 1X embryos, these three common exons are spliced to the fifth and sixth exons, shown in blue, skipping the fourth exon, to produce a male-specific transcript. In 2X embryos, these three exons are spliced to a fourth exon, in red, to produce a female-specific transcript that lacks exons 5 and 6. Although only one male-specific transcript is shown here, a second one is produced by alternative polyadenylation site at the 3′ end. The drawing is not to scale; the introns are actually much longer than the exons.

"deep sequencing" of the transcripts, or **RNA-seq**. RNA-seq is sensitive enough to detect a transcript found in only a few copies per cell, and only a few cells are needed to do an assay, which has greatly increased the detection of alternatively spliced genes.

From an evolutionary perspective, alternative splicing provides an important source of genetic variation. It may be helpful to consider alternative splicing in the same way in which we think about mutations arising during replication, that is, as representing a balance between opposing effects of precision and efficiency. Perhaps a completely precise and reproducible splicing pattern is not possible for the existing spliceosome, so some variation will often occur. If that postulate is true, some alternative splices are simply the result of an occasionally imprecise process and have no other physiological significance. On the other hand, the occasional imprecision gives rise to proteins with amino acid sequences that are slightly or even significantly different from one another, which can then be subjected to different selective pressures.

1.5 Functional RNAs and the gene definition, or the central dogma may be neither a dogma nor central

If alternative splicing challenges our concept of a gene as information for one polypeptide, another class of genes completely demolishes the concept of "one gene, one polypeptide." These are the genes whose final functional product is an RNA molecule rather than a polypeptide, genes classified together as **non-coding RNA (ncRNA)** genes. As may be expected for a category defined by what it does not do (that is, "non-coding"), this is a very diverse group of genes, and their numbers and functions are an active area of investigation.

Well-known exceptions to the central dogma

It has been known for decades that some genes do not code for polypeptides and have RNA itself as their ultimate and functional product. The genes that encode ribosomal RNA and tRNA molecules are long-standing examples. Our gene concept based on "one gene, one polypeptide" could handle such genes because they clearly appeared to be exceptions to the general principle laid out in the central dogma. They were exceptions endowed with their own set of properties (Figure 1.10). For example, compared to mRNA, rRNA and tRNA are stable molecules. In addition, the genes encoding each type of RNA are frequently clustered together in the genome, one or more clusters of rRNA genes and several clusters of tRNA genes, which is distinct from the genetic map in which protein-coding genes are located throughout the genome. Furthermore, transcription of these genes involves different RNA polymerases from conventional protein-coding genes—RNA PolI in the case of the rRNA genes and RNA PolIII in the case of tRNA genes, rather than RNA PolII for mRNA. Most significantly, mRNA genes have distinct expression profiles, whereas rRNA and tRNA genes are usually expressed in all the cells. All these properties are distinctive and familiar for these non-coding RNA genes and different from the properties of protein-coding genes, so investigators who studied rRNA or tRNA genes knew these properties before beginning to study them. In the world of genes, these non-coding RNA genes existed within their own cultures with their own set of rules.

But then even more genes were found that had RNA for the final functional product. The first few of these new non-coding RNA genes, such as the *TERC* gene encoding telomerase RNA and the *Xist* gene involved in X inactivation, were wonderful but exotic—different from both tRNA and rRNA genes, as well as from the main class of protein-coding genes, and certainly worth studying as expansions of the gene concept, but not considered to be "typical" of genes in the genome. In addition, each of these two genes affects a very specific biological process and thus could easily be placed in context as "one more exception" to the gene concept. Our definition of a gene as the information needed to make a polypeptide was being stretched by the discovery of these new RNA genes, but was not yet broken.

However, analysis of genomes during the last decade or so has shown that the "one gene, one polypeptide" principle was naïve, and perhaps even misleading about the definition of the gene. We now know that there are thousands of different genes in the genome of

rRNA and tRNA genes

Ribosomal RNA (rRNA) and transfer RNA (tRNA) genes:
- produce stable RNA products
- are clustered in the genome
- are transcribed by RNA POl I or RNAPol III
- are transcribed in all cells

Protein-coding (mRNA) genes

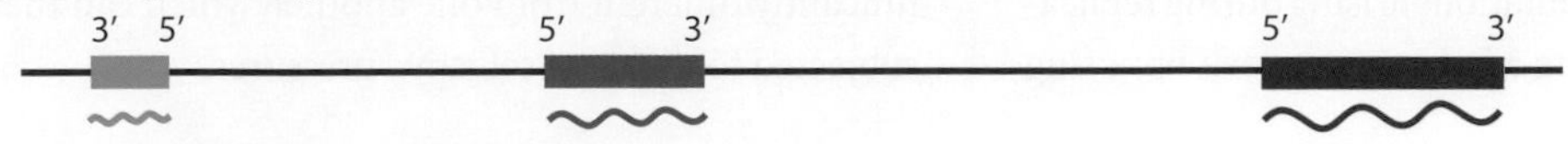

Protein coding (mRNA) genes:
- produce unstable RNA products
- are scattered throughout the genome
- are transcribed by RNA pol II
- have characteristic expression profiles

Figure 1.10 A comparison of rRNA and tRNA with protein-coding (mRNA) genes. Ribosomal RNA genes produce stable RNA products, are clustered head to tail in the genome, are transcribed by RNA polymerase I, and are expressed in all cells. A similar pattern is seen for many tRNA genes, which are also found in clusters and are transcribed by RNA polymerase III. By contrast, mRNAs transcribed from protein-coding genes are unstable; the genes are not clustered in the genome and can be in either orientation with respect to each other; and transcription is done by RNA polymerase II. In addition, the genes are only transcribed in certain cells and at certain times, which results in a specific expression profile for each gene.

Table 1.2 Transcripts and genes in humans, worms, and flies

		Humans		Worms		Flies	
		Number	Size (kb)	Number	Size	Number	Size
mRNAs		20,007	86,560	21,192	34,437	13,940	35,970
Total ncRNAs		22,154	17,770	41,466	2611	21,155	3279
	miRNAs	1756	162	221	20	236	22
	tRNAs	624	47	609	45	314	22
	snoRNAs	1521	168	141	16	287	34
	snRNAs	1944	210	114	14	47	7
	lncRNAs	10,840	10,581	233	184	852	868

From Gerstein et al. 2014, Extended Data Table 1

multicellular organisms whose final product is its RNA. Some data for the genomes of humans, worms, and flies are summarized in Table 1.2.

Literature Link. Gerstein et al. 2014, *Nature* 512: 445–8, with supplemental data

The numbers in this table are not up to date, since new examples are found frequently for these genomes as well as for many other genomes, but this summarizes the overall picture. For example, the human genome has about 20,000 protein-coding genes, which can produce about 100,000 different polypeptides when alternative splicing of exons is considered; this is the category of genes that most geneticists have studied for the past century, and they fit the central dogma. By comparison, there are more than 22,000 genes in the human genome whose final product is an RNA rather than a polypeptide—that is, more than the number of protein-coding genes. This high number of non-coding RNA genes is not unusual to humans, since both worms and flies (and many other organisms not included in the table) also have more non-coding RNA genes than protein-coding genes.

These non-coding RNA genes have been classified into different categories on the basis of their functions when these are known, or on the basis of the lengths of their transcripts when the functions are not known. Among the ncRNA genes with known functions are the tRNA genes; the small nucleolar RNA (snoRNA) genes that target the chemical modifications made to tRNAs, rRNAs, and other small RNAs; and the snRNA genes that are part of the spliceosome complex. The microRNA (miRNA) genes and the long non-coding RNA lncRNA) genes merit further discussion.

MicroRNA genes

So far as we can determine right now, the human genome contains 1,756 miRNA genes, which are many more than the 230 or so that are found in either flies or worms; the significance of this difference in the number of miRNA genes is not clear but, generally speaking, genomes of vertebrates have many more miRNA genes than do invertebrates. *Arabidopsis* is also thought to have about 250 miRNA genes, so the increased number of miRNA genes may be a phenomenon of vertebrate genomes. The properties of genes that encode the regulatory miRNAs are summarized in Figure 1.11. These miRNAs are different in many ways from the RNA-encoding genes described previously and similar to protein-coding genes, which is important as we think about how to define a gene.

First, unlike the rRNA and tRNA genes, the miRNA genes are scattered at hundreds of sites throughout the genome; an investigator never knows where the next one will be found. Some miRNA genes are found in clusters, others sit by themselves, while still others sit in the introns of protein-coding genes. Like protein-coding genes but unlike rRNA and tRNA genes, a cluster of adjacent miRNA genes can be found in either orientation with respect to each other; that is, neighboring genes can use different DNA strands as their template strand.

Second, their RNA products are not large (like rRNA and Xist) or stable (like many other examples of functional RNAs). The miRNAs are very small, with a final product of 22 nucleotides, much smaller even than any mRNA. Furthermore, they are synthesized and degraded, with half-lives that are comparable with those of mRNAs, and different from the other non-coding RNAs that had been found to that point. The dynamics of miRNA synthesis and degradation are important

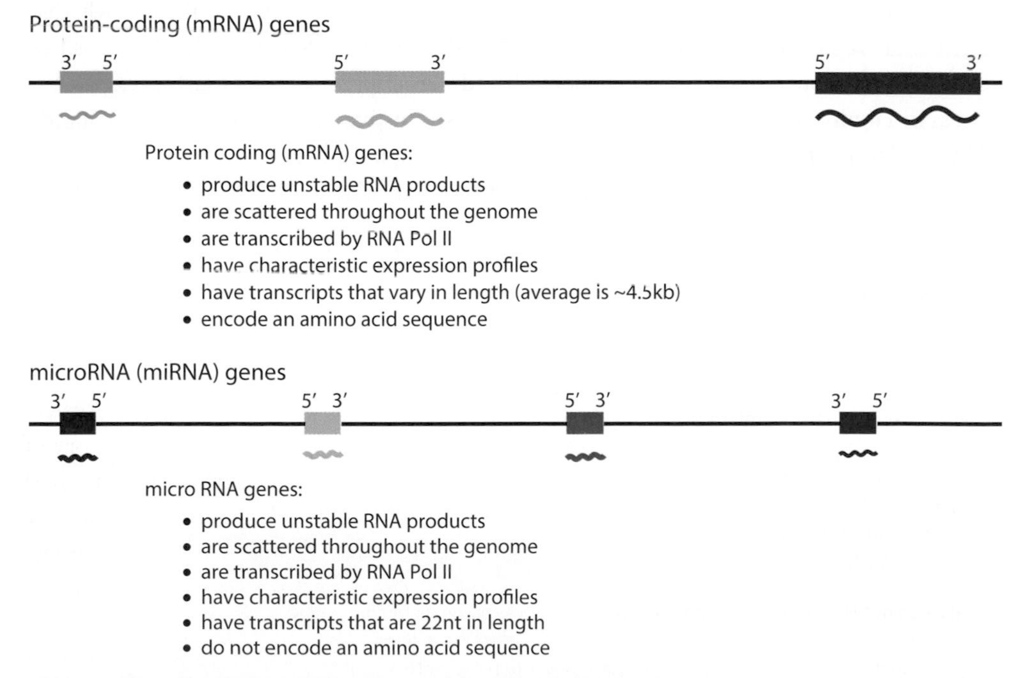

Figure 1.11 A comparison of protein-coding (mRNA) and miRNA genes. Like mRNA, miRNAs are often unstable; the genes encoding each class of RNA are scattered throughout the genome rather than clustered together, they are transcribed by RNA polymerase II, and they have characteristic expression profiles, since not all the transcripts are produced in every cell. The defining differences between mRNA and miRNAs are that miRNAs are a constant length and very short and that miRNAs are not translated into a polypeptide.

given their roles as regulators of mRNA translation and stability, so it is probably not surprising to find that miRNAs are similar to mRNAs in this regard.

Third, these genes are transcribed by RNA polymerase II, the same polymerase as mRNA, rather than by RNA polymerase III, which transcribes the genes encoding tRNA and other small RNAs, or by RNA polymerase I, which transcribes the larger rRNAs.

Fourth and most significantly for our gene definition, miRNA genes have a specific molecular activity, which is highly regulated in different tissues and at different times; in other words, like protein-coding genes but unlike the other RNA genes considered previously, miRNA genes have distinct transcriptional profiles and are not expressed in all cells or at all times. This regulation means that they affect a wide-ranging and ever-expanding set of biological functions. Their specific molecular activity is to target an mRNA molecule for degradation and/or to block its translation, as summarized in Figure 1.12. However, because their interactions with their target mRNA occur by base pairing, miRNAs are involved in many different biological activities. A given miRNA can regulate the expression of a few genes, of a few dozen genes, or of a hundred other genes; most of their targets have not yet been defined. It is no longer remarkable to find that a gene or a biological process involves a miRNA in its regulation. In fact, miRNAs are so widespread that it is an open question whether any biological process in a multicellular organism does not involve miRNA regulation. If we apply the term used to describe mutant versions of protein-coding genes, as discussed in Section 6.2, mutations in miRNA genes are highly **pleiotropic**.

The first few miRNA genes in animals were identified by mutations in *C. elegans*, the same process (and even in the same genetic screens) that led to the discovery of protein-coding genes. However, once it became clear that miRNAs constitute an evolutionarily widespread type of regulatory molecule, the vast majority of miRNA genes have been identified by transcriptional assays of the genome such as microarrays and RNAseq. Thus, with miRNA genes, we realized that the genome contained hundreds of genetic elements with important biological functions and mutant phenotypes that are comparable to protein-coding genes except in one property: their transcript is not translated into a polypeptide. In terms of our gene definition, not only can one gene encode the information to make many polypeptides; one gene may encode the information to make no polypeptide.

Pervasive transcription

But the discovery of miRNA genes was only the beginning of the challenges that genome annotation presents to our gene concept. Further annotation has shown that genomes are "pervasively transcribed," that is, that transcription is occurring at most locations throughout the genome during all stages of development. Many of these newly discovered transcripts could be detected only when sensitive genome-based transcriptional assays were developed.

We can illustrate this change in our understanding another way. Suppose that you were to ask a well-informed molecular geneticist: "What fraction of the

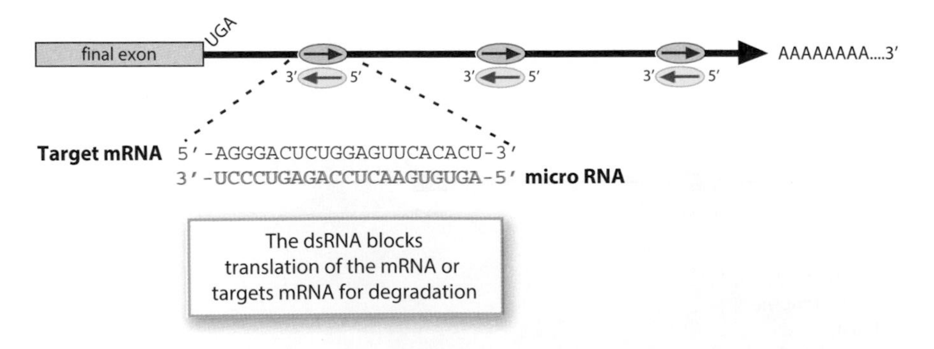

Figure 1.12 The interactions between miRNAs and their target mRNAs. The miRNAs base-pair by complementarity to sequences within the mRNAs; these target sequences in animals are found in the 3′UTR, as shown here, whereas target sequences can be found at different locations within the mRNAs of plants. The diagram shows three target locations for the miRNA within the 3′UTR, but this varies, and most target mRNAs have only a single complementary sequence. The dsRNA hybrid prevents translation of the mRNA or targets it for degradation, or both; in any case, the miRNA reduces the expression of its target mRNA.

genome is transcribed?" That is, transcription is highly regulated, such that no cell is ever making all of the transcripts that are found in the genome, but it may be reasonable to wonder what fraction of the genome is transcribed at some time in the life of the organism and in some cell. If this question were asked some fifteen years ago, around the time when the sequence of the human genome was completed, an acceptable answer would probably have estimated that about 5 percent of the human genome is transcribed, perhaps 40 percent of the genomes of flies and worms are transcribed, and a higher percentage of the genome in yeast, possibly 60–70 percent, is transcribed. We knew that most of the DNA sequence of the genome was not devoted to protein-coding genes, particularly for multicellular organisms, but we assumed that this also implied that most of the genome was transcriptionally silent.

But transcriptional assays since that time have clearly shown that these answers significantly underestimated the amount of transcription in every case. The best evidence now indicates that at least 70 percent of the human genome is transcribed in at least some cells at some level and at some time. The number of protein-coding genes, or the fraction of the genome given to introns and exons, has not changed that much, so the estimated numbers given above would still be fairly accurate if we were referring only to protein-coding genes. But the number of non-coding transcripts has increased dramatically, so that the "typical transcript" from an animal's genome is much more likely to be non-coding than to be coding for a polypeptide. The gene concept that we held so dear for so long and that has been so central to our thinking does not seem to include a high percentage of what the genome actually encodes.

Before considering some of these non-coding RNAs in a little more detail, a few comments should be made about pervasive transcription. First, most of these recently discovered transcripts are very short, definitely shorter than 1 kilobase (kb) and often much shorter. The numbers in Table 1.2 indicate that an "average" mRNA in humans is about 4.6 kb in length, since there are about 20,000 mRNAs that make up 86,500 kb; in contrast, taken collectively, the 22,154 ncRNAs make up only 17,700 kb or an average of about 0.8 kb in length. Second, many of these transcripts are present in only a few copies or in only a few cells. Since even more sensitive assays are finding still more transcripts, and not all cell types in most organisms have been thoroughly tested, many more ncRNAs will undoubtedly be discovered. Some ideas about this cellular specificity are discussed below. Third, as will be discussed, we do not know whether all of these transcripts are functional or, in most cases, what those functions might be.

One example of pervasive transcription is the finding that most core promoters in eukaryotes show bidirectional transcription (Figure 1.13). That is, once a chromatin state has been set up that allows RNA PolII to bind at the promoter region, each strand of the DNA in the promoter region is initially used as a template for transcription. In many cases only the transcript corresponding to the sense strand (the one included in the central dogma) is elongated. The so-called upstream antisense transcripts, those made from the other strand and not

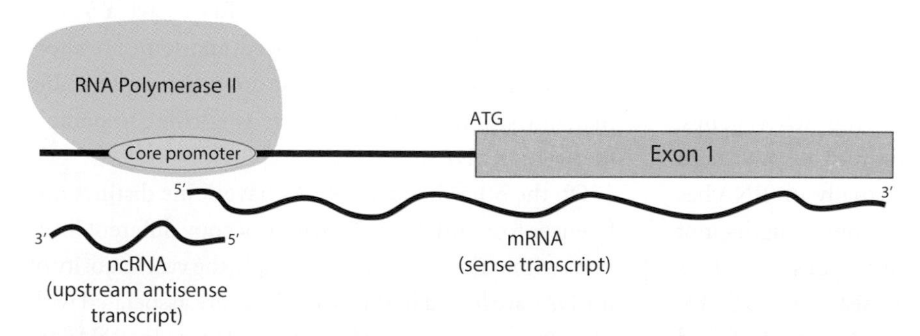

Figure 1.13 Bidirectional transcription at promoters. The canonical view of transcription initiation was that RNA polymerase bound to the promoter and transcription proceeded only in one direction, toward the protein-coding region of the gene, using only one strand of the DNA as the template. It is now recognized that, once the chromatin structure allows a promoter region to be accessible to RNA polymerase, transcription occurs in both directions, that is, both strands are used as a template. Only the transcript toward the protein-coding region of the gene is elongated, by mechanisms that are not yet fully understood. Upstream antisense transcripts are a common class of ncRNAs.

elongated into mRNAs, are among the most common category of non-coding transcripts in the genome. These upstream antisense RNAs arising during transcription initiation may simply be another example in which a molecular process cannot be completely precise, like replication and splicing; once the core promoter region is accessible for transcription, RNA polymerase simply makes transcripts in both directions, without regard to possible fates or functions for those transcripts. Some are elongated into mRNAs, while others might be non-functional transcriptional noise. Of course, also like the variants that arise during replication and splicing, transcripts made as "transcriptional noise" from an imprecise process can provide some genetic variation necessary for natural selection to occur and for new functions to arise.

Long non-coding RNA

The data in Table 1.2 indicate that, on the basis of the annotation at that time, the human genome had 10,840 long non-coding RNA (lncRNA) genes, and both the *Drosophila* and the *C. elegans* genomes are also rich in lncRNA genes. A widely used definition of long ncRNAs is that these are transcripts longer than 200 bases that do not encode a polypeptide. But both of these features in the definition, "long" and "non-coding," are somewhat arbitrary. By setting the cut-off for length at 200 nucleotides, these RNAs are distinguished from small ncRNAs with known functions such as miRNA, snoRNA, and snRNA. Both Xist (which is more than 16 kb in length in humans) and TERC (451 nucleotides in humans), discussed earlier, are considered to be lncRNAs; and they are not only quite different in length from each other, they are different in length from other lncRNAs. It is not at all clear that the length of the lncRNAs should be used as a defining feature in discussing their activities, but that works for now.

In addition, like other non-coding RNA genes, lncRNAs lack a sustained open reading frame that could be translated into amino acids. But virtually any RNA has possible opening reading frames of some length—not even tRNAs and miRNAs consist entirely of stop codons in all three frames—and there are consistent hints that at least some transcripts classified as lncRNA may be found at ribosomes and might produce oligopeptides. But, for now at least, lncRNAs are defined as being not translated. We also note in passing that the sum of the numbers of "comparable ncRNAs" in Table 1.2 is smaller than the total of "annotated ncRNAs," indicating that there are many annotated ncRNAs that cannot be easily assigned to any category. While lncRNAs are treated as a category of non-coding RNAs as we think about transcripts made by the genome, it is almost certainly the case that this is an operational and temporary classification, until more is learned about the many different ones.

Literature Link. Quinn and Chang, 2016, *Nature Review Genetics* 17: 47–62

Many more putative lncRNA genes have been found since the information in Table 1.2 was compiled, both in these organisms and in every eukaryote whose genome has been examined; one estimate arising from pooling data from different sources is that there may be as many as 58,000 loci in the human genome encoding lncRNA. But, no matter how many different loci exist and are included as lncRNAs, lncRNA genes are exceptionally numerous and diverse. On the other hand, even as many more lncRNAs are being discovered in many more genomes, one feature apparent in the numbers in Table 1.2 continues to be true: humans, and vertebrate genomes in general, have many more lncRNA genes that do invertebrate genomes. In fact, it has been noted that the number and diversity of lncRNA genes in the genome correlates more closely with an organism's perceived phylogenetic complexity than does the number of protein-coding genes.

In many ways, lncRNAs are similar to mRNAs, as summarized in Figure 1.14. They are transcribed by RNA polymerase II, and the chromatin structure surrounding their promoters is similar to that of protein-coding genes. (Chromatin structure is discussed in Chapter 2.) Like mRNAs, most lncRNA have a 5′ cap and a 3′ polyA tail; many are spliced, and some are alternatively spliced. Although some are much more stable than mRNAs, most lncRNAs have a high life comparable to that of mRNAs.

On the other hand, lncRNAs have some distinct differences from mRNAs. First, they occupy different locations within the cell. Not surprisingly, the vast majority of mRNAs are found in the cytosol and are associated with ribosomes; in contrast, a high percentage of lncRNAs are found in the nucleus and are associated with chromatin, although many are found in the cytosol or in both compartments. In addition, lncRNAs are more evolutionarily divergent than mRNAs. When related species are compared, lncRNAs have much less sequence identity

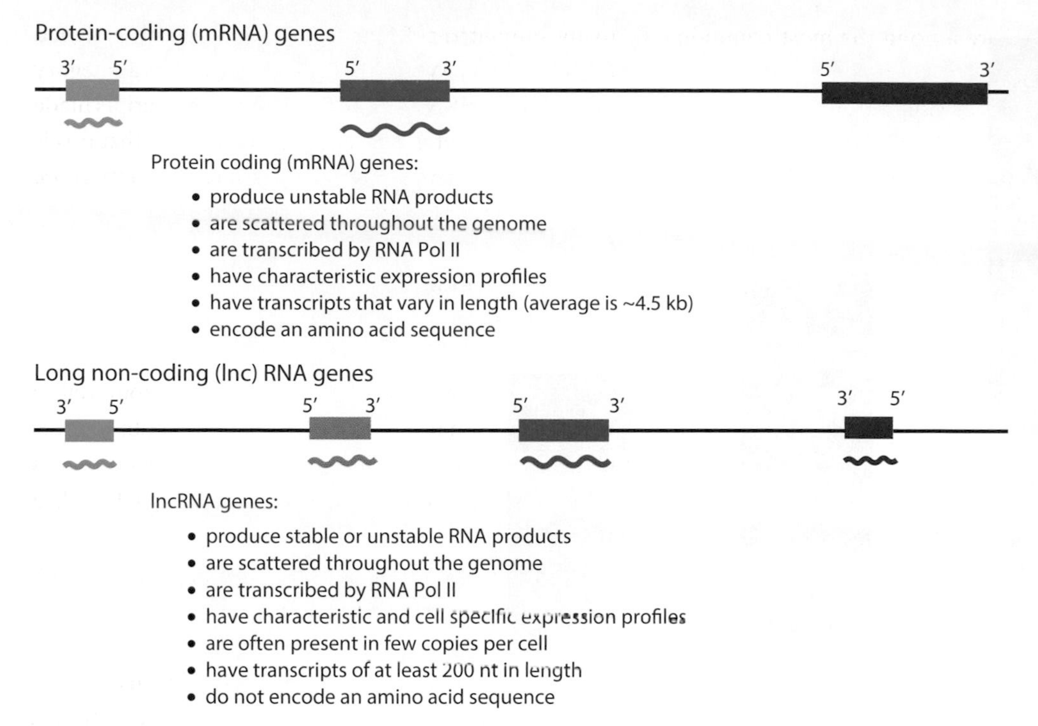

Figure 1.14 A comparison of protein-coding (mRNA) and long non-coding RNA genes. Like mRNA, lncRNAs are often unstable; the genes encoding each class of RNA are scattered throughout the genome rather than clustered together, are transcribed by RNA polymerase II, and have characteristic expression profiles, since not all the transcripts are produced in every cell; lncRNAs are often expressed more specifically and at lower levels than mRNA. The defining differences between mRNA and lncRNAs are that lncRNAs are not translated into a polypeptide and are shorter than a typical mRNA.

than mRNAs, suggesting that there is less selective pressure on their primary sequence than there is for mRNAs. However, the sequences of lncRNAs are more highly conserved than genome sequences predicted to be not under selection, so clearly there is some selective pressure on the nucleotide sequences of lncRNAs.

Most notably, lncRNAs, miRNAs, and mRNAs differ in their cellular specificity. Data for fourteen different human cell types were extracted from online sources and are summarized in Figure 1.15. A majority of mRNAs are found in all cell types, and less than 10 percent are specific to a single type of cell. By contrast, of 6,500 human lncRNA genes analysed in detail, approximately 20 percent are found in only cell type, while only about 10 percent are found in all fourteen different cell types. (The remainder are found in more than one but fewer than all cell types, but in every case the lncRNAs were expressed in fewer cell types than either mRNAs or miRNAs.) The miRNAs are intermediate between the cellular specificity of lncRNA and the widespread expression of mRNAs.

To summarize their properties, lncRNAs are expressed at low levels, their sequences are not highly conserved, they are often found associated with chromatin, and many are expressed in only one or two cell types. This information leads to the hypothesis that lncRNAs are part of a rapidly evolving system of gene regulation, often or even primarily affecting chromatin structure. A few examples with known functions fit this hypothesis. A lncRNA known as HOTAIR is important in establishing the chromosome conformation to allow expression of the *Hox* gene cluster; other lncRNAs are known or thought to have roles in chromosome conformation and the three-dimensional organization of chromatin in the nucleus, including imprinting (discussed in the case study in Chapter 2). In addition, some lncRNAs arise from transcription of enhancer sequences within cis-regulatory modules and appear to play necessary roles in the pre-initiation of transcription.

New cellular functions for lncRNAs are found often, and not all of these involve an interaction with chromatin, or even a nuclear location. While there are numerous

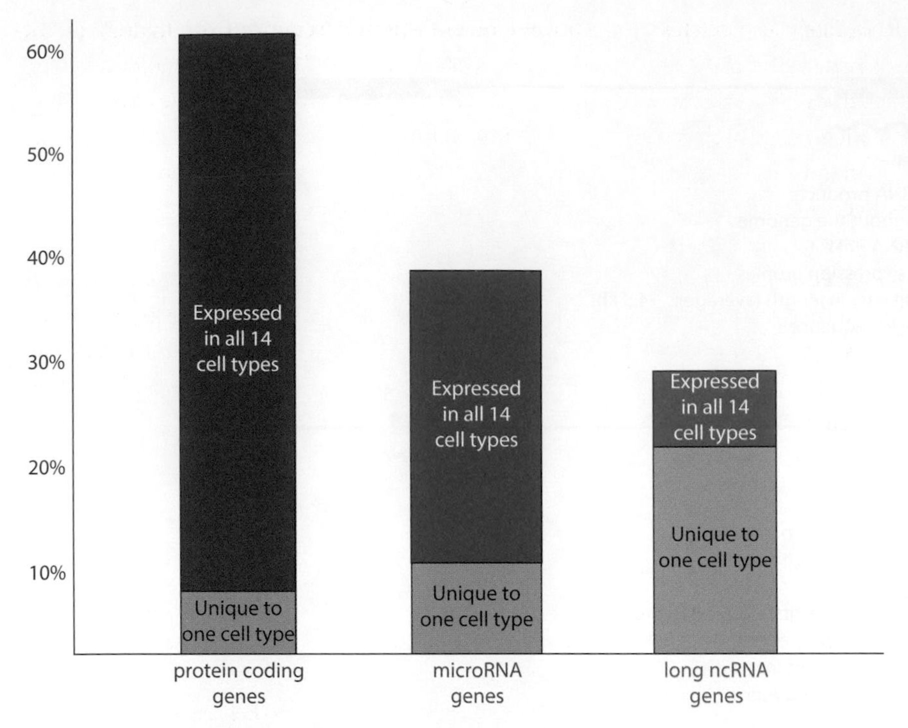

Figure 1.15 Cell-specific expression of lncRNAs. The expressions of several thousand different mRNAs, miRNAs, and lncRNAS were examined in fourteen different human cell types; the data for those specific to one cell type and expressed in all cell types are summarized here. mRNAs are more likely to be expressed in all cell types than either miRNAs or lncRNAs, and relatively few of them are unique to one cell type. In contrast, lncRNAs are much more likely to be unique to one cell type, and relatively few are expressed in all cell types tested. The data are extracted and summarized from information in Gerstein et al. 2014 (see Literature Link) and from the ENCODE and Genecodes websites.

examples that indicate that some lncRNAs are important in diverse cellular functions, it is by no means clear than most lncRNAs have any function. We can reflect briefly on the question of assigning functions by drawing comparisons with protein-coding genes. The functions or activities of protein-coding genes have been assigned by several different methods. Two of these are the tissue and time of expression and an evolutionary similarity to a gene or protein whose function is known in another organism. Neither of these works very well for assigning a function to an lncRNA. As noted above, many lncRNAs are expressed in only one or a few cell types, making it difficult to draw inferences about possible functions on the basis of their expression. In addition, the nucleotide sequences of lncRNAs are not highly conserved; since they are not translated, natural selection for function is probably not based primarily on their sequences. It has been suggested that their functions may depend on secondary structures (arising from intrastrand base pairing, similar to tRNAs and rRNAs) or on short regions of interactions with other macromolecules (similar to base pairing between miRNAs and their target mRNAs), but these are more complicated to study and there are not yet very many data to support this view. It has also been suggested that their functions may be related to their locations in the genome, such as their proximity to enhancers, promoters, and other chromosomal features. This seems like a promising idea, at least for some of the many lncRNAs that have been found.

For protein-coding genes, the most productive method of genetic analysis has usually been to look at mutants; this approach will make up most of what we discuss in the rest of the book. But this is also somewhat more problematic for lncRNA genes than for protein-coding genes. First, unlike for protein-coding genes, we cannot make missense or nonsense mutations in the sequence and expect these mutations to affect their functions in different ways. Thus, making a mutant for lncRNAs genes has often meant creating a deletion for the gene. But, since many lncRNA genes overlap with each other or with other known functional elements in the genome, a deletion of the lncRNA likely also deletes more than one function. (Tool Box 8.1 suggests a method based on dCas9 that can be used specifically to increase or eliminate their transcription, which might help to circumvent this issue.) On the other hand, deletions for protein-coding genes, which have been a standard tool for geneticists for more than seventy years, were probably also affecting unrecognized lncRNAs, which was not considered in the analysis or in the interpretation of the effects of the deletion. In any event, assigning functions for lncRNA genes and determining how to improve upon our operational definition of them are current and future projects for geneticists, as we still try to define the gene.

Revisiting the definition of the gene

With so many genes whose final product is RNA being added to the collection of protein-coding genes studied for so long, we are still making significant adjustments to our definition of genes, more than a century after they were named. One definition put forth as it became apparent that non-coding RNAs were common concluded that a gene is "a union of genomic sequences encoding a coherent set of potentially overlapping functional products."

Literature Link. Gerstein et al. 2007, *Genome Research* 17: 669–81

Another recent attempt reviewed the history of the gene concept and discussed the many challenges associated with an inclusive definition. Those authors propose the definition that "a gene is a DNA sequence (whose component segments do not necessarily need to be physically contiguous) that specifies one or more sequence-related RNAs or proteins that are both evoked by genetic regulatory networks and participate as elements in genetic regulatory networks, often with indirect effects, or as outputs of genetic regulatory networks, the latter yielding more direct phenotypic effects."

Literature Link. Portin and Wilkins 2017, *Genetics* 205: 1353–64

Somehow these do not seem to be definitions that will resonate in our minds like "one gene, one polypeptide," even if they are more inclusive and accurate. By design, neither of these definitions includes aspects of inheritance, which was part of the original definition of a gene, although perhaps that is implied by the reference to genomic or DNA sequences. But maybe we can take a bit of comfort in the history of those who came before us, and realize that we can still study genes and their activities even if we cannot fully define them.

1.6 Genetic analysis

Our attempts in this chapter to define a gene have been a limited success at best. We are able to describe many of its essential properties—its inheritance, location, mutation, and so on. We are able to define its essential nature—a DNA sequence embedded in a genome of other DNA sequences. It is more challenging and rewarding to attempt to connect these two concepts, and to explain the essential properties of a gene in terms of its essential nature.

But we do not always need to define something with precision in order to use it; most people at the gym probably cannot name the exact muscles they are exercising. In a similar way, our inability to define a gene affects our ability to do genetic analysis insofar as we have to consider more possibilities for its functions, but it does not significantly impair it. Genetic analysis uses the properties of genes that we have encountered in this chapter as an approach to solve or dissect a complicated biological problem. In particular, genetic analysis relies on the principle that different alleles of a gene will have different activities, which are observable as different phenotypes.

Let's imagine a biological problem and apply genetic analysis to it. Here is our biological question: What is the mechanism by which insects are attracted to and fly toward a light bulb, the process of phototaxis? This definitely qualifies as a complicated biological problem. Consider a few of the processes that we know have to occur in order for this familiar behavior to happen, as summarized in Figure 1.16.

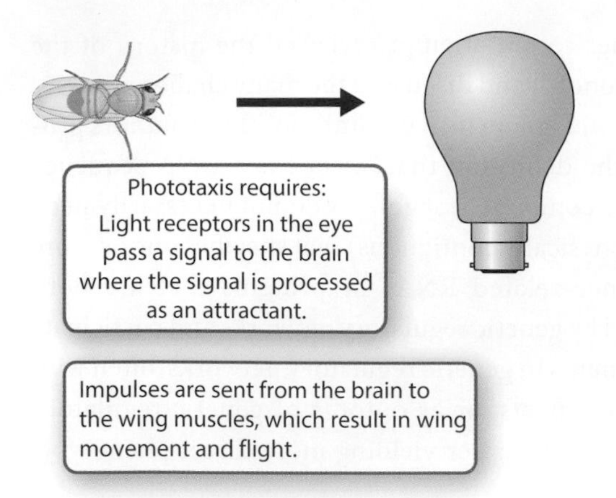

Figure 1.16 An example of a complicated biological question. A familiar but complicated biological question is how insects are attracted to a light source in the process of phototaxis. A few of the biological activities involved in proper phototaxis are summarized here.

- The insect has to have light-sensing genes.
- The input from the light-sensing genes has to be integrated in the brain with other sensory inputs, so that directionality and orientation are maintained during flight.
- The signal has to be processed as an attractant rather than a repellent.
- Nerve impulses have to be transmitted to the appropriate muscles.
- The muscles have to contract and relax, causing wings to beat.
- The air currents have to be detected so directional flight is possible.

As we reflect on this, it is remarkable that any insect ever makes its way to the light—except that anyone who has left a window open on a summer evening knows how reliably this occurs.

The investigator wants to identify and characterize all the component parts by which this process occurs. How should she begin? One approach could start with grinding up or dissecting insect eyes, brains, muscles, and wings, and characterizing all of the RNA and proteins that are present during the response, as well as the changes in them. There would be thousands of molecules, most of them having little to do with phototaxis. Although we may eventually want to know the identity and functions of these molecules, this might not be the most productive way to begin the analysis of the insect's response to light.

A **genetic approach** is different, as illustrated in Figure 1.17. Fundamentally, genetic analysis begins by finding insects that cannot fly to the light normally—that is, they carry a mutation affecting one of the many genes that involve the light response. Then one has to determine the nature of the altered response—the mutant phenotype. From there, the investigator finds the gene, determines its product, attempts to infer its function, and closely examines the mutant phenotype. The expression pattern and the activity of the gene are determined. In order to understand the light response thoroughly, other genes that also affect the response must be found, and their interactions with each other have to be determined. If all that is done carefully, it is possible that at least some aspect of the light response—for example, the process by which the receptor cells in the eye send a signal to the brain—will be understood. One can imagine that the entire process, or any other complicated biological process, could be understood this way.

In fact, a genetic approach similar to this one was carried out by Seymour Benzer and his students in the early 1970s, and identified genes that are needed for the response to UV light. One of the mutants defined a gene known as *sevenless*, since mutants lack the R7 photoreceptor needed for UV light response. Subsequent analysis of the *sevenless* gene involving many other geneticists over many more years learned that *sevenless* encodes a receptor tyrosine kinase, proteins needed for many of the fundamental intracellular signaling pathways in animals; genes in these pathways are among the most frequently mutated in human cancers. Benzer and his students were principally interested in using *Drosophila* mutants to address questions about complicated behaviors. They met that goal, but the fly and its genes were able to tell us more than that.

Genetic analysis using mutations is a cornerstone research approach, and we will introduce many different examples in this book. In many cases, the genetic analysis was addressing one biological question, and the gene shed light on even more questions than the investigators imagined. Table 1.3 shows some examples of biological questions that were originally addressed using mutants, and another biological process that was unexpectedly studied. The list in Table 1.3 includes those found elsewhere in the book, but these are a very small subset of the many thousands of examples of mutant analysis. So, even if we cannot write down a definition that includes every gene, we can use their properties.

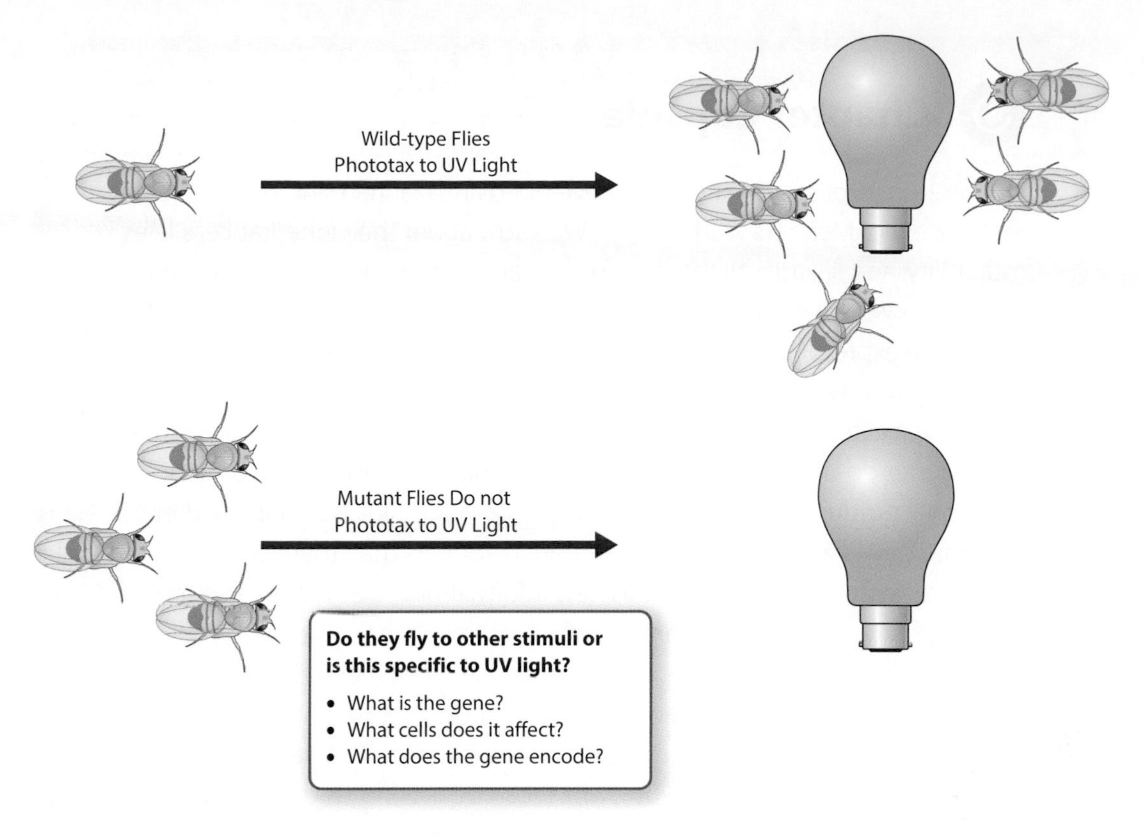

Figure 1.17 A genetic approach to phototaxis. One experimental method to the study of a biological question is to identify mutants that fail in the process, in this case phototaxis to UV light. Once mutants are identified, a number of follow-up questions are important to address, some of which are summarized here.

Table 1.3 Examples of genetic analysis based on mutant screens

Basis for Mutant Screen	Organism	Other Processes Affected	Discussed
phototaxis	*Drosophila*	Intracellular signaling, including cancers	Chapter 1
Homeotic transformation	*Drosophila*	Chromatin structure, silencing and activation	Chapter 2
Eye color variegation	*Drosophila*	Heterochromatin formation	Chapter 2
Growth on sucrose	*S. cerevisiae*	Protein trafficking	Chapter 3
Egg laying and vulval morphology	*C. elegans*	Intracellular signaling, including cancers	Chapter 3
Sex reversal	*Drosophila*	Alternative splicing	Chapter 3
Growth in dark	*Arabidopsis*	Plant hormone signaling	Chapter 3
Feeding	planarians	Regeneration, intracellular signaling	Chapter 3 case study
Embryonic segmentation	*Drosophila*	Many, including intracellular signaling, translational regulation, enhancer sequences, and others	Chapter 4 case study; Chapter 11
Cell division	*S. cerevisiae*	Cell cycle, checkpoints, and cancer	Chapter 6
Eye pigmentation	*Drosophila*	Intercellular transport	Chapter 6
Wing shape	*Drosophila*	Pyrimidine biosynthesis	Chapter 6
Dauer larvae formation	*C. elegans*	Signal transduction, including aging and cancer	Chapter 6 case study; Chapter 11 case study; Chapter 12 case study
Benomyl sensitivity	*S. cerevisiae*	Cell division, mitotic spindle	Chapter 7; Chapter 10 case study

Chapter Capsule

The properties of genes have been developed and clarified since the rediscovery of Mendel's work in 1900. We know about their inheritance, alleles, mutability, and linkage; all these properties are important for the way in which geneticists work with genes. The processes of how genes are expressed and how gene expression leads to phenotypes are broadly understood, but many new insights have arisen as genomes are analysed. While many genes fall into the familiar category of encoding the information to make a polypeptide, many other genes do not encode polypeptides and have RNA molecules as their functional products. In fact, in the genomes of multicellular organisms, there are more genes with RNA as their functional products than there are genes that encode polypeptides. With this new information, it has become even more difficult to propose a definition of a gene that is inclusive of all types, describes the various functions, incorporates the properties of inheritance, and is simple. Nonetheless, even without having a comprehensive definition of all genes, genetic analysis is done by finding or creating variant forms with different activities and phenotypes, which can be used to study many different complicated biological processes.

Additional Reading

Carlson, E. A. 2004. *Mendel's legacy: The origins of classical genetics*. Cold Spring Harbor Press: Cold Spring Harbor, NY.

van Dijk, P. J., and T. H. Noel Ellis. 2016. The full breadth of Mendel's genetics. *Genetics* 204: 1327–36.

Lederberg, J. 1994. The transformation of genetics by DNA: An anniversary celebration of Avery, MacLeod, and McCarty (1944). *Genetics* 136: 423–6.

Strauss, B. S. 2016. Biochemical genetics and molecular biology: The contributions of George Beadle and Edward Tatum. *Genetics* 2013: 13–20.

CHAPTER 2

Genomes, chromosomes, and epigenetics

TOPIC SUMMARY

The genetic information for the biological processes that a species carries out is contained within the DNA sequence of its genome. The DNA sequences of many genomes having been completed, the next steps in genomic analysis involve annotation of the sequence, that is, identifying the functional elements within that sequence. These functional elements include such sequenced-based information as transcripts, splice variants, and binding sites for proteins. While most of the attention in molecular genetics has been focused on the genes and their DNA sequences, the DNA in eukaryotic cells is packaged as chromatin fibers into chromosomes. Changes in chromosome structure are important in regulating transcription, recombination, replication, and many other biological processes but, because these important changes occur locally and dynamically, they have only recently been studied on a genome-wide scale. Thus, in addition to the DNA sequence itself, the functional elements to be annotated include structural modifications in chromatin. These modifications result in heritable changes in the gene function without corresponding changes in the DNA sequence; and they are known as epigenetic changes. Epigenetics has a long history in genetic analysis; with contemporary experimental tools, we are beginning to understand the common molecular basis for some epigenetic phenomena.

INTRODUCTION

The DNA sequence of a species' genome is often called the blueprint for that species. Implicit in this metaphor is the idea that one may be able to construct or reconstruct an organism from knowing the DNA sequence of its genome. However, while an architectural blueprint provides all of the information a builder needs to construct a home, including where to place the walls, doors, electrical wires, pipes, and so on, the ordered list of the nucleotides A, C, G, and T is not a complete blueprint for a geneticist to build a species; additional information about the structure and function of different parts of the sequence must be understood too. Discovering and assigning this additional important information is called annotating the genome. The **annotation** of a genome continues long after the DNA sequence has been obtained, and might be considered as always being in progress.

In this chapter we will discuss some of what has been learned from genome annotation projects. The formal annotation of the human genome is referred to as the ENCODE (**Enc**yclopedia **o**f **D**NA **E**lements) project. Related annotation projects for the model organisms *Drosophila melanogaster*, *Caenorhabditis elegans*, and mice are referred to as the modENCODE projects, and similar projects are underway for many other species without the ENCODE name.

Literature Link. http://www.genome.gov/ and http://www.modencode.org

These projects synthesize the data from many hundreds of computational and molecular experiments in order to determine fundamental information such as the correct structure of each transcript, its alternative splicing isoforms, and its expression patterns. These results form the essential scaffold for assembling all of the information about genes and other DNA sequences as well as their functions.

But the information content of a genome also includes the chromatin structure that packages DNA into chromosomes and so goes beyond the DNA sequence. Heritable changes in chromatin structure that result in functional differences within cells are referred to as **epigenetic changes**, and are also an important part of a genome annotation project. The correlations of changes in transcription, replication, and recombination with changes in chromatin structure, including the combination of post-transcriptional histone modifications comprising the histone code, are providing a rich but complex image of the molecular output of the genomic blueprint.

2.1 Genomes, chromosomes, and epigenetics: An overview

While it is appropriate to consider and analyse a gene in terms of its DNA sequence, we also need to recognize that the larger genome where each gene resides includes much more functional information than we can extract solely from inspection of the DNA sequence. For example, as discussed in Chapter 1, eukaryotic genomes are pervasively transcribed; thus, in order for us to understand the functions encoded in the genome, all the transcribed sequences need to be identified. But, as we now know, only some of these transcripts are also translated into polypeptides, and the majority of transcripts do not code for proteins. It is thus important to distinguish the sequences expressed as mRNA and, so, polypeptides from those expressed as functional RNA and never translated.

Even if all the expressed sequences are identified and classified, many functional elements in the genome appear not to be transcribed; this statement should be read with caution, since new transcripts are discovered frequently. Functional elements that are not transcribed include sites important for DNA replication, regions with structural roles in organizing chromatin, regions regulating recombination, at least some regions involved in regulating transcription, and many other functions that we may not yet recognize at all, let alone be able to identify solely from the DNA sequence.

Moreover, the DNA molecule is packaged into chromosomes; chromosomes, rather than naked DNA, provide the context for the expression, replication, rearrangement, and organization of the genome. It has been recognized for some time that the physical structure of the chromosome affects many cellular functions, but it has only occasionally been possible to attribute a change in function to a change in chromosome structure and organization. In order to understand the functions that are encoded in the genome, we need both to recognize the contribution that DNA sequences make to determining those functions and to determine the effect that changes in chromatin structure and organization have upon them.

Assigning functions to genomic sequences is given the general name of annotation; annotation is an ongoing project for every species whose genome has been sequenced. Some of the functions and features to be annotated are summarized in Figure 2.1 using the *Drosophila* genome as an example and are discussed throughout this chapter.

It has long been recognized that changes in the structure of chromosomes affect changes in genetic functions (such as gene expression and replication) without altering the underlying DNA sequence. Heritable changes in gene function that occur without a corresponding change in the DNA sequence are known as epigenetic changes. Often the change in gene function is recognized whenever the gene is moved to a different position in the genome, or whenever the surrounding chromosome context has been changed. The original recognition of such a position effect was found with the *white* gene in *Drosophila*. When the X chromosome is rearranged such that the wild-type w^+ gene is

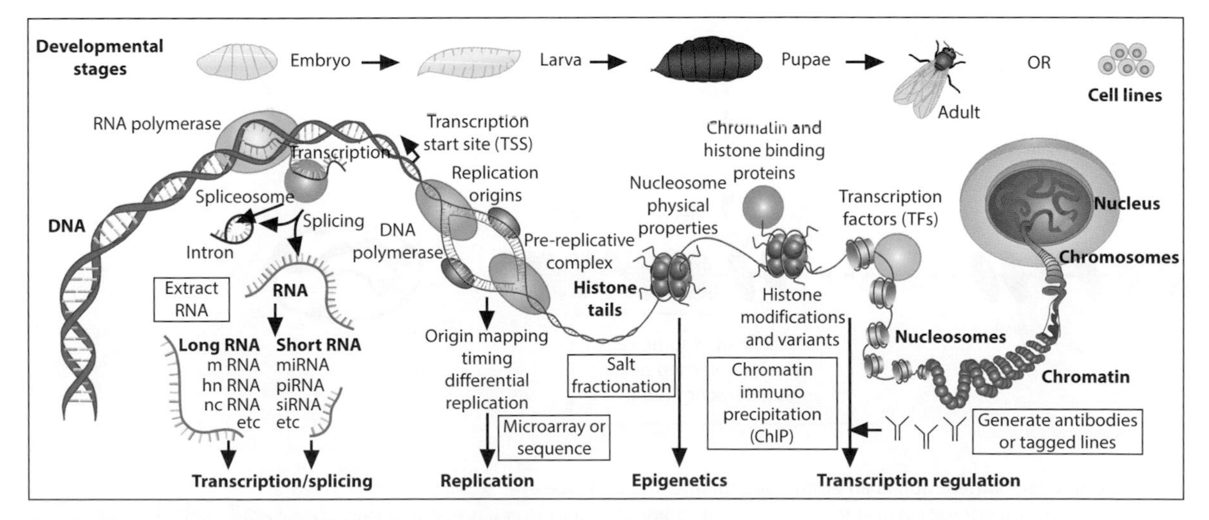

Figure 2.1 The *Drosophila melanogaster* modENCODE project, summarized. Genome annotation projects have similar goals and use similar methods regardless of the organism. The features to be annotation are in boldface at the bottom, some of the methods are displayed in boxes, and the molecular features such as mRNA, miRNA, etc. are in plain text. Redrawn from *PLoS Biology* 2011 9:e1001046, and used with permission.

placed next to heterochromatin, the gene is transcriptionally inactivated in some cells and remains inactive throughout subsequent mitotic divisions (Figure 2.2); as such, it is a heritable change. Similar phenomena have been observed for many organisms in which a transgene has been integrated into the genome; the transgene may be silenced or expressed differently because of the chromatin neighborhood where it has inserted rather than because of any changes in its own regulatory region.

Epigenetics has a long history in genetic analysis. In *An Introduction to Genetics*, their widely used genetics textbook from 1939 (published by W. B. Saunders), Sturtevant and Beadle devoted a chapter to position effect variegation, particularly to cases involving the *w* gene, since these examples are easy to recognize. Although they did not use the term "epigenetics" (coined by Waddington in 1942), Sturtevant and Beadle were describing an example of epigenetic change. Another widely used genetics textbook from the premolecular biology era of genetics, *General Genetics*, by Srb and Owen (Freeman, San Francisco 1952, with a second edition by Srb, Owen, and Edgar in 1965), also devoted part of a chapter to epigenetics. These authors used a definition similar to the contemporary one, and even some examples that are familiar to a modern student of epigenetics. Clearly the importance of epigenetic regulation was recognized long before our recent interest in its mechanisms.

X-chromosome inactivation as an example of epigenetics

A well-known example of an epigenetic change is X-chromosome inactivation in mammals, most strikingly observed with calico cats (Figure 2.3). X inactivation is part of the dosage compensation process in mammals that equalizes the expression of X-linked genes between the sexes. In X inactivation, one of the two X chromosomes in female mammals is transcriptionally inactivated in each cell independently, early on in embryogenesis (although a few genes are transcribed from both X chromosomes), and remains inactive throughout subsequent mitotic divisions.

X inactivation illustrates several features common to other epigenetic phenomena, all summarized in Figure 2.4.

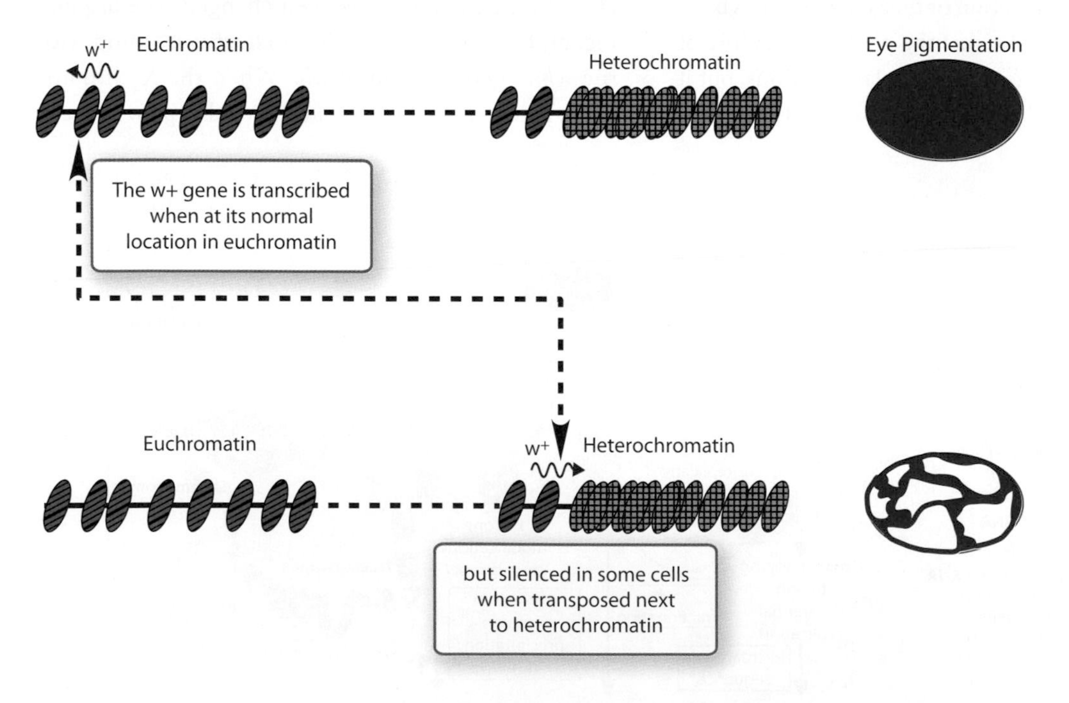

Figure 2.2 Position effect variegation is an epigenetic change. The wild-type eye color in *Drosophila* is red, as shown at the top. If the *w+* allele is moved next to heterochromatin through an inversion, it becomes transcriptionally inactivated in some cells of the eye, which results in a variegated white-eye phenotype, with patches of red and white pigmentation.

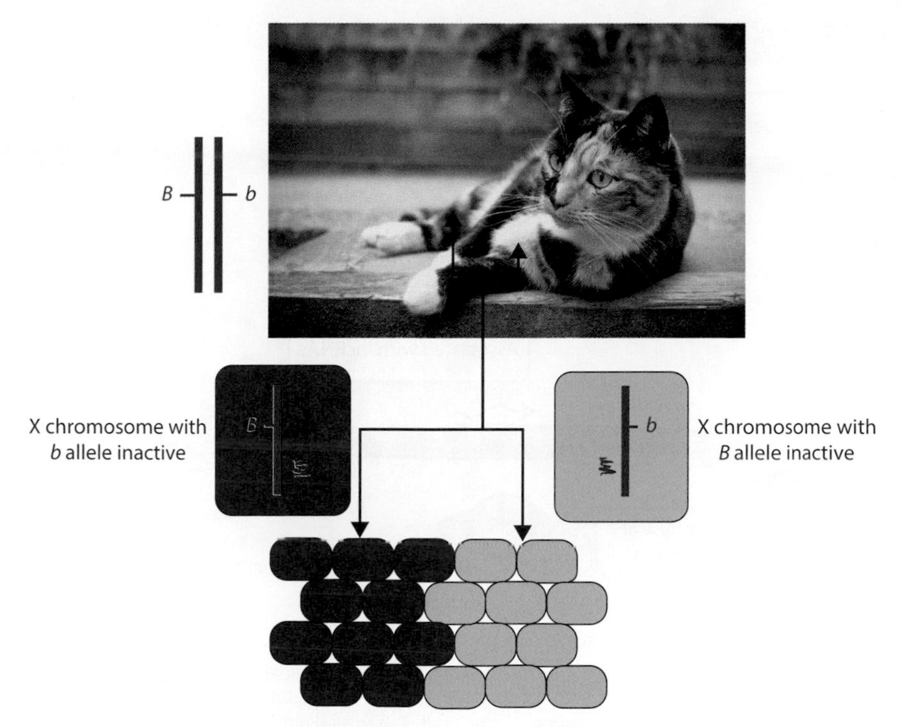

Figure 2.3 X-inactivation in calico cats. Calico cats have patches of orange and black fur owing to X-chromosome inactivation, as seen in the photograph. The responsible X-linked gene has two alleles, represented as *B* and *b*. In some cells, the X chromosome with the *b* allele is inactive, which results in a cell that produces black fur. In other cells, the X chromosome with the *B* allele is inactive, which results in a cell that produces orange fur. The inactive X chromosome is cytologically recognizable as a highly condensed Barr body. The inactive and active states are mitotically stable, so that, when the cell divides, its daughter cells retain the same pigmentation. This produces patches of black and orange fur.

Literature Link. Almouzni and Cedar 2016, *Cold Spring Harbor Perspectives in Biology* 8: a019372; Moris et al. 2016, *Nature Reviews Genetics* 17: 693–703

- First, regions of the chromosomes (in the case of X inactivation, an entire chromosome) often encompassing multiple genes with unrelated functions are affected.
- Second, the inactivated state is stably transmitted from cell to cell during mitosis, as a form of cellular memory. X inactivation is particularly stable and the chromosome remains inactive for hundreds of cell divisions. Other forms of epigenetic regulation may persist over much shorter time scales.
- Third, as in other epigenetic examples, the DNA sequences of the active and inactive X chromosomes are unaltered; the epigenetic mark is reversed during oogenesis for X inactivation, so that either of the mother's X chromosomes can be expressed in the offspring.
- Fourth, the epigenetic state is often observable from the physical state of the chromatin when appropriate assays are available. For X inactivation, the state of the chromatin can be seen cytologically, using dyes for nucleic acids, while chromatin changes with other epigenetic changes may require more specific assays. One of the key goals of genome annotation is to identify the chromatin marks associated with epigenetic changes, even if those marks are not readily visible.
- Finally, X inactivation also involves the functions of several lncRNA molecules, including Xist, Tsix, and at least three others; some other examples of epigenetics also involve lncRNAs, but it is not yet clear whether the involvement of lncRNAs should also be considered as a defining feature of epigenetic changes. (Even in the diverse world of lncRNAs, Xist and Tsix are much longer and more stable than most others.)

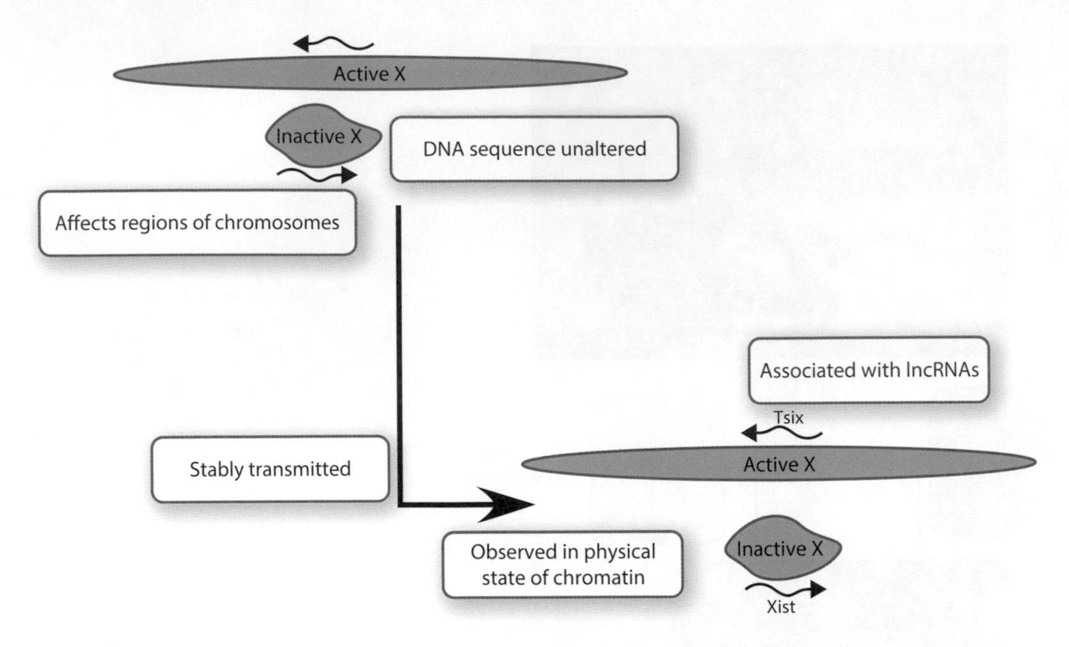

Figure 2.4 Features of epigenetics. X-chromosome inactivation is diagrammed to represent the common features found in epigenetic phenomena. Epigenetics affects regions of the chromosome but does not alter the DNA sequence itself, the effect is stably transmitted during mitotic divisions, and is usually observed as changes in the physical state of the chromatin. X-chromosome inactivation, and many other examples of epigenetics, is associated with several different long non-coding RNA molecules, which may be another feature common to all examples of epigenetics.

During the history of genetic analysis, the importance of epigenetic changes might have been understudied because stable changes in the DNA sequence (that is, mutations) were studied instead; epigenetic changes sometimes seemed to resemble the inheritance of acquired traits, which violated key principles of basic Mendelian genetics. In addition, with the exception of a few isolated examples in which the effect is long-lasting and widespread, such as X inactivation producing calico cats, epigenetic changes were often hard to investigate. Many epigenetic changes are transient or occur in only some cells. As a result, common principles underlying epigenetic changes were difficult to elucidate; in fact there appeared not to be a single mechanism, or even a few mechanisms that unified epigenetic examples. Some of these issues have been resolved through our ability to analyse complete genomes, and with assays that look precisely at chromatin structure on a genome-wide basis. But from these contemporary methods we may yet learn that the broad category called "epigenetics" includes many functionally and molecularly different subcategories.

Accessing genome annotation data

Publications from genome annotation projects often include an abundance of information in every figure and table. The data arise from the synthesis of many different experimental methods, each with its own details. These are generated by individual laboratories working on a particular genome or on a particular question across many different genomes, and are supplied to one of several different large databases, such as those maintained for each model organism. All the critically important but detailed information can sometimes make it difficult for the occasional reader to see a larger picture of the genome structure and function. As we try to emphasize the broader lessons from genome annotation, in this chapter we will look first at the functions associated with DNA and RNA sequences, which are primarily related to gene structure and expression. We will then look at functions associated with chromatin, and then point to a few examples that unify these approaches. Information in this chapter comes from some of the ENCODE-related papers, often from tables in the extended data of those papers, and from searching genome databases on specific topics.

Literature Link. ENCODE Project Consortium 2011, *PLoS Genetics* 9: e1001046

2.2 Gene organization and expression

In discussing the history of the gene concept in Chapter 1, we did not elaborate on one of the most significant and familiar features of genes: the somewhat abstract properties of genes are explained by the biochemical properties of DNA. We cannot describe the findings of molecular genetics in a few paragraphs. But here are some of the fundamentals.

Chromosome size and the spacing of genes

A chromosome consists of one double-stranded DNA molecule, extended from end to end (Figure 2.5). In *C. elegans*, each of the six chromosomes is about 17 million base pairs (Mb) in length. In *Drosophila melanogaster*, the chromosomes range in size from about 22.4 Mb for the X chromosome to about 52.5 Mb for chromosome 3, not including the microchromosome 4 (1.4 Mb). The shortest human chromosome is about 56 Mb in length, the longest about 260 Mb.

Within that long DNA molecule, genes occupy specific locations with the distance between them measured in base pairs or, more frequently, in kilobase pairs (kb), as shown in Figure 2.6 for the 70 kb region of chromosome I from *C. elegans*. Any given gene is encoded on one of the two DNA strands, but neighboring genes could be on the same DNA strand and thus oriented in the same direction, or they could be on the opposite DNA strand and thus oriented in the opposite direction. The region shown in Figure 2.6 has genes in both orientations. Our original vision of the arrangement of genes on the DNA molecule placed them in a linear, non-overlapping arrangement of discrete blocks, separated by intergenic regions of various lengths, with no genes. In fact we now recognize that overlapping transcripts are remarkably common in the genomes of multicellular organisms, and that the regions between the protein-coding genes are often rich in transcripts as well. Both of these features are also evident in the region of the *C. elegans* genome shown in Figure 2.6. Thus, while it is can be helpful to refer to the region between two protein-coding genes as an intergenic region, truly intergenic regions may not be meaningful.

Core promoters and initiation

The regulation of gene expression in eukaryotes occurs principally through the regulation of transcription

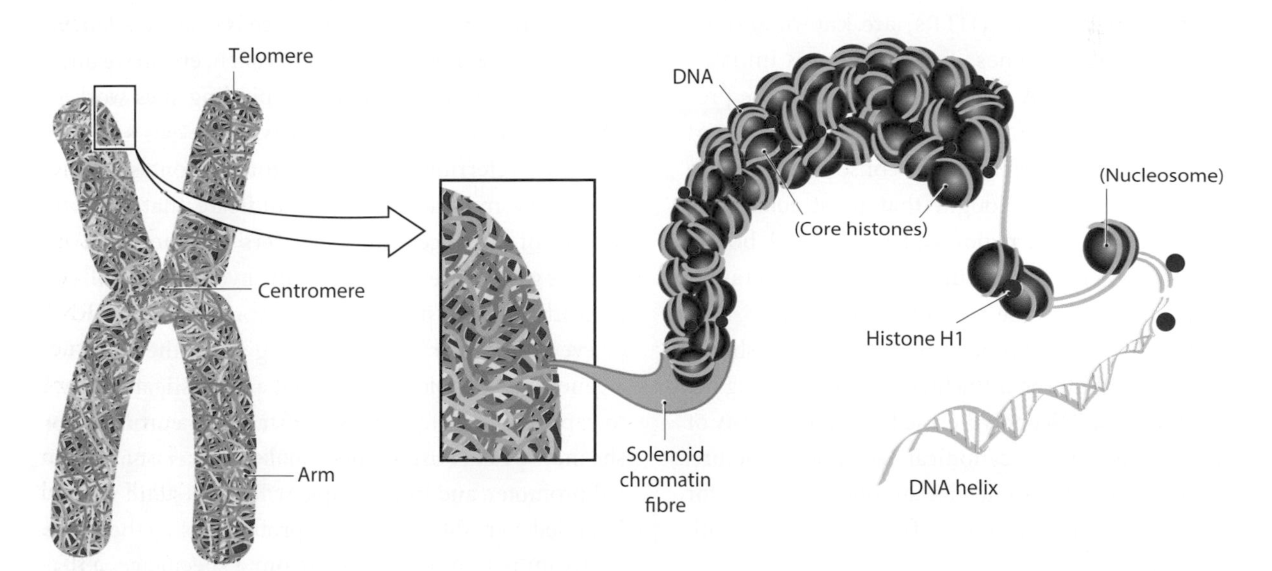

Figure 2.5 Nucleosomes and the structure of chromosomes. When chromatin is extracted under conditions with low salt, a "beads on a string" structure is seen, as diagrammed here and shown in Figure 2.10. The bead consists of two molecules each of H2A, H2B, H3, and H4, while the string is DNA. These beads from the 10 nm fiber are then compacted into the 30 nm fiber by histone H1. The 30 nm fiber is then further compacted into higher order structures. From Meneely et al., *Genetics: Genes, genomes and evolution* (Oxford University Press, 2017), adapted from Figure 3.6.

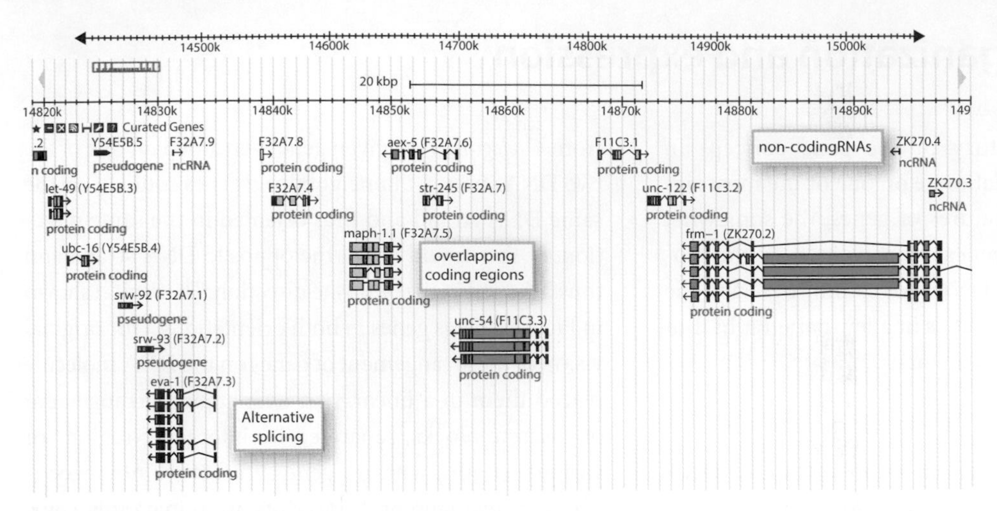

Figure 2.6 Transcripts found in a region of a eukaryotic genome. A region of 70 kb from chromosome I in *C. elegans* is shown as an example. Note that either DNA strand can be used as the template; in worms, these are shown by the transcripts in different colors, but this convention is not used in all organisms. Transcripts from different genes often overlap with each other, either on the same strand but different reading frames or on different strands. Most genes are alternatively spliced. Furthermore, many non-coding RNAs are encoded, of different lengths.

initiation, although many other regulatory mechanisms are known and are important for some genes. A schematic of the structure of a protein-coding gene is shown in Figure 2.7. A protein-coding gene usually begins with an ATG (methionine) start codon, with the A designated as +1. Transcription begins upstream of this start codon, often about thirty–forty bases upstream, but much longer untranslated regions (UTRs) are known and are significant for some genes. Transcription is initiated by the binding of RNA polymerase II to the core promoter region, which includes the transcriptional start site (TSS) and a number of other conserved sequence features. It was once thought that most eukaryotic genes have an AT-rich region called the TATA box in the core promoter, so the subunit of RNA polymerase II that binds first at the core promoter was named TATA-binding protein or TBP; more recent research has shown that, while TBP binding is the first step in recruiting the other subunits of RNA polymerase II, only a minority of genes in humans have a canonical TATA box structure. Core promoters vary somewhat among genes but, for the most part, the core promoter for one gene will work equally well for another gene.

Pre-initiation of transcription and cis-regulatory modules

Specific regulation of transcription—that is, turning on or off the transcription of a specific gene at the proper time and in the proper tissues—is controlled by DNA sequences not located at the core promoter, as shown in Figure 2.7. The **cis regulatory module** (CRM) has one or several DNA sequences of about 8–15 bp (base pairs) that are sequence-specific binding sites for transcription factor proteins. If the binding of the transcription factor stimulates transcription, the sequence is called an **enhancer**; if the binding of the transcription factor reduces transcription, the sequence is called a **silencer**. Enhancers are more common that silencers, so we often assume that transcription factor binding sites work as enhancers, unless contradictory evidence exists.

We are referring to the regulatory region as the cis-regulatory module (CRM) to indicate that the collection of enhancers and silencers together regulate specific gene expression, but many investigators will use a shorthand term such as "enhancer region," "enhancer," or even "promoter" when speaking about the CRM for a gene. For example, suppose that an investigator wants to express a reporter gene specifically in neurons; he or she may speak of using a neuronal enhancer or a neuronal promoter, and most people will understand what is intended. In reality, he or she is probably using the entire upstream region or the CRM from a specific gene that is expressed only in neurons to drive the transcription of their reporter gene. Since the goal of the experiment does not involve distinguishing among the individual transcription factor binding sequences within the CRM, it is more convenient to refer to the entire region as the enhancer or the promoter.

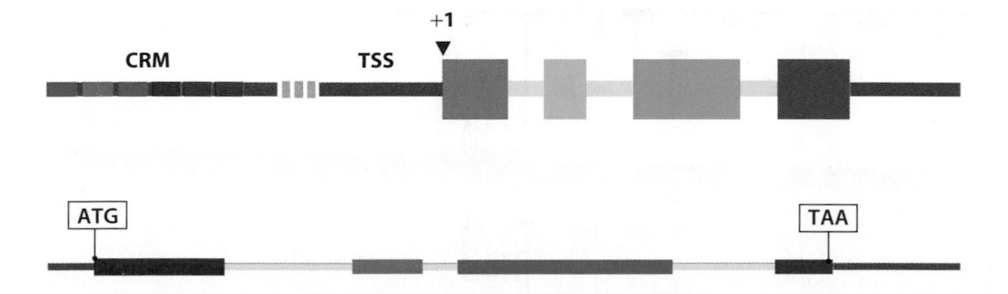


```
aaaaccgacgagaaaacaacATGAAATCTATTATCATTTTATCAGTTTTCTTAGTTTCTTCCGCGTT
GGCTGCCAGTTCTGATAAGGATACTGAATCAAAAGTGGCCAGATATTTGAAATCCTGTGGACTTATC
ACTTCATTGAAGACTACTAAATGTCTTAAGgtgggtttataaaattgttcaagcaggcttcaattg
tcatgattgctcatgcagttcttatttggtggaatttcaaacatggagatttatcacagtagtcct
ttttgttgtacttgctgcatttcgaactatttccgtcattttaataaaaaaacacttttagAGAA
CCGATGAAGTAGATAAGGAGCTCAAAGCTCTCGTAAAGAGCAAAGACAAGGATCTCTCAAAAGTCAA
GGCTTCGTCCAAAGACACCATGgtgggtaacatcagtgatattaaaaacttctcaaatgtgtttcag
GAATGCATTAAAGACCTGAAATCTAAACCACTCGAAGAACATCACAAGGAGACTCTCGACTTCTGTG
GTCGTCCTCTTTTCCAGGACGAGTTCAAGGATTGCAGCAAGAAATTGCAGGCATTGAAGAAAAGTGA
TAAGGATGCTGCAAAGTGTTTGGAAGATTTTAAATCAGAGACAAAGAAGTCGACAAAGGACACTTGC
ACATTTTTGAAGGGCTCGAAAGATTGTATCAAGGCTCATATTAAGAAAGAGTGCGGGGATGATAAGT
TGCAGGGATGGCAGAAGgtaagcggatgatgtcttagaaatcggtttaggcttcggcgcaggcctta
aacttagatgcttaggcttaggcttagagagcctcaaactttcattttcatttccaccaacttttct
aaatttctattttcagTTTGCTCAAGACTCGTTCAAAGAGAACGAATGCGAGAAGGCAATTGGAGC
AAAATGGAACTAActtcattaaatgttcttcgatttcttgttccttgttacagtaaccctgcaaaac
tgcaataattaatttaaaagtctttaaaattccggttgagtacccattcctccctcctacattcttc
atttcttcaaagggtttctctccaggtaccaataaaattttaacgcggtca
```

Figure 2.7 A representative eukaryotic gene. This particular gene is from *C. elegans* and is known as D1086.3; its molecular function and mutant phenotype are not known. The gene has four exons, shown in different color boxes at the top schematic and with the corresponding DNA sequences shown below. The exons are separated by introns, shown in orange. Untranslated regions upstream downstream of the coding region are in gray. Upstream of the ATG at +1 is the core promoter with the transcriptional start site (TSS). Further upstream is the cis regulatory module (CRM), a collection of short DNA sequences that are binding sites for various transcription factor proteins. The core promoter and the CRM are shown in the schematic gene but not in the DNA sequence.

A gene may have more than one CRM that directs its specific expression, and each CRM will almost certainly have more than one enhancer sequence (and may have silencer sequences as well). For most eukaryotic genes, CRMs are located at variable distances upstream of the core promoter, although examples are known in which a CRM is located downstream of the stop codon or within an intron.

In the context of genome annotation, identifying enhancer sequences and CRMs for genes is an important goal. However, enhancer sequences are difficult to determine by computational assays alone, that is, by scanning the genome for a specific sequence where a transcription factor binds. A number of challenges must be considered when attempting to find enhancer sequences and CRMs using computational methods. First, the preferred binding sites are not known for all of the transcription factors encoded in a genome. In addition, for most transcription factors, the binding site or enhancer sequence is not an exact sequence but a consensus among many related sequences in which the bases at some positions are invariant and others can vary. A few examples are shown in Figure 2.8. Furthermore, since enhancers and CRMs can be found at variable locations with respect to the coding region of the gene, the variability in sequences and their location means that many candidate enhancer sequences are identified but only some are actually used. These considerations combine to make a formidable problem in computational biology; we don't know all the sites to look for, we can't predict where in the genome we should be looking, and the sequences at the sites that we do know about can vary. Computational methods to detect enhancer sequences are regularly being improved, but the standard method is to determine directly which sites are occupied by a transcription factor using chromatin immunoprecipitation (ChIP). Descriptions of ChIP and other assays commonly used in genome annotation are listed in Tool Box 2.1, and some more detailed information is posted in the online resource centre.

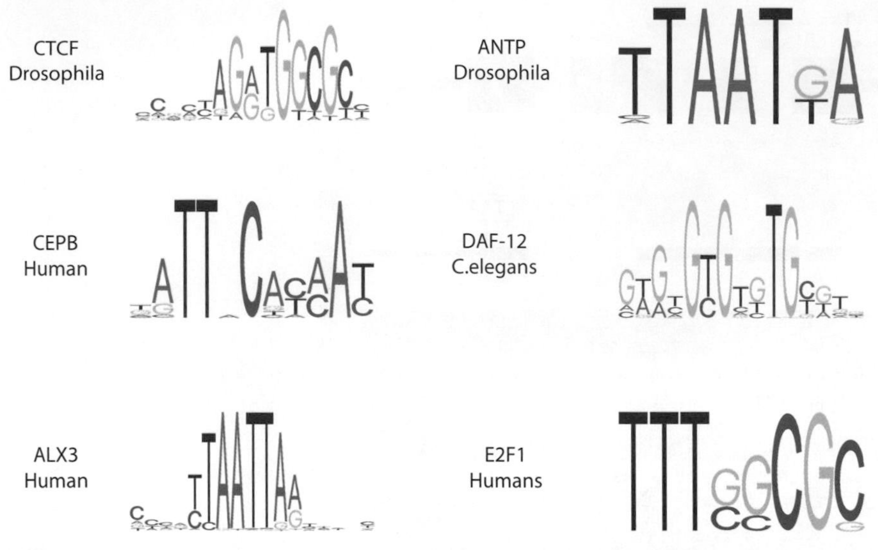

Figure 2.8 Consensus binding sequences for some transcription factor proteins. The size of the letter in the binding sequence represents the level of conservation of that nucleotide at that position, and the presence of more than one letter indicates the degree to which different bases are found at that location in known binding sites. Data compiled by the Jaspar Database of transcription factor binding sites, found at http://129.177.120.189/cgi-bin/jaspar2010/jaspar_db.pl.

TOOL BOX 2.1 Techniques for genome annotation

The tools and techniques used for genome annotation include many of the ones common in genetics and molecular biology labs working on single genes, although adapted for high throughput results. This tool box summarizes some of those tools; more details are available in the online resource centre. New methods are regularly being developed while older methods are being modified, so this is an overview.

Determining sequences of DNA and RNA

DNA sequencing. The most fundamental tool for genome analysis, and the one that makes all of the other methods possible, is DNA sequencing. The first eukaryotic genomes to be sequenced, those of *Saccharomyces cerevisiae* and *C. elegans*, were done primarily by teams of investigators using Sanger dideoxy sequencing, which produces a sequence run of about 700 bases from a single pass through a DNA template molecule. While still widely used for individual genes and small regions, Sanger dideoxy sequencing has largely been supplanted by other methods when entire genomes are sequenced. These methods are variously known as **next-generation sequencing**, deep sequencing, or massively parallel sequencing. Two commonly used techniques are also called 454 or Illumina sequencing, after the companies that market them. These produce sequence runs of 70-300 bases, depending on the method, but each DNA template molecule is sequenced multiple times (deep sequencing), and many DNA templates are sequenced at the same time (massively parallel sequencing). Recently, methods have been developed to use the smallest amount of DNA possible for a template, including the DNA from a single cell. Single-cell sequencing is particularly useful for transcript analysis by RNA-seq.

RNA-seq. Transcript analysis for genome annotation is done primarily by RNA-seq. The name implies that RNA is being sequenced directly, but this is a bit misleading. RNA is isolated in bulk, purified to remove rRNA or any other RNA that does not have a polyA tail, and reverse transcribed into cDNA. The cDNA is then sequenced by a massively parallel method, and the sequences are computationally mapped onto the sequenced genome.

Microarray analysis. While microarrays are discussed in the online resource centre, transcript analysis by microarrays has been largely replaced by RNA-seq for genome annotation.

Determining chromatin structure

DNAse hypersensitive sites or HSS. An early and still common method for characterizing the dynamic structure of chromatin involved the use of nucleases such as DNase I or micrococcal nuclease, which cleave naked double-stranded DNA in a sequence-independent fashion. When chromatin is treated with these enzymes at low concentrations, some regions of the DNA sequence are reproducibly a hundred-fold more sensitive to digestion than other regions. These nuclease **hypersensitive sites** are often found in clusters in regions upstream of active genes, and are not seen at the corresponding site when the gene is not being actively transcribed. Hypersensitive sites arise in these regions from the displacement of histones, and have been widely used to indicate upstream regulatory regions or other regions of "open" or accessible chromatin.

Formaldehyde-Assisted Isolation of Regulatory Elements or FAIRE. Like HSS, FAIRE identifies regions of open chromatin in the genome; FAIRE preserves the DNA sequence at these regions, while HSS digests the DNA. In FAIRE, chromatin is cross-linked with formaldehyde in vivo and sheared, then subjected to phenol-chloroform extract to remove the proteins. The remaining DNA is then sequenced, and the sequence is computationally mapped to the genome.

Chromatin immunoprecipitation or ChIP. ChIP is widely used to determine the locations in the genome at which particular proteins are found. Immunoprecipitation involves the preparation of an antibody against a target protein that is part of a macromolecular complex. For ChIP, the macromolecular complex is chromatin and the antibody is prepared against a target antigen protein such as a transcription factor, a modified histone, a subunit of RNA polymerase, or some other protein of interest. Chromatin is isolated from the cell, stabilized by a cross-linking agent such as formaldehyde to preserve the contacts between DNA and proteins, and sheared so that only the small region of DNA associated with the complex is preserved. The antibody is added to precipitate and isolate the target protein antigen, as well as any DNA sequences and proteins complexed with it. The cross-links are reversed and the DNA sequences that were complexed with the target protein are isolated and sequenced. These DNA sequences are computationally mapped onto the sequenced genome to find the binding sites for the protein in the genome. In order to avoid the need for a different antibody for every chromatin-associated protein, particularly transcription factors, the gene that encodes the protein of interest can also be fused in vitro with the coding region of a gene that encodes a protein for which an antibody already exists. This is known as **epitope tagging**. Commonly used epitopes included the hemagglutinin (HA) polypeptide from the influenza virus or GFP.

Chromosome conformation capture techniques: 3C, 4C, 5C, HiC, and others. These methods are used to determine the likely three-dimensional structure of chromatin in the nucleus; thus they find sites that may not be close to each other in the linear DNA molecule but are in close proximity in the higher order structure of chromatin. Like other methods, chromatin is cross-linked in vivo with formaldehyde, but is then digested with a restriction endonuclease to remove the regions not associated with a cross-link, for example the DNA sequences that have been looped out. The digested chromatin is then ligated together to connect the sequences that were in proximity in the nucleus without the (digested) region between them. The cross-links are reversed, and the ligated DNA molecules are then sequenced. The methods differ in how many interactions are analysed at once. 3C determines the interaction between two specific loci in the genome, 4C determines all of the interactions involving one specific locus, while 5C and HiC determine the interactions among many or all loci in the genome simultaneously.

Since all the somatic cells of an organism have the same DNA sequence, the cell specificity of gene expression does not come from the presence or absence of a CRM. Rather, the explanation for why a particular gene is transcribed in one set of cells and not in another is that the transcription factor proteins that will bind at the CRM and regulate transcription are present in some cells and not others. One can then reasonably ask how a particular constellation of transcription factor proteins came to be expressed in one group of cells; the answer, again, is that the presence of those transcription factor proteins in those cells is the outcome of other transcription factors that regulate their expression. In short, a specific set of transcription factors are produced in a given cell type; those transcription factors regulate the expression of other genes, some of which themselves encode transcription factors, and so on. Each layer of transcription factor regulation is coordinated with the others. The coordination of transcription in an organism becomes a network of interacting transcription factors, a topic we explore in Chapter 12.

The transcription factor proteins bind to the appropriate sequences in the CRM, but typically the CRM is not close enough to the core promoter on the linear DNA molecule to allow direct interaction between transcription factor proteins and RNA polymerase. But the nucleus' solution to this lies in a protein complex known as Mediator; the Mediator complex consists of more than twenty polypeptide subunits (the number varies in different species) and has the function of

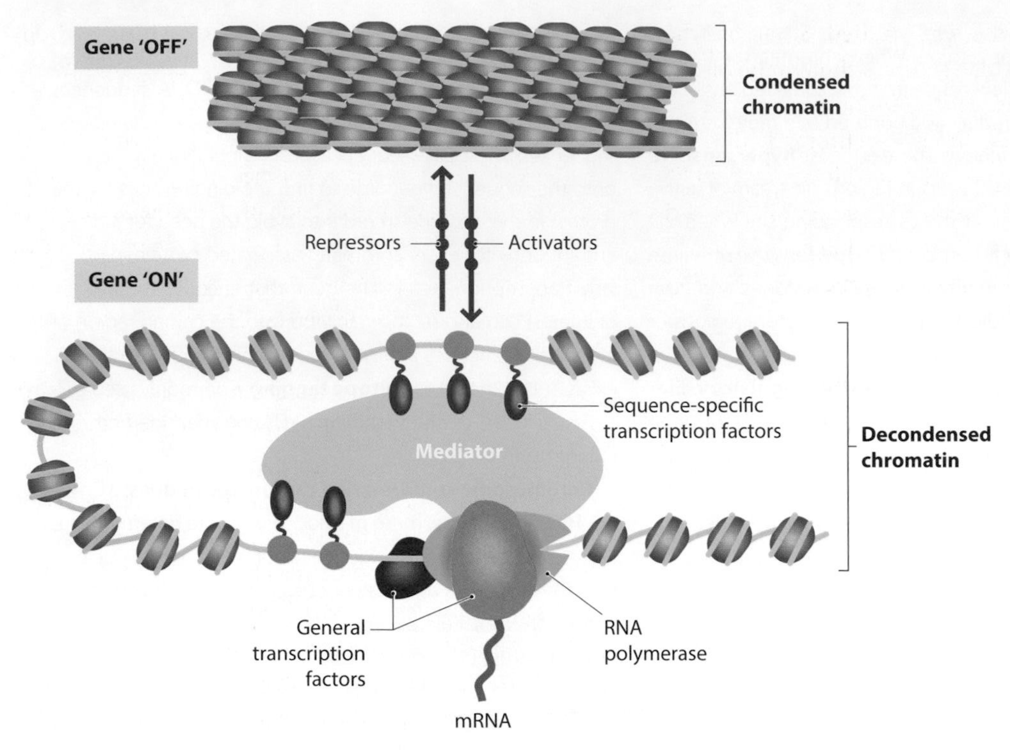

Figure 2.9 The role of the Mediator complex. Chromatin is looped to bring the transcription factors into interactions with the preinitiation transcription complex, with the Mediator complex forming the bridge. The decondensed or open chromatin is available for transcription. From Malik and Roeder, 2010, The metazoan Mediator co-activator complex as an integrative hub of transcriptional regulation, *Nature Reviews Genetics* 11: 761–72.

looping out the chromatin between the CRM and the core promoter to regulate specific transcription. One face of Mediator subunits interacts with transcription factor proteins while the other face interacts with RNA polymerase, as shown in Figure 2.9. The Mediator, sometimes referred to as a coactivator of transcription, is needed for any regulated transcription by RNA polymerase II, but also does not vary much between cell types, and many of its subunits are conserved among eukaryotes. If an investigator is interested in how the transcription of a specific gene is controlled, attention is usually directed toward the transcription factors and the CRM rather than the core promoter and the Mediator.

2.3 Chromatin structure

In the context of the eukaryotic cell, DNA molecules are found in nucleic acid–protein complexes broadly referred to as **chromatin**. As noted at the beginning of the chapter, the functions of a genome are determined not only by the DNA sequences of its genes but also by the chromatin environment in which the gene is located; in other words, changes in chromatin structure affect the function of each gene. There are hundreds of different chromatin-associated proteins, some of which are found in every cell at all times, while others—such as transcription factors—are highly variable in their association with DNA.

At a first approximation, the chromatin proteins found in every cell at all times do not associate with DNA in a sequence-specific fashion, as transcription factors do. The best understood of these proteins that associate with DNA and comprise chromatin are the **histones**. Because the histone proteins are abundant and found in every cell, and because they are among the highly conserved proteins among eukaryotes, they have been exceptionally well studied. Some of what these studies have revealed is that, although the histone proteins themselves are relatively constant in different cell types and at different times, the chromatin that they make up is not at all constant. Considerable variability from cell to cell and from time to time arises from variability in the histones, not in the amino acid sequences of the proteins themselves but in chemical modifications made to them after translation.

Histones and the nucleosome core particle

When chromatin is extracted from cells under conditions of low ionic strength and the extract is viewed by electron microscopy, the resulting structure is frequently described as "beads on a string," as shown in Figure 2.10. The "string" is DNA; there is one double-stranded DNA molecule per chromosome. The "beads" are the nucleosome core particles of histone proteins with DNA wrapped around them.

A nucleosome particle consists of two molecules of each of the four core histones, H2A, H2B, H3, and H4, plus the DNA; some use the term "nucleosome" to refer to the core protein particle alone, without the DNA. The core histones are small proteins, ranging in size from 102 amino acids for H4 to 145 amino acids for H2A; the sequences are rich in lysines, so the histones have a strong positive charge. Histones are among the most highly conserved proteins evolutionarily; there are only two amino acid differences in the H4 sequence between grass and a cow eating it. In fact, the nucleosome core particle can be reconstituted in vitro using histones from different species, so the high sequence conservation results in very similar structures as well.

The structure of the nucleosome has been solved at high resolution and is diagrammed in Figure 2.11. DNA is wrapped around the outside of the protein core particle, with 146 bp that make 1.67 turns. Core particles are separated or linked by a somewhat variable length of DNA, often approximately 60 bp. The nucleosomes form the 10 nanometer (nm) fiber, a figure that refers to its width. The 10 nm fiber has a packing ratio of about 5–10, meaning that the overall length of naked DNA is reduced to about 10–20 percent of its length when it is complexed with the core nucleosome particle. To give some sense of scale, a single mammalian gene that is 10 kb in length will be wrapped around approximately 50 nm particles. However, the length of the linker DNA, and thus the spacing between adjacent nucleosomes, varies in different cells and at different times. This is diagrammed in Figure 2.12, which illustrates the difference between open chromatin regions that are accessible to transcription factors, RNA polymerase, and so on and closed chromatin regions that are refractory to binding by other proteins.

Linker DNA is associated with the histone H1 or one of its many variants, the most variable of the histone proteins. Nucleosomes are compacted with H1 into a 30 nm fiber, which results in an additional packing ratio of about 50; in other words, the physical length of the 30 nm fiber is about 2 percent of the length of the DNA double helix. The 30 nm fiber is further compacted by

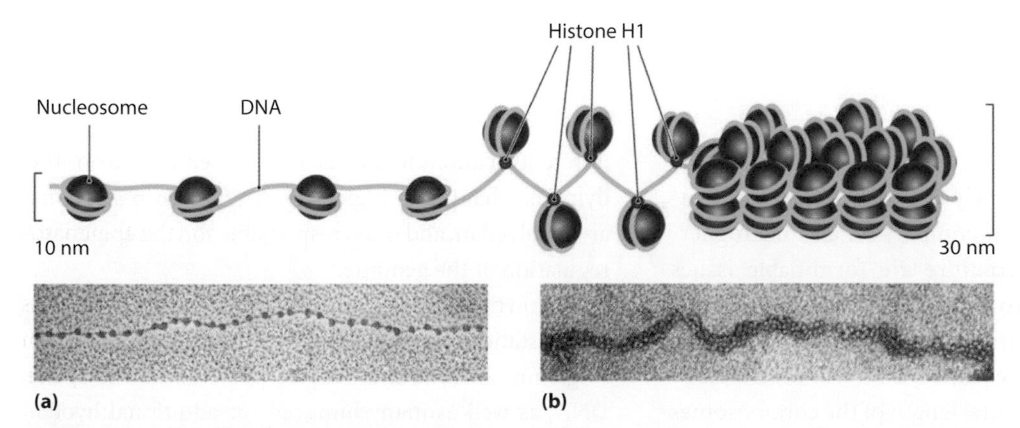

Figure 2.10 The nucleosome core particle. The "beads on a string" structure is seen by electron microscopy when chromatin is extracted at low salt conditions. The beads are the nucleosome core particles, while the string is double stranded DNA. Each core particle has two copies of the four core histones, with 146 base pairs of DNA wrapped around it. More compacted chromatin is shown on the right, with compaction to the 30 nm fiber regulated by the histone H1 or its variants. From Craig et al., *Molecular biology: Principles of genome function*, 2nd edn. (Oxford University Press, 2014).

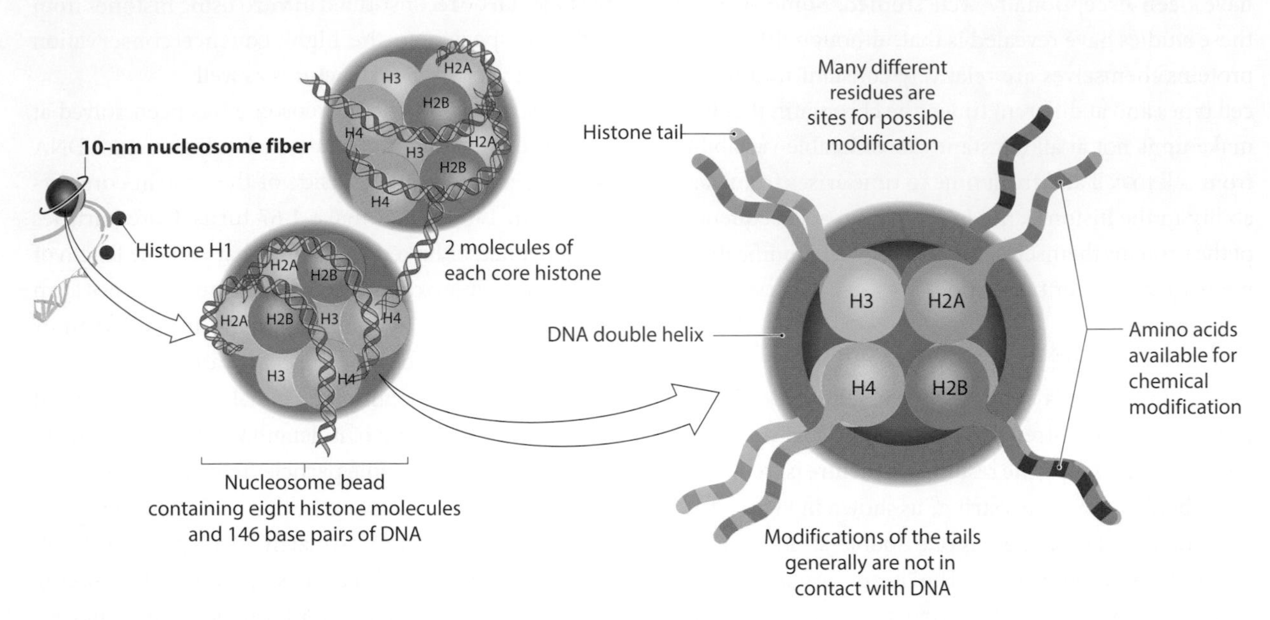

Figure 2.11 The structure of the nucleosome, showing the histone tails. The four histone proteins are arranged around the core, with the N-terminal regions exposed. Although these are the N-termimal regions, they are traditionally called tails. Note that H2B and H3 have longer N-terminal tails that H2A and H4, and more modifications have been found for H2B and H3. From Meneely et al., *Genetics: Genes, genomes and evolution* (Oxford University Press, 2017), Figure 12.21.

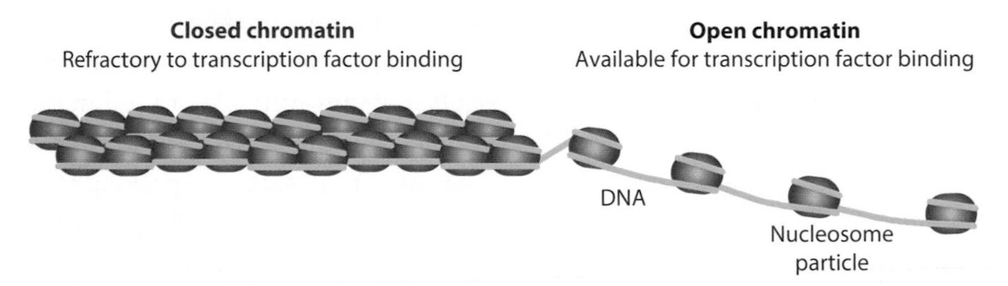

Figure 2.12 Chromatin can be open or closed, depending on how tightly packed the nucleosomes are. In the open state, the DNA sequence is more exposed, so transcription factors can bind more readily. In the closed conformation, transcription factors cannot easily access the DNA sequence. From Meneely et al., *Genetics: Genes, genomes and evolution* (Oxford University Press, 2017), Figure 12.20.

more dynamic and variable loops, a much less defined structure known collectively as a higher order chromatin structure. Chromatin compaction and the higher-order chromosome structure are formidable issues both for the nucleus to solve and for investigators to decipher. A human cell has approximately 1.8 meters of DNA sequence in the standard B form described by Watson and Crick. The total length of the chromosomes at mitosis is about 120 microns, so the overall compaction is more than 10,000-fold. The packing is orderly but also dynamic, so that the higher order structure changes during the cell cycle, differs between cells, varies within different locations in the nucleus of a cell, and so on. The dynamic changes in higher order chromatin structure are involved in, and often responsible for, the epigenetic regulation of the genome.

Within the core particle there are more than 120 direct electrostatic interactions between the core histones in a single nucleosome and the phosphate backbone of the DNA, as well as many hundreds of additional hydrogen bonds mediated by the solvent water, among other interactions. The direct interactions between the histone proteins and the DNA involve particular residues on the histones but almost any location in the DNA sequence;

that is, the electrostatic interactions are with the backbone on the outside of the double helix rather than with the nucleotide bases on the inside. Nucleosomes are capable of sliding along the DNA molecule in an ATP-dependent process, which has the effect of opening or covering particular short regions of DNA sequence in the linker regions. Nucleosome sliding and nucleosome positioning on the DNA molecule are part of the general process known as ATP-dependent **chromatin remodeling**, which affects gene expression, recombination, and replication. Chromatin remodeling has been difficult to elucidate, in part because it is often highly transient: changes in the positions are seen on the scale of milliseconds. Some genes and proteins involved in chromatin remodeling have been found by genetic assays and are described in Sections 2.4 and 2.5.

Histone modifications and a histone code

As Figure 2.11 shows, the N termini of each of the core histones are not directly in contact with the DNA backbone. Being exposed, the residues in the N-terminal tails are subject to many different post-translation modifications, including methylation, acetylation, phosphorylation, ubiquitination, SUMOylation, and ADP-ribosylation, as summarized in Figure 2.13. The enzymes that carry out these reactions have names that indicate their functions: histone methyltransferases, histone deacetylases, and so on. Most of the modified amino acids are lysines, but at least two specific serines and five arginines can also be modified. With so many possible modifications and so many different residues that can be modified, the number of combinations of possible histone structures is staggering. The sites and the types of modifications are shown in Figure 2.14.

For example, histone H3 has a lysine residue (abbreviated K) at positions 4, 9, 14, 18, 23, 27, 56, and 79. Each of these lysines can have from zero to three methyl groups attached to them post-translationally, symbolized as H3K9me2 or H3K9me3, although not all possible modifications have been observed to occur in vivo. (Methylation of H3K14 has not been observed, for example, although it can be acetylated.) Thus, only on the basis of the methylation state of lysines in the tail of H3, with seven lysines known to be modified and up to four modification states for each, any nucleosome could have more than twenty-five different structures. Any of these modifications might have functional significance, although in many cases a possible functional significance is unknown.

As an aside, the estimate that there are more than twenty-five different structures arising from methylation

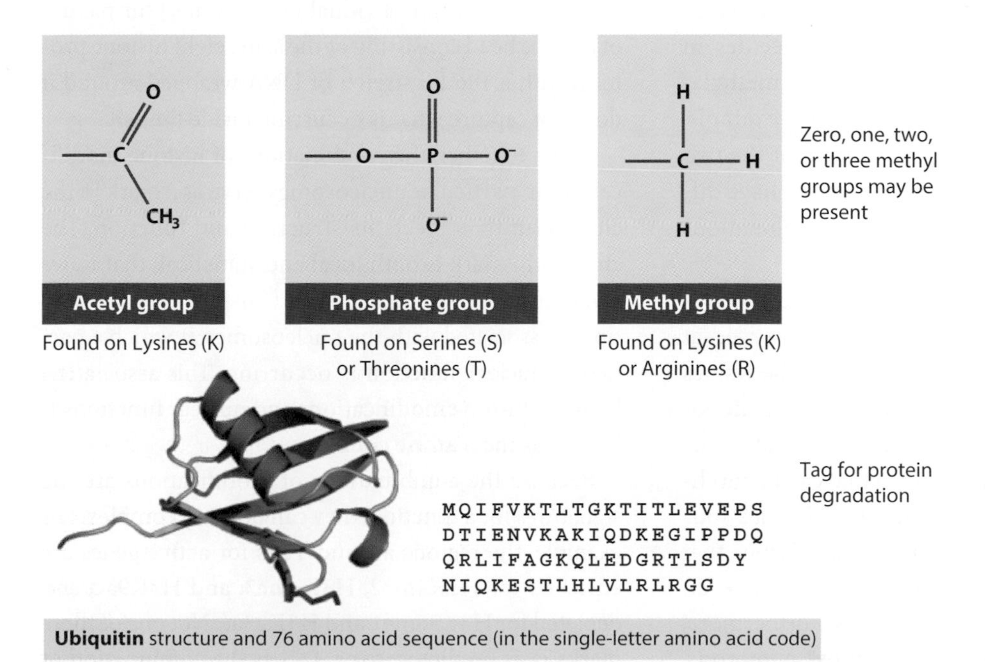

Figure 2.13 The covalent modifications made on histones. The sites for some of the modifications to the histone tails are shown. None of these positions is predicted to be in contact with the phosphate backbone of the DNA molecule, as shown in Figure 2.12. From Meneely et al., *Genetics: Genes, genomes and evolution* (Oxford University Press, 2017), Figure 12.23.

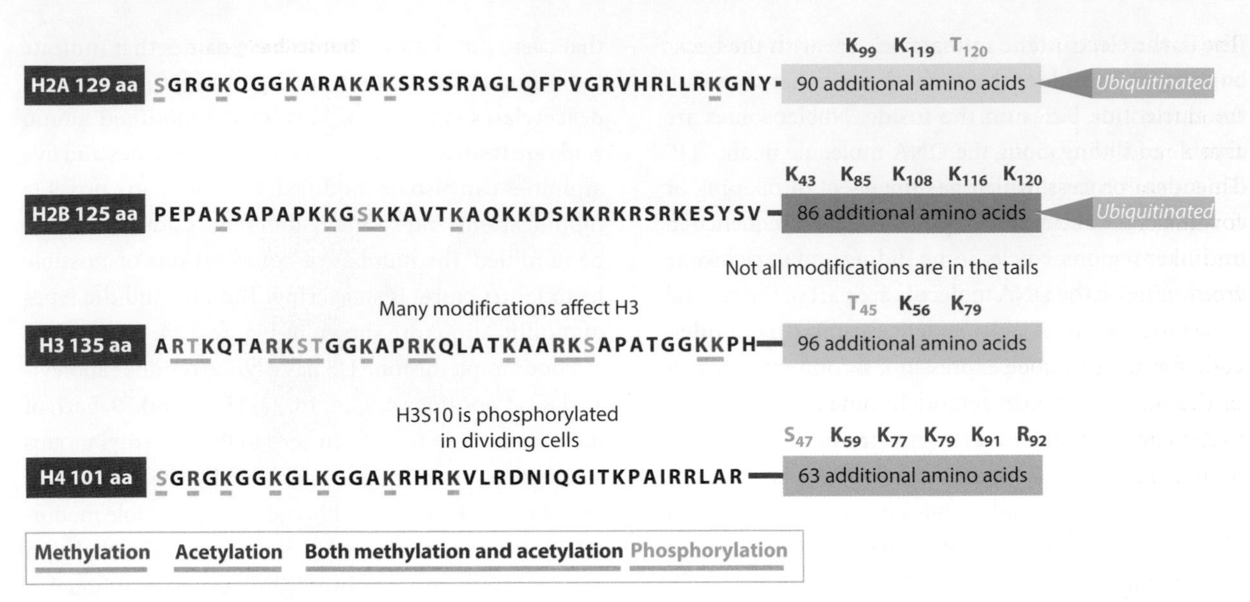

Figure 2.14 Summary of histone modifications. The locations of amino acids that are modified in specific histone tails are indicated. From Meneely et al., *Genetics: Genes, genomes and evolution* (Oxford University Press, 2017), Figure 12.22.

of lysines assumes that a given H3 molecule in a nucleosome has only a single methylation state, so that an H3 molecule is not methylated at both H3K4 and H3K27, for example. Although this seems likely, it is not known whether the assumption is a valid one; if more than one lysine on a single H3 molecule can be methylated, the number of possible structures greatly increases. This further assumes that the two H3 molecules in the core nucleosome particle have the same methylation state, such that both are H3K4me2, for example. Some evidence suggests that this is true, but if the two H3 molecules within a single nucleosome can have different methylation states, the number of combinations increases even more dramatically.

Each of the eight lysines in H3 can also be acetylated—although, again, acetylation has not been observed at all the possible sites. Thus, considering only the lysines found in H3, there are more than forty possible states or structures for each nucleosome core particle. Moreover, lysines are not the only residue in H3 that can be modified to affect nucleosome structure. H3 also has four arginines that can be methylated and two serines that can be phosphorylated, and this contributes to an even greater number of possible combinations. Furthermore, although H3 modifications are the most numerous, the other core histones are also subject to some of the same modifications, as summarized in Figure 2.14. Both H2A and H2B have two lysines known to be acetylated and one serine that can be phosphorylated; H4 has five lysines that can be either acetylated or methylated, one arginine that can be methylated, and one serine that can be phosphorylated. Considering all the known modifications (and assuming that only one modification is found per nucleosome), there are at least fifty possible different structures for each individual nucleosome. Our picture of a stable bead consisting of the same eight histone proteins with a 146 bp stretch of DNA wrapped around it does not capture what is occurring inside the cell.

Taken together, the combinations of histone modifications in particular nucleosomes serve as a mark in the chromatin that affects its structure and function. The chromatin mark is both local and statistical; that is, we can determine whether a particular histone modification is associated with the nucleosomes at which a particular nuclear function is occurring. This association between histone modifications and nuclear functions is known as the **histone code**.

Because the combinations of modifications are the signatures for a function, they can be fairly complex. For example, the regions around TSSs for active genes are enriched for H3K4me2, H3K4me3, and H3K9ac, and depleted for H3K36me1 and H3K23ac. Not one of these marks is as predictive for a TSS as the combination of them, and not every nucleosome core particle at the TSS will have these marks; but, at a genomic level, this combination of modifications is found commonly at a TSS.

The data from genome annotations are often presented in the form of heat maps, with red indicating the likelihood that a modification is present and blue indicating that a modification is absent, as shown in Figure 2.15. Thus a search for this combination of modifications, or for this code, could identify active TSSs in many cells and in many organisms, even if these are not known from other experiments.

We are currently far from deciphering the histone code for any genome in any cell type, except for a few of the many functions in the genome affected by the particular combination of histone modifications found in that region, as summarized in Figure 2.16.

Histone modifications are targets for other chromatin-associated proteins

The mechanisms by which a histone code affects the structure of chromatin and the functions of the genes embedded there are not completely known. Since lysine methylation and acetylation are predicted to alter the charge on the histone, it seemed reasonable to expect that the modifications affect chromatin functions by altering the electrostatic interactions with DNA. However, this is probably not the principal effect of the histone modifications. For one thing, the modifications affect residues at the N-terminus of the proteins, which do not interact directly with the DNA molecule. Thus changes in the charges at these locations might not have much of an effect on the electrostatic interactions with DNA. Instead, modified histones probably form the recognition targets for other chromatin-associated proteins that affect chromatin structure and function.

A few specific interactions between a histone modification and another protein can be used for illustration of a large and expanding set of examples; others are included in Sections 2.4 and 2.5. For example, chromatin-associated proteins with the TUDOR domain recognize methylated lysines and can bind differentially to mono-, di-, or tri-methylated lysines. (The Tudor domain is a specific region of fifty amino acids that is widespread and was named after the *Drosophila* gene in which it was first identified.) The mammalian protein JMJD2A has two TUDOR domains and interacts with H3K4me3 and H4K20me. The human protein 53BP1 with a single TUDOR domain recognizes H3K79me. Thus, while each TUDOR-domain protein is associated with chromatin on the basis of its interaction with methylated lysines

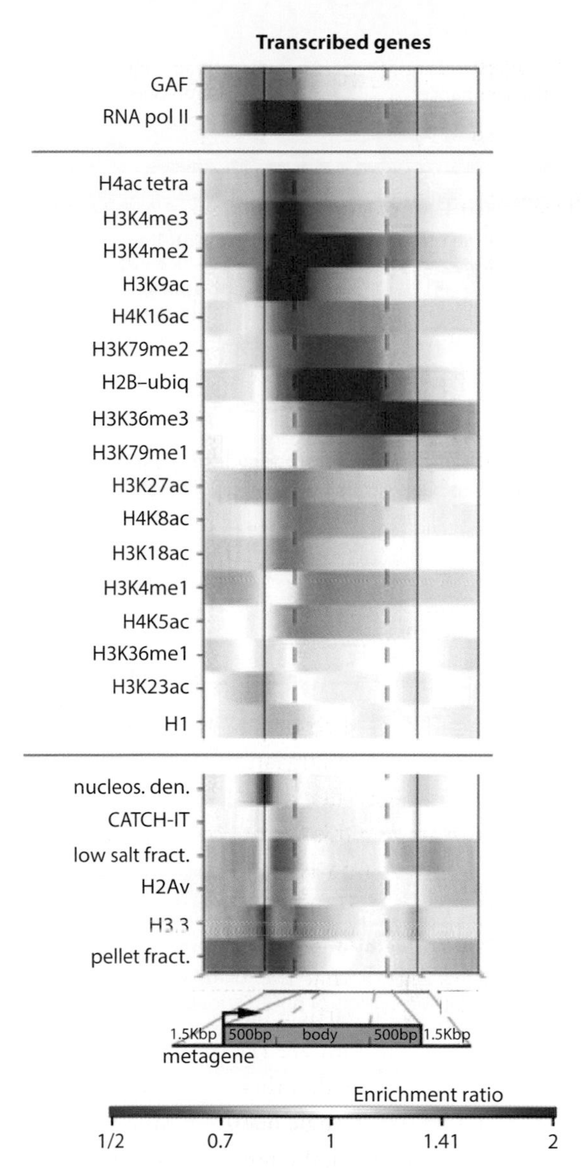

Figure 2.15. Histone modifications and gene structure in *Drosophila*. A generic gene, or "metagene," is shown at base of the diagram. The relative enrichment (or deficit) of different chromatin-associated proteins is shown in a heat map, as marked by the scale, red indicating modifications present and blue indicating modifications that are absent. Different modifications are indicated along the left side of the diagram. Modified from the *Drosophila* modENCODE site and used with permission.

in H3, these proteins do not recognize the same methylated lysines or interact with nucleosomes in precisely the same way.

But other modifications beside lysine methylation are also important epigenetic marks. The phosphorylation of serine at position 10 in H3 (H3S10P) has been recognized as an indicator of mitotic nuclei in eukaryotes for more than fifteen years; this phosphorylation

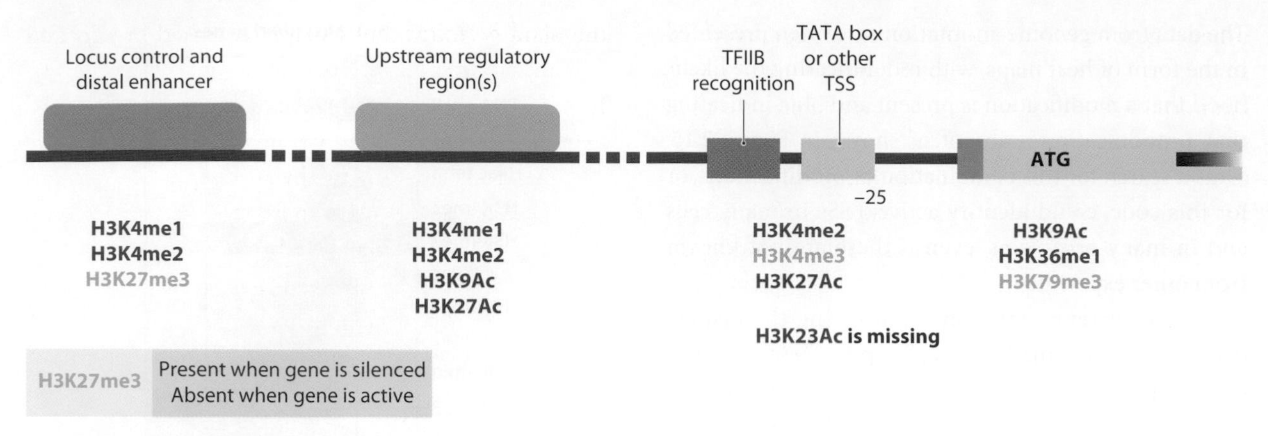

Figure 2.16 An example of a histone code for a typical gene from *Drosophila*. Different combinations of histone modifications are associated with actively transcribed genes. The modifications that are commonly found at each of these locations are indicated, although any individual gene may not have all of these. From Meneely et al., *Genetics: Genes, genomes and evolution* (Oxford University Press, 2017), Figure 12.24.

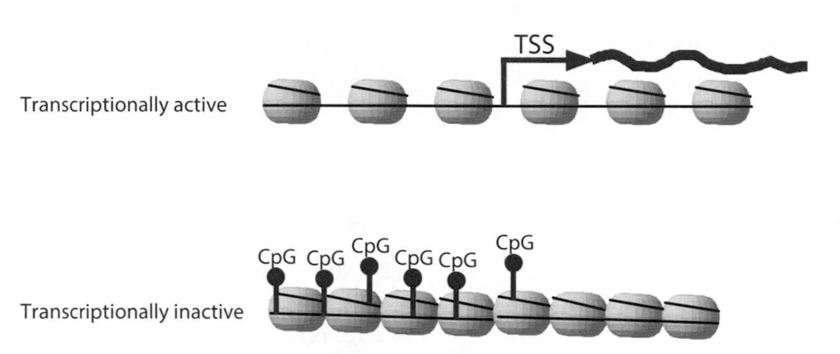

Figure 2.17 DNA methylation and chromatin state. In addition to the covalent modifications on histones, DNA can be methylated in many organisms. In vertebrates, this methylation occurs on clusters of CpG dinucleotides found in the core promoter region of transcriptionally inactive genes.

is carried out by the enzyme Aurora B kinase. Aurora B kinase is also associated with the microtubules in the spindle and is needed for the proper orientation of chromosomes on the mitotic spindle. In mammalian chromosomes, the phosphorylation of H3S10 begins near the centromere, as prophase I begins and spreads along the length of the chromosome; this phosphorylation sequence corresponds to the compaction of the chromosome that occurs during prophase I and is necessary for proper segregation during anaphase I. In the fission yeast *S. pombe*, a mutation in the H3S10 residue preventing its phosphorylation results in multiple mitotic defects and a failure in chromosome segregation. Thus the action of Aurora B kinase on H3S10 and the interaction between H3S10 and the spindle are needed for proper mitotic chromosome segregation.

DNA cytosine methylation is a regulatory mechanism in many species

Chromatin remodeling and histone modifications are closely related, albeit distinct processes.

Before discussing genes affecting chromatin structure in more depth, we should note that methylation also occurs in DNA, whereby cytosines can be methylated to form 5′-methylcytosine, as diagrammed in Figure 2.17. DNA methylation can also be an important aspect of chromatin structure and epigenetic regulation.

Literature Link. Rivera and Ren 2013, *Cell* 155: 39–55

Cytosine methylation probably evolved in the precursor to bacteria and eukaryotes, but has been evolutionarily lost in budding yeast, *C. elegans*, and certain other organisms. Cytosine methylation occurs only in specific sequence contexts: only CpG methylation is common among vertebrates, but in plants cytosines are methylated in the context of the sequences CpG, CpHpG, and CpHpH, where the H can be A, T, or C. Partly for historical reasons, the methylation site in vertebrates is usually referred to as CpG; the notation includes the phosphate between the cytosine and the guanine to indicate that these are dinucleotides rather than a base pair, a convention that we will continue here, although this is not such a common notation in plants.

Cytosine methylation is mediated by DNA methyltransferases, different enzymes having specific methylation context specificities. Both CpG and CpHpG methylation can be maintained during replication, because the symmetric nature of the sites means that both DNA strands are methylated at each site. CpHpH methylation only affects one DNA strand, however, and so de novo methylation is required after every round of DNA replication.

Not every CpG, CpHpG, or CpHpH cytosine is methylated, however. The regulation of which sites are methylated apparently depends on the actions of small RNAs, both in plants and in the mammalian genome. With every generation, DNA methylation patterns are erased in mammalian primordial germ cells but are reestablished in embryos. In plants, however, DNA methylation is not completely erased, and so there is true DNA methylation-dependent epigenetic inheritance. CpG and CpHpG methylation is maintained in the plant germline, but CpHpH is erased in part and re-established in the embryo. These developmental patterns of methylation are regulated in part by differential expression of DNA methyltransferases, as well as by enzymes that actively demethylate DNA.

DNA methylation plays an important role in several epigenetic phenomena. As a broad and general rule, DNA methylation is involved in long-term silencing, which is perhaps analogous to locking the chromatin into an off position. While cytosine methylation can occur at different places in the genome, it is particularly abundant at locations comprising repetitive DNA sequences—including centromeres, which retain their structure through every cell division. In plants and in the germline of mammals, cytosine methylation is also important for repressing transposable element transposition. In addition, DNA methylation of promoters inhibits gene expression, so the regulation of methylation levels can be an important mechanism for regulating gene expression. The inactive X chromosome, for instance, is much more heavily methylated on CpGs than is the active X chromosome. Among vertebrates, the presence of a CpG island—that is, of a cluster of CpG dinucleotides—is commonly found upstream of a gene and is one of the genomic indicators for the locations of genes.

DNA methylation is also important for establishing and maintaining genomic imprinting, which results in the expression of only the maternal or the paternal copy of specific genes. Case Study 2.1 describes an example of genomic imprinting in mammals.

2.4 Polycomb, Trithorax, and chromatin remodeling

Changes in chromatin structure play widespread and crucial roles in many biological processes. As will be often noted in this book, some of the genes and proteins involved in these processes, and thus some of the mechanisms for the processes themselves, were found through analysis of the phenotypes of certain mutants. Also as in many other examples of genetic analysis with mutants, the first clues came from *Drosophila* genetics before the macromolecular structures of genes and chromosomes were known. Here and in Section 2.5 we discuss two of these types of mutant analysis and some of what they have shown us.

The *Drosophila Hox* complex and the body plan

The homeotic gene or *Hox* gene complex is a cluster of genes that encoded transcription factors with a shared DNA-binding domain referred to as the homeodomain. This evolutionarily conserved cluster of genes is responsible for the correct organization of the animal body plan, as was originally recognized, because mutations in these genes produce homeotic transformations; that is, a normal body part (such as a foreleg) forms in a location where it normally does not form (such as on the head, replacing an antenna), as shown in Figure 2.18. The organization and the evolution of individual *Hox* genes and gene clusters are important and fascinating biological topics, but the key feature for our purposes is that each *Hox* gene is expressed in a particular set of cells, and thus provides the developmental identity to that group of cells. The homeotic transformations—such as the one known as *Antennapedia* (*Antp*), whose phenotype with the foreleg and the antennae was just described—arise because the genes are being expressed ectopically, in regions of the animal that do not normally express that particular *Hox* gene, or because the genes are not being expressed at all. The patterns of expression in *Drosophila* and in humans are summarized in Figure 2.19.

The *Hox* genes were found by the homeotic transformations that occurred in *Drosophila*, in which normal *Hox* gene expression more or less corresponds to particular body segments. Two of the most obvious morphological structures that form from particular body segments in insects are the wings, which arise from the second thoracic segment or T2 in a dipteran fly like *Drosophila*, and the legs, one pair of which arises from each of the three thoracic segments. Among the three pairs of legs, the foreleg in males (which arises from the first thoracic segment or T1) is distinctive because it has an easily seen set of bristles known as the sex comb; thus, if another segment is homeotically transformed into T1, additional forelegs can form, and the resulting fly (which usually does not survive) has extra sex combs or "polycombs."

Polycomb

The earliest mutant with an extra sex comb or Polycomb phenotype was found in the 1940s, and a number of mutants with similar phenotypes, both dominant and recessive, were recovered over the years and assigned to different genes on the basis of their map locations. During his landmark studies in the 1970s that defined the role of *Hox* genes in all animals and were recognized through a Nobel Prize, E. B. Lewis interpreted these extra sex combs or Polycomb mutant phenotypes in terms of their effects on the *Hox* genes, an effect that was subsequently verified once the *Hox* genes were cloned.

Literature Link. Lewis 1978, *Nature* 276: 565–70

His interpretation provided a unified explanation for a diverse group of mutant phenotypes. In Lewis's interpretation, the *Hox* genes are normally expressed in

A

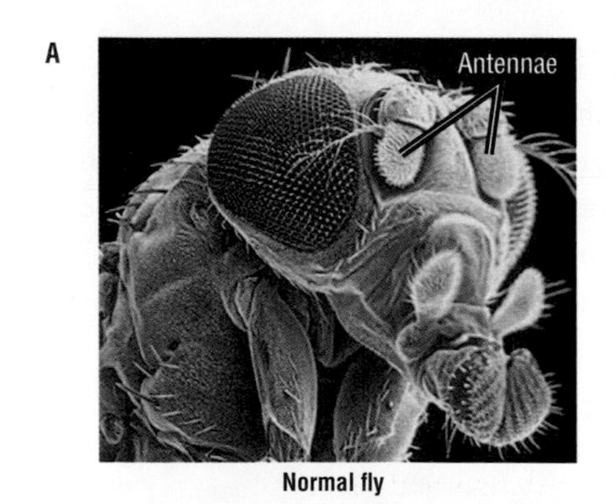

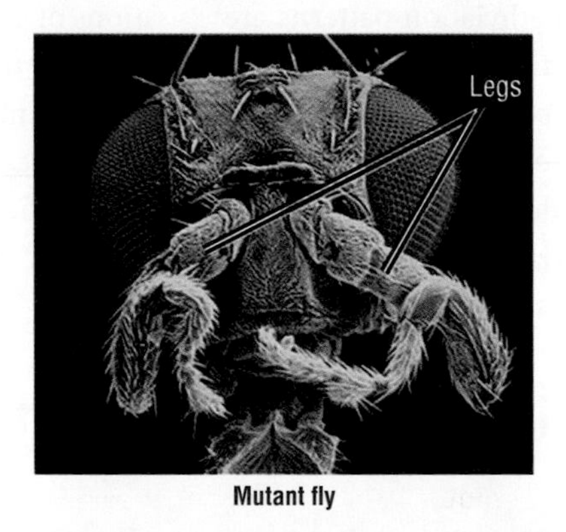

B

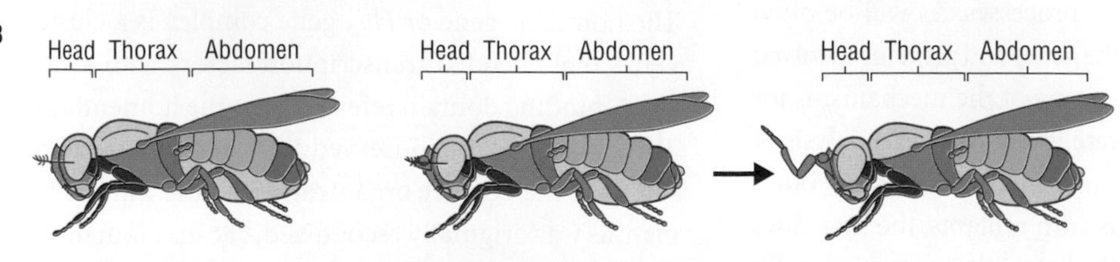

Figure 2.18 Homeotic mutations result in the replacement of one body part with another. Panel A. The heads of a wild type (on the left) and an Antennapedia mutant fly (on the right), showing the homeotic transformation from antenna to legs in the mutant fly. Panel B. The Antennapedia mutant phenotype arises from changes in the CRM of the *Antp* (*Antennapedia*) gene, which changes its pattern of expression but not its function. Modified from Meneely et al., *Genetics: Genes, genomes and evolution* (Oxford University Press, 2017), Figure 12.14.

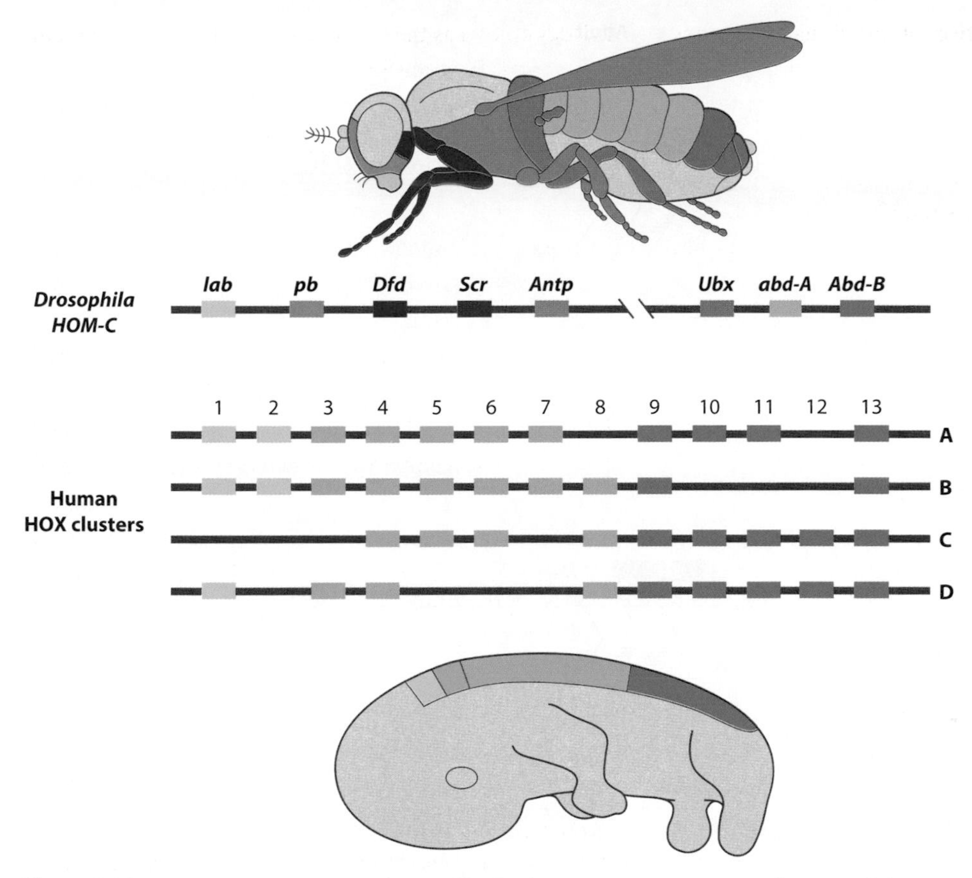

Figure 2.19 ***Hox*** **gene organization and animal body plans.** *Hox* genes are expressed in regional domains along the antero-posterior axis of animals and control the identity of the structures that form in these regions. Note that both *Drosophila melanogaster* and humans have similar *Hox* genes arranged on the chromosome and expressed on the body axis in similar patterns, as shown by the corresponding colors. From Meneely et al., *Genetics: Genes, genomes and evolution* (Oxford University Press, 2017), Figure 12.13.

particular body segments and repressed or silenced in other segments. The Polycomb mutants fail to repress or silence one or more *Hox* genes; in particular, a segment in which the *Hox* gene *Scr* (which is broadly responsible for T1 structures) is normally repressed is now expressing *Scr*, thus making forelegs and additional sex combs, as summarized in Figure 2.20. One mutant was named *Polycomb* (*Pc*), and the other genes with similar mutant phenotypes are referred to as the Polycomb group or PcG. The genes in the PcG are molecularly diverse so they are not a gene family but rather a set of genes that affect related functions. These are genes needed to repress or silence the expression of the *Hox* genes.

Subsequent studies in flies and other organisms have shown that the PcG are evolutionarily conserved among eukaryotes and involved in the repression of gene expression at hundreds, if not thousands of locations in the genome, in addition to the *Hox* gene complex. They carry out this repression through their effects on the chromatin structure, both through ATP-dependent chromatin remodeling and through histone modifications, or through a combination of these. PcG genes in humans are important for many different functions, including cancer biology, X inactivation, stem cell maintenance and regeneration, and so on.

Literature Link. Schuenttengruber et al. 2017, *Cell* 171: 34–56

Nearly all multicellular organisms have two Polycomb Repressor Complexes (PRC), although many organisms have additional complexes or variations in the component subunits of these two complexes, as summarized in Figure 2.21. For clarity, we will consider the role of PRC2 before discussing the role of PRC1. PRC2 has four core subunits, one of which is a histone methyltransferase

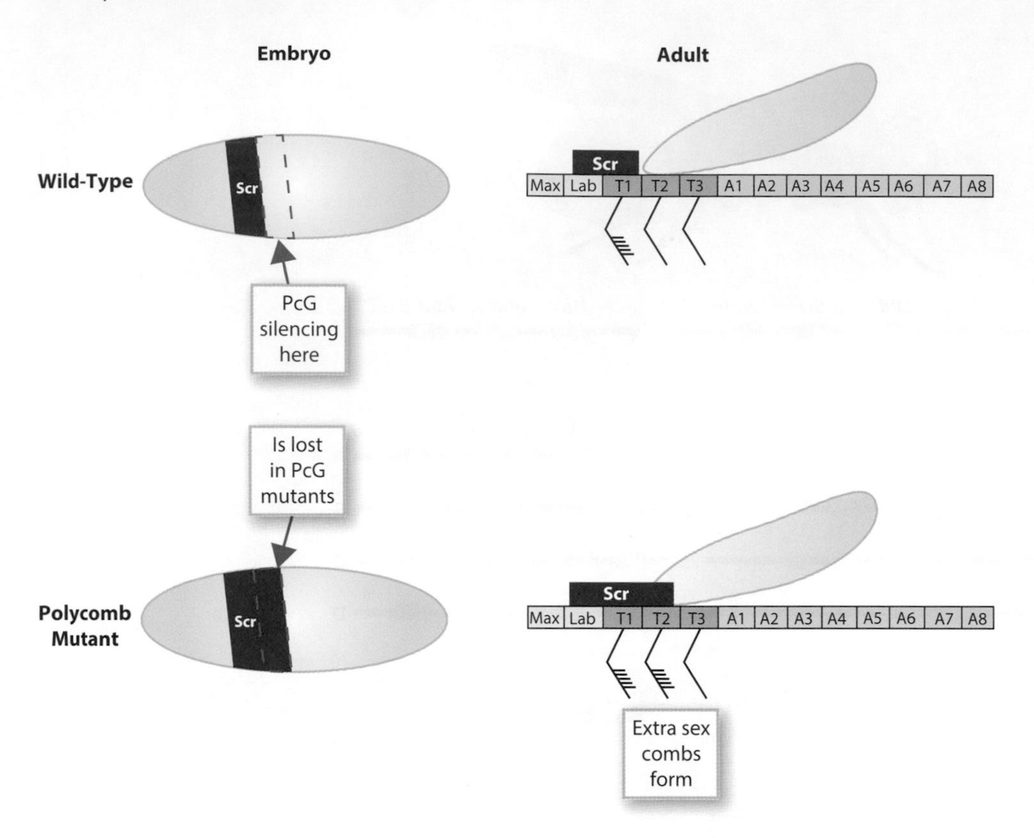

Figure 2.20 The role of Polycomb in gene silencing. The original Polycomb complex genes (PcG) were identified because of their failure to silence the *Hox* genes in particular segments. As diagrammed here, the *Hox* gene *Scr* is normally expressed in a relatively narrow band in the anterior third of the embryo; in wild-type flies, this expression pattern corresponds to the labial (lab) and first thoracic (T1) segments. T1 produces the foreleg, which in males has a sex comb. In PcG mutants, the expression of *Scr* extends more posteriorly into a region that normally has *Scr* silenced. In flies, this region includes the second thoracic segment (T2), which now forms a foreleg with a sex comb rather than a normal second leg. This is the Polycomb mutant phenotype, but the PcG silence many hundreds of genes in addition to the *Hox* genes.

that methylates H3K27; the number of other subunits in PRC2 varies among organisms but is usually about 3–6 proteins. It appears that PRC2 recognizes and binds to DNA at sites for repression, and methylates H3K27 at those sites.

Once H3K27 methylation is carried out by PRC2, H3K27me3 is a binding recognition target for PRC1, as shown in Figure 2.21. PRC1 in *Drosophila* is encoded by four genes, while the corresponding PRC1 in mammals has ten or more proteins. PRC1 binds to H3K27me3, and is responsible for ubiquitination of H2AK119, since one subunit of PRC1 is a ubiquitin ligase; thus it interacts with one histone modification and carries out a different histone modification. The gene altered in the original *Polycomb* mutant encodes the subunit that binds to the H3K27me3. It has been suggested that H2A is ubiquitinated on adjacent nucleosomes by PRC1, and that this ubiquitinated region occludes RNA polymerase and represses the nearby genes, as shown on the right in Figure 2.21.

The postulate that PRC2 initiates repression while PRC1 maintains repression is consistent with the subunit structures of the two complexes. The conserved subunits of PRC1 do not include any proteins that are likely to bind DNA, although some of the additional proteins found in the PRC1 of other organisms might be DNA binding proteins. On the other hand, one of the four core proteins in PRC2 is a DNA-binding protein that would allow PRC2 to bind to DNA prior to and independently of PRC1. In *Drosophila* and *Arabidopsis*, PRC2 appears to bind to specific sequences, known as Polycomb response elements (PREs); in vertebrates, the binding appears to involve CpG islands rather than specific sequences. Because the subunit structures of PRC1

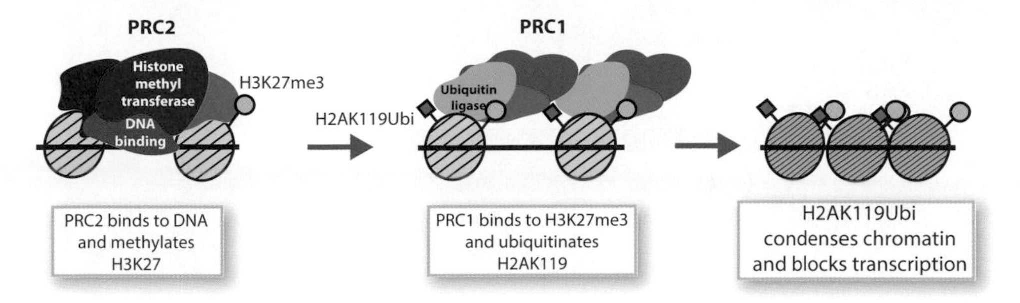

Figure 2.21 The mechanism of PcG silencing. Organisms have at least two Polycomb Repressor Complexes (PRCs), known as PRC1 and PRC2. While each complex is shown with four protein subunits, there are more than four subunits in some organisms, and some organisms have more than two PRCs. PRC2 binds to DNA, either to specific sequences or to CpG islands, depending on the species, and one subunit is a DNA binding protein. Another subunit of PRC2 is a histone methyl transferase, which methylates H3K27 to form H3K27me3, represented here by the small line with the yellow ball at the end. H327me3 is the target for the binding of PRC1, so the activity of PRC2 allows PRC1 to bind. One subunit of PRC1 is a ubiquitin ligase that targets H2AK119, shown here as a small line with a blue box at the end. It is thought that ubiquitination of H2AK119 condenses chromatin and blocks transcription by preventing RNA polymerase from binding.

and PRC2 are somewhat variable among organisms, other biochemical functions have also been associated with these complexes and are probably important in some organisms—or in many. These additional activities include both methyl transferases and demethylases for other H3 residues, DNA and RNA binding, many protein interactions domains, and nuclear scaffold association. The model for repression summarized in Figure 2.21 is almost certainly an oversimplification, but the role of PcG complexes in both local and generalized repression of transcription seems clear.

Trithorax

Around the time when Lewis was interpreting the PcG mutants in terms of repressing *Hox* gene activity, mutants with a different type of phenotype were also found and interpreted in terms of their effects on the *Hox* genes. These mutants appeared to be necessary for the expression of certain *Hox* genes in different segments and were called the Trithorax group of genes, TrG. The TrG genes and the PcG genes act antagonistically, at least at the level of mutant phenotypes, and in some cases at the biochemical level as well. The TrG genes activate the *Hox* genes or maintain the *Hox* genes in an active state throughout mitotic divisions, while the PcG genes repress *Hox* gene expression, as summarized in Figure 2.22.

As might be expected of a group of genes that are needed to activate transcription and to maintain an active state, the TrG genes are very heterogeneous in their functions. They are also very numerous, although many of the TrG genes are evolutionarily conserved among eukaryotes. *Drosophila* has at least twenty-four different genes that are considered to encode subunits of the TrG complexes; one of these genes, *mod(mdg4)*, encodes more than twenty different proteins just by itself, by alternative splicing, so the number of Trithorax-related proteins is quite large. Mammals have approximately forty different TrG genes, some of which are alternatively spliced, so, again, there is a large number of TrG activating proteins.

Broadly speaking, the TrG genes can be divided into two or perhaps three types of complexes, although there are many variations of this generalization, which leads to our imprecise wording about the number of types of complexes.

Literature Link. Schuenttengruber et al. 2017, *Cell* 171: 34–56

The genes in the COMPASS and COMPASS-like complexes include ones that encode histone methyltransferases, which methylate H3K4; the association of H3K4me3 with active genes depicted in Figure 2.16 is the outcome of these histone methyltransferases. The original *Trithorax* mutation in *Drosophila* altered one of the genes encoding a histone methyltransferase, and thus H3K4 methylation. Other proteins often found in COMPASS and COMPASS-like complexes include some that interact with modified histones, others that bind to DNA, and still others that methylate or demethylate histones; one histone demethylase has H3K27 as

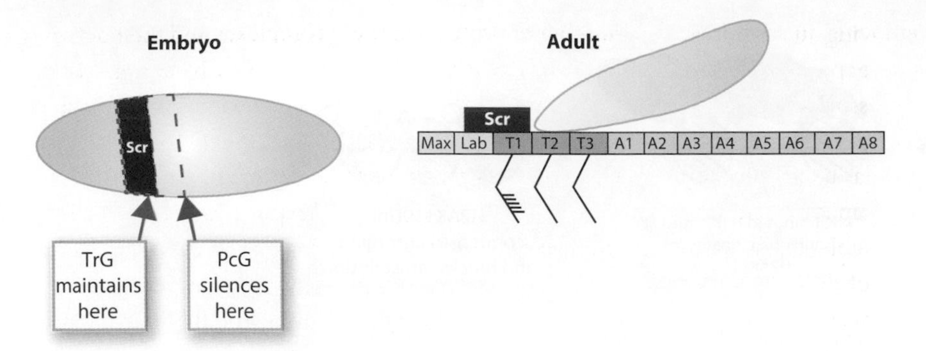

Figure 2.22 PcG and TrG work in opposition to each other. The Trithorax complex genes (TrG) are involved in activating gene expression particular locations and maintaining the active chromatin state. The effects are summarized here for the *Hox* gene *Scr*, as are the effects shown for PcG in Figure 2.20; but hundreds of different genes are affected by TrG. The appropriate location for *Scr* expression in the embryo and adult arises from the combined action of the TrG and PcG complexes.

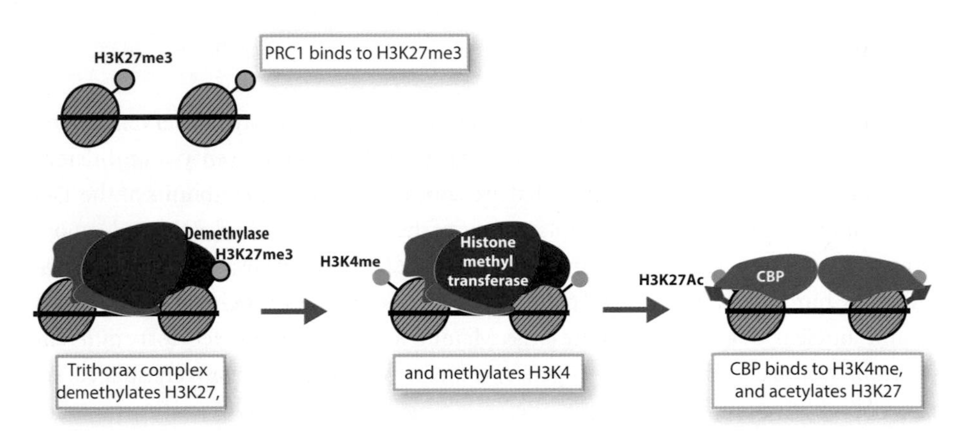

Figure 2.23 The mechanism of TrG activation. The Trithorax complex genes (TrG) activate the expression of genes and maintain them in an active chromatin conformation. The antagonistic activity of the PcG (shown in Figure 2.21) binds to H3K27me3 to silence genes. TrG genes affect the H3K27me target so that PRC cannot bind and silence the genes. Included in the TrG is a histone demethylase that removes the methyl group on H3K27. A histone methyl transferase then methylates H3K4. The CBP protein, which has histone acetyl transferase activity, binds to H3K4me and acetylates H3K27. Since H3K27 is acetylated, the PcG proteins cannot bind to silence expression, so the genes remain active. Many TrG proteins in several different complexes are known, but relatively few have been characterized in detail, so this is an overview.

its target, and so it removes the methyl groups attached by the PRC2 and recognized by PRC1, as summarized in Figure 2.23.

H3K4me3, which arises from the activities of the COMPASS and COMPASS-like TrG complexes, is itself the binding target for CREB-binding protein or CBP, as suggested in Figure 2.23. CBP is a histone acetyltransferase that acetylates H3K27and acts as a coactivator, interacting physically with many different transcription factor proteins, at least forty in humans. One model is that one TrG complex methylates H3K4 and demethylates H3K27, undoing the activity of PRC2. Another TrG complex including CBP then interacts with H3K4me3 and acetylates H3K27, thereby preventing PRC2 from methylating it again. Thus the gene is activated and kept active by these TrG complexes that act on the recognition site for the PcG. There are numerous other proteins in these complexes whose activities are not being considered in these diagrammatic models.

A third type of TrG complex, the SWI/SNF complexes of TrG genes, include some subunits that interact with histones, some subunits that bind to DNA, but, most importantly, a subunit that carries out ATP-dependent chromatin remodeling. Thus this complex functions, primarily or in part, by sliding or removing nucleosomes, opening and closing chromatin for other proteins to bind to DNA or to interact with chromatin-associated proteins.

The PcG and TrG genes were originally identified because of their effects on the chromatin surrounding the *Hox* genes, which made their complicated mutant phenotypes somewhat more interpretable. These complexes affect hundreds or thousands of different genes in every eukaryote in which this has been studied. Our description is accurate but somewhat simplified, and the number of proteins in these complexes and their activities are very large. The principles can be summarized as follows. The TrG genes are needed to activate the expression of genes and to keep genes actively expressed, while the PcG genes are needed to silence or repress the expression of genes and to keep genes repressed. The complexes work in opposition to each other but exert their effects by making or removing covalent modifications of histones and by remodeling the nucleosome positions on the surrounding DNA.

2.5 Positive effect variegation and heterochromatin

X-chromosome inactivation in mammals is probably the most familiar example of an epigenetic modification, while PcG and TrG mediate some of the most widespread epigenetic modifications in a genome. But neither of these can lay claim to being the first example of epigenetic modification to be described. That honor goes to a phenomenon known as **position effect variegation** (PEV); like so many examples of genetic analysis, it begins with white-eyed flies.

Heterochromatin and euchromatin

Let's begin by describing the cytological properties of a eukaryotic genome, such as the one for *Drosophila*. Chromosomes stain readily with DNA dyes; the name "chromosome" is derived from the DNA-staining properties. Cytologists using DNA dyes nearly a century ago to stain nuclei recognized that different regions of a chromosome stain differently. The intensely staining condensed regions were termed **heterochromatin**, while the less intensely stained regions were termed **euchromatin**. Most protein-coding genes lie in euchromatin.

Some regions of heterochromatin, such as centromeres and telomeres, are heterochromatic at all times (as seen in Figure 2.24); these regions are known as **constitutive heterochromatin**. The Y chromosome in most animals and chromosome 4 in *D. melanogaster* are other examples of constitutive heterochromatin. Other regions of chromosomes are heterochromatic only at particular times or in particular cells; these regions are known as **facultative heterochromatin**. About a third of the *D. melanogaster* genome is heterochromatic in at least some cell types, or at least some of the time. Relatively few protein-coding genes lie in heterochromatin (although some do), and heterochromatin is enriched with repetitive DNA sequences. Heterochromatin is refractory to transcription and recombination and includes the last regions of the genome to replicate, presumably because of its condensed state. As will be discussed below, the distinction between euchromatin and heterochromatin, which was based on DNA staining many years ago, turns out to be an oversimplification of the chromatin structure, but it is still as useful one.

Literature Link. Allshire and Madhani 2018, *Nature Reviews Molecular Cell Biology* 19: 229–43

white mottled-4

In the 1920s and 1930s, X-rays were found to be an effective means to generate new mutant alleles of genes; H. J. Muller, who pioneered the use of X-rays as a mutagenic agent in *Drosophila*, was awarded the Nobel Prize for this work in 1946. Many of the mutations produced by X-rays are chromosomal rearrangements, small or large, in which a piece of the chromosome has been deleted, duplicated, inverted, or translocated to a different chromosome. The *w* gene was a convenient target for studying the effects of radiation, since the phenotype is easy to recognize even in individual cells of the eye.

A mutation known as *white mottled-4* (w^{m4}), induced by Muller, inverted the region of the X chromosome such that the w^{+} gene, which normally lies in a euchromatic region, now lies adjacent to or within centromeric heterochromatin, as shown in Figure 2.25. The structure

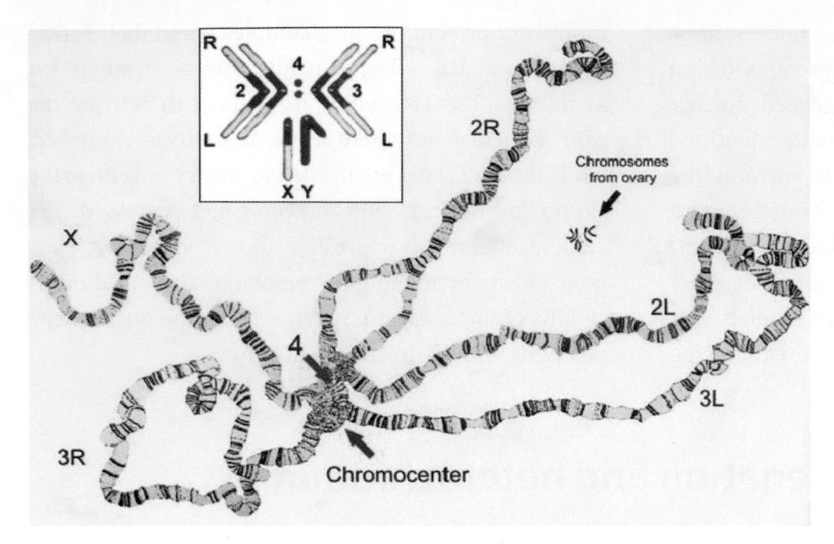

Drawing of polytene chromosomes modified from TS Painter, 1934, J. Hered 25: 465-476.

Figure 2.24 **Heterochromatin in *Drosophila*.** This classical drawing of *Drosophila* polytene chromosomes by Painter has served as the reference for all subsequent cytological studies. Heterochromatin is shown as the dark areas in the inset drawing at the top. Much of the chromocenter is also heterochromatic, since the centromeres are found there in preparation for polytene chromosomes. This image is in the public domain.

of the w^+ gene itself was not altered. Heterochromatin is associated with silenced genes, and the juxtaposition of the *w* gene next to heterochromatin results in the silencing of the normally active w^+ allele.

However, silencing of w^+ does not occur in every cell, so the eyes have a spotted or variegated pattern of red (with an active w^+ gene) and white (with an inactive w^+ gene) facets, as shown in Figure 2.25. Variegation of w^+ expression was postulated to arise because proximity to heterochromatin led to its being inactivated as heterochromatin "spread" over the gene. Once the expression of the gene was inactivated, it continued to be inactivated in subsequent mitotic divisions so PEV is a heritable epigenetic effect. Subsequent studies have generally supported some model of heterochromatin spreading to explain PEV, and have shown that the effect can spread over several contiguous genes, as expected for other epigenetic effects; spreading may not occur linearly but, in general, the genes moved closest to heterochromatin are the most commonly silenced. The implication is that some boundary or barrier between euchromatin and heterochromatin has been altered or compromised, although the nature of this postulated barrier is not known.

Literature Link. Elgin and Reuter 2013, *Cold Spring Harbor Perspectives in Biology* 5: a017780

Suppressors and enhancers of variegation

The easily visible phenotype of w^{m4} made it feasible to find dominant mutations in other genes that affected variegation, and hence heterochromatin formation. These were called suppressors or enhancers of variegation, abbreviated *Su(var)* and *E(var)*; suppressor and enhancer mutations were briefly introduced in Chapter 1 and will be discussed in more detail in Chapter 10. The gene names here can be a little confusing, so it is important to be clear that what is being suppressed or enhanced is the *variegation* of w^+ expression. As shown in Figure 2.26, a suppressor of variegation has more red facets in the eye, and so less heterochromatin formation; thus, in general, the wild-type functions of *Su(var)* genes are to *produce* heterochromatin. An enhancer of variegation has more white facets in the eye and more heterochromatin formation, so the wild-type functions of *E(var)* genes are to *reduce* the amount or the spreading of heterochromatin. Hundreds of different *Su(var)* and *E(var)* mutations have been found in various screens, defining at least 150 different loci, of which about thirty have been studied in some molecular detail. (Since the mutations are dominant and often lethal when homozygous, the definition of separate loci came from mapping experiments rather than from complementation tests.) Unlike most gene names in *Drosophila* and other

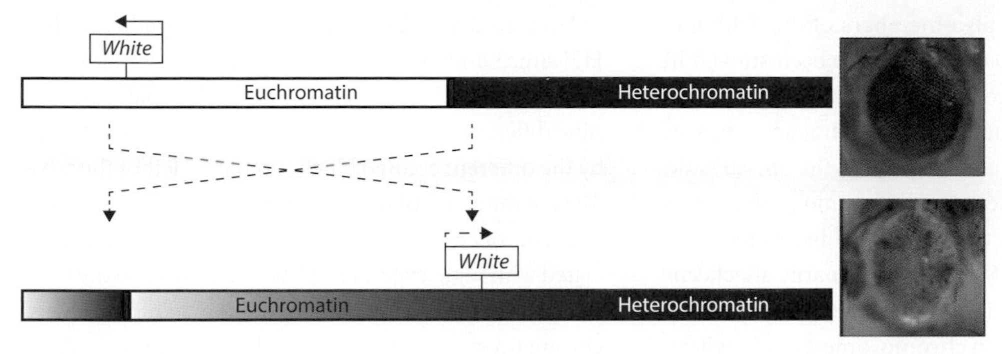

Figure 2.25 ***w^{m4}* and position effect variegation.** As summarized in this drawing and shown in the eye phenotypes at the right, the w^{m4} mutation is an inversion that moves the *w+* locus next to heterochromatin, which affects its expression. In some eye cells, the *w+* gene remains actively transcribed and the cell is red. In other cells, the *w+* is silenced by its proximity to heterochromatin and the cell is white. This image is from Elgin and Reuter, Position-effect variegation, heterochromatin formation, and gene silencing in *Drosophila*, *Cold Spring Harbor Perspectives in Biology*, 2013, 5: a017780, and used by permission.

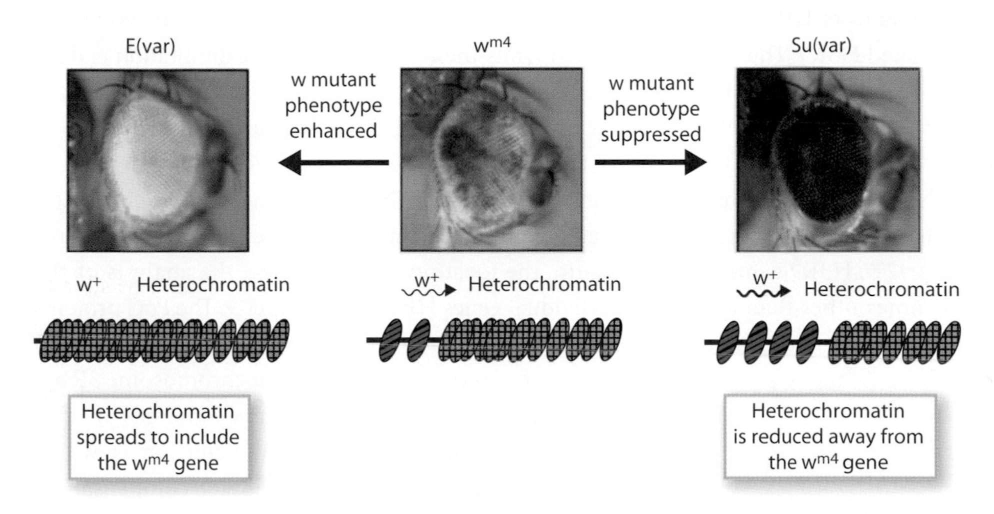

Figure 2.26 Enhancers and suppressors of variegation. The w^{m4} variegated phenotype is shown in the center image. These flies were treated with mutagens, and *E(var)* and *Su(var)* mutants in which the w^{m4} phenotype was enhanced (that is, more white cells) or suppressed (that is, fewer white cells and more red cells) were found. These have been interpreted as mutants that affect the amount and spreading of heterochromatin, as summarized in the drawings at the bottom. The images of eyes are from Elgin and Reuter, Position-effect variegation, heterochromatin formation, and gene silencing in *Drosophila*, *Cold Spring Harbor Perspectives in Biology*, 2013, 5: a017780, and used by permission.

organisms, the gene names are based on the order of isolation and map positions such that the name *Su(var)2-5* refers to the fifth locus on chromosome 2 to be found.

The thirty best-studied loci encode proteins with many different effects on chromatin; most of them are found in the genomes of other eukaryotes, which indicates that the results with heterochromatin formation in *Drosophila* are likely to be generally applicable. *Su(var)2-5*, for example, encodes a protein known as HP1a, a necessary structural component of centromeric heterochromatin; HP1a interacts with many other proteins and has two protein domains, referred to as chromo domains, found in many other chromosomal proteins. About ten of the loci, most of them *Su(var)* genes, have dose-sensitive effects on variegation and encode proteins that seem to be structural components of heterochromatin, including HP1a. Other functions encoded by *Su(var)* genes include histone methyltransferases and demethylases, histone acetylases and deacetylases, DNA binding, phosphatases, and others. Almost none of the *Su(var)* genes is part of the *Polycomb* group, although both types of proteins are involved in gene silencing. In contrast, many of the *E(var)* genes are also members of the Trithorax group; relatively few *E(var)* genes have been studied in molecular detail, but those few that have been studied include transcription factors and general activators of transcription. In fact, the majority of the *Su(var)* and *E(var)* genes have not been studied in molecular detail, so many additional functions could be discovered.

Both *Su(var)2-5* and *Su(var)3-9* primarily affect centromeric heterochromatin, with lesser effects on heterochromatin at telomeres, on chromosome 4, or elsewhere. The two genes work together for pericentric heterochromatin formation. *Su(var)3-9* encodes a histone methyltransferase, one of several in the *Drosophila* genome that are responsible for the di- and tri-methylation of H3K9. H3K9me2 and H3K9me3 are targets for binding by HP1a, and the chromo domains of HP1a are the interaction domains with methylated H3K9. Thus, once the product of *Su(var)3-9* di- or tri-methylates H3K9, the product of *Su(var)2-5* HP1a binds to the methylated histone to form heterochromatin. In the absence of the functions of either of these genes, centromeric heterochromatin does not form, so variegation is suppressed. In the absence of *Su(var)2-5*, H3K9 is methylated in regions of the chromosomes other than at the centromere, so the role of HP1a may be to confine the activity of *Su(var)3-9* to the pericentric regions by some unknown mechanism.

Types of heterochromatin and chromatin states

The recognition that *Su(var) 2-5* and *Su(var)3-9* affect pericentric heterochromatin, but not the formation of heterochromatin in other regions of the genome, means that there must be different types of heterochromatin. Since most of the *Su(var)* and *E(var)* mutants were isolated using w^{m4}, mutants affecting heterochromatin formation other than at centromeres were not readily recovered, although some have been found in other mutant screens. Much more information about types of heterochromatin has come from genome annotation experiments that determine the locations of various histone modifications and chromosomal proteins.

Relevant data are summarized in Figure 2.27. Each of the columns in the table in Figure 2.27 Part A represents the relative presence of a particular histone modification or chromosomal protein, red indicating a high concentration and blue indicating an absence. By computationally clustering modifications that are found together, such as H3K4me3 and H3K9ac, or not found together, such as H3K4me3 and H3K27me3, into the same row, different types of heterochromatin and states of chromatin emerge; nine different states of chromatin are found, as indicated by the different colored blocks on the far left of the table. The combination of modifications just described is characteristic of chromatin state 1 (in red), which is also associated with high levels of RNA polymerase II, so these are actively transcribed genes. Note the blue block labeled chromatin state 7, characterized by high concentrations of HP1a, *Su(var)3-9*, H3K9me2, and H3Kme3; since *Su(var) 3-9* carries out di- and tri-methylation of H3K9, and HP1a binds to H3K9me2 and H3K9me3, it is not surprising that these proteins and modifications cluster together as a group. In fact they are the only members of that group, and other proteins and modifications are absent. Thus the large-scale genome annotation experiments are finding the same results that are revealed by the analysis of individual genes and mutants.

Part B of Figure 2.27 shows the distribution of these various states in the *Drosophila* genome. Blue chromatin state 7 is found almost entirely at pericentric heterochromatin, the location expected from the analysis of the individual genes HP1 and *Su(var)3-9*. The PcG proteins and modifications cluster together at black chromatin state 6; the largest black regions on chromosome 3R are the *Hox* gene cluster. The X chromosome has, overall, different chromatin states from the autosomes, which may be related to the process of dosage compensation. These chromatin states were not distinguished cytologically, but genome annotation is confirming and greatly extending the results from working with individual genes.

While these experiments are based on *Drosophila*, all these marks and chromatin states are likely to be found in other eukaryotes. Not all organisms have easily recognized heterochromatin, but all of these histone modifications are common, and most or all of the heterochromatin-associated proteins are also found in other eukaryotes. (Apparently HP1a itself has been lost in some evolutionarily lineages and its functions have been assumed by other proteins.) Heterochromatin consists of more than histone modifications and specific proteins; there are both DNA sequence elements and lncRNAs that are also preferentially associated with heterochromatin, but the annotation projects are providing a detailed picture of a long-standing cytological question.

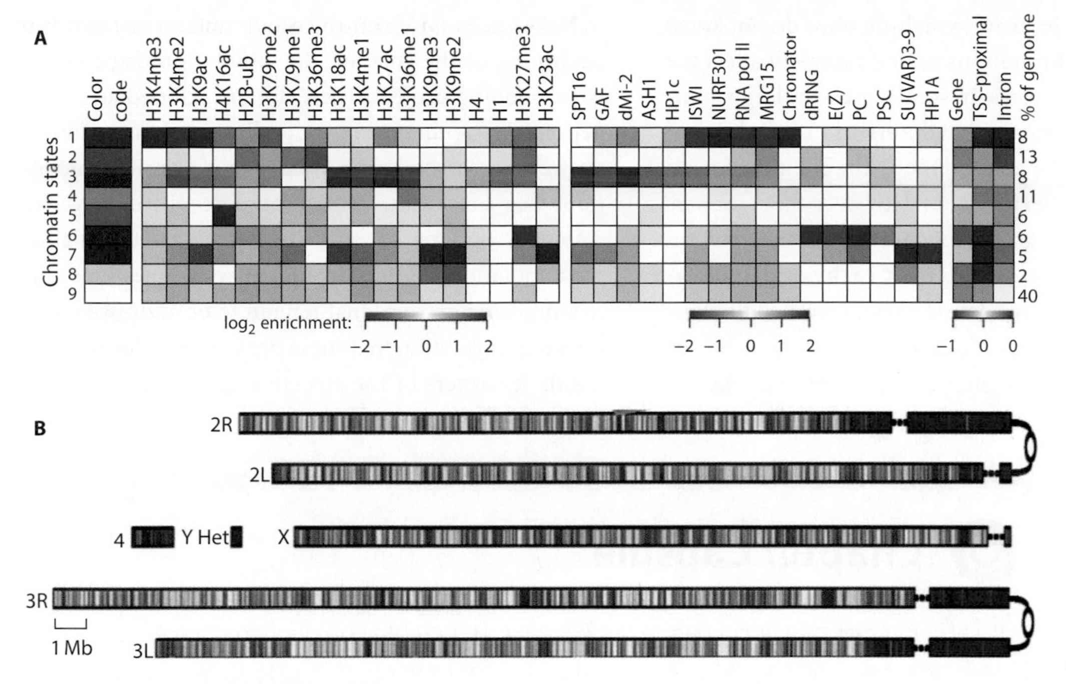

Figure 2.27 Chromatin states in *Drosophila*. Each column in the table in Panel A summarizes the presence of a particular chromosomal protein, including modified histones, red marking high levels of the protein and blue marking its absence. This information was computationally grouped, and seven different combinations of chromosomal proteins emerged; these seven chromatin states are shown as the blocks of colors on the far left of the table. Note, for example, that chromatin state 1 has high levels of RNA pol II, so these are actively transcribed regions. The active chromatin has high levels of H3Kme and H3K27ac, but an absence of H3K27me3, as expected by the activities of the Trithorax complex genes. Chromatin state 7 is centromeric heterochromatin. Panel B summarizes the locations of each chromatin state on *Drosophila* chromosomes. While these data are based on *Drosophila*, similar results are seen for other organisms. This image is from Elgin and Reuter, Position-effect variegation, heterochromatin formation, and gene silencing in *Drosophila*, *Cold Spring Harbor Perspectives in Biology*, 2013, 5: a017780, and used by permission.

2.6 Summary and limitations of genome annotation

There is little doubt that large-scale genome annotation projects are providing more details about the functions in the genome than any single gene approach could. In this regard, these projects have the capacity to change our fundamental understanding of many aspects of molecular biology. The ability to see functions such as replication origins and transcription factor binding sites as they occur across the genome provides a dynamic and integrative perspective that we have not had previously. Results from these genome annotation projects could rewrite molecular biology at every level of knowledge and experimentation.

Nonetheless, genome annotation projects have some limitations. The first of these is that all of the assays other than DNA sequencing itself are typically, snapshots at fixed times, and thus have a limited ability to describe changes as they occur in biological time. Biological time is the key concept here—some important biological changes occur in milliseconds, others in minutes or hours, and still others over days. Experimental techniques cannot yet capture this level of dynamism. In many cases we may not even know what biological time scale is the most important one to analyse.

Other limitations come from the sheer number of methods and data. Every assay method has its own weaknesses and sources of error, and every experimental trial has its own shortcomings. We often realize these limitations for experiments that are carried out by individual investigators, so we can structure control experiments for them. This is much more difficult to do for large-scale projects, for which we often do not know how the inherent limitations of one assay will affect the results from other assays. In some cases, the large number of data usually means that the errors in an individual experiment are compensated for by the additional trials of the same kind of experiment. However, the error bars arising from the integration of data from many different methods are much less clear. Thus, each scientist drawing on the data must validate the results related to his or her gene of interest. In addition, because there can be so many data, an investigator can feel that he or she can appreciate very little of it.

Nevertheless, we return to our statement from the beginning of the chapter: the annotation of genomes is an ongoing project. These are encyclopedias of the genome. The best encyclopedias provide enough answers to satisfy some queries, but also raise questions that cause us to search more deeply, using other resources and approaches. This is the unstated goal of the annotation projects: to answer some questions and to stimulate new ones that remain to be addressed. It is not an exaggeration that these projects are illuminating the dark corners of the structure and function of the genome.

Chapter Capsule

The ability to obtain and analyse the DNA sequence of an organism's genome has fundamentally altered our understanding of many biological processes. But the DNA sequence of the genome cannot always be interpreted to yield the most significant functional components, and not all the important functions are encoded into the DNA sequence directly. Genome annotation, that is, defining as many of the functional elements of a genome, is always an ongoing process. Many of these functions are related to gene expression. In addition to determining all the transcripts encoded in the genome and the splice variants for those transcripts, genome annotation identifies the binding locations for the transcription factor proteins that regulate expression. Genome annotation also includes an analysis of chromatin states and their effects on gene expression through epigenetic modifications.

Additional Reading

Craig, N. L. et al. 2014. *Molecular biology: Principles of genome function*, 2nd edn. Oxford University Press: Oxford.

The ENCODE Project Consortium. 2011. A user's guide to the encyclopedia of DNA elements (ENCODE). *PLoS Biology* 9:e1001046.

Goldman, A. D., and L. F. Landweber. 2016. What is a genome? *PLoS Genetics* 12: e1006181.

Case Study 2.1

Genomic imprinting

We learn in introductory genetics courses that either parent can contribute either allele without affecting the phenotype of the offspring: *Aa* and *aA* genotypes have the same phenotype. While this is true for most genes, there are exceptions. This case study describes one of the best known examples in which *Aa* and *aA* are not equivalent phenotypes, and the phenotype of the heterozygote depends on which parent contributes each chromosome. Examples in which the phenotypes are different depending on which parent contributes which allele are known as **parent-of-origin effects**, **genomic imprinting**, or simply **imprinting**.

The first example of genomic imprinting to be described was for sex determination in mealybugs. All mealybug embryos are diploid, with both sets of chromosomes initially euchromatic. However, by about the seventh mitotic division in the embryos, differences emerge between the parental genomes. In about half of the embryos, all the chromosomes remain euchromatic and transcriptionally active; these embryos develop as females. In the other half of the embryos, the chromosomes that came from the father—the paternal chromosomes—become heterochromatic and are transcriptionally inactivated. These embryos develop as males. The underlying mechanism in this phenomenon is not well understood, but it is a clear case of epigenetics: regions of the chromosome (in this case, an entire set of chromosomes) encoding genes with diverse functions are affected; the chromatin state is stably transmitted and cytologically visible; and the underlying DNA sequences have not changed. When the male undergoes spermatogenesis, the imprint it inherited from its father is erased or reset, so that half of the embryos in the next generation will have an active and euchromatic paternal genome and will become females. The difference between this example and others cases of epigenetics discussed in the chapter is that there is a relationship between silencing and the parent of origin, which is not typical in other epigenetic cases.

Although we did not discuss it in the chapter, a parent-of-origin effect is also found in X-chromosome inactivation. As described in Section 2.1, X-inactivation in placental mammals is random in the cells that give rise to the embryo itself; this random inactivation is why calico cats, even from the same litter, have different patterns of coat pigmentation. But only the cells of the embryo proper have random inactivation of one of the X chromosomes. In the cells that give rise to the extraembryonic structures, the paternal X-chromosome is inactivated and the maternal X-chromosome is active. In fact, in marsupials, the maternal X-chromosome is always the active chromosome, while the paternal X is inactive in both the embryo and the extraembryonic structures. Thus, like the mealybugs, our X chromosomes have retained a mark that indicates their parent of origin, and the maternal and paternal chromosomes are not equivalent.

Thinking through the Experiment

1. How is genomic imprinting an example of epigenetics?
2. Genomic imprinting tells us that an allele coming from the sperm may not be the same as an allele coming from the ovum. What are some of the structural or cytological features of chromosomes that are different between during spermatogenesis and oogenesis and that would be candidates for being involved in genomic imprinting?

3. Of the molecular mechanisms that contribute to other epigenetic examples, which might be best candidates for thinking about the molecular mechanisms of genomic imprinting? Included among these molecular mechanisms are histone modifications, chromatin remodeling, some specific chromosomal proteins, DNA methylation, and possibly long ncRNAs.

Genomic imprinting affects regions of the chromosomes

Mealybug sex determination and X-chromosome inactivation are not isolated examples of genomic imprinting. It has been estimated that 1–2 percent of the protein-coding genes in the human and mouse genomes exhibit parent-of-origin expression differences. More than 80 percent of these genes are clustered into one of sixteen chromosomal regions where two or more genes with distinct functions are imprinted, typically (but not always) with genes on the same homologue that are silenced or activated together. In some regions, the expression of genes on the paternal chromosome is silenced, whereas others have the maternal alleles silenced. As with other epigenetic examples, the effects are exerted on regions of the chromosome rather than on genes themselves, but the phenotypic effects are observed with individual genes; the imprinted genes within a region differ in their functions and in the time and place of expression. Most of these genes are imprinted in both mice and humans, and most if not all of the imprinted genes have remained syntenic between the species, with little variation in the gene order or arrangement on the chromosome.

Thinking through the Experiment

4. What might be the significance of the observation that the genes and chromosomal regions showing genomic imprinting in humans are syntenic and imprinted in mice as well?
5. Genomic imprinting is traditionally considered in terms of which allele is silenced rather than in terms of which allele is activated. On the basis of what we know now about transcription in the genome, why was the emphasis on silencing prescient?

Among the imprinted regions that have been studied, it appears that some of the principles of imprinting are similar to other epigenetic phenomena. Silencing is established, the silencing imprint is maintained throughout mitosis, and the imprinting is reset or erased upon gametogenesis. Resetting or erasing the imprint is probably the least studied aspect of the process, so our attention will be given to the earlier steps. We will describe one particularly well-studied imprinted region on chromosome 15 in humans, corresponding to a syntenic region on chromosome 7 in mice. The arrangement of the genes in this region is diagrammed in Figure CS2.1 Part A. As with many other examples of genetic analysis, the unusual properties of this region of the genome were recognized from mutant phenotypes. In this case, mutations in the region give rise to two distinct genetic syndromes in humans, Prader–Willi syndrome (PWS) and Angelman syndrome (AS).

Prader–Willi and Angelman syndromes are rare genetic disorders that map to the same region of the chromosome

While the properties of the PWS-AS region of chromosome 15 were recognized from children with disorders, we will first describe transcription of genes in this region as it occurs in unaffected individuals. The region is about 2 MB (megabase) and has a number of genes, six of which are shown in Figure CS2.1. Note that the region immediately to the right of *SNRPN* and surrounding *IPW* also has a number of genes encoding non-coding RNA (ncRNA); *IPW* is itself a non-coding RNA gene. Several of these non-coding RNAs are of the small nucleolar RNA class (snoRNA), and multiple copies of genes encoding several of these snoRNAs are

A. The PWS-AS imprinted region

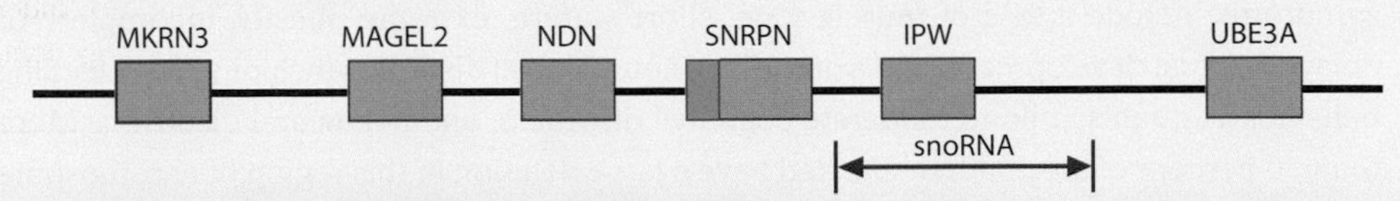

B. Imprinted transcription at the PWS-AS imprinted region

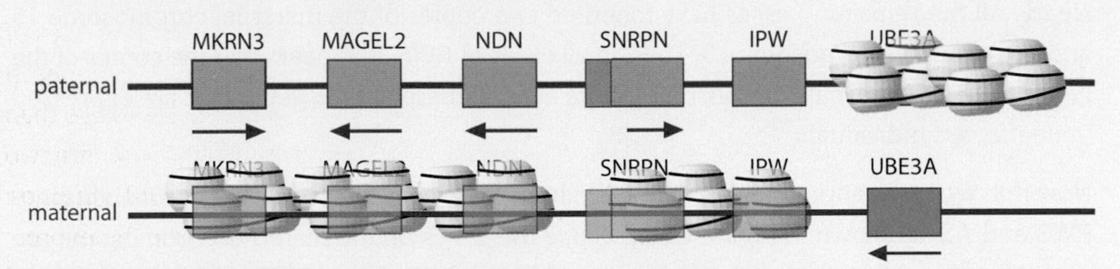

Figure CS2.1 Imprinting at the PWS-AS region on chromosome 15. A region of approximately 2 mb on chromosome 15 has distinct parent-of-origin effects; this is known as the PWS-AS region for the human syndromes that have been mapped to this location. Panel A shows a partial map of the region, with five protein-coding genes and one long non-coding RNA gene (*IPW*) shown. The map is not to scale. There is also a cluster of other non-coding RNA genes, including the snoRNA genes, as indicated by the bar surrounding *IPW*. The pink, once reproduced region immediately to the left of the SNRNP gene is the imprinting center, which is discussed in more detail below. Panel B shows the transcription pattern of the PWS-AS region as it occurs in unaffected individuals. Alleles of the genes on the paternal (blue) homologue are transcribed, with the direction of transcription indicated by the arrow, whereas these alleles are transcriptionally silenced on the maternal (red) homologue. The *UBE3A* gene is an important exception, since it is transcribed from the maternal homologue but silenced on the paternal homologue.

found in this region. snoRNAs have a range of functions in regulating gene expression, often by chemically modifying other RNAs, including tRNAs and rRNAs, but also by being processed to yield microRNAs. It has been widely speculated, with some evidence, that these snoRNA genes are important for the establishment or the maintenance of the imprinted state. An imprinted region on human chromosome 14 also includes two snoRNA genes, lending support to the hypothesis that these genes are important in imprinting. In general, the biochemical functions of the protein-coding genes in the PWS-AS region have not been implicated in the unique imprinted properties. Most of these genes are transcribed in many different tissues, particularly in the developing brain and in the central nervous system, which helps explain the neurological symptoms in affected children.

Genes in this region show distinctive expression patterns in unaffected individuals, as displayed in Figure CS2.1 Part B. Nearly all of the genes, including the non-coding RNA genes, are transcriptionally silenced on the *maternal* chromosome. The significant exception is the *UBE3A* gene, which is transcriptionally silenced on the *paternal* chromosome. Thus, all of the genes have parent-of-origin expression in unaffected individuals. This has significant implications for the two genetic disorders, PWS and AS.

Thinking through the Experiment

6. What is the significance of the fact that the biochemical functions of the genes in the region are not related to the mechanism by which imprinting occurs? Speculate (briefly) about how the biochemical functions of the genes might contribute to the neurological phenotypes seen in children with these syndromes. This connection between gene functions and phenotypes is discussed in more depth in Chapter 6, so this is asking you to think ahead about some of the topics that will be taken up in that chapter.

PWS is an autosomal dominant disorder that affects about 1 in 20,000 newborns. The symptoms include a loss of muscle tone, short stature, extreme obesity, incomplete or delayed sexual development, and a number of neurological disorders including poor feeding behavior as infants, mild to moderate cognitive disorders, and behavioral disorders. More than 70 percent of the children affected have a large deletion in this region of chromosome 15; most importantly, the affected children have inherited this deletion from their father. Nearly all the remaining cases have inherited two copies of the maternal chromosome 15 and no paternal chromosome 15. Thus, in all cases of PWS, it appears that the copies of the genes from the paternal chromosome 15 are absent; these are the genes that are expressed in unaffected individuals.

Now that we know about the transcription pattern in this region, we have some insights into PWS and AS, as shown in Figure CS2.2. Since the genes on the maternal chromosome are normally silenced, these large deletions arising from the father in PWS have do not express most of the genes in this region on either homologue; out of the genes shown in Figure CS2.2, only *UBE3A* is transcribed.

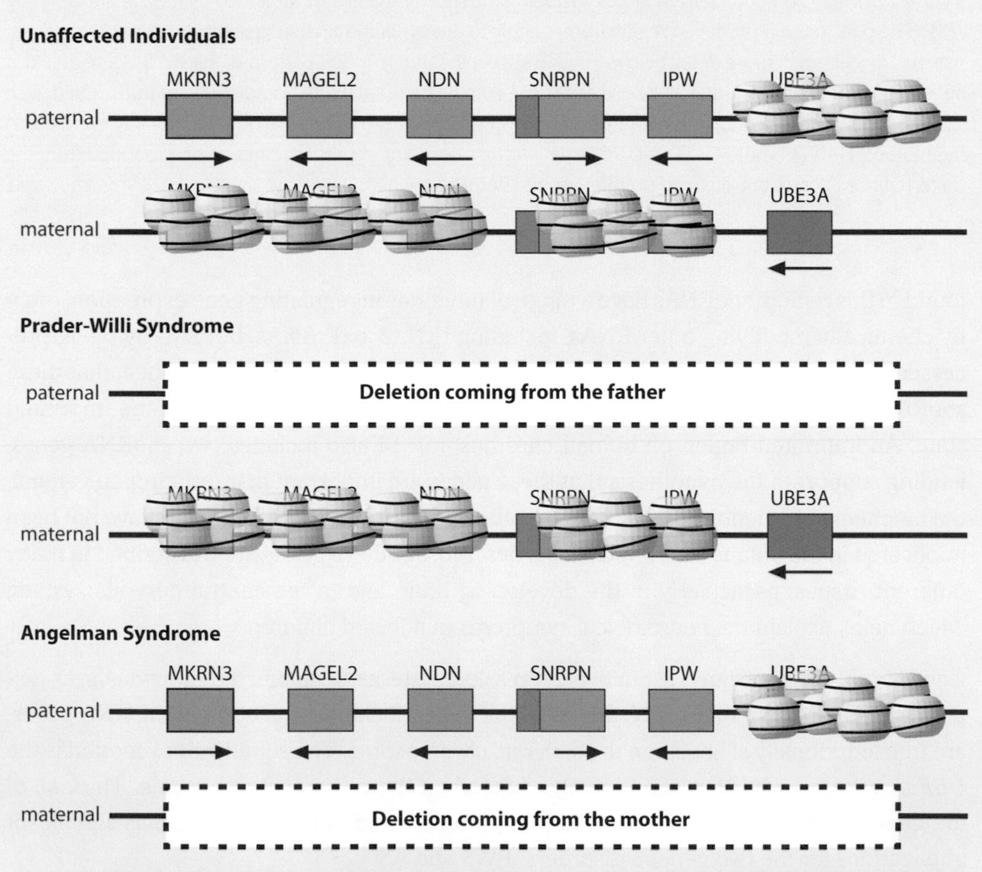

Figure CS2.2 The origins of Prader–Willi syndrome and Angelman syndrome. The parent-of-origin transcription pattern of the chromosomes in unaffected individuals is shown at the top. Because the region shows genomic imprinting, the effects of the same deletion depend on which parent contributed the gamete with the deletion homologue. If the deletion is on the paternal homologue, as shown in the middle, most of the genes that are normally expressed from this region are missing, so PWS arises. If the deletion is on the maternal chromosome, as shown at the bottom, only the *UBE3A* gene is not expressed and AS arises. Mutations that eliminate the function only of the *UBE3A* gene also cause AS, which shows the importance of this specific gene.

When the same large deletions are inherited from mother, as shown at the bottom of Figure CS2.2, a different dominant disorder, known as AS, is seen. AS, too, shows neurological symptoms with severe cognitive disability, impaired and delayed speech, ataxia, and behavioral disorders with an easily excitable personality. In AS cases, the key gene is *UBE3A*, which is silenced as usual from the paternal chromosome and deleted from the maternal chromosome. In fact, some cases of AS have a mutation that inactivates *UBE3A* on the maternal chromosome rather than a deletion of the entire region. *UBE3A* encodes a ubiquitin ligase, which targets a number of proteins for degradation, so its effects could be widespread.

Microdeletions identify distinct imprinting centers for PWS and AS

The original cases of PWS and AS had fairly large and cytological detectable deletions on chromosome 15, which initially suggested that these may be different phenotypes arising from the same mutation. When it became possible to detect very small deletions that affect only a few kilobases, it was clear that the PWS and the AS have distinct genetic causes. Some cases of PWS are due to deletions that remove only a portion of the region but that still have no expression of the paternal alleles. This small region is referred to as the imprinting center (IC) or the imprinting control element (ICE), which indicates that it is necessary for the establishment and maintenance of the imprinted state.

Thinking through the Experiment

7. Describe how the smaller deletions in the region could provide the experimental evidence that PWS and AS arise from different parts of the same chromosomal region. This is a naturally occurring example of a general laboratory procedure known as deletion mapping, discussed in Chapter 4.

In particular, as shown in Figure CS2.3, analysis of microdeletions revealed that a region of about 4.3 kb immediately upstream of the *SNRPN* gene is needed to establish and maintain the imprinted state on the paternal chromosome. Deletions that removed this region prevented expression of the paternal alleles and resulted in PWS, whereas deletions that affected other regions did not affect expression of other paternal alleles. This region is referred to as the smallest region of overlap for PWS (SRO-PWS) and includes the region upstream of *SNRPN*, as well as its promoter and the 5′ portion of the gene. *SNRPN* encodes a small nuclear ribonucleoprotein N, a protein involved in mRNA splicing. Thus its role is also expected to have a general effect on gene expression, although it seems likely that it is the location of this gene rather than its molecular function that contributes to imprinted expression. Some aspect of the chromosome, or something expressed from this region of this chromosome, is regulating the transcription of other genes over a broad domain of the chromosome.

Similar microdeletions identified a region that gives rise to AS on the maternal chromosome separable from the *UBE3A* gene, as summarized in Figure CS2.3. The smallest region of overlap for AS (SRO-AS) is shorter than 1 kb and located about 35 kb upstream of the SRO-PWS. Mutations that affect this region allow transcription of alleles on both homologues, and the maternally inherited alleles are not silenced. (The clinical diagnosis of AS uses the information that *UBE3*, the gene normally expressed from the maternal chromosome, is not being transcribed.)

Thus, as shown in Figure CS2.3, the imprinting center has two distinct loci. A widely accepted model for the role of these two centers is that the SRO-PWS is needed for the establishment

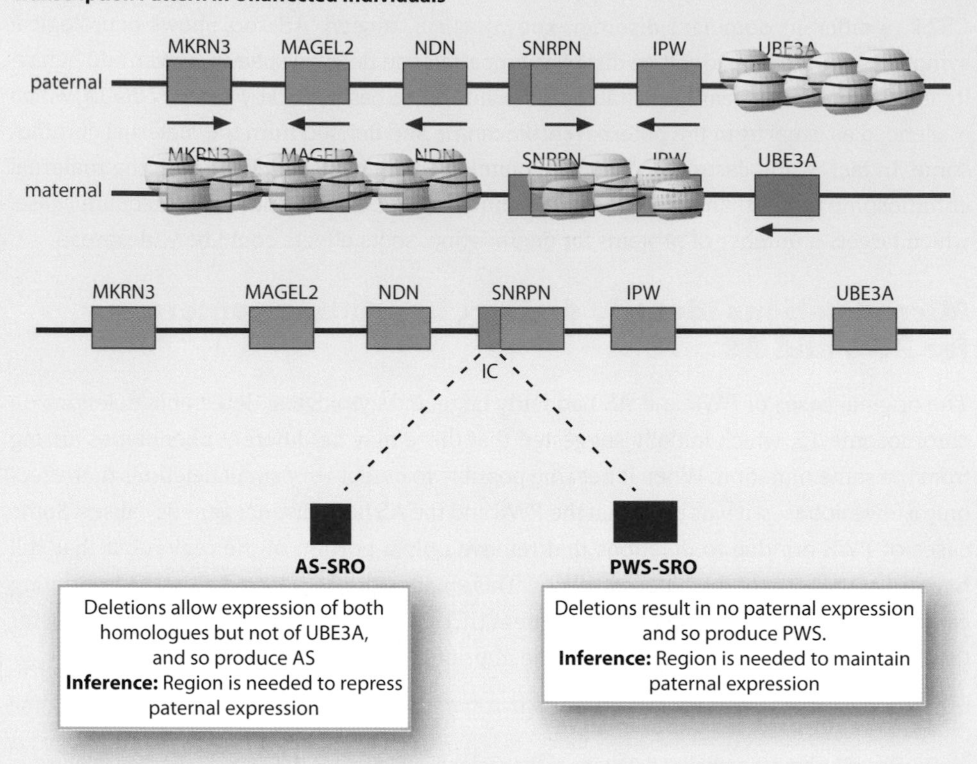

Figure CS2.3 Small deletions defined a bipartite imprinting center. While the original diagnosis of the syndromes depended on large deletions that are missing the entire region, parent-of-origin transcription was also found with much smaller deletions. These deletions defined the region in purple known as the imprinting center (IC), which overlaps the core promoter region and part of the first exon of the *SNRPN* gene. Still smaller deletions of the IC region that result in PWS share a region of overlap of about 4.3 kb, as shown in blue. This is known as the PWS- smallest region of overlap or PWS-SRO. These deletions of the PWS-SRO have no expression of paternally inherited alleles, so some aspect of this region is inferred to be necessary to maintain paternal expression of these genes. Similarly, smaller deletions of IC that result in AS also share a region of overlap of less than 1 kb, the AS-SRO. This is a distinct location, slightly upstream from the PWS-SRO. Deletions of the AS-SRO allow expression of most of the genes from both homologues, but have no expression of the *UBE3A* gene, which explains the AS phenotype. Thus some aspect of this region is inferred to be necessary to silence paternal expression.

and maintenance of the active paternal chromosome since, in its absence, none of the genes that are normally expressed from the paternal chromosome is expressed, and *UBE3A* is expressed from both homologues. The SRO-AS is apparently a negative regulator of the SRO-PWS that prevents expression of these genes on the maternal chromosome. The mechanisms by which these activities occur are not known. Transcripts arise from the SRO-PWS on both homologues, so the mechanism by which the SRO-AS repressed the maternal chromosomes is not clear. The structure of the chromatin in this region has been well described, but the functional relationship among the different structural features is not yet clear.

Chromatin structure at the imprinting center

The assays to examine chromatin structure described in Tool Box 2.1—such as DNase hypersensitive sites, DNA methylation, ChIP for histone modifications, and transcription factor binding sites—have all been directed to an examination of the imprinting center. The experiments have looked primarily at the chromatin state in brain cells, since most of the genes are active in the brain and many of the effects of PWS and AS are neurological. The (silenced) maternal homologue and the (active) paternal homologue show some clear differences. For example, there are four DNase hypersensitive sites that are reproducibly present at the imprinting center on the paternal chromosome but not present on the maternal chromosome in humans; a fifth site is variable, and present in some assays but not all. Thus the chromatin structure on the paternal chromosome is more open than the chromatin structure on the maternal chromosome. Several of these hypersensitive sites are conserved between humans and mice.

The differences between maternal and paternal chromosomes are also seen in the histone modifications that are present. The imprinting center on the paternal chromosome is marked by the presence of H3K4me3 and acetylated H4, which are absent from the maternal chromosome. On the other hand, the same region on the maternal chromosome has H3K9me3, which is absent from the paternal chromosome; CpG islands are also methylated in the DNA sequence of the maternal chromosome but not in the paternal ones.

Several widely used transcription factors are also found to be bound at the imprinting center on the paternal chromosome but not on the maternal one. These include E2F, CTCF, NRF1, YY1, and Sp1, which bind to many sites on human chromosomes, including other imprinted loci in both mice and humans. The functions of these transcription factors at these sites are not clear, and it is worth remembering that the presence of a bound transcription factor does not guarantee that is has a functional role. The paternal chromosome has a more open chromatin structure, as indicated by the DNase hypersensitive sites, so it might not be surprising that it is more accessible to transcription factor binding. Nonetheless, all these assays indicate the range of chromatin differences found at the imprinting center on the two homologues.

Thinking through the Experiment

8. Which of the modifications and changes in the chromatin state have been frequently associated with gene silencing in other examples of epigenetics, and which have been associated with gene activation? Does this pattern hold for imprinted genes?

Genomic imprinting and evolution

The evolution and functions of imprinting in mammals are a matter of some debate. Because the arrangement and structure of imprinted regions tends to be conserved among humans, mice and other placental mammals, and at least in part among marsupials, it is natural to assume that arrangement of the genes and the structure of imprinted region are functionally significant. It may also be, however, that parent-of-origin effects in this region arose in some common ancestor in which they played a role. The unique properties of the region could have imposed an evolutionary constraint that has not allowed much divergence in the arrangement over time, even if the original role of the imprint is no longer crucial. The mechanisms by which an imprint is established and maintained are also not yet clear. Different imprinted regions have generally similar structures, including the presence of numerous ncRNAs, these particular histone modifications, DNA methylation on the silenced chromosome, and, to the extent that it has been examined, even binding of some of the same transcription factors. None of these features is unique to imprinted regions, however, and even the combination of these factors is found at other places in the genome. Genomic imprinting, which has puzzled geneticists for decades, still holds on to many of its secrets.

References

Barlow, D. P. 2011. Genomic imprinting: A mammalian epigenetic discovery model. *Annual Review of Genetics* 45: 379–403.

Nicholls, R. D., and Knepper, J. L. 2001. Genome organization, function, and imprinting in Prader-Willi and Angelman Syndromes. *Annual Review of Genomics and Human Genetics* 2: 153–75.

Rodriguez-Jato, S. et al. 2013. Regulatory elements associated with paternally expressed genes in the imprinted Angelman/Pradi–Willi syndrome domain. *PLoS One* 8: e52390.

Wu, M.-Y. et al. 2012. An unexpected function of the Prader–Willi syndrome imprinting center in maternal imprinting in mice. *PLoS One* 7: e34348.

3

CHAPTER 3

Model organisms and their genomes

TOPIC SUMMARY

The life cycles and genetics of some of the best-studied model organisms are reviewed. The models include the yeast *Saccharomyces cerevisiae*, the nematode *Caenorhabditis elegans*, the fruit fly *Drosophila melanogaster*, the flowering plant *Arabidopsis thaliana*, and the mouse *Mus musculus*. We explore the advantages and potential limitations of each model organism, providing an example of a biological question that has been investigated in each one.

INTRODUCTION

Any lover of nature is amazed at diversity of life on earth. More than 1.75 million species have been named but, since most species are microscopic or live in unexplored habitats, these named species are only a fraction of the number of species alive. Most of the latter have not been studied in any detail, either in nature or in the lab, so it is impossible to imagine that we could ever understand them all.

Although much about the behavior, development, natural history, and physiology of all these species will remain unknown to us, the principles of their genetics are known. As described in Chapter 1, the fundamental principles that Mendel found for traits in garden peas have been extended to many other eukaryotes and, with some occasional adjustments, will undoubtedly apply to all others. We have confidence that this assertion is true because all species evolve from common ancestors. The point is made in a well-known statement by Jacques Monod, co-discoverer of the *lac* operon and mRNA and one of the founding fathers of molecular biology, in 1954: "What is true for *E. coli* must be true for elephants." What we learn from one organism provides the starting point for understanding others.

We assert this confidently, but also humbly: the humility is needed because species are different and even a casual observer can tell many of them apart. *E. coli* is a bacterium and an elephant is a eukaryote; we now know many fundamental aspects of genetics and biochemistry that are not true for both. Like many of the ideas described in Chapter 1, Monod's point is helpful as far as it goes. Monod's statement—which he surely made hyperbolically—does something of a disservice both to *E. coli* and to elephants. Each of these is a fantastically complex organism, worthy of study on its own merits, without the need to justify the study through some broader explanation. As biologists, we would like to investigate thoroughly *E. coli* and elephants and all the species in between. But that is simply not possible.

Instead of spreading our research efforts among thousands of different species, geneticists have focused on a handful of **model organisms**. The choice to study these few species among the many possible ones was often made for fairly weak reasons. Yet, having made the choice, geneticists have investigated these few model organisms thoroughly for decades; nearly all of the tools of genetic analysis have been developed from and applied to the study of these few model organisms. These tools include genetic maps, mutations, cloned genes, gene expression patterns, and interaction networks. Significantly for contemporary genetic analysis, the genomes of these model organisms were among the first to be sequenced and are the best annotated ones apart from the human genome. All the tools that will be developed and used in the chapters to come are in place.

From these model organisms we are able to make intellectual extensions to all others. Our confidence that all other eukaryotes, including those we have not yet studied, will follow these same fundamental principles comes not from the brashness of

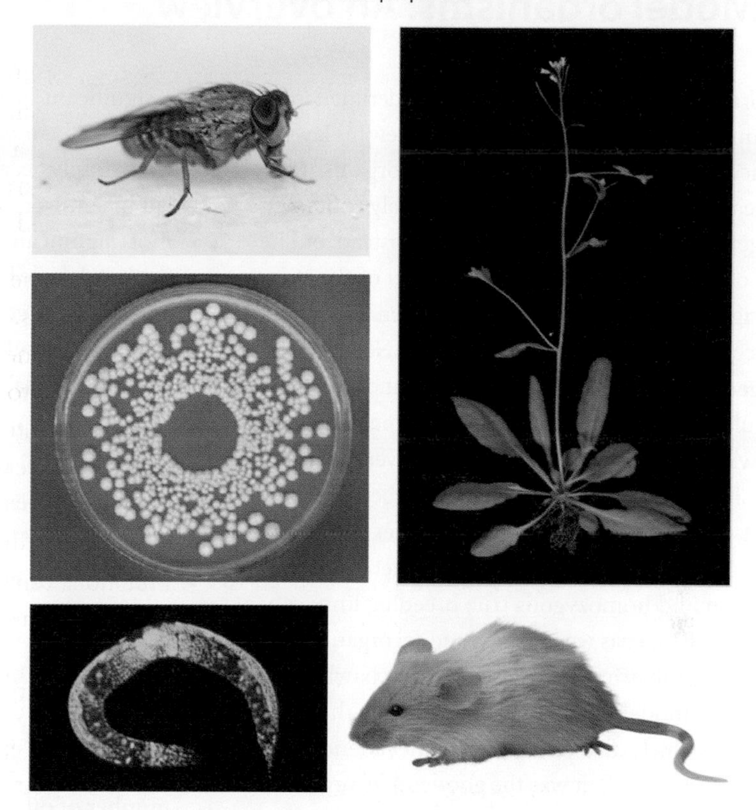

Figure 3.1 The five model organisms widely discussed in this book. The two row shows the budding yeast *Saccharomyces cerevisiae*, the nematode worm *Caenorhabditis elegans*, and the fruit fly *Drosophila melanogaster*. The bottom row has the flowering plant *Arabidopsis thaliana* and the mouse *Mus musculus*. 3.1a: Redrawn from Mogana Das Murtey and Patchamuthu Ramasamy, 2016, Sample preparations for scanning electron microscopy: Life Sciences, in *Modern Electron Microscopy in Physical and Life Sciences*, edited by Milos Janecek and Robert Kral, IntechOpen, DOI: 10.5772/61720. 3.1b: Redrawn from iStock.com/HeitiPaves. 3.1c: Redrawn from iStock.com/nechaev-kon. 3.1d: Redrawn from Salicyna / CC-BY-SA-3.0. 3.1e: Redrawn from iStock.com/GlobalP.

knowing too little but from the humility of knowing how little we truly understand, even about these well-studied models. Perhaps we can restate what Monod said from a perspective he gained from more than sixty years of investigation. If it is true for *Saccharomyces cerevisiae*, *Caenorhabdits elegans*, *Drosophila melanogastor*, *Arabidopsis thaliana*, and *Mus musculus*, it is probably true for elephants as well. (On the other hand, every organism, including elephants, has its own unique features, so we cannot know everything about any organism until we study it individually.) Among our models there have been a yeast, a nematode, an insect, a flowering plant, and a rodent, as shown in Figure 3.1. We have asked these organisms to serve as stand-ins for all of the diverse eukaryotes that we encounter, including humans. We have placed a lot of responsibility on these unassuming creatures, and they have responded by yielding a wealth of insights.

3.1 Model organisms: An overview

The importance of model organisms was underlined from the beginning of genetic analysis. Mendel drew his laws of inheritance from experiments with a single organism, the garden pea *Pisum sativa*. His garden probably contained other vegetables that he might have used instead, but he reported only on peas. An intriguing aspect of the story of the origins of genetics comes from Mendel's correspondence with the distinguished botanist Naegeli, who encouraged him to apply his principles to another plant, in particular hawkweeds. Naegeli's recommendation to study hawkweed is a mistake dreaded by every research mentor, since his suggestion may have been among the worst possible choices for follow-up experiments. Hawkweeds have an unusual parthenogenetic mode of reproduction and homozygous true breeding lines cannot be established. Peas were a good model organism for understanding the principles of inheritance; hawkweeds were a terrible model, which may have set back the study of genetics by thirty-five years. But neither Mendel nor Naegeli knew which plant was the good model organism and which was the poor choice.

Like peas and hawkweeds, many different organisms have been used for genetic analysis. Some have been like peas, and were used to establish the foundations of modern molecular genetics. Others, like hawkweeds, seemed like a good idea at the time but, for one reason or another, little research of general importance has arisen from them. What makes the difference between an experimental organism that becomes a valued model and one that does not? The answer is often complex. Sometimes it lies in some properties of the biology of the organism itself, or in the personalities of and relationships among its early investigators, or even in the willingness of funding sources to invest in a different organism.

Let's look at that question a little more closely. ***What makes an experimental organism a good model?*** A brief comparison of the model organisms used for genetic analysis and described in this chapter generates the following list of relevant features.

- It is easy to maintain in the lab. Ideally, it should not be difficult or dangerous to work with and should not require any peculiar growth conditions. If it is capable of living outside the lab, it should not present a significant health or environmental hazard. Many biologically important organisms failed on this criterion, simply because they are so difficult to grow in the lab in sufficient quantity for research. But one of the most significant breakthroughs in genetic analysis in the last decade has come from our ability to sequence and study the genomes of organisms in natural populations, particularly bacteria, even if they cannot be grown in the laboratory.
- Homozygous true-breeding lines can be established. Not all model organisms have true breeding lines, but having them certainly makes the analysis much simpler.
- Mutations can be induced and mutant phenotypes can be recognized easily. This includes the ability to maintain and analyse mutations whose effects are lethal to the organism.
- The organism produces a large number of offspring in a relatively short period of time. Because a large number of offspring allows the investigator to screen for mutations easily, this feature is also helpful if the organism is comparatively small.
- Genes can be cloned by standard procedures, and cloned genes can be reintroduced into the organism. "Standard procedures" are those that have been developed for other experimental organisms, although all procedures have to be altered for each research application. We discuss reintroducing genes later on in this chapter. In a more contemporary sense, having a substantial amount of genome sequence or an ongoing genome project helps.
- The organism attracts enough young investigators to sustain the analysis beyond the founding generation. This usually requires a complex combination of intriguing biological properties, compelling and persuasive personalities, and available funding and permanent employment.

Many experimental organisms have these qualities yet have not become widely used models for genetics. Thus there is probably an ill-defined component that could be considered happenstance—having the right combination of people at the same place, for example. If we were able to create a computer simulation of twentieth-century genetic research and run it many times, it seems likely that some of the model organisms we use now would not become established and that others, which are not widely used, would emerge as new models. Although such an imagined simulation might change the order in which some key findings occurred, it is unlikely to result in long-term differences in our understanding of genetics. The model organisms have served us very well, but much of what we have learned would probably have been discovered eventually no matter what models were used. It is partly because the model organisms have served so well that contemporary genetics is able to study many new models. Later in the chapter we return to this question when we consider the role of model organisms in an era when genome sequencing is routine.

Having defined some of the characteristics that make an experimental organism a good model, it is appropriate to ask another question. ***What do we want the model organisms to model?*** We can answer this question in different ways, as we think about what our principal model organisms have and, in some cases, have not been able to tell us. First, we can expect that a combination of model organisms can illustrate the principles of biology that are common to all eukaryotes. For instance, all eukaryotes replicate and repair DNA by very similar mechanisms, and the processes and transcription and translation are largely the same. All organisms have many of the same metabolic processes—such as cellular respiration, energy production, and the anabolism and catabolism of macromolecules. All eukaryotes have very highly similar processes of cell division, both mitosis and meiosis. All eukaryotes have their DNA packaged in chromosomes. Many more examples of these common biological processes could be given. Since all organisms carry out these processes, it makes sense to study them in a model organism in which the experiments might be simpler.

Yet, even within this group, there are important individual differences among the species. Consider the structure of chromosomes, which all eukaryotes have and we might expect that any of our organisms could appropriately "model" for us. Within our group of five model organisms, key components of chromosomes are not all the same. As will be discussed, the centromeres of *Saccharomyces cerevisiae* are structurally different from those of other eukaryotes, and *C. elegans* does not have localized centromeres at all. While the centromeres of *Drosophila melanogaster* chromosomes are similar to those of most other eukaryotes, its chromosomes do not have conventional telomeres, as those of other eukaryotes do. Since all these organisms evolved and occupy particular natural environments before they are introduced into the lab, they all have differences from the "standard issue eukaryote"—if there were such an organism. This is why it has been important to have multiple model organisms.

In addition to telling us about the biological processes common to eukaryotes, the model organisms also give us information about their particular group. Thus we need to consider what processes can or cannot be studied easily in each of the model organisms. While yeast is ideal for understanding processes that occur within a cell, it grows primarily as a unicellular organism and is clearly not the source for information about organs and tissues. Among these organisms, only *Arabidopsis* could be used to study photosynthesis, only *Drosophila* could be used to study flight, only mice could be used to study fur coloration, and so on. Each of them can provide information about the taxonomic group that they belong to, and we need to consider to what extent these findings might be more general to other groups. But these organisms allow us to raise questions about processes in other organisms in ways that can be investigated; they give us a place to start.

As we compare these model organisms to others from their taxonomic group, we also recognize quickly that each of these organisms is a single species, and every species has distinct properties. *C. elegans* is not the same as *C. briggsae*, *Drosophila melanogaster* is distinguishable from *Drosophila virilis*, and so on for each of the others. So we can begin to study what makes each species distinct from its close relatives. More recently, this has led to an even more fascinating and individualized biology, as we recognize that the "wild-type" version of the organism in laboratories often has distinct traits, behaviors, morphologies, growth habits, and so on from those of that same organism as it lives in its natural habitat; these differences between wild-type versions and wild versions are reflected in differences between their genomes. Some of the differences between the wild-type laboratory organism and the wild organism in nature have been described in a series of essays.

Literature Link. https://elifesciences.org/collections/8de90445/the-natural-history-of-model-organisms

In short, we ask model organisms to tell us all they can tell about biological processes, and we can learn this at many different levels—those processes common to eukaryotes, those found among particular to taxonomic groups, and even those that make each species unique.

Model organisms and human biology

Perhaps you are thinking that we have overlooked the most obvious answer to the question of why we study model organisms: we want to be able to apply the insight gained from them to human biology and disease. While it may be true that we want to believe that our research will have an important application to humans, the more realistic attitude is that findings with model organisms will provide insight into some of these other properties, which may be relevant to human biology and medicine. In addition to serious ethical and moral issues, humans as subjects of experimental research fail on nearly all the properties of model organisms listed above. Our major advantages are that mutations can be recognized (although the genetic basis for phenotypic differences is often difficult to ascertain) and that young investigators continue to move into the research field. All the other desirable characteristics list here for an experimental organism have presented challenges for human genetics, some of which we will describe throughout the book.

We are certainly unlikely to discover universal biological properties from research with humans, and even mammal-specific and primate-specific properties are probably best studied in other species. It is undoubtedly true, however, that many properties that are found in other species are recognized as being universal or assume much greater significance when they are found to occur in humans. For example, microRNAs were initially described in plants and nematodes and long non-coding RNAs were seen in other genomes first, but both gained much more attention when they were also found in humans and connected to human genetic variation or diseases. Nonetheless, as an increasing number of individual genomes are sequenced and annotated, it becomes possible to think about the species-specific properties of humans, or what biological features make us human. Even before Linnaeus produced the standards that we still use, biological organisms were being classified and compared. Genomics has allowed comparative analysis to reach a new and exciting level of research. Biologists trained before the arrival of molecular biology may have held classes and seminars in subjects such as comparative anatomy, where they examined the skeletons and muscle structure of different vertebrates, or comparative physiology, where they examined respiration or excretion to determine both the common and the distinctive characteristics of each group. We are currently in the age of comparative genomics, when each new genome is understood in light of the others that have been sequenced and analysed. One can even imagine that comparative genomics will be the foundation of a new way to think about biology as a whole.

Our discussion of what can be learned from different model organisms becomes most helpful when put into the context of the comparative analysis of genomes. We are beginning to identify genomic characteristics and genes that are common to all eukaryotes, as well as ones are that are common to multicellular organisms, animals, insects, mammals, and so on. As the genomes of other primates are studied, we can also begin to recognize a few genetic properties that are peculiar, if not unique, to humans. Remarkably and somewhat humblingly, we have many fewer unique genetic properties, particularly when compared to our two best studied extinct relatives, Neanderthals and Denisovans. The uniqueness of humans apparently lies not in specific, novel genes but in more subtle differences, such as the expansion or loss of members of gene families, changes in a few amino acids with many different proteins, and, possibly most provocatively, changes in our non-coding RNA genes.

In the meantime, we describe the properties of five model organisms that are shown in Figure 3.1. Certainly other model organisms could be introduced and discussed, since these five are far from being the only species that contribute to modern genetics. A summary of many of the key features of these model organisms appears in Table 3.1.

The most important information in Table 3.1 is the link to the genome database for each organism. These websites provide the most accurate, comprehensive, and up-to-date information on the model organisms and on some of the related species whose genomes have also been sequenced. These databases also offer links to methods or tutorials about working with the organism, background information, references, and much more—which is essential for planning even the simplest experiments. The journal *Genetics* regularly publishes toolbox and research resource review articles for all these organisms. Geneticists working with one of these model organisms access these databases regularly, and no one would consider planning a project without exploring them first.

Table 3.1 A comparison of selected model organisms

Species	*Saccharomyces cerevisae*	*Caenorhabditis elegans*	*Drosophila melanogaster*	*Arabidopsis thaliana*	*Mus musculus*
Common Names	"yeast". Budding yeast, baker's or brewers yeast	"the worm"	fruit fly	Arabidopsis	mouse
Genome Size	16 chromosomes, 6,607 protein coding genes, 12 Mb	Six chromosomes, approximately 20,400 protein-coding genes, 105 Mbases	Four chromosomes, the 4th with very few genes. Approximately 15,000 protein coding genes. 140 Mb	Five chromosomes, approximately 27,000 protein coding genes. 135 Mb	20 chromosomes. Approximately 22,000 protein coding genes. 3,000 Mb, or 3 billion bases
Initial Attractiveness as a Research Organism	Industrial applications for brewing. Single celled eukaryote that could be grown in large quantities. Can be grown as either haploids or diploids.	Defined cell lineages. Organs and systems including the nervous system have few cells with highly defined connections. Self-fertilizing hermaphrodites produce homozygous strains, with males arising to allow crosses.	Easily grown in lab with easy to score morphological differences. More than a century of genetic research with vast number of mutant lines, rearrangements, etc. available.	Flowering plant. Short life cycle. Small size allows many individuals to be grown in a small space. Naturally self-fertilizing, and produces many seeds. Small genome for a flowering plant. Easiest plant for making transgenics.	Mammals, raised by hobbyists. Small size and fast life cycle.
Method to Make Transgenics	Transformation with plasmids. Homologous recombination with artificial chromosomes	Injection with dsDNA, maintained as an extrachromosomal multicopy array	Injection with P transposable element, which integrates randomly in genome.	*Agrobacterium tumefaciens*-mediated transformation by floral dip.	Embryonic stem cells with targeted gene insertion and disruptions
Mutant Phenotypes Used for Common Markers	Nutritional auxotrophs, drug resistance	Morphological, body shape and movement	Morphological, including eye color, wing shape, body color, etc.	Morphological, including leaf hairs, flower and seed pod characters, etc.; flowering time; hormone responses	Coat colors

(Continued...)

Species	*Saccharomyces cerevisae*	*Caenorhabditis elegans*	*Drosophila melanogaster*	*Arabidopsis thaliana*	*Mus musculus*
Particular Advantages	Ideal for studying nearly every intracellular property of eukaryotic cells. Particular contributions have come in the cell cycle and cell division, meiosis, intracellular secretory mechanisms and protein trafficking, chromosome structure, signal transduction, eukaryotic transcription and replication, and many more. First eukaryotic genome to be sequenced.	Cellular simplicity contributed to embryology, programmed cell death (apoptosis), neurobiology and behavior, most signal transduction pathways. First multicellular genome to be sequenced. Discovery of microRNAs and RNAi in animals.	Long history of genetic research, provides insights into nearly all topics in animal genetics. Polytene chromosomes are easily visible.	Ideal for studying the biology of flowering plants. Multicellularity arose independently in plants and animals; comparisons to the animal models provide a better understanding of the requirements for multicellularity and the make-up of the earliest common unicellular ancestor. Key for delineating the mechanisms of RNA silencing and other epigenetic phenomena.	Compared to humans in nearly all areas of mammalian biology, including immunology, neurobiology, embryology, cancer biology, etc.
Limitations and Peculiarities	Extracellular signaling is limited. Very few introns. Centromeres are unusually small. May lack microRNAs as a means of regulating gene expression.	No established tissue culture. Targeted integration of introduced DNA is difficult. Many genes are in polygenic transcripts with trans-spliced leader sequences	No recombination in males. Targeted gene disruptions are difficult.	No homologous recombination; successful targeted gene disruptions so infrequent that they are rarely attempted. Many seemingly functionally redundant genes.	
Genome Database	The Saccharomyces Genome Database http://www.yeastgenome.org/	Wormbase http://www.wormbase.org/	Flybase http://flybase.bio.indiana.edu/	TAIR—The Arabidopsis Information Resource http://www.arabidopsis.org	Mouse Genome Informatics http://www.informatics.jax.org/

3.2 *Saccharomyces cerevisiae*, or the budding yeast

Geneticists who work with the budding yeast *Saccharomyces cerevisiae* can take a justifiable pride in their ability to manipulate this organism in order to study nearly any process in a eukaryotic cell. Some of the organism's properties are shown in Table 3.1. For fundamental intracellular properties that occur in all eukaryotes, including meiosis and mitosis, DNA replication, the cell cycle, protein trafficking, and many more, the initial comparisons are nearly always made with what was first described in yeast. Although industrial applications for yeast genetics go back for centuries, the use of yeast as a laboratory research organism (apart from brewing) began in the 1930s and greatly increased during the 1950s.

Literature Link. Duina, Miller, and Keeney 2014, *Genetics* 197: 33–48

Yeast cells can be grown in liquid culture or on agar plates with defined growth requirements, as shown in Figure 3.2. Since the nutritional requirements are defined, mutations that cannot synthesize some key molecules are easy to identify and use. These mutations are known as **auxotrophs**. For example, wild-type or prototrophic yeast can produce their own uracil (see **prototroph**). By growing cells in the absence of added uracil—that is, by starving them for uracil—it is possible to find mutations that have a defect in the uracil biosynthetic pathway. These mutants cannot grow unless uracil is added to the growth media, so they are uracil auxotrophs. Auxotrophic *ura3* mutations are useful as genetic markers. Auxotrophic mutations provide a powerful selection scheme since only cells with the desired genotype—uracil prototrophs, for example—will grow and divide.

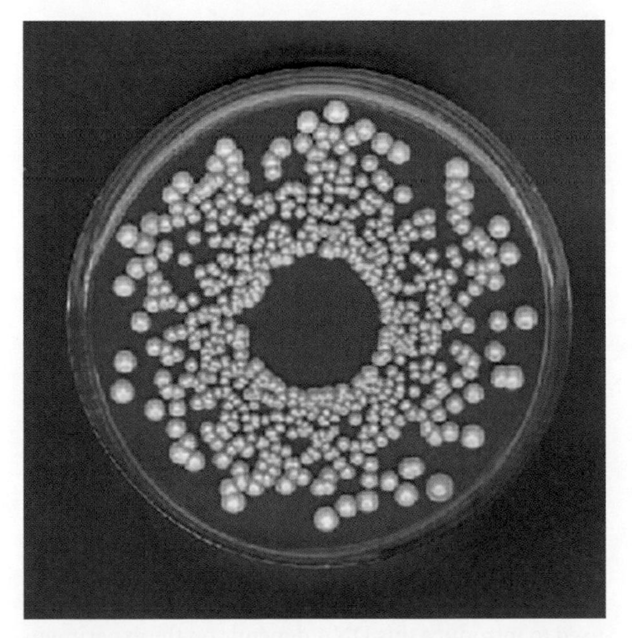

Figure 3.2 ***Saccharomyces cerevisiae*, as seen by most geneticists.** Yeast cells grown as colonies on an agar plate with defined nutrients, which allows easy use of auxotrophic mutations.

Yeast forms colonies, each consisting of thousands of cells under the normal growth conditions on agar plates in the lab, and much of the analysis is done by examining colonies of genetically identical cells rather than individual cells. In this, yeast can be treated much like *E. coli* or other bacteria. Growth rates depend on environmental conditions, of course, but a typical doubling time is less than two hours, so a few cells on an agar plate can produce easily recognized colonies in less than forty-eight hours when grown at a temperature of 30 centigrade. Tens of thousands of yeast cells can be grown on a standard petri dish.

Yeast is often grown as a haploid

Nearly all eukaryotic organisms have both a haploid and a diploid phase of their life cycle. Haploid cells have one copy of each chromosome in the genome; diploids have two copies. Unlike the other organisms we will discuss, yeast is often grown and maintained as a haploid. By examining haploids, issues arising from dominance are avoided, which makes recessive mutations easy to spot. The life cycle of budding yeast is shown in Figure 3.3.

Haploid yeast cells have one of two possible mating types, termed **a** and **α**. Haploid cells of one mating type can fuse or mate with cells of the other mating type to form an **a**/**α** diploid. A diploid strain divides itself by mitotic division and can be propagated indefinitely. Under conditions of starvation (or other stimuli), the diploid initiates meiosis, which is similar to that of cells in the germline of animals. Chromosome pairing, synapsis, and recombination occur to produce four haploid products. These four products, a **tetrad**, remain together in a structure called **ascus**, so all the products of one

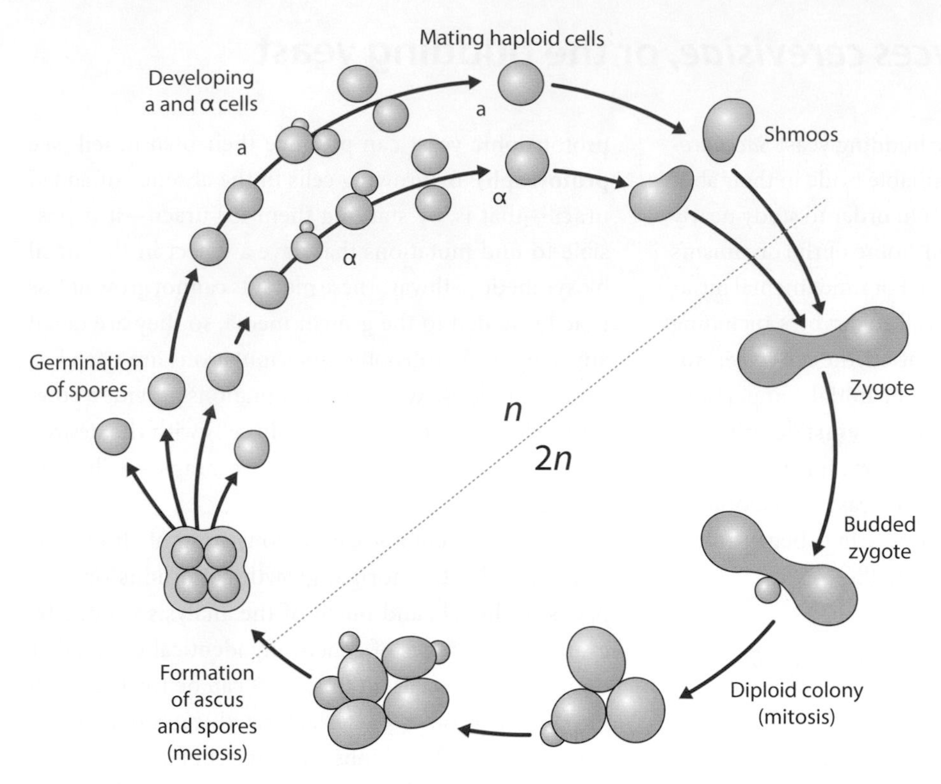

Figure 3.3 The life cycle of budding yeast. Haploid cells can be either of two mating types, a or α. These haploids can fuse into a diploid cell. The diploid cell reproduces mitotically by budding. Under starvation or other stimuli, the diploid cell enters meiosis and forms four haploid cells known as ascospores. These four spores, two of mating type a and two of mating type α, are found together in a sac called an ascus. The ascospores can be germinated and grown as haploid cells, which also reproduce mitotically, by budding.

meiotic division can be recovered together, as shown in Figure 3.4. (Saccharomyces has unordered tetrads because the four haploid ascospores can be found in any orientation and the ones arising from sister chromatids need not lie adjacent to each other. Some other fungi, for example *Neurospora crassa*, have ordered tetrads, since the haploid ascospores are located in the ascus in the order of the sister chromatids during meiosis.) Tetrad analysis has allowed the genetic mapping and analysis of many of the events of meiosis that occur in all eukaryotes, particularly those that involve gene conversion.

The processes of cell division—the cell cycle, mitosis, chromosome segregation, cytokinesis, and so on—have been thoroughly studied through genetic analysis in yeast. One unusual but useful feature of yeast chromosomes deserves a particular mention. Chromosomes are positioned on the spindle by the attachment of the centromere to the spindle fiber. In yeast, a single spindle fiber attaches at the centromere, and the centromere is an unusually small and well-defined sequence—about 125 bp in yeast, by comparison with millions of base pairs of repetitive sequences in most other organisms, as shown in Figure 3.5. Despite the difference in the length and composition of the underlying DNA sequence, the protein complex comprising the kinetochore that forms at the centromere is composed of similar proteins, in yeast and in other organisms. The small size of the yeast centromere sequence has allowed the construction of some useful cloning vectors, as described in what follows.

Transformation in yeast involves naturally occurring plasmids

DNA can be introduced into yeast cells by transformation, using either electroporation or osmotic shock. The ease with which transformation occurs has been one of the key advantages of yeast genetics, since foreign DNA can be readily cloned into yeast cells. The form that is used to introduce this DNA depends on the experimental application. Many of the transformation vectors are engineered so as to be capable of growth in either *E. coli*

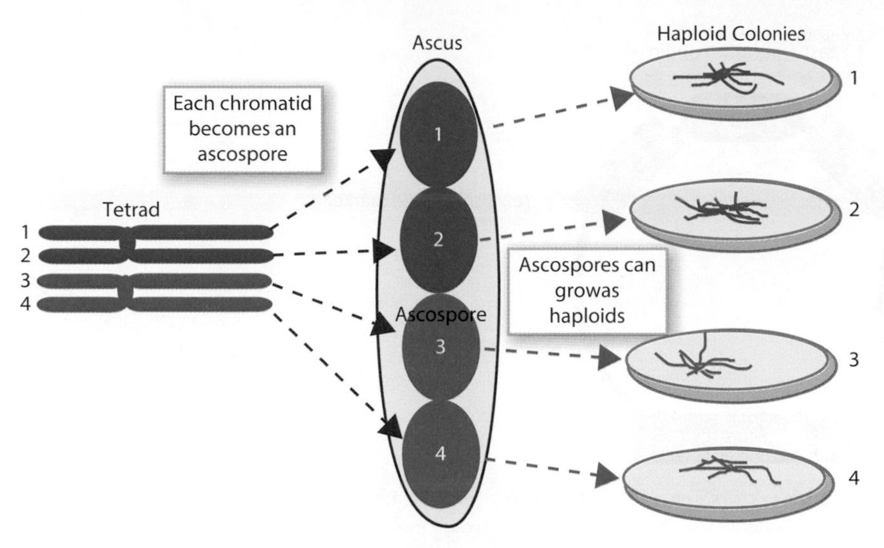

Figure 3.4 Tetrads. In ascomycete fungi such as *Saccharomyces*, *Schizosaccharomyces*, *Neurospora*, and *Aspergillus*, the four products of a single meiotic division can be found together in an ascus. In *Saccharomyces*, the four haploid products can be oriented in any formation in the ascus, although the drawing depicts the ascospores in the same order as the sister chromatids for clarity.

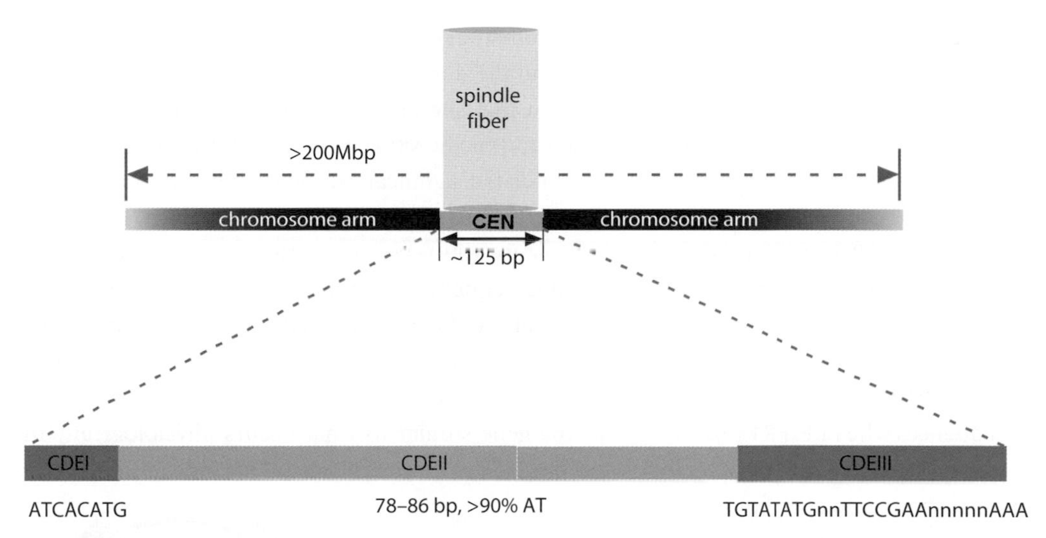

Figure 3.5 Centromere sequence and structure in *Saccharomyces*. The top diagram shows the centromere in the context of a chromosome. A single spindle fiber attaches to the centromere in *Saccharomyces*, unlike in other organisms, which have multiple spindle fibers per centromere; and the centromere is only 125 base pairs in length, much smaller than the centromeres in other organisms. The lower diagram shows the organization of the centromere. There are three sequence blocks, CDEI, CDEII, and CDEIII. CDEI and CDEIII have sequences that are conserved among the centromeres on different chromosomes. CDEII is slightly variable in length and sequence across different chromosomes, but is more than 90 percent AT in overall composition.

or yeast and have selectable markers appropriate for each host, as diagrammed in Figure 3.6, as well as replication origins from both bacteria and yeast. Growth in bacteria provides a sufficient quantity of DNA to conduct the experiments in yeast. These transformation vectors are called **shuttle vectors** because they can be moved back and forth between yeast and bacteria.

Yeast cells have a naturally occurring plasmid referred to as the 2 μm plasmid, which in its native form is about 6.3 kb in length; the shuttle vector diagrammed in Figure 3.6 has sequences from the 2 μm plasmid, including the origin of replication. The 2 μm plasmid can exist in up to twenty copies in the cell and segregates at random during mitosis; due to its high

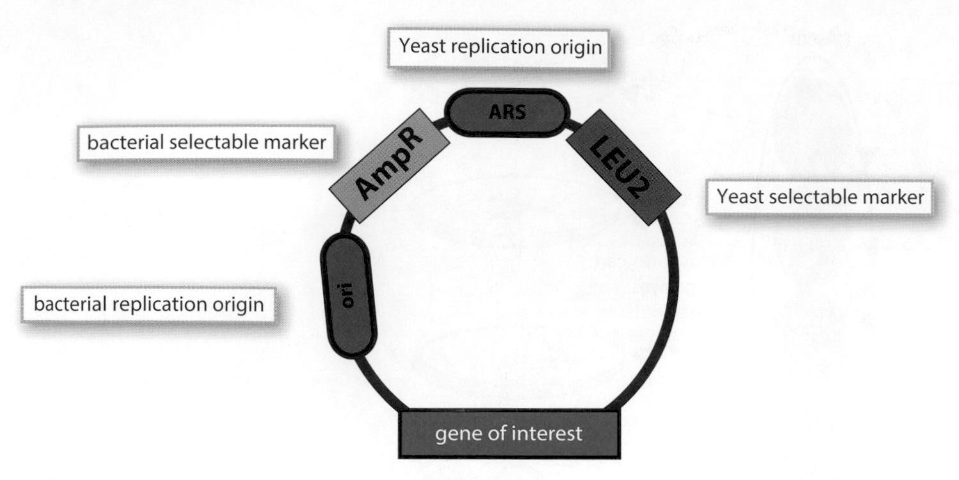

Figure 3.6 The general structure of yeast shuttle vectors. These plasmids can be grown in either *E. coli* or yeast and have both an origin of replication and a selectable marker appropriate to each organism. The actual organization of these elements, as well as the selectable markers used, vary with the different vectors. By growing the plasmid in bacteria, large quantities can be obtained for transformation into yeast. Genes cloned into a plasmid of this type will be present at high copy number in the yeast cell.

copy number, both daughter cells receive copies of the plasmid.

Genes cloned into a plasmid vector derived from the 2 μm plasmid such as the one diagrammed in Figure 3.6 are present at a high copy number in the cell, which is useful for expressing a gene at high level. This property will be discussed again in Chapter 10. In addition to its value as a high copy cloning vector, the 2 μm plasmid has another special importance for genetic analysis: the naturally occurring 2 μm plasmid encodes a site-specific recombination enzyme known as FLP; the target sites for FLP are the FRT repeat sequences. The FLP/FRT recombination system will be mentioned again in Chapter 7 as a means of targeting insertions into specific sites in the genome.

Some artificially engineered plasmids have other uses as vectors. Plasmids with a bacterial origin of replication but lacking a yeast replication origin can be grown stably only in bacteria, but can be transformed into yeast; such as plasmid, referred to as a yeast integrative plasmid (YIp), is diagrammed in Figure 3.7. Yeast has an extremely high rate of meiotic recombination, much higher than any of the other experimental organisms we will discuss. Thus, at detectable frequency, DNA introduced into yeast is integrated into the genome by a double crossover, as shown in Figure 3.8. Because sequence homology between the introduced donor DNA and the recipient host DNA directs the integration, homologous recombination targets the donor sequences to a specific site, and the donor sequences can replace the genomic copy at that site in the recipient. The ability to target genes to specific sites in the genome using integrative plasmids is a significant strength of yeast genetics. This topic is discussed in more detail in Chapter 7.

Because the yeast centromere consists of a short, defined sequence, plasmids containing a centromere (CEN plasmids), diagrammed in Figure 3.9, have also been useful for genetic analysis, particularly for experiments that require a copy number and expression level of the gene similar to what occurs physiologically in

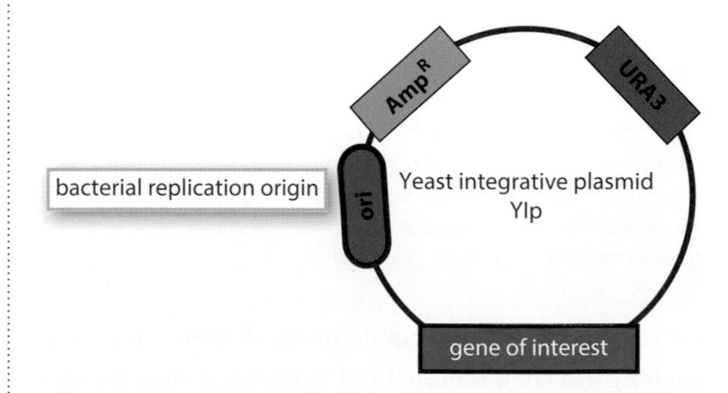

Figure 3.7 Yeast integrative plasmids, a YIp. This plasmid has a bacterial origin of replication (ori) and a selectable marker for growth in bacteria, but lacks a yeast origin of replication. Thus, when transformed into yeast cells, this plasmid cannot replicate and will be lost unless crossing over between the plasmid and the chromosome occurs, as diagrammed in Figure 3.8.

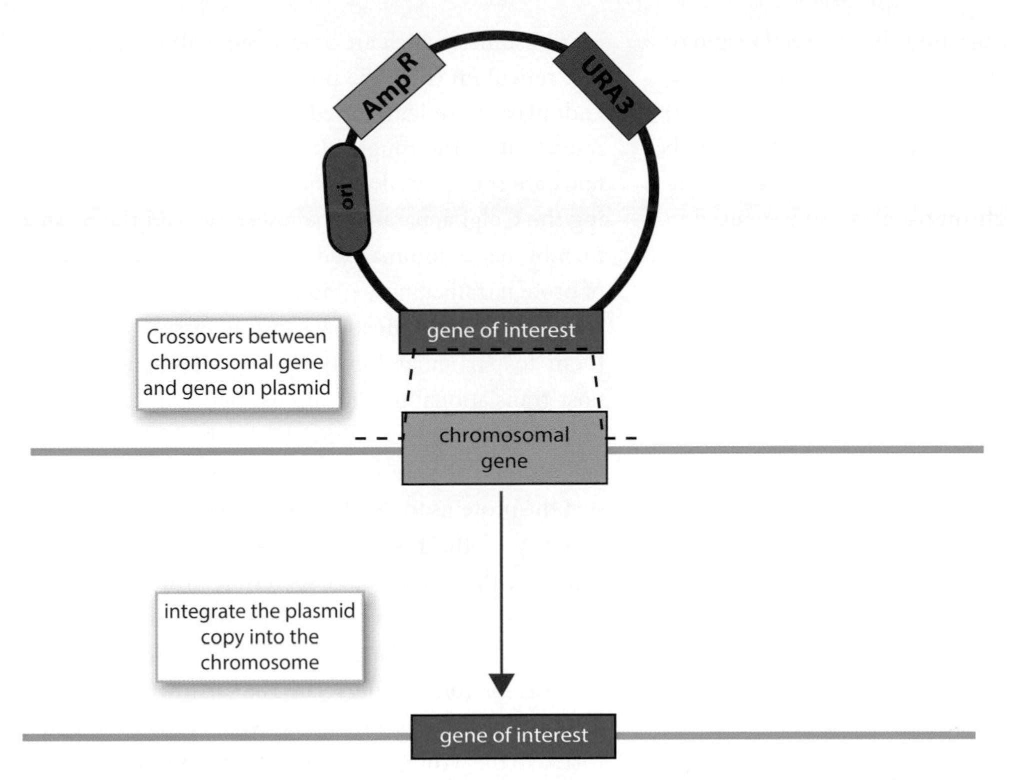

Figure 3.8 Crossing over between the YIp and the chromosome. Any DNA introduced into yeast cells that has sequence similarity to the chromosome can be integrated by homologous recombination. As shown by the dashed line, the introduced DNA on the plasmid replaces the chromosomal copy by a double crossover.

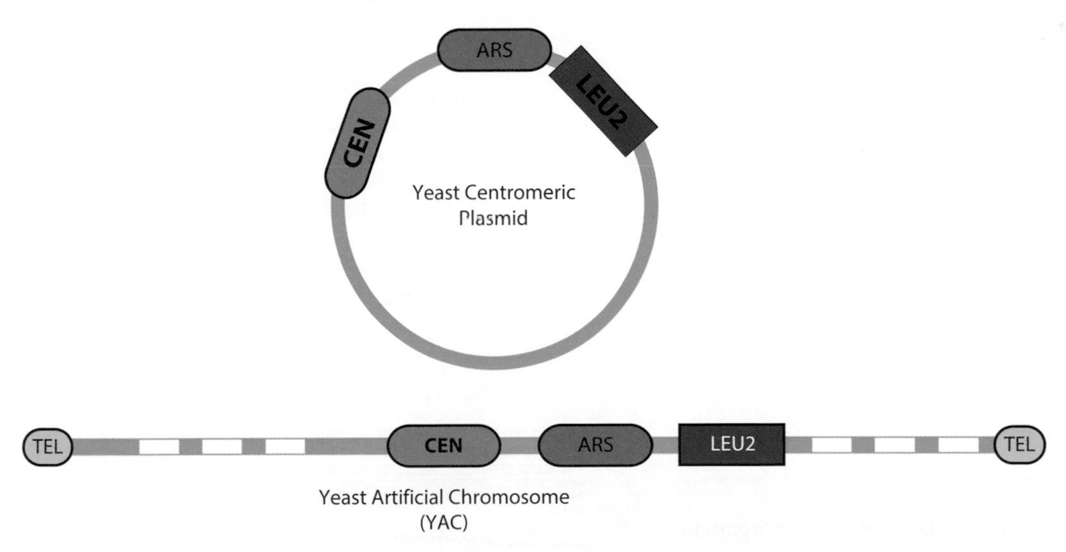

Figure 3.9 A yeast centromere (CEN) plasmid and a yeast artificial chromosome (YAC). A centromere (CEN) plasmid includes the centromere on the plasmid; as a result, the plasmid segregates mitotically as do the regular chromosomes and does not accumulate to high copy number. The centromere plasmid forms the basis for the yeast artificial chromosome, shown below. The linear chromosome includes telomere sequences at the termini, as well as the elements of CEN plasmid. A YAC includes 200 kb or more of other DNA, which makes YACs widely used as cloning vectors.

yeast. The circular CEN plasmids have a yeast origin of replication and a centromere, so they replicate autonomously but do not accumulate a high copy number; they also segregate normally during meiosis, although with higher rates of loss. Yeast centromere plasmids are in effect small circular chromosomes and were used in many of the early experiments on chromosome structure. Yeast centromere plasmids are used in yeast two-hybrid assays, as described in Chapter 11.

An important modification to the centromere plasmid was a linear version of it known as a **yeast artificial chromosome (YAC)**, also shown in Figure 3.9. To make a circular plasmid into a linear chromosome required more than simply opening the circle; telomeres were added to each end to make the YAC stable mitotically and meiotically. In addition to the telomeres, YACs have an origin of replication and a centromere, so they are stable chromosomes present at a single copy. YACs have been an important experimental vector because they can consist of more than a million base pairs of additional DNA that can be derived from any source. YACs are used to clone large fragments of the genomes of multicellular organisms, including humans.

Protein trafficking is one example of a fundamental process in eukaryotic cells studied by genetic analysis in yeast

Yeast has contributed so much to our understanding of eukaryotic molecular biology that it is almost impossible to single out one example for further discussion; and many examples appear throughout the book. A very brief one will be used here to show how the awesome power of yeast genetics has been applied to understanding a key property of eukaryotic cells. Randy Schekman shared the Nobel Prize in 2013 for this work.

Literature Link. Schekman 1992, *Current Opinion in Cell Biology* 4: 587–92; Mellman and Emr 2013, *Journal of Cell Biology* 203: 559–61

Eukaryotic cells are characterized by cellular organelles with different functions. In order to carry out these functions, proteins must be appropriately localized to their correct organelle; the particular constellation of proteins is characteristic for the organelle. Regardless of their ultimate destination in the cell, proteins are made on ribosomes, which are associated with the endoplasmic reticulum (ER); the presence of ribosomes on the endoplasmic reticulum led electron microscopists to refer to it as the rough ER. From the rough ER, proteins are transported to other parts of the cell, including the Golgi apparatus, the lysosome, and the plasma membrane, as summarized in Figure 3.10. The process of protein trafficking is fundamental to eukaryotic cells. Different proteins must have characteristics that target them to particular locations—signal sequences and post-translational modifications such as glycosylation, for example. But there must also be the "traffic controller" proteins that recognize the molecular signals and sort the proteins to the proper compartment.

Many of the traffic controller proteins—or, more properly, the sorting and secretion proteins that are common to all eukaryotic cells—were identified by genetic analysis in *Saccharomyces cerevisiae*. Different investigators, often using related but distinct approaches, have looked for yeast mutants in which proteins are missorted to the wrong cellular compartment. For example, one powerful genetic selection diagrammed in Figure 3.11 exploited the nature of sorting of a fusion protein made by attaching part of the N-terminus of carboxypeptidase Y (CPY) to the enzyme invertase. The fusion protein is diagrammed in Figure 3.11 A.

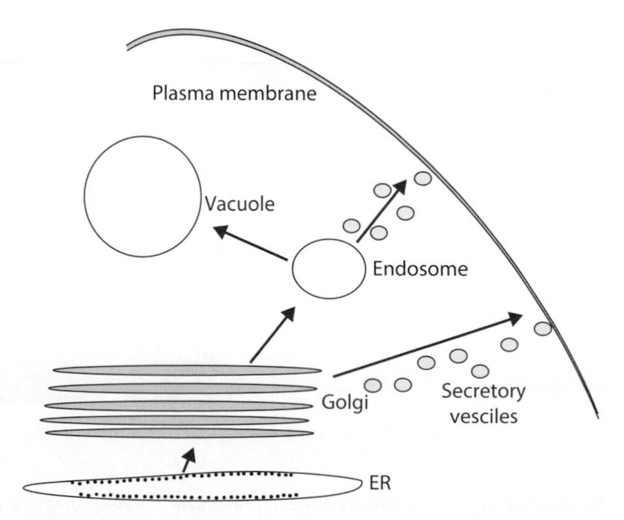

Figure 3.10 A schematic diagram of protein trafficking in eukaryotic cells. The yeast vacuole is the functional equivalent of the lysosome in other cells. Proteins from the rough ER move to the Golgi, where they are sorted to further processing in the endosome or in the vacuole, or secreted to the plasma membrane.

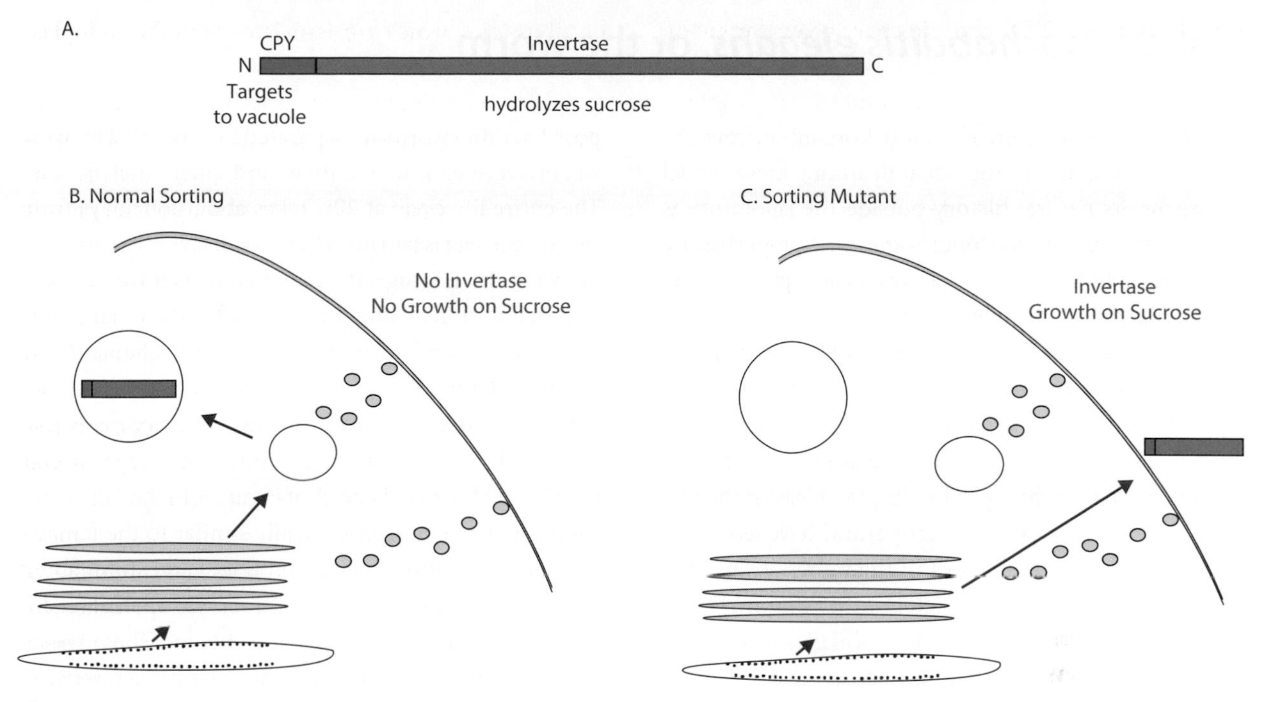

Figure 3.11 One method of identifying genes involved in protein trafficking. In Part A, a fusion protein between carboxypeptidase Y (CPY) and invertase is shown. CPY is normally found in the vacuole, while invertase is normally secreted from cells and breaks down sucrose. The coding region for the N terminus of the CPY protein is fused to the coding region of invertase. As shown in Part B, CPY targets this fusion protein to the vacuole. As a result of this location, invertase is not secreted from the cell and the cell cannot grow on sucrose. The scheme used to find protein-sorting mutations is shown in Part C. A mutation that mis-sorts the fusion protein will send it to the plasma membrane rather than to the vacuole. As a result of the mutation in protein trafficking, the fusion protein is secreted from the cell and the cell is able to grow on sucrose.

Carboxypeptidase Y is normally found in the vacuole in yeast cells, the *Saccharomyces* equivalent of the lysosome. Its enzymatic role is unimportant for this mutant selection; the important feature is its localization. CPY is synthesized as a longer inactive precursor protein (preproCPY) in the lumen of the ER. Here a signal sequence is cleaved from it to yield proCPY, which is then glycosylated and moved via transport vesicles to the Golgi. In the Golgi, the proCPY is sorted away from proteins that are destined for the cell surface and is instead transported to the vacuole.

The portion of the fusion protein that provided the genetic selection comes from invertase. Invertase is the enzyme that hydrolyses sucrose to glucose and fructose, providing a carbon source that yeast cells can utilize. An organism lacking invertase cannot grow on sucrose, because it cannot metabolize the sucrose to generate glucose or fructose. Invertase is secreted from the cells, so its localization is completely different from that of CPY. However, when the N-terminus from preproCPY is attached to invertase, invertase is sorted according the CPY localization and ends up in the vacuole, as shown in Figure 3.11 B. Invertase is still enzymatically active in this fusion protein, but the cells cannot grow on sucrose because invertase is not being secreted from the cell and sucrose is not hydrolysed. However, a mutation that *fails* to sort the CPY-invertase fusion protein to the vacuole but instead sends the protein to the cell surface will secrete invertase and will be *able* to use sucrose as a carbon source, as shown in Figure 3.11 C.

Hundreds of such mutations that affect protein sorting have been found by yeast genetic selections and screens like this one. Through the procedures described in Chapter 4, these mutations have been assigned to different genes by complementation tests; more than forty such vacuole protein sorting or *vps* genes are known in yeast, nearly all of which are also found in all other eukaryotes. In this genetic analysis, yeast served as a model for universal eukaryotic properties of protein trafficking.

3.3 *Caenorhabditis elegans*, or the worm

C. elegans was employed as a model organism from the beginning of its use in the lab and, among these model organisms, its natural history outside the laboratory is probably the least understood. Some of its properties are shown in Table 3.1. In the lab, it lives on agar plates spread with *E. coli*, crawling along the surface eating bacteria, as shown in Figure 3.12. Worms are about 1 mm long and are barely visible without a microscope; nearly all genetic experiments are done using a dissecting microscope, and worms are transferred manually from one plate to another.

Worms are morphologically simple, at least at the resolution of a dissecting microscope, and have few obvious anatomical features. Most of the mutations used as genetic markers in *C. elegans* affect either the movement or the morphology of the worm; typical genetic markers include uncoordinated (*unc*) mutations that affect the motion of the worm on the surface of the agar plant and dumpy (*dpy*) mutations that result in a short fat worm.

C. elegans has two sexes, but no females

The life cycle of *C. elegans* consists of an egg stage, four larval (or juvenile) stages termed L1 through L4, and an adult, as diagrammed in Figure 3.13. The stages of the post-hatching worm are separated by molts at which the worm secretes a new cuticle and sheds the old one. The entire life cycle at 20C takes about four days from the time an egg is laid until that worm lays eggs. An alternative to the L3 stage, the dauer larva, is formed under stress conditions. Mutations that affect dauer larva formation are described in the case study for Chapter 6 and in Chapter 12.

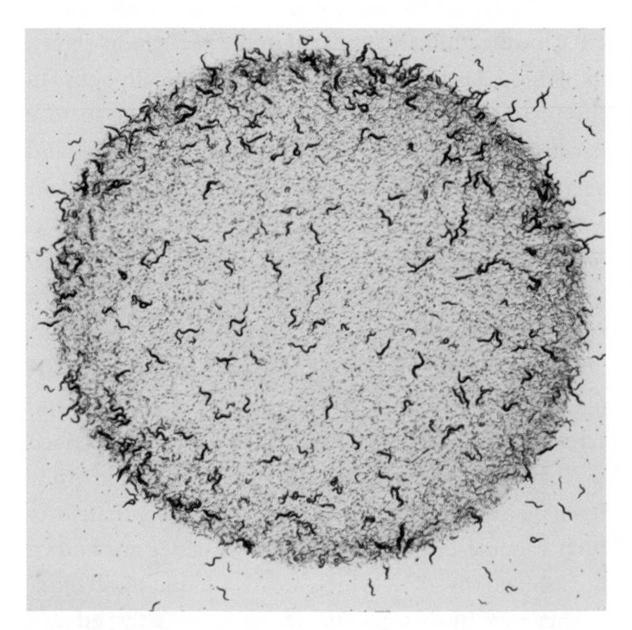

Figure 3.12 ***C. elegans*** **in the lab.** ***C. elegans*** **crawls on the surface of agar plates, eating bacteria that have been spread as a lawn.** Most manipulations are done using a wire pick while viewing the worm through a dissecting microscope.

One of the distinctive features of *C. elegans* is that it has self-fertilizing hermaphrodites rather than true females, as shown in Panel A of Figure 3.14. An hermaphrodite, which is morphologically similar to the females of related nematode species, produces and stores about 300 sperm at the beginning of meiosis. Spermatogenesis is then shut off and oogenesis begins. These sperm will fertilize an ovum from the same germline internally, if no sperm from males are present. The capacity for self-fertilization makes it easy to maintain homozygous strains, and even mutations with severe effects on movement and morphology (such as paralysed mutants) can be maintained as homozygous hermaphrodites. The number of sperm produced by hermaphrodite determines the number of self-progeny, so a wild-type hermaphrodite produces about 300 offspring.

A male is shown in Panel B of Figure 3.14. Sex is determined by the number of X chromosomes. Hermaphrodites have a pair of X chromosomes (abbreviated XX), whereas males have a single X chromosome and no Y chromosome (abbreviated X0). Loss of an X chromosome during meiosis in the hermaphrodite produces a nullo-X gamete, which results in a male when fertilized by a gamete with the usual one X chromosome. Males arise spontaneously during hermaphrodite self-fertilization at a frequency of about 0.2 percent.

Two modes of reproduction are possible, as shown in Figure 3.15: self-fertilization by hermaphrodites and cross-fertilization between males and hermaphrodites. Nearly all the offspring arising from self-fertilization in an hermaphrodite are themselves hermaphrodites. In order to perform a genetic cross, males are mated to hermaphrodites. Cross-fertilization of an hermaphrodite by a male produces 50 percent males and 50 percent hermaphrodites among the cross progeny, and male strains can be easily created and maintained by mating

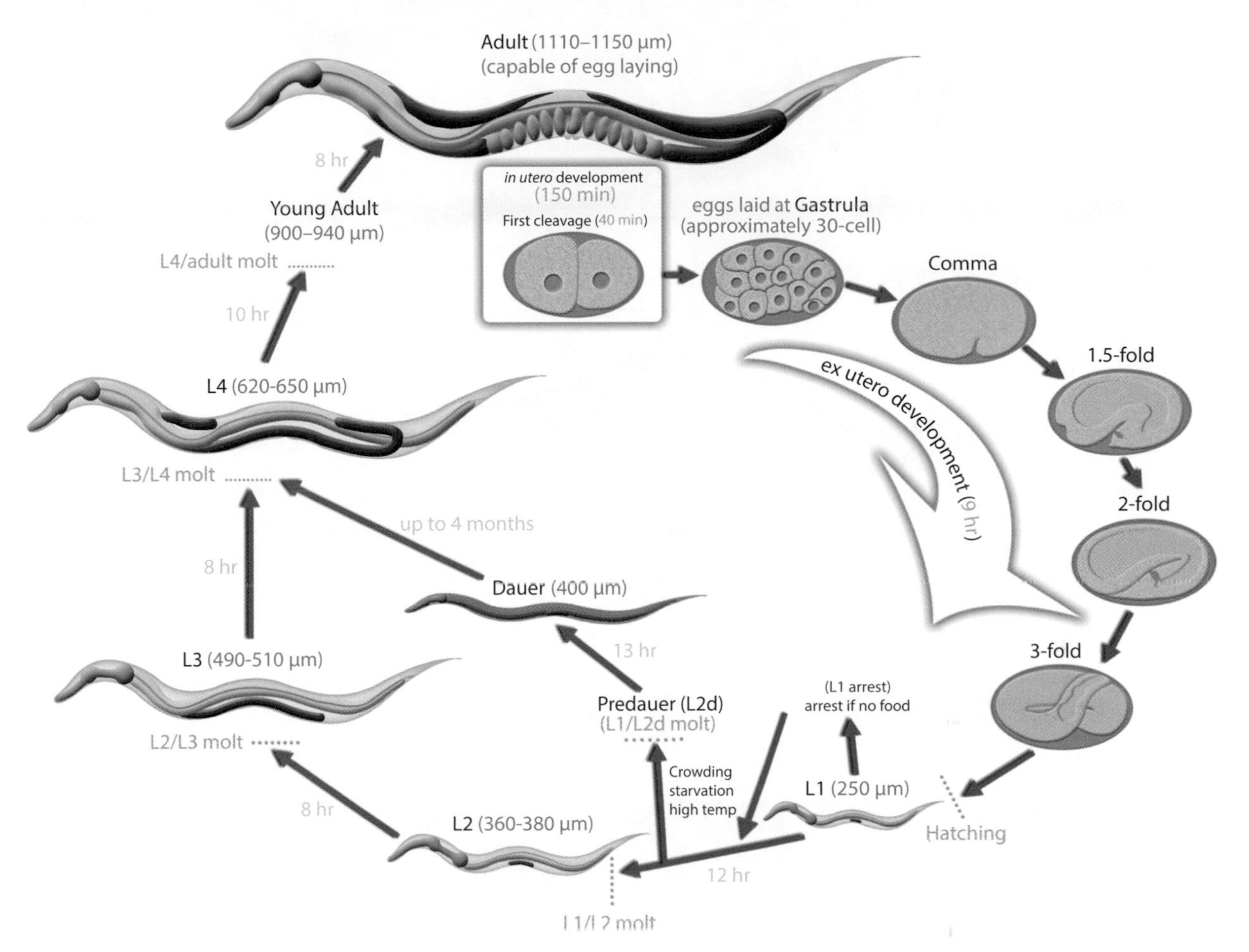

Figure 3.13 The life cycle of *C. elegans*. The entire life cycle from the time an egg is laid until that worm lays eggs is about four days at 20 C. The cycle consists of an egg, four larval stages (L1 through L4), separated by molts when the cuticle is replaced, and finally an adult. The dauer larva is an alternative to the L3 stage that forms under stress conditions. Used by permission from Wormatlas.org.

with hermaphrodites. Although sperm from a male outcompete self-sperm from an hermaphrodite during a mating by an unknown mechanism, for all practical purposes the offspring of a male-by-hermaphrodite mating include both cross-progeny and self-progeny. Males and hermaphrodites are readily distinguished, the elaborate fan-like tail in the male being the feature that is most easily recognized. The sexes also differ in size, movement, and behavior.

Nematodes have precisely defined cell lineages

One of the unusual features of *C. elegans*, as of all nematodes, is that the number of cells is small and to a large extent invariant between individuals. This feature was noted by classical embryologists such as Theodor Boveri more than a century ago; Boveri was able to track the early cell divisions in the embryo of the parasitic nematode *Ascaris*. *C. elegans* is transparent, so that all the internal structures and cells can be observed in live worms by light microscopy. All the cell lineages from fertilization to adulthood have been traced in *C. elegans*, so the complete ancestry of every cell at every stage is known. The cell lineage of the early embryo is shown in Figure 3.16. The precision in the location and timings of the divisions is impressive; an observer can watch a cell divide, take a break for a few minutes, and come back in time to see the next cell division occur on schedule. The relatively few variable cell fates involve discrete "decisions" between only a few cells; if one cell adopts cell fate and lineage pattern A, its partner cell adopts cell fate and lineage pattern B, as illustrated in Figure 3.17 for the anchor cell (AC) and the ventral uterine cells (VU). Because these cell signaling events are so precisely delimited, many of them have been thoroughly studied by genetic analysis.

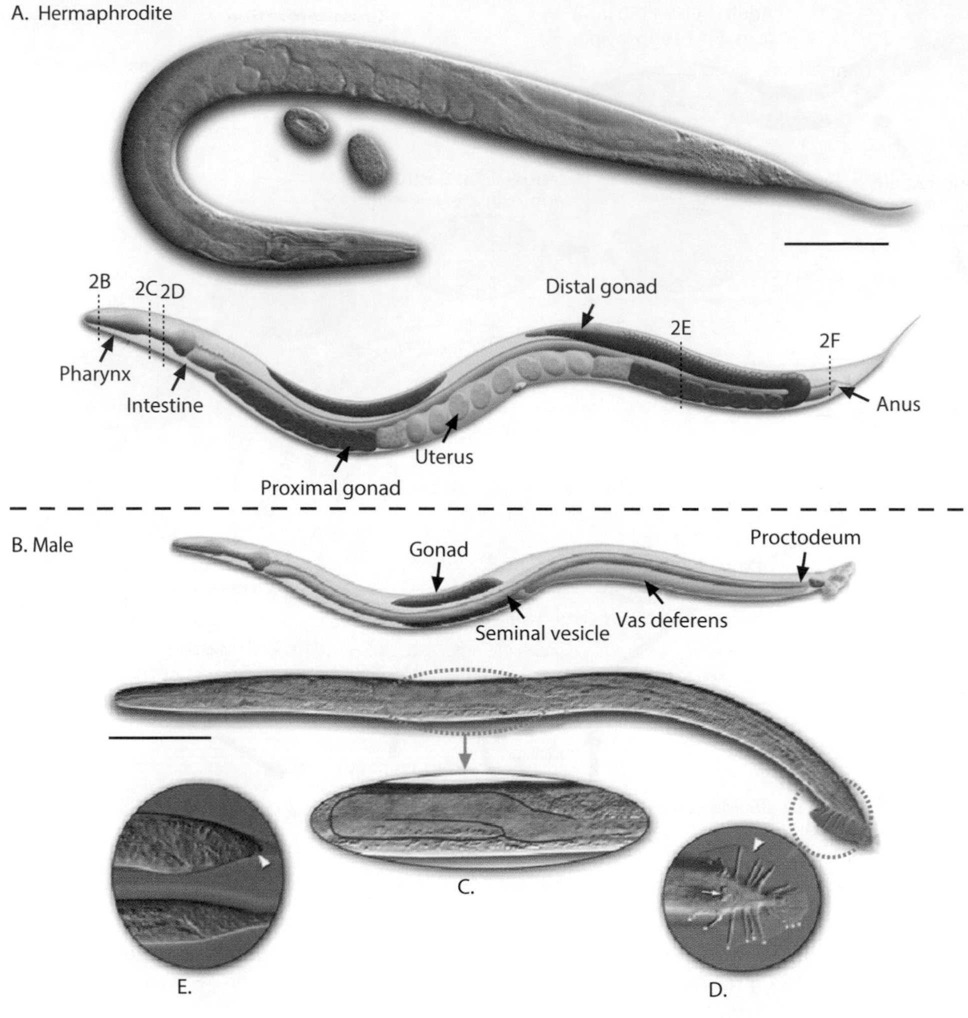

Figure 3.14 Sex differences in *C. elegans*. Worms have two sexes, hermaphrodites (Panel A) and males (Panel B). Hermaphrodites have two X chromosomes (XX), whereas males have one X chromosome and no other sex chromosome (X0). Hermaphrodites have a gonad with two arms centered on the uterus and vulva in the middle of the animal. The male gonad has one arm that begins in midbody and extends to the posterior. Males have a elaborate fan-shaped tail used in mating. Used by permission from Wormatlas.org.

The precise cell *lineages* also result in precise cell *locations*, since there is relatively little cell migration. Thus the intercellular connections have also been described with precision and reconstructed by electron microscopy. For example, there are exactly 302 neurons in the adult hermaphrodite, and the complete wiring diagram with all of the connections has been compiled from electron micrographs. The roles of individual cells have been determined by genetic analysis—that is, finding a mutation in which the cell does not form—and by laser ablation. Typically, if a cell is destroyed by a laser beam, no other cell replaces it and the worm will lack the function of that cell. For those cells whose functions are replaced, the pool of cells that are capable of adjusting their fate and cell division pattern is known. This inability to replace missing cells has been a particularly useful quality for studying cell interactions and neurobiology. Laser ablation of a certain sensory neuron can result in the failure to respond to a specific environment signal such as some class of chemical, or to a gentle touch.

It was noted during observations of the cell lineage pattern that some cells are produced and then die. These were among the first detailed descriptions of programmed cell deaths that occur during metazoan development. Genetic, cellular, and molecular analysis of this cell death process in *C. elegans* formed the foundation

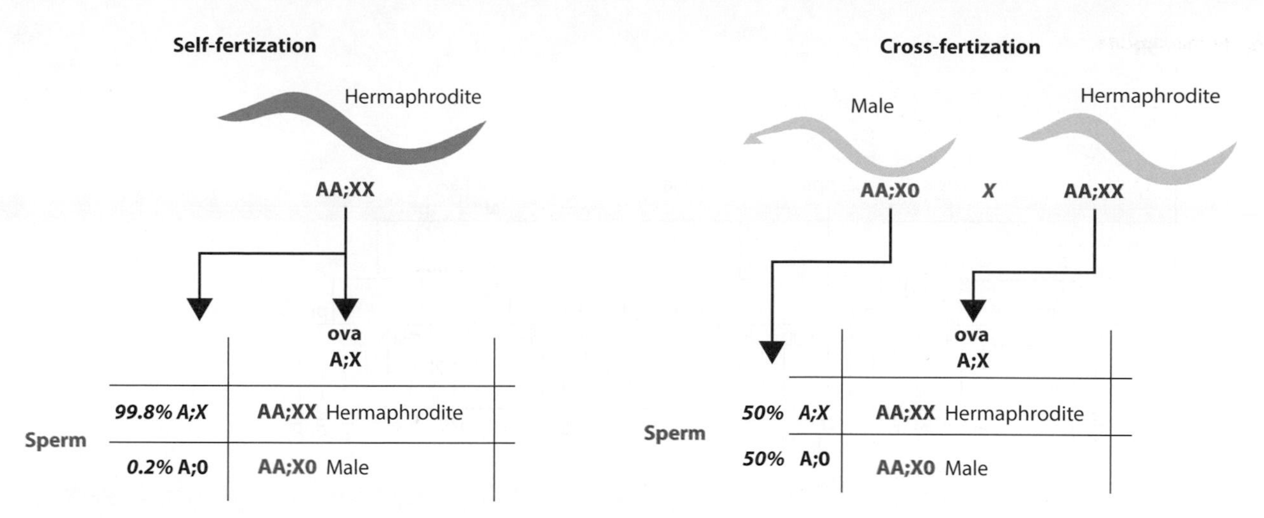

Figure 3.15 Modes of reproduction in *C. elegans*. *C. elegans* can reproduce either by self-fertilization in a hermaphrodite or by cross-fertilization between a male and a hermaphrodite. A haploid set of five autosomes is indicated by A, and the X chromosome by X. A diploid hermaphrodite has two X chromosomes (AA;XX), whereas a male has only one X chromosome (AA;X0). A hermaphrodite produces both sperm and ova that can fertilize each other internally. Approximately 99.8 percent of the gametes have an X chromosome, so most of the offspring of self-fertilization are hermaphrodites. About 0.2 percent of the games lack an X chromosome, which is symbolized by A;0. When a gamete without an X chromosome fertilizes a game with an X chromosome, the offspring is AA;X0, which is male. For cross-fertilization, a male can mate with a hermaphrodite. Since half of the sperm produced by a male have an X chromosome (symbolized A;X) and half lack an X chromosome (symbolized A;0), half of the offspring of cross-fertilization are hermaphrodites and half are males.

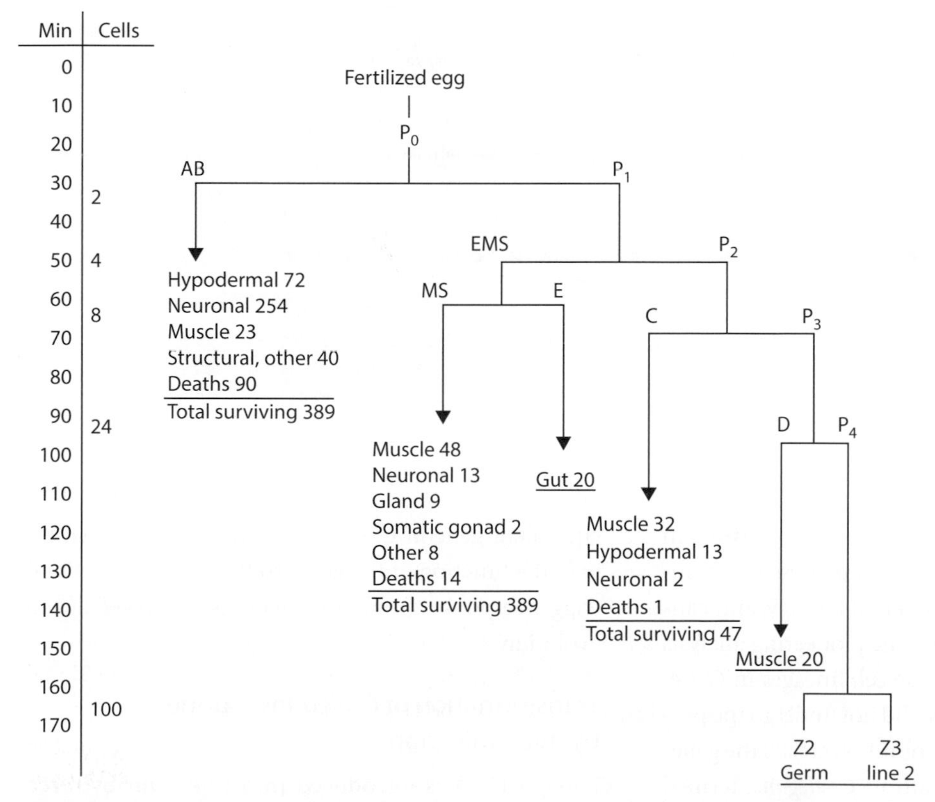

Figure 3.16 The early embryonic lineage of *C. elegans*. The lineage of every cell in *C. elegans* has been determined by observing cell divisions in living worms by light microscopy. The P cells are stem or blast cells, which divide to produce another P cell and the founder cell (such as AB, EMS, and so on) for a particular cell type or lineage; cell divisions of the founder cells are not shown, but the total number and types of cells derived from each founder cell are summarized. Redrawn from the Wormbook.org.

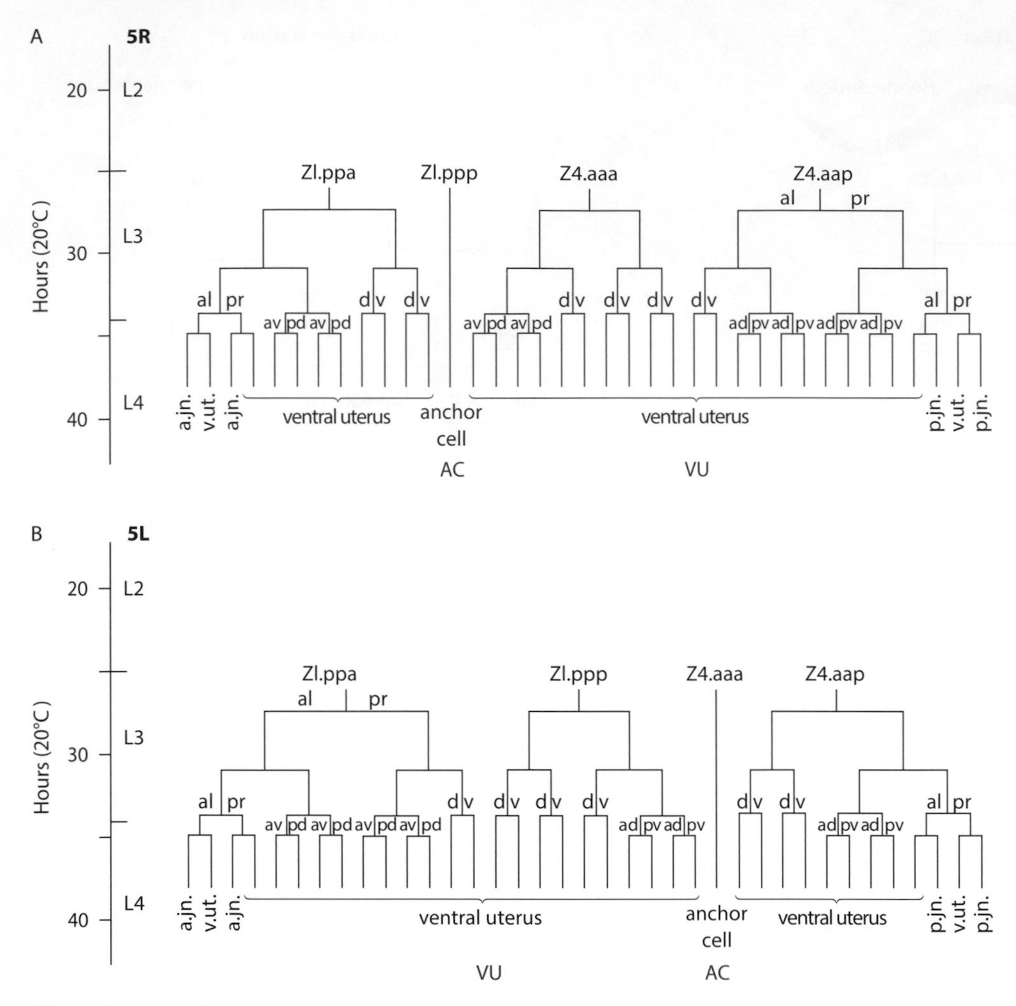

Figure 3.17 Cell decisions in *C. elegans*. Most cell lineages and fates are invariant between individuals in *C. elegans*. Even for those few variable cell lineages, the variation is strictly limited. This example shows the cell lineages of four cells, called Z1paa, Z1ppp, Z4aaa, and Z4aap, which produce the anchor cell (AC) and the ventral uterus (VU). The anchor cell can be derived from either Z1ppp (at the top as pattern 5R) or from Z4aaa (at the bottom as pattern 5L). The cell that does not give us to the anchor cell produces the ventral uterus. In either pattern, Z1paa and Z4aap do not change their cell lineage. This regulation occurs during the L3 stage, the lineages occurring during the L4 stage, as shown in the time scale at the left of the figure. Redrawn from the Wormbook.org.

for understanding the broader process of **apoptosis** in other animals. Apoptosis is a central process in immunology, neurobiology, developmental biology, and cancer biology, among other fields. The pioneering analysis was made possible by the precise cell lineages in *C. elegans*, where mutant worms that did not undergo the programmed cell deaths could be found. Nearly all the genes identified as affecting cell death in *C. elegans*, termed the *ced* genes, have one or more orthologous genes in humans, often carrying out a very similar cellular function. In fact some of the mammalian genes involved in apoptosis are capable of functionally replacing the equivalent gene in *C. elegans*, which indicates that not only the function of the gene itself but also its ability to trigger responses from other genes is conserved during evolution.

Transformation of *C. elegans* is done by microinjection

Foreign DNA is introduced into *C. elegans* by direct injection of linear DNA into the gonad using a needle, as illustrated in Figure 3.18. The injected DNA forms a complex linear array, usually a head-to-tail concatamer of 50 to 200 copies. Somewhat remarkably, this DNA

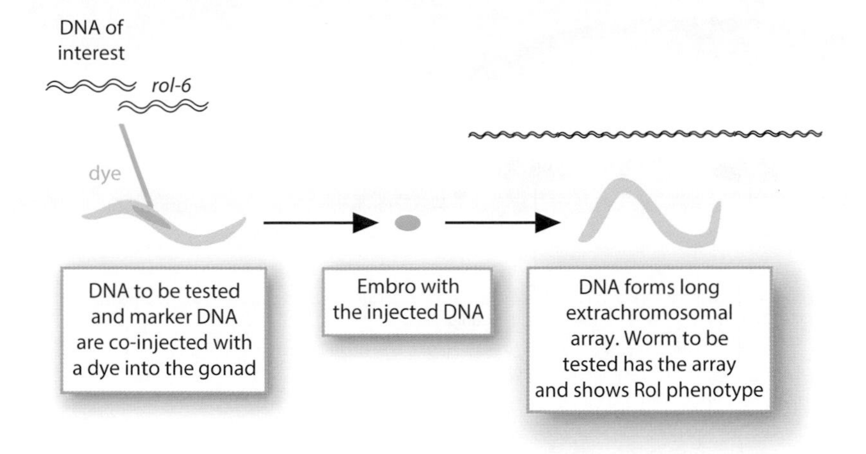

Figure 3.18 Transgenic worms and microinjection. Transgenic worms are produced by injecting DNA directly into the gonad. As shown in the figure, the DNA mix to be injected includes a dominant mutant allele of the *rol-6* gene, which causes the worm to move in a corkscrew motion. The DNA is often, but not always, co-injected with a dye, shown here in pink. The resulting egg grows into a worm with rolling movement. The DNA has been made into a complex array of both the *rol-6* gene and the other DNA, with as many as 200 copies in a long linear array. Other transgenic markers, such as ones encoding fluorescent proteins, are often used rather than *rol-6*.

array is maintained stably during both mitotic and meiotic divisions, although it can be lost at unpredictable and sometimes high frequencies.

Maintenance of the injected extrachromosomal array is a function of an unusual property of nematode chromosomes. Unlike the situation in most other eukaryotes, nematode chromosomes do not have localized centromeres. Instead, microtubules from the mitotic spindle can attach anywhere along the length of the chromosome. In classical cytology, chromosomes like this, without localized centromeres and spindle attachment, are termed **holocentric**. The proteins associated with the kinetochore in other organisms are also found in *C. elegans*, but they localize along the length of the chromosome rather than being targeted to one specific region at the centromere.

Thus, when DNA is injected into the gonad of a *C. elegans* hermaphrodite in a transformation experiment, it forms a complex repeat sequence array similar to the complex repeat sequences found at the centromeres of nearly every other eukaryote. The repeated DNA array is then transmitted during cell division. In order to determine that a particular offspring of an injected hermaphrodite has inherited the extrachromosomal array, some additional genetic marker is usually included in the injection mixture. Widely used markers include a dominant mutant allele of the collagen gene *rol-6* that causes the worm to move in a corkscrew rolling motion, or a fluorescent protein such as dsRed or one of its many derivatives, as shown in Figure 3.18.

Extrachromosomal arrays of injected DNA work very well for many types of transgenic experiments, as described in subsequent chapters. However, other experiments such as targeted gene replacements and disruptions require that the DNA integrate into the chromosome; hence many investigators integrate the array into the chromosomes by introducing double-stranded DNA breaks. Integration is at random locations, so one limitation of working with *C. elegans* continues to be that targeted gene replacements cannot be done efficiently.

Mutations and the cell lineage patterns are used together for genetic analysis in *C. elegans*

Many examples from genetic analysis in *C. elegans* are used throughout the book. In particular, dauer larva formation is discussed in Chapters 6, 11, and 12 and sex determination in worms is also used as example in Chapters 11 and 12. In this section we describe the development of the vulva to illustrate how the cell

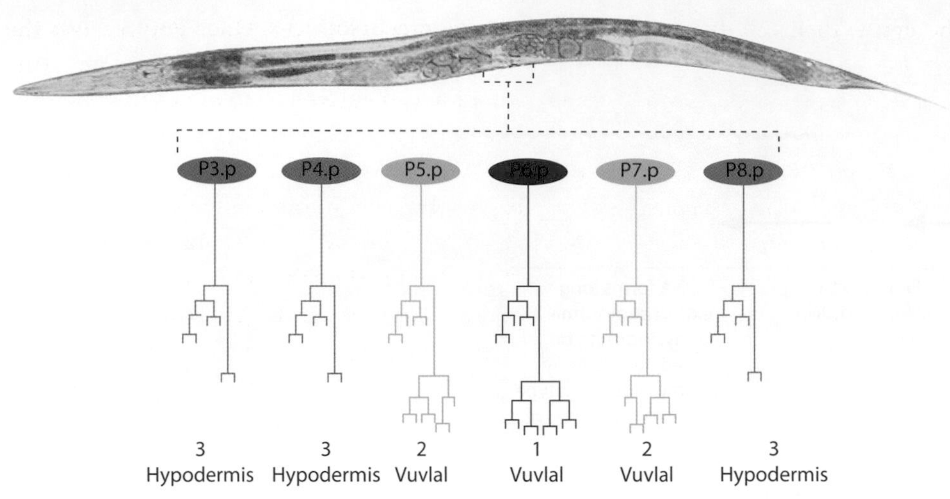

Figure 3.19 The cell lineages that produce the vulva in *C. elegans* hermaphrodites. As described in the text, the vulva and surrounding structures arise from the cell divisions of three cells, known as P5p, P6p, and P7p or the vulva precursor cells (VPC). In wild-type worms, P6p responds to a signal from the anchor cell, which would lie directly above it in this diagram, and divides to produce eight cells; this is known as primary lineage (1°). The flanking cells P5p and P7p produce seven cells each, by the secondary lineage (2°). P3p, P4p, and P8p form, each, part of the hypodermis in wild-type worms by the tertiary lineage (3°) shown. These cells can produce the primary or secondary lineages in various mutants or with different experimental manipulations. Redrawn from various sources, including Sternberg, "Vulval Development," in the Wormbook.org.

lineages and mutations were used together in a genetic analysis of a signal transduction pathway.

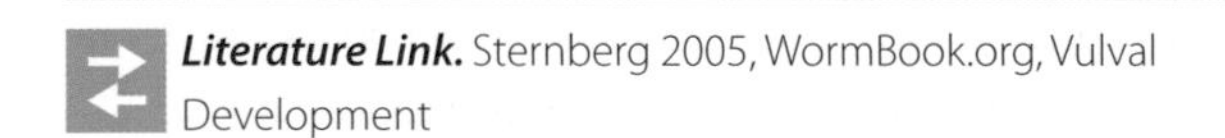
Literature Link. Sternberg 2005, WormBook.org, Vulval Development

The vulva in the adult hermaphrodite has twenty-two cells in precisely defined locations arising from invariant cell lineages. Three cells, called P5p, P6p, and P7p, undergo a series of cell divisions during the L4 stage to produce the vulva in wild type, as shown in Figure 3.19. Two different patterns of cell lineages occur: the one called the primary lineage, which generates the eight cells produced by P6p, and the one called secondary lineage, which generates the seven cells produced by P5p and P7p. The neighboring cells known as P3p and P4p (on the more anterior side) and P8p (on the posterior side) do not participate in vulva formation in the wild-type worm, but may regulate their cell lineage pattern to produce vulva cells in mutant worms,

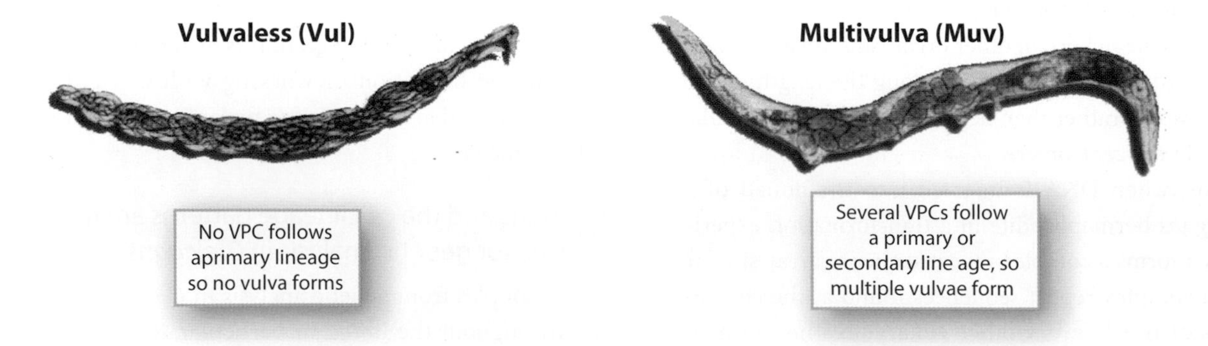

Figure 3.20 Mutant phenotypes from the vulva lineages. Two principal classes of mutations were found that affected the vulva lineages in Figure 3.19. Mutants in which none of the vulva precursor cells (VPC) adopt the primary lineage are vulvaless or Vul. Mutants in which more than one of the Pnp cells have adopted a primary lineage result in multiple partial vulvae, a phenotype known as multivulva or Muv.

so they are included in the figure; their cell division pattern is called the tertiary lineage. Collectively, the cells P3p through P8p are called vulva precursor cells—VPCs. The anchor cell—which does not form part of the vulva—lies adjacent to P6p. When the anchor cell is destroyed using a laser, a vulva does not form and none of the cells adopts the cell lineage pattern that they do in wild type. This provides evidence that the anchor cell sends a signal to the VPCs to trigger vulva formation. In particular, other laser ablations have shown that the signal between the anchor cell and P6p is critical for normal vulval development.

Many different mutations that affect the lineages of the vulva were found among mutant worms that were defective in egg-laying. Broadly speaking, these mutants had one of two phenotypes, as shown in Figure 3.20. Some do not form a vulva at all, a phenotype known as Vulvaless (Vul, Figure 3.20 A). By recognizing the role of the anchor cell in relation to P6p, some of these mutants are easy to understand. In some, the anchor cell itself does not form, or it lies in an unusual location. In others, P6p and the other VPCs do not form normally. Thus these do not form a vulva because one of the component parts, either the signaling anchor cell or the receiving VPC, is absent.

Other Vulvaless mutants have both the anchor cell and the vulva precursor cells, but still do not form a vulva. Subsequent analysis, by procedures described in subsequent chapters, showed that for some of these mutations the defect lies in the anchor cell, which seems not to send a required signal. In others mutations the defect lies in the VPCs, which behave as if they are not receiving the signal or not passing it on within the cell (i.e. "transducing" the signal) properly.

Another class of mutant phenotype is represented by worms in which more than one of the VPCs follows the primary lineage that only P6p does in wild type. These mutants have multiple partial vulvae, a phenotype known as Muv (Figure 3.20 B). At first blush, these mutants are not so easy to explain. Some of the mutations cause transduction of the intracellular signal in the absence of the signal from the anchor cell. Cells such as P4p or P8p, which normally do not receive the signal from the anchor cell and thus do not respond by producing vulva lineages, are now responding constitutively. Another group of mutations affects a different signal transduction process among vulva precursor cells. Once P6p receives the signal from the anchor cell, it signals its neighboring cells and inhibits them from responding to the anchor cell signal; this is a process of **lateral inhibition**. The normal response of P5p, for example, arises from a composite made up of the amount of signal it receives from the anchor cell and the amount of inhibitory signal from P6p. If the inhibitory signal is removed by mutation, P5p or P7p can also follow the primary lineage. By combining their knowledge of the cell lineages with the mutant phenotypes, the investigators were able to recognize how the cells signal to each other during vulval formation.

These mutations identified two different cellular pathways involved in human cancers

The analysis of this complex biological question relied on knowing the cell lineages in worms and on the ability to find mutations that affected these cells and their interactions. The entire analysis was done without knowledge of the molecular identity of the signals and without initial knowledge of how many different extracellular signal molecules were involved. By having genes that affect the processes—that is, mutants that failed in some parts of the differentiation but not in other parts—it was possible to learn the developmental and cellular logic of the signal transduction process. Once the genes were cloned, by processes similar to those described in Chapter 5, the molecular functions of these signal transduction pathways were revealed: the signal from the anchor cell is a molecule similar to the epidermal growth factor, and the receptor on P6p is a protein in the class of epidermal growth factor receptors. The Vulvaless mutations that failed in the intracellular signal transduction process were even more interesting. These genes defined the components of the Ras/Raf signal transduction pathway, a highly conserved signal transduction pathway in animals. Interestingly, some of the individual genes and proteins, including Ras and Raf, had been identified in mammals as oncogenes, since many tumors have mutations in one of these genes. The pathway was learned from genes, and the connection to cancer biology was made when the genes were cloned.

(To be historically more complete, at about the time when the role of Ras pathway was recognized as affecting the vulva lineages in worms, the role of Ras in affecting the phototactic response in *Drosophila* described in Chapter 1 was also being recognized. The two genetic analyses, together, connected the signal transduction pathway to the defects in the cancer cells.)

The lateral inhibition response depends on a different signal transduction pathway. The genes in this pathway were first described by mutations in *Drosophila melanogaster*, including the gene *Notch*, after which the pathway is now named. Again, the signal transduction pathway among these cells was inferred from the phenotypes of the mutant worms and from the interactions among the mutations, the analysis described in Chapter 10 and 11. Later on some human tumors and other genetic diseases were found to have mutations in the *Notch* pathway.

Through the genetic analysis of a very specific phenotype, the ability of a worm to make a vulva and lay eggs, two signal transduction pathways that are used in all animals were uncovered. These are further examples of universal properties of eukaryotes being revealed by the model organism. The equivalent signal transduction pathways in humans govern entirely different developmental processes rather than vulva formation. The pathway is the same, but evolution has "tinkered" with its use and applied it to different developmental processes.

3.4 *Drosophila melanogaster*, or the fly

No other experimental organism has such a long and continuous history of research that involves so many geneticists for so many years. The first mutation in flies, found in 1908, was *white* eyes, the mutation that persuaded scientists that genes were part of chromosomes. The first paper published in the journal *Genetics* was devoted to research on *Drosophila* and, in fact, more than half of the pages in the first two issues of *Genetics* were devoted to Bridges's classic work *Non-Disjunction as Proof of the Chromosome Theory of Heredity*. A comparison of some of the properties of *Drosophila* with other model organisms is shown in Table 3.1.

This history of research has generated a huge catalog of mutants, chromosome rearrangements, and other tools of genetic analysis in *Drosophila*. Reducing genetic research in *Drosophila melanogaster* to a few paragraphs is a daunting task; after all, a series of books entitled *The Genetics and Biology of Drosophila* ran into twelve volumes and was compiled before most molecular analysis or any genomic analysis had been done.

Drosophila melanogaster is raised in glass or plastic vials, originally in pint milk bottles with mashed bananas, but now with commercially available fly food that includes corn meal, agar, and yeast. At 24 C, the life cycle takes about ten days, as shown in Figure 3.21, with complete metamorphosis from crawling larvae to flying adults. There are three distinct larval stages called **instars**, followed by a period of pupariation; as pupae, most of the larval cells and organs degenerate, to be replaced in the adult fly. The third instar larvae have groups of undifferentiated cells termed the **imaginal discs**; these separate pockets of cells do not degenerate but will differentiate to give rise to the mouth parts, antennae, eyes, legs, wings, halteres (the modified hind wing), and the genitalia in adults, as

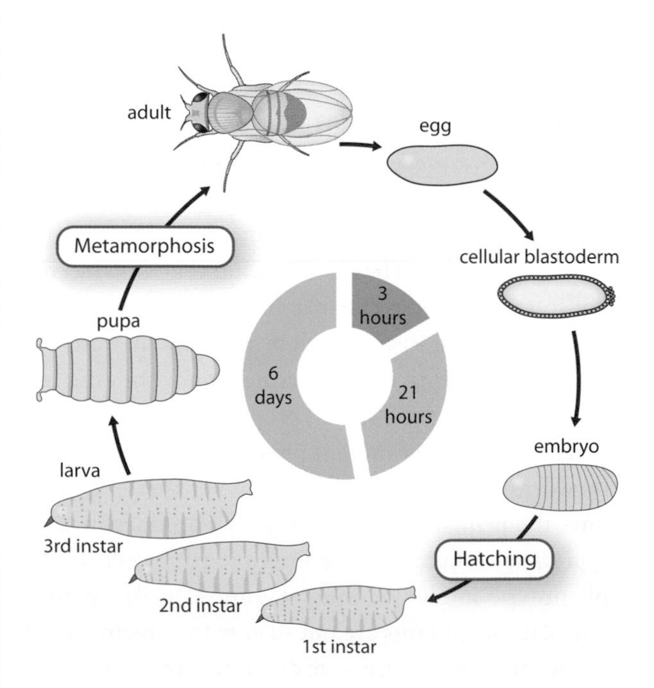

Figure 3.21 The life cycle of *Drosophila melanogaster*. The entire life cycle of *Drosophila* melanogaster can be completed in slightly more than ten days, depending on the growth temperature. The three larval instars crawl and feed on their own. During pupation, most of the larval cells die. Many adult structures arise from imaginal discs, as shown in Figure 3.22.

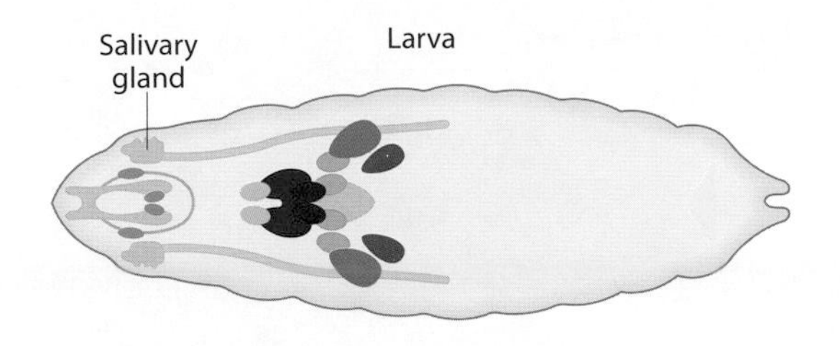

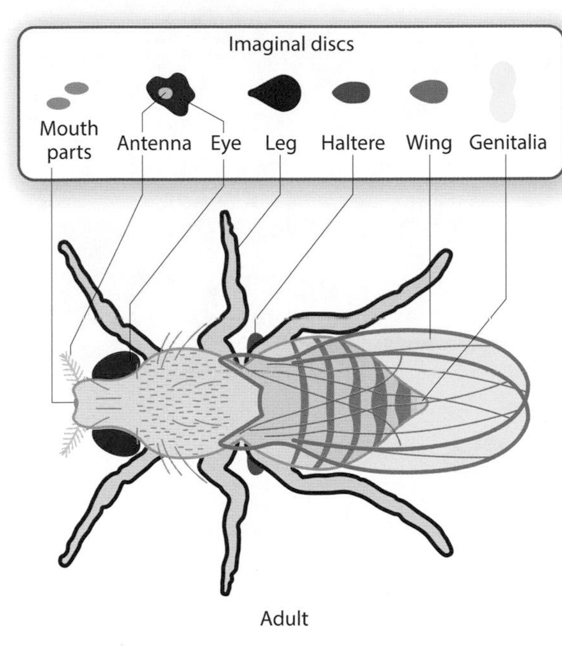

Figure 3.22 The imaginal discs in the *Drosophila* larva. Theses pockets of cells form no structures in the larva but differentiate during the pupa stage to form the adult structures shown.

shown in Figure 3.22. When an adult female emerges from the pupa case or ecloses, she is unable to mate for about twelve hours. These virgin females are used for crosses, and it is common to hear fly geneticists talk about "collecting virgins." Mated females can store sperm for several days and continue to mate with other males, so virgin females ensure that the proper cross has been done.

Flies have an intricate external anatomy, with many bristles and hairs, pigmentation differences in the body and the eye, structures such as wing veins, sex combs, antennae, and more. Seemingly all the anatomical features have corresponding mutant phenotypes—forked bristles, multiple wing hairs, black body, white eyes, crossveinless wings, and so on. Mutant strains are maintained by matings, often involving strains that have many defined mutations throughout the genome.

Literature Link. Mohr et al. 2014, *Genetics* 197: 1–18

Drosophila melanogaster has some peculiar features

Because it is such a well-studied model organism, it is tempting to think of *Drosophila melanogaster* as typical and of other organisms as strange. Nonetheless, some aspects of *Drosophila* biology are unusual; more precisely, the biology of dipteran insects (of which *Drosophila* is a representative), affects the way that *Drosophila melanogaster* is used as a model organism. First, unlike other model organisms, *Drosophila* has no recombination in males. In nearly all animals, the sexes have a different number and distribution of crossovers, and genetic maps are typically based on recombination in females. Dipteran insects are extreme in this regard. In *Drosophila*, males segregate their chromosomes at meiosis without chiasmata and recombination, and all crossing over occurs in females. The absence of recombination in males has been exploited by geneticists to maintain particular configurations of alleles.

Second, in certain tissues such as the larval salivary glands, *Drosophila* chromosomes replicate ten or eleven times, but the DNA strands do not separate. This results in very thick chromosomes known as **polytene chromosomes**, as shown in Figure 3.23. Polytene chromosomes have a distinctive and consistent pattern of light and dark bands, which are easily visible under the light microscope. The banding pattern of polytene chromosomes allowed Painter, Bridges, and others to construct detailed structural maps of the chromosomes with each of the five chromosome arms divided into twenty segments and the bands in each arm designated by a combination of letters and numbers. The break points of chromosome rearrangements and the genes could be located with respect to these bands, and the position of the gene on the chromosome can be described by different types of coordinates. For example, the *forked* bristle locus is found at position 1-56.7 on the genetic map (that is, on chromosome 1 and 56.7 map units from the reference point at the left tip) and at position 15F4-7 on the cytological map of polytene chromosomes.

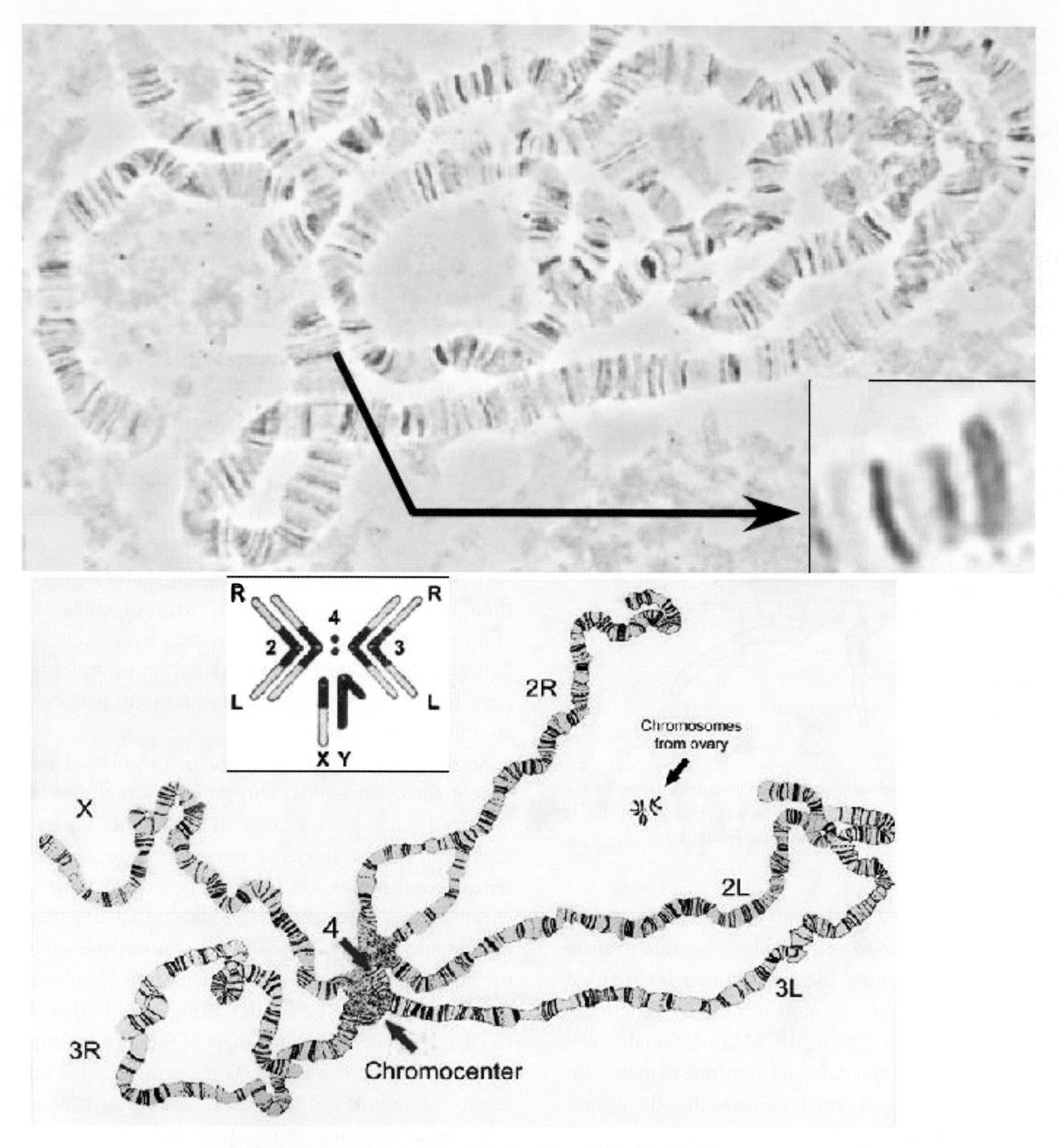

Figure 3.23 Polytene chromosomes in *Drosophila*. In this classic photo and diagram from Painter in 1934, the chromosomes from larval salivary glands were stained with a DNA dye, to illustrate the banding pattern. Each chromosome arm is labeled. Polyteny occurs when the DNA replicates but the sister chromatids do not separate. This image is in the public domain.

Transformation of *Drosophila* uses a transposable element

One of the most significant advances for *Drosophila* genetics in the molecular age has been the use of P elements as a vector for transformation. P elements are DNA transposable elements, and a complete P element of about 2.9 kb in length encodes a transposase that acts on the 31 bp inverted repeats at its ends so as to catalyse transposition of the full element, as shown in Figure 3.24. Like other transposable elements, most P elements

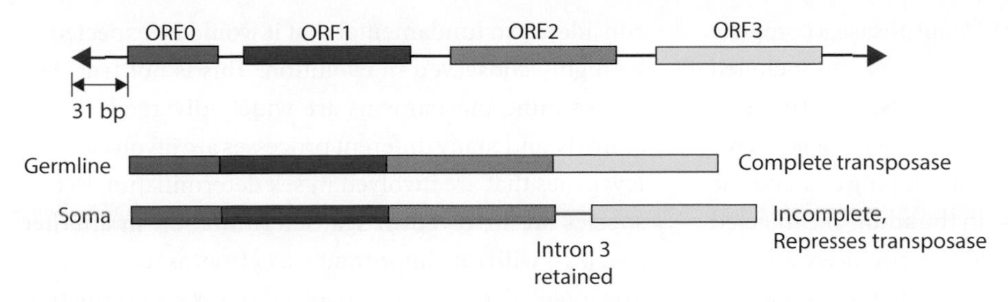

Figure 3.24 P elements in *Drosophila*. A complete P element has four open reading frames, referred to as ORF0 through ORF 3, and inverted repeats at the ends, as indicated by the red arrowheads. In the germline, the introns are removed and the four exons are spliced together to make active transposase. In the somatic cells, the third intron separating ORF2 from ORF3 is not removed, generating an inactive transposase. In fact these product represses transposition. Thus, unlike most transposable elements, the P element is a DNA transposable element that is mobile only in the germline.

are defective and cannot transpose autonomously, because either the transposase or the ends of the element have become mutated. However, an active transposase encoded at any site in the genome can catalyse the transposition of any element with the appropriate repeats at the ends, regardless of what that target element encodes.

P elements were apparently absent from the original laboratory strains of *Drosophila melanogaster* but were present in fruit flies in the wild. *Drosophila* geneticists learned to capture wild male flies and mate them to laboratory females. The F_1 offspring of this cross are normal but have a number of defects in their germlines that appear in subsequent generations. This syndrome of genetic alterations, known as **hybrid dysgenesis,** includes chromosome breakage, male recombination, sterility, and, most importantly, new mutations.

During the 1930s, hybrid dysgenesis became a preferred mechanism for producing new mutations at high frequency. The cross had to involve wild males and laboratory females, so paternal factors (hence the name P elements) were clearly implicated. The mechanism for hybrid dysgenesis was not resolved until the late 1970s, although the existence of a transposable element similar to what Barbara McClintock had shown in maize was one long-standing hypothesis. This hypothesis proved to be true, and hybrid dysgenesis is caused by the sudden introduction and mobilization of P transposable elements.

The laboratory strains that lacked P transposable elements also lacked the ability to regulate their transposition, so P elements were able to move at a high frequency in laboratory strains. The unusual feature that the effects are restricted to the germline is explained by alternative splicing of the transposase gene encoded by the P element, as shown in Figure 3.24. The splicing pattern that yields a functional transposase occurs in the germline, whereas the splicing pattern in somatic cells results in an incomplete and non-functional transposase.

Rubin and Spradling harnessed the effects of the P element as a transformation vector by replacing the transposase gene with other DNA sequences; they retained the terminal inverted repeats as target sequences, as shown in Figure 3.25. Since these engineered elements

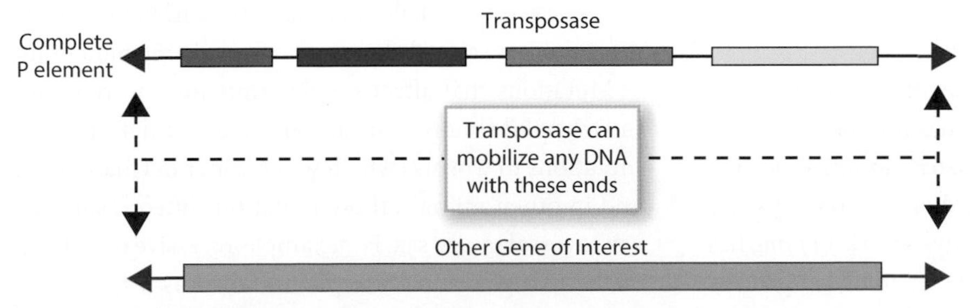

Figure 3.25 P elements as transformation vectors. Any active transposase from a P element can catalyse movement of DNA between the P element ends. For transgenic experiments, the transposase exons are removed from the P element and replaced by a gene to be transformed into the fly, shown here in purple. The inverted repeats are retained at the ends. When this is injected into a fly, it cannot move on its own but can be moved by active transposase from a different P element. Integration into the genome occurs at random sites.

cannot move without a source of transposase, a complete P element encoding the transposase was also included in the transformation mixture. This DNA mixture of a complete P element and an engineered element is micro-injected into the pole cells of the developing embryo; the pole cells become the germline in the adult. The injected fly has active P transposase and the engineered P element in its germline, so it is analogous to the F_1 offspring of a dysgenic cross. The P elements, both the complete element and the engineered element, insert at random throughout the genome in the offspring of injected fly. Once inserted, the P element is inherited along with the rest of the chromosome. Thus, while the insertions result in stable inheritance, they occur at random locations, and targeted insertions to particular sites are rare. If the transposase gene is subsequently outcrossed and removed from the chromosome, the engineered element becomes a stable integrant in the genome. (The concept of outcrossing is discussed more completely in Chapter 4 in our discussion of mapping newly arisen mutations.)

Many modifications to this basic procedure of P element transformation have been introduced, but the fundamental strategy has remained the same. The most significant modification has been the production of *Drosophila* strains with the P transposase integrated at known specific sites into chromosomes. The transposase does not have the terminal repeats and cannot move once inserted; these are referred to as "wings clipped" elements. With integrated transposase, there is no need to inject a complete P element along with the engineered transformation vector. In addition, since the location of the "wings clipped" element in the genome is known, outcrossing the newly arisen, transformed line in order to eliminate transposase and subsequent transposition is simplified.

Sex determination in *Drosophila* involves a pathway of alternative splicing

Drosophila provides many examples of advanced genetic analysis, a few of which will arise in this book. Sex determination in *Drosophila melanogaster* was introduced in Chapter 1 and will be described here in some greater depth to illustrate how genetic analysis in *Drosophila* has provided insights into an important biological process in nearly all multicellular organisms.

Nearly all animals have some form of sexual dimorphism, so sex determination is a truly fundamental biological process. Naively, sex determination could be considered so fundamental that it would be expected to be highly conserved in evolution. This is not true. Sex determining mechanisms are widely diverged among animals, and many different processes are involved. Very few genes that are involved in sex determination in one species are involved in sex determination in another species—with one important exception, as we will note. The molecular processes involved in sex determination in *Drosophila* are used in all other eukaryotes; they are just not involved in sex determination.

In *Drosophila*, sex is determined by the number of X chromosomes, so that embryos with one X chromosome become males and those with two X chromosomes develop as females. The question of how a *Drosophila* embryo "counts" the number of X chromosomes is itself an outstanding example of the application of genetic analysis that we can only quickly summarize here. The X-linked gene known as *Sex-lethal (Sxl)* is the key regulatory switch.

Literature Link. Cline and Meyer 1996, *Annual Review of Genetics* 30: 637–702

Sex-lethal is among the first genes to be expressed in *Drosophila* embryos and, because it is X-linked, its level of expression is initially different between those embryos with one X chromosome and two X chromosomes. The level of *Sex lethal* expression from two X chromosomes is enough to trigger female differentiation, whereas the level of expression from one X chromosome is not enough to trigger female differentiation and males result, as shown in Figure 3.26. Above a threshold, *Sxl* activates its own expression, so that this relatively slight difference in expression between 1X and 2X flies becomes amplified through auto-activation, as illustrated for 2X flies in Figure 3.26. *Sex lethal* activation sets the switch, but what are its targets for female sex differentiation? Potential target genes were already known among existing mutations.

Mutations that affect sex determination were found in *Drosophila* many years ago and, like sex-determining mutations in worms (which we consider in Chapter 11) and in other animals, these mutations often result in a **sex reversal** in one sex. For example, recessive mutations in the two genes in *Drosophila* known as *tra* and *tra2* transform 2X flies into the male differentiation pathway. (The 2X pseudomales are not fertile in *Drosophila*, since the germline is not transformed.) Conversely, 1X *tra* or *tra2* mutant animals are phenotypically normal males.

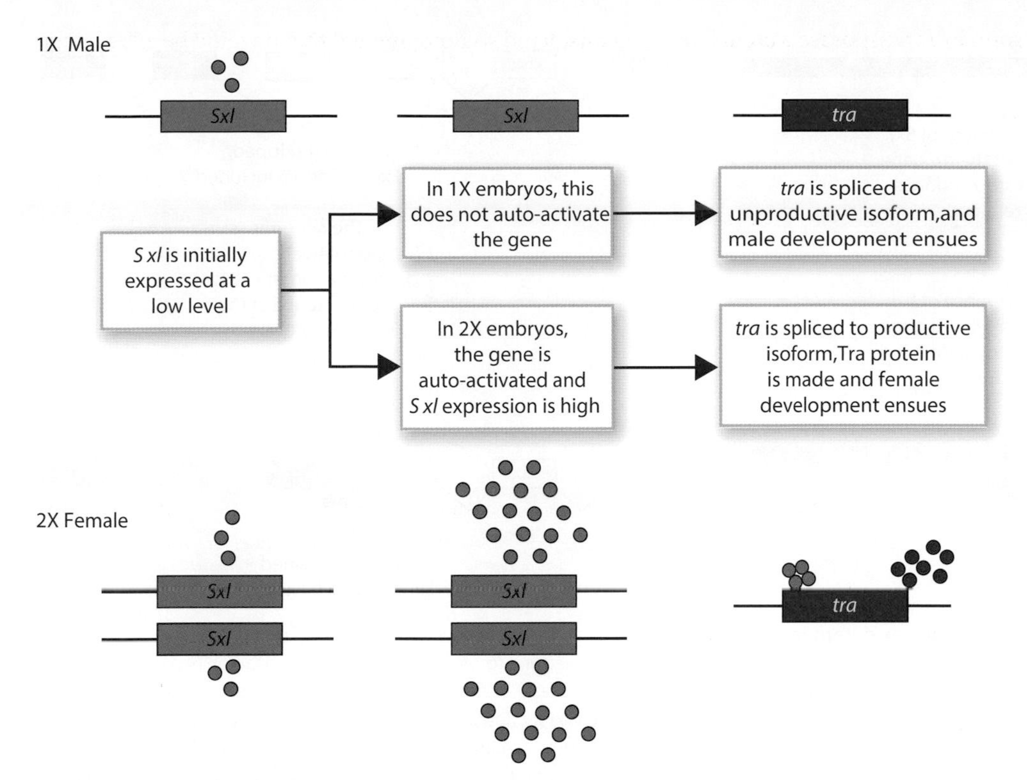

Figure 3.26 Sex determination and the number of copies of the *Sxl* gene in *Drosophila*. Sex determination in *Drosophila* depends on the number of X chromosomes; flies with one X chromosome are males, whereas flies with two X chromosomes are females. The key gene that initiates the sex determination pathway is the X-linked gene *Sex-lethal (Sxl)*. *Sxl* is expressed very early in the embryo, and because the gene is X-linked, the expression is higher in 2X embryos than in 1X embryos. In 2X embryos, the expression of *Sxl* is high enough for the gene toauto-activate and regulate its own expression. This auto-activation amplifies a twofold difference in the level of *Sxl* expression into a much larger difference between 1X and 2X embryos. The *tra* gene is a direct target of *Sxl* activation; as shown here, this activation only occurs in 2X flies, where *Sxl* lethal activity is high. As a result of *tra* expression, female differentiation occurs. In the absence of *tra* expression in 1X flies, male differentiation occurs.

Since the absence of *tra* and *tra2* functions results in 2X animals becoming males with no obvious effect on 1X differentiation, these two genes are inferred to be needed for normal female development, but not for normal male development in the somatic tissues. The genes exert their effects on all somatic tissues. The *tra* gene is a target of *Sxl* regulation, as summarized in Figure 3.26 and in Figure 3.27. The *tra2* gene is not a direct target of *Sxl* regulation but works with the Tra protein to regulate genes further downstream.

Sex-lethal encodes a protein that is involved in RNA splicing, and its effects can be explained by its regulation of alternative splicing. The *Sxl* protein has numerous RNA targets, including its own transcript and that of the *tra* gene. The first two exons of the *tra* gene are diagrammed in Figure 3.27. The *tra* gene is transcribed in both 1X and 2X flies, but the RNA splicing pattern of the second exon is different in the two sexes. In the absence of Sxl protein in 1X flies, as shown in the top half of Figure 3.27, an upstream splice site in exon 2 is used; this portion of exon 2 includes a stop codon, so the protein produced in males is truncated at this position and non-functional. In the presence of Sxl protein in 2X flies, as shown in the bottom half of Figure 3.27, a more downstream splice occurs in exon 2 of the *tra* gene. This splice site is located after the stop codon in exon 2, so that the transcript produces a Tra protein that is full-length and functional in females. The regulation of splicing does not stop with Sxl protein, however. The Tra protein is itself a splicing factor, which also regulates alternative splicing of its downstream target genes.

Mutations in another gene, known as *doublesex (dsx)*, have a different effect than the mutations in the *tra* and *tra2* genes. Rather than sex reversal, both 1X and 2X flies

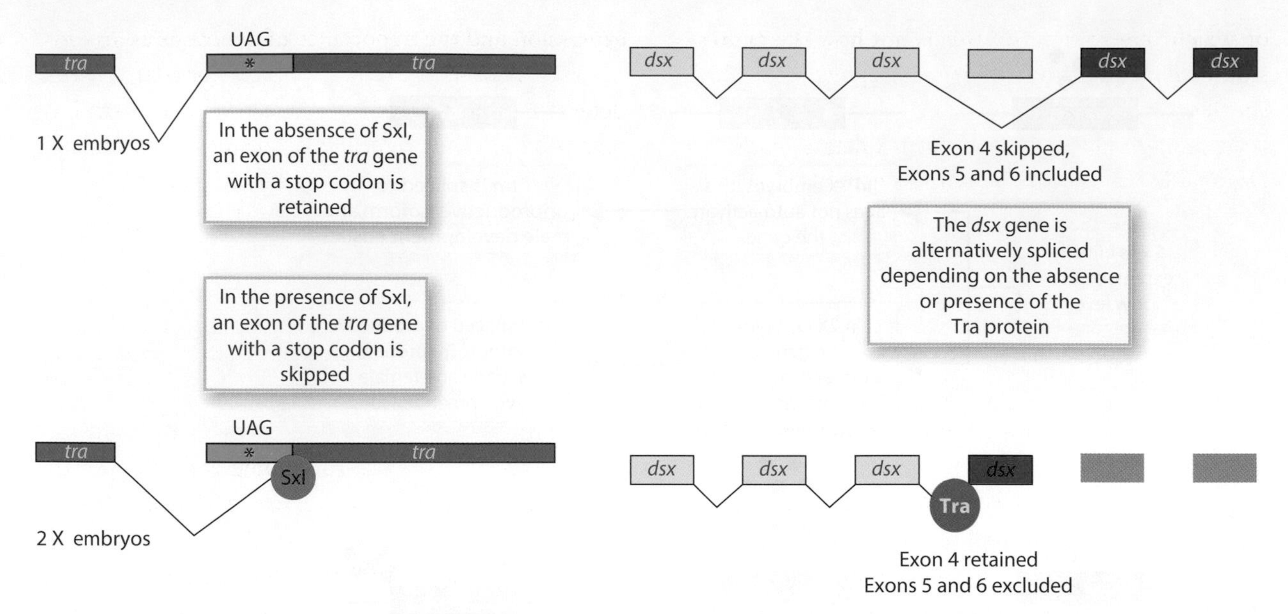

Figure 3.27 Alternative splicing and sex determination in *Drosophila*. The *tra* gene is a direct target of *Sxl* regulation, as summarized in Figure 3.26, but the mechanism of regulation was apparent only when *Sxl* was cloned and found to be an RNA splicing factor. Two different splice acceptor sites are present in the second exon of the *tra* gene. In the absence of *Sxl*, that is, in 1X flies, the more upstream splice acceptor is used, as shown in the top figure. This sequence of exon 2 includes a UAG stop codon, so no functional Tra protein is made. In the presence of *Sxl* protein, which occurs in 2X flies, this upstream splice acceptor site is blocked and a more downstream splice acceptor site is used, as shown in the lower half. The downstream splice acceptor results in exon 2, which does not have the stop codon, so a full-length functional Tra protein is made. Only the first two exons of the *tra* gene are shown, for clarity; and, even in the presence of Sxl protein, some of the transcript with the stop codon is made. The *tra* gene also encodes a splicing factor; and the Tra protein directly regulates the splicing of the *doublesex (dsx)* gene, as shown on the right. In this case, the alternative splice of the *dsx* gene involves the fourth exon. In the absence of the Tra protein in 1X flies, the fourth exon is skipped and the first three exons of the *dsx* gene are spliced to the fifth and downstream sixth exon. In the presence of the Tra protein in 2X flies, the splice is directed from the third exon to the fourth exon and neither of the downstream exons are used.

in *dsx* mutants become intersexual, having some aspects of somatic differentiation of both sexes. The 1X and 2X intersexes are nearly identical in phenotype, indicating that *dsx* is involved in sexual differentiation in both males and females. The difference between the male and the female versions of the Dsx protein, again, lies in the pattern of its splicing, as summarized in the right half of Figure 3.27. The presence of the Tra protein, which is only found in females because of its own splicing, results in the female-specific version, whereas the absence of the Tra protein results in the male-specific version. The role of the *tra2* gene is to function with Tra in regulating this splice; but it does not encode a splicing factor itself.

Sex determination in *Drosophila* involves the recruitment of general cellular functions to a sex-specific role

Although *Sxl* and *tra* have orthologs that are involved in alternative splicing, these orthologs are not involved in sex determination in other species. Alternative splicing is not the key mechanism for controlling sex determination other than in *Drosophila* species and in a few dipteran insects. The *doublesex* gene on the other hand encodes a transcription factor known as DMRT (the "d" standing for "*doublesex*"). These are evolutionarily conserved transcription factors that are involved in sex determination, or more specifically in male sex determination, testes formation, or spermatogenesis in many other species, including worms and mammals, and even in species without distinct sex chromosomes. One inference is that DMRT proteins such as Dsx are key regulators in an evolutionarily ancient pathway for sex determination, but they are located in the pathway downstream of the initial signal that triggers sex differentiation. Species vary widely in how the expression of the *dmrt* gene is regulated, probably depending on the specific environmental and evolutionary pressures at work in each species; in *Drosophila*, the regulation occurs at the level of splicing, which is triggered by the number of sex chromosomes. But this is not how the expression of *dmrt* genes is necessarily regulated in other species.

In sum, the genetic analysis of sex determination in *Drosophila* has modeled both the general molecular role of alternative splicing as a means to regulate gene expression and the importance of *doublesex* as an evolutionarily conserved transcription factor related to sex determination. Analysis of the model organism has revealed both the molecular details of a fundamental step in gene expression and the evolutionary history of key developmental process in animals.

3.5 *Arabidopsis thaliana*, sometimes known as "the weed"

Plant scientists who study the model flowering plant *Arabidopsis thaliana* are no strangers to the questions: "Why do you work on a weed? Why not corn or tomatoes or soybeans?" The assumption is that agronomic crops should be studied because of the easily comprehended relationship to crop improvement, environmental stewardship, and beautiful garden flowers. In fact, prior to the widespread use of *Arabidopsis* thirty years ago, the central plant genetic models included both food and horticultural crops—peas, maize, tomatoes, barley, petunias, and snapdragons.

Like *C. elegans*, *Arabidopsis thaliana* was an intentional and considered choice as a model organism from the beginning of its use; some of its main attributes are summarized in Table 3.1. In keeping with the properties of ideal model organisms, *Arabidopsis* is small, has a generation time of about six weeks, thereby allowing 8–9 generations a year, and produces thousands of offspring per parent plant (Figure 3.28). *Arabidopsis* is also naturally self-pollinating, which allows for easy production of homozygotes, but can be manually crossed so that new genetic variants can be bred, as diagrammed in Figure 3.29. In addition, the *Arabidopsis* genome is small for plants, about 125 Mb, and it was the third multicellular eukaryote, after *C. elegans* and *Drosophila melanogaster*, to be completely sequenced. Its genome is thought

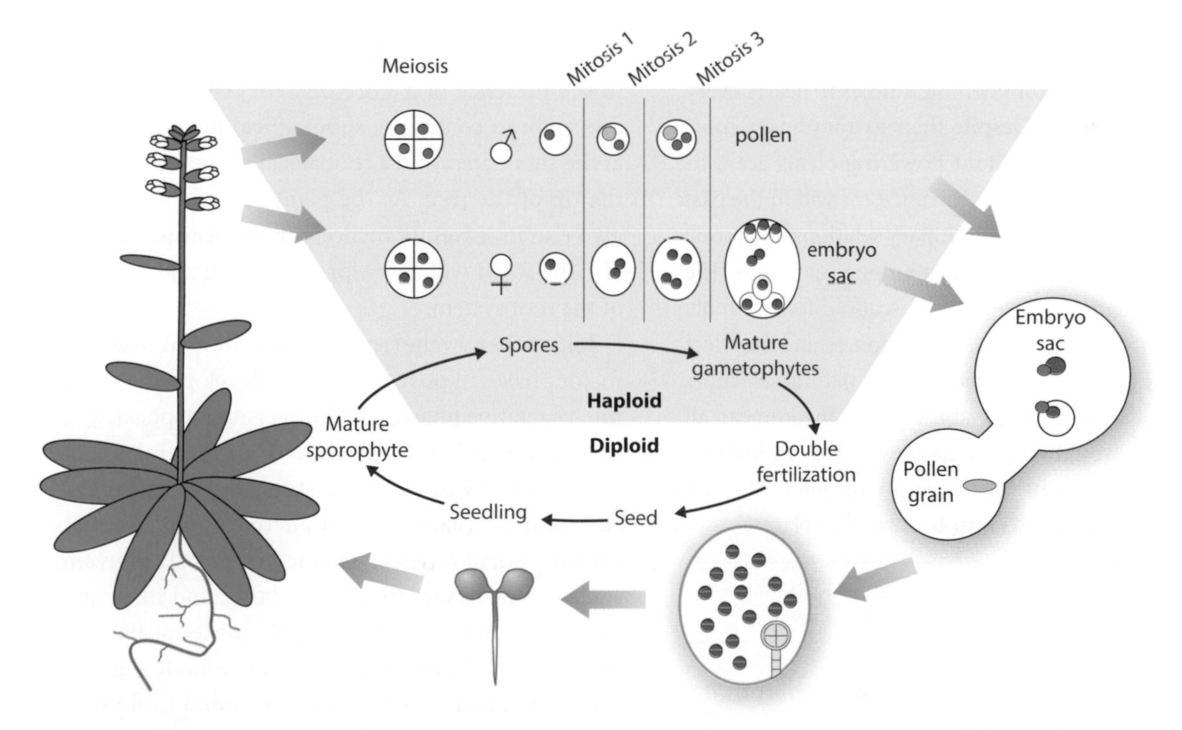

Figure 3.28 The life-cycle of *Aradopsis thaliana*. *Arabidopsis* has a life cycle similar to that of other flowering plants, but much shorter. The complete life cycle requires about six weeks, and thousands of seeds are produced per plant. The haploid phase of the life cycle is shaded in light blue and the diploid phase in white.

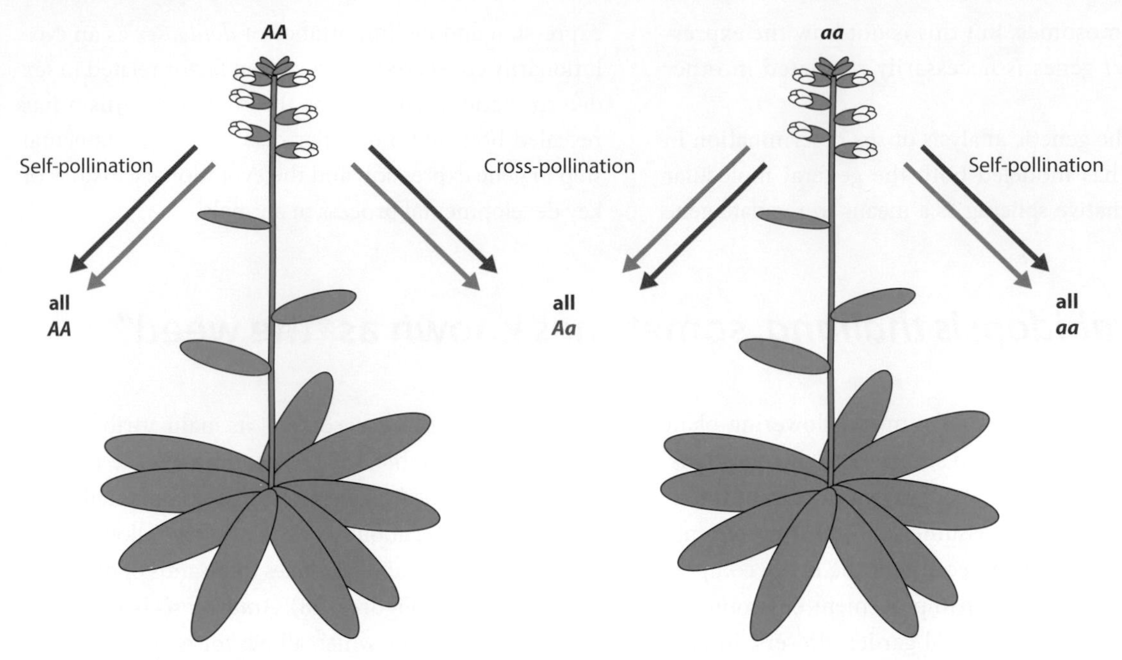

Figure 3.29 Cross-pollination and self-pollination in *Arabidopsis*. Each flower produces both pollen and eggs, and the plant is capable of both self-pollination and cross-pollination, leading to fertilization. Multiple generations of self-fertilization, as in *C. elegans*, produce lines that are homozygous at all loci. Cross-pollination is needed to perform matings.

to encode approximately 27,000 protein-coding genes, also fewer than most other plants. Furthermore, since *Arabidopsis* has a natural worldwide distribution, it is well suited for studies of natural diversity.

The *Arabidopsis* genome has in fact become the primary reference point for genomic studies in nearly every other flowering plant. Despite their extraordinary morphological diversity, all extant flowering plants are estimated to have shared a common ancestor within the past 150 million years, so the evolutionary relationship among flowering plants is similar to that among mammals. The *Arabidopsis* genome is used as the scaffold for assembly and comparison of nearly all plant genomes, and the effect of evolution, environmental adaptation, natural selection, and domestication throughout the plant kingdom can all be observed through comparisons with the *Arabidopsis* genome. Although it may seem like an insignificant weed, for geneticists *Arabidopsis* now towers other plants.

Literature Link. Wu et al. 2015, *Genetics* 200: 35–44; Woodward and Bartel 2018, *Genetics* 208: 1337–49

The life cycle of *Arabidopsis* is shorter than that of many flowering plants

Arabidopsis has a six-week life cycle from germination to mature seeds. The seed itself contains a dormant embryo consisting of two cotyledons or seed leaves that act as an energy store for the young seedling, two immature leaf primordia, a hypocotyl, and a root. In contrast to animal embryos, the plant embryo does not have all the organs of the final adult. Instead, plants display indeterminate tip growth. This is primarily due to the presence of a store of stem cells in meristems. The embryo contains a shoot apical meristem (SAM) at the shoot apex and a root apical meristem (RAM) at the tip of the root. As the plant grows, the SAM will give rise to leaves, branches, inflorescences, and flowers. The RAM will allow for the growth and branching of the root system.

Like other higher plants, Arabidopsis has three major phases of post-embryonic development: a juvenile vegetative phase, an adult vegetative phase, and a reproductive phase (Figure 3.30). The juvenile vegetative phase consists of the first few leaves and is characterized by a lack of reproductive competence and certain morphological characteristics. The juvenile stage can last years in woody plants, and likely plays an important role in delaying the onset of flowering and hence reproduction, since in the adult vegetative phase the plant is sensitized to respond to flowering signals. Complex signaling pathways and regulation by small non-coding RNAs result in major shifts in gene expression that control the transitions between these three phases.

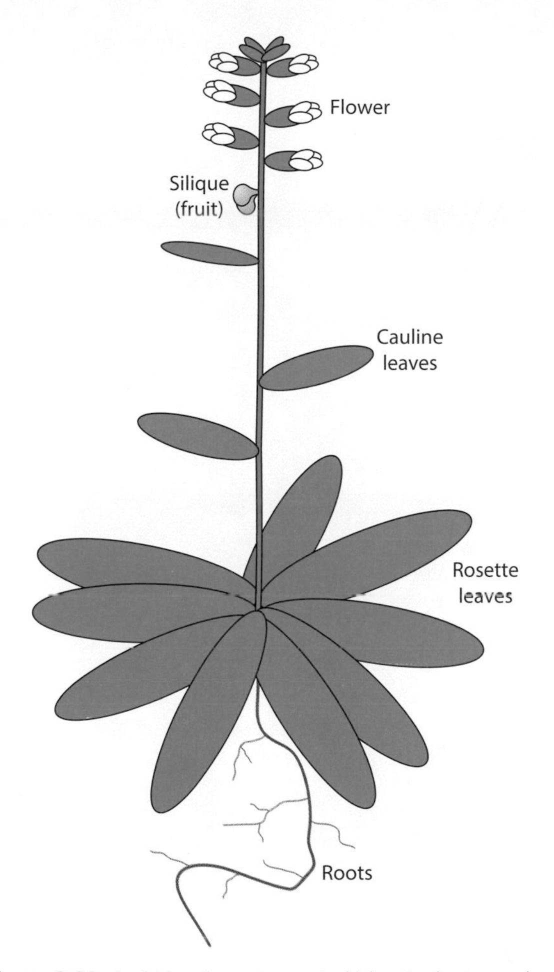

Figure 3.30 ***Arabidopsis* anatomy.** Arabidopsis plants are about 45 cm tall at maturity and consist of leaves (rosette leaves at the base and cauline leaves on the inflorescence stem), a stem, flowers, fruits called siliques that contain seeds, and roots.

The first flowers of the wild-type plants open within three or four weeks of germination. A diagram of a flower is shown in Figure 3.31. Each flower has three symmetrical whorls of organs—four leaf-like structures called sepals, four petals, six stamens, and one pistil. Some of the earliest work on *Arabidopsis* development studied genes regulating floral development, and many mutants are known that regulate the identity and structure of the floral organs. The structure of the flower is such that the plants are naturally self-pollinating, which is advantageous to genetic studies in the same way in which self-fertilization is important to those who study *C. elegans*. Cross-pollinations are made by carefully emasculating flowers before the stamens release their pollen and then manually pollinating with pollen from another flower. After fertilization, the pistil develops into a specialized seedpod called a silique, which may hold 50–100 seeds.

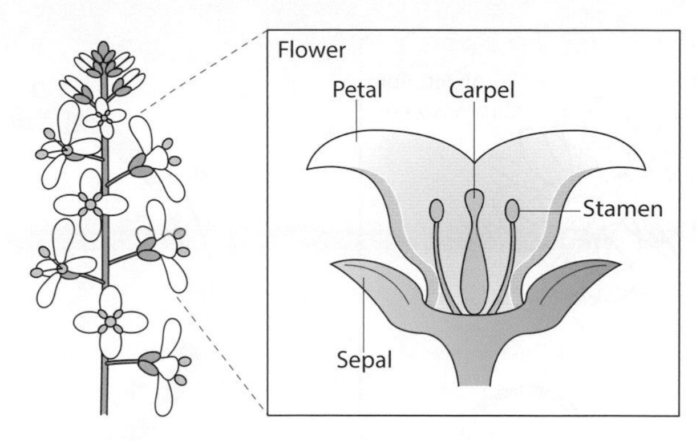

Figure 3.31 ***Arabidopsis* flower.** The *Arabidopsis* flower consists of four concentric radial whorls of organs: four sepals, four petals, six stamens, and one pistil made up of two carpels.

Transformation of Arabidopsis is done using *Agrobacterium tumefaciens*

In addition to its other advantages over many plant models, *Arabidopsis* can be readily transformed using the crown gall-causing bacterium *Agrobacterium tumefaciens. Agrobacterium* is a natural plant pathogen that infects a wide range of dicots by inducing the formation of a tumor or gall at a wound site with the aid of a tumor-inducing (Ti) plasmid. A 20 kb region of the plasmid called T-DNA is stably transferred into the genome of the host, as diagrammed in Figure 3.32. Within this T-DNA are genes encoding biosynthetic enzymes for making the plant growth hormones auxin and cytokinen, as well as a gene for synthesizing opines, amino acid derivatives that the bacteria use as a carbon and nitrogen source. High expression of these hormones directs the formation of a tumor from the infected cell, all of which produces food for the bacteria, allowing them to multiply and flourish. Generally, these kinds of tumors are fairly innocuous to the plant, and hundred-year old trees sometimes have huge crown galls.

The process of transformation is shown in Figure 3.33. Insertion of T-DNA into the plant genome is directed by 25 bp repeats, the right and left borders (RB and LB) flanking the sequence to be inserted. The Ti plasmid also encodes the *vir* genes, which encode virulence factors required for the transfer but are not themselves transferred. This naturally occurring gene transfer process forms the basis for making transgenic plants. The hormone and opine synthesis genes are replaced by any DNA of interest as well as by a selectable marker that encodes antibiotic or herbicide resistance, and the re-engineered Ti plasmid is transformed into *Agrobacterium*. When *Agrobacterium* infects the plants, the DNA between the

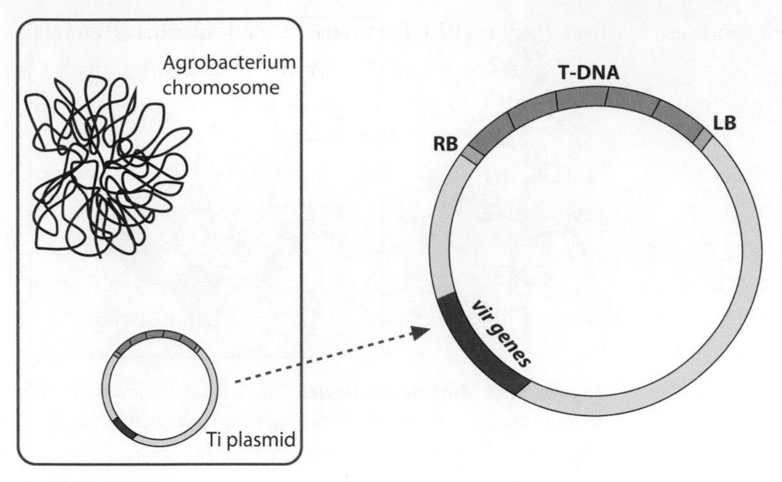

Figure 3.32 The Ti plasmid of *Agrobacterium tumefaciens*. Transformation in *Arabidopsis* takes advantage of the soil bacterium *A. tumefaciens*, which produces a tumour known as a crown gall. A complete Ti plasmid has T-DNA, which encodes the genes required for tumour formation, flanked by 25 bp repeats known as the right and left borders (RB and LB), shown here in green. Upon infection, the Ti plasmid is transferred to the plant cell, and the T-DNA randomly integrates into the plant genome. The plasmid also encodes the virulence genes (*vir*) that encode functions needed for infection and transfer of the T-DNA into the plant DNA.

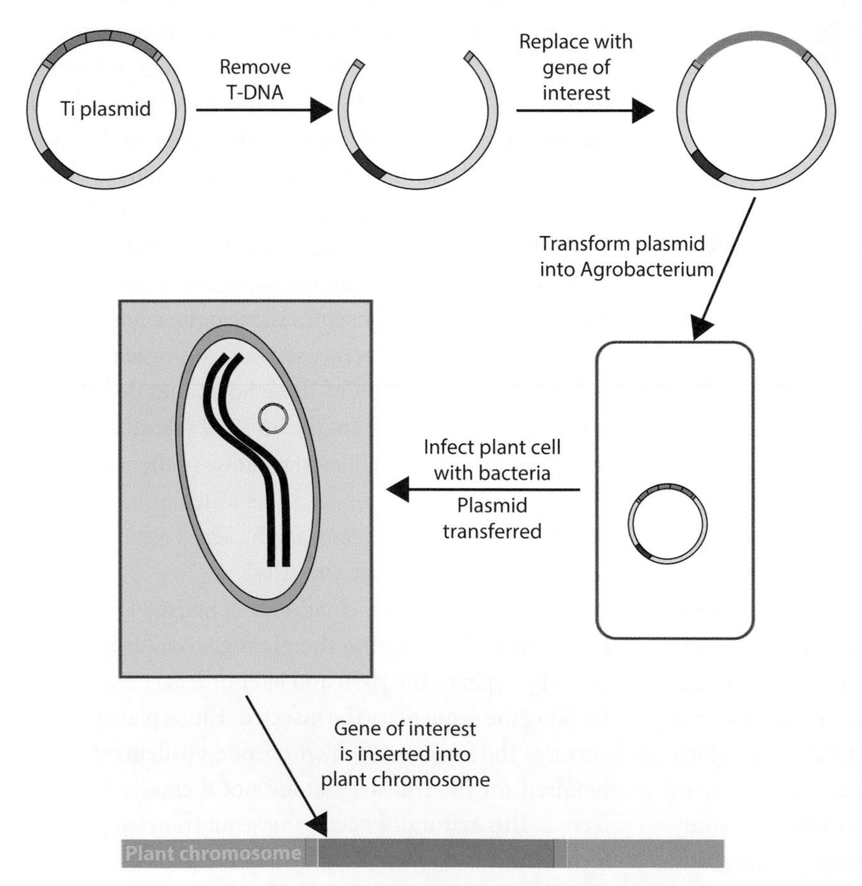

Figure 3.33 Transformation of *Arabidopsis* using modified Ti plasmids. The T-DNA is removed from the Ti plasmid in vitro and replaced with the gene of interest, to be integrated into the plant DNA. The modified plasmid is transformed into the bacteria, which are then used to infect the plant. In *Arabidopsis*, the infection is done by dipping the flowers into a solution containing the bacteria. The bacteria transfer the Ti plasmid to the plant cell, where the region flanked by the RB and LB is integrated into the plant chromosome. Integration occurs at random. In *Arabidopsis*, the infected flowers produce transgenic seeds containing the T-DNA.

RB and LB sequences is inserted into the plant's genome. Because insertion occurs at random sites, multiple transgenic lines are usually examined to control for any effects of insertion site or position effects on the expression of the T-DNA genes. The random insertion of T-DNA is analogous to the random insertion by P elements used for making transgenic *Drosophila*. To date, targeting insertions of the sort used in yeast and mice cannot be done efficiently in *Arabidopsis* and other higher plants.

Although *Agrobacterium* is an effective transformation vector for many plants, *Arabidopsis* is particularly readily transformed by simply briefly dipping the immature inflorescences in a dilute *Agrobacterium* suspension. Since *Agrobacterium* is infecting the egg cells with the immature flower, some of the seeds collected from a floral dip transformed plant will be transgenic. The selectable marker is used to identify the transgenic plants from those that have not been transformed.

The brassinosteroid hormone biosynthetic and signaling pathways have been delineated by genetic analysis in *Arabidopsis*

Plants, like animals, have a number of hormones—key molecules produced at one location and acting on another that are required for proper growth and development. In addition to auxin and cytokinen, other plant hormones include gibberellins, ethylene, abscisic acid (ABA), and brassinosteroids, with wide-ranging and overlapping roles.

Literature Link. Yang et al. 2011, *Plant Cell* 4: 588–600

The last of the plant hormones to be identified were the brassinosteroids in 1979, originally isolated from the pollen of the plant *Brassica napus*. As their name suggests, brassinosteroids are steroid hormones structurally similar to the insect-molting hormone ecdysone and to the mammalian hormones testosterone, estrogen, progesterone, and corticosteroids. Despite the similarity, plants do not have receptors for animal steroid hormones. (Molecular biologists have taken advantage of this to make estrogen-inducible promoter constructs for use in transgenic plants.)

Brassinosteroids have since been found in all flowering plants tested, as well as in at least one alga. They come in many forms, but the most active endogenous form is brassinolide (BL). They have defined roles in cell elongation, vascular differentiation, senescence (the active process of plant death), repression of photomorphogenesis (light-regulated growth) in the dark, and the modulation of stress responses. Brassinosteroids regulate these processes via crosstalk with other hormones, particularly auxin. Recent work has shown that the global gene expression patterns of BL- and auxin-inducible genes are significantly overlapping, which helps explain their synergistic interaction in certain plant processes.

Mutations demonstrated the presence of brassinosteroid pathways in *Arabidopsis*

Despite various studies demonstrating the effect of exogenous BL on plant growth, convincing evidence about the roles of brassionostreids as natural hormones governing plant growth came with the identification of the first brassinosteroid mutants in *Arabidopsis* in 1996. Brassinosteroid biosynthetic mutants were first identified in screens for seedlings defective in skotomorphogenesis or etiolation (growth habit of plants grown in the dark). In wild-type plants, dark-grown seedlings are phenotypically distinct from those grown in the light; they are characterized by an elongated hypocotyl, closed, hooked cotyledons, and an absence of chlorophyll. Researchers interested in the regulation of this process screened for mutants that displayed normal light-grown growth (short hypocotyl, open cotyledons, and chlorophyll) even in the dark. The phenotypes of some of these mutants are shown in Figure 3.34 A. When two of these mutants, *cpd* (*constitutive photomorphogenesis*) and *det2* (*de-etiolated 2*), were cloned, they were found to be mutated in genes predicted to encode brassinosteroid biosynthetic enzymes, as shown in the diagram of the signaling pathway in Figure 3.34 B. Treating the plants with exogenous BL restored a wild-type dark-grown phenotype, which suggested that brassinosteroids repressed photomorphogenesis in the dark. When grown in the light, the mutants displayed many other phenotypes, including severe dwarfism, dark-green leaves, delayed senescence, and poor fertility.

At the same time, other scientists were carrying out screens specifically looking for the brassinolide receptor. While BL had been shown to enhance shoot elongation, it had also been shown to inhibit root elongation in *Arabidopsis* seedlings grown on BL-containing artificial media. The mutant *bri1* (*brassinosteroid insensitive 1*)

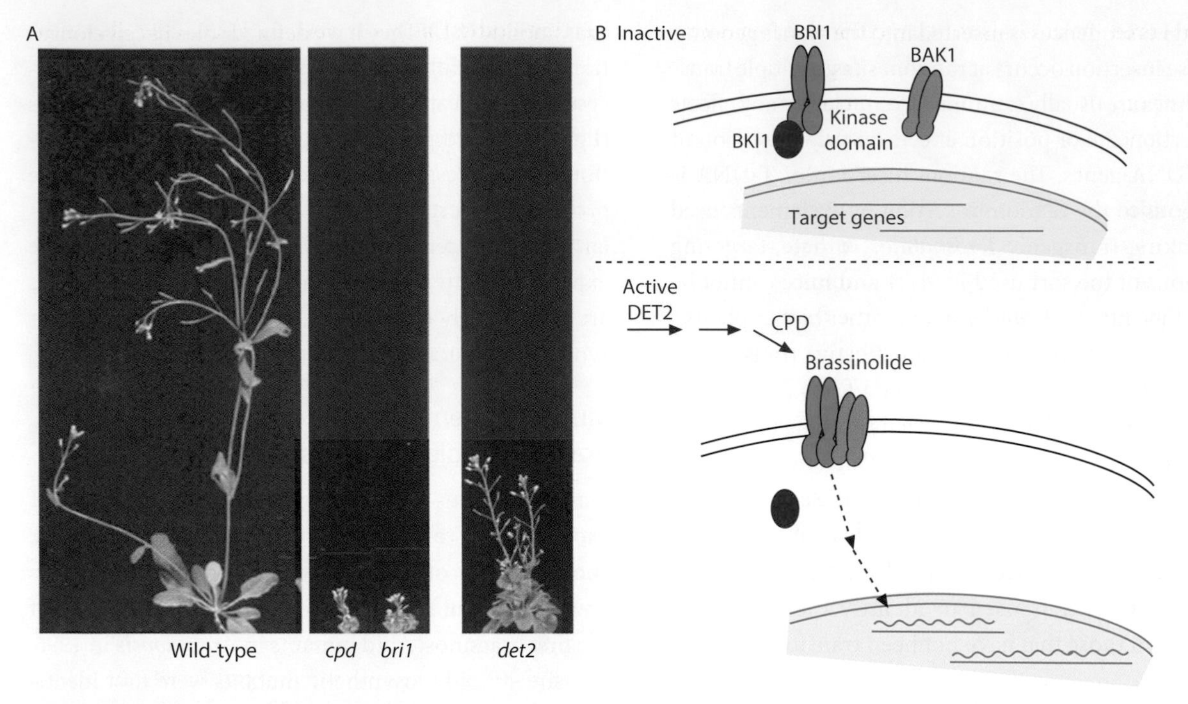

Figure 3.34 The brassinosteroid signaling pathway in *Arabidopsis*. The biological significance of brassinosteroids was proven through the use of mutants in *Arabidopsis*. (A) The phenotype of some of the dwarf mutants involved in hormone synthesis and signaling. The phenotypes of the three mutants described in the text (*cpd*, *det2*, and *bri1*) are shown. (B) In the absence of the active hormone brassinolide, the receptor BRI1 is in an inactive complex with the inhibitor BKI1, and the target genes are not transcribed. The genes *DET2* and *CPD* encode two of the enzymes involved in the synthesis of the active hormone brassinolide. The gene *BRI1* encodes the receptor itself. Brassinolide binds to the BRI1 receptor, releasing it from a complex with BKI1 and allowing BRI1 to form a complex with BAK1. In a series of signalling steps, not all of which are known, this triggers the transcription of the target genes. In *det2* and *cpd* mutants, little or no hormone is made, which results in the dwarf phenotype shown in panel A. In a *bri1* mutant, no receptor is made and so the pathway is not active.

was identified as a mutant insensitive to the root growth-inhibiting effects of BL. Soil-grown *bri1* knockout alleles look indistinguishable from *cpd*. When it was cloned, it was shown to encode a leucine-rich repeat (LRR) receptor-like kinase. BRI1 is localized to the plasma membrane, and BL directly interacts with BRI1 via the extracellular LRR domain, activating the kinase and setting off downstream signal transduction cascade (Figure 3.34 B). This is in contrast with steroid receptors in animal that bind to steroids in the cytoplasm before translocating to the nucleus to act as a transcription factor.

The identification of *cpd*, *det-2*, and *bri1* and their pleiotropic phenotypes convinced many plant scientists that brassinosteroids really were required for growth and development. Subsequent screens for plants that display the severe dwarfism of these mutants have identified most of the enzymes in the brassinosteroiod biosynthetic pathway and almost an entire signal transduction pathway from BRI1 to the downstream transcription factors. This pathway, together with some of the supporting genetic and molecular evidence, will be revisited in Chapter 10.

3.6 *Mus musculus*, or the mouse

Even non-geneticists probably have yeast in the house, nematodes in the gardens, and fruit flies near fruit stands. There may also be *Arabidopsis* growing uninvited in our gardens or along our sidewalks. But these organisms were not raised as pets. Alone among the model organisms, mice owe their standing to their role as house pets, particularly among "mouse fanciers," hobbyists who, from the early days of the twentieth century, bred the animals for coat color and behavioral differences. Other mammals such as dogs and cows also have important and long-standing relationships with humans, and many of them also provide useful examples in Mendelian genetics. Nearly all these other mammals are the objects of genome projects, but none of them has the history of genetic experimentation that comes with the mouse. Among domesticated mammals, mice are among the smallest in size and have the fastest life cycle. Although many other mammals have provided important insights into the fundamentals of human genetics, none has been such a good model for human genetics as the mouse. Some of the properties of mice as model organisms for genetic analysis are shown in Table 3.1.

Literature Link. Peters et al. 2007, *Nature Reviews Genetics* 8: 58–69

Mice have the typical life cycle of placental mammals. Gestation takes about twenty-one days in mice, as compared to about thirty-eight weeks in humans, and puberty is reached in about six weeks, as opposed to twelve or more years in humans. Thus a mouse life cycle is slightly more than two months, from fertilization of an ovum until that fertilized egg can produce its own offspring.

Mouse development is described in more detail in Chapter 7 in the context of transgenic mice and gene targeting, a technique that was recognized with the Nobel Prize in 2007. For the purposes of this chapter, the important aspects are that mice can be genetically manipulated either through controlled crosses between males and females or through the alteration of single cells. Embryonic stem cells from mice can be grown in culture, and genes can be introduced into these cells. Transformed cells are selected that have integrated the gene stably into the chromosome by a double crossover, a process similar to yeast integration. These transformed embryonic cells are then used to produce an adult mouse that has the introduced DNA as a targeted integration. The introduced gene can replace or "knock out" the function of the chromosomal gene; other manipulations of mouse genes are described in detail in Chapter 7.

It is estimated that mice and humans diverged from a common ancestor approximately 75 million years ago. The mouse haploid genome is approximately 3×10^9 base pairs, or about the same as in humans. The number of protein-coding genes is about 22,000, also about the same as in humans (although the number is regularly revised in both mice and humans). One estimate is that more than 99 percent of genes in the mouse have at least one orthologue in humans, although different gene families have been expanded in each lineage. By one count, there are 118 genes in the mouse for which humans have no orthologue; many of these genes are found in other mammals as well as in the mouse, so they could have been lost during primate evolution. There is extensive preservation of the linkage arrangement of genes as well, so that 80 percent or more of the mice genes are found at the same relative chromosomal location as their human orthologues. These conserved linkage arrangements are known as syntenic blocks (see **synteny**). The value to understanding diseases of having a well-studied genetic model that is so closely related to humans cannot be overstated.

Because mice were raised as pets, variations in coat color were noticed and bred. The most widely used coat colors are white, black, and brown, although genetic modifications in pattern and intensity of color are known, as shown in Figure 3.35. More than 75 genes that affect coat color are known. A classic reference book with many color pictures, *The Coat Colors of Mice*, by W. K. Silvers, has been made available electronically by the Jackson Laboratory.

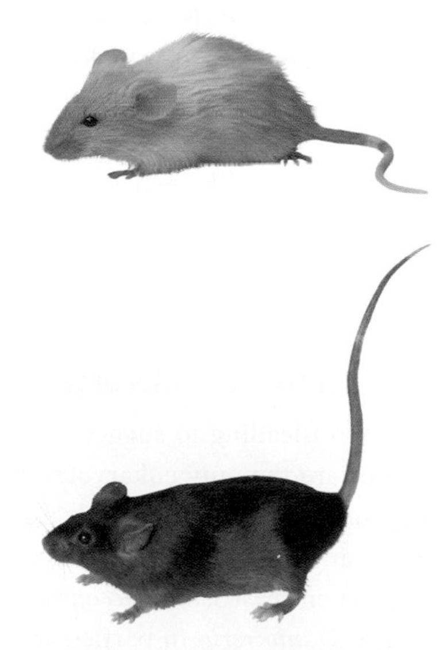

Figure 3.35 Different coat colors in mice. Mutations in more than seventy-five different genes that affect coat color have been identified in the mouse. The most commonly used colors are black, brown, and white.

 ***Literature Link.*** http://www.informatics.jax.org/wksilvers

Other genetic markers of the mouse that are used are morphological features such as eye color and size, tail structure, coat texture, and others. *Mouse Genetics* by Lee Silver, which presents a fascinating historical record of mouse genetic research, can also be accessed from the website of the Jackson Laboratory. Each of these books is integrated with the current mouse genome database, so that a curious reader can easily compare the classic mouse fancier mutations with contemporary genomic analysis of these same genes.

3.7 Model organisms: An overview

We have barely begun to describe genetic analysis in these five model organisms, so we refer again to Table 3.1, especially to the sections on the particular advantages and peculiarities of each model organism. In every case, the procedure to make transgenic organisms arises directly from some aspect of the biology of the organism. Similarly, some biological questions are more readily explored using genetic analysis in one model organism rather than in another; often but not always, the advantages and limitations of the organism as a model arise directly from some of its biological characteristics.

One often overlooked feature shared by these model organisms is the active and interactive research communities working with them. The importance of a community of scientists who share reagents, strains, ideas, methods, and criticisms cannot be overstated. Many excellent genetic models have not quite risen to the prominence warranted by their biological properties because other researchers did not adopt them readily; in a few cases, rivalries among the researchers hampered progress on the organism. The extensive genome databases referenced in Table 3.1 and in the online supporting material are a testimony to the collaborations and energy of the researchers who study these model organisms.

A few other widely used model organisms

It would be very misleading to suggest that these five model organisms are the only eukaryotes that geneticists use. Other organisms are frequently used, too, and deserve more than the passing mention that can be made here; the fission yeast *Schizosaccharomyces pombe* and the zebrafish *Danio rerio* in particular have made significant contributions to genetic research and were omitted only because of space. A few of these organisms are listed in Table 3.2, with examples of their utility and with links for their genome databases. Both the organisms and the examples of the biological questions they can be used to answer are selective rather than comprehensive. The list in Table 3.2 does not include some organisms that are researched principally because they are closely related to one of the five most widely used organisms, such as other *Drosophila* or *Caenorhabditis* species. Table 3.2 also does not include any of several domesticated mammals whose genomes have been sequenced, such as cats, cows, pigs, rats, and so on. Many mutations have been identified in these organisms and serve as models for human biology and disease. Mutations are known and the genome sequence is available for many plants of agricultural significance as well. The genomes of many other organisms have been sequenced, including most of the higher apes, although these animals are not genetic models in the same sense as the others, since they are usually not raised in the lab and few mutations have been studied in them.

Are more model organisms needed?

We have emphasized that these model organisms act as stand-ins for the many thousands of eukaryotic species, so it is appropriate to evaluate their role as representatives and to ask whether more representatives are needed. Again, the answer depends in part on what we expect the model organisms to model.

If our goal is to determine the common underlying molecular biology of eukaryotes, then it seems at least possible that no additional model organisms will be needed. Further, if our goal is to have model organisms that tell us about *human biology*, an argument could be made that no additional model organisms are needed.

Table 3.2 Some other eukaryotic model organisms widely used for genetic analysis

Species	Type of Organism	Example of Biological Questions	Genome Database
Schizosaccharomyces pombe	Fission yeast	Cell cycle, chromosome structure	***Schizosaccharomyces pombe*** GeneDB
Aspergillus nidulans	Fungus	Meiosis	Aspergillus Comparative Database
Neurospora crassa	Fungus	Meiosis	***Neurospora crassa*** Database
Tetrahymena thermophile	Ciliated protozoa	Chromosome structure, nuclear differentiation	Tetrahymena Genome Database
Paramecium tetraurelia	Ciliated protozoa	Excitable cells and ion channels	Paramecium DB
Dictyostelium discoideum	Slime mold	Signaling and morphogenesis	dictyBase
Chlamydomonas reinhardtii	Green alga	Flagella and motility, chloroplast development	ChlamyDB
Physcomitrella patens	Moss	Evolution of plant development	JGI ***P. patens*** subsp. ***patens***
Zea mays	Maize	Transposons, epigenetics, plant development	MaizeGDB
Oryza sativa	Rice	Plant development, abiotic stress, model monocot	OryzaBase Database; TIGR Rice Genome Annotation
Populus trichocarpa	Poplar	Woody plant growth and development, secondary growth	JGI ***P. trichocarpa***
Danio rerio	Zebrafish	Vertebrate development, behavior	ZFIN
Canis familiaris	Dog	Behavior, disease, large mammal development	NHGRI Dog Genome Project

We have the human genome sequenced and annotated, we have powerful tools for genetic analysis in mice for comparative studies, we can use zebrafish as a model for analysis of some complex questions in neurobiology, and we have comparative genomic sequences from higher apes, primates, and other mammals, and even other *Homo* species such as Neanderthals and Denisovans. In addition, we have genome information from people from many dozens of different populations and geographical origins, a rich trove of our evolutionary history over the past millennia, and we have the individual genomes of thousands of different people; perhaps this is enough information for understanding much of human biology.

However, understanding the common biological properties of eukaryotes and applying these principles to humans are only two of the reasons why investigators rely on model organisms. If our goal for research with model organisms is to understand biology—that is, to understand the evolutionary processes that shape the living world—then these model organisms are far from enough. Certain phylogenetic groups are very well represented, such as ascomycetes fungi, insects, and mammals, while other taxa are nearly absent.

One can study some of these other species, contemplate some of their peculiarities, and dream about the evolutionary grandeur that we have not yet viewed.

Genetic analysis in non-model organisms

However, we might not need to continue to speculate about interesting biological questions for many other species. Genome sequencing has become faster and less expensive, which has obviated the need to culture the organisms in a lab—one of the most important limitations encountered for many species and one of our primary advantages shared by model organisms. Genomes can be sequenced from single specimens, even from fragments or preserved samples. With well-annotated genome sequences, much more comparative genomic analysis is possible and evolutionary relationships and changes can be inferred. In Tool Box 3.1 we describe

TOOL BOX 3.1 Model organisms and morphological diversity: Seahorses, think zebra(fish)

What do we want our model organisms to model? One of the many applications of model organisms is that they provide a foundation for investigating and testing ideas about the origins of the morphological diversity that arises during evolution. It is often relatively easy for us to distinguish between two related species, for instance to know that one bird is a cardinal while another is a blue jay. Most of their genes are highly similar or nearly identical to one another, yet we can tell the birds apart at a glance. Because the species are different but their genomes are highly similar, we can focus our attention on the parts of the genome that vary in order to understand the genetic and molecular origins of the variation.

In this box, we describe how the zebrafish *Danio rerio* can be used to analyse morphological diversification of other teleost fish. The teleost fish include more than 95 percent of all extant fish species, and nearly all the species of fish familiar to us; as any angler, customer at a fish market, or visitor of an aquarium knows, the teleost fish vary widely in size, habitat, behavior, and morphology. Some of the most common variations are in the size, number, and shape of the fins. Fins in fish are the evolutionary homologues of limbs in tetrapods, and the developmental genetics for vertebrate limb formation has been well studied. The same genes and processes responsible for forelimb and hindlimb development in reptiles, amphibians, birds, and mammals are also involved in fin bud development in zebrafish and other teleosts. The homologues of our forelimbs (our arms) are the pectoral fins, while the hindlimbs (our legs) are equivalent to the pelvic fins. These are shown in the diagram in Figure B3.1.

Pelvic fins are believed to function primarily in vertical motion in the water, as well as in stability during turning and stopping. Some fish lineages, including seahorses, pufferfish, and some sticklebacks, lack pelvic fins. The loss of pelvic fins has occurred by independent evolutionary processes in these three lineages, as shown by the comparison of their genomes with zebrafish. The positions of the limb buds along the body axis of vertebrates are regulated by the *Hox* gene clusters. In particular, the gene *hoxd9* is associated with the development of the hindlimb bud, such that the loss of function of *hoxd9* in mice results in a failure of the hindlimb buds to form. Fish have two *hoxd* gene clusters. In pufferfish, the expression of the *hoxd9a* gene is altered; thus, in a broad sense, pufferfish lack pelvic fins because of a change in the positional information about where the hindlimb bud forms. Sticklebacks have *hoxd9a* expressed in its normal position during embryogenesis, but the pelvic fin bud does not grow.

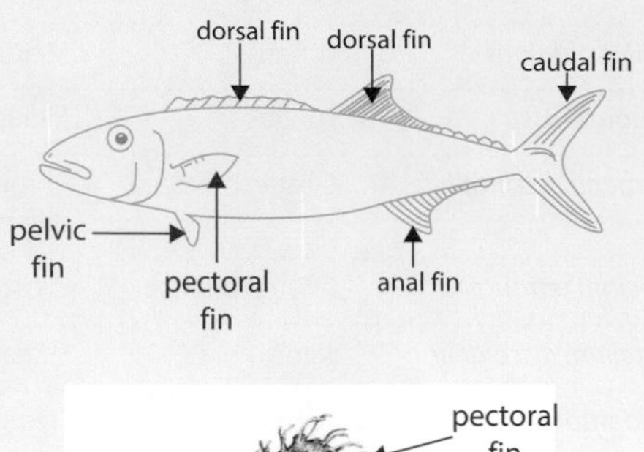

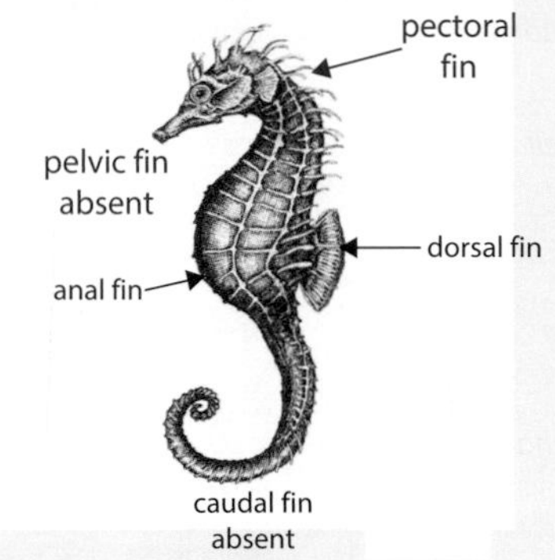

Figure B3.1 The anatomy of fish. The generalized anatomy of a teleost fish is shown at the top, with the fins labeled. The pectoral fins arise from the forelimb buds, while the pelvic fin is derived from the hindlimb bud. The other fins are outgrowths from the body itself and are not derived from limb buds. A diagram of a seahorse is shown below. Note that the seahorse has a pectoral fin (derived from the forelimb bud), but the pelvic fin is absent.

Outgrowth of the limb buds is affected by the transcription factor Pitx1, which in mammals is expressed in the developing hands and feet, among other locations (including the pituitary gland, after which the gene is named). In the sticklebacks lacking the pelvic fin, an enhancer sequence specific to the hindlimb in regulatory region of the *pitx1* gene has been altered, so that, while *pitx1* is expressed and functions normally in other locations, it is not expressed in the pelvic fin bud, and the bud does not mature.

The evolutionary change that results in the loss of the pelvic fin in seahorses is different yet. Seahorses have a number of distinct morphological features, as shown in Figure B3.1, such that at a casual glance they barely resemble a typical fish. The lack of pelvic fins probably contributes to their unusual swimming posture: seahorses appear to swim upright rather than horizontally. Analysis of the seahorse genome, published in 2016, showed that the genome lacks the gene *tbx4*. This is not a loss of gene function or a change in gene expression as seen with many other genes such as *hoxd9a* or *pitx1*, but rather a deletion of the gene itself, as demonstrated by syntenic alignments with the genomes of other fish. At the corresponding locations in their genomes, seahorses encode the genes that flank *tbx4* in other fish, but lack any detectable coding or regulatory regions of *tbx4*. The gene encodes the transcription factor protein Tbx4, which is required for the development of the hindlimb in birds and mammals; mutations in *tbx4* result in the loss of hindlimbs or in changes in hindlimb development in mice and humans. The expression of *tbx4* is regulated by *hoxd9*, so the failure of hindlimb formation occurs because a target of *hoxd9* regulation is missing. *tbx4* expression may also be a target of *pitx1*, or each of them may be separate targets of *hoxd9* regulation. In brief, among the three lineages that lack a pelvic fin, pufferfish have changes in the positional information about where the fin bud forms, sticklebacks have specific changes in the expression pattern of a key transcription factor that regulates bud growth, and seahorses lack one of the primary target genes that control fin bud differentiation.

In birds and mammals, *tbx4* has pleiotropic effects, so that, in addition to the formation of the hindlimb, the gene is also expressed in the allantois, in the developing lung, in the esophagus, in the hind gut, and in the heart. All these tissues (or their equivalents) develop normally in seahorses despite the absence of the *tbx4* gene, so it is apparently not necessarily for the development of any of these other tissues and organs. Is *tbx4* still needed, then, for pelvic fin formation? In order to answer this question, the investigators turned to zebrafish. Zebrafish have a functional *tbx4* gene, which is expressed in the pelvic fin bud. Using CRISPR-Cas9 (as will be discussed in Chapter 8), the investigators created a four-base-pair deletion in the coding region of *tbx4* gene in zebrafish, which knocks out its function. The mutant zebrafish lacked a pelvic fin, just like seahorses, but developed normally in other tissues and were viable. Thus an evolutionary question about morphological diversity that arises from the analysis of the seahorse genome could be directly answered by a genetic analysis in the model organism zebrafish.

Literature

Lin, Q. et al. 2016. The sea horse genome and the evolution of its specialized morphology. *Nature* 540: 395–9.

how the model organism zebrafish has been used for insights into the evolution of one of the most unusual fish, the seahorse.

But genome sequences, and even cloned genes, provide only some of the important tools of genetic analysis and do not meet all the criteria for a model organism presented earlier in the chapter. Another criterion is the availability of mutations, which until recently has usually been a most difficult criterion for potential research organisms to fulfill. While it is unrealistic to hope that any other organisms will have as many genetic mutations available as one of the main models, other approaches to finding mutant phenotypes offer considerable promise here as well. In Chapter 8 we describe how **genome editing** using CRISPR-Cas9 could make it possible to make directed mutations for any DNA sequence in any organism. In Chapter 9 we describe genome-wide screens involving **RNA interference (RNAi).** RNAi allows investigators to knock down the function of a specific gene and observe the effects without the need to handle or maintain mutant strains. While it is not quite the same as having a mutation, RNAi provides a powerful and usually rapid alternative for organisms in which mutations are not readily available but whose genome sequences are known.

We close this chapter on model organisms by describing one example of what genetic analysis might increasingly look like. Case Study 3.1 discusses how RNAi has been used to study genes involved in regeneration in planaria. Planaria are from the phylum Platyheminthes, an understudied group, and are famous for their ability to regenerate entire organisms from small dissected pieces, a fascinating biological phenomenon whose mechanism is not thoroughly studied. Genetic analysis looks like a perfect method to approach this question, except that mutations were not available for planaria and none of the current model organisms is a close relative evolutionarily. But, as the case study shows, with sequenced genomes, genome editing, and RNAi or related methods, any organism can become a model organism for experimental genetics.

Chapter Capsule

Five eukaryotic species have been used as the experimental model organisms by which much genetic analysis has been done. These five organisms, *Saccharomyces cerevisiae*, *Caenorhabitis elegans*, *Drosophila melanogaster*, *Arabidopsis thaliana*, and *Mus musculus*, are easy to grow in the lab and have large and expanding storehouses of mutants. Complete genome sequences are available for all of them, in easily accessible databases. Most of the remainder of the book will use examples from these five organisms. However, these five species represent only some of the organisms that have been used for genetic analysis. As sequenced genomes are completed for more species, tools such as RNAi and genome editing open up the possibility of genetic analysis for many more organisms and for many more biological questions.

Additional reading

In addition to the literature links and genome databases for individual organisms, the following essays discuss the use of model organisms:

Aitman, T. J. et al. 2011. The future of model organisms in human disease research. *Nature Review of Genetics* 12: 575–82.

Bonini, N. M., and S. L. Berger. 2017. The sustained impact of model organisms—in genetics and epigenetics. *Genetics* 205: 1–4.

Brenner, S. 2003. Humanity as a model system. *Science* 302: 533.

Davis, R. H. 2004. The age of model organisms. *Nature Review of Genetics* 5: 69–76.

Case Study 3.1

Regeneration in planaria

The model organisms discussed throughout this book have been have used to investigate and illuminate many different biological processes. As valuable as these organisms are, they cannot possibly represent the full range of interesting biological questions, and other organisms can prove to be more useful for investigating these processes. In this case study we describe a genetic approach to the study of regeneration in planaria, for which no mutations are available.

Thinking through the Experiment

1. Think about a biological process in some organism that has piqued your curiosity, possibly from a visit to a zoo, an aquarium, an arboretum, or a science museum.
    a. Could this biological process be studied using one of our five primary model organisms or not?
    b. If yes, outline a strategy by which one of the model organisms could provide the foundation for understanding the genetic and molecular basis for this process.
    c. If no, what might be an experimental approach to studying this process and what molecular, genomic, and genetic tools would be needed?

Regeneration in planaria has been studied for more than a century

Planaria are flatworms, members of the phylum Platyhelminthes. The overall anatomy of a planarian is shown in Figure CS3.1. Planaria have a nervous system, muscles, gland cells, epithelium, and a gut with a single opening. They live in fresh water, feed on small animals or animal debris, and average about 10 mm long, depending on the species (or about ten times the length of *C. elegans* and four times the length of *Drosophila melanogaster*).

There is a long history of research with planaria, much of it centered on their ability to regenerate new organisms after dissection or wounding; in fact T. H. Morgan studied regeneration in planaria before working on *Drosophila*. The only proliferating cells in the adult are the neoblasts, which are constantly dividing and are the equivalent of a totipotent stem cell population. Upon wounding, a blastema forms of neoblast cells covered by epidermis, and the neoblast cells proliferate and differentiate, as shown in Figure CS3.2; the neoblast cells are equivalent to stem cells in vertebrates. Dissected pieces from any part of the animal are capable regenerating entire animals with all of the organ systems, patterning in the correct positions and with the proper structures. Many physical manipulations such as transplantations and surgeries have been performed during the past century to investigate this amazing regenerative capacity, but the genes and molecules were unknown until Newmark, Sanchez Alvarado, and others began working with *Schmidtea mediterranea*, a diploid species with both sexual and asexual reproduction, and applying experimental tools derived from work in model organisms.

Thinking through the Experiment

2. One of the many instructive findings from physical manipulations was that cells from any part of the animal were capable of regenerating an entire animal. The regenerated animal showed a normal body pattern, with a proper anterior–posterior access, dorsal and ventral sides, and even left–right orientation.

a. How is this similar to the regeneration process that occurs during wound healing in mammals such as humans?
b. How is this different from what occurs during wound healing and regeneration in mammals?

Genes important for body plan regeneration were identified using RNAi

In order to identify and characterize genes involved in regeneration, a process known as RNA interference or RNAi was used. RNAi screens will be described in more detail in Chapter 9; but, for now, the important aspect is that a double-stranded RNA (dsRNA) corresponding to a gene is introduced into the organism, whereupon it reduces or eliminates the expression of that gene. As a result, the organism exhibits the phenotype of a mutation in the gene, although the gene itself is unaffected. Since regeneration occurs from the neoblasts, RNAi was applied to a group of genes expressed in neoblasts. The regeneration

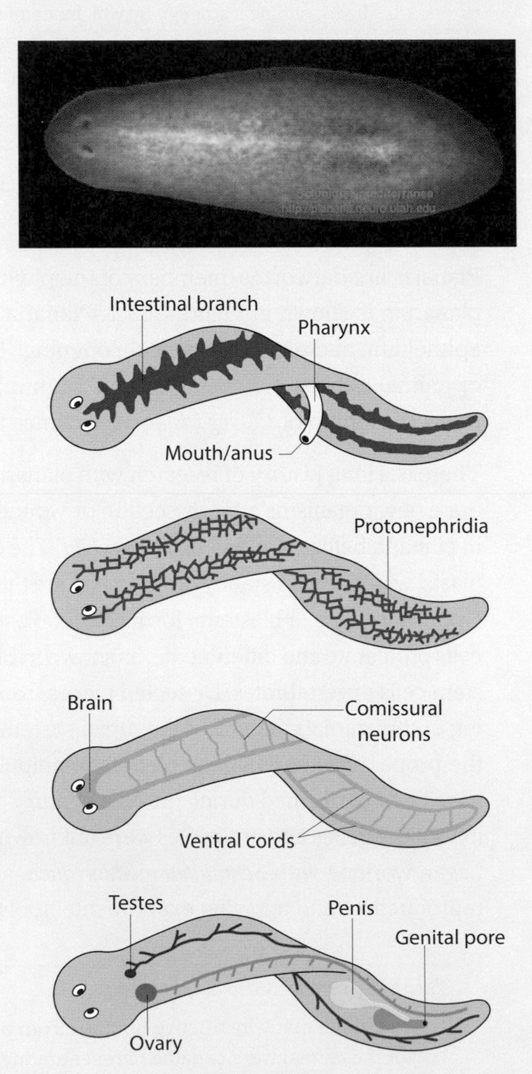

Figure CS3.1 The anatomy of the planarian *Scmidtea mediterranea*. This image is in the public domain.

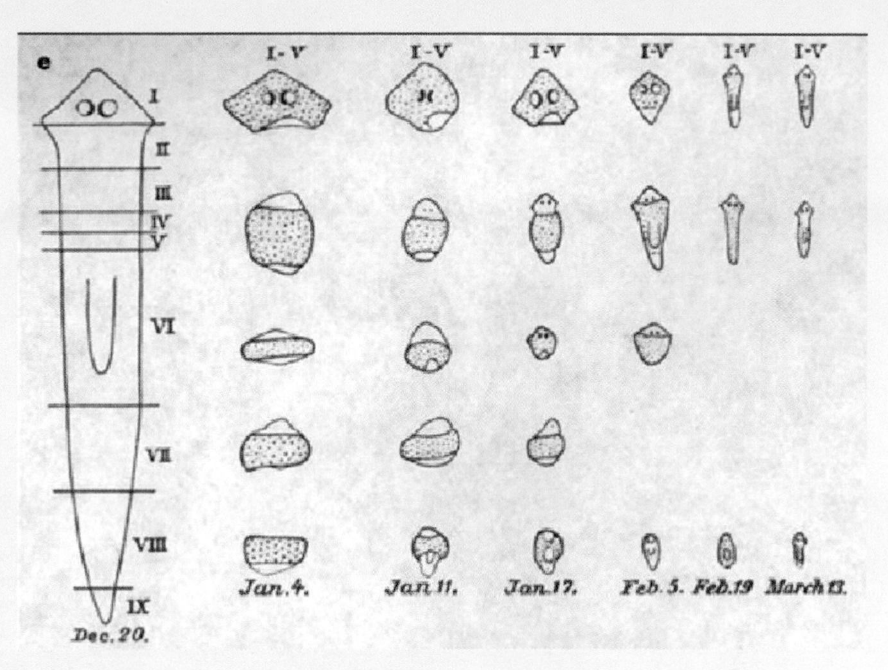

Figure CS3.2 Regeneration in planaria. These drawings are from a paper by T. H. Mongan published in 1898; they show the ability of planaria to regenerate from different fragments.

assay is shown in Figure CS3.3. A dsRNA from a gene expressed in neoblasts was mixed with a liver extract and fed to the animal; in some experiments, after specific genes were identified, the dsRNA was injected rather than fed. After feeding, the head and the tail were removed from the planarian and regeneration was allowed to occur for nine days. The process of feeding dsRNA and dissection was repeated on these "regenerated" organisms and regeneration ability was assessed.

Thinking through the Experiment

3. Genes expressed in the neoblasts were tested as candidate genes for being involved in regeneration.
   a. Describe some of the molecular processes that are necessary for regeneration and would be identified by this approach but might not be specifically involved in the *patterning* of tissues in the regenerating blastema.
   b. Speculate about some of the phenotypes that might be found for a gene that affected some aspect of the patterning process during regeneration.
   c. On the basis of your answer to Question 3b, what is one of the most significant advantages of using a genetic approach (including RNAi) for the analysis a biological process like this?

In the initial experiments, RNAi assays were performed for 1,065 neoblast-expressed genes from more than 50,000 dissected planaria, and a collection of 240 different genes were found that affected regeneration. The regeneration defects were classified into eleven different morphological categories, including an inability to regenerate at all and alterations in the shape and growth of the blastema. One of the main advantages of genetic approaches like this is that little prior knowledge of the process is required, and unexpected results can be obtained. This was true for the regeneration process. For example, as illustrated in Figure CS3.3, some genes affected only regeneration by the tail blastema and not the head blastema, which suggests that regeneration in the two ends of the animal involve some different genes.

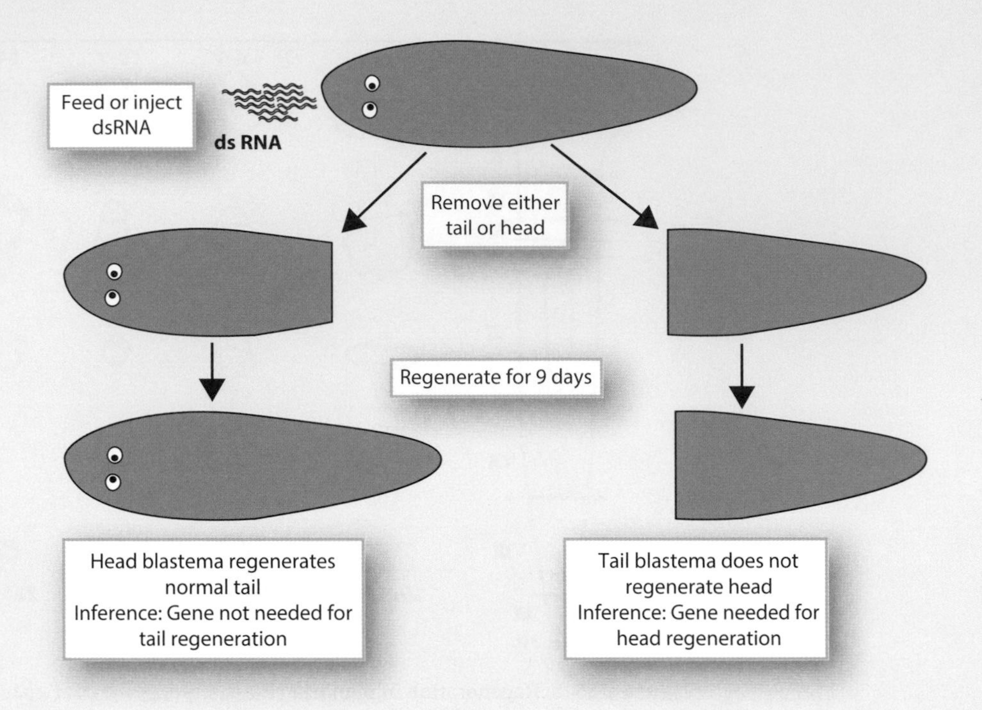

Figure CS3.3 Assaying genes for their effect on regeneration using RNAi. A preparation corresponding to an individual gene is fed to or injected into the planarian. The dsRNA knocks down expression of the gene by RNAi. Either the head or the tail is removed, and the remaining cells are allowed to regenerate for nine days. The process is then repeated. In the hypothetical example shown, the gene affects the ability of the tall blastema to regenerate a normal head but does not affect the ability of the head blastema to regenerate a normal tail.

These experiments implicated β-catenin and other genes in the Wnt pathway, a highly conserved signal transduction pathway in animals, as being responsible for the anterior–posterior patterning.

 ***Literature Link.*** Gurley et al. 2008, *Science* 319: 323–7; Petersen and Reddien 2008, *Science* 319: 327–30.

For example, if the regenerating fragment is injected with dsRNA corresponding to the β-catenin gene, the fragment produces head structures (and expresses some head-specific genes) on both the anterior and the posterior faces. The genes of the Wnt pathway, including β-catenin, are expressed in a gradient with the highest levels of expression at the tail and very little expression in the anterior region or the head. Recent work has focused on how neoblasts interpret and respond to this Wnt pathway gradient, and its relationship to other body patterns.

 ***Literature Link.*** Oderberg et al. 2017, *Current Biology* 27: 733–42; Stuckemann et al. 2017, *Developmental Cell* 40: 248–63

Thinking through the Experiment

4. Explain the result in which both the anterior and posterior faces of the regenerating fragment produce head structures when injected with β-catenin dsRNA, in light of the Wnt gradient.
5. What molecular tools are needed to show that β-catenin and other genes in the Wnt pathway are expressed in a gradient?

Pharynx Regeneration is a model for organ-specific repair and regeneration

Regeneration of an entire body pattern is remarkable and dramatic, but planaria (like many eukaryotes) also have the capacity to replace only specific organs and structures that have been lost. For example, damaged skin or liver cells are replaced efficiently and specifically in humans; in contrast, damage to heart or spinal cord cells is not readily replaced. When the pharynx is removed in planaria, only the pharynx is regenerated rather than an entire organism or head. This suggests that the organism can somehow "sense" what is missing, and regulate the regeneration process accordingly. One hypothesis for how this occurs is that populations of neoblast cells are heterogeneous in their gene expression and differentiation capacity. Like other types of stem cells, some cells are totipotent and can regenerate all structures, while others can regenerate only some structures. This idea has been investigated for pharynx regeneration by Adler et al.

Literature Link. Adler et al. 2014, eLife 3: e00238

The pharynx is the only opening to the digestive system and consists of neurons, muscles, secretory gland cells, and epithelial cells, all of which can be regenerated in the appropriate positions with normal functions. However, there are no neoblasts in the pharynx. When the pharynx is removed, mesenchymal neoblast cells in the vicinity of the pharynx begin to adopt pharyngeal fates and accumulate at the site of the wound or lost pharynx; the cellular reprogramming to become pharyngeal cells appears to begin before the cells accumulate at the wound site or as they do so.

Rather than perform surgical removal of the pharynx, Adler et al. used a process they refer to as "chemical amputation." Animals briefly soaked in sodium azide extrude their pharynx, which is then dislodged from the rest of the animal by gentle agitation. Other organs such as the gastrointestinal system do not appear to be affected by this treatment, and the regeneration of the pharynx after azide treatment is the same as regeneration by surgery, as measured by both cellular and molecular markers. Because the pharynx is essential for feeding, feeding behavior can be used as an assay for the presence or absence of a functional pharynx during amputation and regeneration.

Thinking through the Experiment

6. What would be some advantages of a chemical amputation procedure over a surgical one?
7. Outline how an RNAi screen for genes needed for pharyngeal functions could be done, and speculate about what phenotypes might be observed.

Once the chemical amputation is done, or any other type of wound occurs, two waves of cell proliferation are observed. Within a few hours, cell division increases throughout the body as assayed by the presence of histone H3S10 phosphorylation, a marker for cell division described in Section 2.3. Within twenty-four hours, mitosis is occurring at high rates in the region of the pharynx, that is, the wound site, but not throughout the body. Within three days, pharynx-specific cellular and molecular markers are seen as the pharynx begins to regenerate, and pharynx regeneration is complete within seven days.

These waves of proliferation helped define the times when genes needed for the pharynx-specific program would be active and might be informative targets for RNAi. By expression profiles compiled at different times, 718 candidate genes were identified as being active in the first twenty-four hours after amputation, of which 274 were tested by RNAi; eighty-two additional genes active at forty-eight and seventy-two hours were also isolated, out of a total of 356 genes tested in the RNAi assay. A subset of these genes were also tested by in situ hybridization and found to be expressed around the regenerating pharynx but not expressed prior to amputation.

The planaria were fed dsRNA from each of these 352 genes, chemical amputation was performed, and pharynx regeneration was assessed; feeding behavior after amputation was used as a measure of feeding behavior. When this assay was used in combination with cellular and molecular markers for regeneration, twenty genes needed for the regeneration of a function pharynx were found and assigned to different functional categories. Eight genes were needed for neoblast proliferation and mitosis overall, and were not pharynx-specific; in addition, four genes appeared to be required for food sensing or other pharyngeal functions, but not for regeneration, since RNAi knockdowns of these gene regenerated a nearly normal pharynx but the animals did not feed. Neither of these categories was the gene of the greatest interest for understanding pharyngeal-specific regeneration. RNAi knockdown of the remaining eight genes, which are referred to as the specific effectors, produced some pharyngeal tissues, showed a nearly normal rate of mitosis overall, and could regenerate other tissues but did not produce a complete pharynx. These genes were the ones most likely to be involved in the specific regeneration of the pharynx.

The most severe phenotype among the specific effector genes arose from RNAi knockdown of the gene *FoxA1*, which encodes a forkhead type transcription factor. In *FoxA* (RNAi) animals, the dividing cells that normally comprise the regenerating pharynx are a disorganized mass with no discernible pattern, which suggests that *FoxA* is needed for the specification of pharynx regeneration. Previous studies had shown that *FoxA* is expressed in the developing, regenerating, and mature pharynx, so its expression pattern is consistent with its being a master regulator of pharynx development and function. Importantly, regeneration of other organs occurs normally in *FoxA* (RNAi) animals, including the Wnt-dependent anterior–posterior body pattern as well as nervous system and intestinal regeneration. In addition, forkhead transcription factor proteins such as FoxA have been shown to be important for foregut development in many different animals, including nematodes and mammals.

Thinking through the Experiment

8. Thinking carefully about transcription factors and animal evolution, what is implied by the fact that a particular type of transcription factor (such as FoxA) is found to be needed for the differentiation or specification of the same organs or tissues (such as the pharynx or foregut) in many animals?
9. Recall the original hypothesis that populations of neoblasts may be heterogeneous in what organs they can regenerate. How could *FoxA* expression be used to test this hypothesis, and what might be the outcome if this hypothesis is true?

Since *FoxA* appears to encode a transcription factor needed for the specification and organization during pharynx regeneration, its expression could be used as an indicator to test the hypothesis about heterogeneity among neoblast cells. By using an antibody to determine which cells are expressing FoxA, the investigators found that the population of dividing cells are in fact heterogeneous, some expressing FoxA and others not. By some process yet unknown, after pharynx amputation the FoxA-expressing cells divide more rapidly than the rest. Once expressed, FoxA stimulates the transcription of other genes that are necessary for the regeneration and patterning of the pharynx.

The papers studying regeneration in planaria are worth reading for what they tell us about regeneration, wound healing, and embryonic patterning in other organisms. However, our purpose in this case study has been to illustrate how genetic analysis can be done in organisms like planaria, in which fascinating and sometimes extraordinary biological processes occur, but no mutants have yet been found. The chapter asks this question: Are more model organisms needed? With genome sequences and tools such as RNAi and CRISPR, the answer may be "no." But a more nuanced response is that, with such tools, nearly every organism can become a model for studying some biological process. The path for the genetic analysis of remarkable biological questions has never been more inviting.

References

Adler, C. E. 2014. Selective amputation of the pharynx identifies a FoxA-dependent regeneration program in planaria. eLife 3: e00238.

Gurley, K. A. et al. 2008. Β-catenin defines head versus tail identity during planarian regeneration and homeostasis. *Science* 319: 323–7.

Newmark, P. A., and A. Sanchez Alvaredo, 2002. Not your father's planarian: A classic model enters the era of functional genomics. *Nature Reviews of Genetics* 3: 210–19.

Oderberg, I. M. et al. 2017. Landmarks in existing tissue at wounds are utilized to generate pattern in regenerating tissue. *Current Biology* 27: 733–42.

Petersen, C. P., and P. W. Reddien. 2008. *Smed β-catenin-1* is required for anterioposterior blastema polarity in planarian regeneration. *Science* 319: 327–30.

Reddien, P. W et al. 2005. Identification of genes needed for regeneration, stem cell function, and tissue homeostasis by systematic gene perturbation in planaria. *Developmental Cell* 8: 635–49.

Reddien, P. W., and A. Sanchez Alvarado. 2004. Fundaementals of planarian regeneration. *Annual Review of Cell and Developmental Biology* 20: 725–57.

Sanchez Alvarardo, A. 2004 Planarians. *Current Biology* 14: R737–8.

Stuckemann, T. et al. 2017. Antagonistic self-organizing patterning systems control maintenance and regeneration of the anteroposterior axis in planarians. *Developmental Cell* 40: 248–63.

Unit II

Mutants and Phenotypes

4

CHAPTER 4

Identifying and classifying mutants

TOPIC SUMMARY

The process of genetic analysis often begins with a mutant phenotype. The great power of mutant analysis is that the investigator does not need advance knowledge about what to expect when the biological process is disrupted. The process of mutant analysis can be summarized as:

- finding mutations, often using some mutagenic agent to increase their frequency;
- characterizing the phenotype of the mutation, including examining its phenotype under different conditions;
- assigning the mutations to genes using complementation tests;
- classifying the type of defect in each mutation on the basis of its effect on the function. While most mutations are recessive and cause a loss or a reduction in the function, some interesting mutations are dominant and arise from other alterations in a gene.

Many of the details for this overall process depend on the organism and the mutant phenotypes, so general principles of mutant analysis will be surveyed. While we present mutant analysis as a linear process, the actual laboratory procedure is rarely stepwise. Many of the interpretations in the remainder of the book depend on having a well-characterized set of mutants.

INTRODUCTION

Biological processes are the outcomes of millennia of evolution. Nobel Laureate Max Delbrück is reported by Hood and Galas to have said: "Any living cell carries with it the experiences of a billion years of experimentation by its ancestors."

Literature Link. Hood and Galas 2003, *Nature* 421: 444–8

We cannot trace the long evolutionary history that determines how an organism has come to employ different genes and gene products in specific ways to execute a certain biological process. On the other hand, it does not take any special training in biology to appreciate the complexity and glory of a living organism.

In order to understand the intricacies of an organism, we want to analyse how it carries out its functions. Without knowing all the turns and dead ends of the evolutionary history, we cannot deduce this underlying biological logic from first principles. How, then, can we unravel the strategy of normal biological processes? For geneticists, we can infer the history by examining specimens that fail to carry out the process normally—that is, specimens that have a mutant phenotype. **Mutant analysis**, the topic of this chapter, is the study of failed biological processes. Something in the normal development, physiology, or behavior of the organism is different or goes awry, and we can see a mutant phenotype as a consequence. From that failure we infer what must occur normally: genetic analysis is about learning from mistakes.

Literature Link. Markovetz, 2010, *PLoS Computational Biology* 6: e1000655

Not every mistake, mutant, or variant is equally useful for genetic analysis, however. We need to know what gene has been mutated and how it has been changed. We need to find other mutations in the same gene and compare their effects. We need to use that mutation to lay the foundation for many other types of analysis. A mutant is fundamental to many of the other tools in the rest of the book. Finding a mutant is often the first and crucial step in untangling the complex strategies of biological processes. Genetic analysis with mutants is recognizing the beauty of failures.

4.1 Finding mutations: An overview

How does one begin to study a complicated biological process? Imagine working with one of the model organisms described in Chapter 3. You watch a fruit fly buzzing around its vial, tracking your movements with those big red eyes: the process is more intricate and subtle than you thought. Or you see a worm sliding gracefully on the surface of the Petri dish, attracted to some chemicals and repelled by others. You see a plant blooming in season but sitting dormant at other times. What could you do to understand these or any other biological processes?

Almost all geneticists who work with model organisms will answer this question with the same mantra: find a mutant. Do you want to know what determines the eye color in a fruit fly? Find a mutant with a different eye color. Do you want to understand how the worm moves so gracefully? Find a mutant that cannot move. Do you want to know why plants bloom at certain seasons but not others? Do you want to understand any biological process? Find a mutant that cannot carry it out properly.

It may seem paradoxical that the strategy for analysing a normal biological process begins with observing its malfunctions. Yet that is what mutant analysis involves: disrupting the process in some way and asking what happens. In other words, breaking it in a limited and controlled way to see what happens. Or, to repeat the geneticist's mantra, "find a mutant."

This chapter describes the procedure by which geneticists find and characterize mutants. The mutated gene and the mutant phenotype provide essential tools for every other part of genetic analysis, and topics in subsequent chapters in the book typically assume that the investigator has a mutated gene to work with. Furthermore, we assume not only that the investigator has such a mutant but also that the mutant has been characterized—that is, that the investigator has an answer to such questions as:

- Is the mutation recessive or dominant?
- Does the mutant phenotype change in different growth conditions?
- Does the mutation affect more than one biological process?
- Does the mutation reduce or eliminate the function of the gene? Or does it increase its function, or introduce a new one?
- How many other genes with similar mutant phenotypes are there in the genome?

Genetic analysis can begin with a mutant organism, with a functional gene, or with a genome sequence

It is sometimes helpful to distinguish between two approaches to genetic analysis called "forward genetics" and "reverse genetics." In the contemporary era, when molecular analysis and genome sequencing are readily available, there is little experimental distinction between these approaches and it seems unlikely that an investigator will recognize when an experiment has made the transition from one approach to the other. Nonetheless, the distinction is helpful for our descriptions in this chapter and in Chapter 5, so we will define these terms and illustrate them with the vestigial wing phenotype in Figure 4.1. **Forward genetic analysis** or, more simply and commonly, **genetic analysis** means that the investigator found or generated an organism with a mutant phenotype and worked toward identifying and characterizing the gene that is responsible. "Forward" means that the experimental analysis has moved from phenotypes in wild-type and mutant organisms as the starting point to the sequence of the gene in wild-type and mutant genomes. **Reverse genetic analysis** means that the investigator has identified a cloned gene or DNA sequence and works toward identifying the

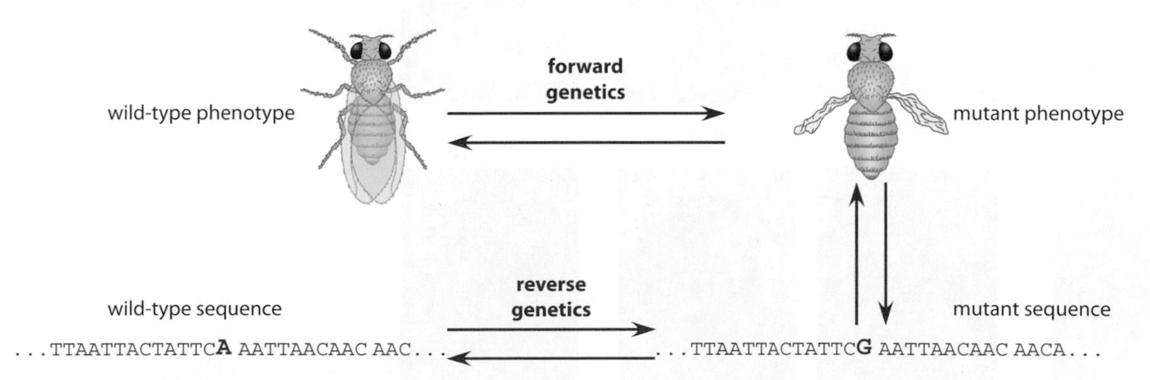

Figure 4.1 Forward and reverse genetic analysis. Genetic analysis connects a phenotype with a gene or DNA sequence. Forward genetic analysis, usually just called genetic analysis, indicated with red arrows, begins with a wild-type phenotype and continues with a mutant phenotype, then progresses to a DNA sequence in the mutant and in the wild type. In the reverse genetic analysis, represented by the blue arrows, the DNA sequence for the gene is known but the mutant phenotype needs to be determined. Reverse genetic analysis is described in Chapter 5. The mutant phenotype and the DNA sequence are from the *vestigial wing* or *vg* gene in *Drosophila*.

mutant phenotype or toward understanding the biological process that the gene effects. This chapter deals with forward genetic analysis, hereafter referred to as genetic analysis. Chapter 5 addresses reverse genetic analysis. Chapter 9 describes some combined approaches to genetic analysis in which the sequence of the genome is the starting point to finding both mutant phenotypes and individual genes.

The greatest advantage of forward genetic analysis is that the investigator allows the organism to provide insights into the biological process. The investigator does not need to understand the biochemistry or the physiology of the underlying process in advance, and often will know little about the molecules that are involved. The investigator does not need to know in advance how many genes are involved, or anything about their functions or relationships, although that often helps. In fact the investigator does not even need to know what alternative phenotypes exist for genes involved in the process. The mutant phenotypes yielded by the organism will reveal those things. The mutations can occur with any gene in the genome. Genetic analysis that begins by finding mutants is an extremely powerful first step for nearly any biological problem.

Reverse genetic analysis has the advantage that the molecular function of the gene under investigation could already be known. Often reverse genetic analysis begins when a gene has been identified in one organism, in the fruit fly for instance, and the investigator wants to know what the equivalent gene does in another organism, say, the mouse. Or perhaps the gene is known to be expressed in the material under study—cells, or stage in a given life cycle. Thus the gene is presumed to be important. Reverse genetic analysis requires more knowledge to start from than forward genetic analysis does, but the two produce similar information in the end.

Let's consider a simple example of forward genetic analysis. A wild-type fruit fly has brick-red eyes. How does this eye color come about? What other colors are possible? Does the biochemistry of eye pigmentation allow flies to have blue eyes, brown eyes, or black eyes? Does eye color have any relationship to the colors of other tissues in the fly? How can we analyse this complex biological problem?

Remember our mantra: find a mutant. To do this, a geneticist will look for flies that have eyes of any color other than the wild-type brick red. Some of the phenotypes that might be found are shown in Figure 4.2. These mutant flies will be analysed through mating with flies of defined genotypes, in order for us to determine how many genes are involved and what has happened to those genes to give rise to this mutant eye color. The initial mutant hunt is repeated, and more flies with mutant eye colors will be found and analysed through

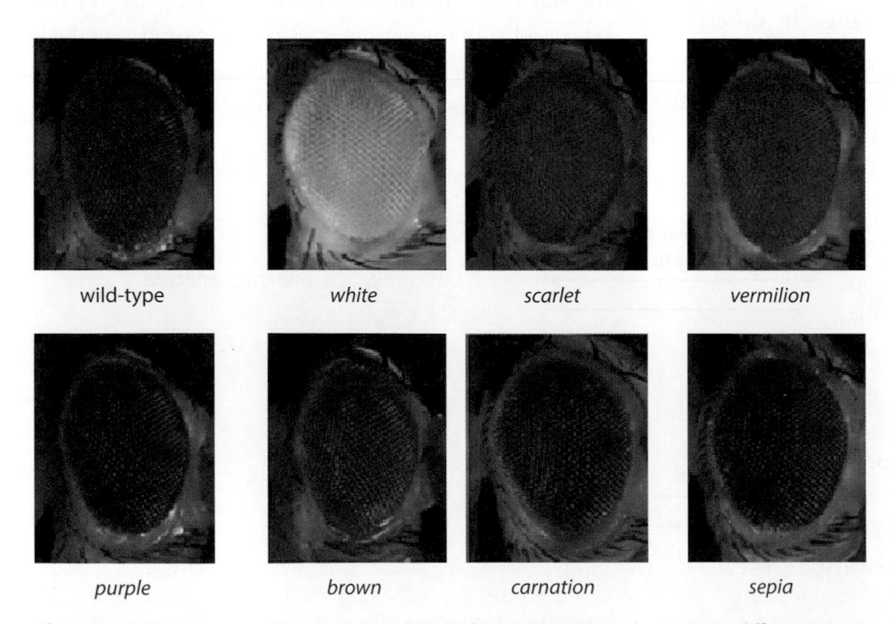

Figure 4.2 Mutant eye colors in *Drosophila melanogaster*. More than twenty different genes have effects on eye color, of which a few are shown here; the gene name is under the picture. Images are from Flybase (flybase.org) and used with permission.

this scheme of crosses. If the investigator persists long enough, many of the genes that affect eye color will be found. Then we can begin to answer questions about how the color of the eye is determined. In reality, investigators will address many of these questions as they are collecting the mutants; the analysis of the first mutations to be found will help determine what scheme of crosses can be used to identify and analyse additional mutations. But it is important to recognize some of the questions that were *not* necessary for the initial analysis.

- The investigators did not ask what colors could be found, or whether they could find a particular color. Instead they asked what colors are actually found in mutant flies.
- The investigators did not ask what molecular or biochemical process was being affected. The different color could occur because the gene for one or more pigments is not being transcribed, because some enzyme involved in pigmentation synthesis is not being made properly, because the pigment granules are not being transported into the cells, or as a result of a host of other molecular and biochemical processes. That understanding comes later, but it is not needed in advance.
- The investigators did not have to ask what other tissues use the same pigmentation process as the eye. After finding an eye-color mutant, they might see, for example, that the body color was not affected but the color of an organ called the Malpighian tubules is affected by these same mutants. Thus, on the basis of this mutant, the genetic control of eye color is related to pigmentation in the Malpighian tubules but distinct from body color.

In one sense, genetic analysis is simple: start anywhere and find a mutant. In fact the lure (and lore) of genetics is such that there are many stories about geneticists whose research progress seemed to be stalled, so they simply mutagenized their organism, found mutants with unexpected phenotypes, and jump-started a new project. But, after finding one new mutant, the investigator begins to ask several questions. "What can I do to produce more mutants with similar phenotypes?" "What do I do with my mutant once I have found it?" "How do I know what has gone wrong in this mutant?" Genetic analysis can help to answer all of these questions.

4.2 Producing mutations

Mutations in single genes arise spontaneously as errors in DNA replication and repair, among other processes, but these natural events occur too infrequently for a geneticist who wants to find many mutations as quickly as possible. Naturally occurring variation is particularly important in human genetics for which mutations cannot be induced at will, and natural variation is increasingly used in other species as well. In this section we focus on the topic of inducing mutations in laboratory species; once the mutant has been found, whether in the lab or in nature, many of the follow-up steps are similar, at least in rationale if not in actual procedures. For laboratory species, a geneticist will use some other reagent—a **mutagen**—to induce mutations at a much higher rate than they occur spontaneously. There is a strong and vital interaction between the fields of mutation research and environmental health, in large part because mutagenic chemicals in one organism are often mutagenic or carcinogenic in another. Thousands of different mutagens have been identified, of which a smaller number have been used in the laboratory. Some of these mutagens and their primary mechanism of action are summarized in Table 4.1. Our attention will be on a few that are commonly used in our experimental model organisms discussed in Chapter 3.

Chemical mutagens modify or replace nucleotide bases

Chemical mutagens often produce a known but limited spectrum of single gene mutations. They can act by chemically modifying the nucleotide base, by being incorporated into the DNA during replication instead of a normal nucleotide, or by intercalating between the bases on a DNA strand. In the first two cases, the altered nucleotide base is more likely than the original base to mispair during replication; intercalating agents often produce frameshift insertions or deletions (**indels**).

Table 4.1 Widely used mutagens in model organisms

Class	Mutagen	Primary effect	Type of mutation	Administration	Organisms
Chemical	ethylmethane sulfonate (EMS)	Alkylates purines	G to A transitions	Soaking, feeding	*Drosophila, C. elegans, Arabidopsis*
	Nitrosourea (ENU)	Modifies thymidine	Transitions	Direct application to testes	Mice
	Acridine orange	Intercalation between bases	Frameshifts		bacteria
Radiation	UV light	Pyrimidine dimers	Transitions, tranversions	Irradiation	many. Sometimes combined with psolaren to potentiate the effect.
	X-rays, gamma rays	Chromosome breaks	Chromosome rearrangements	Whole organism irradiation	Many
	Fast neutrons	Chromosome breaks	Deletions	Irradiation of seeds	Plants
Insertions	Transposable elements	Random insertion or excision	Single gene disruptions and small deletions	Mutator strains	Many
	T-DNA	Random insertion	Single gene disruptions	Floral dip	*Arabidopsis*

The widely used mutagen ethyl methane sulfonate (EMS) is an example of a chemical mutagen that modifies the nucleotide base. EMS attaches ethyl groups to guanine or sometimes to adenine nucleotides, which creates a higher probability that the modified guanine will mispair with thymidine rather than with its normal cytosine (or that the modified adenine mispairs with cytosine), as shown in Figure 4.3 A. Thus, if the investigator uses EMS as a mutagen, most of the mutations recovered will affect a single base, although other types of events such as chromosome breaks and small deletions are known to occur occasionally, particularly when higher dosages of mutagen are used. The common mutation induced by EMS is a transition (one purine replaced by the other purine and one pyrimidine replaced by the other pyrimidine). EMS is usually administered to the organism by ingestion, either by mixing the mutagen with the food or by soaking the organism in a solution contained the mutagen.

Since EMS is not a particularly powerful mutagen in mammals (which is an advantage from the point of view of the investigators' safety), the alkylating agent N-ethyl N-nitrosourea or ENU is widely used in mice for a similar effect. ENU acts primarily on thymines, resulting in transitions of A/T to G/C or transversions of A/T to T/A. ENU is often directly applied to the testes of the male mouse and is absorbed through the skin.

In contrast to chemical mutagens like EMS and ENU, which modify bases, base analogs such as bromo-deoxyuridine (BrdU) and 2-amino purine (2-AP) are incorporated into DNA during replication in place of a normal nucleotide base; once incorporated, these bases are more likely to mispair than their naturally occurring analogs (Figure 4.3 B). Neither of these mutagens is widely used in eukaryotes, in part because the proofreading functions of DNA repair enzymes render them relatively ineffective.

Radiation induces chromosome breaks and structural rearrangements

The first mutagen to be demonstrated in a model organism was X-rays; for this H. J. Muller was awarded a Nobel Prize in 1946. Most of the mutations induced by ionizing radiation like X-rays and gamma rays and by fast neutrons in plants are double-stranded chromosome breaks. These breaks result in chromosomal rearrangements such as deletions, duplications, and translocations, deletions of varying size being the most common (or the most commonly used) outcome. The effects of high energy radiation on chromosome structures are summarized in Figure 4.4. In most uses of X-rays and gamma rays, the entire organism is irradiated with doses of 1,000 rads or more, or more than twenty times the dose of a typical X-ray used in medical or dental diagnosis; fast neutrons are used by exposing seeds. The mutagenic

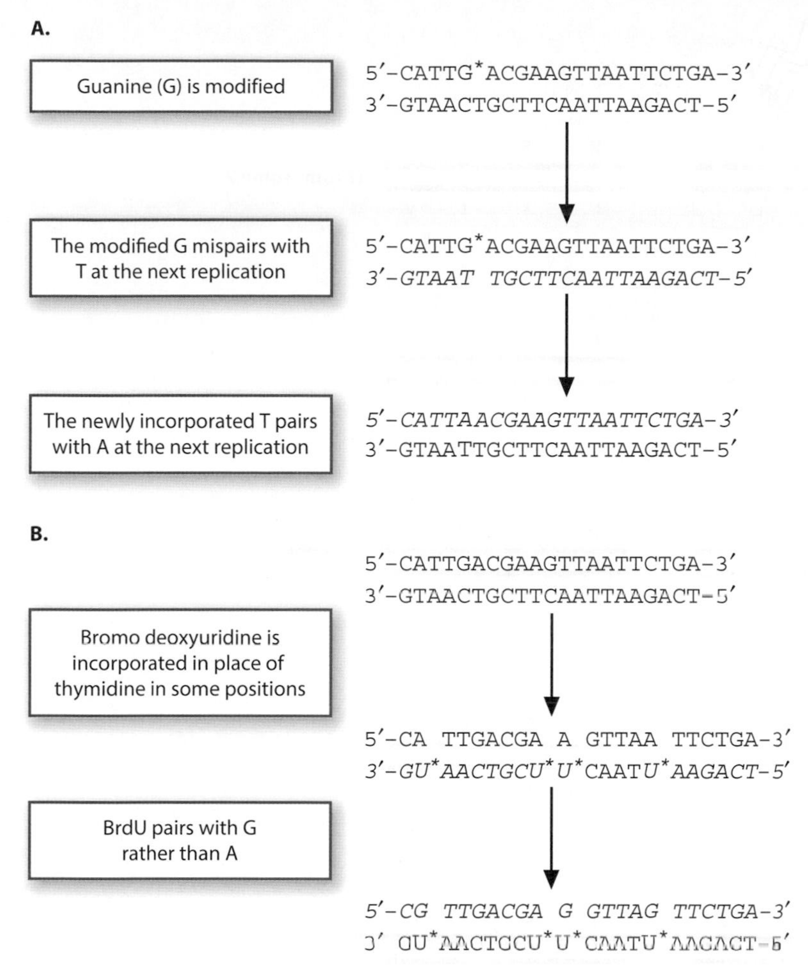

Figure 4.3 The effects of chemical mutagens. Two different effects of chemical mutagens are shown. In Panel A, the mutagen modifies one of the nucleotide bases, in this case a guanine. The modified base is likely to mispair in the next rounds of DNA replication, causing a transition from a G:C base pair to an A:T base pair. Alkylating agents such as EMS and ENU are examples of mutagens of this type. In Panel B, the chemical is structurally similar to one of the nucleotide bases and is incorporated in place of the normal base, in this case, bromo-deoxy-uridine instead of thymine. The base analogue is more likely to mispair than the normal base in subsequent rounds of DNA replication, which results in a transition.

process is complex and the induction of mutations does not follow a simple dose–response curve, in part because the repair mechanisms vary in different organisms, and perhaps even in different cell types.

Chromosome rearrangements often involve multiple break points on the same chromosome or on different ones, which makes them somewhat difficult to characterize. Although we often do not know precisely what a rearrangement has done to the chromosome, such rearrangements provide useful genetic tools for varying gene dosage or suppressing the crossing over. We will discuss some of these uses in Sections 4.4 and 4.6.

An entirely different type of mutation arises from ultraviolet radiation. The preponderance of effects from ultraviolets (UVs) is the induction of pyrimidine dimers, that is, a co-valent cross-link between adjacent pyrimidines on the same DNA strand. The intrastrand cross-link affects base pairing, giving rise to mutations. Because UV damage is quite effectively repaired (fortunately for our health), UV radiation has had limited utility for generating mutations in eukaryotes. Although widely studied as a common environmental mutagen, UV radiation is not often used as a laboratory mutagen in the search for new mutations.

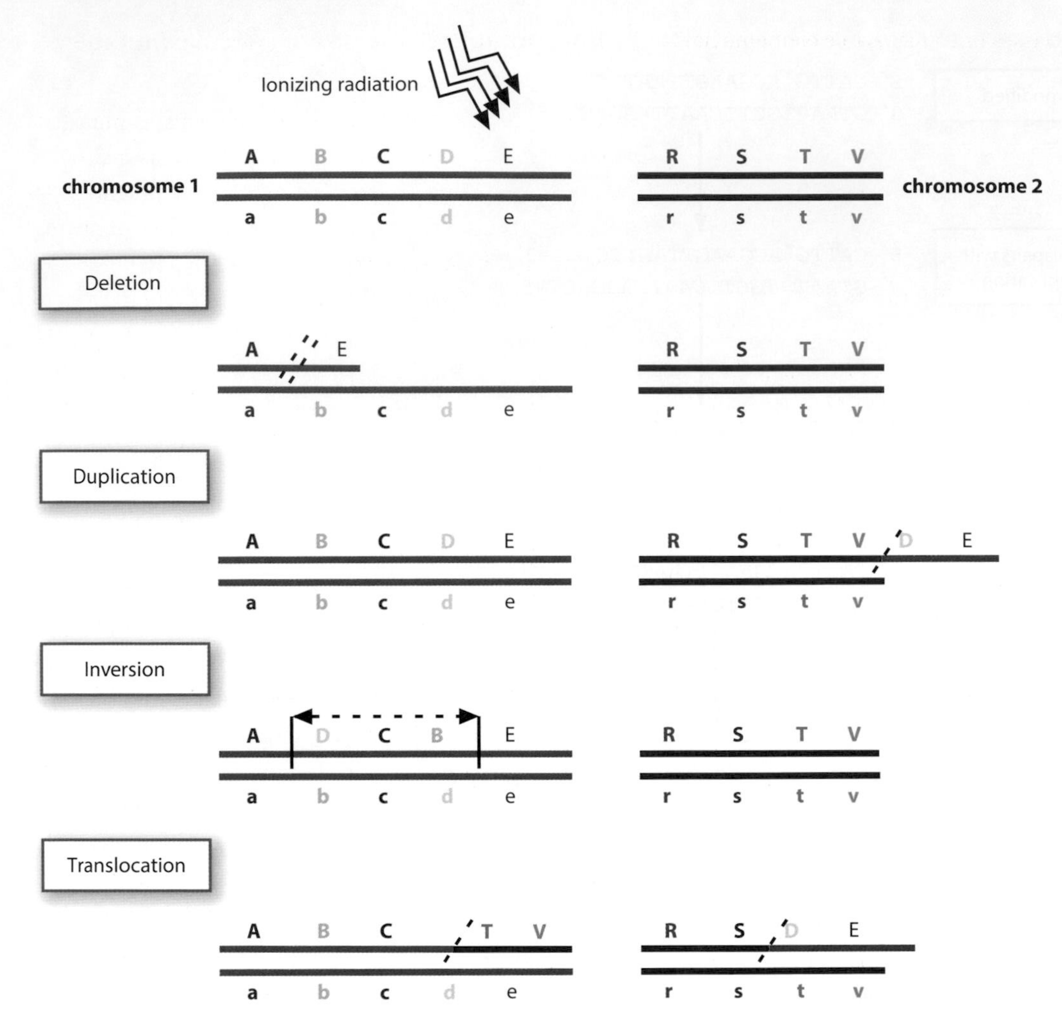

Figure 4.4 The effects of ionizing radiation. X-rays and gamma rays induce double-stranded breaks in the DNA. The mutagenic effects are complex and can occur for a single gene or part of a chromosome. Two different chromosomes, in blue and in green, are shown, with genetic regions indicated by upper and lower case letters. Four different types of chromosome rearrangements are illustrated. A deletion removes part of the chromosome with the subsequent refusion of the two broken ends. The deletion shown here results in a chromosome that lacks (or is deficient for) the genes B, C, and D. Such rearrangements are also called deficiencies. A duplication has an additional region of a chromosome, which could be on located on the same or a different chromosome, as shown here. The organism has three (or possibly four, if the duplication is homozygous viable) copies of a region of a chromosome, which might not have the same alleles. An inversion flips a segment of the chromosome, so that the order of genes is reversed when the ends are reattached. The inversion shown here affects the genes B, C, and D. With a translocation, shown here as a reciprocal translocation involving two non-homologous chromosomes, regions are moved from one chromosome to another, non-homologous chromosome. In the example shown, the D and E genes from the blue chromosome have been moved to the green chromosome, and the T and V genes from the green chromosome have been moved to the blue chromosome. Close analysis of the end points of chromosome rearrangements has found that the break events often involve multiple structural changes, such as small deletions or duplications at junctions of inversions and translocations.

Insertional mutagens are among the most useful laboratory mutagens

The mutagens discussed so far exert their effects via changes to the DNA that are not completely repaired. But other widely used mutagenic agents are the naturally occurring insertional mutagens such as transposable elements and T-DNA insertions, introduced in the previous chapters. These insert complete DNA sequences into novel locations in the genome, thereby altering the activities of nearby genes.

Many different classes of transposable elements have been studied, and no species of prokaryotes or eukaryotes is free of them. The first element to be recognized, the *Ac/Ds* transposable element in maize, is illustrated in Figure 4.5. This element was found by Barbara McClintock, who was awarded a Nobel Prize in 1983 for her studies (these began in the 1930s); the insertion of a version of an *Ac/Ds* element was probably also responsible for Mendel's wrinkled peas. Transposable elements, both active and inactive, comprise a sizeable percentage of the genome sequence of multicellular organisms. The use of transposable elements as laboratory mutagens began in *Drosophila* at about the same time as McClintock's experimental studies, although the mutagenic effect was not attributed to transposable elements until decades later. Regulating the movement of transposable elements, particularly that of the P element in *Drosophila melanogaster*, has given rise to some of the most valuable tools in modern genetic analysis. For now, we will limit our description to the use of transposable elements as mutagens, with P elements as our primary example.

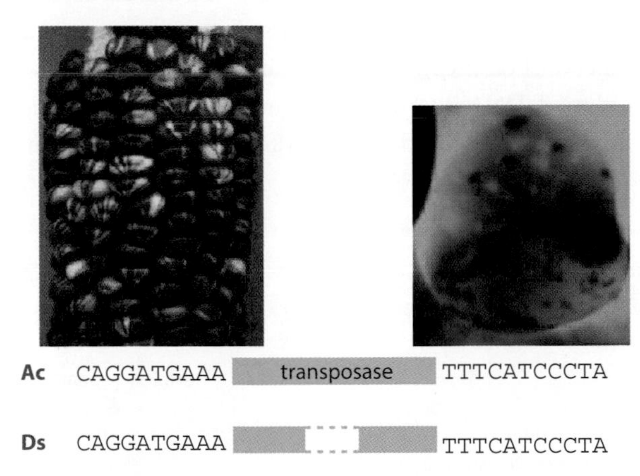

Figure 4.5 The *Ac/Ds* transposable element in maize. The photograph on the left shows the element inserted into the *P* or *pigmentation* gene; the photograph on the right shows a kernel with the element inserted into the *bronze* gene. The lack of pigmentation in each figure is due to inactivation of the gene through insertion of a *Ac/Ds* element. Classically described as a "two-element system," molecular analysis has revealed that *Ac* and *Ds* are two versions of the same element. As diagrammed below the figure, *Ac* is a complete element, with eleven base pair repeats at each end and an intact gene encoding transposase. *Ds* has the same the eleven base pair repeats but has deletions that eliminate part of the transposase gene and its function. The functional transposase encoded by *Ac* elements can move *Ds* elements, since the terminal repeats are the same.

Transposable elements are effective mutagens for inducing single gene mutations

The effectiveness of transposable elements as mutagens rests on two properties: first, they move as a discrete sequence unit; and second, every gene is a potential target for transposable element insertion. The frequency with which a transposable element moves and produces a mutation depends on the type of element and the genetic make-up of the organism. The genomes of many laboratory strains are full of quiescent transposable elements that rarely move, if ever. In other strains, transposable element-induced mutation will arise in a particular gene, once in every few thousand gametes. These strains are known generally as **mutators**, owing to their high rate of mutations.

Mutator strains in which a transposable element is especially active are usually not very fertile because the regulation of the rate of transposition of a mutagenic element requires the same balance as other mutagenic processes: movement occurs frequently enough for it to be possible to use the transposable element as a laboratory mutagen, but not so often that the overall fitness of the organism is affected. Since transposable elements are found in every genome, organisms have different mechanisms, often involving microRNAs, to repress the rate at which they move; mutator strains often lack one or more of these repressive mechanisms. Most active transposable elements (which, again, are a small minority of all the elements in a genome) move in all the cells of the organism and induce both somatic and germline mutations. One of the advantages of P elements, however, is that they move only in the germline and not in somatic cells, as discussed in Section 3.4. Thus, even *Drosophila* strains in which P elements are very active in the germline are relatively healthy and show few somatic effects. The effects are seen in their offspring.

The means by which an investigator induces transposable elements to move depends on the type of element and on the organism. With P elements, regulating the movement was originally done by particular genetic crosses between wild-caught flies (which had the transposable element) and laboratory strains (which at the time did not have the element and also lacked the repressive mechanisms). In the 1930s, *Drosophila* geneticists isolated many of their "spontaneous" mutations by using crosses that mobilized P elements, not realizing why the mutagenic effect was occurring. Many of the original mutant alleles of classic *Drosophila* genes are the result of P element insertions.

Contemporary *Drosophila* geneticists do not rely on naturally occurring versions of P elements. Instead they have modified P elements in vitro and transformed these back into flies to control the movement of P elements, as discussed in Section 3.4. The mobility of a DNA-based transposable element such as a P element depends on two components: an enzyme known as transposase, encoded by the P element itself; and its substrate, the terminal repeat sequences of the P element. If either component is defective, the element cannot move. However, functional transposase encoded on one P element can mobilize many other P elements, so long as their terminal repeats are intact. McClintock considered *Ac* and *Ds* to be separate elements; only many decades later, when their DNA sequences were characterized, was it recognized that *Ac* elements encode a complete transposase gene while *Ds* elements have deletions in the transposable gene, as shown in Figure 4.5. *Drosophila* geneticists have created strains that allow them to control P element movement, as diagrammed in Figure 4.6. A gene that encodes transposase under the control of an inducible promoter can be inserted into the genome; the inserted gene has no terminal repeats, so the modified transposon cannot move. Other strains have transposons with the terminal repeats (which are short and relatively easy to place on a cloned construct) but do not have the active transposase. When the two strains are crossed, the F_1 has both an active transposase and terminal repeats on which it can act, so the cloned construct with the terminal repeats can be moved about the genome.

The mutagenic effect of a transposable element arises from either or both of two activities. First, the element can be inserted into a new location within a particular gene, thereby disrupting the gene or its expression. Second, an existing transposable element can excise imprecisely from its location within a gene. Imprecise excision induces a mutation by deleting a few nucleotides surrounding the insertion/excision site. Both effects are known to occur frequently, and both have been used to produce mutations within a gene.

Insertional mutagenesis in *Arabidopsis* offers many of the same advantages as transposon mutagenesis, but does not use a plant transposable element. The basis for insertional mutagenesis is the plant pathogen *Agrobacterium tumefaciens* which, as discussed in Section 3.5, inserts T-DNA at random into the plant genome. Investigators replace the sequences between the right and the left borders with any sequence, which the bacterium will insert into the plant genome upon infection. Because the sequence inserts into the genome at random locations, insertions are often within a gene and disrupt its normal function. Like transposable elements, T-DNA insertions in exons often result in a complete loss of gene expression and function. Because *Agrobacterium* infection is an easy procedure, T-DNA insertions are commonly used mutagens in *Arabidopsis*, second only to EMS in its use as a mutagen.

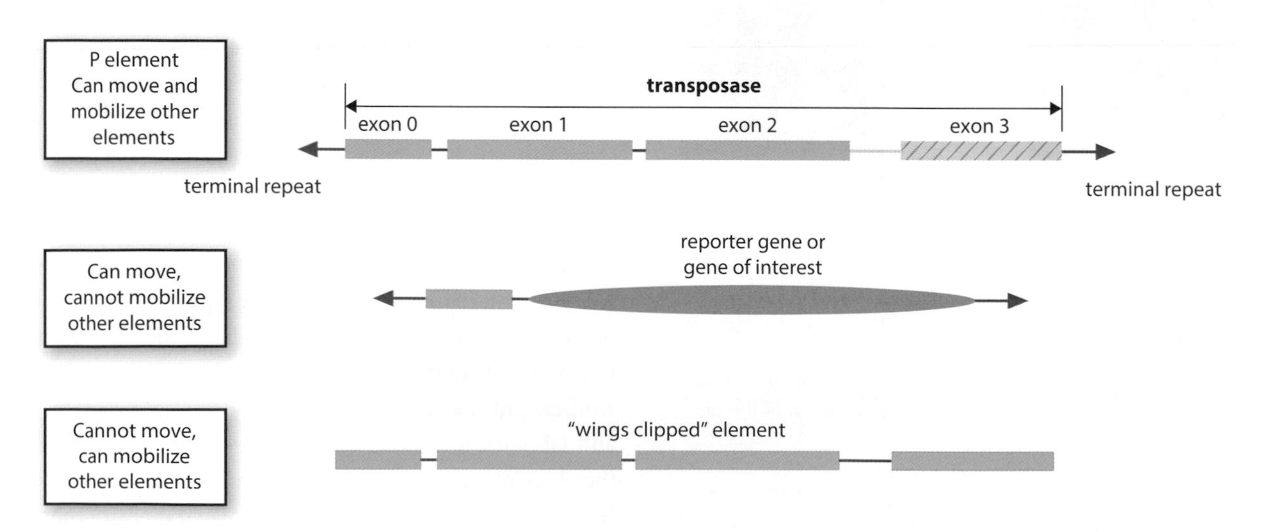

Figure 4.6 The P element in *Drosophila* with some experimental derivatives. An intact P element, shown at the top, has a transposase encoded by four exons. The last exon (shown in green) is alternatively spliced, and a complete transposase is made only in the germline. P elements have been genetically engineered, two examples being drawn below. The element in the middle has the transposase replaced by a reporter gene or other gene of interest. It could not move on its own but can be mobilized by an element with complete transposase. The "wings clipped" element depicted at the bottom encodes functional transposase but lacks the terminal repeats needed for transposition. A "wings clipped" element cannot itself move, but can move elements with terminal repeats when both elements are expressed in the same genome.

One reason why transposable elements and insertional mutagenesis became methods of choice for many applications is that they provided a means to clone a gene, that is, to identify its corresponding DNA sequence, in an era before genome sequences were known. The investigator could use the known sequence of the transposable element to find the gene that had been disrupted with that transposable element, then work from that insertion to obtain the rest of the DNA sequence of the gene. This method of transposon tagging is rarely necessary nowadays, but was once the standard practice for cloning a gene. Another reason for their utility as mutagens is that transposable elements and insertional mutagenesis do not expose the investigator to a potentially hazardous chemical or procedure.

A summary of mutagenic effects

Let's summarize this section. By choosing one type of mutagen from Table 4.1 in preference to another, the investigator has the means to induce mutations of different types. If the primary goal is to obtain single gene events, either chemical mutagens or insertional elements are effective mutagens. If the primary goal is to induce mutations that affect multiple genes, ionizing radiation is usually the preferred mutagen.

A key question for any mutagenesis is the potency of the mutagen and the appropriate dosage or treatment. The investigator wants to find many mutations, so it helpful to use the highest mutation rate possible. This usually means that mutations are being induced in many different genes in the genome. On the other hand, if too many mutations are induced in the genome, the organism is often rendered sterile or unable to grow. Because mutations at many different locations in the genome are induced in any mutagenesis scheme, the newly recovered mutant probably has several mutations in addition to those in the gene of interest. Thus the investigator has to cross the mutant strain to a wild-type strain with no known mutations, in order to replace the genome in the mutant with unmutagenized chromosomes. This process, called **outcrossing**, can take several generations of careful mating to the wild type; a general scheme is shown in Figure 4.7. Thus inducing many mutations in the genome increases the chances of finding a mutant

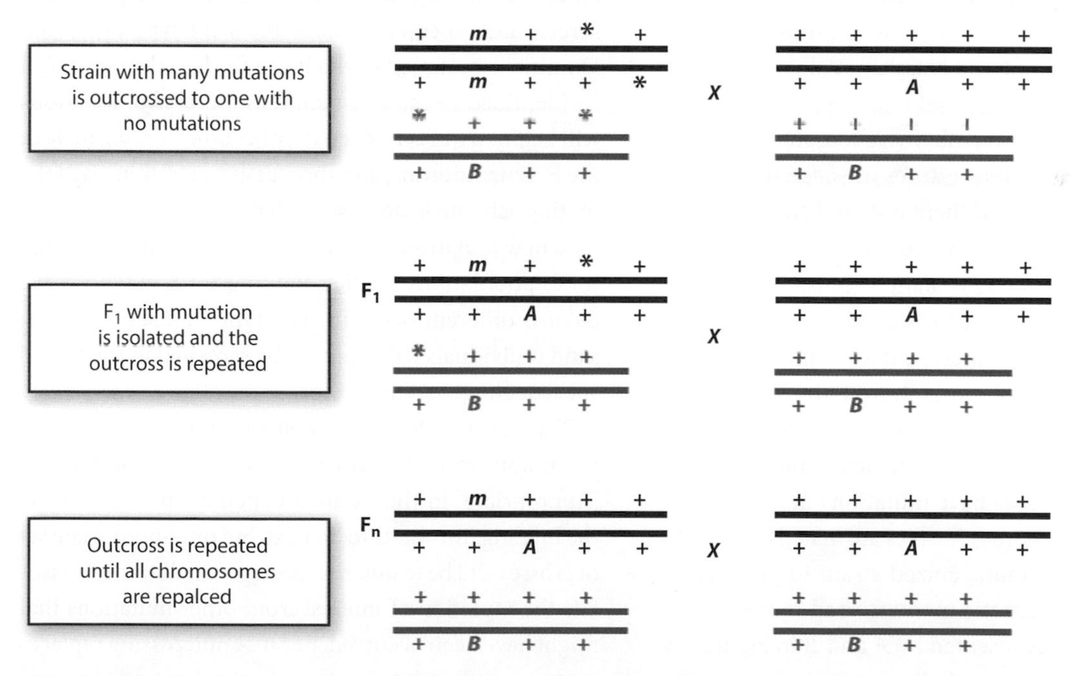

Figure 4.7 Outcrossing for the purpose of eliminating unlinked mutations. In a conventional mutant screen, many mutations (represented by red stars) are introduced into the genome; they can affect the phenotype of the mutant desired, represented by *m*. Genetic markers represented here by A and B are used to monitor the mutagenized and non-mutagenized chromosomes. The goal of the outcrossing scheme is to replace every region of the chromosomes that has been mutagenized, except for the newly arisen mutation with the corresponding region from an non-mutagenized chromosome. A simple outcrossing scheme is shown, and more elaborate ones are likely to be used. In each generation, the mutagenized chromosomes, which have no other genetic markers in this example, are replaced by chromosomes that were not mutagenized, which in this example have the genetic markers. Each round of outcrossing replaces one or more region of the chromosomes and removes some of the additional mutations. Outcrossing is often done as part of the procedure for mapping mutations.

phenotype; but usually it also increases the amount of effort needed to isolate the new mutation of interest from any other mutations induced by the mutagenesis process.

4.3 Finding mutations

Having discussed the ways that an investigator can induce mutations, we turn to some of methods that are used to isolate mutations. The process of finding new mutations among populations of mutagenized organisms is referred to as a mutant search or **mutant screen**. If some method is used to favor the recovery of one type of mutants, or if only one type of mutant can survive and grow, the search is called a **mutant selection**. Mutant screens and selections might be either simple or complicated, and sometimes they have the beauty and self-expression that allows the investigator to find, with minimum effort, exactly what she is searching for. No general description can adequately capture that process, and every issue of *Genetics* or its companion journal *G3* has at least one publication about a genetic screen. Since we cannot describe all of these, we use the case study in this chapter to provide some details about one of the most successful genetic screens. Christiane Nusslein-Volhard and Eric Wieschaus collaborated to find mutations affecting segmentation in *Drosophila* embryos, a mutant screen that displays some of the best features of successful genetic analysis; this work earned them a shared Nobel Prize in 1995. We will compare what was done in their screen with our general description of mutant screens.

While the case study provides an actual example, a simplified hypothetical scheme of a mutant search intended to find eye-color mutations is drawn in Figure 4.8. A population of flies with wild-type eyes is mutagenized with EMS or with a P element. The gametes of these mutagenized flies have mutations in many genes throughout their genome. The flies are mated to a genetically identifiable unmutagenized strain to produce an F_1 generation. The genetic markers used in the unmutagenized parent, represented by *A* and *B* in Figure 4.8, play the important role of distinguishing the mutagenized and unmutagenized chromosomes in subsequent generations.

The F_1 generation will be heterozygous for each of the newly induced mutations, which are almost certainly recessive. In order to find homozygotes for a mutation of interest, the F_1 generation could then be mated among itself, to produce an F_2 generation and then an F_3 generation in which the mutagenized chromosome can be made homozygous. Many other mating schemes could also be used with the same goal of identifying homozygotes for the mutation. The F_3 generation might then be mated to produce an F_4 generation that is homozygous for the chromosomes that have been mutagenized; many variations of the scheme can be introduced along the way. (It is somewhat easier to make mutagenized chromosomes homozygous in self-fertilizing organisms such as *Arabidopsis* and *C. elegans* than in outbreeding organisms such as flies and mice, and even easier in organisms such as yeast that grow as haploids.)

Because most mutations are recessive, all or nearly all of the flies in the F_1 generation of our hypothetical screen are expected to have the wild-type eye color. Dominant mutations, which are far less common, can be identified in the F_1 generation. Recessive mutations will begin to be seen when they become homozygous in the F_2 generation and in subsequent generations, and the mating scheme is done such that flies are homozygous for a new mutation. The investigator can screen through these populations of flies, looking for any whose eye color is different from the wild type, and several thousand individuals will typically be screened in the F_2 and subsequent generations.

The newly induced eye-color mutant is then mated to an appropriate strain for further characterization. "Appropriate" in this context depends on what exactly the investigator wants to do next, but one appropriate set of crosses will be to outcross mutagenized strain and isolate the mutation of interest from other mutations that might have been occurred. Because outcrossing replaces mutagenized with unmutagenized genes and chromosomes, the outcrossed individuals often grow better, are more fertile, or have a slightly different phenotype from that of the original mutant strain. The process of outcrossing may also yield other information that is useful for subsequent analysis, such as dominance and linkage.

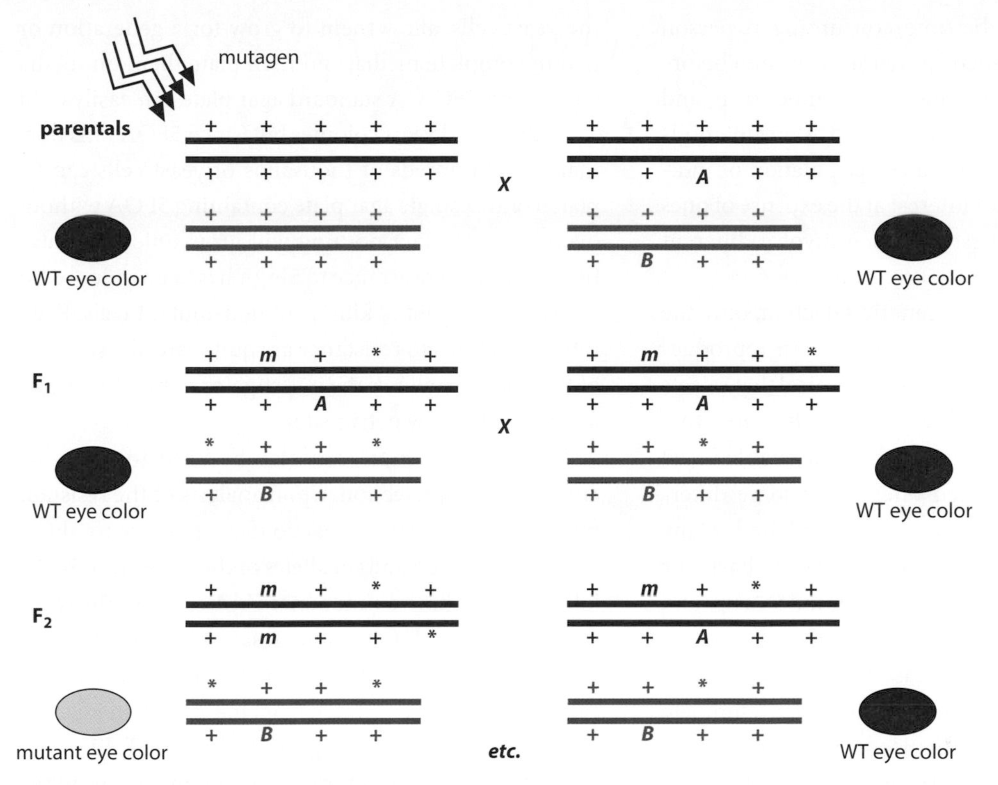

Figure 4.8 A simple scheme for finding eye-color mutations. In this scheme, a wild-type fly is treated with a mutagen and mated to a fly with wild-type eye color but with other genetic markers, represented by *A* and *B* on two different chromosomes. Because most mutations are recessive, the F_1 flies are expected to have wild-type eyes but will be heterozygous for different mutations affecting eye color. In this diagram, the eye-color mutation, represented by *m*, is on the same chromosome as the A genetic marker but on the other homologue. The F_1 flies are mated among themselves to produce an F_2 generation, some of which have a mutant eye color, shown here in pink. Other mutations that do not affect the eye color are represented by stars (*) and will be outcrossed as in Figure 4.7. This scheme would be laborious, and much more clever schemes involving fewer F_1 matings are more likely to be done.

But, as a general rule, outcrossing newly arisen mutations is both necessary and tedious.

"Genetic tricks" use well-established principles of genetics

Imagine that an eye-color mutation has been recovered and outcrossed, so that it is likely to be the only mutation remaining in the strain. An investigator could stop the mutagenesis scheme there, characterize this one mutation and its corresponding gene, and have an idea about a single step in the eye pigmentation process in *Drosophila*. Historically, many geneticists in different model organisms did just this: stopped after a single mutation and characterized it. It is uncommon nowadays to stop after a single mutation, and most mutant screens will be repeated to generate many mutations. As we discuss in Chapter 9, genome-wide mutant screens are done to identify *all* the genes that affect a particular biological process. While the overall idea of finding a mutant seems simple, complications do arise during this process, and geneticists have developed methods to deal with these complications. These methods are sometimes referred to as "genetic tricks," but that term suggests that something is being done by sleight of hand. The strategies are in fact rooted in our understanding of basic genetics, and most of them have been used in one form or another for many decades. We will summarize a few of these briefly.

Genetic selections can be used to search mutagenized strains more efficiently

Even with the most potent mutagens and clever screens, it is not unusual to screen thousands or tens of thousands of individuals in a typical mutant hunt. Each of these candidate mutant lines needs to be outcrossed and categorized, which can be time-consuming. A person does not have to apply such a scheme very often before wishing for a strategy to reduce the number of F_1 and F_2 individuals that have to be examined. Many methods have been developed to enrich the population of individuals with a mutation of interest at the expense of ones with no mutation, the best of which involve different kinds of a **genetic selection**.

In the strongest type of a genetic selection, only the mutation of interest is capable of growth or reproduction. It is analogous to a weed killer chemical that selects against a weed and lets the flowers grow. Of course, that requires that the weeds (that is, the undesirable ones) and the flowers (the desired ones) have some different biological properties that can be exploited. Many different biological properties can form the basis for a selection scheme. For example, antibiotic, drug, and nutritional selections have been widely used in bacterial and eukaryotic genetics. In the case study, a genetic selection with lethal mutations was done in order to eliminate flies that did not carry mutagenized chromosomes.

An example of a selection from yeast is illustrated in Figure 4.9. Wild-type yeast cells cannot grow in the presence of the chemical 5FOA. To find mutations that confer resistance to 5FOA, the investigator can mutagenize the yeast cells, allow them to grow for a generation or two in complete media, and then plate them on media containing 5FOA. A standard agar plate can easily hold a few thousand yeast colonies but, since 5FOA kills normal cells, hundreds of thousands of yeast cells can be plated onto a single agar plate containing 5FOA without overgrowth. In fact nothing will grow unless a mutation conferring resistance to 5FOA has been induced, so the selective agent is killing all non-mutant cells. Even if the mutations to resistance are quite rare, the strength of the selection allows the investigator to find them easily with only a few petri plates.

This same example can also be used to illustrate another type of selection. Upon analysis of the resistant mutants, the mutations that confer resistance to 5FOA are all found to be mutant alleles of the gene *ura3*. 5FOA kills cells that have a functional *URA3+* gene. The wild-type product of URA3 is necessary for uracil biosynthesis, so uracil auxotrophy provides another means of selection. If uracil is omitted from the growth media, any *ura3* mutant cells cannot grow and only URA3+ cells will be found. Thus it is possible to find mutations in the *ura3* gene relatively simply.

These two conditions make it possible to set up a **counter-selection** for either URA3+ or *ura3* mutant

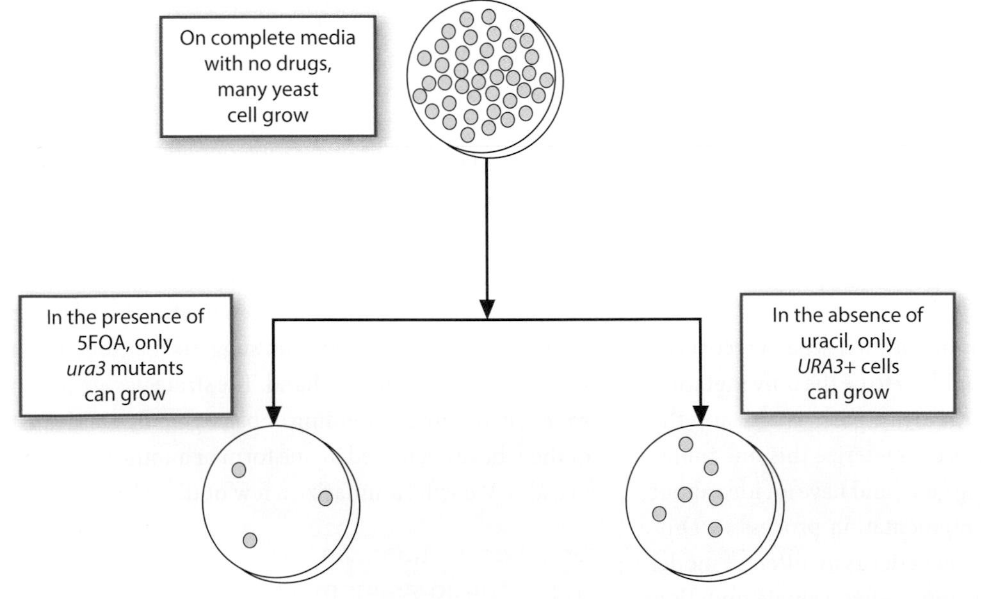

Figure 4.9 Selection and counter-selection in yeast. Yeast cells grow on complete media. When cells are replica-plated onto plates containing the toxin 5FOA, only a few yeast cells grow. 5FOA is metabolized by normal cells into a toxic product, in a process that requires the uracil biosynthetic enzymes. The only cells that can grow on 5FOA are auxotrophic mutants for uracil biosynthesis, particularly *ura3* mutants. Thus, 5FOA can be used to select for *ura3* mutants. On the other hand, when normal cells are replica-plated onto plates lacking uracil, only cells with a functional *URA3* gene can grow. Thus it is possible to select from either *ura3* mutants or the wild type.

cells. On media with 5FOA (but with all other nutrients), *URA3+* cells will be killed and only *ura3* mutant cells will grow. On media lacking uracil, the *ura3* mutant cells cannot grow and only the *URA3+* cells will be found. Nutritional and drug selections of this type are commonly used in yeast genetic analysis but are not limited to yeast; an example of drug selections involving mouse ES cells is described in Chapter 7.

Selection schemes in multicellular organisms can also be based on nutrition, drug and hormone sensitivity and resistance; and they respond to a chemical stimulus, among many possibilities. But we can make a more general example of selection by thinking of the *URA3* gene as a **selectable marker** for a particular DNA region. For example, suppose that we construct a plasmid that includes the *URA3* wild-type gene and a copy of another gene that we are interested in, as shown in Figure 4.10. The yeast cell without the plasmid has a mutated *ura3* gene and is a uracil auxotroph. Populations of cells are transformed with the plasmid, and the cells are grown in the absence of uracil. The only cells that will grow are those that have taken up the plasmid and express the *URA3+* gene; but these cells will also have our gene of interest, which we have not directly selected for. Thus, by using uracil auxotrophy as the basis for our selection, we have also been able to select for the presence of our gene of interest without having to monitor its effect or phenotype directly.

Genetic selections can also be done using linked lethal mutations

This manner of use of selectable markers can be extended to many other types of genes beside nutritional markers; the general principle is that the phenotype of one gene serves to monitor the presence or absence of certain alleles of linked genes whose phenotypes are not actively being examined. This principle is widely used in multicellular organisms such as *Drosophila* and *C. elegans* by employing lethal mutations as selectable markers on the unmutagenized chromosome. A general setup is diagrammed in Figure 4.11, and a more specific example is used in the case study's genetic screen. In our hypothetical eye-color mutant hunt, a quarter of the flies are homozygous for the parental non-mutagenized chromosome and are of no interest to the screen; in fact their presence makes it more difficult to find the mutations with eye-color defects. Suppose that, before the mutant hunt has begun, a recessive lethal mutation is crossed onto the marked chromosome in the non-mutagenized parent, as in Figure 4.11. Now, all of those offspring who inherited only the non-mutagenized chromosome will die, making it easier to spot the ones with eye-color defects. One of the most common means to reduce the number of offspring that have to be examined is to include a lethal mutation or some other selectable marker in the mating scheme.

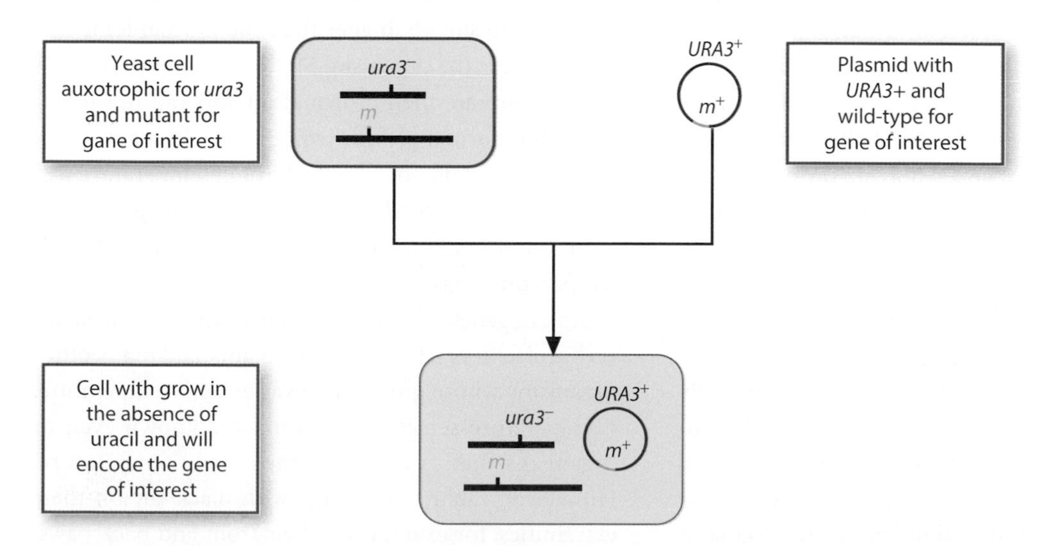

Figure 4.10 Selectable linked markers on plasmids. Cells are also mutant for the gene of interest *m* as well as *ura3⁻*, so they cannot grow in the absence of added uracil. A plasmid that has the wild-type *URA3⁺* and the wild-type *m⁺* gene is transformed into *ura3⁻ m⁻* cells. Selection for the ability to grow in the absence of added uracil also selects for the inheritance of the *m⁺* allele, regardless of its own phenotype.

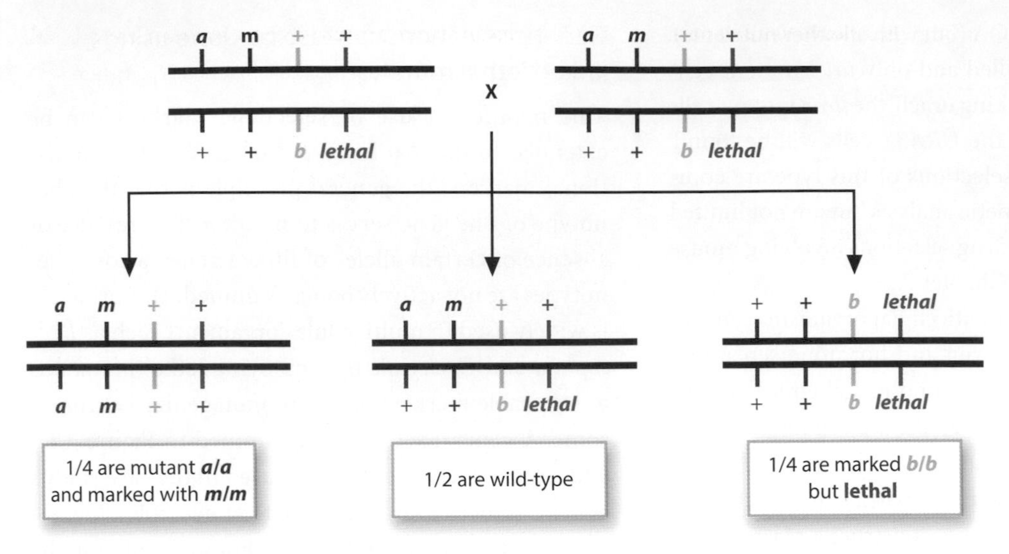

Figure 4.11 Selection with linked lethal mutations. A similar principle of selection as depicted in Figure 4.10 is used for selecting or screening for linked mutations on chromosomes in diploids. A heterozygote is shown that has the *a* marker closely linked to the *m* mutation of interest. The other homologue has the *b* marker and lethal mutation with a phenotype unrelated to the *m* mutant phenotype. Since all the alleles depicted are recessive, the heterozygote has a wild-type phenotype. When two heterozygotes are mated, the *b/b* offspring die because of the linked lethal mutation, the *a/a* offspring are also homozygous for the mutation of interest *m*, and the heterozygotes are of the wild type. The offspring with the *a* mutant phenotype are examined for the effects of the mutant *m*, while the wild-type heterozygotes can be used to maintain the strain.

Lethal mutations can still be maintained in culture

In the previous example, a lethal mutation was used as a selectable marker, to reduce the number of individuals that need to be grown and examined. But, rather than thinking of a lethal mutation as a selectable marker, suppose that the mutation of interest is itself lethal when homozygous. This is not uncommon, since many genes are essential and thus lethal when mutated; these essential genes may be the ones of the greatest interest. For example, the segmentation mutations described in the case study are all lethal when homozygous.

Two general approaches have been used to maintain stock (or strains) with lethal mutations. The first approach is to employ growth conditions, especially temperature regimens, under which recessive homozygotes can survive. These are known as **conditional mutations**, with the specific example of **temperature-sensitive mutations**. The second approach is to maintain the mutations as heterozygotes; this approach is known as **balancing** the mutation. Both approaches are widely used; let's explore each of these in turn.

Conditional mutations

Temperature-sensitive conditional mutations were first employed with the bacteriophage T4, and have been widely used since then. As a standard practice, it is often helpful to conduct the mutant hunt at a growth temperature slightly higher than the normal temperature used for the organism. Mutants found at the high temperature are then grown at a temperature slightly below the normal growth temperature. In many cases, mutations that do not grow at the high temperature will grow and exhibit a wild- or nearly wild-type phenotype at the lower temperature, as shown in Figure 4.12. Such temperature-sensitive conditional mutations have been a staple of genetic analysis not only in viruses and bacteria but also in yeast, worms, flies, plants, and many other organisms whose growth rate varies with temperature. (Temperature-sensitive mutations are known even in organisms that can regulate their body temperature. Himalayan rabbits are white, with black fur on their extremities, for example on their front and back paws, the tips of their ears, and the end of their nose, where the body temperature is cooler. At very low temperatures, the rabbits are all black. The allele that produces the

Permissive temperature **Restrictive temperature**

wild-type mutant

Figure 4.12 Temperature-sensitive phenotypes. Mutations with a temperature-sensitive phenotype can be maintained as homozygotes by being cultured at the permissive temperature, usually a lower temperature. At the permissive temperature, a wild-type phenotype is observed, while the mutant phenotype is seen at the restrictive temperature. Temperature-sensitive phenotypes are particularly useful for maintaining and analysing lethal mutations.

black coat color is temperature-sensitive and inactive at higher body temperatures.)

The temperature at which the mutant phenotype is seen is termed the **restrictive temperature**. The temperature at which the wild-type phenotype is seen is called the **permissive temperature**. Often these temperatures differ only by a few degrees; the permissive temperature for *C. elegans* is 15 C, whereas the most widely used restrictive temperature is 25 C. Not all temperature-sensitive mutations are lethal, although this is the use we are considering. The key is that the mutant phenotype is seen under a specific set of growth conditions, in this case temperature.

In most cases the restrictive temperature is the high temperature, so that the mutations actually have a heat-sensitive phenotype. For this reason, when the term "temperature-sensitive" mutation is used with no further explanation, as above, it can be assumed that the restrictive temperature is the high temperature and the permissive temperature is the low temperature. Such temperature-sensitive mutations are often designated by the notation "*ts*." Exceptional cases have been described in which the restrictive temperature is the lower temperature and the permissive temperature is the normal growth temperature for the organism. These mutations are called "cold-sensitive," designated by "*cs*."

It is easy to imagine how a temperature-sensitive phenotype might arise by recalling that the three-dimensional shape of a protein depends on the temperature and that the function of a product depends on its three-dimensional shape. Thus a mutation that slightly destabilizes the structure of the protein may have no effect on the function until the protein is placed under higher temperature conditions—not so high that the native protein is destabilized but high enough for an already slightly destabilized protein to become non-functional. With a non-functional protein, a mutant phenotype can occur. Under lower permissive temperature conditions the protein is stable and functional, so a wild-type phenotype is observed. Although this is an easy way to think about temperature-sensitive mutations, this explanation has not been verified for most temperature-sensitive mutations, and it is clearly an oversimplification for some of them. Nonetheless, it provides an appropriate intellectual framework for thinking about this kind of mutations.

Temperature-sensitive mutations are introduced as a means to maintain a mutation, for example a lethal mutation, as a homozygote with no effect on the phenotype. In Chapter 6 we discuss how temperature-sensitive mutations can be used to estimate the time at which the action of a gene is needed; this is described in Section 6.3 and in that chapter's case study.

Balanced heterozygotes

Since most mutations are recessive, lethal mutations in diploid organisms can also be maintained as heterozygotes; the homozygous lethal mutant individuals of interest are produced when two heterozygotes are mated. An example was introduced in Figure 4.11. These are known as **balanced heterozygotes**, and a generalized genotype for such a state—a balanced heterozygous state—is diagrammed in Figure 4.11. However, generations of geneticists have greatly modified this simplified example, and the case study presents a more realistic use of balanced heterozygotes.

The greatly simplified example in Figure 4.11 illustrates several important features of using balanced heterozygotes. Notice that the effectiveness of this method will be affected by crossing over between the visible marker *a* and the mutation of interest *m*—or, in population genetics terms, by the length of the haplotype on which *a* and *m* are found. When crossing over occurs between the marker mutation *a* and the mutation of interest *m*, the mutation of interest would be lost. In our simple diagram, recombination would limit this approach to an interval of a few map units.

The early *Drosophila* geneticists developed a set of **balancer chromosomes** to circumvent the problem with recombination and extend the intervals over which lethal mutations can be maintained as heterozygotes. Many chromosome rearrangements, particularly inversions, act as **crossover suppressors**, which may be because the homologous chromosomes do not form any synapsis along their length, because the gametes arising from a crossover with the rearranged chromosomes give rise to non-viable offspring, or for some combination of these effects. Rearrangements that serve as crossover suppressors are the basis for balancer chromosomes, as described in more detail in Tool Box 4.1.

In addition to crossover suppression meant to extend the length of balanced interval or the haplotype, each of the homologous chromosomes must be marked with appropriate visible markers; ideally, these would have phenotypes that can be scored very early in the course of development. *Drosophila* geneticists refined this further by including a dominant visible marker that is itself homozygous lethal (such as Curly wings) on the

TOOL BOX 4.1 Balancer chromosomes in *Drosophila*

Many mutations recovered in a genetic screen have to be maintained in heterozygous strains because they are non-viable as homozygotes. As discussed in Section 4.3, these mutations are maintained as balanced heterozygotes with easily visible markers both nearby on the same homolog and on other homologs, so that the homozygote with the mutation of interest can be recognized. The linkage arrangements of the visible marker mutations have to be retained during subsequent generations, which means that either the visible makers must be very closely linked to the mutation of interest or crossing over needs to be reduced or prevented in the region.

The *Drosophila* geneticists H. J. Muller and Alfred Sturtevant found a solution to this problem more than ninety years by constructing the first of many **balancer chromosomes**. Balancer chromosomes are commonly used in *Drosophila* genetics, so this Box focuses on flies, but balancer chromosomes have also been used for mutant screens in *C. elegans* and in mice.

Properties of balancer chromosomes

A good balancer chromosome has two essential characteristics. First, it must greatly reduce or eliminate crossing over in a region of the genome being investigated. With crossover suppression between the homologs, the inheritance of the balanced mutation of interest on the non-rearranged chromosome can be tracked even if the visible marker is not normally nearby on the standard genetic map. Second, individuals homozygous for the balancer need to be distinguishable from heterozygotes, which in turn need to be distinguishable from the homozygous non-balancer mutants.

Crossover suppression

Crossover suppression has been achieved with the use of inversions or other complex chromosomal rearrangements. Cytological observations in maize showed that an inversion heterozygote can form a loop structure upon meiotic synapsis, as shown in Figure B4.1; such inversion loops are assumed to occur in other organisms with less easily visible chromosomes. A crossover that occurs within the loop will produce gametes with duplication and deletion chromosomes, which may include chromosomes that lack a centromere (known as acentric fragments) and ones that have two centromeres (dicentric fragments), which fail to complete meiosis normally and thus do not contribute to viable gametes. Since only the non-crossover chromatids will appear in viable gametes and in the subsequent offspring, an inversion effectively reduces or eliminates crossing over.

The most widely used balancer chromosomes in *Drosophila* have multiple inversions induced by repeated irradiation over many years, and these multiply inverted chromosomes eliminate crossing over in a large region of the chromosome. In fact, most of the euchromatic portion of the *Drosophila melanogaster* genome can be balanced by one of three balancer chromosomes, one for the X and one for each of the two autosomes. The ability to use only three different balancer chromosomes to screen the genome reflects both the years of genetics research using *Drosophila* and the small number of chromosomes. The case study in this chapter describes the use of balancer chromosomes to identify and maintain mutations that affect segmentation in *Drosophila* embryos, mutations that are lethal when homozygous.

Distinguishing marker mutations

The second necessary characteristic for a balancer chromosome is the ability to distinguish the balanced heterozygote, which is the genotype of interest for strain maintenance, from either of the two homozygotes. In *Drosophila* balancer chromosomes, this is usually accomplished with a dominant marker on the inversion chromosome. Some of the dominant markers that have been used are themselves homozygous and lethal, or the inversion also has other recessive lethal mutations present on it. For example, one balancer chromosome described in the case study is marked with the dominant marker Curly wings as well as with several recessive lethal mutations. An example is shown

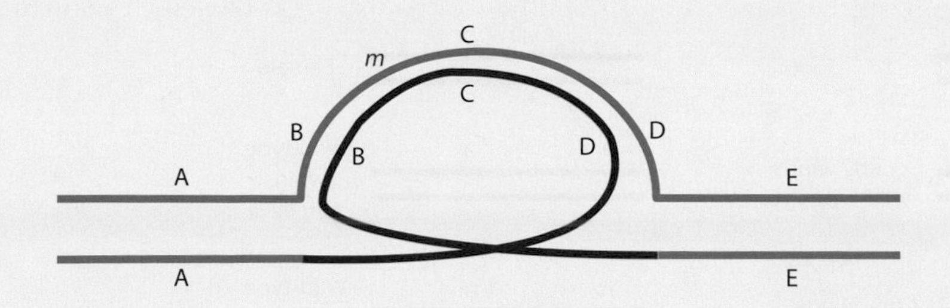

Gametes produced

Non-crossover

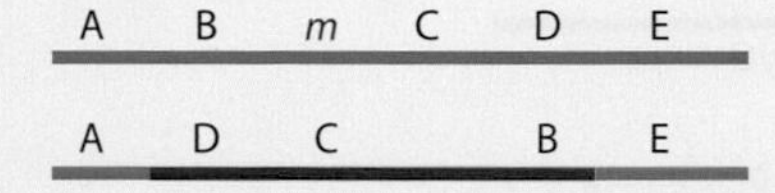

Crossover

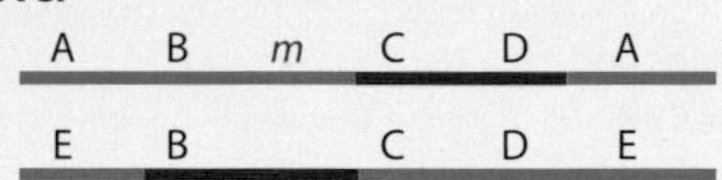

duplicated for A, deleted for E

duplicated for E, deleted for A

Figure B4.1 Inversions suppress crossing over. Two homologous chromosomes are shown, with the upper case letters indicating regions of the chromosome, which could be genes. The non-rearranged chromosome has the order ABCDE, the mutation of interest *m* on this homolog being situated between the markers B and C. The inverted chromosome has the order ADCBE, with the region of the inversion shown in red. In order for synapsis of each region to occur between the homologs, one needs to form a loop, as diagrammed here. While such loops have been observed in some plants with large chromosomes, the chromosomes may fail to synapse altogether rather than loop, but the outcome is the same. If a crossover occurs in the inverted region, such as between *m* and C here, the crossover gametes will have duplications and deletions for some regions of the chromosome, as shown. If the inversion is large or is adjacent to the centromere, the duplications and deletions chromosomes could be dicentric or acentric, both of which are meiotic and unstable. Gametes with such duplications and deletions either fail in meiosis or produce inviable offspring. Thus, crossing over is suppressed in the sense that only non-crossover gametes can produce viable offspring.

in Figure B4.2. The balanced heterozyote has Curly wings and is easily identified for stock maintenance. The homozygotes for the balancer inversion die because of the recessive lethal mutations. The offspring that do not have Curly wings will be ones with the mutation being studied. Such a system is particularly effective, since *Drosophila* is maintained in culture tubes by random mass matings rather than by matings of an individual pair of flies. Therefore, if the mutation of interest is lethal, which is a standard use of balancer chromosomes, the two homozygous classes will die and only the balanced heterozygote will survive. The balanced mutant line can be maintained in culture by simply dumping the contents of one vial into a fresh culture vial, without any need to separate males and females or perform individual crosses. The only surviving flies are those that are heterozygous for the balancer and for the mutation of interest.

A more recent innovation has been to insert the GFP gene or a modified derivative of GFP as a reporter gene onto the balancer inversion. The fluorescence arising from GFP can be identified in early embryos long before any adult visible phenotype such as Curly wings could be spotted. Even more importantly, the embryos can be sorted automatically by fluorescence in a flow cytometer, and the three genotypic classes easily identified and separated from one another at a very early stage. Since balancer chromosomes are used to maintain lethal mutation stocks, it is very helpful to be able to distinguish the inviable embryos arising from the effects of the mutation from those arising from the effects of the balancer chromosome.

The balancer chromosomes in *C. elegans* and in mice are less widely used, in part because genome-wide mutant screens (as discussed in Chapter 9) have obviated some of the need to maintain lethal mutations like this, and in part because these organisms have more chromosomes so many more balancer chromosomes would be needed. Nonetheless,

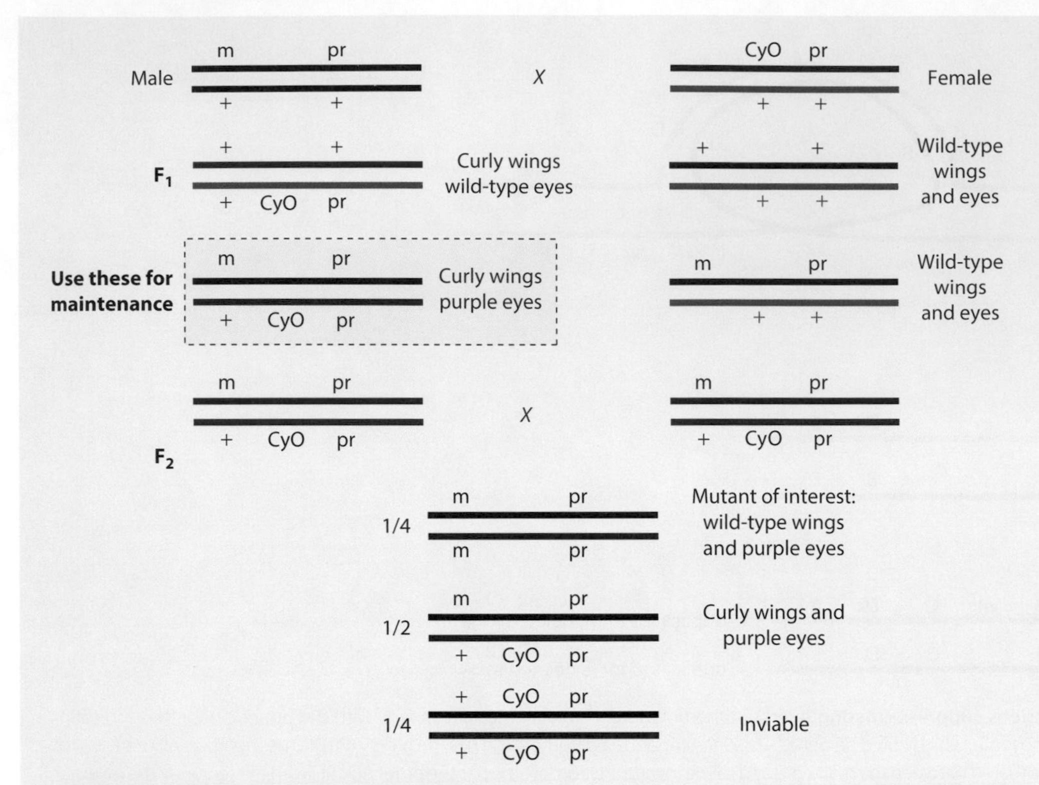

Figure B4.2 Using a balancer chromosome. One widely used balancer chromosome in *Drosophila* is called CyO. This chromosome has multiple inversions that suppress recombination, the dominant marker Curly wings (*Cy*) and the recessive marker purple eyes (*pr*), among other genetic markers. A heterozygous male that has the mutation of interest (*m*) linked to the *pr* gene is mated to the balancer heterozygous female. Since males do not recombine in *Drosophila*, the *m* and *pr* will remain linked in this cross. Four classes of F_1 progeny will result, as shown. The ones of interest for further analysis and maintenance are recognized because they have Curly wings and purple eyes. These flies are mated to each other to produce the F_2 offspring, as shown. The ones with purple eyes and normal wings are also homozygous for the mutation of interest. Even if *m* is only loosely linked to *pr*, the two genes always segregate together because of the CyO balancer. The flies with Curly wings and purple eyes are like the parents and can be used to maintain the stock. The ones that are homozygous for the CyO chromosome are inviable because Cy/Cy homozygotes die.

these have the same properties as the balancer chromosomes that have been standard tools in *Drosophila* genetics for decades.

Literature

Hentges, K. E., and M. J. Justice. 2004. Checks and balancers: Balancer chromosomes to facilitate genome annotation. *Trends in Genetics* 20: 252–9.

balancer chromosome, so that homozygotes for the balancer chromosome do not survive; other lethal mutations can also be included on the balancer chromosome. It is not necessary to have the balancer chromosome be homozygous lethal so long as it can be readily recognized, but this does make the analysis simpler. The case study illustrates this use.

Although the use of balancer chromosomes is most advanced in *Drosophila*, similar methods for maintaining lethal mutations as heterozygotes have been used in other organisms, including worms and mice. Most of these methods bear some resemblance to the more sophisticated methods used in *Drosophila*. Rather than a dominant morphological marker such as Curly wings, an integrated copy of a gene with a fluorescent product, such as dsRed or GFP, can also be used.

Conditional mutations and balancer chromosomes are, both, methods that can be used to maintain other lethal mutations in culture so that their phenotypes can be analysed. Questions and strategies for the analysis of mutant phenotype are the topic of Chapter 6.

4.4 Mapping genes

Once a mutant with a disruption in a particular biological process is found, the next critical step to identify which gene has been mutated. Only when the mutated gene has been identified is the investigator able to begin analysis of the function of the normal gene.

Identifying which gene has been mutated typically involves a procedure of **mapping** the mutation to a particular location on a chromosome. The mutation is compared to other mutations, which are similar in phenotype and map position, in order to determine whether these mutations are alleles of the same gene, a process known as complementation testing (see **complementation test**). Often mapping and complementation are done as a coordinated effort, so that only one allele of a gene has to be carefully mapped and only mutations that map to the same general location are tested for complementation against the first allele. The mutation screen can be set up in such a way—with balancer chromosomes, for instance—that only genes that map onto a particular region of the genome are recovered, so that more detailed mapping may not be needed before complementation testing is done. Other times, depending on the number of mutants and the amount of time required to map them, complementation tests are done before mapping, so that only one allele has to be mapped for each gene. For the sake of completeness we discuss the process of mapping in our description mutant screens, but mapping experiments usually involve screening and scoring a large number of individuals and can be tedious to carry out, so our description might not represent how the experiments are actually done.

Mutations are usually mapped using recombination

The most widely used procedure for assigning a mutation to a gene is to assign it to a genetic locus, that is, to map the mutation to a location on a chromosome using segregation and recombination. Mapping mutations involves basic principles of classical genetics, and a reminder of the expected results is summarized in Figure 4.13.

Suppose that *m*, the newly identified mutation, is being mapped with respect to a known genetic marker *a* whose map position has been established; for the sake of this example, both *m* and *a* are recessive to wild type, and both are alive and fertile. Our notation *a* refers in general to genetic variants whose known locations can be used in mapping. The marker *a* need not be an allele of a gene and, in fact, in most contemporary mapping experiments, the genetic markers are molecular variants such as single nucleotide polymorphisms or sequence-length polymorphisms rather than phenotypic variants such as white eyes or uncoordinated movement. The logic of the procedure is the same no matter what type of variant is used for a mapping marker; the only differences are the methods used to assess the genotype and the phenotype. If *a* and the allele of interest *m* are unlinked, they will segregate independently of each other, as shown on the left in Figure 4.13. If *a* and the allele of interest *m* are linked, they will not segregate independently of each other, the precise frequency of *m a* and wild *m*+ *a*+ gametes depending on the map distance (p) between the genes; if *m* is closely linked to *a*, p will be small, so different marker variants can be progressively used to find ones most closely linked to *m*. The challenge in these experiments is to find enough markers with known map locations and to have the patience to test them all, but this work has been simplified by having sequenced genomes with many polymorphisms that often can be scored at once.

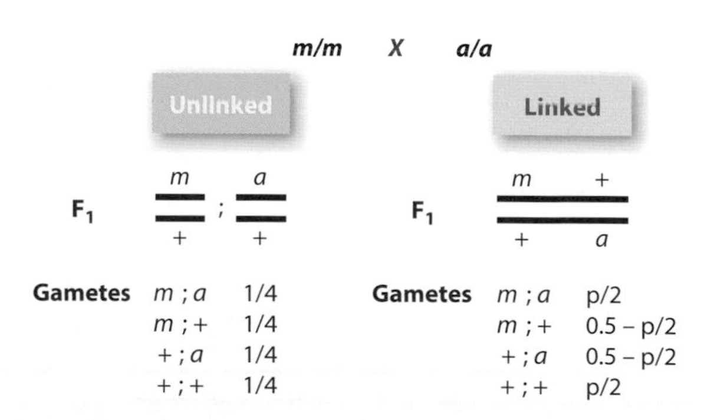

Figure 4.13 Determining linkage for a mutation of interest. Mapping a newly arisen mutation of interest relies on the segregation ratios observed for unlinked and linked genes. If the mutation of interest *m* is not linked to the marker *a*, then heterozygotes for *m* and *a* will produce four types of gametes in equal frequency. If *m* is linked to *a*, then the parental (that is, non-crossover) gametes will be much more frequent than the recombinant gametes; the exact frequency depends on the map distance p between the genes but if p is small, almost no recombinant gametes will be observed.

In fact the effort devoted to mapping the gene depends in part on the overall goal of the project. For monitoring segregation in a subsequent generation, or for determining which mutants should be tested for complementation, it is often sufficient to map the gene with respect to some nearby markers, but not necessary to map it to the closest flanking markers. In order to clone the gene on the basis of its map position, a process described in Chapter 5, it is helpful to map it to the smallest feasible interval, so that fewer candidate genes needed to be sequenced or surveyed.

In some cases it has proved to be easier to identify the location of the newly arisen mutation in the genome by sequencing the genomes (or at least the exons) of the mutant individual and by comparing the result with the sequence of the wild-type genome. In Chapter 5 we describe the first example for which this was done with a human disease trait; the genes for several hundred rare human disease traits have now been located by directly sequencing the entire genome or the exons of an affected individual, and similar but often simpler procedures can be applied on many other organisms. Thus the need to identify a map position on the basis of recombination is obviated, at least for situations in which the goal of mapping is to find the corresponding gene.

A mutation can be mapped using deletions and duplications

When a mutant has been mapped to a particular chromosomal region, it is sometimes easier to determine its more precise position using chromosome rearrangements such as deletions and duplications, at least in organisms such as *Drosophila* and *C. elegans*, where many deletions and duplications are known. (Many more duplications and deletions have been found and analysed in *Drosophila* than in any other model genetic organism, an advantage arising from a century of genetic analysis.)

Figure 4.14 summarizes the structures of deletion and duplication chromosomes schematically. A **deletion** chromosome (also known as a deficiency chromosome) is a chromosome in which all the genes are missing in a contiguous portion of it; it is often drawn as a chromosome with a gap or a hole in it, although in reality the two end points are fused together to make a new and shorter chromosome. A **duplication** chromosome has an additional fragment of itself elsewhere in the genome, usually with the wild-type alleles for each of the duplicate genes. In effect, deletion chromosomes have the recessive mutant alleles for every gene in a region, while duplication chromosomes have the dominant wild-type alleles for every gene in a region.

A deletion almost always has to be maintained as a heterozygote (often with a balancer chromosome), because homozygous deletions that affect more than a few genes are usually lethal. Balanced this way, the deletion strain will have only one copy of every gene that is missing from the deleted region, the copy on the other homologous chromosome; in other words, it is hemizygous for those genes. In Figure 4.14 it lacks the wild-type alleles for genes *c*, *d*, and *e* but has the wild-type alleles for *a*, *b*, and *f*. A duplication strain will have three

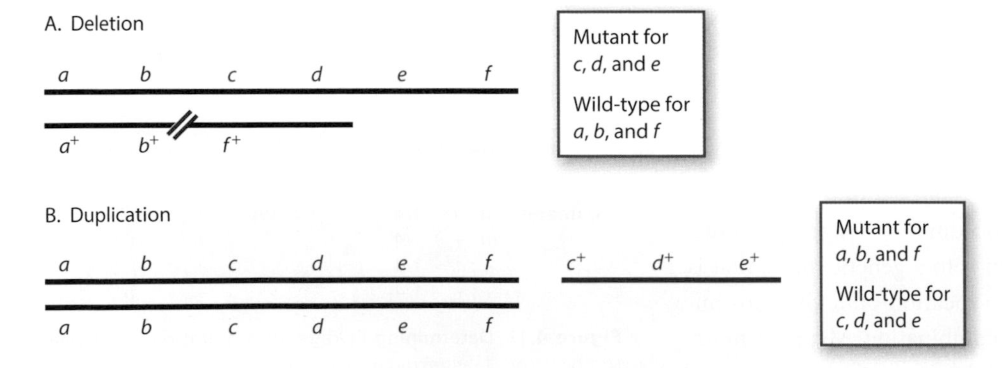

Figure 4.14 The structures of deletions and duplications. The deletion chromosome shown in Panel A lacks the genes (or regions) ***c***, ***d***, and ***e***, as shown in red, and can be considered as having a mutant allele for these three genes. The diploid shown will exhibit a mutant phenotype for genes ***c***, ***d***, and ***e***, but will have a wild-type phenotype for genes ***a***, ***b***, and ***f***, which are heterozygous. The duplication shown in Panel B has the wild-type alleles for genes ***c***, ***d***, and ***e***. The diploid shown will have a mutant phenotype for the genes ***a***, ***b***, and ***f***, which are homozygous for the recessive mutant alleles, but will have the wild-type phenotype for genes ***c***, ***d***, and ***e***. The duplication is shown as free and unattached, but is more likely to be attached to a chromosome. Deletions and duplications have respectively one or three copies of these genes, a genotype known as segmental aneuploidy.

(or possibly four) copies of every gene present within the duplicated region; two of these copies are the ones present on the normal chromosomes, which might be recessive mutant alleles with the additional wild-type copy (or copies) present on the duplication, as shown in Figure 4.14. When compared to deletions, duplications are nearly always alive and fertile as heterozygotes (that is, with three total copies of the region) and often alive as homozygotes (with four total copies of the region). Both deletions and duplications are called **segmental aneuploids**, being monosomic (the deletion having one copy) or trisomic (the duplication having three copies) for a particular defined segment of a chromosome.

Segmental aneuploids provide a convenient means to map a gene within the duplicated or deleted chromosomal segment, and, for many regions of the genomes, they are available from the genetic stock center for different model organisms. The procedure for mapping using a series of deletions is shown in Figure 4.15; for reasons to be described in Section 4.6, deletions are more formally called deficiencies and are denoted by *Df*, so we use these terms interchangeably. In a simple form of deletion mapping, a heterozygote is made with the different recessive markers on one homologue and the deletion with otherwise wild-type alleles on the other homologue, as shown in Figure 4.15. This is mated with a strain that has at least one of the recessive makers (*a* in our figure) and the mutant *m* of interest. The offspring from this cross that exhibit the wild-type phenotype for *a* (which have also inherited the deletion chromosome) are then examined for the *m* phenotype. If the progeny with wild-type phenotype for *a* exhibit an *m* mutant phenotype—whatever

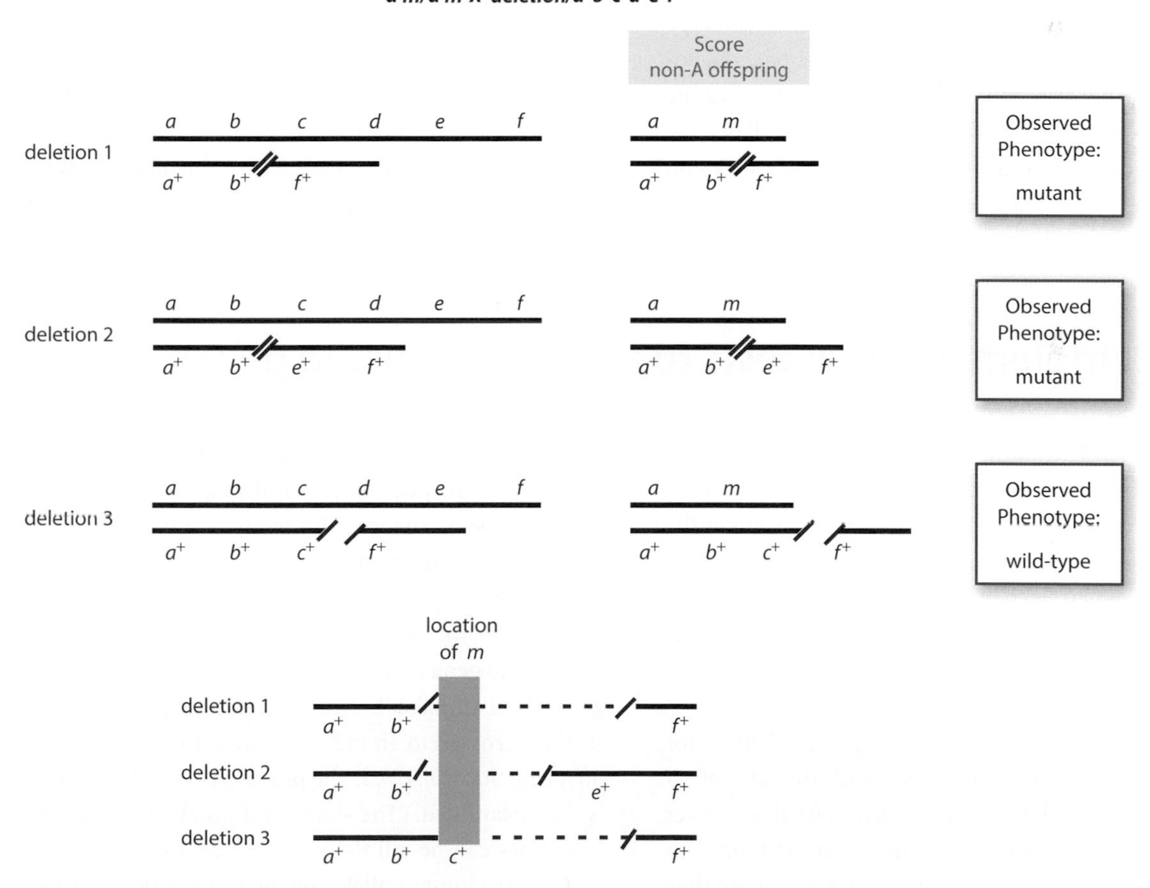

Figure 4.15 Mapping with deletions. The mutation ***m*** is mapped by using three different deletions, referred to as deletions 1, 2, and 3. The deletions remove different but overlapping segments of the chromosome. The mutation is linked to a marker known *not* to be deleted by the deletion, in this case the marker allele ***a***. A deletion strain, typically maintained as a heterozygote, as shown here, is mated to the maker strain, and the F_1 progeny are examined. As shown here, only the non-A progeny are examined, since these are the ones that inherited the deletion. In the results illustrated, the mutant ***m*** is deleted by deletions 1 and 2 but not by deletion 3. This places ***m*** mutation in the region shown by the green box, approximately the same region as the ***c*** gene. A complementation test would be used to determine if the mutation ***m*** is an allele of the ***c*** gene.

the *m* mutant phenotype might be—then the deficiency chromosome must not carry the wild-type allele of the *m* locus, and the *m* locus lies within the deleted region, as shown for deletions 1 and 2 in Figure 4.15. The deficiency chromosome fails to complement the mutation *m*. If, on the other hand, the progeny that are of the wild type for the *a* phenotype do not exhibit an *m* mutant phenotype, the deficiency chromosome must have the wild-type allele of the *m* gene and the *m* locus does not lie in the deleted region, as shown for deletion 3. With a set of partially overlapping deletions, the *m* locus can often be positioned using relatively few crosses.

The procedure for mapping with duplications is conceptually similar, except that the duplication strain will provide the wild-type function for the mutated gene and will complement the mutation rather than fail to complement it, as happens with deletions. This can be described in terms of genotype and phenotype: *m/m* has a mutant phenotype; whereas *m/m; Dp(m+)* includes the wild-type allele on the duplication, and so it has a wild-type phenotype. If we test a series of overlapping duplications, some will provide the wild-type function and some will not; the *m* locus can be positioned with the chromosomal region as result. Often duplications and deficiencies can be used together, to define the position most carefully.

As will be described in Section 4.6, duplications and deletions are also used to classify the type of mutant that is being characterized. Duplications and deletions allow the investigator to vary the number of wild-type (and mutant) copies of the gene—which, as we shall see further, is a key aspect of determining what type of mutant allele is being studied.

Let's summarize where we are in our hypothetical process of identifying genes that affect a biological process. Again, the case study can be read in parallel for an actual example of this procedure. We have collected a number of mutations with a range of phenotypes, all of them affecting the biological process that we are analysing. The mutant strains have been outcrossed sufficiently, so that we have confidence that the mutant phenotypes are due to a single mutation rather than to multiple different mutations in the same strain. We have mapped the mutations with respect to the location of other known genes, using either visible markers or segmental aneuploids. However, we do not yet know how many different genes are represented by the mutations we have collected, we have not yet named of these genes, and we have not yet worked out how the mutation affects the function of the gene and the biological process. Each of these steps will be considered in turn.

4.5 Complementation tests to assign mutants to genes

In a successful mutant screen, many mutants with similar phenotypes are found. These mutants could be different alleles in the same gene—and thus affect the same molecular function—or could be mutant alleles in different genes that have related functions but encode different molecules. The goal of complementation testing is to determine how many different genes have been identified by mutations.

The complementation test was explicitly defined for the first time by Seymour Benzer, with the help of *rII* mutations in the bacteriophage T4, and used to answer the following question: are these two mutations with similar phenotypes alleles of the same gene, or are they alleles of two different genes? The only requirement for a complementation test is that both mutations be recessive to wild type. The procedure for a complementation test is shown in Figure 4.16.

Let's call *m1* and *m2* two different mutations with similar mutant phenotypes and map locations. Thus there may be one gene at this locus, and *m1* and *m2* are two different mutant alleles of the same gene. Alternatively, there may be two genes with related functions in this same region of the chromosome, *m1* being a mutation in one gene and *m2* being a mutation in the other gene. To distinguish between these possibilities, an *m1* strain is crossed to an *m2* strain so as to make a hybrid offspring that is *m1/m2*. The phenotype of this F_1 hybrid is then examined. (The shorthand notation for writing genotypes can be a little confusing at this stage. If *m1* and *m2* are mutant alleles of the same gene, then *m1/m2* is a proper way to write the genotype. If *m1* and *m2* are alleles of two different genes, the genotype would be written *m1/+; m2/+* or $m1\ m2^+/m1^+\ m2$. The process and the results are easier to describe than to notate formally.)

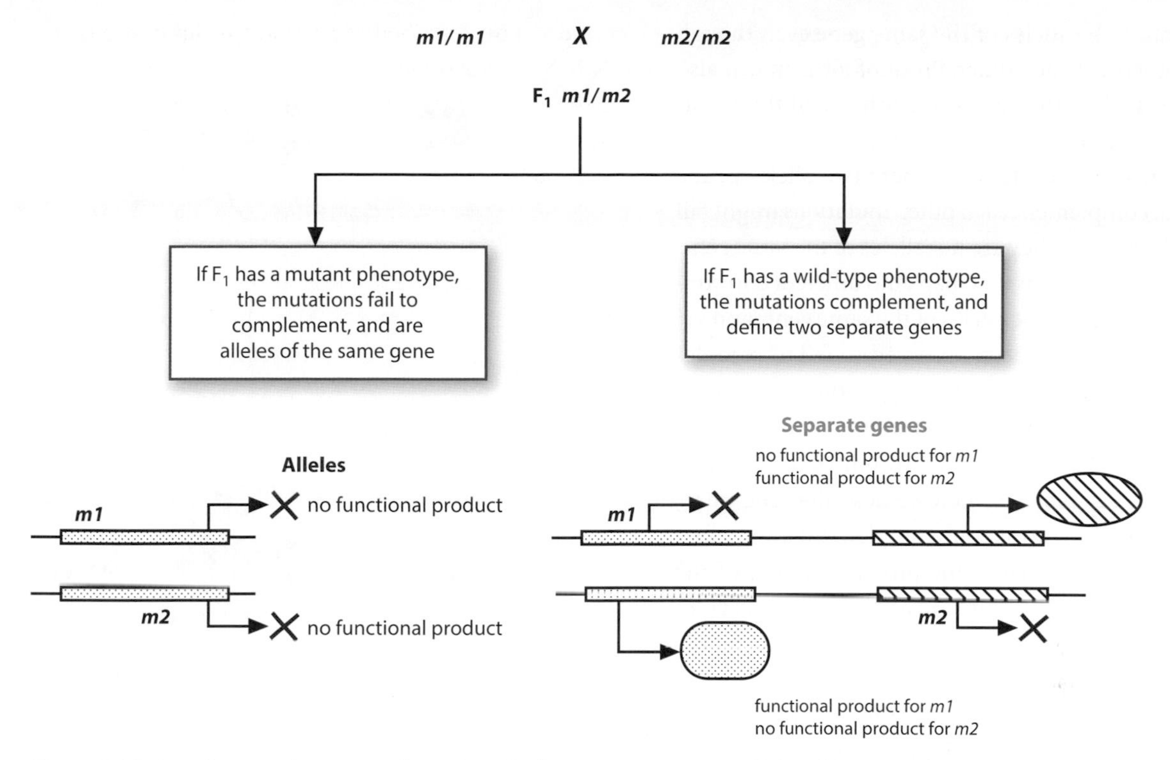

Figure 4.16 Complementation tests to determine whether two mutations are alleles of the same gene. Both ***m1*** and ***m2*** are recessive mutations. They are crossed to each other to produce an F_1 with each mutation. If they are two different alleles of the same gene, as shown on the left, the F_1 heterozygote between them will not make a functional product and will have a mutant phenotype. The mutations fail to complement. Conversely, if ***m1*** and ***m2*** are alleles of two different genes, as shown on the right, the ***m1*** mutant strain will provide the wild-type ***m2*** function and the ***m2*** mutant strain will provide the ***m1*** wild-type function. The heterozygote will have a wild-type phenotype, and the mutations complement.

According to interpretation of the complementation test, if the *m1/m2* hybrid has a *mutant* phenotype, *m1* and *m2 fail to complement* and thus are alleles of the *same gene*. On the other hand, if the *m1/m2* hybrid has a *wild-type* phenotype, *m1* and *m2 complement* each other and thus are alleles of *different* genes. In effect, if the two mutations are alleles of two different genes, the *m1* strain is capable of providing the wild-type function that the *m2* strain lacks, and the *m2* strain is capable of providing the wild-type function that *m1* lacks—the *m1* mutant completes or complements the defect in *m2* and vice versa. Mutations that fail to complement each other are assigned to the same complementation group.

Complementation tests define a gene but complementation tests are not perfect

Complementation tests are generally simple to carry out and are usually considered a necessary part of a mutant screen, but they are not a perfect method to determine whether two mutations are alleles of the same gene. The few situations in which complementation tests give ambiguous results can themselves be informative.

In a few cases, two mutations that are alleles of the same gene can complement each other owing to some unusual property of the protein encoded by the gene. This phenomenon is known as **intragenic complementation**. Recall that the essence of a complementation test lies in the ability of one mutant strain to provide a molecular function missing in the other mutant strain. Thus a protein with two independent functional domains can sometimes be mutated in one domain without affecting the function of the other domain. If the *m1* and *m2* affect different domains, they might complement each other, since each is providing one mutant and one functional domain for the protein. Intragenic complementation has also been seen when the protein functions as a multimer. In cases involving intragenic complementation, the proof of allelism often comes from a third mutation, which fails to complement either of the other two. If both *m1* and *m2* fail to complement *m3*, then *m1* and *m2* are inferred to be alleles of the same gene even though they complement each other. Proof of allelism can also be obtained when the gene is identified and the mutations are sequenced.

In contrast to the situation where two alleles of the same gene complement each other, mutations might fail to do so even when they are not alleles of the same gene. For example, mutations that map onto different chromosomes clearly cannot be alleles of the same gene; and yet they have sometimes been observed to fail to complement each other. This describes a phenomenon known as **non-allelic non-complementation**, in which mutations in two different genes fail to complement. This is discussed in Chapter 10 when we describe gene interactions. Whereas intragenic complementation often indicates something about the protein product of the gene, non-allelic non-complementation often provides information about the cellular process affected by the two genes. Nonetheless, with these warnings in mind, complementation tests are usually the standard method for defining allelism and for determining whether two mutations affect the same gene and the same function.

Complementation tests can be used to screen for new mutant alleles of a gene

Complementation tests are so fundamental to our definition of genes that they can themselves be used as an experimental tool in a screen or selection to find new alleles of a gene that the investigator wants to study in more detail. After all, one mutation in a gene may be interesting, but many more mutations in the same gene are much more informative in understanding the biological process.

For example, suppose that we have a mutation in a gene known as *vestigial* that results in vestigial wings in *Drosophila* and that we want more alleles of this same gene. Of course we could continue to mutagenize the flies, collecting mutations that affect wing shape, mapping them, and performing complementation tests to determine whether they are alleles of the *vestigial* gene that we most want to study. But, once the investigator has decided to study one particular gene in more detail, a mutagenesis procedure known as **F_1 screen** or non-complementation screen can be performed to specifically identify alleles of this one gene. The strategy is diagrammed in Figure 4.17. In this case, the mutagenized flies are mated immediately to a strain that carries a known mutation in the gene, the *vestigial* wing mutation in this example. The F_1 offspring are examined for wing defects, in particular for vestigial wings. Most of the F_1 offspring will have wild-type eye wings, because they don't have a mutation in the *vestigial* gene. A few of the F_1 offspring will have vestigial wings, however, because a new mutation in the *vestigial* gene has been induced in the mutagenized male parent. That is, the new mutation induced in the male will fail to complement an existing mutation inherited from the female and will then be a new allele of the same gene.

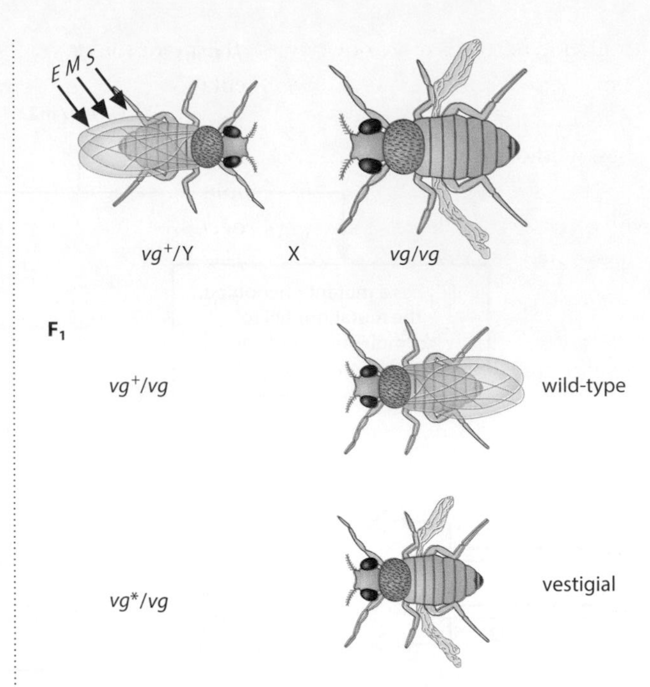

Figure 4.17 An F_1 or non-complementation screen, illustrated here for the vestigial wings gene in *Drosophila*. A wild-type male is treated with a mutagen and mated to a female that is homozygous for the *vg* mutant allele. The parental female could be homozygous, as shown here, or could be heterozygous with a balancer chromosome. Most of the F_1 flies will have wild-type wings, since they have inherited a wild-type ***vg***$^+$ allele from the father. A few of the F_1 progeny will have a newly induced mutation in the vg gene, shown here in red, as ***vg****, which will fail to complement the ***vg*** allele from the mother and this will result in a mutant fly.

One significant advantage of such a screen is that the mutations are screened in the F_1 generation rather than in the F_2 or F_3 generation, which saves the time and effort of breeding and examining a second generation that includes few mutations of interest. In addition, mutations with only a subtle effect can be detected, because they fail to complement the original mutation. By placing a mutation with a weak phenotype in *trans* to a mutation with a strong phenotype, the mutant phenotype of the weak mutation becomes more obvious. Furthermore, the new mutations found by a non-complementation screen do not need to be mapped. In short, an F_1 screen is one of the best methods to find new alleles of any gene of interest.

The case study describes the combination of complementation tests and mapping procedures used for genes affecting segmentation in the *Drosophila* embryo. Because 577 mutations were placed into 137 genes by these methods, efficient and informed procedures for mapping and complementation were crucial.

The total number of genes that could be found by a mutagenesis procedure can be estimated by statistical methods

As the mutants from a large collection are assigned to genes using complementation tests, an investigator will want to estimate how many additional genes can be found from a particular mutant hunt. That is, she will want to know whether it is worthwhile to continue the same screen or selection to look for more mutants, or whether the procedure will simply provide yet more mutant alleles of genes that have already been identified. The question of whether one has reached **saturation** for a screen is a relatively complicated statistical problem. Some aspects of this problem are discussed in Tool Box 4.2. Briefly stated, a key consideration in the estimate is the fraction of genes that have only one identified mutant allele and the fraction that have more than one mutant allele. From this it is possible to make a rough estimate of the number of genes that remain to be identified, that is, that have no mutant alleles yet. The case study describes how the *Drosophila* geneticists estimated that they were nearing saturation for their mutagenesis scheme.

TOOL BOX 4.2 Estimating the number of genes

A mutant screen is considered to be saturated when all the genes that *can* be identified by that particular screen *have* been found. We cannot know the total number of genes that could have been found, so the genes that have been identified will inevitably represent a sample of the ones that could be found. Estimating the number of genes that can be found by a particular mutagenesis procedure, that is, estimating to what extent the screen is saturated, is an inherently statistical concept. As mutations are isolated and assigned to genes, the number of mutant alleles per gene is the most useful parameter in estimating how many more genes could be found—in short, whether a newly isolated mutation will identify a new gene or simply another allele of a previously known gene. If newly isolated mutations are very likely to identify only alleles of existing genes and not new genes, the mutant screen can reasonably be discontinued, since the screen is likely to be saturated.

We can express this situation mathematically. Suppose that there are N genes that can be found in a genetic screen. The probability that the first mutation will identify a new gene is 1, since no genes have been found before. The probability that the second mutation is an allele of the same gene as the first one is 1/N, whereas the probability that it will be a different gene is (1-1/N). If N = 100, that is, if there are 100 genes that could be found in this screen, it is very likely (99 percent in fact) that the first two mutations will define different genes; if N = 5 and only five genes could be found by this screen, then it is possible (20 percent) that the mutations are alleles of the same gene. We can express this more generally if we define **h** as the number of genes that have been identified previously when each mutation is found. With each new mutation, the probability of identifying a new gene is then (N-h)/N: the number of genes that could be found minus the number of genes that have been found, divided by the number of genes that could be found.

Let's illustrate with some numbers. When the first mutation is isolated, **h** = 0, so the probability of finding a new gene is (N-0)/N or 1, regardless of the size of N. After the first gene has been found, **h** = 1; if the second mutation identifies a different gene from the first one, then **h** = 2. In other words, as the number of genes identified increases, **h** approaches N, so N-h approaches zero: most mutations will be alleles of genes that have already been found. Although we do not know N, we can estimate it by using **h** and the change in **h** as more genes are identified. This is the statistics behind the concept of saturation.

Before considering the statistics of this situation more fully and the assumptions that have been made, we need to compare them to the results of a mutagenesis scheme. Every mutation can be identified as an allele of some gene, and mutations are being induced at random with respect to the genes. Thus mutations can be found in any gene without an intentional bias introduced by the investigator. The assumption is that each gene is being mutated at the same rate and that the mutations are equally likely to be recognized and retained by the investigator.

All genetic data indicate that this assumption is an oversimplification that can result in a significant distortion in the estimate of gene number. Consider the data from mutations on chromosome 2 that affect *Drosophila* segmentation as tabulated in Table CS4.1. Even without much background about frequency distributions, it seems very unlikely that the genes have the same probability of being mutated when one gene is hit eighteen times, another one is hit seventeen times and a third is hit fifteen times and yet thirteen genes are hit only once. Intuitively it seems clear that the genes hit so much more often must somehow present a better target for mutagenesis.

Apparently high rates of mutability could happen because genes are not physically the same size, because some mutations are easier to recognize and retain, or because of some property of the mutagen (among other possible explanations). Differential rates of mutability have frequently been observed, but the underlying explanation has usually not been pursued.

Saturation and the *Drosophila* mutagenesis screen

Having examined the underlying assumptions that went into our model, what did the investigators observe and how did they determine whether saturation was nearby? They plotted the number of new mutations found and the number of new genes found as a function of the number of mutagenized lines. This plot is reproduced in Figure CS4.4. In other words, they are showing the change in **N-h**, the number of total genes (which is not known) minus the number of hit genes, as a function of the number of mutations or tosses. The data show that roughly 6,000 chromosomes were scored and the number of mutations increased linearly with the number of chromosome tested. The number of genes increases until about 1,500 chromosomes have been tested and about seventy-five mutations defining fifty genes have been found. Then the number of genes increases much more slowly than the number of mutations, such that among the final 1,500 chromosomes almost no new genes were found. This indicates that nearly all genes that could be found by this method have probably been found.

Gene names are usually but not always based on mutant phenotypes

As the investigator is identifying mutations, the mutations will probably have a laboratory designation that only the investigator can understand or recall. Once the mutations have been assigned to a gene, the gene has to be named and the mutation is given an allele designation according to the conventions for that organism. It is probably safe to say that no other topic in genetics has more potential for confusion than gene names and allele designations.

The potential for confusion arises from many sources. First, different organisms have different systems of nomenclature. Gene names in *Drosophila* consist of one to four letters and rarely involve a number (e.g., *w, ptc, Antp*), whereas gene names in *C. elegans* are usually three letters long and always include a number (e.g., *tra-1, dpy-10*). In *Arabidopsis*, genes names are usually three letters but can vary from two to six letters. They may include a number, or even two numbers (such as *bri1-1*), of which the second refers to the allele.

Second, many of the nomenclature systems are products of a long history of research in the organism. A gene name that made sense to an investigator working on a particular mutant allele years ago may be quite obscure to a modern investigator. For example, the *Notch* gene in *Drosophila* encodes a receptor involved in cell–cell signaling; orthologous genes are found in nearly all animals. The gene name arises because the first mutation resulted in notches in the wing margin, a relatively insignificant defect in light of all of the other roles of this gene. However, the gene name indicates its history.

Third, a gene name reflects a particular historical viewpoint on the nature of genes. In an era of genetics that pre-dated genome projects, or even molecular biology, genes were recognized solely through mutant alleles. This began with one of the first genes to be named, the *white* gene in *Drosophila*, and continues until today. The gene was called *white* because flies with a mutation in the gene had white eyes. Thus the gene was named after its *mutant* phenotype—in effect after what happens when the normal function is missing or altered. The wild-type function of this gene is to produce red eyes.

This method of naming genes on the basis of their mutant phenotypes is contrary to the way everyday objects are named. For example, when I am driving and want my car to stop, I press a pedal on the floor and the car slows down. The normal function of this pedal and its associated parts is to brake the car's momentum; and we call the pedal the "brake," after its normal function.

But with genes, especially in the earlier days of genetics, the normal function was usually unknown; the best clue to normal function was to see what phenotype resulted when the function was disrupted. Thus, when this particular gene was disrupted, it was observed that the fly had white eyes; so the gene was named *white*. If early geneticists had been asked to name the parts of a car, the pedal that we call the brake might have been instead called "no stopping" or something similar.

Genetic nomenclature has a particular type of confusion for human genes. Sometimes we speak of "disease genes" when we actually mean that a known disease arises when the gene is *mutated*. The cystic fibrosis gene (called CFTR and discussed in more detail in Chapter 5) does not cause cystic fibrosis. The normal function of the gene is to transport chloride ions across the membrane. It is a *mutation* in the CFTR gene that results in cystic fibrosis: the gene is named after its mutant or disease phenotype.

This type of shorthand is rampant, particularly when the news media report that a gene for some behavioral trait such as schizophrenia or Alzheimer's disease has been identified. It must be recalled that the gene is being named (or nicknamed) after the phenotype that arises when the gene is altered. The normal function of the gene is not to cause schizophrenia or Alzheimer's disease any more than the normal function of the *white* gene is to make white eyes. But we may not know the normal function of the gene; we only know what happens when the gene is altered. Thus the gene is named after its mutant phenotype.

The potential for confusion becomes exacerbated by geneticists' own inconsistency in conferring gene names. The inconsistency arises in part from the way a particular gene was first found and studied. Not all genes are now recognized first by their mutant phenotype, especially since biochemical and molecular genetics and genomics are used in addition to traditional mutant analysis to find genes. In humans, the β-globin gene is very well studied. Notice how the gene is named: after its polypeptide product, and not after the diseases that arise from mutations. Hundreds of naturally occurring mutations in the β-globin locus are known in humans. These mutations result in a group of diseases called β-thalassemia, which include sickle cell anemia. But the gene is not called the *thalassemia* gene or *thal-1* or something similar. It is named instead after its normal function: the β-globin gene. The same is true for many genes that encode metabolic enzymes or familiar proteins such as β-tubulin or collagen. It should be noted that, even with a single organism and category of genes, different names can be used. In *C. elegans*, *rol-6* mutations result in a worm with a rolling corkscrew movement; the wild-type *rol-6+* gene encodes a collagen, and this is its mutant phenotype. But this is not the rule; many other collagen genes in *C. elegans* are named *col-*, after their normal protein products. In addition, in an era in which genomes are sequenced and genes are found by computational analysis, many members of gene families have been named on the basis of their predicted functions and of their similarities to genes in other organisms.

So genes are named after their mutant phenotypes most of time but not always, and there is no simple universal rule. It is unfortunately something that must be learned for each gene individually. One small satisfaction is that each gene name records something of the history of its own discovery, so learning the name often allows the investigator the opportunity to learn something of the background and of circumstances in which the gene was identified.

4.6 Classifying mutants

As different mutations that affect a phenotype are collected, mapped, and assigned to genes, the investigator also needs to determine what each mutation has done to the normal function. That is, a mutation disrupts the normal function of a gene. But exactly how does it disrupt the function? Does the mutation completely eliminate the function or merely reduce it? Does it overproduce the normal function of the gene, determining a mutant phenotype? Perhaps it produces the normal product but at the wrong time or in the wrong tissue, thereby causing a mutant phenotype. For most contemporary molecular biologists, these questions may be best answered by a molecular analysis of the DNA sequence and of the expression pattern of the gene product, either RNA or protein. From such analyses one could determine whether the mutation produced a stop codon, resulted in the substitution of an amino acid, altered a splice site, deleted or altered an upstream regulatory sequence, or caused some other change in the expression of the gene. All this molecular information can be helpful for interpreting the action of a gene and for understanding the mutant phenotype.

Geneticists working in the days before molecular analysis faced these same questions about the effects of mutations and developed methods to answer them, at least in part. We discuss the logic used to interpret mutant phenotype in more depth in Chapter 6. In this section we will describe how varying the number of copies of the gene—either of the wild-type allele of the mutant allele—is used to classify alleles as an early step in interpreting the mutant phenotype. Since the methods to vary the number of copies of the gene typically involve duplications and deficiencies, classification of mutant alleles could occur during the processes by which the mutant were isolated, mapped, or outcrossed.

The classical work involving this concept was published by H. J. Muller in 1932, and many of his terms and ideas are still in use, although a few have been clarified or renamed.

Literature Link. Muller 1932, *Proceedings of the 6th International Congress of Genetics*, 1: 213–55 (frequently reprinted)

Our description will draw heavily on that of Muller while attempting to combine his ideas with what we know now about gene functions. This sets a foundation for some concepts developed in Chapter 6.

Two points need to be made at the outset. First, Muller was working in *Drosophila melanogaster*, which is diploid, and our description will also draw on diploids. Similar concepts can be used with organisms that grow as haploids, but the actual procedure might be a bit different. Second, recall that most genes are named and described in terms of the *mutant* phenotype rather than the wild-type function. Most mutations are recessive and reduce or eliminate the function of a gene. Recessive mutations are often termed **loss-of-function mutations**. Thus, when we refer to a strong mutant allele, we are describing an allele that has a more severely mutant phenotype than does a weak mutant allele of the same gene. The inference is that the strong allele is likely to have less of the wild-type function remaining in it than does a weak allele; but the comparison is made with the help of the mutant phenotypes rather than through any molecular knowledge of the function of the gene.

Recessive mutations generally have a loss of or a reduction in the normal function of the gene

In classifying mutations, the first test is one of the simplest: to determine whether the mutation is recessive or dominant to wild type. The investigator usually learns this during isolation outcrossing, mapping, and complementation testing. Most mutations are recessive to wild type, at least when the phenotype is compared at the level of the organism. That is, if m is a mutant allele, the heterozygote $m/+$ has the same wild-type phenotype as the $+/+$ wild-type homozygote. In the context of gene function, this result implies that the organism is producing enough of the activity of the gene that even a single wild-type dose is enough to carry out the normal function. From an evolutionary perspective, dominance of the wild-type allele makes superficial sense: a single mutation in a single gene occurs regularly in both somatic cells and the germline, so there has been selection for a level of gene expression that is insensitive to the loss of a single dose. Exceptions are certainly known in which the heterozygote has a mutant phenotype, so one mutant allele is enough to disable the gene; such genes, called **haplo-insufficient**, are discussed below.

With recessive mutations, the next question asks whether the mutation eliminates the normal function of the gene or whether it reduces it but retains some residual wild-type activity. These possibilities are compared in Figures 4.18 and 4.19. Muller referred to mutations that eliminate the normal function as **amorphic** mutations, the name reflecting that these had no detectable wild-type function or form of the gene. A more common contemporary usage is to refer to these as **null** mutations, and amorphic and null are synonyms. We will employ these terms interchangeably so that the reader becomes familiar with both usages. Note that Muller was not using any information about the type of molecular change to the gene; this was not known to him. A null mutation is illustrated in Figure 4.18.

Mutations that retain some residual normal function but have a reduction in function were referred to **hypomorphic** mutations, "hypo-" reflecting a lower activity by comparison to the wild type. Hypomorphic mutations are

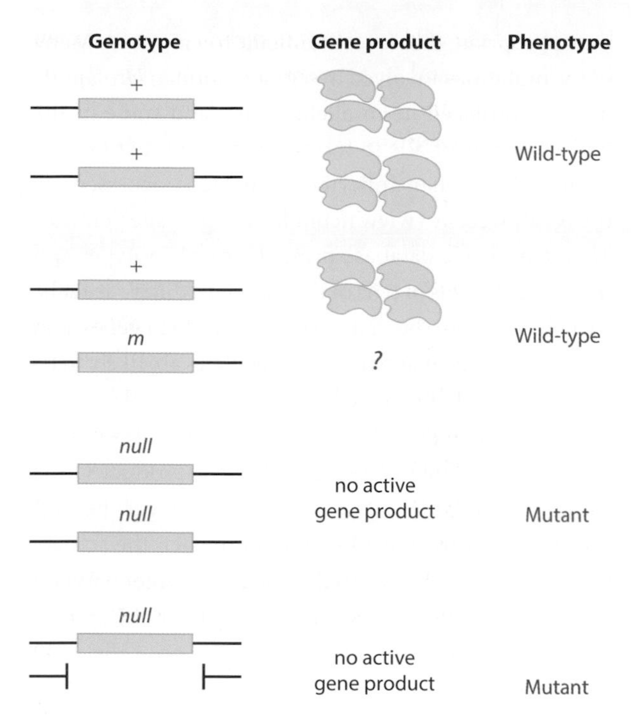

Figure 4.18 Test demonstrating a null or amorphic alleles. An amorphic or null mutation is shown. Each wild-type allele of gene produces an amount of the gene product, shown here as the four green shapes. A heterozygote *m/+* makes only half as much gene product but has a wild-type phenotype since *m* is recessive. A homozygote for the null allele makes no functional gene product and is mutant. When the null allele is placed opposite a deletion or a deficiency of the gene, no change is seen in the phenotype; the null or amorphic allele behaves like a deficiency in this test. Compare this with the results of a hypomorphic mutation in Figure 4.19.

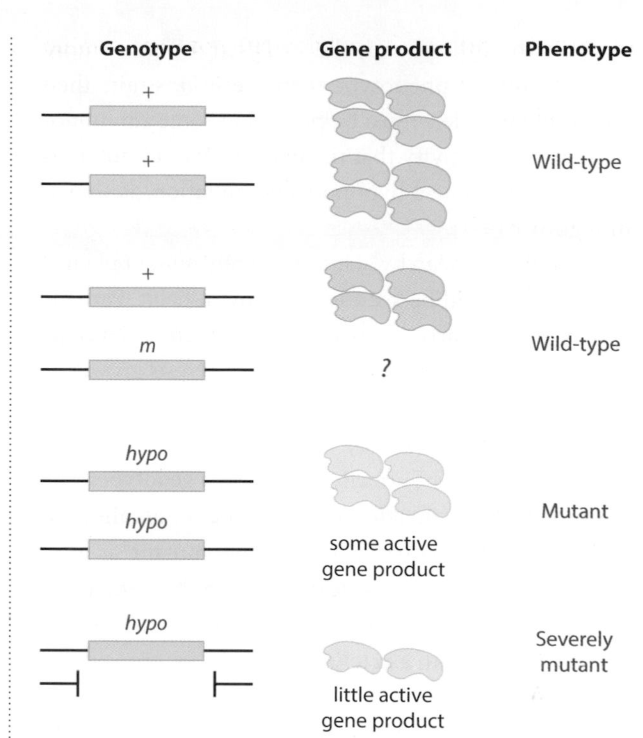

Figure 4.19 Test demonstrating a hypomorphic mutation. As in Figure 4.18, each wild-type allele of gene produces an amount of the gene product, shown here as the four green shapes. In this case, the mutation only reduces the function of the gene but does not eliminate it, either because it reduces the amount of the gene product (from four to two per allele in the figure) or because it reduces the gene's activity (as suggested by the lighter green), or through some combination of these. As a mutant homozygote, *m/m* has a mutant phenotype but retains some functional activity of the gene. This can be inferred when the hypomorphic mutation is placed opposite a deficiency. In this case, the deficiency has less activity than the hypomorphic mutation and the mutant phenotype becomes more severe. Compare this with the results of a null allele in Figure 4.18.

often referred to as "leaky" mutations, but "leaky" can have somewhat different meanings in different contexts, so we prefer to use the older term hypomorphic mutations or hypomorphs. Hypomorphic mutations could arise from a number of different molecular changes, such as a missense mutation or possibly a splicing defect, among other possibilities; but again, Muller used these terms without knowing about molecular changes to the genes. This is illustrated in Figure 4.19, but note that hypomorphic mutations could arise from changes that reduce the *amount* of the gene product being made or the *activity* of the gene product that is made—or some combination of these.

To distinguish amorphs from hypomorphs, Muller devised a simple test using deficiencies that is summarized in the bottom diagrams in Figures 4.18 and 4.19. His strategy follows this logic: a deletion completely lacks the physical genetic locus, so it certainly has no remaining gene activity and must therefore be an amorphic mutation. For Muller, such deletions were usually cytologically visible and deleted multiple adjacent loci, so any other recessive mutation that failed to complement several adjacent loci was termed "deficiency" and assumed to be a deletion even if the molecular event was not cytologically evident. (In a strict usage, a deletion had to be cytologically visible, so "deficiency" was used to encompass genetic events that behaved like deletions even if the physical change was not cytologically visible.)

Since a deficiency lacks the gene and thus all its activities, a mutation that has the same mutant phenotype as a deficiency in genetic tests is also considered to be amorphic. That is, if the *m/Df* strain has the same mutant phenotype as the *m/m* strain, *m* is regarded as an amorphic mutation, since substituting a deficiency for one mutant copy makes no difference in the mutant phenotype of the strain. This is shown in Figure 4.18.

If, on the other hand, the *m/Df* strain has a more severely mutant phenotype than the *m/m* strain, then the *m* mutant allele must be providing some amount of residual gene activity that the deficiency does not; *m* is then defined to be a hypomorphic mutation, as shown in Figure 4.19.

What do we mean by "a more severely mutant phenotype"? This could be seen in the growth habit. Perhaps the *m/m* strain arrests growth at a particular developmental stage, whereas the *m/Df* strain arrests growth at an earlier developmental stage. A more severely mutant phenotype might also be observed from comparing what cells or organs are affected in each genotype; perhaps the *m/Df* strain affects more tissues or organs than the *m/m* homozygote. "More severely mutant" is a bit abstract, but the sense is clear. By using this test, the *m* mutation is defined to be hypomorphic because replacing the *m* allele with a deletion results in a more severely mutant phenotype.

Another test that Muller used for hypomorphs is to increase the dose of the mutant allele rather than to decrease the dose as shown in Figure 4.18 and 4.19; since increasing the mutant dose usually requires more specialized strain constructions, it is not commonly used now. If *m/m* is a hypomorphic mutation with some gene activity, then the segmental trisomic *m/m/m* with three mutant doses should be less severely mutant than *m/m*, and *m/m* should be less severely mutant than *m/Df*. If the phenotype becomes less severe (or more like wild-type) with each additional copy of the mutant allele, *m* is a hypomorphic mutation. On the other hand, increasing or further decreasing the dose of an amorphic mutation will have no effect on the mutant phenotype, since there is no gene activity to begin with.

Null and hypomorphic alleles of a gene can be placed in an allelic series

Both amorphic and hypomorphic mutations provide useful information in understanding the activity of the gene. Having a collection of alleles with different phenotypes can allow the investigator to infer an **allelic series** that ranges from alleles that lack the activity of the gene to alleles that have reductions in the activities of the gene and to alleles that alter the activity of the gene in other ways. In general, the allele with the most severe mutant phenotype is considered to be the null allele, particularly when deletions of the locus are not readily available or the appropriate crosses are difficult to carry out. Many of the mutations involved in segmentation in *Drosophila* have been placed in an allelic series, and some of the best examples of allelic series can be found at OMIM (Online Mendelian Inheritance in Man), accessed via the NCBI website (www.ncbi.nlm.nih.gov/sites/entrez). This remarkable database presents the phenotypes and molecular lesions of tens of thousands of different allelic variants that have been described for human genes, and this often makes it apparent which alleles are likely to be null alleles and which are hypomorphic.

The amorphic or null mutation phenotype reveals the stage or tissue that *first* requires the gene product. Often the most fundamental activity of a gene can be best inferred from an amorphic mutation when the activity is absent; we will return to this idea in Chapter 6. When gene interactions are described in Chapters 11 and 12, we note that it is important to know the type of allele and to use null mutations whenever possible.

On the other hand, the amorphic phenotype for an essential gene cannot reveal functions of the genes that occur later in development. Hypomorphic mutations are useful for observing these other activities of the gene. Hypomorphic mutations may have enough residual wild-type function to accomplish an early stage, but not enough to carry out a later step in development. This may apply not only to the developmental stage but also to tissues that require a particularly high level of gene activity. We return to this concept in Chapter 5, with a few examples of clinically distinct human genetic diseases; molecular analysis has shown that these different phenotypes arise from either null or hypomorphic mutations in the same gene.

Dominant mutants generally arise from overproducing a normal function

Most mutations are recessive to wild type and arise from a complete or partial loss of gene activity. However, mutations that are dominant to wild type are found in some genes, and some human genetic syndromes are primarily recognized from dominant mutants. For example, the form of dwarfism known as achondroplasia is dominant to wild type, as are syndromes such as polydactyly (extra digits), Huntington's disease, and several cancer syndromes. Dominant mutations, or the gene they affect, often have some unusual feature rather than simply reducing the function of the gene, as recessive mutations do. In fact the molecular explanation for dominance is different in each of the examples just listed.

Most dominant mutations are thought of as **gain-of-function mutations**, but a precise definition of "gain of function" can be a bit elusive. Muller divided dominant mutants into three helpful categories, although his terms are not all widely used nowadays. The most frequently encountered type of dominant mutant is the class Muller referred to as **hypermorphic** mutations; many geneticists nowadays call them **overproducers**. These mutants produce the normal gene product, usually in the same cells as the wild-type gene, but they have too much gene product and a mutant phenotype results.

For example, a hypermorphic mutation might eliminate a site or a sequence needed for repression, so that the gene activity is overproduced as a result. The site or sequence needed for repression could occur at any level of gene expression. Hypermorphic mutants could include a mutation that results in unusually high levels of transcription or in the failure to degrade the transcript; thus the mutation affects the level of RNA being produced. In either case, too much gene product is being made and the activity of the gene is overproduced. Hypermorphic mutants could also include a mutation that results in a protein that is not degraded or that fails to interact with a repressor protein such as kinase or a phosphatase; thus the mutation affects the level or the activity of the protein product from the gene. At the level of mutational analysis, all these different types of mutations have roughly the same effect, since the mutant phenotype arises because there is an elevated level of gene activity.

Muller's test for identifying a hypermorphic mutation was, again, to vary the mutant and the wild-type doses of the gene, as shown in Figure 4.20. The test is essentially the same as the one used for amorphs and hypomorphs, but the expected results are different. Suppose that *M* is a dominant mutation, perhaps one that causes notches in the margin of a wing in *Drosophila*. Thus *M*/+ has a more severe mutant phenotype than does +/+, which has no mutant phenotype at all. If *M*/*M* has a more severe mutant phenotype that does *M*/+, for example with even more notches or deeper notches in the wing margin, then *M* is defined to be a hypermorphic mutation; replacing the wild-type activity of the gene (the +allele) with a mutant activity (the *M* allele) results in a more severe mutant phenotype. Furthermore, the partial trisomic *M*/*M*/*M* (if it can be examined) should be even more severely mutant than is *M*/*M*. That is, each additional dose of the mutant allele makes the phenotype more severely mutant, because the mutation is causing an overproduction of the activity of the gene.

A similar and important supplemental test can be applied by decreasing the copies of the wild-type allele through a deletion, as diagrammed at the bottom of Figure 4.20. If *M*/+ is mutant and *M* is a hypermorphic mutation, then *M*/*Df* should be *less severely* mutant than *M*/+; the amount of gene product has been diminished by replacing the wild-type allele with a deletion.

Many dominant oncogenes involved in cancers in vertebrates are the result of hypermorphic mutations, in which the function of the protein often cannot be repressed. For example, the normal cellular and the oncogenic versions of the *src* proteins differ at their C termini. In the normal cellular protein, phosphorylation

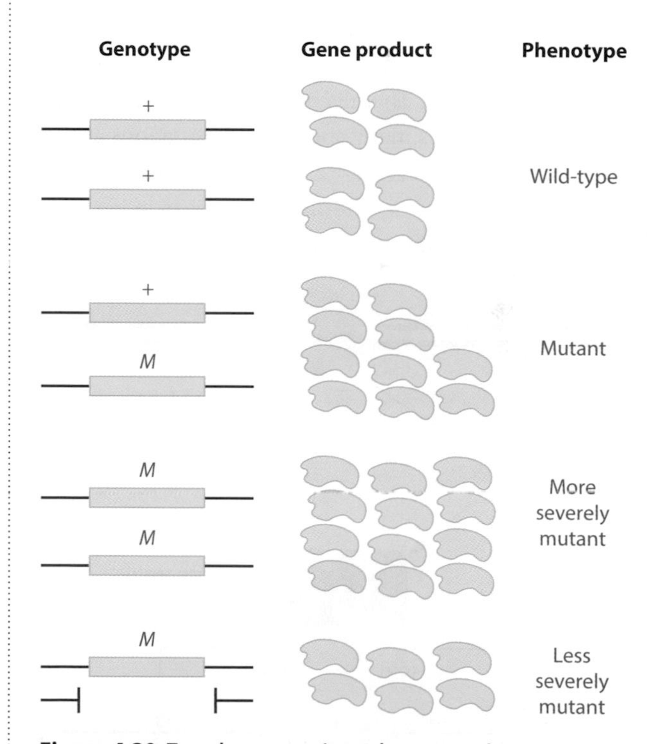

Figure 4.20 Test demonstrating a hypermorphic mutations. As in Figure 4.18, each wild-type allele of gene produces an amount of the gene product, shown here as the four green shapes. A hypermorphic mutation, represented as M, overproduces the gene product or the gene function, represented by six green shapes. The overproduction results in a mutant phenotype in both heterozygotes and homozygotes. A deletion that removes the wild-type copy of the gene may restore the wild-type phenotype, or may make it less severely mutant. Compare this with the results of recessive amorphic and hypomorphic mutations in Figures 4.18 and 4.19.

at a tyrosine residue near the C terminus inactivates the protein. The oncogenic version of the protein lacks this tyrosine and is not inactivated; thus, the activity of the protein is overexpressed and a hypermorphic mutation results.

Another example is found in the *ras* oncogene in humans and in its orthologues in other organisms. The Ras proteins are part of a well-known signaling cascade that is used in many different processes in different organisms; the first few steps of this cascade are summarized in Figure 4.21. The Ras protein binds guanine nucleotide, such that Ras is active when bound with GTP but inactive when bound with GDP, as shown in Figure 4.21. Hydrolysis from GTP to GDP requires interaction with a GAP or GTPase activating protein, the role that the Sos protein plays in Figure 4.21. Mutations that replace the glycine at position 12 in Ras prevent the interaction with the GAP and thus prevent hydrolysis to GDP. As a result, Ras is constitutively bound to GTP and active, so Ras activity is overexpressed and cell proliferation is unregulated; this mutation commonly is found in human cancers. The glycine residue at position 12 is conserved across many species and mutations at this position in other species are also hypermorphic in other species. Cancers are not the only outcome of hypermorphic mutations of this type. Achondroplasia also appears

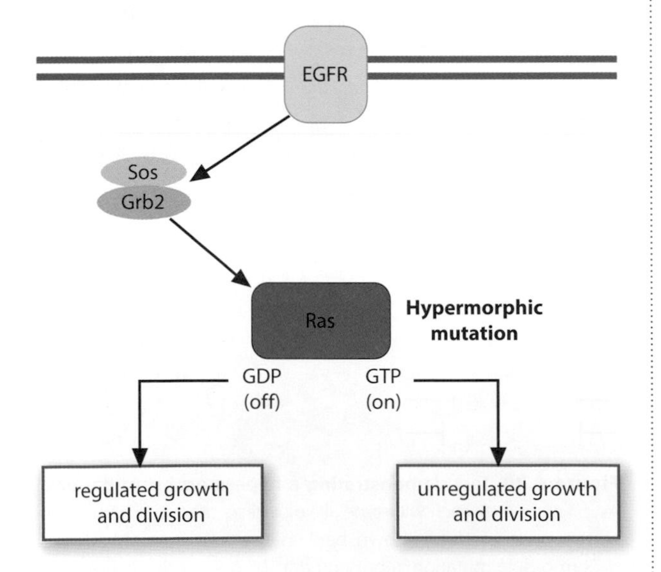

Figure 4.21 Mutations in the *ras* gene are often hypermorphic. The activity of the Ras protein, shown in red, is regulated by the hydrolysis of GTP, which is carried out by proteins such as Sos in combination with Grb2. If the Ras protein is mutated such that it GTP cannot be hydrolysed to GDP, the protein is not repressed and the gene has an overexpression phenotype.

to be the consequence of a constitutive active mutation, in this case a particular missense mutation in the fibroblast growth factor receptor gene FGFR3.

In the examples of dominant mutations discussed so far, the hypermorphic phenotype arises from a mutation that makes the protein insensitive to inactivation by another protein. Hypermorphic mutations can also arise from other forms of overproduction of an active protein. Examples of hypermorphic mutatons are known in which the protein is not degraded and thus persists, in which the gene is overtranscribed because of a mutation in a regulatory region, and in which the gene has become duplicated at its site in the genome. In the case Huntington's disease, a CAG codon that is repeated in tandem more than 40 times results in an unusually stable protein. In all these examples, the overproduction of a normal form of the protein results in a dominant mutant phenotype. By definition, all these overproducer mutations are hypermorphic.

Hypermorphic mutant phenotypes can also be encountered when the expression level of a gene is manipulated experimentally. For example, the coding region of a gene can be placed under a transcriptional regulatory region that expresses it a high level; high copy number plasmids in yeast (discussed in Chapter 3) often express the gene at higher than normal levels. When high expression molecular constructs or plasmids are reintroduced into the organism, the gene is overexpressed and a hypermorphic mutant phenotype is sometimes observed.

Dominant hypermorphic mutations might not exist for all genes because the biological process the gene affects may be insensitive to the dose of that particular gene product. Hypermorphs are analogous to having too much of a good thing. For some good things, such as health and happiness, having too much is not a problem and does not produce an aberrant phenotype. For some other good things, such as money, having too much may be a problem in some circumstances, and an unusual phenotype can arise. For yet other good things, having too much becomes a problem once the accumulation reaches a critical level.

Mutations producing an unexpected function can also be dominant

Muller recognized two other categories of dominant mutants, which he called neomorphs and antimorphs. Although the concepts are still useful, the terms have been largely replaced by other descriptors. Neomorphs are mutations that result in a novel function or novel phenotype (see **neomorphic allele**). As a strict definition of novel phenotypes, this would seem to be uncommon, because it suggests that the gene has somehow altered its nature or acquired some new properties. However, many neomorphic mutations cause the **ectopic** expression of the gene in a cell, in tissue, or at a developmental stage when it is not usually expressed.

With this recognition about ectopic expression, neomorphic mutations are not so rare. The *Drosophila* homeotic mutation *Antennapedia* (illustrated in Figure 2.18) is a well-known example. The Antp protein is normally expressed in the foreleg, and if the locus is deleted, the forelegs fail to develop. In the *Antennapedia* mutant, the Antp protein is expressed in the head region and a foreleg arises at an ectopic location. Thus a phenotype that to Muller looked like a novel function of the gene is due to the normal expression of the gene in a novel tissue, which gives rise to a neomorphic phenotype. One of the common *Antp* neomorphic mutations arises from mutations in the regulatory region, such that the gene is expressed in the wrong tissue.

An example in humans is illustrated by Burkitt's lymphoma, a solid tumor of B lymphocytes; the mutation is summarized in Figure 4.22. The majority of cases of Burkitt's lymphoma are characterized by a specific reciprocal translocation between chromosomes 8 and 14 that fuses the regulatory region of the antibody heavy

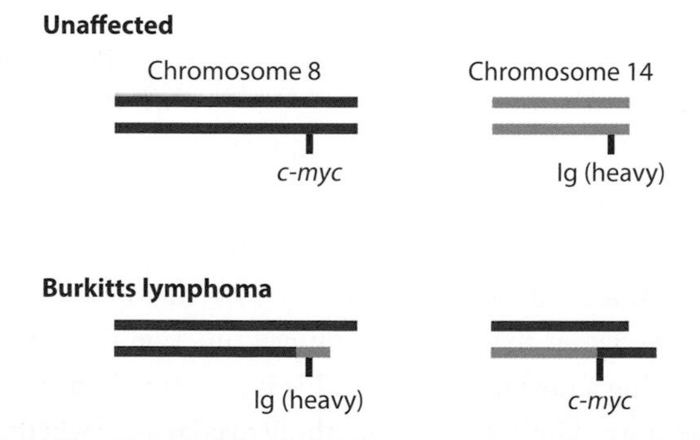

Figure 4.22 Burkittt's lymphoma is an example of ectopic expression. In unaffected individuals, the *c-myc* gene on chromosome 8 is expressed at low levels in many cells, while the immunoglobulin-heavy chain genes (Ig heavy) of chromosome 14 are expressed specifically in the immune system at high levels. A translocation that places the *myc* gene under the regulation of the Ig (heavy) enhancer region results in the *myc* gene being expressed ectopically at high levels. This could be considered an example of a neomorphic mutation of the *myc* gene.

chain genes on chromosome 14 with the protein coding region of the *myc* gene, normally found on chromosome 8. The Myc protein functions in regulation of the cell cycle but is not normally expressed at high levels in B lymphocytes, in contrast to antibody heavy chain genes. The translocation places the *myc* gene under the control of the transcriptional enhancers of the antibody heavy chain gene, and this results in unusually high and unregulated expression in a cell type that does not normally express the Myc protein in this pattern. Other cases of Burkitt's lymphoma are due to translocations that place the *myc* gene under the control of other antibody genes located on chromosomes 2 or 22. In other words, Burkitt's lymphoma is due to a dominant neomorphic mutation in the *myc* gene.

Antimorphic mutations are ones that **antagonize** the wild-type function of the gene; these mutations are most easily thought of as poisons of the normal function. Although they are dominant mutations, antimorphs behave like a loss of function in the gene rather than like a gain of function. For this reason, the term "antimorph" has been largely replaced by a more descriptive synonym, "dominant negative," so we are now talking of **dominant negative** mutations.

A dominant negative mutation is often indicative of a protein product that forms dimers or multimers. One type of dominant negative mutation is illustrated in Figure 4.23. For example, many transcription factor proteins form homodimers or heterodimers (involving two different genes) to stimulate transcription; similar protein interactions occur with many membrane-bound receptors. A dominant negative mutation can arise if mutated subunit continues to interact with the wild-type subunit but the dimer cannot bind to DNA to stimulate transcription. The mutant subunit acts as a dominant mutation that prevents the function of the normal subunit, and the heterozyote has a mutant phenotype.

The *Shaker* gene in *Drosophila* is characterized by antimorphic mutations. The wild-type *Shaker* locus encodes a polypeptide that assembles as a multimeric potassium ion channel protein in the cell membrane. The presence of one defective subunit interferes with normal channel assembly and function, and thus results in the Shaker mutant phenotype. Another example is seen with mutations in some of the collagen genes of *C. elegans*. Collagen proteins form a highly organized bundle; the absence of one copy of the gene (that is, an

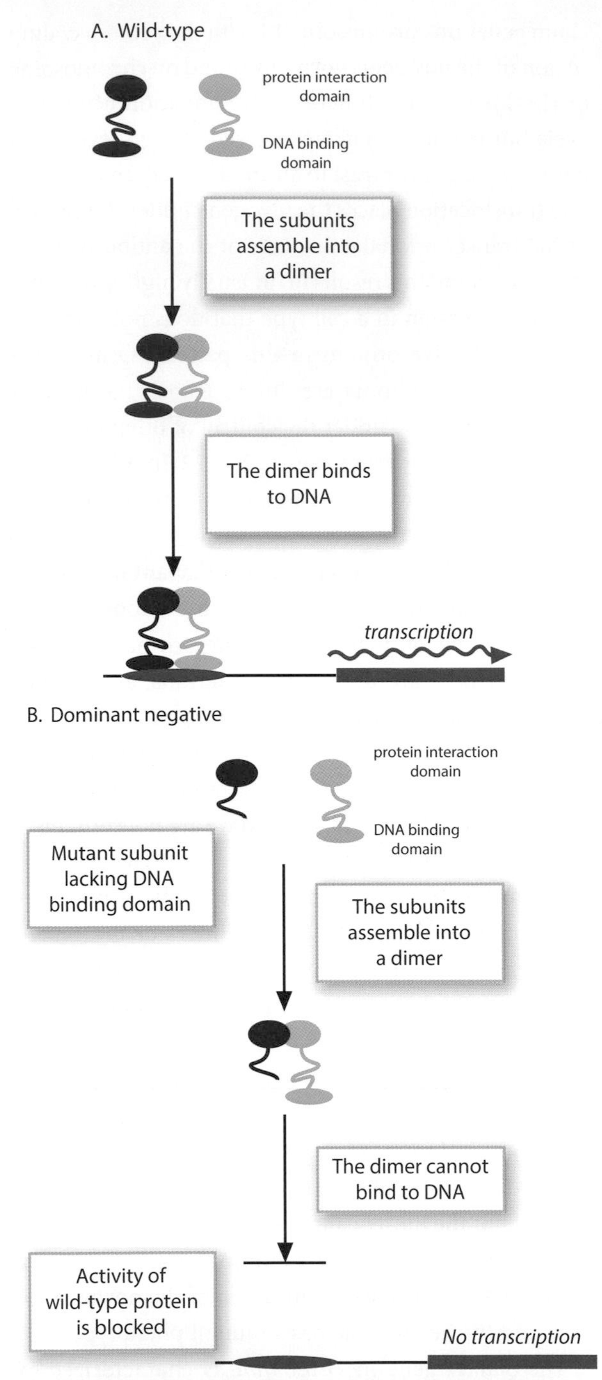

Figure 4.23 Dominant negative mutations or antimorphs. These are mutations that interfere with the normal wild-type function. One type of example is shown here using a hypothetical but common type of transcription factor. The transcription factor protein has two domains, a protein interaction domain and a DNA binding domain. As shown for the wild-type in Panel A, the subunits form a dimer that binds a regulatory region to trigger transcription of the target gene. In a dominant negative mutation (Panel B), the mutant allele encodes a subunit that lacks the DNA-binding domain. Because it can form a normal dimer with the wild-type subunit, the mutant subunit is dominant but the dimer cannot bind DNA or trigger transcription. Thus the mutant allele is an antimorph that "poisons" or prevents the activity of the wild-type allele.

amorphic mutant) is rarely detrimental, because the normal protein is still made from the wild-type allele and assembles normally. Thus, amorphic mutations are recessive. However, a mutated protein can assemble with the wild-type protein and interfere with the normal function of the entire bundle. The antimorphic or dominant negative mutation acts as a loss-of-function mutation because it is interfering with the normal gene product, often in the assembly of a multimeric protein. Curiously, since they antagonize the normal function of the gene, many antimorphic mutations depend on the presence of a wild-type allele for their mutant phenotype; homozygotes for antimorphic mutations may have no mutant phenotype at all or may have a different mutant phenotype.

Dominant negative mutations are generally uncommon among the mutants recovered in a mutant hunt. However, investigators frequently and intentionally produce a dominant negative mutation in vitro by mutating one domain, such as the DNA binding domain, but not the protein interaction sites. Such a mutant has the advantage of behaving like a loss of function or even as a null allele, so it reveals the loss of function mutant phenotype for the gene. At the same time, because the mutated version is dominant, it is not necessary to also eliminate the wild-type copy of the gene to see the effect; in fact, often the mutant effect depends on having the wild-type copy present. Thus, introducing the mutant version into an otherwise wild-type cell or organism is enough to produce a mutant phenotype.

Haplo-insufficient mutants are dominant and define dose-dependent genes

Although dominant gain-of-function mutations such as the ones described above are the exceptions, most mutations reduce or eliminate the normal function of a gene rather than creating a novel function or overexpressing a normal function. This is true simply because there are many ways for something to go wrong, whether with a gene or any other function. Most loss-of-function mutations are recessive because the remaining wild-type allele provides enough gene activity to be able to carry out the normal function. However, for a few genes, the organism requires both wild-type copies for the normal function. This situation is diagrammed in Figure 4.24. A mutation that eliminates gene function in such

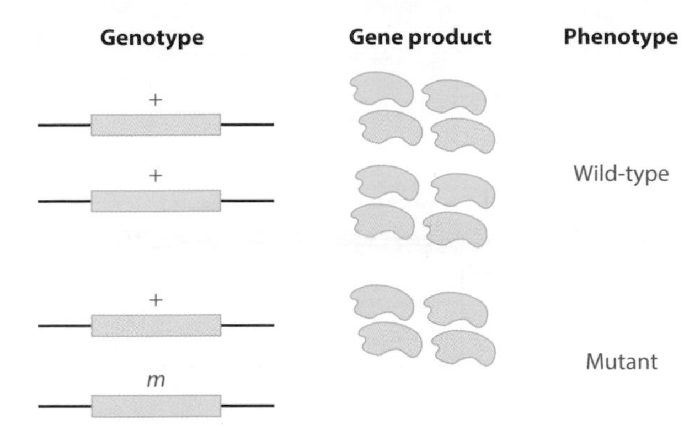

Figure 4.24 Haplo-insufficient mutations. Since most mutations are recessive to wild type, most genes are not dose-sensitive in their activity in the range of one versus two copies. However, some genes are haplo-insufficient and require two wild-type alleles; a null allele that eliminates the activity of the gene, as suggested here, will be dominant.

a dosage-sensitive gene is dominant to wild type, and the gene is considered to be haplo-insufficient; that is, a single functional copy of the gene is inadequate to carry out the normal function. Muller did not encounter any such mutations in *Drosophila*, so he did not name this category.

Not many genes have such a dose-sensitive mutant phenotype, but some important examples are seen with genes implicated in cancers. For example, many cancers have mutations in the gene *p53*; mice that are heterozygous for an inherited mutation in *p53* are predisposed to tumors, particularly lymphomas, so this behaves like a dominant mutation. Why does a mutation that knocks out *p53* function produce a dominant phenotype? Two possible explanations are illustrated in Figure 4.25.

The first idea postulates that the phenotype arises from a combination of germline and somatic mutations in the same gene, a two-hit hypothesis of recessive mutations for the gene. With this hypothesis, the inherited *p53* mutation is recessive, so all the cells in the affected individual are initially *p53*/+ and have normal functions. Under the two-hit hypothesis of cancer, the high rate of tumors arises because the one remaining wild-type allele in these *p53*/+ heterozygous cells becomes knocked out by a somatic mutation; both copies of the *p53* gene have been inactivated and a mutant phenotype arises. If this hypothesis provided the complete explanation, then the tumor cells would be predicted to be homozygous mutant, with one inherited mutation and one somatic mutation. Because the inherited mutation and the somatic mutation are independent events, the somatic mutations would be different molecular lesions, and different tumors are also expected to involve different mutations.

An alternative explanation is that the *p53* gene is haplo-insufficient or dose-sensitive. A mouse that is *p53*/+ has a reduced activity of the gene and presents an elevated risk for cancers without the need for a second, somatic mutation. On this hypothesis, one mutation in the gene reduces its activity to a level that is below what is needed for normal function. The mutant "predisposes" the mice to tumors with no second genetic event needed. In effect, the *p53* mutant is dominant but has reduced penetrance, a concept discussed in more detail in Chapter 6. The amount of gene activity has been lowered close to the threshold needed for normal function; in some cells, the activity of the gene falls below this threshold.

Both hypotheses may be true. Upon sequence analysis, many tumors are homozygous for mutations in the *p53* gene, and different types of molecular lesions in *p53* are involved; this supports the first hypothesis, that the mutation is recessive and a second mutation occurs. But in at least some cases, the tumor cells have retained the wild-type *p53* allele and do not have a somatic mutation in the gene. Thus, this supports the hypothesis that the *p53* gene is dose-dependent or haplo-insufficient.

A similar explanation has been posited for cancers arising from mutations in the *PTEN* gene and may be true in other oncogenes as well. Haplo-insufficient mutations differ from hypermorphic or neomorphic mutations in that they are *loss-of-function* mutations rather than overproducing mutations. They are different from antimorphic mutations in that they do not require the presence of a wild-type allele; and, like hypomorphic alleles, haplo-insufficient mutations often produce a more severe phenotype when they are homozygous. Thus haplo-insufficient mutations are similar to null or hypomorphic mutations, except that they are dominant because of the peculiar dosage requirements for that gene.

Systematic analysis of *Drosophila* chromosomes using deletions and duplications suggests that relatively few genes are haplo-insufficient for overall morphological phenotypes; but it does not rule out the possibility that a gene may be dose-sensitive in particular cells. In fact

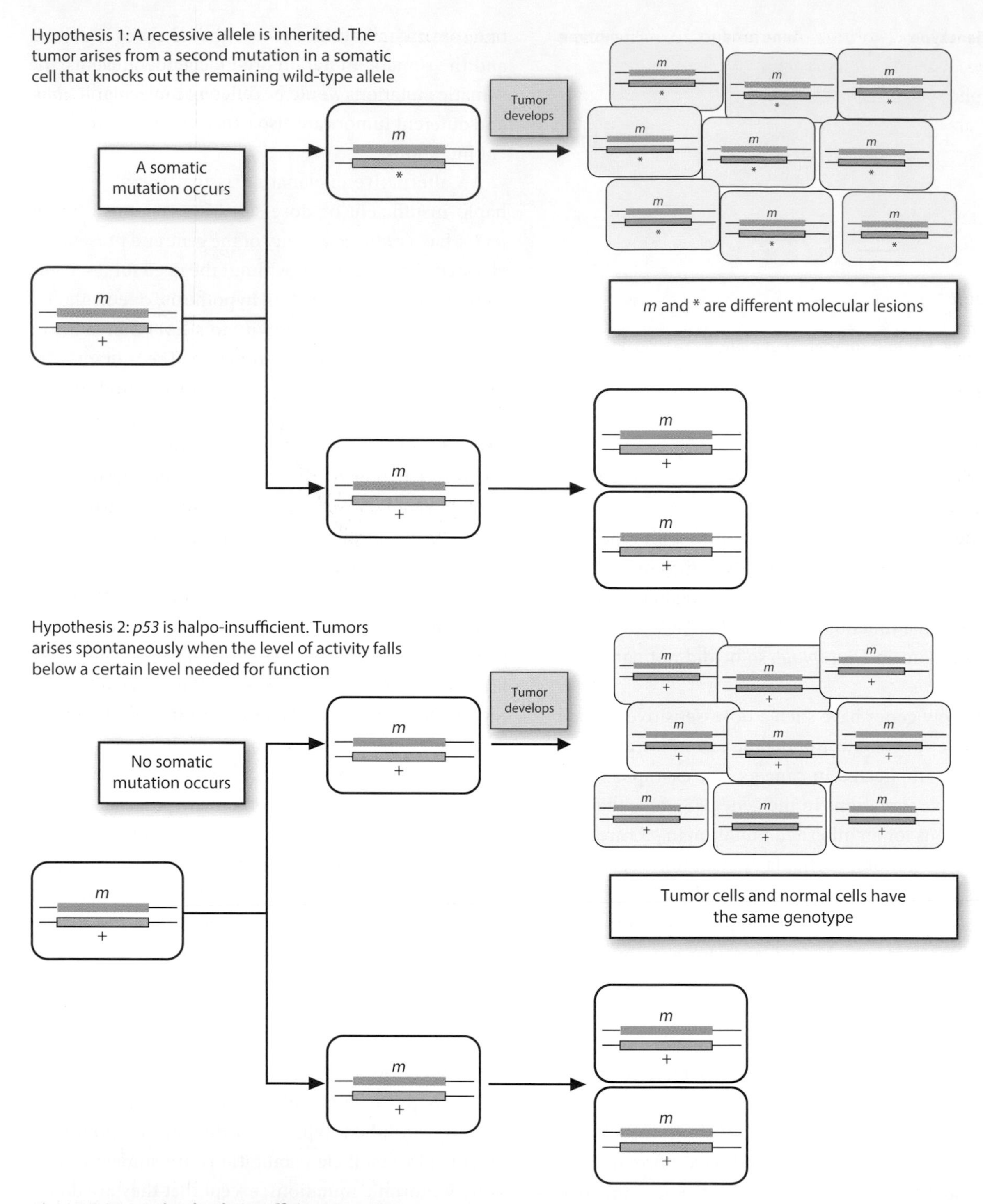

Figure 4.25 ***p53*** **has haplo-insufficient mutations.** Two hypotheses have been proposed for the tumor-causing effects of mutations in the *p53* gene. Under the first hypothesis, *p53* mutations are recessive. An individual who is heterozygous is at risk for tumor formation, since a second, somatic mutation in the *p53* gene will reduce or eliminate the activity of the gene. If this hypothesis is true, tumor cells will have two different molecular lesions in the gene, the inherited one and the somatically acquired one; in fact, different primary tumors may have different mutations. Under the second hypothesis, *p53* is haplo-insufficient and two wild-type alleles are required for normal gene activity. An individual who is heterozygous is at risk for tumor formation, because some cells require higher levels of the protein or because the protein levels may otherwise fall below threshold levels. No additional somatic mutation occurs, so that the tumors have a wild-type allele of *p53*. In mice heterozygous for a *p53* mutation, some the tumors have a wild-type *p53* allele, which supports the second hypothesis.

analysis of hypomorphic mutations in an allelic series indicates that this is almost certainly true. The sensitivity of the phenotype being monitored in the mouse experiments above—that is, cancer occurring in particular tissues—may allow the effects of subtle changes in gene dosage to be observed. This may suggest that genes that are dose-sensitive are more likely to be oncogenic when mutated.

4.7 Summary: Find a mutant

Mutations and mutant phenotypes are foundational tools of classical genetic analysis. Geneticists spend considerable amount of intellectual energy and time in finding and classifying mutants because these have proved to offer one of the best ways to understand biological processes, especially those for which very little prior information is available. Imaginative and productive genetic screens and selections can be an expression of creativity, and even a type of beauty. It is only a slight exaggeration to say that geneticists view the world through the lens of mutant phenotypes and natural variation.

While identifying mutants is the crucial first step, and often the one that is most intriguing, the subsequent steps are also important, if somewhat more laborious. By a series of outcrosses to wild-type, mapping, and complementation tests, individual mutations can be assigned to genes and the number of genes in the process can be estimated. While carrying out these steps, the geneticist will probably learn whether the mutant involves a complete loss of function—that is, is a null mutation—or a reduction of function, or whether it involves some alteration of function that results in a dominant mutant phenotype. Although we have outlined these subsequent steps as if there were a standard procedure and order in which to carry them out, the more precise methods depend on the organism, the phenotype, and even the geneticist. No matter how and in what order they are taken, these subsequent steps are necessary for accurate insights into how a mutant affects the biological process. We return to this discussion in Chapter 6, with some further examples.

Chapter Capsule

Genetic analysis often begins with a mutant screen to find an organism with an altered phenotype for some biological process. Since such mutations do not arise spontaneously at a high frequency, several different mutagens are used to increase the frequency at which they occur. Once mutants have been found, the next steps are to assign them to genes and to determine how each particular mutant affects the activity of the gene. Assigning the mutants to genes helps in understanding how many different genetic activities contribute to the biological processes and whether it is useful to continue to look for more mutations. Most mutants are recessive to wild and arise from a complete or partial loss of the activity of the gene, although uncommon dominant mutations provide different types of insights, once they have been characterized.

Case Study 4.1

Segmentation in *Drosophila* embryos

Many possible examples of the application of mutant analysis could be used as a case study to this chapter, and nearly every monthly issue of *Genetics* or *G3* has a report of a new successful genetic screen. However, few if any mutant screens have been more successful than the isolation of segmentation mutants in *Drosophila* embryos. When Nusslein-Volhard and Wieschaus began the analysis in the late 1970s, no one would have been so bold or prescient as to predict publicly that this process would not only unlock the strategy for embryogenesis in flies but also provide insights for embryology in nearly all other animals. Even more remarkably, no one dreamed that it would provide key insights into the underlying biology of some of the most common forms of cancer. Nusselin-Volhard and Wieschaus and their collaborators combined remarkable knowledge, observation, dedication, and hard work with the power of genetics, in a study that few have matched. The results of their work are well known and intellectually compelling; many students will have some familiarity with terms such as "gap genes," "pair-rule genes," or "segment polarity genes," but may not be familiar with some of the details of how they got to their conclusions. Our goal in this case study is to show the use of the application of genetic analysis in the context of this broader project. The case study forms nearly a parallel chapter, but it is one of the best ways to demonstrate how a successful mutant screen is done.

A summary of early *Drosophila* development

The most visible aspect of an insect embryo is that it has segments—fourteen of them in *Drosophila melanogaster*, although segments in the head are fused and harder to distinguish that those throughout the main body. A picture of a *Drosophila* embryo is shown in Figure CS 4.1. Each segment has a characteristic appearance and location along the body axis, and several give rise to the distinctive structures in the adult. For example, the three thoracic segments, T1, T2, and T3, give rise to the three pairs of legs; in dipteran insects such as *Drosophila*, T2 also produces the wings, whereas T3 produces an evolutionary derivative

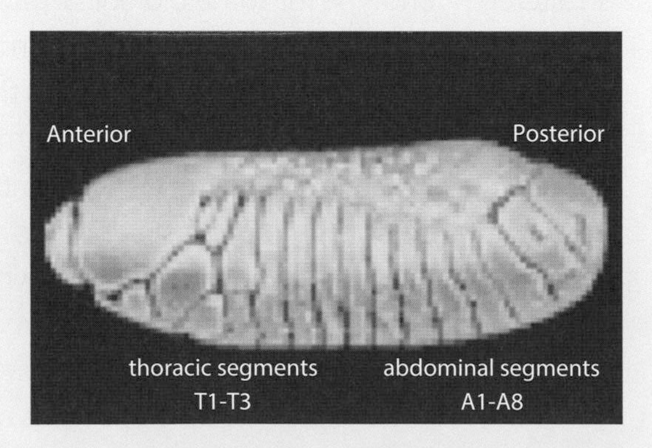

Figure CS4.1 Segmentation in the *Drosophila* embryo. In this scanning electron micrograph, the positions and structures of the segments are shown. There are fourteen segments overall, although the three head segments are fused and hard to recognize.

of the hindwing known as a haltere. As discussed in Chapter 3, in 1978 E. B. Lewis had published a highly influential paper (*Nature* 276: 565–70) that laid out a strategy by which the homeotic genes specified the identities of individual segments just about the time when Nusslein-Volhard and Wieschaus began to search for mutations that affected the pattern of segmentation itself. These are phenotypes of the type discussed in Section 4.1 of the chapter.

When they began, segmentation patterning was widely believed to be connected to the determination of the body axis in the embryo, in particular the anterior–posterior axis; their mutant analysis would confirm and extend this hypothesis. Physical manipulations of cells and nuclei—including transplantation, ligation, and ablation—had been done in *Drosophila* and other insects, but these could provide only the broadest framework of how segments arose in their proper positions, and no genes or molecules had been identified.

It was also clear, from both simple observation and physical manipulations, that the single-cell embryo already had a polarity derived from an underlying polarity in the oocyte. The nucleus of the single-cell *Drosophila* embryo divides rapidly without cytoplasmic division, eight divisions in about an hour, to generate a synticium with roughly 1,000 nuclei in a common cytoplasm. The nuclei migrate to the periphery of the egg and divide four more times before boundaries form between them to make individual cells around the edge of the embryo. Thus, during the early stages before cellularization occurs at the periphery, any gene product (either mRNA or protein) that is free to diffuse in the cytoplasm might affect all the nuclei in the embryo. The progression from the synticial blastoderm with the nuclei in a common cytoplasm to the cellular blastoderm stage with the nuclei arrayed around the periphery of the embryo to a segmented embryo had been thoroughly described by scanning electron microscopy and other methods, but the underlying process of segmentation itself was not understood at all.

Thinking through the Experiment

1. Most animal embryos do not have a synticial stage in which all of the nuclei are in a common cytoplasm. How does the presence of a synticial stage in *Drosophila* affect the processes underlying early development?

These observations that the single-cell embryo has an underlying polarity because of an underlying polarity in the oocyte also affected thinking about some of the mutants that might be found. Some of the mutants affecting the segmentation pattern seemed likely to be **maternal-effect mutations**. The general observation concerning maternal-effect mutations is that the phenotype of the offspring depends on the genotype of the mother; the genotype of the offspring may or may not play a role itself in the phenotype. Maternal-effect mutations had been described in many different animals for different processes; for example, whether the shell of a snail swirls in a clockwise or counter-clockwise direction is determined not by the genotype of the snail itself but by the genotype of the snail's mother. Although molecular explanations for maternal-effect mutations were missing, the overall interpretation was that some molecules needed for the developmental process, particularly for developmental processes that occur very early in embryogenesis, were supplied by the mother in the oocyte and were present before gene expression in the embryo began.

Thinking through the Experiment

2. Maternal-effect mutations are known in many different animals and were long recognized as being important for early embryonic events. What is the underlying developmental or cellular basis for a maternal-effect mutation?
3. A key experiment in thinking about maternal-effect mutations was to mate a wild-type male to a homozygous mutant female. For some genes and mutations, the offspring of such a mating survived and had a wild-type phenotype. For other genes and mutations, the offspring were mutant even though the sperm was providing a wild-type allele. Speculate about why different genes and mutations gave different outcomes in these experiments. Hint: remember that the embryo is heterozygous and that the wild-type allele is provided by the sperm.

Mutagenesis

In order to understand the genetic logic of segmentation, Nusslein-Volhard and Wieschaus set out from the very beginning to find as many genes affecting segmentation as possible. They used EMS as their principal mutagen and concentrated almost exclusively on recessive mutations; every gene is a potential target for EMS-induced mutations, and most of the mutations are single-base changes that result in a loss or reduction of gene function. The mutagenic effect of EMS is discussed here in Section 4.2. The initial mutagenesis strategies of the two researchers made three important assumptions about the genes they might find, all of which proved to be true.

- First, mutations that affect segmentation in embryos are very likely to be lethal, such that the embryo cannot progress to adulthood or become fertile. Therefore Nusslein-Volhard and Wieschaus needed to establish methods to maintain their mutant strains as heterozygote (see Section 4.3 on balanced heterozygotes and Tool Box 4.1).
- Second, the genes would map throughout the genome, so procedures would be needed that allowed them to balance heterozygotes at many different loci.
- Third, some of the mutations would exhibit a maternal effect. While they found maternal-effect mutations in their screen—and some of the genes most important in interpreting the logic of segmentation exhibit a maternal effect—Nusslein-Volhard and Wieschaus focused on the ones that had effects in the larvae, postulating that these mutations would identify genes that interpret maternally deposited signals.

Balancer chromosomes were used to maintain heterozygotes

Three different papers detail the mating schemes that were used to find these mutations, each paper addressing genes from a different chromosome. Nusslein-Volhard, Wieschaus, and their collaborators used different balancer chromosomes for the X chromosome, for chromosome 2, and for chromosome 3. The procedure was approximately the same for each chromosome, but the genetic strains differed. For example, the screen that identified mutations on chromosome 2 used the balancer chromosome CyO, an inversion that includes most of the second chromosome and is marked with the dominant mutation Curly wings (*Cy*), as shown in Figure CS4.2. In the chromosome 2 scheme shown in Figure CS4.2, males homozygous for the mutations cinnabar (*cn*) eyes, brown (*bw*) eyes, and spotted (*sp*) were fed EMS to induce mutations during spermatogenesis. These males were mated to the females that were heterozygous for the CyO inversion and a multiply marked chromosome.

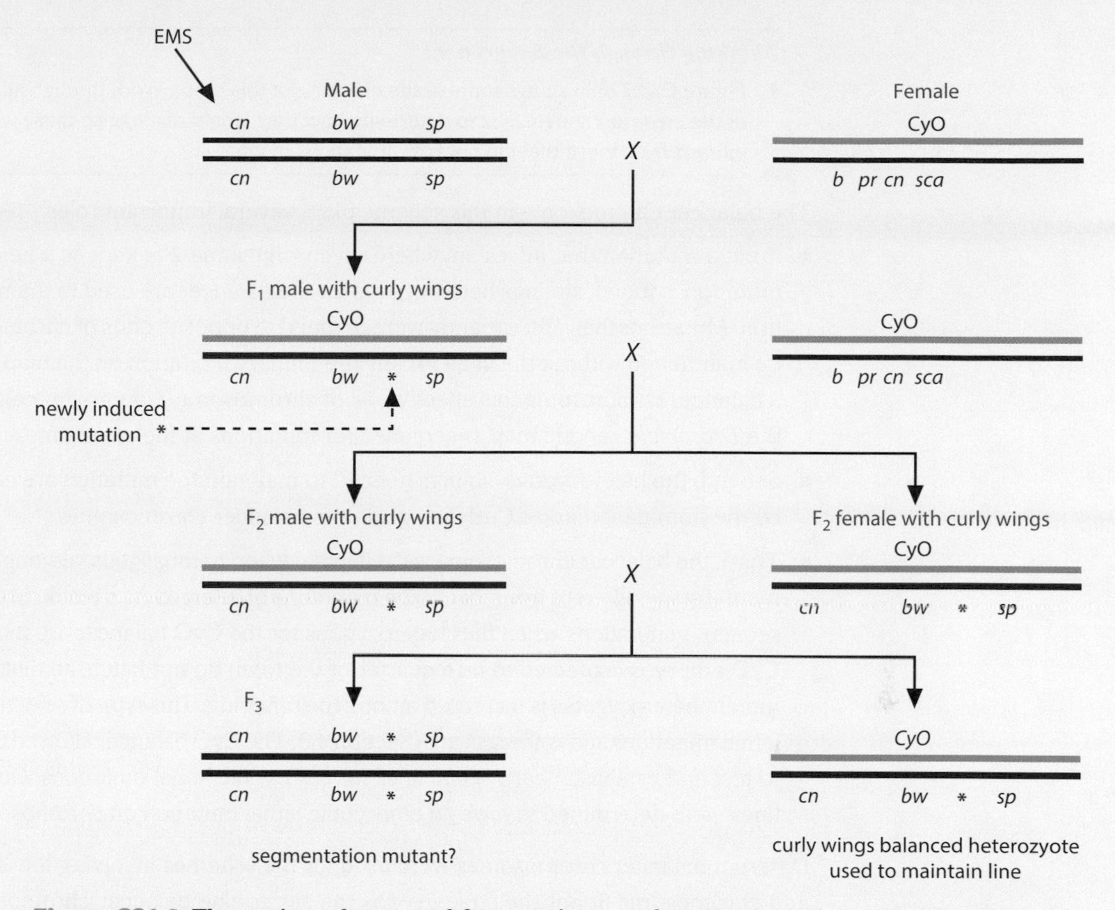

Figure CS4.2 The mating scheme used for mutations on chromosome 2. A male homozygous for the recessive marker mutations *cn*, *bw*, and *sp* was mutagenized with EMS and mated to a female heterozygous for the CyO balancer and *b*, *pr*, *cn*, and *sca* as shown. An individual F_1 male with the CyO chromosome balancing a newly induced mutation on chromosome 2 (although this is not known until the scheme is completed) was then mated to another female with the same genotype as the one in the first generation. The F_2 flies heterozygous for the CyO balancer and the newly induced mutation were mated and their non-Curly F_3 offspring examined the presence of a mutation affecting segmentation. If a mutation of interest was found, sibling flies with the CyO balancer and hence Curly wings were used to maintain the mutant line. Not all possible offspring are shown in each generation, but the various visible marker mutations (and lethal alleles on the CyO balancer) allowed them to keep track of the newly arisen mutation. Similar mating schemes with different balancer chromosomes were used for the X chromosome and for chromosome 3.

The F_1 flies with newly induced mutations balanced over the inversion were recognizable because they have Curly wings. A single F_1 male fly was then backcrossed with a female from the CyO heterozygous strain to produce a balanced F_2 strain with both males and females. At this point, the mutagenized copy of chromosome 2 is heterozygous but is not yet known to have any mutations of interest. By using an unmutagenized F_1 female, the F_2 males will have an unmutagenized X chromosome. Chromosome 3 may still contain newly arisen mutations (and in fact some were found), but these are eliminated in subsequent outcrosses and mapping experiments, as described in Sections 4.3 and 4.4. The subsequent crosses also determined that the mutations were recessive and that each mutant phenotype was due to a single mutation rather than to two mutations acting synergistically. Each F_2 line was mated with its siblings, and the F_3 larvae were examined for mutations.

Thinking through the Experiment

4. Figure CS4.2 only shows some of the offspring of this cross. Work through all of the offspring of the cross in Figure CS4.2 to determine how they would distinguish those with mutations of interest from those that did not have mutations of interest.

The balancer chromosome in this scheme plays several important roles.

- First, a mutation that arises anywhere on chromosome 2 is kept as a heterozygote. If a mutation is found, siblings heterozygous for the balancer are used to maintain the mutation. Mutations that subsequently were mapped to opposite ends of chromosome 2 could be maintained without the need for any preliminary limitation on the map location. With a balancer chromosome this effective, all of chromosome 2, approximately 40 percent of the *Drosophila* genetic map, is screened for mutations at the same time.
- Second, the heterozygous siblings needed to maintain the mutation are easily identified by the dominant marker Curly wings on the balancer chromosome.
- Third, the balancer chromosome is itself lethal when homozygous, although at a developmental stage different from that of the mutations of interest. As a result, in the F_2 and subsequent generations when flies heterozygous for the CyO balancer are mated, the CyO/CyO embryos (expected to be a quarter of the total) do not hatch, so that the fraction of mutant heterozygotes is increased among the offspring. This type of selection used linked lethal mutations and is discussed in Section 4.3. The CyO balancer allowed the researchers to test and establish nearly 4,600 lines for embryonic lethal mutations. Out of these, 321 lines were determined to have an embryonic lethal mutation on chromosome 2.

Different balancer chromosomes were used for the schemes involving the X chromosome and chromosome 3, but the strategy was the same. The balancer chromosomes allowed the researchers to test nearly the entire length of the chromosome as one unit, without mutations being lost through recombination. For the X chromosome, 122 lines with lethal mutations were established; for chromosome 3, 198 mutant lines were established. One of the crucial features of this mutagenesis scheme was the ability to isolate and maintain as many lethal mutations as possible, which can be done only with the balancer chromosomes available in *Drosophila melanogaster*.

Thinking through the Experiment

5. No other organism has the number and variety of balancer chromosomes that are found in *Drosophila melanogaster*. Discuss briefly how the availability of these balancer chromosomes made this screen feasible.
6. Contemporary geneticists working on other model organisms have generally not focused on generating a set of balancer chromosomes as exists in *Drosophila*, although a few balancer chromosomes have been made for other organisms. In your view, why is it that most geneticists no longer think that creating balancer chromosomes is a good use of their time, even as we recognize the importance of these chromosomes in *Drosophila* genetics?

A temperature-sensitive selection against the unmutagenized flies

In order to find as many mutations as they did, Wieschaus, Nusslein-Volhard, and coworkers cultured and examined hundreds of thousands of flies. The amount of effort was formidable, but they were able to use a genetic selection to keep the numbers somewhat more manageable. Genetic selections and temperature-sensitive mutations similar to this are described in Section 4.3. Prior to the work of this team, several dominant temperature-sensitive lethal mutations (referred to as DTS) had been identified by other geneticists. One of these temperature-sensitive lethal mutations was crossed onto each chromosome in the non-inverted homologue in the unmutagenized mother. These temperature-sensitive lethal mutations provided the means for a very useful selective strategy.

It is important to realize that individual flies often need to be mated when finding mutations in *Drosophila*. Matings of individual flies require that the females be virgins; the DTS mutations provide an efficient means to recover virgin females. The F_1 flies were mated and then shifted to the restrictive temperature. At the restrictive temperature, the parents with the dominant temperature-sensitive mutation die, so that no special effort was required in order to collect virgin females at each generation. The temperature-sensitive lethal mutation also ensures that only the balanced heterozygotes are maintained in the F_2 generation, since the siblings with the non-inverted chromosome die. This particular genetic selection was thus used to reduce the number of flies that had to be examined for mutations rather than to select specifically for the mutations of interest. Nonetheless, the selection also allowed the investigators to maintain and examine many more mutant lines than would otherwise have been possible.

Complementation tests

From the extensive mutant hunts, more than 640 different mutations affecting segmentation in *Drosophila* embryos were found, mapping to three different chromosomes. The next step for Nusslein-Volhard, Wieschaus, and co-workers was to separate the mutants by chromosome and by phenotype and to determine the total number of genes that these mutations represent using complementation tests. This is described in Section 4.5.

They simplified the number of complementation tests needed by assuming that mutations in the same gene might have similar phenotypes, and by focusing on those combinations first. For example, among the 321 mutations on chromosome 2, five phenotypic categories were used: anteroposterior defects, dorsoventral defects, holes in the cuticle, general differentiation defects, and head defects. Subgroups were also used within each category. These were subjective phenotypic categories that arose from the investigators' experience and knowledge of mutant *Drosophila* embryos. Since it was not realistic to perform pairwise complementation tests among all 321 lethal mutations expeditiously, separation by categories reduced the labor and time. Nothing other than the mutant phenotype suggested that these were meaningful distinctions, since different mutations from the same gene might have fallen into different categories. The paper is candid about the success and limitations of this approach.

> Several mutations were tested as members of more than one subgroup. In general, this strategy proved useful and in many instances our predictions of allelism were correct. (Needless to say, our abilities in this matter improved great during the course of the experiment.) ... In other cases, the initial classification was less successful. Several mutant phenotypes had more than one prominent feature and what later turned out to be allelic mutants had initially been distributed among two or three different categories. (Nusslein-Volhard et al. 1984, p. 272)

Problematic classifications are described, and it is noted that complementation tests were performed among genes that proved to map close together, in order to be sure that the mutations belonged to separate genes. The results are discussed in more detail below, in the section on estimating the number of genes.

Thinking through the Experiment

7. Classifying mutations by phenotype before doing complementation tests allowed the researchers to conduct this number of tests in a timely manner. But this also makes certain assumptions about the mutations in a gene. What assumption is made? Do you think this assumption was a safe one to make? (The investigators themselves were very cautious about this.)

Map positions

More than 120 genes affecting the segmentation pattern in *Drosophila* embryos were identified in these screens, located throughout the genome. Each of these genes had to be mapped and tested with previously identified genes that mapped in the same location. Both recombination distances and segmental aneuploidy, as described in the Section 4.4, were used to assign these new genes to a chromosomal location: the method used for recombination mapping on chromosome 2 is summarized in Figure CS4.3; similar methods were used for the genes on chromosome 3, and a slightly modified version was used for the genes on the X chromosome. Most genes were also mapped using deficiencies and duplications.

Even upon re-reading these papers after more than three decades, the immediate impression is about the remarkable amount of work that went into finding these mutations, assigning them to genes, and mapping the genes. Nearly every genetic method available in *Drosophila* was used in these experiments. Most current readers will probably skip quickly to the pictures of the mutant phenotypes and to the tables of the genes, since those tell us what the flies reveal about the strategy for segmentation. But the lessons for us in these papers lie in the details of the methods that were used and they constitute an example of how a thorough genetic analysis of a biological process can be.

Saturation and the number of genes

As a result of the complementation tests, the 259 mutations on chromosome 2 defined 61 genes, 48 genes having more than one allele and 13 genes having one allele; 4 mutant lines had separate mutations in two or more different genes and were not pursued. The results are tabulated in Table CS4.1. The number of alleles per gene ranged from one to 18. Similar strategies identified 198 mutations in 45 genes on chromosome 3 and 101 mutations in 20 genes on the X chromosome. Overall for the three chromosomes, the number of mutations per gene ranged from one to twenty, with an average of 4.5 mutant alleles per gene.

With such an impressive collection of mutations, how do investigators know when they are done? At some point, most of the mutations are found to be new alleles of previously identified genes, and they add little new information. This is known as saturation—a situation where all the newly identified mutations are alleles of previously identified genes

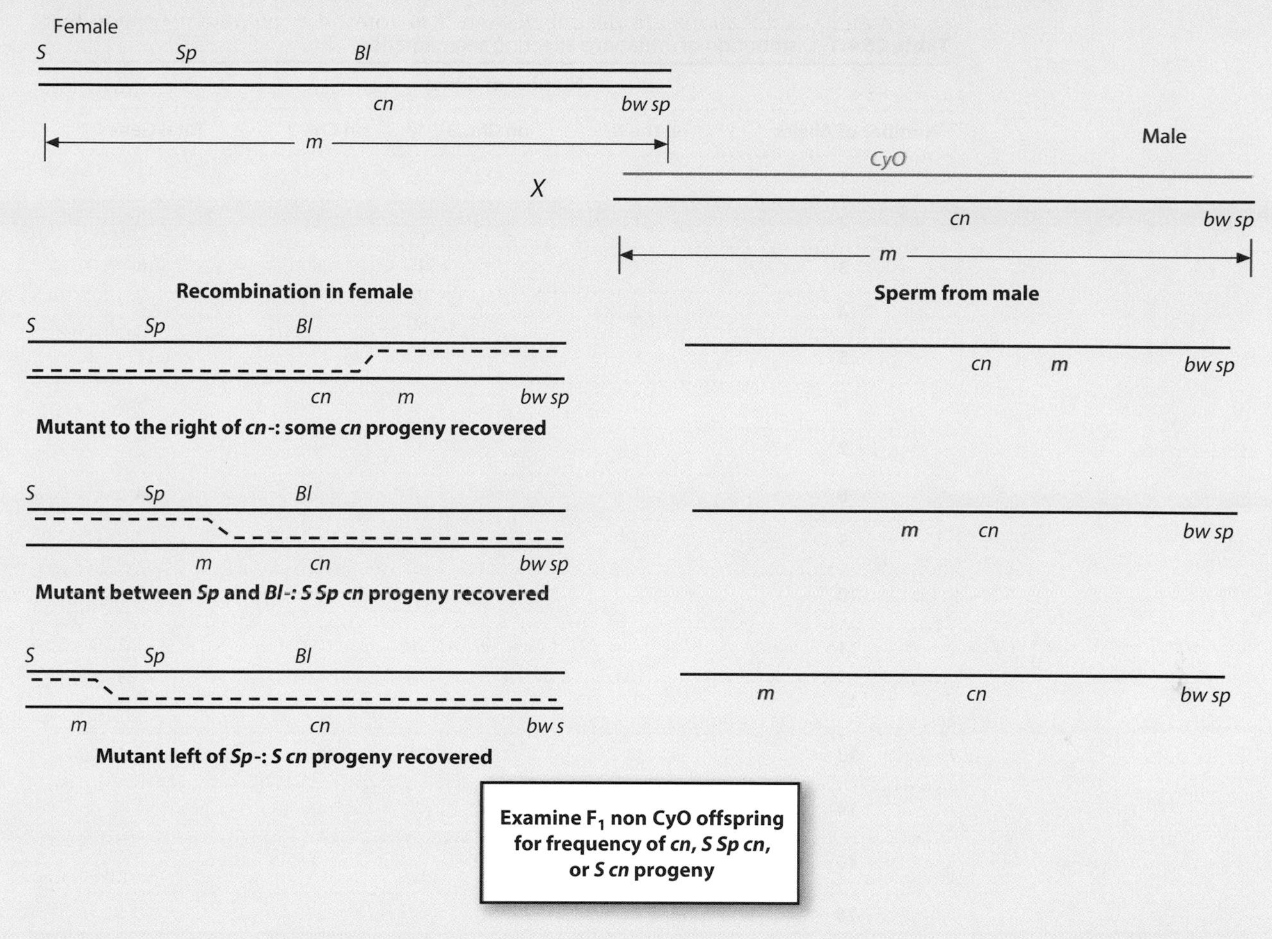

Figure CS4.3 Recombination mapping of a segmentation mutation on chromosome 2. The recessive marker mutations linked to the newly arisen mutation *m* from the mating scheme in Figure CS4.2 were also used for recombination mapping. The location of *m* with respect to the recessive marker mutations *cn*, *bw*, and *sp* and dominant mutations *S*, *Sp*, and *Bl* could be determined by determining which phenotypes arose in the recombinant offspring found in this cross, as well as the ones not found. The surviving offspring were examined, since those arose from a recombinant gamete in the female. For example, if some *cn* offspring but no *S* or *Sp* offspring were found, the mutation must be located to the right of *cn*. Dashed lines indicate the possible recombinant chromosomes in the heterozygous female, all of which have lost the *m* allele since the recombinants survived.

and no new genes can be found by this method. Tool Box 4.2 discusses how to determine when saturation has been reached or approached.

Thinking through the Experiment

8. Define the concept of saturation, as applied to mutant hunts, and *discuss its importance, as well as some of the important assumptions being made.*
9. *While we are using the concept (and the term) "saturation" for mutant hunts, a very similar idea underlies other examples involving sampling from a population. Can you think of* another example that relies on this same concept, and explain what assumptions are made to apply this concept?

Table CS4.1 Distribution of mutations affecting segmentation

	Number of Genes			
Number of Alleles	**on the X**	**on Chr. 2**	**on Chr. 3**	**Total Genes**
1	13	13	13	39
2	1	13	5	19
3	7	7	4	18
4	2	8	6	16
5	1	3	7	11
6	3	5	1	9
7	1	1	1	3
8	1	1	2	4
9	2	5		7
10			1	1
11			1	1
12	1	1		2
13		1		1
14			1	1
15		1	1	2
16				0
17		1		1
18		1		1
19				0
20			1	1

Each row is the number of genes identified with that number of alleles; that is, 13 genes on the X chromosome had one allele, one gene on the X chromosome had two alleles, seven genes had 3 alleles, and so on. In total, 137 genes were identified.

In order to determine whether they were near saturation, the investigators plotted the number of new mutations and the number of new genes as a function of the number of mutagenized lines. This plot is reproduced in Figure CS4.4. The data show that roughly 6,000 chromosomes were tested and the number of mutants increased linearly with the number of chromosome tested; in other words, they were continuing to isolate new mutant alleles. The number of genes identified increased until some 1,800 chromosomes had been tested and about fifty different genes had been identified. Then the number of genes increases much more slowly than the number of mutations, such that among the final 1,500 chromosomes tested almost no new genes were found. This indicates that nearly all genes that could be found by this method have probably been found. The goal of using genetic analysis to find all the genes that affecting segmentation in *Drosophila* was met.

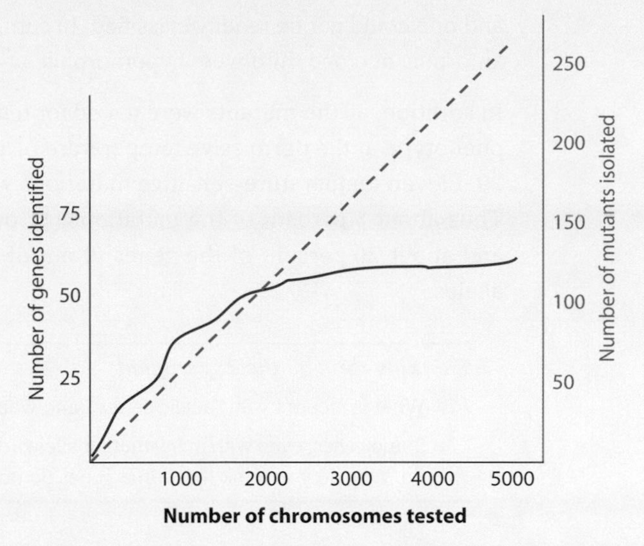

Figure CS4.4 Estimating the number of segmentation genes on chromosome 2. The graph shows the number of chromosomes mutagenized and tested on the X axis with the number of new genes identified (in red) and the number of new mutants isolated (in blue) on the two Y axes. Note that, while the number of mutants continued to increase with each mutagenesis, the number of genes identified did not increase, since the newly isolated mutations were alleles of genes that had been previously found.

Classifying mutants: Types of alleles

Because of the mutagen and the methods used to find them, nearly all the mutations recovered in these screens were recessive: among the more than 20,000 mutations recovered of all types, fewer than forty proved to be dominant mutations. None of the dominant mutants was tested further. Since most of the genes had many recessive mutant alleles, the investigators were able to compare the phenotype of different mutations in the same gene. That is, they could attempt to construct an allelic series for some genes. In addition, by mapping the mutations using deletions, the investigators could also compare the phenotype of a deletion with the phenotype of a mutant allele—Muller's test for hypomorphs and amorphs, as described in Section 4.6.

Thinking through the Experiment

10. While it may seem like an obvious point, the recognition that most mutations are recessive to wild type is significant. What is the biological and evolutionary significance of the fact that most mutations are recessive to wild type?

The information for the genes on chromosome 3 is both informative and typical of the results. Among the thirty-two genes with more than one mutant allele, the mutants have the same phenotype in seventeen cases. Thus no allelic series could be constructed for those seventeen genes. Six of these seventeen genes were tested with deletions, and in all cases the phenotype of the m/Df was the same as phenotype of m/m. Therefore, for these genes, the phenotype arising from these mutations was considered amorphic or null. For fifteen other genes, mutant alleles had different phenotypes, and an allelic series could be constructed for several of these genes. Thirty-five mutations in five of these genes were tested with deletions, with the result that nineteen were hypomorphic, fifteen were amorphic, and one could not be readily classified. In sum, about two thirds of the mutations were amorphic and one third were hypomorphic.

In addition, all the mutants were tested for temperature sensitivity by comparing the phenotype at the permissive temperature of 18 and at the restrictive temperature of 29. Eleven temperature-sensitive mutations were identified for nine different genes. Thus about 5 percent of the mutations (11 out of 198) were temperature-sensitive, and about 20 percent of the genes (9 out of 45) had temperature-sensitive mutant alleles.

Thinking through the Experiment

11. What is meant by an "allelic series," and what can be learned from one?
12. Some genes, even with many mutant alleles, do not exhibit a true allelic series, while others do. Why do you think that some genes do not have an allelic series of mutations?

Three categories of mutant phenotypes were observed

The interpretation of mutant phenotypes is the topic of Chapter 6, but we will include some information now from this screen; the classifications of mutant phenotypes are likely to be the most familiar outcomes of this screen, for many readers. The phrase applied in Chapter 6 to describe Barbara McClintock—having a feeling for the organism—could also be applied to Wieschaus, Nusslein-Volhard, and their collaborators.

The range of mutant phenotypes found among these genes was remarkably diverse. Some mutants appeared to be lacking entire blocks of the embryo, whereas others were lacking particular regions or had some disorganized regions. Nusslein-Volhard, Wieschaus, and collaborators recognized a pattern in the defects and divided the mutant phenotypes into three main categories: the gap genes, the pair rule genes, and the segment polarity genes. The categories are summarized in Figure CS4.5. One category of mutants is lacking large regions of the embryo, which results in embryos that are often much smaller. That is, these mutants have gaps in their segmentation pattern, and the genes are called, collectively, **gap genes**. The inference was that these genes are needed for the organization of broad regions of the embryo.

A second group of genes has probably the most unexpected mutant phenotype: these genes affect particular combinations of segments, often a pair of segments or alternating segments, but typically only seven of the fourteen segments. These genes are collectively called the **pair-rule genes**. The inference from this mutant phenotype is that the broad regions of the embryo defined by the gap genes are further subdivided by the pair-rule genes, so that each segment is specified by a combinatorial code of gap and pair-rule gene expression.

A third group contains genes that affect every segment, often by altering the polarity or orientation of the segment. This group, which has the largest number of genes found, is known as the group of **segment polarity genes**. The inference from this group is that each segment is specified by the expression of some combination of gap and pair rule genes, but the orientation of structures within each segment is specified by the segment polarity genes.

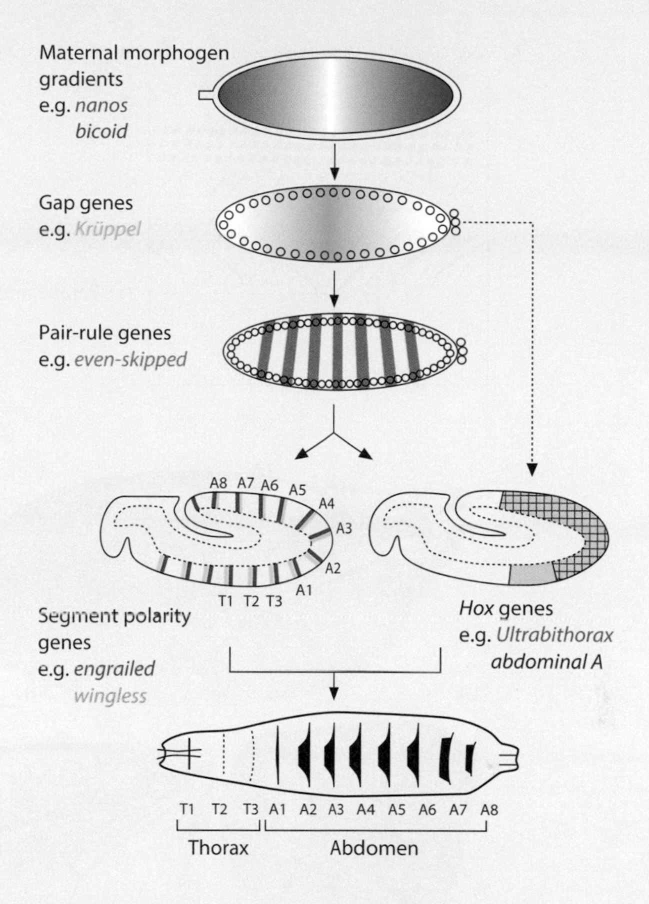

Figure CS4.5 Gap, pair-rule, and segment polarity genes. The mutant phenotypes of the genes were assigned to different categories. The genes *nanos* and *bicoid* act in the mother to establish an anterior–posterior gradient in the egg before fertilization. The anterior–posterior pattern then becomes progressively defined into segment and segment structures by the gap genes, the pair-rule genes, and the segment polarity genes.

Thinking through the Experiment

13. We interpret the normal role of a gene from its mutant phenotype. What would be the normal role of a gene classified as
    a. a gap gene;
    b. a pair rule gene;
    c. a segment polarity gene?
14. Which, if any, of these gene categories might also include some maternal effect mutations? Explain your answer.

The *Patched* gene is an example of a segment polarity gene

One of the segment polarity genes was named *patched* (abbreviated *ptc*) because the mutant phenotype has patches of bristles; to the untrained eye, the mutant phenotype, as shown in Figure CS4.6, looks somewhat like a man who shaved with a dull razor using a poor mirror, so that patches of bristles remained at the unusual places. But the pattern and orientation of the remaining bristles in *patched* mutants reveal that the mutant shows deletions of structures in adjacent segments with mirror image duplications of remaining structures. The mutants have twice the normal number of segment boundaries, with reversed polarity

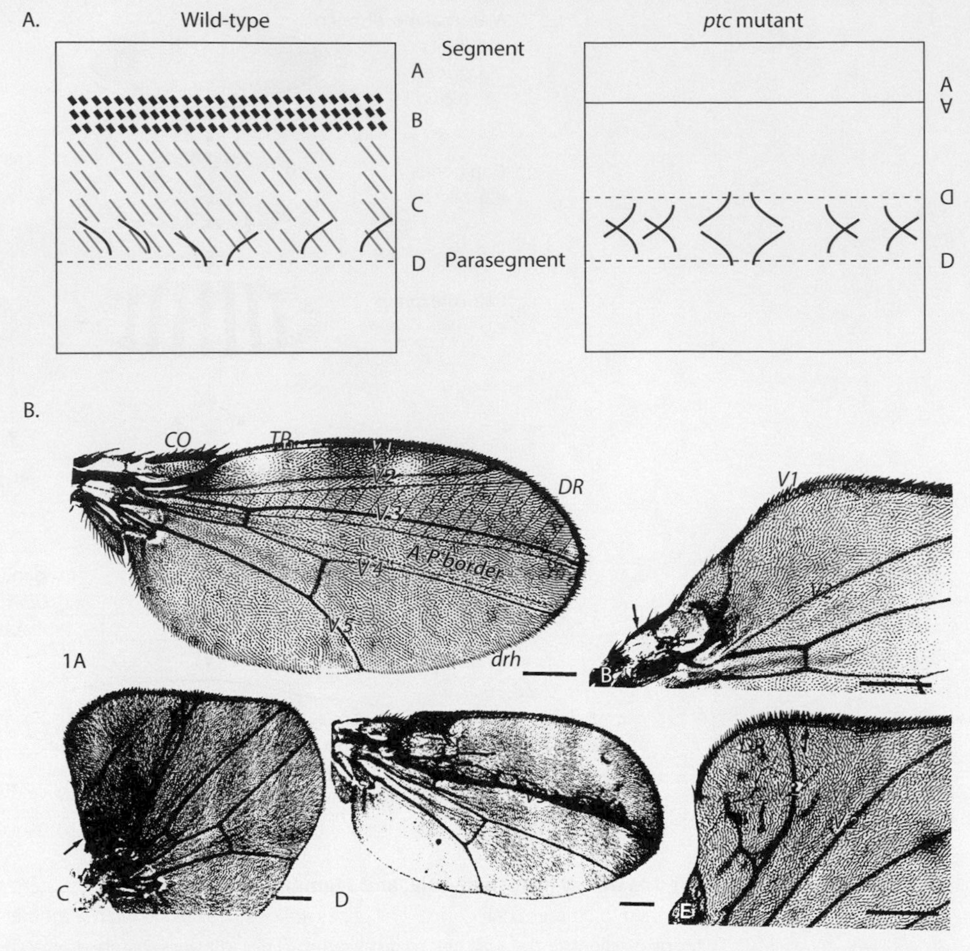

Figure CS4.6 The *patched* mutant phenotypes. The *patched* (*ptc*) gene will be used as an example in subsequent chapters, so its mutant phenotypes are summarized here. In the embryonic segments, diagrammed here for the T3 segment, *patched* is a segment polarity gene, and the mutant is lacking some bristles and hairs while having mirror-image duplicates of others. As discussed, *patched* also affects the patterning of the adult wing, as shown in the drawings at the bottom. The segment polarity effects are redrawn and modified from a diagram by Hooper and Scott (1989) while the wing phenotypes have been redrawn from photographs in Phillips et al. (1990) since the original images are not available; see the References section in this chapter's case study.

around the segmental boundaries. Since *patched* will be used as an example in other parts of the book, we will describe some of the characterization here in a bit more detail.

In the original screens, twelve mutant alleles of *patched* were found, including one temperature-sensitive allele and one other hypomorphic allele that affected fewer structures than the null alleles. The *patched* gene mapped to a location that corresponded to that of a gene called *tufted*, found more than thirty years earlier but not extensively studied. *tufted* mutations are viable and have variable effects on the wing and lesser effects on bristles and structures in the head, eye, and antenna. In the wing, *tufted* alleles cause deletions of some structures such as wing veins, sensory organs, and bristles and duplications of other structures in the anterior compartment of the wing, including other veins and bristles; in severe examples, the wings are much shorter than in the wild type and are shaped liked paddles.

The *tufted* mutations and the *patched* mutations map onto the same location and have some similarities in their effects on bristles, although at different times in development and in different tissues. Given the similarities, it was logical to test whether these were different alleles of the same gene. When complementation tests were done between the *tufted* mutations and the *patched* amorphic mutations, the mutations failed to complement, both for the embryonic lethality and for the wing defects, demonstrating that these are alleles of the same gene. The *tufted* alleles are hypomorphic, whereas the *patched* alleles are more likely to be null; these categories are discussed in Section 4.6. These classes of mutant phenotypes show an advantage to having an allelic series for a gene. If the phenotype of the *tufted* mutations were considered without knowledge of the *patched* mutation phenotype, the important function of the gene in embryo segmentation could be overlooked. Conversely, in the absence of the *tufted* hypomorphic alleles, the effects of the gene on adult tissues would have been missed. The effect on the adult wing is the phenotype used in Chapter 11.

Thinking through the Experiment

15. Explain why *patched* and *tufted* have different phenotypes and what this might tell us about the role of the gene.

In subsequent studies using an antibody to monitor the levels of *ptc* protein, the null alleles had no detectable amounts of the wild-type protein in the wing, whereas a *tufted* hypomorphic allele had greatly reduced amounts of the protein; this was reduced even further in a hypomorphic *tufted*/null *ptc* heterozygote. Thus our description of hypomorphic and amorphic alleles in the chapter holds true in this case. The gene name *tufted* was dropped in favor of the name *patched*, and the original *tufted* alleles were designated as alleles of *ptc*.

Although the original screens for all segmentation genes found twelve alleles of *patched*, other investigators subsequently studying the *patched* gene itself conducted more mutant screens, to find even more alleles. Additional *patched* alleles were also found by an F_1 screen, including some that were induced by P elements; F_1 or non-complementation screens are described in Section 4.5, and P elements as a mutagen are discussed in Section 4.3. The F_1 screens found about a dozen more alleles of *patched*, including viable alleles (similar to the *tufted* alleles) and embryonic lethal alleles. The viable alleles were used to examine the effect of *patched* on the wing and on other adult structures. More than 140 mutant alleles of *patched* have now been identified. We will return to a discussion of *patched* in later chapters.

A triumph for genetic analysis

It may be hard for a contemporary reader to understand the excitement that surrounded this mutant collection when it was published in 1984. (The Nobel Committee recognized the importance of this work by awarding its authors a Nobel Prize for Physiology and Medicine, shared with E. B. Lewis for his interpretation of homeotic mutations.) Here at last were representative alleles of most (if not all) of the genes that affect segmentation in *Drosophila* embryos, a distinctive but mysterious biological process. Now at last we had the genetic tools to analyse this process. The papers on individual genes began appearing quickly afterwards. Each of these 120 genes has its own story to tell, and similar accounts could be made for the mutant analysis of many other biological processes. The *patched* gene will be referred to in subsequent chapters, because most of the tools of genetic analysis were used in studying it.

References

Hooper, J. E., and M. P. Scott. 1989. The Drosophila *patched* gene encodes a putative membrane protein required for segmental patterning. *Cell* 59: 751–65.

Jurgens, G., E. Wieschaus, C. Nusslein-Volhard, and H. Kluding. 1984. Mutations affecting the pattern of the larval cuticle in *Drosophila* melanogaster. *Wilhelm Roux's Archives of Developmental Biology* 193: 283–95.

Nusslein-Volhard, C., and E. Wieschaus. 1980. Mutations affecting segment number and polarity in *Drosophila*. *Nature* 287(5785): 795–801.

Nusslein-Volhard, C., E. Wieschaus, and H. Kluding. 1984. Mutations affecting the pattern of the larval cuticle in *Drosophila* melanogaster. *Wilhelm Roux's Archives of Developmental Biology* 193: 267–82.

Phillips, R. C., I. J. Roberts, P. W. Ingham, and R. S. Whittle. 1990. The *Drosophila* segment polarity gene *patched* is involved in a position-signalling mechanism in imaginal discs. *Development* 110: 105–14.

Wieschaus, E., C. Nusslein-Volhard, and G. Jurgens. 1984. Mutations affecting the pattern of the larval cuticle in *Drosophila* melanogaster. *Wilhelm Roux's Archives of Developmental Biology* 193: 296–307.

5

CHAPTER 5

Connecting phenotypes with DNA sequences

TOPIC SUMMARY

Once mutations have been identified and classified in a gene that affects a biological process, one of the next steps is to connect the mutant phenotype to the corresponding DNA sequence. This process of connecting phenotypes to DNA sequences is colloquially known as "cloning the gene." Every gene that has been cloned has its own story, particularly from the era before genome sequences were widely available. For many genes, the initial approach to cloning was based either on its map position or on its expression pattern. These approaches usually find several candidate genes. Once candidate genes are identified by these approaches, the causative gene for the phenotype can be confirmed by complementation tests or by sequence analysis. As genomic sequencing has progressed, it is now possible to identify the DNA sequence corresponding to a mutant or variant phenotype directly, by sequencing the genomes or the exons of affected individuals. This has been particularly helpful in identifying the causative genes for some rare genetic conditions in humans.

INTRODUCTION

A mutant phenotype cries out for an explanation. How could a fly have legs where its antennae should be? Why does the worm move like a corkscrew rather than according to its usual smooth pattern? What is the explanation for the child with a life-threatening but rare syndrome that occurs in this family? What is the normal function encoded by that gene, and how has the mutation altered it? In order for us to answer these questions and many others, the gene must be cloned—that is, the mutant defect seen in the cell or organism must be connected to the change in a specific gene and to the product of that gene.

One of the powerful features of genetics is the possibility of making a connection between a change in the phenotype and a change in the DNA sequence. This chapter describes some of the ways to find the DNA sequence that corresponds to the gene affecting a variant phenotype. We will see how some of the properties of genes discussed in Chapter 1 and discovered by those geneticists who were working before we even knew the molecular composition of genes have been used to identify a gene's DNA sequence. Many different strategies to clone a gene have been tried and shown to work. Our aim is to explain a few of them in the context of these known properties of genes.

To date, most of our emphasis in genetic analysis—and in this book—has been on the mutants and their phenotypes, and genetics was a lively intellectual field for decades before anyone knew that genes were composed of DNA sequences. But the remarkable reality is that the detailed insights provided by mutants are also explained by the properties and actions of DNA sequences, and that neither mutants nor DNA sequences can be fully analysed unless both are available and studied together. In this chapter we discuss some of the strategies that have been used to connect mutant phenotypes to DNA sequences, primarily in the era before genome sequences were available, but also in the current era. A mutant phenotype can demonstrate what process the underlying gene affects; but it is essential to clone a gene in order to know how it functions at the molecular and biochemical level in the process altered in the mutant.

5.1 Connecting mutants to DNA sequences: An overview

The field of genetics combines several very different concepts under one scientific heading, and touches on or contributes to nearly every other field of biology. In each of the sub-disciplines of genetics, the term "gene" has a slightly different meaning, as summarized in Figure 5.1. These different meanings are part of the challenge of defining a gene with an inclusive and relatively simply description, as discussed in Chapter 1. For example, in Mendelian genetics (also known as classical genetics or transmission genetics), where mutant phenotypes predominate, the gene is often represented by a symbol such as *A* or w^+ and is defined by its pattern of inheritance in a family over a few generations with known parents. Included under this heading are the principles of linkage, allelism, dominance, and so on. If the term "genetics" brings to mind Punnett squares or

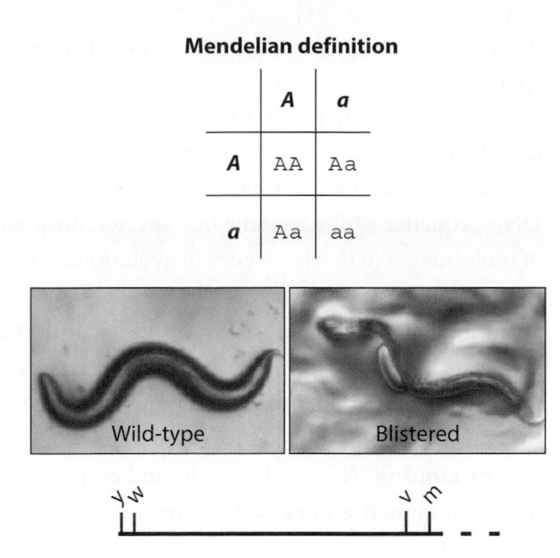

Population definition

$$p^2 + 2pq + q^2 = 1$$

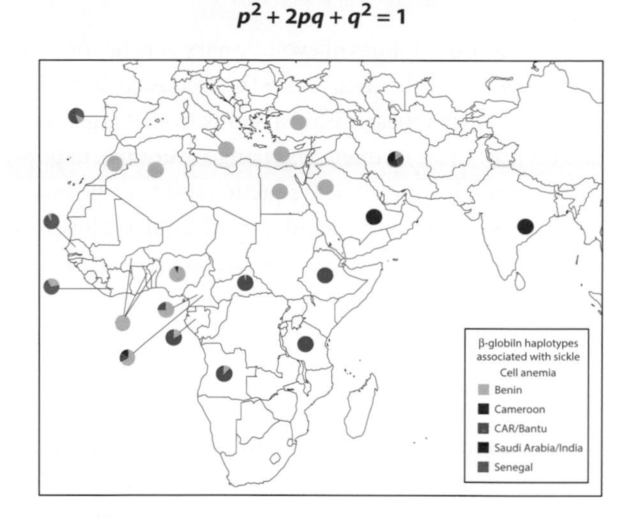

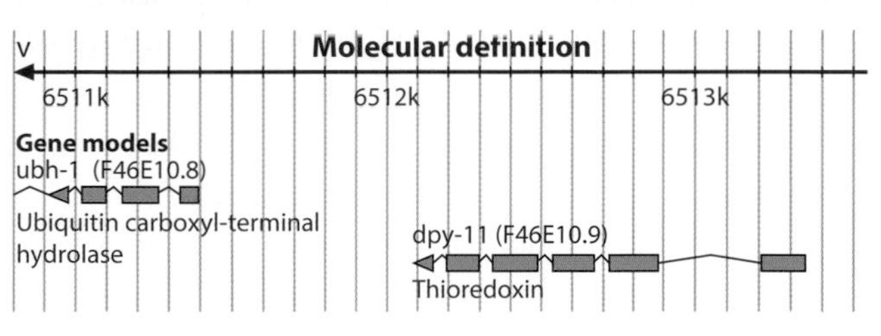


```
>F46E10.9.1 unspliced + UTR
aATGCTGCTCCGATTGCTCGCCGTGCTCGGACTTTTCGCCGTTCCAGTCTCGGGGGGACCTACCAGATCTTCCAAGCTCGTCTTTCTCAATGAGGAAAACTGGACAGATCTTATGAAGG
GAGAGTGGATGATTGAATTgtaagttgtttcattgaaagctccaatttggaacaatgtatatagtttgtacttggtagaattaactactaaggttctcaagcatacactttaaaaacat
tttaattagttacaaaactacaaaattgataactgataaatttttaaaaatccgccgagatgaaaatctcaattaaattccatcttactagtacctatgatatgaaaagatatagagta
gaatcttttcataatttgaattcactgttggttattagaatctttctaaatagctatagaagagttaggtctagatcatttgaaaaaaataacttttaaaattccagCCATGCTCCATG
GTGCCCAGCTTGCAAGGATCTTCAAAAAGCATGGAACGCATTCGCCGACTGGTCTGATGATCTTGGAATCAAGGTTGGAGAAGTCGATGTTACCGTCAATCCAGGACTTTCCGGTAGAT
TCCTTGTCAGAGCTCTCCCAACAATTTATCAgtgagttttgaaagctatttatttctacaaaaatcgaacgtttttttttcagCGTCAAGGATGGTGTCTTCCGTCAATATTCTGGA
GCTCGTGACAAGAATGACTTCATTTCATTCGTCGAGGATAAGAAATACCGTGTGATCGACCCAGTTCCAGACTACAAGCATCCAAACTCTAAGCAgtacgctttttgtggcacagcaa
aaataatattataatttttttcagAATGGCCGTCGTTGCCGTCTTCTTCAAACTTTCGATCTCGGTTCGTCACTTGCACAACCATCTCGTCGAAGATAACCCAATCCCATCATGGGCCA
GCTACGGACTCTTCGCCGGAGTTACTCTTGCTCTCGGATGTCTTCTTGGATTCgtgagtttttaaatagaaatttctgtaaacatcgtgatttccagTTCATCGTTATCATCATCGAC
CAAGTATTCCCTACTGGACCAAGAAAAAGCCAACAAGCCAAGAAAACCGAGAAGAAAGACGCAAACAAAGgtactttaaaatattcggattctgatttctaaaaatttatttattttca
gATTCTGGAACTGAATCTCCGACAAAGAAAAATGGAAATAACAATAATGGCAAAGAAACGAAGAAGACCAAGTAAattctatgcaacttccaaaaaccccaattgtttaccttttgtg
catcttgttattttatttcaaattataatttaaaaacttcaaaatgtttcatatttatcccaaaccagcagtctctgccccaagttttgaattttaaaaaatttgttaattataagta
ttgccgtcccccaagatatatataagtcacttcttcctatgggttcccgtctttttttgtctcttttgttttcaccaatgcttctctccttctgcgcatttcatttcggactctgcaacg
acgacacgtaattaaagcgacatctcttgctgcgagagccagtcggtgctggtcgagtgtgtgttcgctttggcgtcgtcacgccttctttccctgctgcttttcgctcctcattcaat
gcattgatccgtttgtccgtttttttgtcctctgctttctctcaaatatttcacttttttacgaaaaaagttgttaataaatcataattcaattaaaaccttttttgataattcagaaat
atattagcgcattgg

>conceptual translation
MLLRLLAVLGLFAVGVSGGPTRSSKLVFLNEENWTDLMKGEWMIEFHAPWCPACKDLQKAWNAFADWSDDLGIKVGEVDVTVNPGL
SGRFLVTALPTIYHVKDGVFRQYSGARDKNDFISFVEDKKYRVIDPVPDYKHPNSKQMAVVAVFFKLSMSVRDLHNHLVEDKGIPS
WASYGLFAGVTLALGCVLGFFIVIIIDQVFPTGPRKSQQAKKTEKKDAKKDSGTESPTKNGNNNNGKETKKTK*
```

Figure 5.1 The gene, as defined by different approaches. The gene, defined as the Mendelian unit of inheritance, involves wild-type and mutant phenotypes, progeny ratios as predicted from Punnett squares, and linkage and map distances. The gene defined as the segregating unit in populations involves allele and genotype frequencies, as represented by the equation for genotype frequencies at Hardy Weinberg equilibrium and the map of haplotype frequencies. The gene defined in molecular biology includes a set of exons and introns, a DNA sequence, and an inferred RNA and amino acid sequence. The map of haplotype frequencies for sickle cell in Africa is taken from A. Gabriel and J. Przybylski, 2010, Sickle-cell anemia: A look at global haplotype distribution, *Nature Education* 3(3): 2 and used with permission.

linkage maps, the gene is being defined by Mendelian genetics.

In the sub-disciplines of evolutionary genetics or population genetics, the focus is on the frequency at which different alleles appear within a population of organisms and on how these frequencies change over time. In population genetics, the allele frequencies of two alleles are generally symbolized by p and q. In addition to allele frequencies, this field also considers the concepts of mating patterns, selective advantages, migration, genetic drift, and so on. If the term "genetics" recalls Hardy–Weinberg equilibrium, haplotypes, or natural selection, the gene is being defined by population or evolutionary genetics.

Both Mendelian genetics and population genetics were intellectually rich fields of science before Watson and Crick proposed the structure of DNA, and even before DNA had been conclusively shown to be the biochemical constituent underlying the abstract concept of gene. In fact, the "evolutionary synthesis" of the 1930s brought together Mendelian genetics and population genetics to explain the processes of natural selection, drift, and so on in ways that are still used, without any knowledge that the genes under discussion consisted of a DNA sequence. There may be a tendency among some current scientists to overlook the insights and contributions from these pre-molecular fields, but such an attitude would miss much of the beauty and power of genetics.

We can also talk about molecular genetics, in which the gene is defined by its DNA sequence and its RNA and protein products. Included in this sub-discipline are the concepts of cloned gene, gene expression, transcription, translation, replication, and so on. No reader of this book will need to be persuaded that molecular genetics is also a fantastically rich field of science.

However, the great attraction and power of genetics is that all three of these seemingly different views of genes can be integrated. In fact, not only can they be integrated, but they *must* be integrated if we are to understand how we can use genetics to fully analyse a biological problem. An allele can be described and analysed in terms of an eye color variation (a Mendelian phenotype), its frequency in the population (an evolutionary concept), or a DNA sequence (a molecular concept). All these concepts are relevant to our understanding of genetics. It has not always been easy, either for geneticists or for students, to make the intellectual connections between these different aspects of a gene, as our discussion in Chapter 1 illustrated.

We will set aside making the connections to evolutionary and population genetics in this book, at least for the most part. However, it should be recognized that the DNA sequence of the genome of a species, the epitome of molecular genetics, is a record of evolutionary history, so population and evolutionary principles can never be far away. But the topic of this chapter is connecting a mutant phenotype, found in genetic screens such as those described in Chapter 4, with the DNA sequence of the corresponding gene—a process that has been known as **gene cloning**. This is a big topic, and every gene that has been cloned comes with its own history of false leads and unexpected results. No molecular geneticist of a certain age will forget the thrill (and possibly the frustration) of the first gene he or she cloned. They can probably tell you where they were when they read about the cloning of the human globin gene, the cystic fibrosis gene, or the *Drosophila* homeotic genes. Most could tell you about a time when they scratched their head and wondered how some particular DNA sequence (the molecular definition of the gene) could explain some particular mutant phenotype (the classical, Mendelian definition of the gene). One of the most intellectually exciting parts of genetics is bridging this gap between mutant phenotype and DNA sequence.

If each gene has its own cloning story, how can the anthology of all cloned genes be compressed into one chapter? We will take the following approach. Regardless of their molecular composition—their DNA sequences—genes were known, from studies of classical and population genetics, to have certain characteristic and defining properties. As we note in Chapter 1, the analysis of genomes has added to and modified some of these properties, but few of them have changed in fundamental ways. These characteristic properties—such as having different alleles, mapping to a particular location on a chromosome, affecting different tissues, and so on—all can be explained in terms of the molecular biology of the gene. However, these classical properties can also be used as the basis for actually identifying the DNA sequence that corresponds to the gene—that is, for cloning the gene. While many different classical properties have been used as the starting point for cloning a gene, the map position and the expression pattern came to dominate most approaches.

An investigator who wants to clone a gene for further study is faced with at least two different but overlapping challenges, as summarized in Figure 5.2. The first of these is to collect candidate genes, DNA sequences that might correspond to the gene of interest. These candidates are, typically, those to map to a particular location on the chromosome, or those that are expressed in the expected tissues and times. The candidate genes that meet both criteria of map position and expression are studied further. It is possible (although somewhat unlikely) that only a single candidate gene will be found by mapping or expression pattern, however. Thus, other properties of the gene are used to classify or sort through the candidate genes in order to find the one gene of interest, that is, the exact DNA sequence that explains all every other aspect of the gene. The most widely employed of these other properties is **mutation** or allelism, which connects a specific set of alterations in the DNA sequence with a particular change in the phenotype. For some experimental organisms, **complementation**, another property that defines a gene and is discussed in Chapter 4, can also be used to determine which one among the candidates is the actual gene of interest.

This chapter does not seek to provide a comprehensive list of the different ways in which genes have been cloned, nor do we offer an instruction manual that will allow a person to clone a particular gene. In fact some of the cleverest methods used to clone a gene are not included here simply because greater ingenuity was needed once the more general methods we discuss proved to be inadequate. Rather the chapter is an effort to show how all these different properties of genes have provided the insights needed for connecting genes to DNA sequences. Our case study of the human gene implicated in cystic fibrosis, the *CFTR* gene, should be read as an example of how all these different properties were used as experimental methods to ensure that the investigators had found the DNA sequence that could explain all the genetic and molecular properties of cystic fibrosis. It also compares cloning the gene in the pre-genomic era with cloning a similar gene now. We begin our exploration of how genes are cloned by discussing the method most widely used in this era of genomics, namely positional cloning.

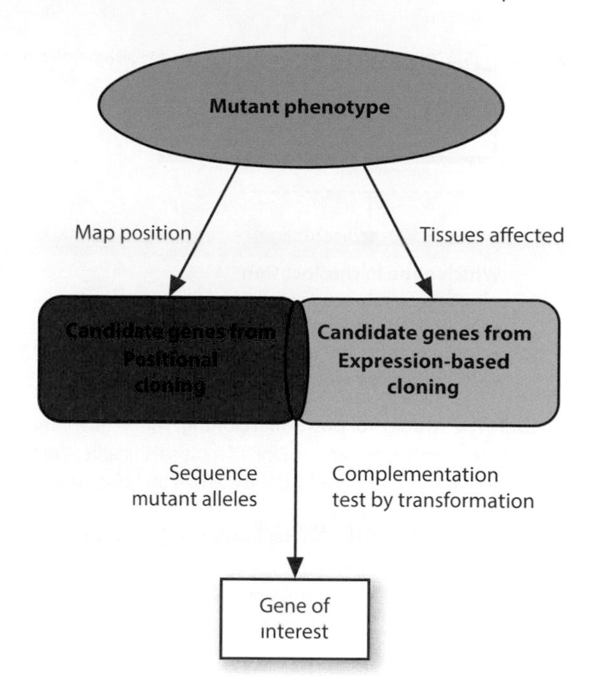

Figure 5.2 Approaches to connecting a mutant phenotype to a DNA sequence. For genes originally defined by a mutant phenotype, the most common approach to finding the corresponding DNA sequence is to rely on the map position, an approach known as positional cloning. Genes known primarily by their pattern of transcription are cloned on the basis of their expression. Both approaches yield many candidate genes, and even combining the information from position and expression might not pinpoint an individual gene. The candidate genes can be narrowed to the single gene of interest by comparing the DNA sequences of different alleles or, in model organisms, by performing a complementation test known as transformation rescue.

5.2 Identifying candidate genes: Map position

A gene maps to a particular locus on a chromosome, and the terms "locus" and "gene" are often treated as synonyms. For example, the *white* gene of *Drosophila melanogaster* maps to the X chromosome at position 1.5, or 1.5 map units from the reference point at the left tip of the chromosome. The fact that a gene can be mapped to a particular locus is the principle that defines the strategy known as **positional cloning**.

Genes were first mapped to a specific location on a chromosome in the early days of *Drosophila* genetics,

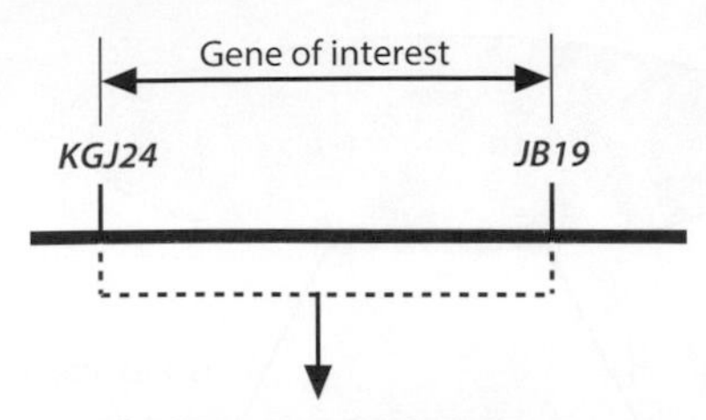

Which gene in this location
- Is expressed in the expected pattern?
- Shows DNA sequence variation?
- Complements the mutant phenotype?
- Has homology with genes with similar effects?

Figure 5.3 The strategy of positional cloning. The gene of interest is mapped between two previously cloned molecular markers, in this hypothetical example, the polymorphisms *KGJ24* and *JB19*. The DNA sequence from the region between the two cloned markers are then examined for genes expressed in the expected time and tissue, for sequence variation between wild-type and mutant alleles, for possible complementation with the wild-type sequence when possible, and for homology with other genes with similar effects.

and genetic maps have been the atlas for geneticists ever since. It is not too surprising, then, that one of the properties most widely used for cloning a gene has been its map position. The underlying logic is simple, and is summarized in Figure 5.3. The gene of interest has been shown (by mapping methods described in Chapter 4) to lie in the genetic interval between two loci, both of which have been placed on the DNA sequence. Therefore, if the region between two loci can also be cloned, the cloned sequence must encode the gene of interest. Positional cloning has been a major impetus for genome-sequencing projects. Indeed, for the early genome projects, whether for bacteria, yeast, or worms, the progression from genetic map to genome project was more gradual than abrupt, making it difficult to determine exactly when the genetic map became the genomic project. Let's examine the two primary requirements for initiating positional cloning: locating a gene between two molecularly defined markers and being able to clone the region between the two markers.

Locating a gene with respect to cloned markers and genes

We will start by discussing the types of molecularly defined flanking markers and the process for recombination mapping. Remember that, for a marker to be used in recombination mapping, it must exist in two different alleles or polymorphisms in the mapping population. The genetic markers that flank that position could be other cloned genes with characterized mutant phenotypes. For many inbred laboratory organisms, such as yeast, flies, worms, or *Arabidopsis*, these are often the most convenient flanking markers to use, especially as more genes have been mapped and cloned. But genes with mutant phenotypes are not the only flanking markers that can be used, and in fact they often are not the best flanking markers to use in genetic mapping.

Regardless of the types of markers used, the mapping of a gene to the region between two cloned markers is done by conventional recombination mapping. The primary goal of this stage of positional cloning is to find cloned markers at the margins of the flanking region as close to each other as possible, so that fewer candidate genes need to be examined. On the other hand, it is more challenging to find closely linked markers and to map genes in relation to them, because recombination events are less frequent within short genetic distances. Much of the effort in genetic mapping has been directed at increasing the resolution of maps, that is, at finding cloned markers separated by the smallest realistic intervals.

An insight that was crucial for efforts in positional cloning was that the mapping markers do not need to be genes. Any type of molecular polymorphism—for example a restriction fragment length polymorphism (RFLP), a single nucleotide polymorphism (SNP), a sequence length polymorphism such as a copy number variation (CNV), or a short insertion or deletion (indel)—works as well as a cloned gene for this purpose.

The procedure for mapping using morphological markers and molecular markers is very similar, although the order of the assays' steps is reversed, as summarized in Figure 5.4. With morphological markers (Figure 5.4, Panel A), the recombinants in the interval are found first, and the phenotype of the gene of interest is then assayed in those recombinants. With molecular markers, the phenotype of the gene of interest is found first, and the recombinants are then assayed. But the interpretation is the same and the position of the gene of interest is determined with respect to known molecular variation.

Cloning the region between two molecular markers

The second of the two requirements for positional cloning is the ability to clone the region between the cloned markers. Cloning the region of interest used to be a laborious process involving chromosome walking in a

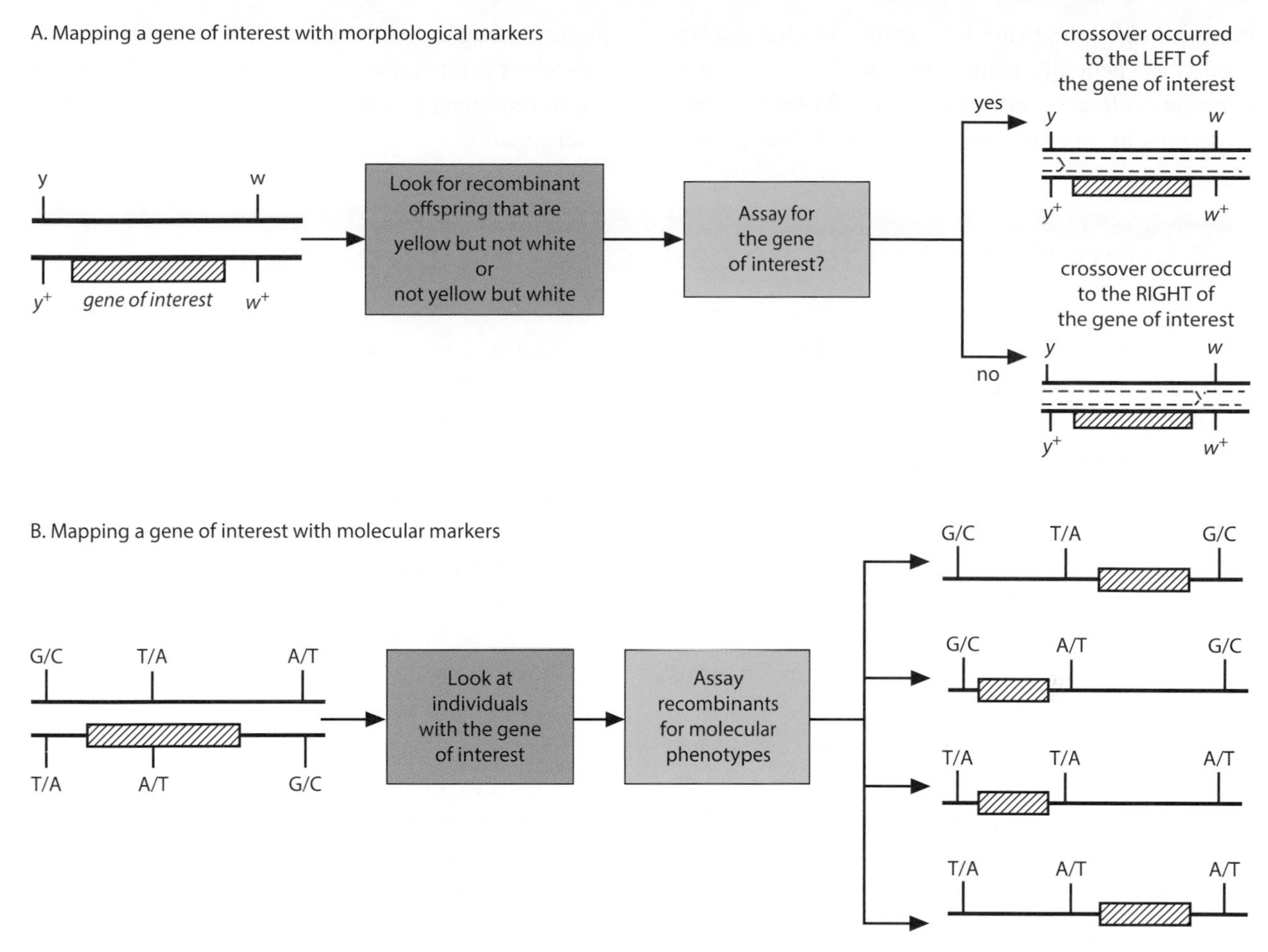

Figure 5.4 **Using molecular polymorphisms as markers for mapping.** Mapping a gene of interest with molecular markers is analogous to mapping with morphological markers such as yellow body and white eyes, although the order of steps is usually reversed. With morphological markers in Panel A, recombinant offspring that have one of the two mapping markers are recovered first and then scored for the gene of interest using the mutant phenotype. With molecular markers such as single nucleotide polymorphisms (in Panel B), individuals within a population with the gene or phenotype of interest (such as a genetic disease) are scored for their flanking markers by molecular assays such as DNA sequencing or restriction digests; thus the phenotype of interest is determined first and the recombinants are scored next. In either case, the location of the gene and crossovers in the region of the genome are determined.

library—a process of using a portion of one cloned fragment as a probe to find overlapping clones, which are then used to find additional overlapping clones. This was a significant focus of the papers describing the cloning of *CFTR*, which is summarized in the case study. Fortunately genome projects now automatically assemble overlapping clones for the entire genome of a species without regard for the genes themselves, so the sequence between the two makers can usually be obtained by downloading the genome onto a computer. No one who has done it longs for the "good old days" of chromosome walking.

Other strategies for obtaining the DNA sequence for part of the chromosome were also used occasionally before genome assembly and sequencing projects were developed. Among these other position-based strategies were the microdissection of *Drosophila* polytene chromosomes and the flow sorting of mammalian chromosomes. In all cases, the underlying goal was to place the gene at a particular location in the genome and to isolate the DNA from that location.

Genome projects have made positional cloning the most widely used strategy for cloning a gene in nearly all well-studied organisms. With genomes that are assembled, sequenced, and annotated, every gene has already been cloned, so the role of the investigator is to find the right one among the many candidates. Genome databases are then used to identify candidate genes in the region, and strategies using the other properties of genes help to determine which candidate gene is the gene of greatest interest. The widely used **genome-wide association studies (GWAS)** for complex traits in humans are in effect positional cloning projects using known molecular markers; the individuals with the phenotype of interest are found first, and the recombinant variants in each individual are then determined.

5.3 Identifying candidate genes: Expression pattern

The second most commonly used method to identify the underlying DNA sequence for an observed phenotype is to use the gene's expression pattern. Either the expression of RNA or that of the protein could be used for the cloning process, but RNA-based cloning is much more common.

Genes can be cloned based on their expression patterns

With what we know about mutant phenotypes, it seems almost trivial to be reminded that a gene has a particular expression pattern. A mutant phenotype can suggest where the gene is likely to be expressed. For example, while phenotypic descriptions of mutations in the *white* gene in *Drosophila* usually focus on eye pigmentation, these mutations may also affect other pigmented structures in the fly, such as the Malpighian tubules. Therefore, whatever the function of the *white* gene, it must at least be expressed in the eye and in the Malpighian tubules. Similarly, it seems reasonable to assume that genes involved in cancer-causing phenotypes will be expressed differently in normal and in tumor cells. The pattern or level of gene expression has provided many different cloning strategies, some of them quite elaborate. These approaches are collectively termed **expression-based cloning**. While many different expression-based cloning strategies have been successful, nearly all current expression-based cloning strategies are based on the pattern of transcription.

RNA-based expression cloning

The fundamental and original procedure for expression cloning using RNA, summarized in Figure 5.5, is to extract total mRNA from a tissue type or developmental stage that is thought to transcribe the gene. The RNA is then reverse-transcribed into cDNA, which is more stable and much easier to work with than RNA. In the past, and still to some extent now, the cDNA from a particular developmental stage or tissue can be used to make a cDNA library that contains all (or nearly all) of the transcripts from that sample. The cDNA library provides many candidate genes that are expressed at that time or in that tissue, so additional screening methods were required to find the particular gene of interest.

More contemporary genomic approaches to expression-based cloning come from the use of microarrays or from RNA-seq. In a two-channel microarray such as the one shown in Figure 5.6, gene-specific oligonucleotides corresponding to tens of thousands of genes are spotted robotically onto a support (such as silicon or glass slide) in a specific pattern, such that each location in the array corresponds to a defined gene. RNA is isolated from one sample of interest, reverse-transcribed, and fluorescently labeled to make labeled cDNA (or cRNA—complementary RNA), which is hybridized to the array. The fluorescence pattern is scanned, and the intensity at each spot is converted to a numerical expression value. In the case of two-channel arrays, two different samples are co-hybridized to the same array. The expression from the two samples is distinguished by labeling them with different fluorophores—Cy3 or Cy5. The results from the different samples are compared to identify differentially expressed genes, with particular interest in transcripts found at different levels in the two samples, as seen in Figure 5.6. Since the gene represented by the oligonucleotide feature at each location is known, the gene whose expression has changed can be cloned. With RNA-seq, RNA is extracted, reverse-transcribed into cDNA, and the cDNA is then sequenced. The sequences are then computationally mapped onto the sequence of the genome to identify genes that are differentially transcribed; no additional cloning is needed.

Expression-based cloning approaches will usually identify a large number of candidate genes—for example, genes expressed differently in tumor cells and normal cells of the same type. Often there are too many

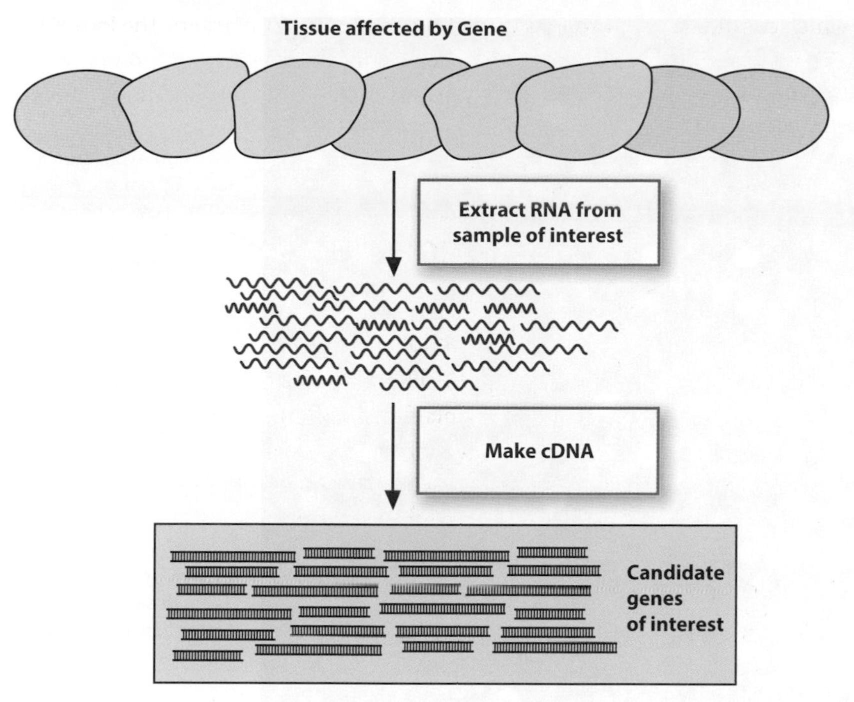

Figure 5.5 **RNA-based cloning methods.** The basis of most expression-based cloning is to extract RNA from a sample in which the gene of interest is thought to be transcribed. The RNA is reverse-transcribed to make a cDNA library or to be labeled for further experiments. Many candidate genes are typically found with transcription patterns similar to those of the gene of interest.

expression differences for it to be possible to clone one individual gene of interest, so follow-up approaches based on map position, sequence variation, or sequence similarity to other genes will be used to limit the candidate pool to just a few genes. Thus gene cloning based on transcriptional profile is among the best initial strategies for compiling a large collection of candidate genes and provides the material for follow-up experiments that use other properties of genes.

An intermediate method: Homology-based cloning

When cloning and the molecular analysis of genes were still novel, many investigators used the inferred **gene homology**—or, more specifically, the nucleotide sequence similarity of genes from a related species—in order to clone genes from another species. For example, when the tubulin gene was cloned from one organism, tubulin genes could be cloned from other organisms on the basis of the nucleotide sequence hybridization between the cloned gene and the other species. This method is no longer widely used, mostly because direct genome sequencing is much faster and accurate. However, homology-based information was helpful in identifying the best candidate for CF, as discussed in the case study. In addition, once the *patched* gene was cloned in *Drosophila*, as we shall see in what follows, homology-based cloning was used to clone the *patched* gene from other insects and eventually from other animals including mammals.

5.4 Evaluating the candidate genes: Complementation and mutation

Both positional cloning and expression-based cloning generate a number of candidate genes that fit the initial criterion of either mapping at the right location on the chromosome or being expressed in the expected pattern. Applying the two criteria together can often limit the number of candidate genes further, and may yield

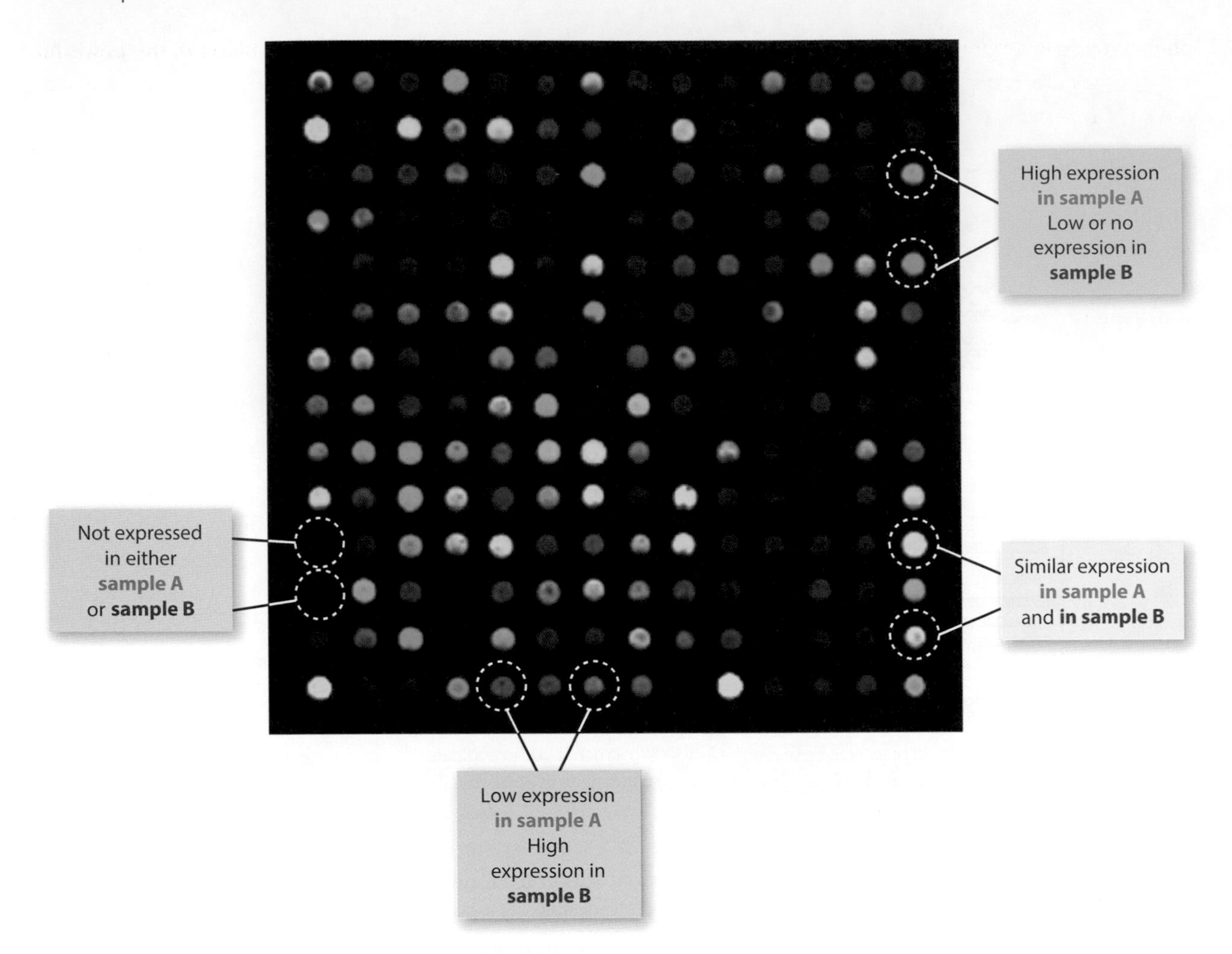

Figure 5.6 Microarrays and expression-based cloning. In a microarray, sequence probes specific to different genes are fixed to a support in a defined array. This array shows 14 × 14 or 196 different features, but hundreds of thousands of features can be loaded onto an individual microarray chip. For a two-channel array such as the one shown here, RNA from sample A has been made into cDNA and labeled in green. RNA from sample B has been made into cDNA and labeled in red. The cDNAs are hybridized to the array. Genes expressed at high levels in one sample but at low levels in the other sample will appear as spots of one color, as indicated here. Genes expressed in both samples at similar levels will appear as yellow spots, and genes not expressed in these samples are black. Since genes occupy known positions on the array, genes expressed in one or the other sample can be readily identified and cloned.

only a single candidate gene of interest. However, even if only a single candidate gene is found, proof that this is the proper gene comes from either or both of two properties of genes: complementation and mutation, as summarized in Figure 5.2. Although mutation or, more generally, sequence variation among the alleles is applicable in more situations and more organisms than is complementation, complementation tends to be more definitive, so we discuss it first.

Complementation testing or transformation rescue can be used in model organisms to identify the best candidate gene

As described in Section 4.5, the functional definition of a gene is the complementation test: two recessive mutations are said to be in different genes if they can complement each other, restoring the wild-type phenotype, and are said to be in the same gene if they cannot complement each other and fail to restore wild-type function to the organism.

Functional complementation can form the basis for identifying which candidate cloned gene corresponds to the one responsible for the mutant phenotype. This functional complementation is often called **transformation rescue**, a name that describes well the nature of the test. That is, if the dominant *wild-type* allele of a gene is introduced into a recessive *mutant* organism that lacks that function, the wild-type copy will provide the normal function lacking in the mutant and will restore it to the wild-type growth, movement, color, or other phenotype. In laboratory jargon, the wild-type allele has *rescued* the mutant phenotype.

While it is most useful to confirm that a candidate sequence is the correct gene, transformation rescue or functional complementation has also been used as the initial strategy for cloning a gene, particularly in an organism such for auxotrophic mutations in the yeast *Saccharomyces cerevisiae*. Recall from Section 4.3 that wild-type alleles can be introduced on a plasmid in yeast. Suppose that the mutant strain is a uracil auxotroph and cannot grow in the absence of additional uracil. A library of plasmids containing fragments of the yeast genome is transformed into the mutant cells, which are then plated onto media lacking uracil. The only cells that can grow to produce colonies are those in which the wild-type gene has been introduced and where it is expressed to make a functional product. Although only a small number of cells acquire the appropriate plasmid, the powerful selection for growth in the absence of uracil ensures that many plasmids can be quickly assayed to identify the correct gene.

This functional complementation strategy has not been widely used as an initial strategy in multicellular organisms. However, transformation rescue has been used frequently to narrow down a list of candidate genes and to *confirm* that one has cloned the correct gene, as summarized in Figure 5.7. For example, imagine that a positional cloning strategy has yielded a collection of four cosmids, each about 40 kb in size, at least one of which must contain the gene. Each cosmid can be introduced into the mutant organism and the transformants tested for rescue of the mutant phenotype. Once these are found, subcloned regions of the cosmid can be used to find the smallest rescuing

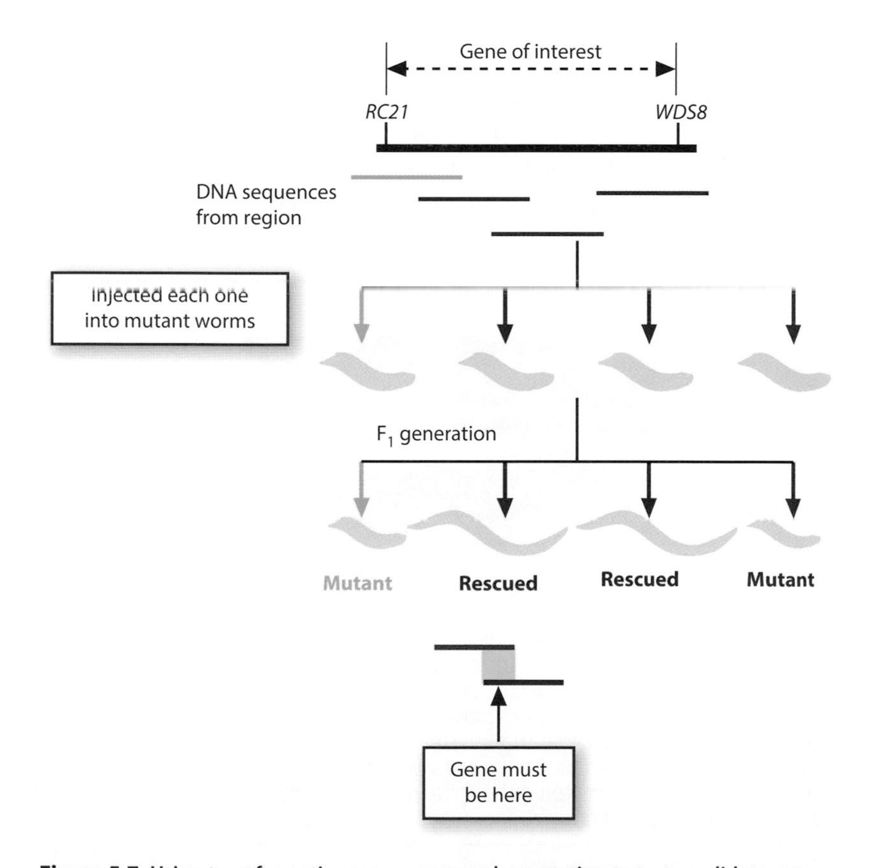

Figure 5.7 Using transformation rescue or complementation to test candidate genes. In this hypothetical example from *C. elegans*, a mutant that results in a short fat phenotype has been mapped between the polymorphisms *RC21* and *WDS8*, a region that is spanned by four overlapping sequences (such as cosmids) shown here in different colors. DNA from each region is microinjected into a mutant worm. Since DNA from both the red and the blue regions rescue or complement the mutant defect and give a phenotypically normal worm in the next generation, the wild-type copy of the gene must therefore lie in the region where these sequences overlap.

fragment. The rescuing fragment is now sequenced from both wild-type and mutant organisms, to find the gene with a molecular lesion in the mutant organism. In this way, complementation is being used to reduce candidate genes to a more manageable number than position alone could manage, and the molecular defect in the mutant allele is used after complementation to confirm that the correct gene has been found. Note that, as above, the mutant phenotype must be recessive to wild type for this to work.

It is worth noting that these complementation strategies are all widely employed in bacteria and none was used first on eukaryotes. Whenever convenient methods have been available to screen or select for the transformants, complementation by transformation rescue has provided a very powerful strategy for cloning genes and for confirming that the cloned gene is the one responsible for the mutant phenotype. Investigators rely on this property when using bacterial or yeast plasmids for many other molecular biology experiments, usually without thinking about complementation tests. That may be the highest praise for the strategy of complementation: it works so well that most investigators use it routinely and rarely think about it.

Mutations, or DNA sequence variations, are the most widely used property to confirm the relationship between the cloned gene and the mutant phenotype

As we noted in Chapter 1, a gene exists in multiple different forms, known as alleles or mutations. The positional cloning strategies discussed in Section 5.2 require alleles with different phenotypes, but the molecular basis of these alleles is not important for the cloning strategy. However, because mutations are so specific to an individual gene—that is, a particular base is changed, or a sequence is inserted or deleted at a specific site, and this results in a particular mutant phenotype—mutations are essential for confirming that the investigator has cloned the correct gene.

To use molecular changes between alleles to confirm that a candidate gene is the right one, the candidate gene is first sequenced from the wild-type strain and from several different mutant strains (Figure 5.8). If the correct gene has been cloned, the sequences will show different changes in each of the mutant alleles—missense mutations, nonsense mutations, splicing mutations, deletions, insertions, and so on. For model organisms that are derived from highly inbred strains

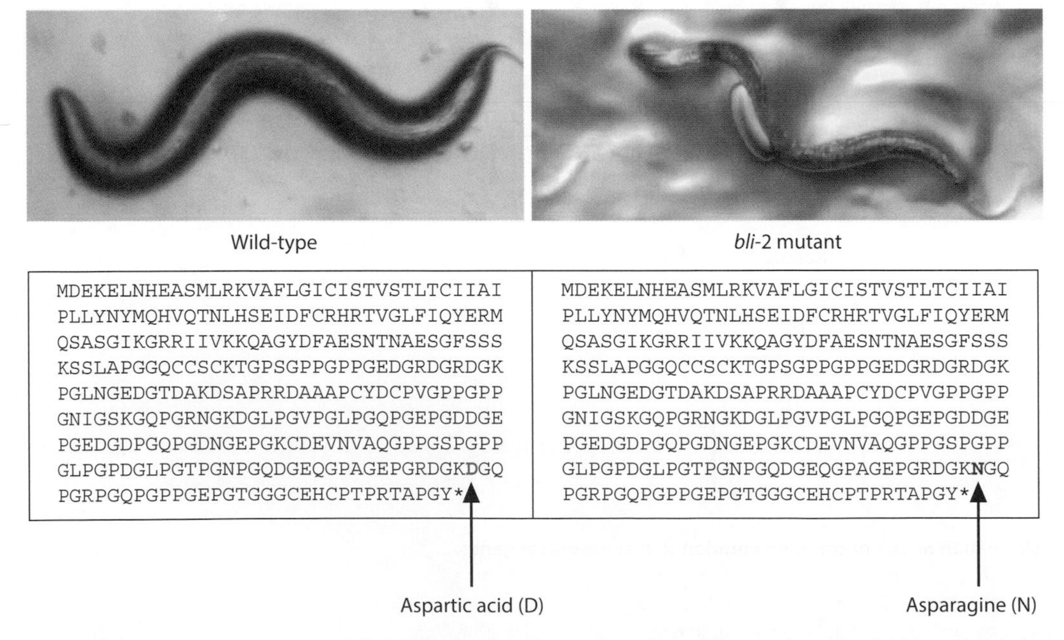

Figure 5.8 Phenotypic variation and mutations. The mutant phenotype of *bli-2*, as shown by the blistering on the cuticle of the worm, arises from an amino acid change in the *bli-2* gene. The wild-type gene has an aspartic acid (D), while the *bli-2* mutant shown has an asparagine (N). The amino acid sequences are taken from Wormbase.org.

such as flies, worms, and *Arabidopsis*, even a single base change in the coding portion of the gene can confirm that the investigator has cloned the correct gene. For organisms that do not have highly inbred laboratory strains, including humans, the investigator needs to be careful to distinguish a mutation that alters the function of a gene from a naturally occurring polymorphism that has little or no functional impact on the gene. For this reason, the case for having cloned the right gene becomes more persuasive with each mutant allele that is sequenced and shown to have a molecular lesion in the gene.

Since every cloned gene has its own history, our description of the cloning strategies has been fairly general. A specific and important example that uses all the principles we have discussed is provided in the case study. Positional cloning, expression pattern, homology to known genes, and sequence variation in affected individuals were all used in cloning *CFTR*, the gene responsible for cystic fibrosis in humans.

5.5 Direct searches for causative mutations: Exome sequencing

For an investigator working with an inbred model organism such as fruit flies, yeast, or worms, there are typically only one or a very few sequence changes between the wild-type organism and the mutant organism isolated by a genetic screen in the laboratory. But this is not the case for humans, because there is no wild-type genome sequence from which all other mutations arise. While there is a reference "human genome" sequence, no individual actually has this sequence, and all of us have hundreds of thousands of sequence changes between our genome and the reference genome.

As DNA sequencing technology became less expensive and faster, it has become possible to find the causative mutation in some laboratory model organisms by directly searching the genomes of mutant individuals for sequence changes. This strategy of direct sequencing to identify the causative mutation (or sequence variation) has also been used to identify the gene for some rare human genetic diseases. The greatest challenge in such studies is not to identify the sequence changes, which is a technical challenge but lies within current capabilities, but, more conceptually, to determine which of the many sequence changes is the one responsible for the genetic syndrome.

To be sure, over the past decade most genes in humans, particularly those involved in complex traits, have been cloned or identified molecularly through GWAS, a strategy based on positional cloning. But GWAS for complex traits relies on common or polymorphic variants in the analysis: rare mutations, which are called "private" (see **private variants**), are very difficult to find through linkage studies alone, since the number of affected individuals is small and the mutant allele is present at very low frequency in the population.

In the United States rare diseases are defined as diseases that affect fewer than 200,000 people; for some genetic disorders, fewer than fifty affected individuals (usually children) have been identified. While any single disease might be classified as "rare," it is estimated that there are more than 7,000 such diseases affecting as many as 25 million people in the United States. Many of these diseases have some genetic component, and a substantial number of them might well be single gene traits. Because the traits are rare and families with more than one affected member are uncommon, the exact mode of inheritance and the genetic contribution to the disorder are often not conclusively known. It is also the case that different disorders may be mistaken for one another, or that genetically related disorders may not be recognized as arising from different alleles of the same gene.

Identifying the altered DNA sequence for a human trait through direct sequencing of the genome of affected individuals faces some challenges that are not encountered in working with model organisms, as summarized by the flowcharts in Figure 5.9. Compared to non-mammalian model organisms, our genomes are larger and have more repeat elements, more numerous and longer introns, and so on. More significantly, much more sequence variation is found among naturally occurring populations. Genomic variation between

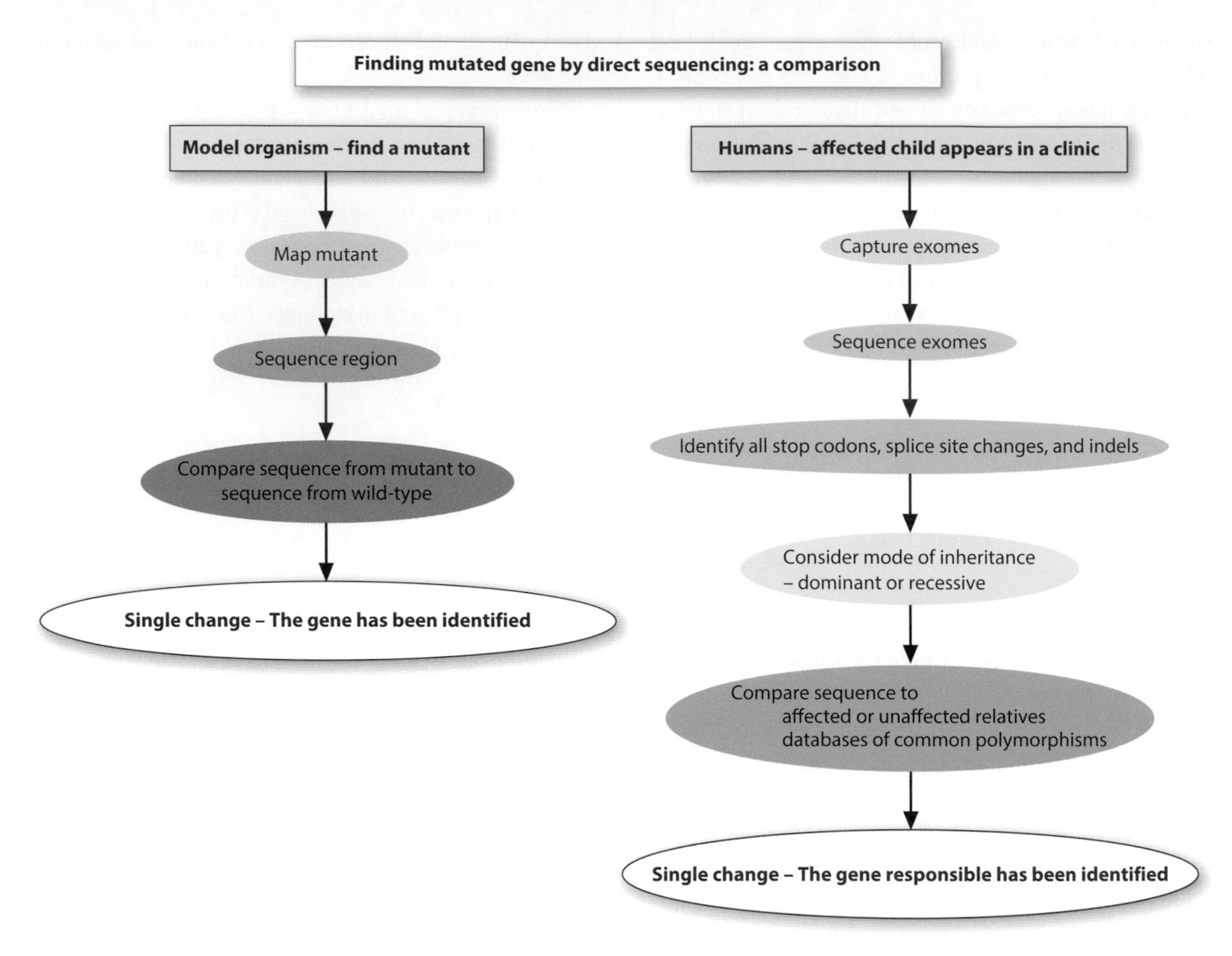

Figure 5.9 A comparison between model organisms and humans for using direct sequencing to identify a causative gene and mutation. Contemporary sequencing methods have made it possible to find mutated genes by directly sequencing the mutant genomes. In model organisms, as shown on the left, a new mutant has typically been mapped, and the sequence at this region of the genome is compared between mutant and wild-type. With direct exome sequencing in humans, all the exons of an affected individual, who is typically a child with a genetic disease, are captured and sequenced. All of the mutations that are likely to cause major changes in the coding capacity of the gene, such as stop codons, are identified. The mode of inheritance of the disease is also considered. These mutations are then filtered by comparing the exon sequences to the databases of known human polymorphisms, which are very unlikely to have any of the disease-related mutations for rare genetic diseases. The exons of other affected individuals are also sequenced to find what mutations are in common among affected individuals but missing in unaffected individuals. This combination of filters often results in one or a few candidate mutations.

unrelated people can be expected at hundreds of thousands or even millions of sites in the genome, most of which have nothing to do with the variation in the trait being analysed. While we have a reference human genome—in fact thousands of individual genomes—none of them could be considered a "wild type" from which other genomes have been derived, as is true in model organisms. Even when the genomes of affected and unaffected individuals can be sequenced, identifying the one causative mutation among all that genetic variation is a formidable task.

Exome sequencing

Although these are enormous challenges to overcome, the analysis of genetic diseases on the basis of direct genome sequencing has progressed rapidly since the first publications appeared in 2009. As the costs of sequencing have decreased, the feasibility of rapid genome sequencing has improved. The initial publications estimated the cost of sequencing at about $50,000 per genome, but further advances in the past few years have reduced that cost to $5,000 or less, or put it below the cost of many other medical procedures in the United

States. Furthermore, improved filtering methods have been developed to identify the best candidate mutations among the vast array of genetic variation.

Initial approaches focused on sequencing the exons of the human genome rather than the entire genome. This is known as **exome sequencing**. Exons comprise about 1 percent of the total genome sequence, so if only exons are sequenced the amount and the complexity of sequence information that is compiled and analysed is greatly reduced. Mutations in exons are likely to be the mutations with the most profound effect on the function of the gene, since they can cause amino acid changes or polypeptide chain termination. On the other hand, by sequencing only exons, any mutations in regulatory regions will be missed; a few published studies have sequenced the complete genomes of affected individuals, primarily for disorders with relatively subtle health effects. Most of the disorders that have been analysed by exome sequencing have severe health consequences, such as childhood mortality, and thus are predicted to have significant alterations in the coding regions of the altered gene. Exome sequencing offers this direct connection between the gene and the syndrome.

The human exome consists of about thirty megabases spread among approximately 180,000 exons, so the first stage in sequencing is to "capture" the exons from among the entire genome of 3 billion base pairs. While different methods have been used, exome capture is usually based on hybridizing fragmented genome DNA to two or more different exon microarrays, specifically constructed to represent all the protein-coding regions in the genome. Hybridizing fragments are extracted, purified, and amplified by PCR for sequencing, each fragment being sequenced repeatedly. With current sequencing methods, more than thirty gigabases (30×10^9) of sequence can be generated from one machine in a ten-day sequencing run. The sequences produced by these methods are very short, typically shorter than 100 bp in length, and are then assembled against the scaffold of the reference human genome.

With this strategy, it is feasible to have thirty to forty-fold average coverage of the exome. While any given exon might be sequenced fewer than thirty times, as few as seven times in some cases, the multiple passes (known as massively parallel sequencing) minimize the impact of sequencing errors that might arise in any single run.

Miller syndrome was among the first rare genetic diseases to be analysed by exome sequencing

To illustrate the process of exome sequencing and its power in identifying the altered gene for rare diseases, let us consider one of the first applications in some detail, and add other examples as illustrative of specific situations.

Literature Link. Ng et al. 2009, Targeted capture and massively parallel sequencing of 12 human exomes, *Nature* 461: 272–6

Ng et al. first did a proof-of-concept study with the rare dominant disorder Freeman–Sheldon syndrome (FSS), which is characterized by hypercontraction of multiple joints. Previous analysis based on positional and expression-based cloning had identified mutations in the myosin heavy chain gene MYH3 as being causative for FSS, so this foreknowledge was used to determine what methods would be needed to find a gene for which no prior information is available. The investigators developed the strategy for exon capture and sequence analysis that has been widely applied in other cases.

Having shown that they could identify the causative mutation of the dominant disorder in FSS by direct sequencing, they then applied their methods to the sequence analysis of Miller syndrome, for which the genetic basis was not known.

Literature Link. Ng et al. 2010, Exome sequencing identifies the cause of a mendelian disorder, *Nature Genetics* 42: 30–5

Miller syndrome is a rare genetic disease characterized by cranio-facial abnormalities. Only about thirty cases are known worldwide, so the exact mode of inheritance is not clear. The inheritance is consistent with the mutation being an autosomal recessive trait, but an autosomal dominant trait with reduced penetrance could not be absolutely ruled out. Thus the analysis had to include both possibilities, which gives insights into the processes needed for other disorders known to be either recessive or dominant. Autosomal recessive disorders are easier to detect because an affected individual has to contain a mutant allele on each homologue; since the mutant alleles are typically not a consequence of the same molecular alteration (except in cases where the parents are descended from a common ancestor, for instance through first-cousin marriages), these individuals are said to be **hetero-allelic** (in classical genetic terms) or **compound heterozygotes** (the term more often used in human genetics), as shown in Figure 5.10. There are only three known families with two affected siblings for Miller syndrome, so these families formed the core group for sequence analysis.

Flow chart diagrams of the filtering process for dominant and recessive inheritance are shown in Figures 5.11 and 5.12. The area of the boxes in these figures is proportional to the number of remaining candidate mutations for the gene as each additional step in the filtering process is taken. Reading the arrows from left to right shows the effects of including additional affected individuals. Reading the arrows pointing downward shows the effects of filtering common polymorphisms. While the numbers are those found for Miller syndrome, they are similar of what has been found in subsequent studies. Since Miller syndrome could be either dominant or recessive, both sets of data are shown.

The first step was to obtain the exome sequences of the two affected individuals in one family; about 164,000 regions were sequenced, representing 27.9 Mbases and about 96 percent of the exons. This amounted to 5.1 gigabases of DNA sequence per individual, in lengths of about seventy-six base pairs, or an average forty-fold coverage of their exons. Many different types of polymorphisms are seen when the two siblings are compared. Because this is a severe disease with profound phenotypic consequences, the authors made the reasonable assumption that the mutation changed the coding function of the gene. Thus mutations that resulted in amino acid substitutions or stop codons (that is, non-synonymous variants, abbreviated NS in the figures), mutations that altered a splice site (SS in the figures), or mutations that inserted or deleted part of the coding region (that is, indels, abbreviated I in the figures), were compared between the siblings.

Now the mode in which the trait is inherited becomes an important part of the analysis. If the disease gene is an autosomal dominant trait, then the siblings only need to share one mutation; in this study each person had 4,680 such single gene variants, of which 3,940 were shared between the two affected siblings, as summarized in Figure 5.11. On the other hand, if the disease is an autosomal recessive trait, then an affected

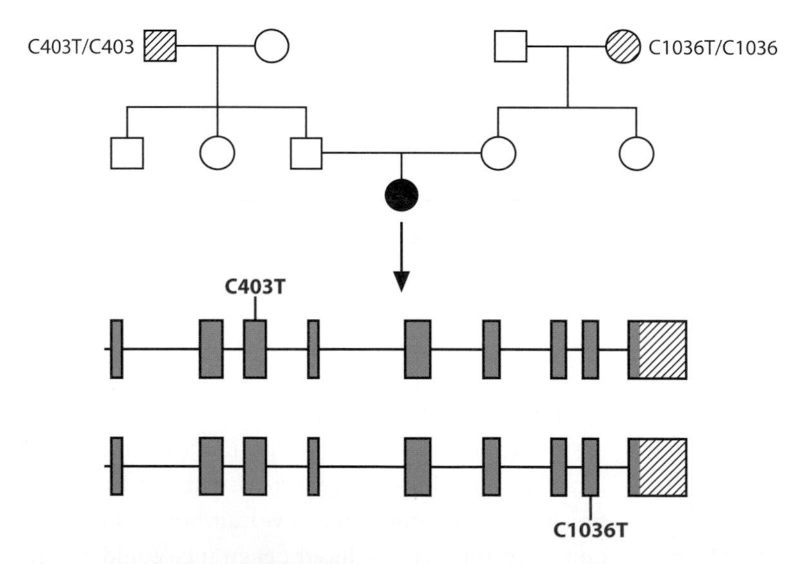

Figure 5.10 A compound heterozygote. For rare recessive genetic traits that arise in the offspring of unrelated individuals, it is unlikely that the two recessive alleles will have the same molecular change. This is known as a compound heterozygote in human genetics, or as heteroallelic in model organisms. The pedigree shows an affected female (the filled circle), for what must be a recessive trait. Two of her grandparents (in this example, her paternal grandfather and her maternal grandmother) must have been heterozygotes in order for her to be affected, but the two grandparents have different molecular lesions in the gene. While the affected female is a homozygote, her two alleles are different molecular events.

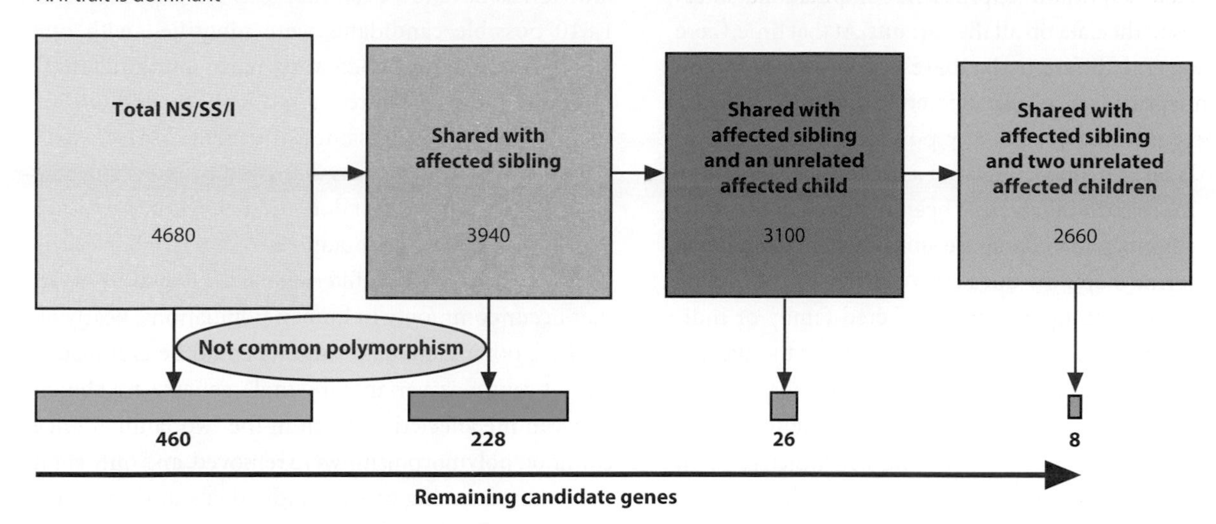

Figure 5.11 A flow chart for filtering the mutations for a dominant disorder. The effects of testing additional individuals are shown horizontally, while the effects of filtering against databases of known polymorphisms are shown vertically. The size of the boxes and the numbers indicate the effectiveness of different filtering methods for candidate mutations for Miller syndrome.

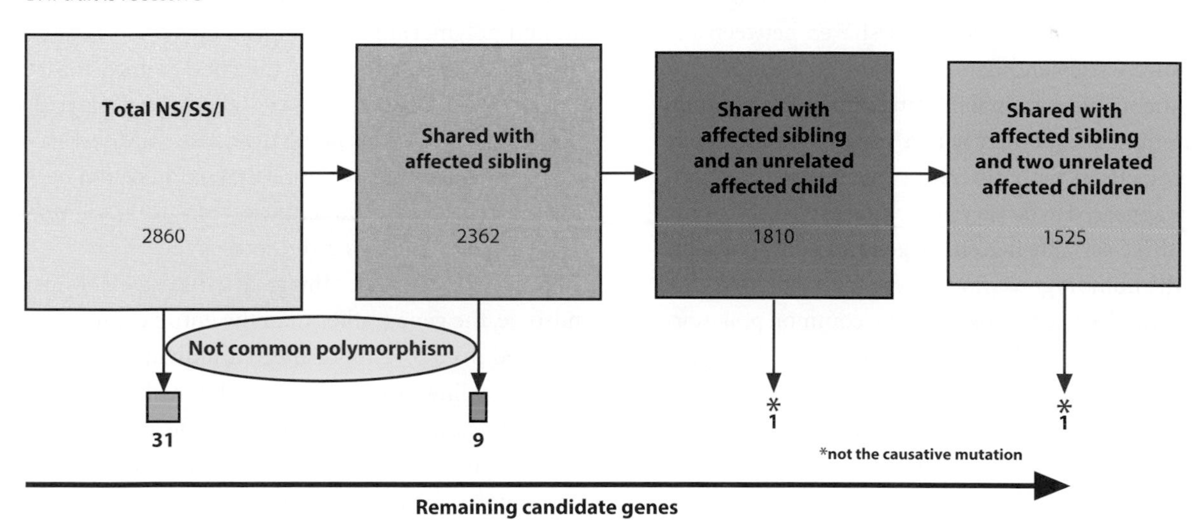

Figure 5.12 A flow chart for filtering the mutations for a recessive disorder. The organization and information are similar to those in Figure 5.11, except that the number and size of boxes indicate the number of remaining candidate genes if Miller syndrome is inherited as a recessive disorder. Only one candidate gene is common to both affected siblings and one affected but unrelated child. Subsequent analysis found that this gene is mutated in all affected children. Studies for other genes have produced different numbers of initial mutations, but roughly the same effectiveness of different filtering methods.

individual has to have mutations in both alleles of the gene; furthermore, these variants would be the same between the affected siblings. Under this model, each person in the study was homozygous for 2,860 variants, of which 2,362 were shared between the two siblings, as summarized in Figure 5.12. This large number of common variants reflects that, on average, siblings have a half of their genome in common; so, while somewhat fewer candidate genes were found, as expected, if the syndrome was recessive, comparing the two affected siblings was not a particularly helpful filtering technique. One of these shared variants is the Miller syndrome disease gene, but there are many other shared variants as well.

The next step was to apply some computational filtering steps to the data on all the variants. At that time, there were several different databases of known common polymorphisms from unaffected individuals; because Miller syndrome is rare, any polymorphism found in these databases can be ruled out as a candidate mutation. Many more individuals have been included in the common polymorphism database since this initial study, so this filtering step now does even a better job of finding the mutations unique to each affected family or individual. Thus all the polymorphisms found in the affected individuals that were also found in these databases could be disregarded as potentially causative mutations. This proved to be a very effective filtering step, which reduced the number of candidate variants by more than a factor of 15 for a dominant trait and by more than a factor of 250 for a recessive trait, as seen by tracing the downward arrows in Figures 5.11 and 5.12. After eliminating common polymorphisms found in these databases, only 228 single gene variants (that is, dominant) and only nine homozygous variants (recessive) shared between the siblings were still candidates.

Another computational filtering step used a program that attempted to predict which variants have severely damaging effects with different protein structure models. This appeared to be very promising as well, since only eighty-three variants in a single gene and only one gene under the homozygous model were shared between the affected children, were absent in the common polymorphism databases, and were predicted to be damaging to the encoded protein. If Miller syndrome is recessive (as it proved to be), this gene was the only remaining candidate, and no other individuals would need to be analysed. However, this strong candidate gene was not implicated when an affected individual from a different family was compared, and in fact a further complicating (but illuminating) issue was encountered. This gene, *DNAH5*, is discussed in what follows.

In addition to these two affected siblings, the exomes of two unrelated individuals also affected by Miller syndrome were sequenced. Figures 5.11 and 5.12 show the result of including these individuals as well, under a different assumption about the mode of inheritance for the mutation. If only a single dominant mutation is required to cause the disease, there were about 3,100 shared variants, of which 2,660 were present among all three affected children from different families (Figure 5.11). If affected individuals have to be homozygous for the mutations, 1,810 possible candidates were identified with one unrelated family and 1,525 were shared among all three affected children (Figure 5.12). Identifying additional children affected by the same rare genetic disease, which is not easy, is important, as discussed below in the case of Kabuki syndrome, but does not help very much with filtering out candidate mutations.

On the other hand, by filtering out the polymorphisms that occur commonly in human populations, nearly all of these other candidate variants could be eliminated. Only twenty-six variants in single genes were shared between the affected children in the two families once common polymorphisms were removed, and only eight were shared among affected individuals in three families (Figure 5.11). If the recessive mode of inheritance is accepted, then only one gene was shared among two separate families. Thus the affected gene could have been identified using the exome sequences of two affected siblings and one affected but unrelated individual, once common polymorphisms were filtered out. Variants in this same gene were found in the child from the third family. The mutations in this gene, called *DHODH*, made it the best candidate for being the cause of Miller syndrome. The inheritance of unrelated mutations in both alleles of this gene in three unrelated families is very strong evidence that this is the correct gene.

In order to confirm that this gene is the cause of Miller syndrome, the gene (rather than the entire exome) was sequenced from four more affected individuals, one of them the sibling of the affected individual in kindred 2 above and three of them the only affected member of their family. All found individuals were compound heterozygotes for mutations in the *DHODH* gene. In addition, the parents of the affected individuals were all found to be heterozygous for *DHODH*, so none of these affected children was the result of a new mutation. In total, eleven different mutations in six different families were found, and none of them was shared except within the same family (although two of them were different mutations in the same glycine codon). None of these mutations in *DHODH* was found among 200 unaffected control individuals.

We noted above that *DHODH* was *not* one of the candidate genes identified in the first family after filtering out predicted deleterious mutations. This seeming discrepancy is due to the fact that one of the mutations found in *DHODH* in the affected siblings was not predicted by the program to be damaging to the function of the protein, although it clearly is. This was a shortcoming of the program used at the time; since then, many improvements have been made in our ability to predict damaging mutations. For example, some current programs include comparisons with orthologous proteins from other species as part of their method for the prediction of damaging mutations, since evolutionarily acceptable mutations are not likely to be damaging enough to cause a severe disease phenotype. But this indicated the importance of including affected but unrelated children in the study.

Some additional information emerged from these studies. Recall that the siblings in the first family were homozygous for a candidate gene that is not found in an affected individual from another family. Both of these affected siblings also have recurrent lung infections. Although these are the only individuals with Miller syndrome who have recurrent lung infections, the small number of affected individuals meant that it was not clear whether recurrent lung infections were part of the same syndrome or an unrelated phenotype, which happened to arise in these individuals. The genetic data are clear: the recurrent lung infections are due to a separate mutation, one found in *DNAH5*, a dynein heavy chain that is expressed in cilia lining our airways. Within this family, both diseases are segregating, and these two siblings happen to be homozygous for both of them. Other individuals with Miller syndrome do not have this additional mutation. Thus the sequence data clarified a diagnosis of this rare disease.

In addition, there had been suggestions, on the basis of the phenotype, that Miller syndrome might be the same as another cranio-facial syndrome, known as Naylor syndrome. However, analysis of the individuals affected by Naylor syndrome found no mutations in the *DHODH* gene, indicating that these clinically similar syndromes are in fact genetically distinct.

Refinements to the procedure used for Miller syndrome have made exome sequencing even more powerful

Exome sequencing was clearly shown to have great promise as a method for the identification of genes with rare traits, such as Miller syndrome. The most effective filtering method proved to be removing all of the common polymorphisms rather than including more affected children or attempting to predict damaging mutations. The databases of common polymorphisms have greatly expanded since then.

Literature Link. MacArthur et al. 2014, *Nature* 508: 469–76

Other modifications designed to enhance the power of exome sequencing are also possible. For example, expression data could also be useful; one expects that the candidate gene will be expressed in the affected tissues at the proper time. Expression data were important in the cloning of the gene for FSS and, as the case study details, for CFTR as well. In addition, these investigators had no linkage information that could have helped them identify the gene, since the syndrome is so rare; many other examples include some linkage data as well. At about the time when this work was carried out on Miller syndrome, another group was attempting to identify the affected gene for familial exudative vitreoretinopathy (FEVR), a dominant retinal disorder that had been mapped to a 40 Mb region with about 300 genes on chromosome 7. The investigators sequenced the exomes from this region from a single individual, in order to find the causative mutation in the gene TSPAN12; this mutation was confirmed by sequencing the gene from additional families with affected individuals. Thus they combined linkage data with exome sequencing, in much the same way as genes are identified in laboratory species.

Even without the subsequent refinements, the achievement of identifying the gene for Miller syndrome is remarkable. The exome sequences of only four individual in three families were necessary to identify the gene, and four more affected individuals were tested for verification. In fact, if the mode of inheritance had been unequivocally known to be autosomal recessive, only the two affected and one unrelated affected individual would have been needed to find the gene.

Other published examples in which exome sequencing has been able to identify the causative gene have yielded a number of different lessons, which are summarized below.

First, relatively few individuals are needed to identify the gene, even if no other information is available. More than three quarters of the disorders to be analysed by exome sequencing so far have been recessive, which makes them easier to identify. For recessive disorders in which there is no family history or linkage data, three or four affected individuals from two or more families have been sufficient to identify the gene, although some investigators have used three (or more) affected individuals from the same family, as well as one or more unaffected individuals from that family; nearly all investigators have confirmed the identification by sequencing the candidate gene from several more individuals, affected but unrelated. For the relatively few dominant disorders that have been analysed (examples are discussed below), most of the analysis has concentrated on multiple members of an affected family. In general, because exome sequence information is not that expensive to obtain, there is no particular impetus to use fewer patients; the emphasis instead has been on improving the filters with a more comprehensive database of common polymorphisms and with better predictions of damaging mutations.

Second, diagnosis can be done more accurately, and this can in turn result in better therapies. In one study, a family had been initially diagnosed with Bartter syndrome, a prenatal renal disorder affecting potassium channels. It was not clear from the diagnosis precisely which version of the syndrome affects this family, since several different clinically related disorders arising from different genes have been described. In fact, the family had none of these versions. Exome sequencing revealed that the family actually had mutations in the gene *SLC26A3*, which affects chloride channels and had been previously implicated in congenital chloride diarrhea. Fortunately, the symptoms for the chloride channel disorders can be partially managed by different therapies, so a proper diagnosis had direct therapeutic impacts: individuals can be offered more appropriate care on the basis of more accurate knowledge of the underlying condition.

The use of genetic data like this, to improve the definition of a disorder, has been termed **reverse phenotyping**, a usage that parallels that of "reverse genetics"; in this case the molecular information is used to clarify the diagnosis of Bartter syndrome. Similar clarifications arose from the analysis of Miller syndrome. The recurrent respiratory infections in the one family were shown to be due to mutations in a different gene, and Naylor syndrome was shown to be genetically distinct from Miller syndrome despite the clinical similarities.

Third, different disorders can be caused by the same gene. This parallels our discussion in Section 4.6 about the classification of mutations in model organisms. A number of examples have identified a gene for one disorder that had previously been found to be causative for a different disorder. Some of these involve dominant and recessive mutations in the same gene. For the gene *PTPN11*, loss-of-function mutations were determined from exome sequencing to result in the recessive skeletal disorder metachondromatosis. This gene had previously been identified as the basis for a different skeletal disorder, known as Noonan syndrome, which is inherited as a dominant trait. The recessive and dominant mutations are summarized in in the amino acid sequence at the top in Figure 5.13. The dominant mutations in this gene, of which many are known, appear to represent a gain of function, while the recessive mutations that result in metochondromatosis cause a loss of function. Thus, different types of mutations in the same gene appear to cause distinct phenotypes, as revealed by exome sequencing.

While Noonan syndrome and metachondromatosis represent the gain-of-function and loss-of-function phenotypes for the same gene, other examples found by exome sequencing show how an allelic series of recessive mutations can also result in clinically distinct disorders, as summarized in the amino acid sequence in the lower half of Figure 5.13. Complete loss-of-function mutations in the gene *HSD174B*, such as stop codons or missense mutations affecting evolutionarily conserved amino acids, had been shown to result in a disorder known as bifunctional protein deficiency (BPD), which is fatal in infancy or early childhood. These mutations are likely to be null alleles. As we have seen in the discussion of null mutations and hypomorphic mutations in Section 4.6, here too exome sequence revealed that the gene also has missense hypomorphic mutations, with a slightly different phenotype. These mutations, which have some residual function, give rise to Perrault syndrome, a disorder with ovarian dysfunction and deafness, in which symptoms are clear only at puberty. The consequences of the gene later in development were not seen previously because children with BPD do not survive.

Similarly, nonsense mutations in the gene *BAG3* were found by exome sequencing to result in dilated cardiomyopathy; missense mutations in the gene had been previously shown to cause a less severe disorder known

A. The PTPN11 Protein from humans

```
MTSRRWFHPNITGVEAENLLLTRGVDGSFLARPSKSNPGDFTLSVRRNGAVTHIKIQNTGDYYDLYGGEKFATLAELVQYYMEHHGQLKEKNGDVIELKYPLNCAD
PTSERWFHGHLSGKEAEKLLTEKGKHGSFLVRESQSHPGDFVLSVRTGDDKGESNDGKSKVTHVMIRCQELKYDVGGGERFDSLTDLVEHYKKNPMVETLGTVLQL
KQPLNTTRINAAEIESRVRELSKLAETTDKVKQGFWEEFETLQQQECKLLYSRKEGQRQENKNKNRYKNILPFDHTRVVLHDGDPNEPVSDYINANIIMPEFETKC
NNSKPKKSYIATQGCLQNTVNDFWRMVFQENSRVIVMTTKEVERGKSKCVKYWPDEYALKEYGVMRVRNVKESAAHDYTLRELKLSKVGQALLQGNTERTVWQYHF
RTWPDHGVPSDPGGVLDFLEEVHHKQESIMDAGPVVVHCSAGIGRTGTFIVIDILIDIIREKGVDCDIDVPKTIQMVRSQRSGMVQTEAQYRFIYMAVQHYIETLQ
RRIEEEQKSKRKGHEYTNIKYSLADQTSGDQSPLPPCTPTPPCAEMREDSARVYENVGLMQQQKSFR
```

B. The HSD17B4 Protein from humans

```
MGSPLRFDGRVVLVTGAGAGLGRAYALAFAERGALVVVNDLGGDFKGVGKGSLAADKVVEEIRRRGGKAVANYDSVEEGEKVVKTALDAFGRIDVVVNNAGILRDR
SFARISDEDWDIIHRVHLRGSFQVTRAAWEHMKKQKYGRIIMTSSASGIYGNFGQANYSAAKLGLLGLANSLAIEGRKSNIHCNTIAPNAGSRMTQTVMPEDLVEA
LKPEYVAPLVLWLCHESCEENGGLFEVGAGWIGKLRWERTLGAIVRQKNHPMTPEAVKANWKKICDFENASKPQSIQESTGSIIEVLSKIDSEGGVSANHTSRATS
TATSGFAGAIGQKLPPFSYAYTELEAIMYALGVGASIKDPKDLKFIYEGSSDFSCLPTFGVIIGQKSMMGGGLAEIPGLSINFAKVLHGEQYLELYKPLPRAGKLK
CEAVVADVLDKGSGVVIIMDVYSYSEKELICHNQFSLFLVGSGGFGGKRTSDKVKVAVAIPNRPPDAVLTDTTSLNQAALYRLSGDWNPLHIDPNFASLAGFDKPI
LHGLCTFGFSARRVLQQFADNDVSRFKAIKARFAKPVYPGQTLQTEMWKEGNRIHFQTKVQETGDIVISNAYVDLAPTSGTSAKTPSEGGKLQSTFVFEEIGRRLK
DIGPEVVKKVNAVFEWHITKGGNIGAKWTIDLKSGSGKVYQGPAKGAADTTIILSDEDFMEVVLGKLDPQKAFFSGRLKARGNIMLSQKLQMILKDYAKL
```

Figure 5.13 Examples of genetic information gained from exome sequencing. Mutations in the *PTPN11* gene (shown above) are responsible for three different clinical syndromes. Mutations that given rise to Noonan syndrome (shown in blue) and Leopard syndrome (shown in green) are all dominant gain of function missense substitutions that replace one amino acid with another. Mutations that give rise to metochondromatosis (shown in red) are all recessive loss of function mutations that result in a stop codon, a deletion, or a frameshift. Only some of the known mutations are shown in each case; all known changes can be found on OMIM. At least fifteen different mutations give rise to Noonan syndrome, at least five mutations give rise to Leopard syndrome, and at least eleven give rise to metachondromatosis. Mutations in the *HSD17B4* gene (shown below) are responsible for two different clinical syndromes. Both syndromes are recessive traits. Mutations at the positions shown in red give rise to BPD, which is fatal in newborns or infancy. All these mutations are either stop codons or in amino acids that are evolutionarily conserved and are likely to be null mutations. The mutation shown in green gives rise to Perrault syndrome, which appears at puberty. This missense mutation is likely to be a hypomorphic mutation, so mutations in this gene show an allelic series. Information extracted from OMIM.

as myofibrillar myopathy. Interestingly, since dilated cardiomyopathy is an autosomal dominant trait, the gene may be haplo-insufficient, another type of mutation described in Chapter 4.

Fourth, similar disorders can arise from distinct genes. Just as different mutations in the same gene can provide clinically distinct disorders, clinically similar disorders can arise from distinct genes, a phenomenon known to human geneticists as **genetic heterogeneity**. (Individuals working with model organisms also recognized that many different genes had similar mutant phenotypes; but, since genetic mapping and complementation testing were relatively easy to do, the phenomenon was not named.) Genetic heterogeneity is often seen with dominant disorders.

Literature Link. Bamshad et al. 2011, *Nature Reviews of Genetics* 12: 745–55

One illustrative example is Kabuki syndrome, a rare facial syndrome with cardiac abnormalities, skeletal defects, delayed growth, and other symptoms affecting a range of organ systems. There are only about 400 cases known, with an estimated frequency of about 1 in 32,000 people. The disorder is likely to be an autosomal dominant, and most of the affected individuals appear to have it from new mutations, since few if any had an affected parent. In line with the condition's arising from new mutations rather than from inherited (but incompletely penetrant) mutations, none of the affected children in the study with exome sequencing had an affected sibling, so the approach of filtering the data with the help of another affected sibling could not work. In this case, the investigators sequenced the exomes of ten unrelated people—seven with European ancestries, two Hispanic, and one Haitian European.

Literature Link. Ng et al. 2010, *Nature Genetics* 42: 790–3

The results were filtered through the same overall strategy used for Miller syndrome, by asking what variants are common and predicted to be damaging among the affected children but absent from the general population. The numbers and areas of the boxes in Figure 5.14 are the candidate genes that remain after common polymorphisms and damaging mutations have already been filtered, so the arrows indicate the effects

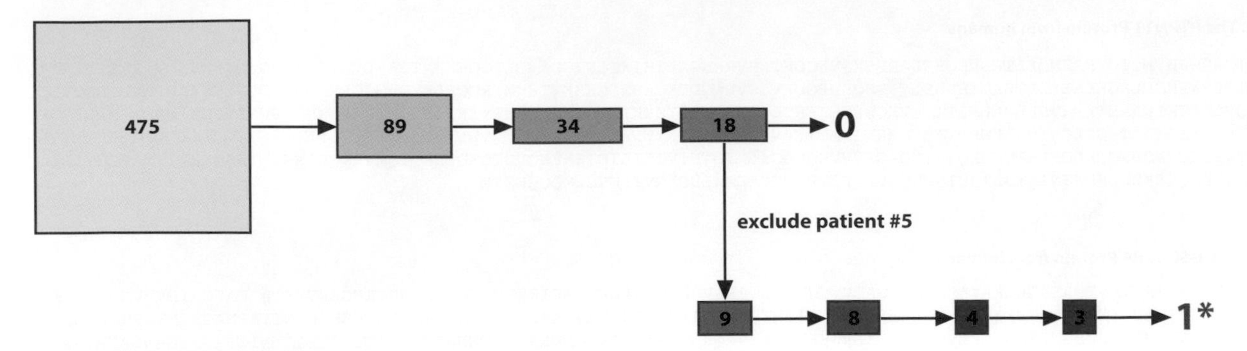

Figure 5.14 Filtering Kabuki syndrome mutations. Kabuki syndrome is a rare dominant disorder. Exomes were sequenced from affected individuals and compared as with Figure 5.11, but only the stop codons, splice site mutations, and indels not found in the databases of common polymorphisms are shown. The horizontal arrows indicate the effects of additional affected individuals. Individuals 1 through 4 had eighteen mutated genes in common that were not also found in unaffected people. Individual #5 did not have a mutation in any of these eighteen genes. However, if individual #5 is excluded from the analysis, all the other affected individuals shared a mutation in this one gene. Kabuki syndrome is genetically heterogeneous, and individual #5 apparently had mutation in a distinct but phenotypically similar gene.

of including additional affected children. However, once this filtering was done, the researchers found that no mutation was common to all ten affected children. They did note that seven of the ten individuals shared a mutation in the same gene, a gene encoding a SET domain protein that encodes as a histone H3 methyl transferase. (Histone methyl transferases are introduced in Chapter 2.) Two of the other three individuals also had a mutation in this same gene, but it had been filtered out when only potentially damaging variants were considered. Thus nine of the original ten individuals had a mutation in the same gene, but the tenth individual did not.

The investigators then sequenced this gene (but not the entire exomes) from forty-three additional patients diagnosed as having Kabuki syndrome, and found a mutation in the same SET domain protein in twenty-six of them. So this gene, now named *KMT2D*, is clearly implicated in the disorder. But why are there some patients who do not have a mutation in this gene? The disorder is genetically heterogeneous; that is, mutations in some other genes produce a phenotype that has similarities to Kabuki syndrome but is genetically distinct from it. In fact a follow-up study by a different group of researchers found that some patients with Kabuki syndrome have dominant mutation in an X-linked gene known now as KDM6A, which is predicted to encode a histone demethylase.

Literature Link. Bokinni 2012, *Journal of Human Genetics* 57: 223–7

After all, with only about 400 cases known, clinicians are unlikely to have seen another case previously, so their diagnosis depends on the description in the literature; the diagnosis of "Kabuki syndrome" might not be subtle enough to capture the different genotypes that are represented, so reverse phenotyping based on genetic data could be useful for distinguishing similar individuals. In general, disorders that are genetically heterogeneous require more affected individuals to be examined. Not all patients with Kabuki syndrome have mutations in either *KMT2D* or *KDM6A*, so additional genes remain to be found.

Exome sequencing is also used for somatic mutations

While our discussion has focused on the use of exome sequencing to identify the gene that has been mutated in some inherited disorders, the most frequent application of exome sequencing is to determine the somatic mutational events that arise during cancer, as shown in Figure 5.15. That is, the exons are captured and sequenced from normal and tumor cells of the same patient, in much the same way as for germline disorders. The genetic variants found in the normal cells of the same individual can be used to filter the tens of thousands of variants found in the tumor cells. Variants found only in the tumor cells and missing in the normal cells must be those that have arisen during the disease progression. Not all of them will be disease-causing—in fact most of them will arise as a consequence of the cancer rather than as a cause, and thus are classified as passenger mutations.

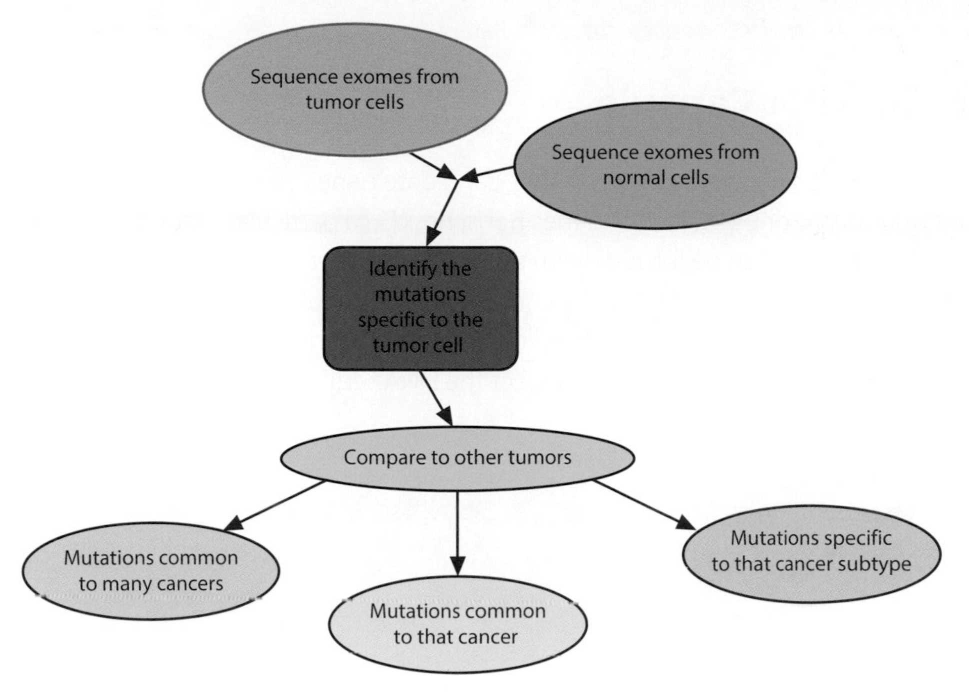

Figure 5.15 Exome sequencing for cancer-causing mutations. The overall strategy of using exome sequence in somatic cells for the Cancer Genome Atlas is diagrammed. Knowledge of the genetic basis for cancers can suggest different prognosis and therapies, as well as some unexpected genetic relationships among cancers once thought to be distinct.

Nonetheless, the complete spectrum of mutations found in various cancer cells can be compared to produce a mutational landscape of different types of cancers. Dozens of different types of cancers have been analysed this way, resulting in the resource known as the Cancer Genome Atlas (http://cancergenome.nih.gov/).

5.6 Summary: Connecting a phenotype to a DNA sequence

The goal of this chapter has been to collate some of the many approaches that have been used to clone genes or to identify the DNA sequence that corresponds to a gene with a particular mutant phenotype. These properties of genes were known during the earliest days of genetics, even before we knew that genes consisted of DNA sequences. As seen with exome sequencing, approaches to connecting mutant phenotypes to DNA sequences continue to be developed, and new techniques will probably be used in the near future. Nonetheless, the classical properties of genes will continue to be the foundations for gene cloning.

Having a cloned gene that corresponds to a mutant phenotype provides one of the essential tools for the analyses described in the rest of the book. We have taken an organizational approach in which cloning the gene often follows upon finding the mutant phenotype, which reflects the historical sequence in which these two essential tools were connected: mutant phenotypes were usually found before the corresponding DNA sequence was identified. From its mutant phenotype, two essential properties of the gene could be determined or inferred: its map position and its expression pattern. These are often the first approaches to connecting the gene to its DNA sequence, but each also yields many candidate genes that are the gene of interest. Thus, other properties of genes, including complementation tests and sequence variation among alleles, are used to confirm that the correct gene had been cloned.

Chapter Capsule

A typical gene-cloning strategy is to identify candidate genes by mapping a mutant phenotype or by looking at genes transcribed in a particular pattern. The candidate genes can be further delimited by complementation tests or by sequencing the DNA from different mutants to find a molecular lesion. In the past few years, direct sequencing of the exons of affected individuals could, with appropriate filtering, be used to connect the DNA sequence to the mutant phenotype.

Additional Reading

Bamshad, M. J. et al. 2011. Exome sequencing as a tool for Mendelian disease gene discovery. *Nature Reviews of Genetics* 12: 745–55.

Gilissen, C. et al. 2011. Unlocking Mendelian disease using exome sequencing. *Genome Biology* 12: 228.

Kiezun, A. et al. 2012. Exome sequencing and the genetic basis of complex traits. *Nature Genetics* 44: 623–30.

Araya, C. L. et al. 2016. Identification of significantly mutated regions across cancer types highlights a rich landscape of functional molecular alterations. *Nature Genetics* 48: 117–25.

Case Study 5.1

Positional cloning of the cystic fibrosis gene in humans

Cystic fibrosis (CF) is one of the most commonly occurring genetic diseases among people of European descent. While the frequency of the disease varies among populations, approximately one in 2,500 Caucasian babies has CF. The principal symptom is an accumulation of mucus in the lungs, which leads to difficulty in breathing and an increased susceptibility to bacterial infections. Most affected individuals also have defects in exocrine secretion in the pancreas; ducts in other organs (such as the vas deferens in the testes) are often affected. Some symptoms of the disease have been familiar for decades, in particular the unusual secretion of chloride ions in sweat. As recently as fifty years ago, few individuals with CF lived past their teens; however, thanks to dramatic developments in diagnosis and treatments, the median life expectancy is now about forty years.

CF is inherited as an autosomal recessive trait. From the frequency of affected children, it is estimated that as many as 4 percent of Europeans are heterozygous for the disease but have no symptoms. Because of its high frequency and simple pattern of inheritance, CF has been the focus of a substantial amount of research. Cloning the gene and characterizing the most common molecular lesion that causes the disease were a *tour de force* of human genetic analysis.

Identifying candidate genes by position

This process, which cloned the gene responsible for CF, is summarized in Figure CS5.1. The gene was first mapped to the long arm of chromosome 7 (7q) by pedigree analysis and subsequently placed in a region of about 1,500 kb, with a large number of polymorphic molecular markers mapped to either side; the locations of the most important of these polymorphisms are shown in Panel A of Figure CS5.1. In a Herculean effort involving some novel cloning strategies that need not concern us here, this entire 1,500 kb region was cloned into a set of forty-nine overlapping phage and cosmid clones.

Thinking through the Experiment

1. Two of the most experimentally demanding aspects of positional cloning are glossed over in the preceding paragraph. While identifying the DNA sequence between the linked polymorphisms is now only historically relevant, the process by which the gene was mapped is still worth considering. Briefly outline the process that would map a human disease gene with respect to polymorphisms using pedigree analysis.

This mapping and cloning process occurred more than a decade before the human genome project was completed, at a time when few details were known about the overall structure of the human genome. Once the DNA sequence of region of the chromosome that included the *CF* gene was delimited by mapped markers and cloned, it was necessary to identify the locations of the genes in the region. This was itself an arduous process. The investigators first identified homology or, more specifically, nucleotide sequence similarity to other mammalian DNA sequences. Sequence similarity was established by hybridization of each

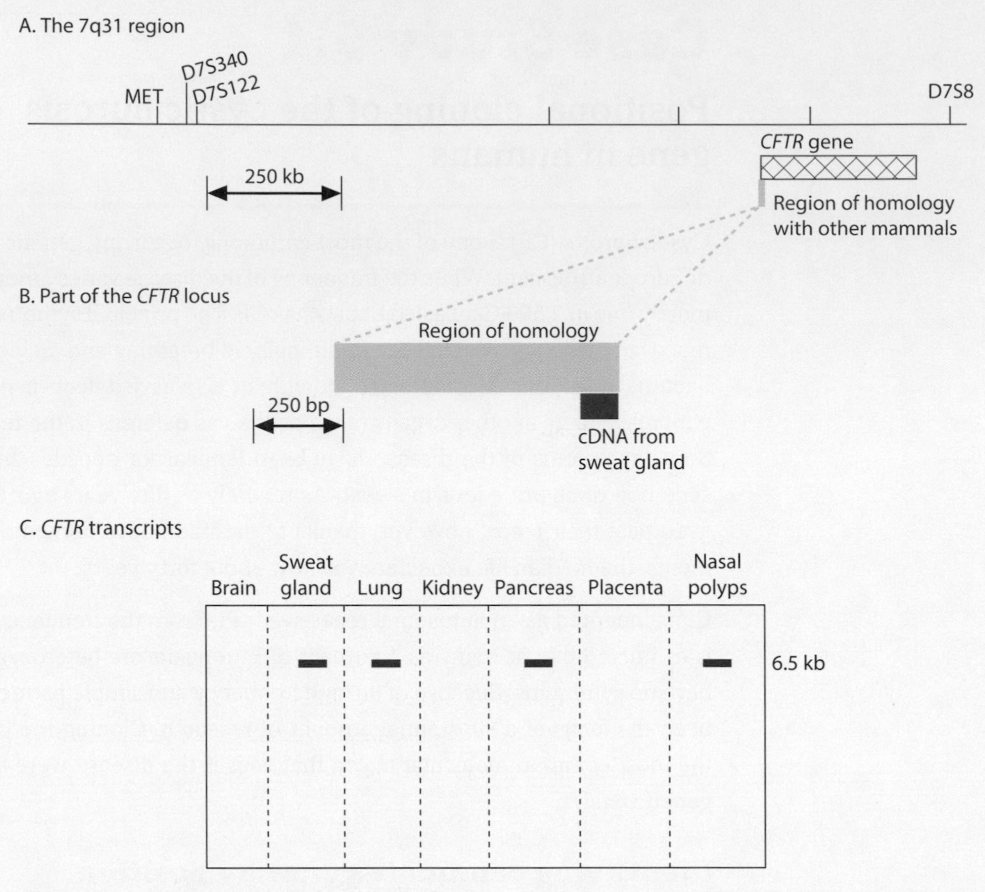

Figure CS5.1 Positional cloning of the human cystic fibrosis gene *CFTR*. A. The polymorphisms closest to the CF gene on chromosome 7q31 as determined by recombination in pedigree analysis are shown, and the region between them was cloned. This region was probed by hybridization for nucleotide sequence similarity to other vertebrates. The region of hybridization and inferred homology found with other mammals is shown in green, with the actual location of the *CFTR* gene included for reference. The scale is 250 kb, so the entire region is approximately 1.5 Mb and the region of homology is approximately 1 kb. B. An expansion of the region of homology, with the small fragment of exon 1 of *CFTR* that was used as a probe indicated in blue. This exon was found by probing a cDNA library from sweat glands of a healthy individual. The scale bar represents 250 bp. C. The fragment of exon 1 was also used to probe transcription of the gene using RNA blots from a variety of human tissues. Only some tissues tested are shown. A transcript is detected in the tissues known to be affected by CF such as sweat glands, lungs, and pancreas, but not in tissues that are not affected by CF, such as brain and kidney. Simplified from Rommens et al., 1989, Identification of the cystic fibrosis gene: Chromosome walking and jumping, *Science* 245: 1059–65.

part of the 1,500 kb region with DNA from other mammals. Because coding sequences are more highly conserved than non-coding sequences, the regions of hybridization with other mammals served as a first approximation of the locations of exons. With candidate exons identified by hybridization, the investigators then located CpG islands (described in Section 2.3) found at the 5′ end of most mammalian genes. These two approaches gave the investigators more insight into the structure of all the candidate genes in this region of the chromosome—which included the *CF* gene, but several other genes as well. One region of homology proved to be derived from the first exon of the *CFTR* gene.

Thinking through the Experiment

2. Again, most of these experimental steps are now supplanted by genome annotation. However, without thinking about the experimental techniques themselves, what assumptions about the structures of mammalian genes were made during this process and how have these been incorporated into gene prediction annotations?

With the locations and overall exon structure of candidate genes identified, the investigators then relied on the disease symptoms to infer the expression of the CF gene. Investigators isolated mRNA and probed cDNA libraries from tissues such as tracheal epithelium, lungs, sweat glands, and pancreas—that is, tissues and organs known to be affected in CF children. They also used cDNA libraries prepared from tissues not expected to transcribe the CF gene, such as the kidney, brain, and placenta, in order to rule out some candidate genes. One sequence probe had the expected pattern of transcription and emerged as the best candidate gene.

From these approaches, the investigators expected to find a candidate gene at the right position on the chromosome and expressed in the appropriate tissues. No one of these methods was guaranteed to identify all the genes in the region. In fact, a careful reading of the cystic fibrosis papers illustrates the limitations of the approaches available at that time, since several false candidates for the CF gene were implicated as well. The correct gene was found only after repeated and exhaustive screening of a cDNA library prepared from the sweat glands of a normal individual.

Thinking through the Experiment

3. What is the significance of "exhaustive screening of a cDNA library prepared from the sweat glands of a normal individual"? In particular, why would exhaustive screening be needed? And why did researchers screen a normal individual rather than one with CF?

The CFTR protein and gene

What could be said about the protein encoded by the candidate gene *CFTR*? Cystic fibrosis is associated with defects in secretion, and suggestions that the gene might encode a membrane channel protein predated its molecular or biochemical analysis. In particular, the secretion of chloride ions mentioned earlier was considered a strong hint that the CF gene may be involved in secretion, either as the ion channel protein itself or as a regulatory protein. When compared with other protein sequences, the predicted CFTR protein has extensive sequence similarity to membrane channel proteins. The inferred amino acid sequence of the candidate CFTR protein supported the conclusion that this could be the gene responsible for CF.

But how could the investigators be certain that they had found the gene that is responsible for CF? Map position alone, even for the carefully mapped CF gene, identified candidates but did not limit the number to only one gene. Homology searches and expression pattern analysis refined the search for candidate genes but still did not pinpoint the exact gene. The inferred amino acid sequence of the protein is consistent with the predicted role of the CFTR protein but, by itself, this would not have persuaded critics that the investigators had found the CF gene. So what property would offer the most unambiguous evidence that the investigators have found the CF gene?

Thinking through the Experiment

4. Anticipate the next step in this procedure.
   a. What experiment would be done?
   b. What individuals would be chosen for this experiment, and why?
   c. What would they expect to find?

Mutations in the *CFTR* gene

For proof that they had cloned the causative gene for cystic fibrosis, the investigators turned to mutations. After all, the gene is defined by the disease that occurs when it is mutated. If investigators had indeed found the gene responsible for a disease, individuals with the disease would have mutations in the candidate gene, and unaffected individuals from families with no history of the disease would usually be homozygous for the wild-type allele. In principle, then, one could simply sequence the candidate gene from an affected individual and from an unaffected individual and look for sequence differences. Natural polymorphisms are expected among humans, so merely the presence of sequence differences would not be enough to identify the gene. However, CF patients should all have sequence differences in the same gene, which is then the best candidate gene.

For a rare genetic disease, an unaffected individual from a family with no history of the disease can reasonably be expected to be homozygous for two wild-type alleles of the gene. Thus any sequence difference in this person can be attributed to natural polymorphisms rather than to causative mutations. This assumption will play an important role in filtering out variations for Miller syndrome and other rare genetic diseases, for example. Sequence differences between the patients affected by the rare disease and an unaffected individual should identify the causative gene.

But unaffected heterozygotes for CF are relatively frequent, so even this conceptually straightforward approach was more challenging. Any person in the population could well have one mutated copy of the gene; in fact, about 4 percent of the overall population are expected to be carriers of causative mutations in the gene. Furthermore, a person might not be recognized as a carrier, because the disease is recessive. A person is known to be a carrier only when he or she has an affected child. Suppose that two individuals, both of whom are heterozygous carriers for CF, marry and have children. The probability of a healthy child is three quarters. Even with two children, the probability is 9/16 that neither child is affected, so the carrier status of the parents remains undetected in many families. In short, undetected carriers for CF are not rare. Thus it may not be informative simply to compare the sequence from the candidate gene from a CF patient with the sequence from an unaffected person; the unaffected person could be a carrier.

Since the investigators could not make an assumption about *unaffected* families, they turned to a more comprehensive analysis of *affected* families. Unaffected parents who have an affected child are obligate heterozygotes and must have one wild-type and one mutant allele of the CF gene. The investigators looked not only at the putative CF gene (*CFTR*) itself but also at nearby molecular polymorphisms that flank the gene to determine which were linked to the inheritance of CF. These polymorphisms allowed them to establish a correlation between the inheritance of other linked molecular markers and the inheritance of CF, which could then be used to determine both wild-type and mutant alleles of the gene. Previous studies had indicated that many cases of CF occur on the same chromosomal background or haplotype, which suggested that many children with CF have inherited the exact same molecular lesion. This suggestion was directly confirmed. Upon sequencing the candidate gene *CFTR* from heterozygous carriers, investigators discovered that 68 percent of the CF chromosomes had a three base-pair deletion that removes a phenylalanine at residue 508 in the predicted CFTR protein, a mutation referred to as ΔF-508. Other samples from European populations have somewhat higher or lower percentages of this specific mutation, but in all cases δF508 is the common mutation associated with CF among people of European ancestry.

Thinking through the Experiment

5. The analysis of the CF gene was a significant template for thinking about many aspects of human genetics. What is the evolutionary or biological significance of the fact that most of the affected children have the same molecular lesion and the same surrounding polymorphisms?
6. The best studied gene associated with a human genetic disease is probably the mutation in β-globin that gives rise to sickle cell anemia. (CF is likely to be second.) Essentially every person affected by sickle cell anemia has the same molecular lesion. However, they do not share the same surrounding polymorphisms on the chromosome. What does this indicate about the evolutionary history of this mutation?

None of the normal chromosomes in obligate heterozygotes has this mutation, and no recombination has been observed between it and the inheritance of CF so inheritance of this mutation completely co-segregates with the inheritance of the disease. This analysis confirms that a mutation in the *CFTR* gene is responsible for CF. In subsequent studies, the *CFTR* gene has been sequenced from other children with the disease, particularly children whose symptoms differ from the classical CF disease phenotype and children from other ethnic backgrounds. These children also have mutations in the *CFTR* gene, although usually not the delta F508 mutation. To date, more than 100 different mutations in the *CFTR* gene have been observed in patients with CF, but delta F508 still comprises about 70 percent of CF mutations, and more than that in some populations.

Thinking through the Experiment

7. The “Allelic Variants” feature on OMIM includes information on many other mutations in the CFTR gene.
   a. Are there other mutations in CFTR that are common in some populations?
   b. What is particularly unusual about the Variant .0005 ARG117HIS?

Cystic fibrosis and positional cloning

In this case study we have seen how investigators used genetic map position, homology to other genes, and expression pattern to identify candidate genes for CF. The analysis of mutant alleles confirmed that the candidate gene *CFTR* is the gene that is actually responsible for the disease.

The cloning of the *CFTR* gene is rightly recognized as one of the landmark events in human genetics. However, a current reader of the original papers from 1989 would probably be struck by how much the availability of the sequence of the human genome has changed laboratory procedures. A comparison of the procedures used in the classic papers and a current similar analysis is presented in Figure CS5.2. Unlike in current papers in human

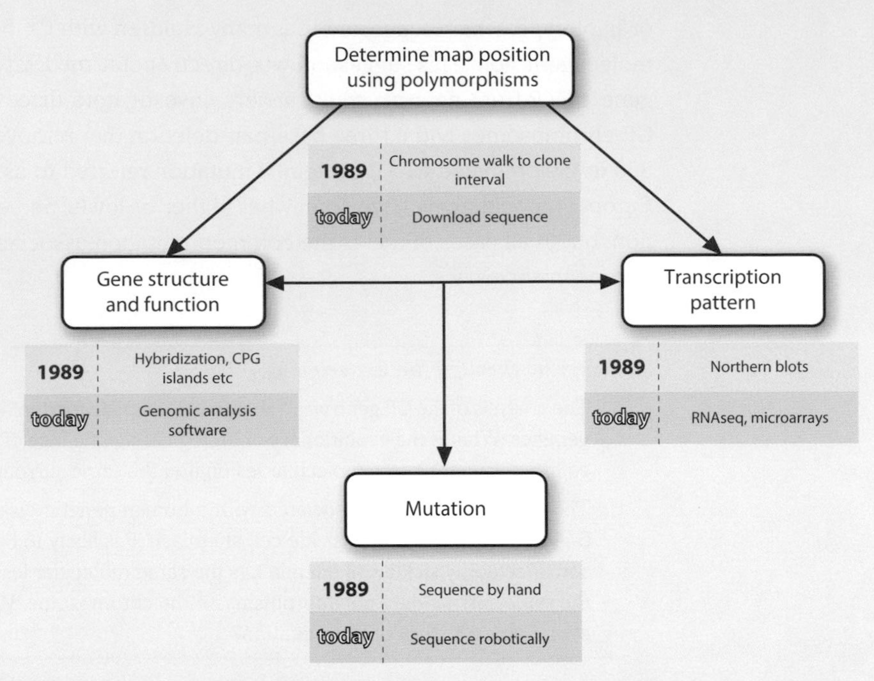

Figure CS5.2 Comparing the cloning of *CFTR* with current procedures. The flow chart shows the strategy and process used to clone *CFTR* and identify it as the causative gene for CF. The experimental procedure involved in carrying out each step is summarized below the box. While the overall strategy for gene cloning in humans has not changed since *CFTR* was cloned, the procedures have been greatly simplified.

genetics, much of the description in the classic papers is devoted to the recombination analysis used to position the gene and to procedures for chromosome walking, designed to clone the entire region.

Even more significantly, positional cloning is now done by searching sequences on the computer, so the entire process of using hybridization and other wet lab procedures to identify the genes in a region has become largely unnecessary. For a current investigator, the entire region of the genome has been sequenced and information on sequence similarity to other mammals is available computationally in a few minutes rather than in months, as it took investigators at that time. Thus the need to do hybridization with the DNA of other mammals to find the highly conserved regions has been eliminated. Likewise, the structure of most human genes, including the locations of CpG islands, splice sites, and stop codons, has been accurately determined, so all these steps are also already completed. Predicted amino acid sequences have been obtained for every known protein-coding gene, and orthologs for many human genes have been identified in other species. In addition, for many human genes, expression patterns have been determined by microarrays or RNAseq, so the protocol for finding mRNAs or cDNAs from the appropriate tissues is also shortened or eliminated. Thus most of the processes that required months of work by a team of highly trained scientists can now be done by a single investigator with minimal experience in a few hours or less.

Thinking through the Experiment

8. The preceding paragraphs may give the mistaken impression that cloning the gene for a human disease is relatively easy. While it is much easier now than then, it is still often difficult to associate a particular genetic disorder with a particular DNA sequence. Summarize some of the features of CFTR cloning that may have made this easier than for many other genes, whether in 1989 or now.

If the genetic times have changed so much, why does the cloning of the *CFTR* gene still deserve a place of honor in a genetics textbook? Several reasons come to mind. First, just as was done for CF, it is still essential to sequence candidate genes from affected individuals (or their parents) to confirm that one has cloned the right gene. That crucial part of the analysis has not changed, and CF shows why it is necessary. Although it is true that sequencing methods and the availability of the data have dramatically changed this part of the analysis, the only true confirmation that one has cloned the right gene comes from identifying mutations that cause the disease.

In addition, a current reader will realize how much more difficult cloning a human gene is when the gene does not cause a well-described syndrome or the disease does not have a simple Mendelian inheritance pattern. Cystic fibrosis is one of the genetic diseases that we know best, both in the clinic and in the research lab; the analysis would have been more difficult for a more obscure disease gene or for a disease with a multifactorial pattern of inheritance. The final reason to describe the cloning of the *CFTR* gene is that the high quality of the research itself—not only the quality of the data but, more importantly, the quality of the logic—has been an underlying guide in the human genome project. The cloning and analysis of the *CFTR* gene set the standard by which the cloning of other human genes and the human genome project itself has been measured.

References

Rommens, J. M. et al. 1989. Identification of the cystic fibrosis gene: Chromosome walking and jumping. *Science* 245: 1059–65.

Riordan, J. R. et al. 1989. Identification of the cystic fibrosis gene: Cloning and characterization of complementary DNA. *Science* 245: 1066–73.

Kerem, B. et al. 1989. Identification of the cystic fibrosis gene: Genetic analysis. *Science* 245: 1073–80.

6

CHAPTER 6

Mutant phenotypes and gene activity

TOPIC SUMMARY

The two primary tools for genetic analysis are to have a mutant phenotype for the gene and to know its DNA sequence, from which we can often infer its molecular function. Together, the mutant phenotype and the DNA sequence can present a picture of the role that the gene plays in the organism—in other words, the gene activity. The mutant phenotype and the DNA sequence, each, contribute insight into the activity of the gene that the other does not provide as clearly. This chapter focuses on the interpretations of mutant phenotypes and on what these tell us about the activity of the gene.

INTRODUCTION

The keys to genetic analysis as described so far are to find both a mutant and the corresponding DNA sequence from wild-type and mutant individuals. That is, a common first step in using genetic approaches to dissect a biological process is to identify a mutant organism that does not carry out the process normally. As discussed in Chapter 4, this is a particularly powerful strategy when very little is known about the underlying molecular or biochemical steps. In fact, identifying and characterizing an informative mutation may provide the opening necessary for breaking down the molecular and cellular biology of a biological process. Many decades of insightful genetic analysis unfolded with very little knowledge of the molecular mechanisms involved; in fact, much was learned before it was recognized that genes were composed of DNA, or before the central dogma of molecular biology was posited. The logic of that type of genetic analysis—interpreting mutant phenotypes appropriately—forms the basis for this chapter.

Nonetheless, we live and work in an era in which much attention is focused on the second of the keys to genetic analysis, that is, on findinging the gene and its corresponding DNA sequence. We discussed this topic in Chapter 5. Every current investigator who finds a mutant or variant phenotype, whether in the lab, the field, or the clinic, also immediately thinks of finding the gene's DNA sequence, which is crucial for understanding its molecular and biochemical functions. Likewise, once we know the DNA sequence for a gene, we want to know what happens to the organism when that sequence is altered, a topic discussed more fully in Chapters 7, 8, and 9.

But we should not overlook the insights that can be gained from thinking carefully about mutant phenotypes themselves. A key concept in analysing mutant phenotypes is the somewhat abstract notion of **gene activity**. The activity of a gene is an inference about the role that the gene plays in the biological processes of the organism. Gene activity certainly can encompass important molecular information about the pattern of gene expression, the biochemical function of the gene product, and the metabolic and cellular pathways affected by the gene, and an investigator would be foolish not to try to learn this type of information in addition to making inferences from the mutant phenotype. But gene activity offers a broader perspective than the molecular and biochemical functions might provide. We will illustrate this perspective with some of the examples discussed in this chapter.

Much of genetics can be summarized as the process of connecting genotypes and phenotypes, as summarized in Figure 6.1. Phenotypes can be easy to see—plants have flowers of different colors, some animals have striped coats while others have solid coats, worms form blisters on their cuticles, a particular stimulus provokes a particular behavioral response, and so on. That is why geneticists typically began

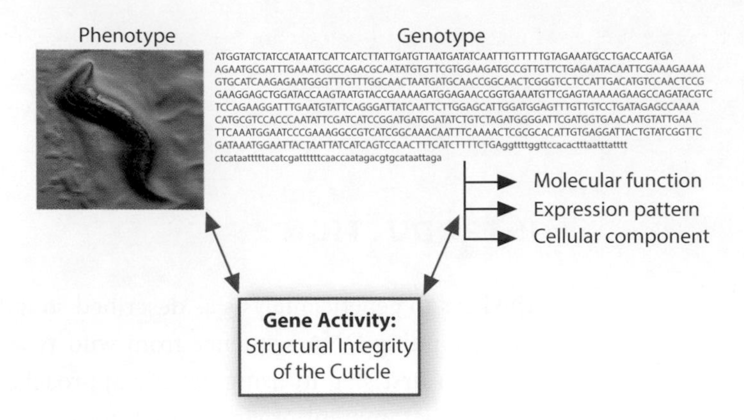

Figure 6.1 Gene activity. The concept of gene activity combines observations from multiple types of experiments designed to infer the functions that a gene plays in the organism. In this example, the mutant phenotype shows worms with blisters on the surface. This suggests that the gene has some function in the structural integrity of the nematode cuticle. The genotype, in this case the DNA sequence of the gene, provides information about the molecular function of the protein, the time and location when the gene is expressed, and possibly where the protein is localized; these reveal that the gene encodes a collagen that is expressed at a specific time in the life of the worm and is localized to the extracellular matrix. Together, the approaches provide insights into the process of cuticle formation.

with phenotypes. With molecular biology and genomic tools, genotypes have also become much easier to determine. We know quite a lot about the mechanisms that connect genotypes and phenotypes—the central dogma of transcription and translation, the biochemical activities of proteins, and the signaling pathways within and between cells, for example. From knowing these mechanisms in detail, we might be able to form a logical connection between a genotype and a phenotype. But more often a gap remains in our understanding; it is not altogether apparent how a mutation in a gene that encodes a protein with a particular biochemical function results in a particular change in the phenotype. Or, conversely, it may not be apparent how a particular phenotype can be explained by the known molecular functions of a gene. This is the gap that the concept of gene activity attempts to fill.

6.1 Overview of mutant phenotypes and gene activity

One of the best known descriptions of the way in which a geneticist thinks about research problems was evoked by Evelyn Fox Keller in the title of her biography of the renowned maize geneticist Barbara McClintock; McClintock was described as having "a feeling for the organism." In the Preface to that book, Fox Keller quotes McClintock as saying: "The important thing is to develop the capacity to see one kernel that is different, and make that understandable. If [something] doesn't fit, there's a reason, and you want to find out what it is."

Genetic analysis is one powerful strategy "to find out what it is," but we should not skip over the preceding sentence in McClintock's statement. A feeling for the organism, understanding mutant phenotypes, or seeing the one that is different depends to a large extent on recognizing what happens in the wild-type organism: this can be very broad or very specific. We might ask ourselves what usually occurs during growth and development, and when during the life cycle each event happens. What is the normal behavior in response to a stimulus?

What is a typical number of offspring or gametes, and how long does it take to produce that number? These questions are posed at the organismal level, but a similar set could be posed if we can watch what happens at the cellular or subcellular level. These are often the types of observations that we make nowadays using microscopy techniques, molecular and biochemical assays, and so on.

In short, then, in order to recognize and characterize a mutant phenotype accurately and in detail, we also need to study the wild-type phenotypes in detail. The "feeling," for McClintock and for others who examine mutant phenotypes, is not based primarily on intuitive leaps of imagination, as sometimes occurred among nineteenth-century naturalists, but on a history of experiences with and observations of the normal organism.

It may be helpful to recall the genetic screen that yielded a large collection of mutants affecting segmentation in *Drosophila* embryos, as discussed in the case study for Chapter 4. It was a demanding and, sometimes, no doubt tedious task to collect more than 640 mutants, to maintain all the mutant strains, and to carry out the crosses in order to assign each of these to a gene. Any person reading those papers carefully is impressed by the amount of work involved. But perhaps even more impressive, and certainly better remembered, is the effort and creativity that came from interpreting the disparate mutant phenotypes for all of those genes and from realizing that these mutant phenotypes could be grouped into three general categories: gap genes, pair rule genes, and segment polarity genes. Many of us looked at the published pictures of the cuticle patterns and were not sure what we were looking at. Only investigators who were intimately with the details of *Drosophila* segmentation patterns would have made those interpretations and insights into the mutant phenotypes.

It is in this context that we can discuss gene activity. Gene activity is a broad term to describe the role that the gene (or, more often, the gene product) plays in the biological processes that occur in normal individuals under a particular set of environmental conditions. It may encompass, or it may come to encompass, the expression pattern of a gene and how that expression pattern is regulated, the molecular and biochemical function of the gene product, and the cellular pathways that use that gene product. Gene activity is often defined progressively, so that, as we come to know more of the details about the expression pattern, the biochemical and molecular functions, and the cellular pathways associated with a particular gene, we can refine what we mean by its activity. But often the earliest definition of a gene's activity comes from its mutant phenotype; that is, from the capacity to see the one kernel, cell, embryonic segment, or organism that is different from the wild type.

White eyes and pigmentation

Let's consider two examples by way of illustration. The first gene to have its activity recognized was likely the *white* gene in *Drosophila*; mutants have white eyes rather than the brick red eyes of the wild type, as shown in Figure 6.2. Recognizing this phenotype requires no specialized knowledge of fruit flies and no particular feeling for the organism, since the white-eyed flies are very easy to spot. Even a casual observer would conclude that the activity of this gene involves eye pigmentation. A slightly more subtle interpretation, although one familiar to genetics students, is that the *normal* activity of this gene involves eye pigmentation and that the mutant lacks that normal activity. Nonetheless, the gene is needed for eye pigmentation.

Literature Link. Flybase.org

A somewhat closer observation of wild-type and mutant flies shows that the eyes are not the only organ whose pigmentation is affected by the *white* gene; in *w/w* mutant flies, the Malpighian tubules in larvae and the testes sheath (in *w/Y* males) are also lacking in pigmentation. So the mutant phenotype tells us that the same cellular and molecular process in the adult eye is likely also being used for pigmentation in these other tissues and at these other times. On the other hand, the cuticle in white mutant flies has normal pigmentation (to give one example), so that must involve a different cellular and molecular process, which w^+ does not affect. We don't know from the mutant phenotype alone precisely what cellular or molecular process is affected by *w*, but we can refine our understanding of the activity of the gene and its mutant phenotype when we learn its molecular function.

Our intuition about flies might have led us to think (incorrectly, as it turns out) that w^+ is needed to make the pigments, in which case we could predict that its gene product is an enzyme involved in pigmentation biosynthesis. However, we can refine our thinking about the

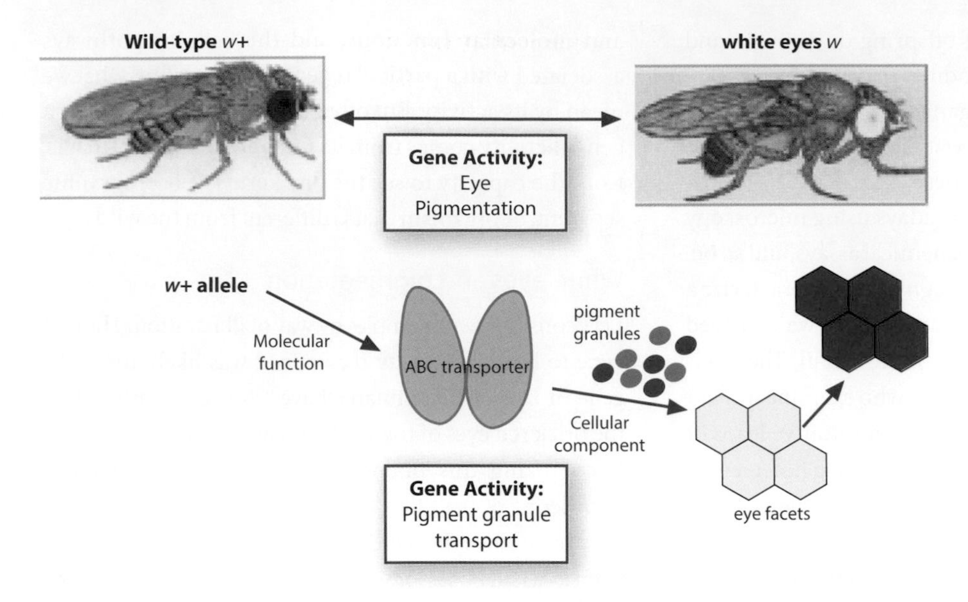

Figure 6.2 Gene activity for the *w* gene in *Drosophila*. The mutant phenotype shows that the gene functions in eye pigmentation but does not provide information about how pigmentation is altered; the gene could affect the synthesis of pigment granules or any of several other processes in eye pigmentation. The molecular analysis shows that the gene encodes a transport protein of the ABC transporter class, indicating that the role of the gene is to transport pigment granules into the eye facets. While the gene encodes a transport protein, it would not be clear from the gene sequence alone what macromolecule or substance is being transported. The combination of approaches gives a more complete insight into the function of the gene.

activity of the gene once we learn that the gene encodes an ABC transporter protein; thus w^+ does not make the pigment granules, but it in effect encodes the biochemical "truck" that delivers them. This biochemical function explains the mutant phenotype, and the mutant phenotype helped us recognize that the same transporter proteins are involved in other tissues.

Similar examples are found for every organism, especially once we recognize that inferring the normal activity from mutant phenotypes usually means that we are asking what happens when the activity is lacking. (This can cause particular confusion when references are made to genes associated with human diseases and conditions. There are probably no genes in humans whose normal activity results in schizophrenia, but there are an undefined number of genes for which an altered or mutant version *contributes* to the condition.) The yeast cell cannot grow in the absence of added leucine, the worm is paralysed, the pea is wrinkled, and so on—in each case, we can use the mutant phenotype as the first approximation to describe the activity of the gene before we know the biochemical functions or cellular pathways that the gene affects.

Rudimentary wings and pyrimidine biosynthesis

On the other hand, we can sometimes be misled when gene activities cannot always be so easily inferred from mutant phenotypes. Let's use another example, also from the classical literature in *Drosophila*. Suppose that you find a fly that belies its name and is unable to fly. Upon examination of the mutant phenotype, you see that the wings are severely truncated, the wing veins don't form normally, and the wing remnants that do form sometimes have blister-like structures on them. These were among the original observations of the mutant phenotype of the gene *rudimentary (r)*, as shown in Figure 6.3. The inference is that the activity of the gene is needed for normal wing formation. Maybe this is a gene that tells the fly how to make a wing.

 ***Literature Link.*** Flybase.org

However, closer observation and additional mutations in the gene refined that description of the gene activity. First and significantly, the original *r* mutants (such as the r^9 allele, whose phenotype was described above and is

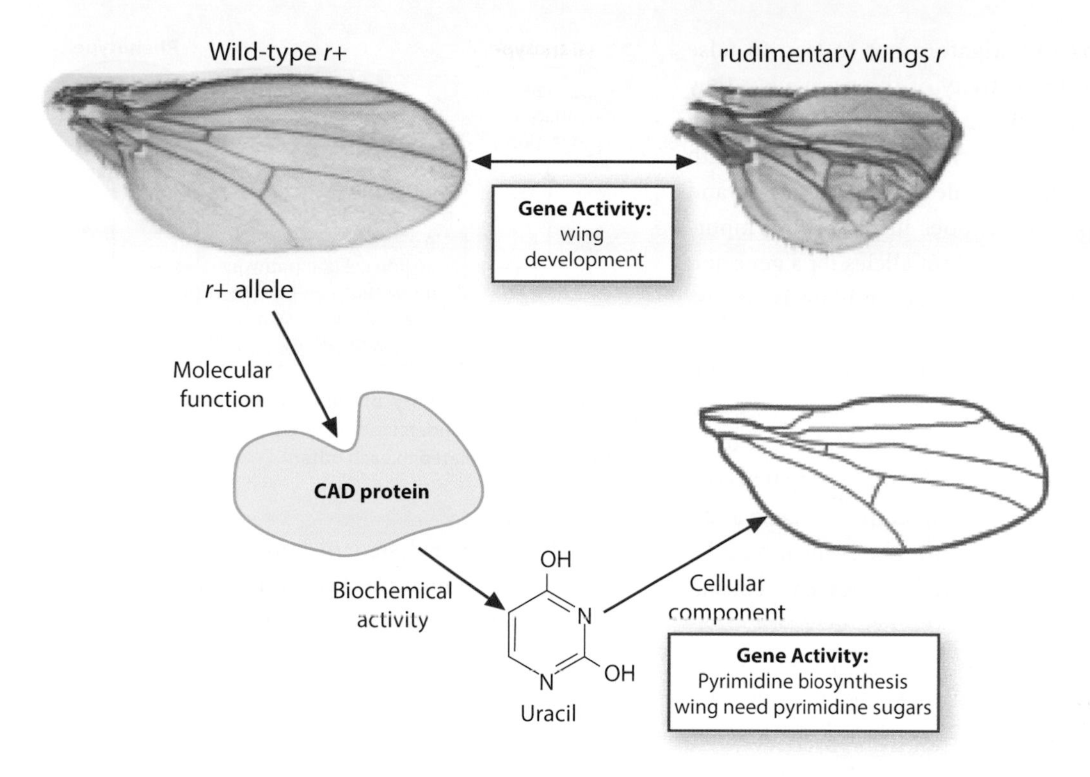

Figure 6.3 Gene activity for the *rudimentary wings* (*r*) in *Drosophila*. The mutant phenotype, shown here, suggests that the role of the *r* gene is related to wing development. In fact the molecular analysis indicates that the gene encodes the *Drosophila* CAD protein, which is needed for pyrimidine biosynthesis. The combination of the results illustrates the role that pyrimidine sugars play in the growth and structure of the wing, an insight that neither could have provided just by itself. Reproduced from Rawls, 2006, Analysis of pyrimidine catabolism in *Drosophila melanogaster* using epistatic interactions with mutations of pyrimidine biosynthesis and β-alanine metabolism, *Genetics* 172: 1665–74 and used with permission.

similar to the one shown in Figure 6.3) were found to be hypomorphic. As we noted in Chapter 4, the best inferences about normal gene activity are made by examining null alleles, although hypomorphs can also be helpful, as this example will illustrate. Other alleles of the *r* gene, and likely also null alleles, had much more severe phenotypes, including female sterility, malformed egg formation, poor yolk deposition, and occasionally lethality. (The various *r* alleles have a complex complementation pattern, which has been the focus of many papers but goes beyond our discussion here.) In addition, when the nutrients in the media were changed, the mutant phenotypes became more pronounced. This suggests that activity of the r^+ gene has something to do with the general metabolic requirements for growth rather than with wing specification itself; wing formation may require a particularly high level of these nutrients, whatever they are.

When the gene was cloned, that interpretation proved to be a more accurate representation of the mutant phenotype. *rudimentary* encodes the CAD protein in *Drosophila*, a multifunctional protein with domains that catalyse the initial steps in pyrimidine biosynthesis in eukaryotes. Clearly all cells require pyrimidines, so we might have expected that mutations in the gene would be lethal (as some *rudimentary* mutants are). However, standard fly food includes enough pyrimidines that the flies can sometimes survive without extensive de novo pyrimidine biosynthesis; *r* hypomorphs made just enough pyrimidine in addition to what was in their food to become adult, and even to reproduce. But they cannot make the components of the wings; wing formation requires extensive pyrimidine biosynthesis, since insect wings are rich in polysaccharides derived from UDP-galactose, a pyrimidine sugar.

The phenotypes of the hypomorphic alleles revealed the process that requires the most extensive activity of the gene, but not the primary function of the gene. Additional alleles as well as changes in growth conditions helped investigators not to leap to false conclusions about the activity of the *r* gene; but they could not determine what the actual activity was until the gene's sequence was determined. Our example with *rudimentary* is a cautionary tale about gene activity and mutant phenotypes, but it does illustrate the importance of having multiple mutant alleles for a gene and knowing what type of alleles each mutant is, as discussed in Chapter 4.

One of the main themes of this book is that geneticists routinely try to make connections between genotypes and phenotypes and have learned how to address some of the challenges associated with making that connection. For example, in order to connect a gene and a phenotype, we will want to know about the molecular biology of the gene and its function, a topic we considered in Chapter 5. We want to know the nature of the mutation and its effect on the gene, a topic we considered in Chapter 4. We want to know about any possible interactions with other genes, a topic to be introduced in Chapter 10. Some of the other information that we would like to have in order to connect genotypes and phenotypes falls outside the topics of this book. Among those would be the effects of the environment, including drought, nutrition, and (for a plant) soil conditions. If the organism is an animal, particularly a mammal, we would want to know whether maternal and paternal nurturing might have affected the phenotype. We might want to know about the genetic contributions of each parent, particularly if the gene shows evidence of imprinting or a maternal effect. All these affect our ability to understand how the genotype produces the phenotype; so, if we really want to answer the question about genotypes and phenotypes, we would want to know as much as possible about all of these.

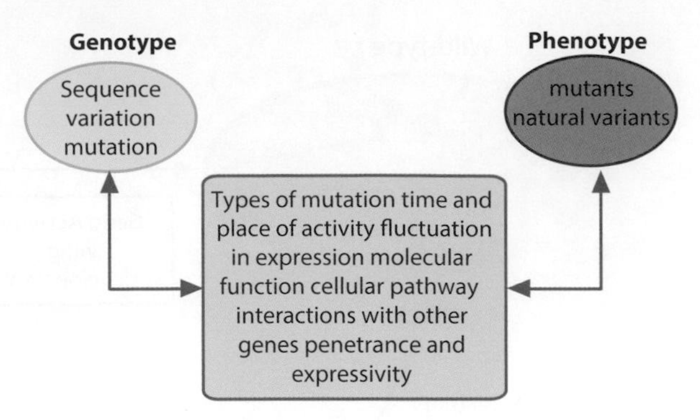

Figure 6.4 A summary of some of the information to be considered when understanding how a particular genotype and phenotype are related to each other.

With our current state of knowledge and technology, we often know the two end points of the connections between genotypes and phenotypes for our model organisms, as summarized in Figure 6.4. We know genotypes and genetic variation, because DNA sequencing has become easier, cheaper, and faster, and we have many tools to introduce changes into the DNA sequence, both random changes and directed edits. We know phenotypes, particularly mutant phenotypes, from genetic screens, gene disruptions, and RNAi screens, which will be described in subsequent chapters. Even for humans in whom targeted gene mutations cannot be made (although we will discuss this further in Chapter 8), we know genotypes from DNA sequencing and genome-wide association studies, and we know diverse phenotypes simply by looking around, so we can infer the connections. Much of what remains for us is to fill in that large poorly charted territory in the middle. How exactly does this phenotype arise from this genotype? That is the topic that we will now go on to explore in more detail.

6.2 Interpreting mutant phenotypes

Generations of geneticists have thought and written about interpreting mutant phenotypes and inferring the normal activity of the gene from these phenotypes. It is informative that many of these geneticists made such interpretations without knowing the molecular nature of genes or the processes of gene expression, which form such an important link when we think about gene activity today. Yet, even without knowing this important link, geneticists from previous generations were able to connect genotypes and phenotypes. Upon rereading some of these early studies with our present understanding of genes and genomes, some of the ideas and interpretations remain remarkably insightful; it is worth considering some of these insights in light of what we have subsequently learned from our now huge collections of mutants.

A classic work on interpreting mutant phenotypes from the era preceding molecular biology was written by the Swiss geneticist Ernst Hadorn in 1955; an English translation by Urusla Mittwoch entitled *Developmental Genetics and Lethal Factors* appeared in 1961. A pdf of this work is available.

Literature Link. https://www.ncbi.nlm.nih.gov/pmc/articles/PMC2973062

Many of the concepts that Hadorn introduced or discussed in terms of "lethal factors"—what we would now call essential genes—set up a framework for thinking about mutant phenotypes for other genes. An **essential gene** is one whose function is necessary for the growth and reproduction of the organism; an individual that is unable to survive and reproduce has a mutation (or is homozygous for a mutation) in an essential gene. The book surveys a wide range of mutant phenotypes in many different animals (and a few plants), attempting to synthesize the effects of mutants in some coherent interpretations. We will briefly review some of these ideas using Hadorn's terminology, although not all of it has been widely adopted. In the next section we will discuss a more familiar example that applied these same concepts.

Mutant phenotypes tend to affect particular developmental stages and tissues

Since Hadorn was addressing lethal effects of mutations, he focused attention on the time at which the effects of the mutant gene became evident. He referred to this as *the phase of gene activity*. He noted that lethal mutations in a variety of different species arrest development at or near certain transitions in the life cycle—hatching, molts, pupation, and so on. He called the stage of arrest the effective lethal phase, which we could generalize to the **effective mutant phase**. For many different mutations, the effective mutant phase corresponded to a "boundary" in the life cycle or to a transition from one stage of the life cycle to the next, so these were also referred to as **boundary mutations**. These concepts are diagrammed in Figure 6.5, using the results with genome-wide RNAi screens on the life cycle of *C. elegans*, (described in Chapter 9); similar results have been seen in many other organisms and mutant screens.

Many more mutations and mutant phenotypes have been analysed since Hadorn surveyed lethal phenotypes in the 1950s, but the concept of a boundary mutation is still worth thinking about. For example, many screens for

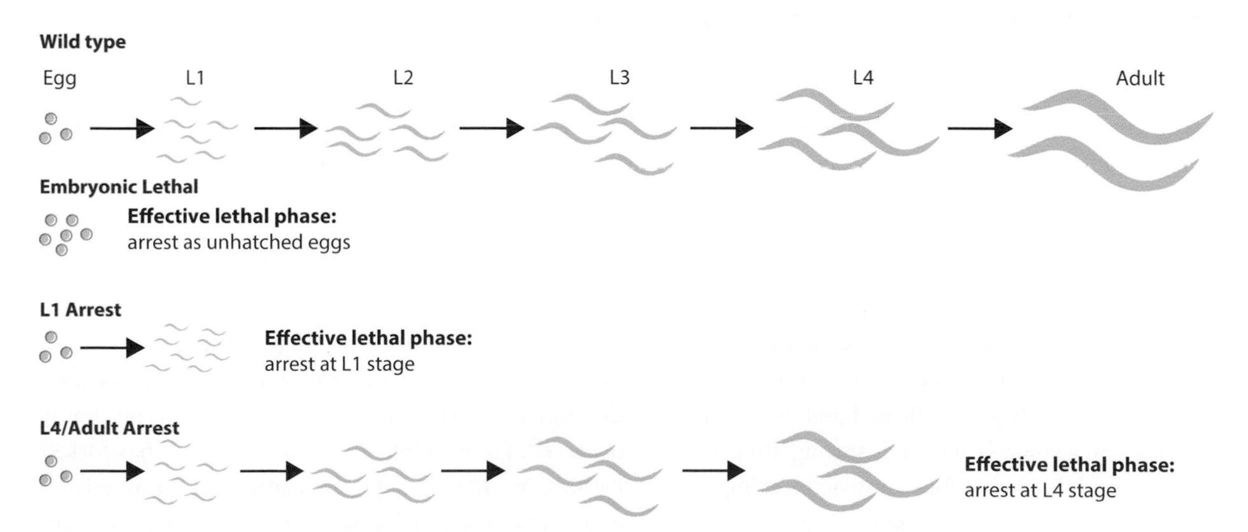

Figure 6.5 Effective lethal phases and boundary effects. Mutations in many different genes have effects at the same developmental stage, which is referred to as a boundary effect or as the effective lethal phase. This is diagrammed here, with a summary of results from RNAi and mutant screens in worms, in which arrests at the egg stage and at the L1 stage are common but arrests at the L2 and L3 stage are much less common. These phenotypes may reflect the level of overall physiological and morphological activity that is needed to complete that stage and progress to the next one. A similar logic might be extended to particular cell types rather than particular developmental stages.

lethal mutations as well as genome-wide RNAi screens in *C. elegans* found that the first larval-stage molt is the effective lethal phase for many genes; the affected worms hatch from the egg but arrest development during the first larval stage. Similar effects have been seen with lethal mutations in many other genetic screens in many organisms. There are particular stages of the life cycle during which many different mutations cause an arrest, and other stages during which relatively few mutants arrest. The identification of mutants affecting segmentation in *Drosophila*, described in the case study for Chapter 4, also illustrates the importance of this point. As noted above, the ability to recognize and classify a very diverse group of mutants into three phenotypic categories on the basis of their effective mutant phase—gap genes, pair-rule genes, and segment polarity genes—was among the key elements in interpreting the results of these screens.

How should the effective mutant phase or the boundary effect be understood? One possibility is that the effective mutant phase represents an observer bias from our tendency to assign diverse things into the same few categories. That is, we classify mutants as being arrested at particular stages because these stages are easy to observe and we like to put things into simple categories. The mutant may actually arrest at subtly different points within a stage or a life cycle, but we cannot distinguish these by the assay we are using. This was certainly the case with the *Drosophila* segmentation mutants; although the category was called (for example) "pair-rule genes" in order to emphasize some fundamental similarities in the mutant phenotypes, the differences in phenotype among various pair-rule genes also offered a critical insight into understanding the activities of these genes. It was important to recognize that the mutants fell into distinct categories, and also that there were differences among mutants assigned to the same category.

Although an observer bias is probably important, a more fundamental and biologically based interpretation of boundary mutations is also likely. Certain life cycle stages involve many complicated and interacting physiological and cellular events: cells are migrating into new locations and dividing, gene expression is changing (some genes being activated and others being repressed), metabolism is changing, and so on. A mutation affecting any part of these complicated events might be expected to result in an arrest or a mutant phenotype at this same stage. Effective mutant phases then represent physiologically or developmentally critical stages in a life cycle. This is the interpretation Hadorn favored.

While Hadorn wrote primarily about developmental stages, we might be able to extend this concept of physiologically critical stages to thinking about the effects of mutations on different tissues and cells. For example, many mutations in animals exhibit a phenotype in the nervous system. This is not too surprising when we consider the cellular and physiological complexity of the nervous system. Neurons extend lengthy cytoplasmic processes and secrete and receive signals from other cells, so any mutation that affects cell shape, secretion, or intracellular transport is likely to exhibit effects on the nervous system. These are not "neuronal genes" in the sense that the activity of the gene is found only in the nervous system. A gene that has general activity throughout the body could appear to be affecting the nervous system preferentially, because the proper functioning of the nervous system is a highly demanding cellular process. The effect on the nervous system of a particular mutation may reflect the underlying physiology of neurons rather than the fact that the principal focus of gene activity is the neuron.

It may be helpful to think again about the lessons from the mutant phenotypes of the *rudimentary* gene, discussed above; the effective mutant phase affecting wing development and seen with the hypomorphic mutations in *rudimentary* revealed a physiologically intensive time for pyrimidine biosynthesis rather than some specific activity for wing development. Recognizing that some tissues or organs are affected by many different mutants can help interpret the mutant phenotypes and focus on the phenotypes that might be more specific.

Terminal phenotypes might not indicate the time or location of gene activity

We should also note that the effective lethal or mutant phase refers to the **terminal phenotype**. Even for mutations that are not lethal, the mutant phenotype is often described in terms of the terminal phenotype that is observed. For example, an adult mutant fly has forked bristles, an adult mutant worm cannot move properly, an adult mutant mouse has small eyes, the mature mutant plant produces wrinkled peas; all these relate to the terminal phenotypes. But often a phenotype observed at earlier times is more instructive in helping us understand the activity of the gene.

Hadorn used the term **phenocritical phase** to describe the time at which the earliest mutant defects are observed. This term is not in common use, but the underlying concept is quite important, so we should discuss it at some length. The terminal phenotype is the outcome of a number of different failed or abnormal cellular processes, and it can be difficult to know which defects are the most informative in understanding the function of the gene. The earliest mutant phenotype, the phenocritcal phase, is often particularly helpful in interpreting the terminal mutant phenotype. Consider, for example, a neurological mutation resulting in abnormal movement. The terminal phenotype may be that the adult animal moves poorly or is even immobile. But the phenocritical phase—the phase when the earliest defects are detected—could be much earlier in development, perhaps when a particular subset of neurons fails to extend axons properly and the mutant phenotype is more subtle and specific.

We can certainly recognize parallels with many human diseases, including cancers. The phenotype when the disease is fully manifested is not necessarily the most informative for understanding the initial focus of the disease or the genes that affect it. The cause of the disease is often more easily understood by examining the first symptoms to appear.

Identifying the earliest phenocritical phase also depends on the type of assays that are being used to observe the phenotypes. For example, a molecular marker for a particular cell or subcellular structure, or the expression pattern of a reporter gene, might identify a phenocritical phase that is much earlier and informative than one seen by watching the cells during later development. Thus it is important to study the phenotype by as many different assays as possible, in order to recognize details of the differences between wild-type and mutant organisms.

Hadorn also recognized that a complicated phenotype may be determined by a range of factors that may exhibit different phenocritcal phases. That is, the mutant effect may be seen in some tissues or organs earlier than it is seen in others. This difference in timing may be due to the fact that the gene is active at different time in the different tissues or that one tissue needs to develop normally for another tissue to develop via a signaling process. The distinction between the phenocritical phases in different tissues can also be extremely helpful in understanding a mutant phenotype, an effect that we will revisit below.

Most genes have pleiotropic effects

Most mutant phenotypes are a complicated mixture of cells and tissues that develop normally and ones that develop abnormally. In "Evolution in Mendelian Populations," his classic paper from 1931 that helped to lay the foundation for the modern evolutionary synthesis, Sewall Wright wrote that "each character is affected by many genes and each gene affects many characters." This observation defines the property known as **pleiotropy**. Pleiotropy means that a mutation in the gene has more than one phenotype; our white-eyed fly that began the chapter lacked pigmentation not only in the eyes but also in the Malphighian tubules. Mutant phenotypes from all sources and organisms have completely supported Wright's perspective, and we recognize now that pleiotropy is the rule for gene activity; it is unusual to find a gene that affects only a single phenotype and has no other effects.

Let's consider an example: *Sonic hedgehog* (*Shh*) mutations in the mouse. *Shh* encodes a secreted protein that is involved in many inductive interactions that pattern the mouse embryo; orthologs are found in many animals, including *Drosophila*. (In fact, as we will discuss in Chapter 11, the *Drosophila* ortholog *hedgehog* encodes the ligand for the receptor encoded by the *patched* gene.) Mutations in *Shh* are highly pleiotropic, causing defects in the development of the heart, limbs, eyes, ribs, skin, hair, and many other organs. The terminal phenotype is too complicated to be analysed fully enough for us to understand the normal activity of the *Shh* gene. However, the earliest defects in the *Shh* mutants are in the establishment of structures along the dorsal midline of the embryo, including the notochord and the dorsal floor plate. The recognition that *Shh* affects the development of these structures, which then induce the formation of other organs, helps explain the complexity of the mutant phenotype.

Pleiotropy arises from any of several different sources, as illustrated in Figure 6.6. First and commonly, the activity of the gene is required in more than one cell type. This is the simplest explanation for the pleiotropy seen with white eyes; the activity of the w^+ gene is needed for pigment granule transport both in the eye facets and in the cells of the Malpighian tubules.

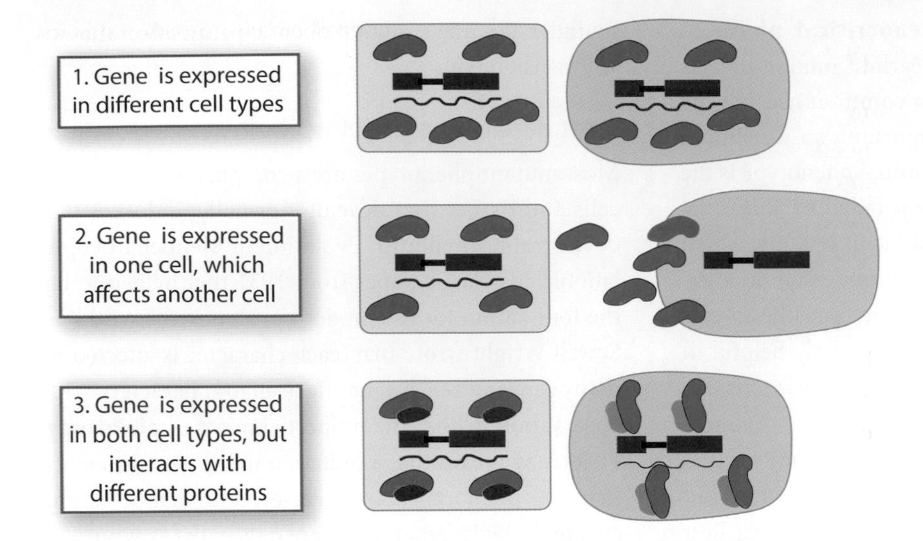

Figure 6.6 Sources of pleiotropy. Mutations in most genes are pleiotropic, with several different underlying explanations. Most commonly, the gene is expressed in more than one cell type, so the mutation shows effects in more than one cell type. Alternatively, the gene may be expressed in one cell type, but that cell signals to another cell; thus a mutation in the gene affects not only the cell in which the gene is expressed but also the cell that receives signals from it. A third possibility is that the protein is expressed in both cell types but interacts with different proteins that are cell-type specific.

Second, the activity of the gene is needed in one cell type, and the absence of the activity in that cell has effects in turn on other interacting cells. Thus, even if the gene is not active in some cells, it affects their structure or function because the gene affects other cells in the same organ. *Shh* is expressed in many cells, but some of the pleiotropic effects in *Shh* mutants arises from the failure of the notochord and dorsal floor plate to develop, since these structures induce the formation of many other cellular structures. Thus, even cells or organs that do not themselves express *Shh* are affected by its absence.

Third, as discussed more fully in Chapter 12, a gene (or, more likely, the gene product) interacts with other genes (and gene products) to carry out its functions. These other gene products can have many different functions, and may vary between cells. The variety of interactions that a gene or a gene product makes is also a significant source of pleiotropy; in fact, as will be discussed in Chapter 12, recognizing the types of interactions that a gene forms can be predictive for it being pleiotropic.

Somewhat less commonly, the gene may encode a protein with more than a single function, or even with distinct functions in different cells. For example, cytochrome C, a protein needed for electron transport and energy production in nearly all organisms, is also involved in apoptosis in many animals. Thus cytochrome C mutants identified on the basis of their apoptotic effect also have defects in electron transport, and vice versa. Similarly, a small heat shock protein that protects cells from environmental stress is expressed at high and constitutive levels in the eye where it comprises alpha-crystallin of the lens.

Literature Link. Constance 2015, *Frontiers in Genetics* 6: 211

This last effect—in which a protein, often an evolutionarily ancient protein such as cytochrome C, has acquired a second and distinct function—is referred to as moonlighting. It is thought that most of the enzymes of the glycolytic cycle and of the TCA cycle have secondary or moonlighting functions in some species. Since gene and protein moonlighting represent a distinctive type of molecular evolution, they are not always defined as pleiotropy and are not illustrated in Figure 6.6; however, they fall within the broad definition of a gene that plays multiple roles, or a mutant that shows multiple different phenotypes, so they are included here.

The mutant phenotype may not reflect when and where the gene is expressed

One of the key questions in observing the effects of a mutation in different parts of the body is whether the gene is active in that tissue or organ or whether the mutant phenotype is the result of effects in other tissues or organs, that is, the cellular focus of the activity of the gene. We suggested this above, when discussing *Shh*. In short, does the activity of a gene work primarily or exclusively within the cell in which the gene product is made, or does it involve something that works between (or even outside of) cells? A gene whose activity occurs within a cell is referred to as **cell-autonomous**, whereas a gene whose activity works between or outside of cells is referred to as **cell-non-autonomous**.

It may be helpful to think of a few specific examples of gene functions that illustrate these terms, although these are not the only functions that meet these descriptions. These concepts are summarized in Figure 6.7 with some examples. A gene that encodes a transcription factor almost certainly will be cell-autonomous in its activity; transcription factors work within the cells in which they are made and do not diffuse between cells. A gene that encodes a ligand in an extracellular signaling pathway will almost certainly be cell-non-autonomous, since the gene product is secreted outside the cell and works on cells at a distance from its place of origin. There are many variations and exceptions to these broad generalities, but determining the cellular focus of the activity of the gene is still a key to interpreting its mutant phenotype.

It is possible to know if a gene's activity is cell-autonomous or not without knowing the molecular function of the gene. Knowing whether the activity of a gene is cell-autonomous or cell-non-autonomous typically requires us either to express the wild-type gene specifically in some cells but not others in a mutant organism or to remove the expression of the gene in some cells but not others in an otherwise wild-type organism. In other words, the organism is a **mosaic** consisting of cells with different genotypes. These strategies are summarized in Figure 6.8; one particular method is described in Tool Box 6.1.

The key experiment is to express the wild-type gene in only some cells, while the other cells have the mutant gene. We then ask if the wild-type and mutant *activities* of the gene correlate at a cellular level with the wild-type and mutant *phenotypes* of the gene. An example is

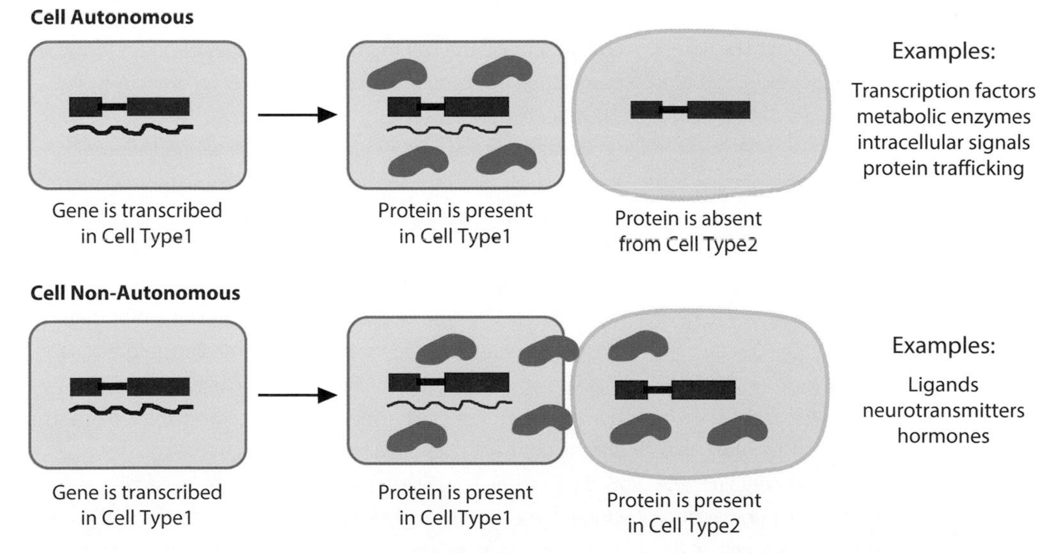

Figure 6.7 Cell-autonomous and cell-non-autonomous gene activity. Genes whose functions work within the cell in which they are transcribed and translated are referred to as cell-autonomous. Among the many examples are genes that encode transcription factors, enzymes for most metabolic processes, and proteins involved in signal transduction. Genes whose functions are needed in cells other than the one in which they are expressed are referred to as cell-non-autonomous. Among the examples are most of the secreted proteins, such as ligands and neurotransmitters. The cellular autonomy of the activity of a gene provides another insight into its function in the organism.

TOOL BOX 6.1 Mosaic analysis

As the chapter discusses, the time and cellular focus of gene activity can often be inferred from mutant phenotypes arising from the gene. The yeast cell cycle presented in Section 6.3 and the case study illustrate how temperature-shift experiments have been used to analyse the *time* of gene activity, or when the function of the gene is needed. In Tool Box 6.1 we discuss how mosaic analysis can be used to analyse the *location* or the cellular focus of gene activity. The examples of mosaics that are the simplest to analyse are the ones in which the activity of the gene of interest is cell-autonomous and works with each cell, so these will be in our focus.

The underlying strategy for mosaic analysis is to construct and analyse an organism with cells of two different genotypes; some cells will have the wild-type or functional allele of a gene, while other cells have the mutant or non-functional allele. As current technology allows investigators to introduce a cloned gene back into an organism, mosaic analysis is most often performed by adding back wild-type activity to some cells of an organism that is mutant overall, as discussed in the chapter.

Before cloned genes and transgenic methods were widely available, when the activity of genes was recognized almost solely from their mutant phenotypes, a different strategy was used. The overall approach in this type of mosaic analysis is to remove the wild-type function of the gene in particular cells and to observe their phenotypes. If the cells continue to show a wild-type phenotype despite the absence of a wild-type allele of the gene, the activity of the gene is assumed not to be necessary in those cells. On the other hand, if the cells show a mutant phenotype, the activity of the gene is assumed to be needed in those cells.

The easiest strategy for removing the wild-type function of a gene from a cell is to begin with an organism that is heterozygous; if the wild-type allele is dominant, as is usually the case, the heterozygous organism shows a wild-type phenotype. Thus only one wild-type allele must be removed for the cell to have a mutant genotype.

The methods of removing the wild-type allele were usually based on the occurrence of some random event during mitosis, such as spontaneous loss of a chromosomal fragment (in *C. elegans*), or an X chromosome (in *Drosophila*) that contained the wild-type allele. As a consequence of relying on random events, different mosaic organisms have mutant patches of cells in different locations on their bodies, depending on where and when the random event occurred. Since the tissues that were mosaic varied, it was important to have recognizable cellular phenotypes to determine the genotypes of individual cells and tissues. These cellular phenotypes, which in *Drosophila* were often bristle structure or cuticle pigmentation, were not themselves the object of study. Rather they were markers for the genotype of individual cells.

The wild-type allele of the gene of interest would then be genetically linked to the wild-type allele of these markers; or, more commonly, the investigator would use markers that happened to map close by the gene of interest. The investigator first scores the population of mosaic organisms to identify cells showing the loss of wild-type activity for the marker gene in the cells of interest. For example, if the gene of interest affects structures in the wing, cellular makers that were mosaic in the wing were found and analysed; mosaic individuals in which the wings were entirely of the wild type or entirely mutant for the cellular marker would be ignored. Thus the effects of the gene of interest can be tested independently of its own phenotype.

While other methods were used to produce mosaic individuals in *Drosophila*, the most common method has been to use somatic crossing over, also known as mitotic recombination, to construct the mosaic flies. The method is summarized in Figure B6.1 and discussed below.

Suppose that the goal of the experiment was to determine whether the gene *m* (*miniature wings*) is required only in particular parts of the wing or it is needed throughout the wing. At the beginning of the experiment, a population of flies is heterozygous for the gene of interest *m* and one or more cell-autonomous markers. A typical configuration is shown in Figure B6.1 which has the cell marker *forked bristles* (*f*) on the same homologue as *m* and a different cell marker *yellow body (y)* on the other homologue.

Somatic crossing over is induced by low-level irradiation with X-rays, usually applied at the larval or pupa stages. The irradiation induces double-stranded DNA breaks, and in the repair of these breaks chromatids are exchanged between the mitotic homologues; the exact mechanism by which this occurs is not known. In Figure B6.1, a hypothetical cell is shown in which somatic crossing over as occurred to the right of *f* marker; note that, in this hypothetical cell, each chromosome has two chromatids but, unlike in mitotic cells of non-irradiated organisms, the sister chromatids have different alleles. When such a cell divides and its daughter cells receive a sister chromatid from each homologue, some of the daughter cells will be homozygous for one of the two markers; others will be homozygous for the other marker. (Other daughter cells will be heterozygous for the two markers, and will not be recognized as being different from cells in which no somatic crossing over occurred.) In our example, some daughter cells will be homozygous for *f* and others will be homozygous for *y*. As these cells continue to divide, they

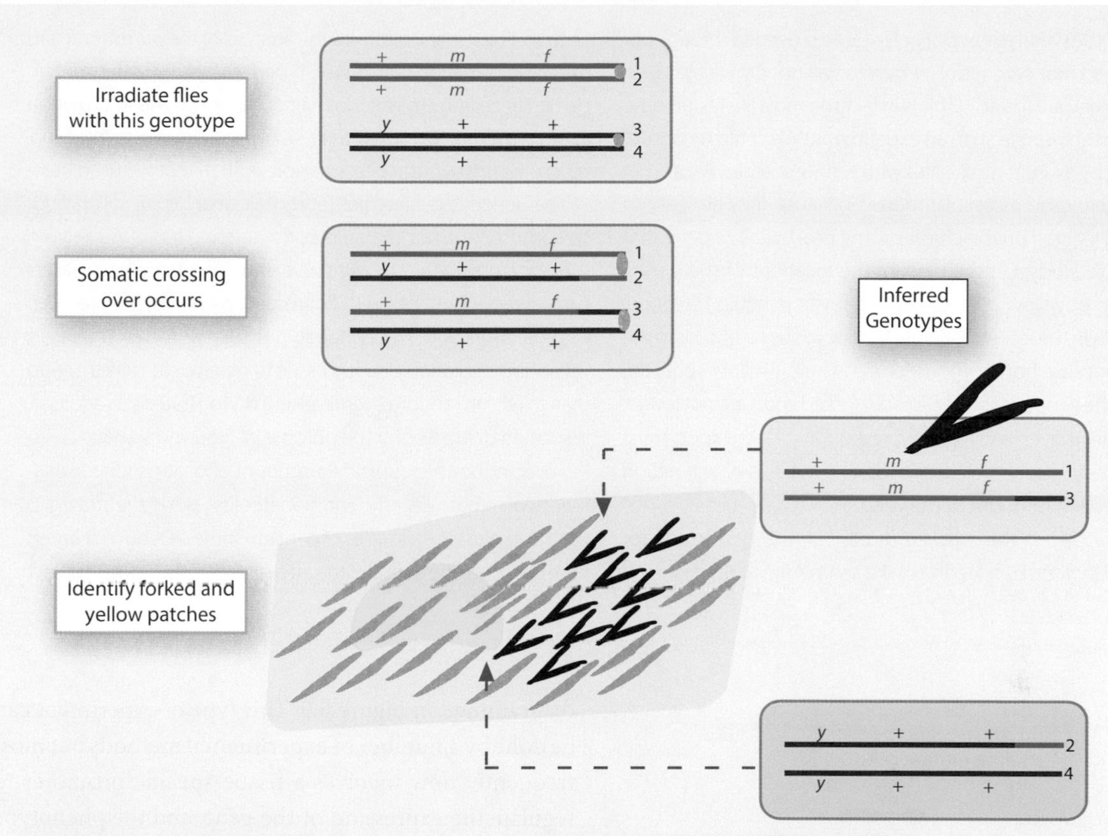

Figure B6.1 Generating mosaics by somatic crossing over in *Drosophila*. A fly that is heterozygous for the gene of interest (designated *m*) is irradiated to induce somatic crossing over. Note that the mutant *m* allele is genetically linked to the easily scored marker for forked bristles (*f*); the other homologue has another easily scored marker, for yellow body (*y*). Irradiation induces somatic crossing. The cell illustrated here has sister chromatids that are not identical; this is a hypothetical situation to illustrate the effect, and may not reflect events in the actual cells. When such as cell undergoes mitosis, its daughter cells will inherit either chromatids 1+4 or 2+3 (which would be wild type and not recognized) or chromatids 1+3 or 2+4. During mitosis, when sister chromatids separate, cells with sister chromatids 1+3 will have forked bristles, while cells with sister chromatids 2+4 will have a yellow body. The cells with forked bristles will also be homozygous for the mutant allele of the gene of interest, and are the ones of interest to score the effect of the mutation in these cells. The presence of cells with a yellow body—that is, with twin spots of yellow and forked—is an important control.

produce mitotic clones or patches of cells with their same genotype, just as happens during mitosis normally. Since patches of both types of marked cells, forked or yellow, are found, these are known as **twin spots**, and are diagnostic for the occurrence of somatic crossing over and for a mosaic fly.

Because the entire larva or pupa has been irradiated, twin spots could be found in any part of the body. Their size and shape will depend on the time when somatic crossing over occurred and on the orientation and number of the subsequent cell divisions; larger clones of cells arise from more cell divisions. Since populations of pupae or larvae are irradiated rather than single flies, some of the flies will have no twin spots and some may have died from the irradiation. In others, the twin spots will be located in parts of the fly that are not of particular interest to the investigator; in our example with *miniature wings*, twin spots on the head or abdomen would not be of interest. Nonetheless, as the investigator sorts through the population of mosaic flies, some will have twin spots in the part of the fly under study.

Look again at our example. The patches with forked bristles would also be homozygous for *m*, and the presence of homozygous *m/m* cells in the wing of a heterozygous fly can be recognized by its bristles without knowing what the effect of *m* will be in these cells. Suppose that some flies have patches of forked bristles in the proximal part of the wing, near the body, and that such flies also have wings reduced in size. We could infer that the wild-type activity of the *m* gene is needed in these cells for the normal outgrowth of the wing. Suppose, by contrast, that some flies have patches of forked bristles in the distal parts of the wing, and that these wings are nearly normal in size; we could infer that the wild-type activity of *m* gene is not needed in these cells.

What if the wild-type activity of *m* gene is needed throughout the wing, so that patches of *m/m* cells will not grow regardless of where they are found? (This is in fact the most likely outcome if *m* had been used in such an experiment.) We could recognize this result by examining the flies with yellow patches. Recall that *y* and *f* were used as cellular makers because they are easy to score and located on the chromosome nearby the *m* gene that we want to study. On the basis of the locations of the genes, twin spots of yellow cells and forked bristles should be found. Just as cells in the forked spots are homozygous for *m*, cells in the yellow spots are homozygous for *m*+. Thus, if yellow spots are found in the wings or mosaic flies but forked spots are not found, we can infer that cells with the corresponding forked spots must not be able to grow—in other words, that the wild-type function of *m* is needed throughout the wing.

We introduced the *patched* mutant in the case study for Chapter 4; *patched* (*ptc*) is one of the segment polarity genes. In Section 11.2 we will return to the analysis of the *ptc* mutant and use the information that its wild-type activity is necessary only in particular cells in the anterior half of the wing. This information came in part from a mosaic analysis of *ptc* in the *Drosophila* wing, much like the hypothetical example with *m* that we used here.

The underlying strategies for mosaic analysis are still in wild-spread use, but not many investigators of *Drosophila* rely on somatic crossing to carry out these experiments. Cellular makers such as reporter genes with fluorescence phenotypes are often easier to work with than phenotypes such as forked bristles or yellow pigmentation, and it is easier to manipulate genes in vitro than to rely on random events arising during mitosis. In addition, irradiation of the flies for the purpose of inducing somatic crossing over introduced substantial amount of death, so the experiments were not very efficient. Nonetheless, while the strategies used to express the activities of genes in some cells but not others have changed, the goal and the interpretation have not changed.

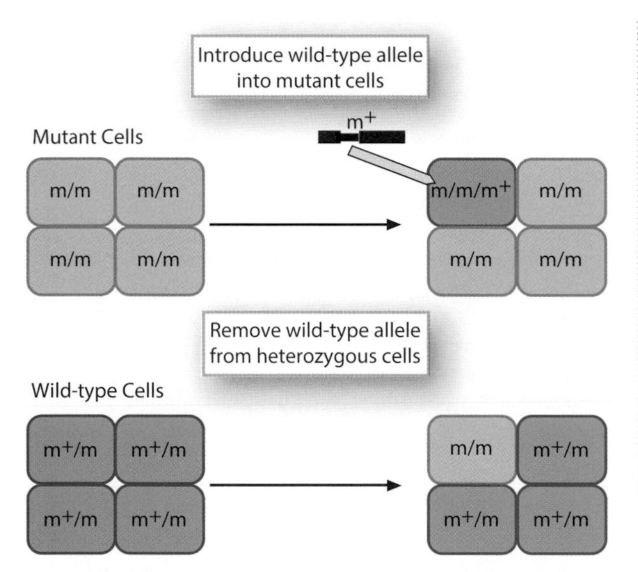

Figure 6.8 Strategies for recognizing whether a gene is cell autonomous. Two general strategies are used to determine whether the activity of a gene is cell-autonomous or not, although the actual experiments can be done in several different ways. In either case, a mosaic organism is made in which some cells are mutant and others are of the wild type, and the phenotype is observed from the mutant and from the wild-type cells. If wild-type activity is observed in cells with the wild-type gene, or if the mutant phenotype is observed in cells with the mutant activity of the gene, the gene is inferred to be cell-autonomous. In the first strategy, the organism is mutant and lacks the wild-type function of the gene in all its cells. The wild-type function can be introduced into some cells, via a transgene for example, and the phenotype can be observed. In the second strategy, the organism has a wild-type phenotype and the wild-type allele is removed from some cells; this is most often done with heterozygotes, so that only one wild-type allele has to be removed.

diagrammed in Figure 6.9. This type of experiment can be done by a number of experimental methods but most frequently now involves a tissue-specific promoter to regulate the expression of the gene, and the phenotype may be the expression of a reporter gene rather than some aspect of the organism; but the interpretation is the same. If the cells with the wild-type activity of the gene have a wild-type phenotype but surrounding cells with a mutant activity have a mutant phenotype, the mutation is said to be cell-autonomous. If, on the other hand, the genotypically mutant cells show the wild-type phenotype of their neighboring cells, the mutation is considered to be cell-non-autonomous.

While mosaic analysis can either introduce the wild-type activity (via a wild-type allele) into cells that lack it or remove the wild-type activity from cells that previously had it, most approaches nowadays involve introducing a wild-type activity (or allele) of the gene into an organism that lacks it. The introduced gene may be expressed only in particular tissues or at particular times, or its expression could be regulated by the investigator through external stimuli such as heat shock or light flashes, as is done with optigenetic methods.

Suppose, as suggested by Figure 6.9, that we find a worm that is paralysed and we identify and clone the wild-type copy of the gene that has been mutated. We want to know whether the activity of this gene is needed in the muscles or in the nervous system.

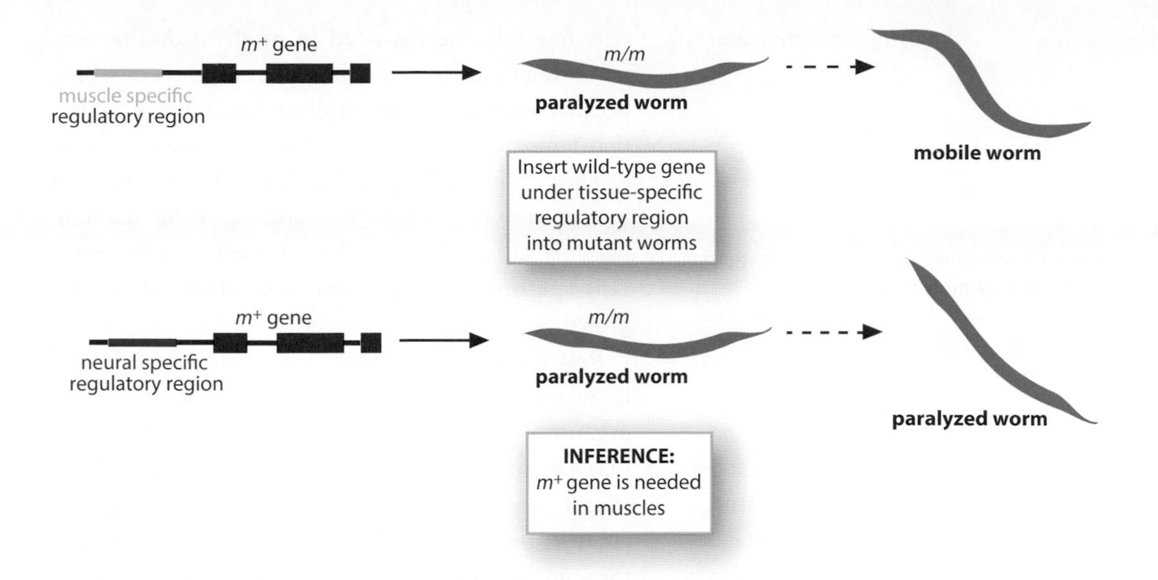

Figure 6.9 The cellular locus of gene activity using transgenes. One strategy for determining the cells that require the activity of a gene consists in using different transgenes. This is illustrated for paralysed worms, which are *m/m*. In one strain, the wild-type m^+ allele is provided on a transgene under the regulation of a muscle-specific regulatory region, whereas in another strain the wild-type m^+ allele is provided on a transgene under the regulation of a neural-specific regulatory region. If the m^+ gene is required in muscles, as illustrated here, the strain with expression with muscle-specific regulation will have a wild-type phenotype, whereas the strain with expression with neural-specific regulation will have a mutant phenotype.

We could place the wild-type copy of the gene under the control of a regulatory region that is active only in the muscles, and introduce this into a mutant worm. If the worm is now able to move, we infer that the gene is active in the muscles. A similar set of experiments could be done for the nervous system, and to determine whether the activity of the gene is cell-autonomous or cell-non-autonomous.

Originally mosaic analysis was done entirely with mutant phenotypes, since no gene had yet been cloned and effective methods to reintroduce a gene into an organism or cells did not exist. In these older approaches, the wild-type activity of the gene (as determined by its wild-type allele) was removed in certain cells or tissues and the effect on the phenotype was observed. This use of mosaic analysis in *Drosophila* is described in Tool Box 6.1. This method resulted in substantial cell (and organismal) death and has largely been supplanted by molecular manipulations of the expression of particular genes.

By whatever method mosaic organisms are made, the inferences are similar. If removing the wild-type allele from a particular cell results in a mutant phenotype, the activity of the gene is needed in that cell; similarly, if supplying the wild-type activity of the gene to particular cells results in a wild-type phenotype, the activity of the gene is needed within that cell. If removing the wild-type allele does not result in a mutant phenotype, the activity of the gene is not needed in that cell.

6.3 Conditional mutations and the time of gene activity

One of the most commonly used strategies for analysing a mutant phenotype and to infer the activity of a gene has been to use conditional mutations. The concept of a conditional mutation was introduced in Chapter 4 as a means to maintain lethal mutants as homozygotes, but we want to expand their use here. In its broadest sense, a conditional mutation is one whose mutant phenotype is observed only under certain environmental conditions. Under other environmental conditions, the mutation has a wild-type phenotype. With this definition, many types of mutations are conditional, including auxotrophic mutations and drug-resistant mutations.

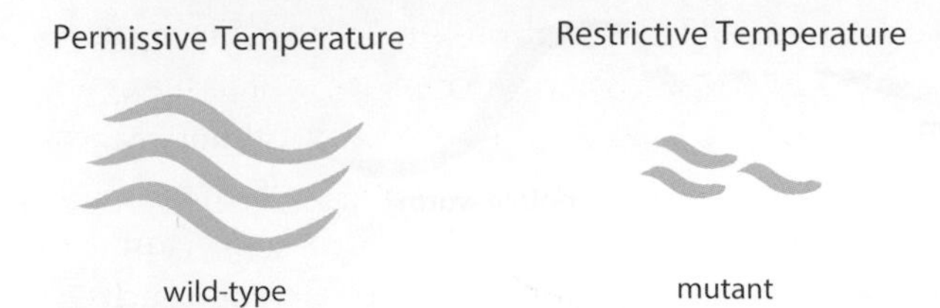

Figure 6.10 Temperature-sensitive conditional mutations, illustrated here with the dumpy morphology in worms. At the low or permissive temperature, the genotypically mutant worm has normal body shape. At the restrictive or high temperature, the genotypically mutant worm is dumpy.

For the purposes of analysing gene activity, the most useful type of conditional mutant is a **temperature-sensitive mutation**. A temperature-sensitive mutation is one that shows a mutant phenotype at one growth temperature but a wild-type phenotype at a different growth temperature. We illustrate this in Figure 6.10 with a dumpy mutant phenotype in the body shape of worms. Most temperature-sensitive mutations are heat-sensitive: the mutant phenotype is seen at the high temperature and the wild-type phenotype is seen at the low temperature. A few temperature-sensitive mutations are **cold-sensitive** and show the mutant phenotype at low temperature. However, since heat-sensitive mutations are by far the more common category, most investigators use the term "temperature-sensitive" to refer to mutations that show a mutant phenotype at high temperature, and "cold-sensitive" to refer to mutations that show a mutant phenotype at low temperature. In standard notation, temperature-sensitive mutations are often indicated with "*ts*," while cold-sensitive mutations are notated with "*cs*."

A few other terms need to be defined. The temperature at which the mutant phenotype is observed is called **restrictive temperature**. This reflects the fact that temperature-sensitive mutations have been used extensively in working with lethal mutations, and this temperature restricts the organism's growth. The temperature at which the wild-type phenotype is seen is called **permissive temperature**, since it permits normal growth of the organism.

Restrictive and permissive temperatures are defined in relation to the normal growth conditions of the organism. For example, *C. elegans* is normally grown in the laboratory at 20 C. Thus the restrictive temperature for temperature-sensitive mutations in worms is about 25 or 26 C, near the top of the temperature range for wild-type worms. The permissive temperature for a worm temperature-sensitive mutation is usually 15 C. *Saccharomyces cerevisiae* is normally grown at 30 C, so the restrictive temperature is 36 C or 37 C and the permissive temperature is 22 C or 23 C. In most cases, the molecular or biochemical explanation for temperature-sensitivity is not known. However, the simplest and most likely explanation is that a missense mutation in the gene resulted in an amino acid substitution that has made the protein less stable. At the slightly elevated restrictive temperature, the thermolabile mutant protein unfolds while the wild-type protein maintains its normal conformation. This explanation has been demonstrated to be correct for some genes and proteins, and is assumed to be appropriate for many others without a direct demonstration. For example, the dumpy phenotype illustrated in Figure 6.10 corresponds to a missense mutation in the gene, as shown in Figure 6.11.

>DPY-7 Wild-type
MEKPSSGANHVAKATVSLSIASVLILGAVLTMLSIQLDEAHERLQNRMGSFKFVARNIWHDIVLVKSNGRIKRQYGGYGS
DSAQSDNQQCTSCVQLRCPP**G**PIGPPGVSGEPGMDGANGRPGKPGLDGLDVPLDPEPAFPCVICPAGPPGTRGPQGEVGR
PGQTGESGHPGLPGRPGKPGRVGDAGPQGEPGEQGEPGIKGPPGDDSIGGTGIKGPPGPPGPRGPKGPPGSNGLPSQNSG
PPGPIGEMGPPGPPGPRGEPGPPGPFGPPGDSGEPGGHCPSSCGVQEIVAPSVSELDTNDEPEKPARGGYSGGGYGKK

>DPY-7 (tsmutant)
MEKPSSGANHVAKATVSLSIASVLILGAVLTMLSIQLDEAHERLQNRMGSFKFVARNIWHDIVLVKSNGRIKRQYGGYGS
DSAQSDNQQCTSCVQLRCPP**R**PIGPPGVSGEPGMDGANGRPGKPGLDGLDVPLDPEPAFPCVICPAGPPGTRGPQGEVGR
PGQTGESGHPGLPGRPGKPGRVGDAGPQGEPGEQGEPGIKGPPGDDSIGGTGIKGPPGPPGPRGPKGPPGSNGLPSQNSG
PPGPIGEMGPPGPPGPRGEPGPPGPFGPPGDSGEPGGHCPSSCGVQEIVAPSVSELDTNDEPEKPARGGYSGGGYGKK

Figure 6.11 Temperature-sensitive mutations are often missense mutations. A common and reasonable inference is that a temperature-sensitive mutation makes a polypeptide product that is less stable at elevated temperatures. A specific example is shown here, in which the temperature-sensitive allele for the *dpy* gene in worms has an arginine (R) rather than the glycine (G) found in the wild-type allele.

Temperature-shift experiments can determine the approximate time of gene activity

One way in which temperature-sensitive mutations have been widely used is as a means for working with lethal mutations; at the permissive temperature, the mutant cells or organism develop normally, so an otherwise lethal mutation can be maintained in culture as a homozygote. In addition to maintaining lethal mutations, temperature-sensitive mutations have also been widely used to estimate the *time* of gene activity. The method involves doing a **temperature shift**—that is, growing the organism at the permissive temperature for part of its life cycle and at the restrictive temperature for another part of its life cycle. Both a **shift up** from the permissive temperature to the restrictive temperature and a **shift down** from the restrictive temperature to the permissive temperature provide informative data. The shift up is thought of as blocking gene activity, while the shift down is thought of as restoring gene activity. Typically, only one temperature shift is done for any set of experiments so that once shifted to a growth temperature, the organism remains at that temperature for the rest of its life cycle.

We can illustrate with a hypothetical but realistic example from *C. elegans*, beginning with the shift-up experiments. The example is diagrammed in Figure 6.12; the temperature-sensitive mutant phenotype is such that the worm is Dumpy in shape, and the gene is abbreviated as *dpy*. A worm that is homozygous mutant for a temperature-sensitive *dpy* mutation (abbreviated *ts*; so the mutant is *ts/ts*) is allowed to lay eggs at the permissive temperature. The eggs will develop and hatch into larvae at the permissive temperature and the worms have a normal morphology. If the mutant strain had been grown at the restrictive temperature, all the worms would be Dumpy, since the gene's activity would be blocked.

At various times, some of the eggs and larvae are shifted up to the restrictive temperature, and the phe notype of the adult worms that arise from these eggs and larvae is observed. If the worms are shifted to the restrictive temperature *before* the time when activity of

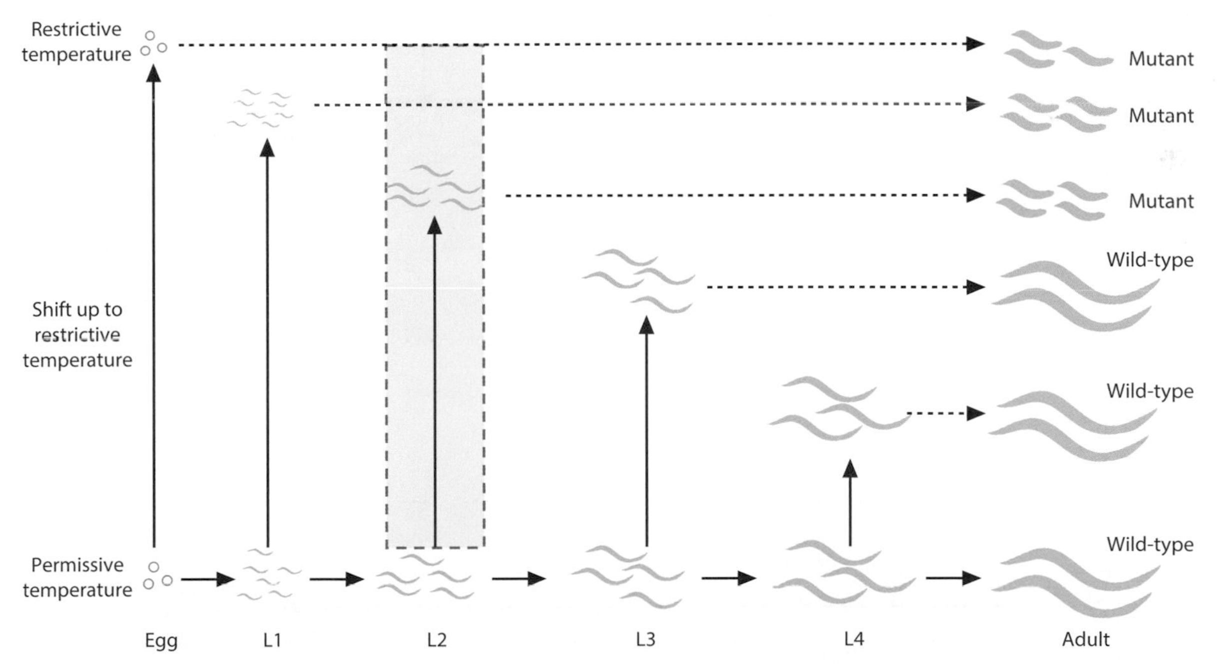

Figure 6.12 Shift-up experiments. The goal of temperature-shift experiments is to determine when the activity of the gene is needed. In this example, all the worms are homozygous for a temperature-sensitive *dumpy* mutation. A population of worms is grown at the permissive temperature. At various times during the life cycle, some of the worms are shifted up to the restrictive temperature, as indicated by the upwards red arrows. Once shifted up, the worms are maintained at the restrictive temperature, and the phenotype is observed in adults. The diagram illustrates that worms shifted up the egg stage or at the L1 or L2 stage become dumpy mutant adults. Worms shifted up at the L3 or L4 stage become wild-type adults. The inference is that the activity of the gene is needed after the egg and L1 stages and before the L3 and L4 stages; that is, probably during the L2 stage, as shown by the green boxed area.

the gene is needed, the *mutant* phenotype will be seen and the adult worms are Dumpy. Since the activity of the gene is assumed to be impaired at the restrictive temperature, we infer from this result that the activity of the gene must have been needed during the time when the worm is at the restrictive temperature. On the other hand, if the worms are shifted to the restrictive temperature *after* the time when the activity of the gene is needed, the *wild-type* phenotype will be seen. We infer that the activity of the gene has occurred normally at the permissive temperature and was concluded before the shift up was done.

Imagine that the activity of our hypothetical gene is needed only during the second larval stage, as suggested by Figure 6.12. If the worms are shifted to the restrictive temperature during embryogenesis in the egg or at the first larval stage, the worm will experience the second larval stage at the restrictive temperature and the activity of the gene will be impaired. Mutant worms arise. If the worms are shifted to the restrictive temperature at the third or fourth larval stage, the worms will have experienced the second larval stage at the permissive temperature at which the gene functions normally. Although they are growing now at the restrictive temperature, the worms will be wild type in morphology. By doing shift-up experiments at different times of the life cycle, the time at which the gene is active can be inferred.

We can now perform the same type of analysis with the same temperature-sensitive mutant using shift-down experiments. This experiment is illustrated in Figure 6.13. The *ts/ts* mutant worm is allowed to lay its eggs at the restrictive temperature; the eggs and larvae are shifted to the permissive temperature at different times after egg-laying. Worms that are shifted to the permissive temperature during embryogenesis or the first larval stage will go through the second larval stage at the permissive temperature, and will therefore be of the wild type. The gene that affects morphology has functioned normally. Worms that are shifted to the permissive temperature at the third or fourth larval stages will have

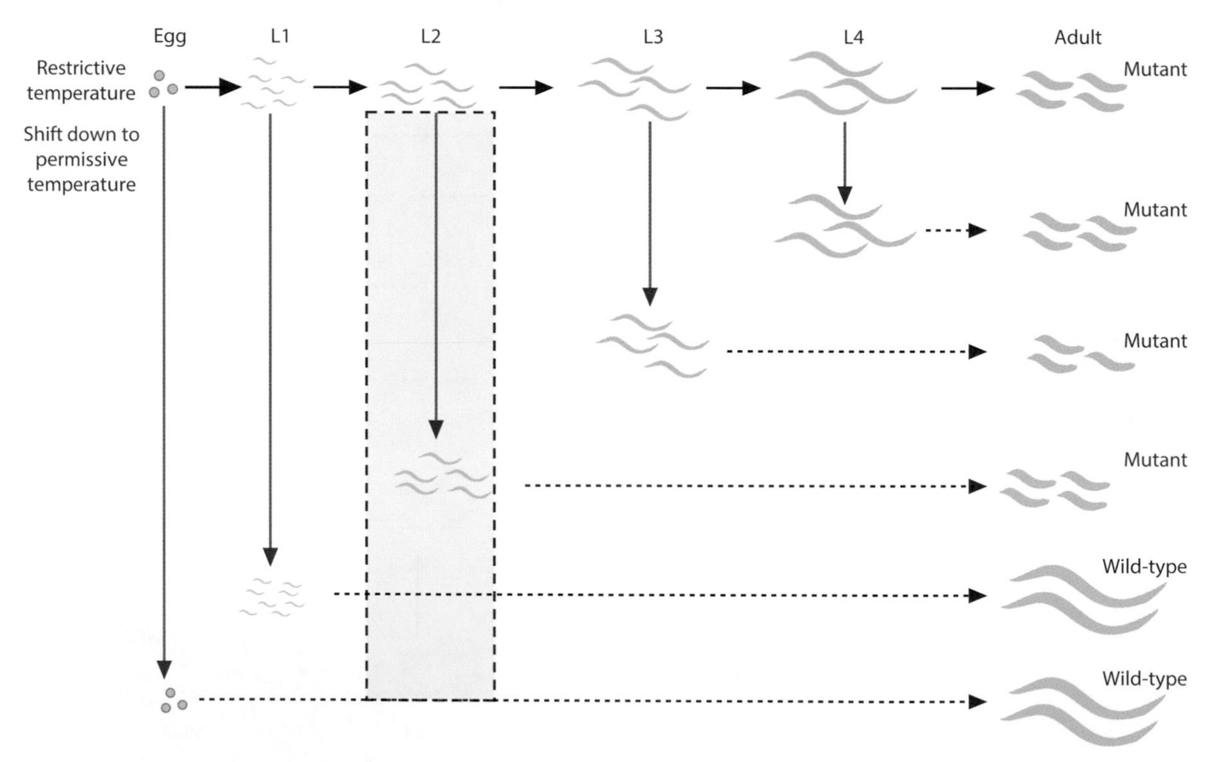

Figure 6.13 Shift-down experiments. As with the shift-up experiments in Figure 6.12, the goal of shift-down experiments is to determine the time when the activity of the gene is needed. A population of worms that is homozygous for a temperature-sensitive *dumpy* mutation is grown at the restrictive temperature. At different stages of the life cycle, worms are shifted down to the permissive temperature, as indicated by the downwards blue arrows. Once shifted down, the worms are maintained at the permissive temperature and the adult phenotype is observed. The diagram illustrates that worms shifted down before the L2 stage become phenotypically wild-type adults, whereas worms shifted during or after the L2 stage become phenotypically mutant adults. The inference is that the gene is active after L1 stage and during the L2 stage, as shown in the green box.

already experienced the critical second larval stage at the restrictive temperature and will therefore be mutant because the activity of the gene has been blocked.

The reciprocal shift experiments, the shift up and the shift down, have each identified the second larval stage as the stage of gene activity. The results of the shift-up and shift-down experiments are synthesized in Figure 6.14. The second larval stage has been variously termed **t-crit**, which stands for critical time, **stage of execution**, or more commonly **temperature-sensitive period (TSP)**. If a population of the temperature-sensitive mutants can be synchronized in development well enough, it may be possible to define the TSP even more narrowly, by shifting the worms up or down at specific times during the second larval stage. However, in practice, it is often more useful to think of the TSP broadly or as an approximate time of gene activity rather than to attempt to pinpoint it very precisely.

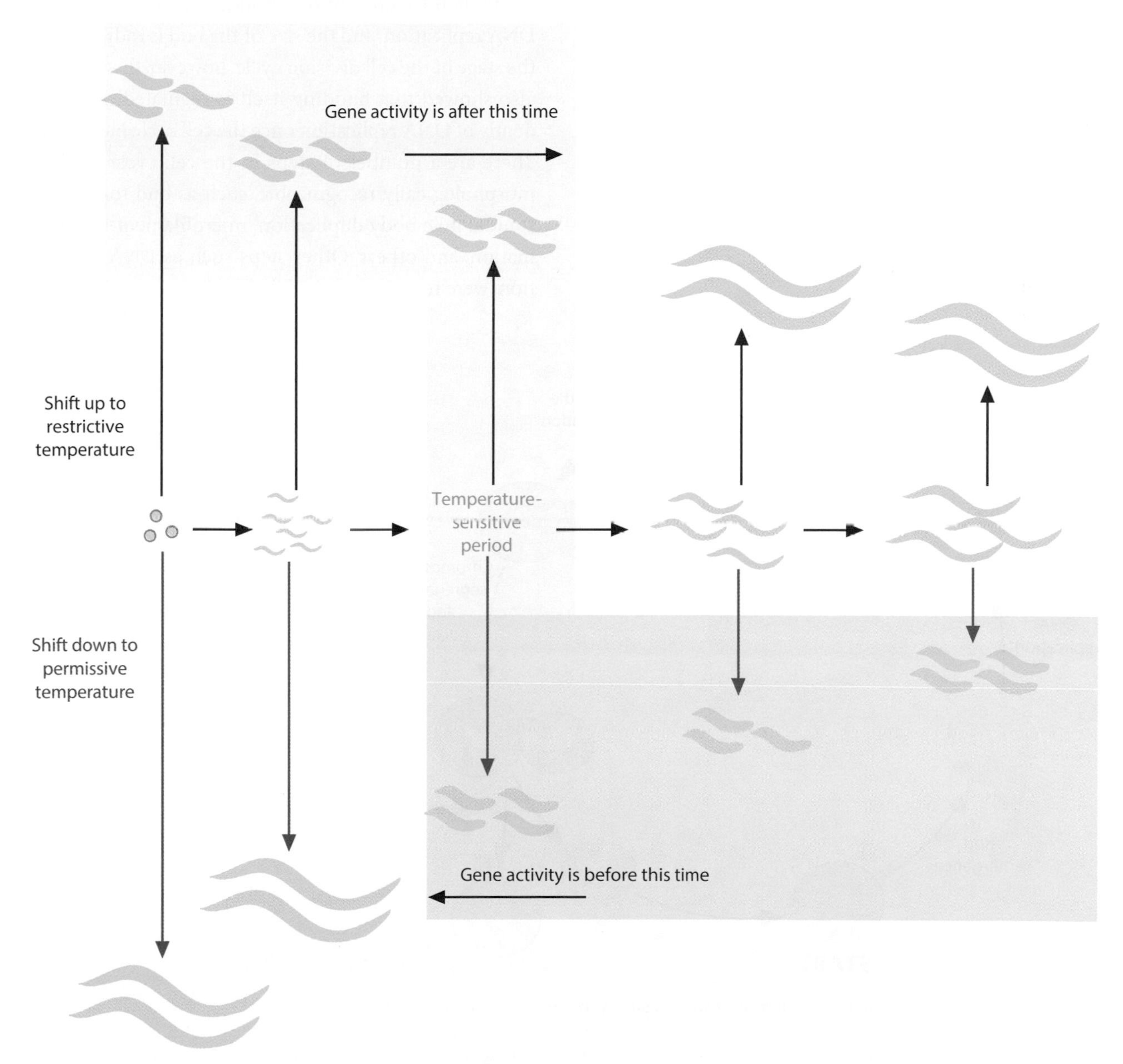

Figure 6.14 The temperature-sensitive period. The results shown in Figures 6.12 and 6.13 are combined here. If the temperature-sensitive period is experienced at the restrictive temperature, the mutant phenotype will be observed. Shifts up to the restrictive temperature before the L2 stage result in a mutant phenotype, so the activity of the gene is required during or after this stage. Shifts down to the permissive temperature after the L2 stage result in a mutant phenotype, so the activity of the gene is required before or during this stage. The region of overlap is shown in the green box as the temperature-sensitive period.

The cell cycle in yeast was analysed using temperature-sensitive mutations

The analysis of the yeast cell cycle is a venerated example of the uses of temperature-sensitive mutants in analysing and answering a significant biological question. This example shows how the general concepts on interpreting phenotypes introduced in Section 6.2 were used in a specific case. In the decades since Lee Hartwell and his students began this analysis, the eukaryotic cell cycle has become a fundamental and familiar topic in molecular and cell biology courses, and it dominates our thinking about cancer cell therapy. Modern students might learn the cell cycle as a cascade of phosphorylation steps—that is, as a biochemical pathway—and might not realize that it began with the analysis of temperature-sensitive mutations in yeast. For this and other work on the cell cycle, Hartwell shared the Nobel Prize in Physiology and Medicine in 2001. In an autobiographical sketch, he mentions that his thinking about the yeast mutants was influenced by his contact with Robert Edgar, who pioneered the use of temperature-sensitive mutations in his analysis of the bacteriophage T4 assembly.

Literature Link. Hartwell, Leland H. et al. 1974, *Science* 183: 46–51

The yeast *Saccharomyces cerevisiae* is called budding yeast because a cell divides by producing a bud, as shown in Figure 6.15. Like mitosis, budding is correlated with DNA replication, and the size of the bud is indicative of the stage of the cell division cycle; however, this analysis also showed that budding itself is regulated independently of DNA replication once the cell cycle has begun. There are a number of steps in the cell cycle that are morphologically recognizable, such as bud formation, spindle pole body duplication, microfilament ring formation, and others. Other steps such as DNA replication, were recognizable by biochemical assays. Hartwell

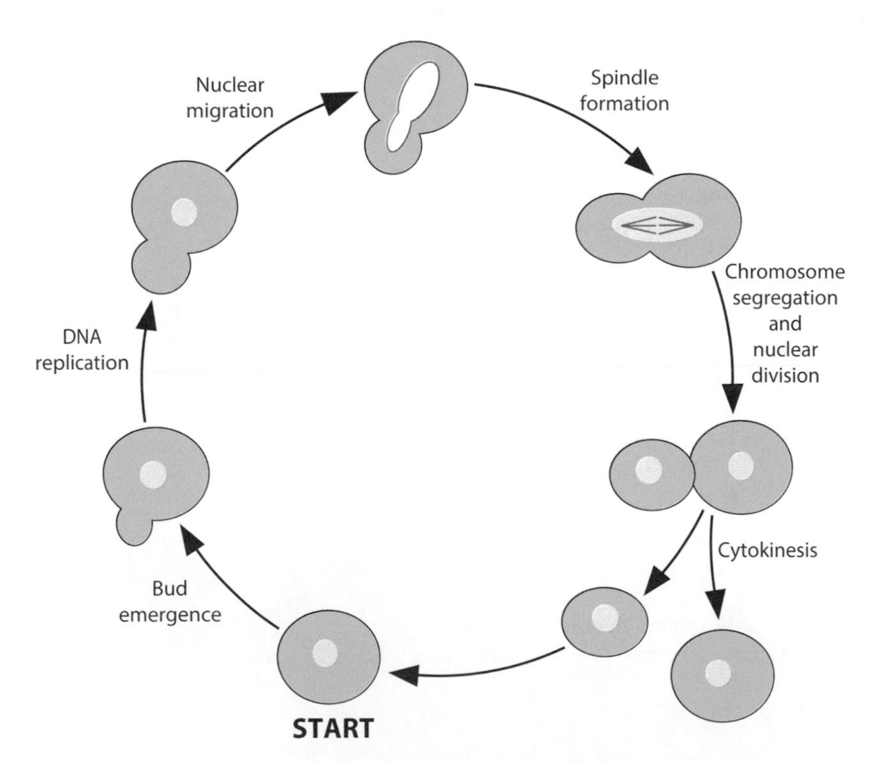

Figure 6.15 The cell cycle of the budding yeast *Saccharomyces cerevisiae*. Some of the key morphological features of the cell cycle are depicted, including the corresponding cellular and molecular processes that could be used as landmark events. Temperature-sensitive mutations that arrested the yeast cell cycle at different points, identifying the *cell division cycle* or *cdc* genes, were used to understand the genetic, molecular, and biochemical regulation of the cell cycle of nearly all eukaryotic cells. A shift to the restrictive temperature before the critical stage results in a cell cycle arrest with a diagnostic phenotype. A shift to the restrictive temperature after the critical stage allows the cell to complete one cell cycle before arresting at the next cycle.

referred to these recognizable steps as **landmark events**, which could be used as specific markers of the general phenotype. He and his students isolated 150 mutations in thirty-five different genes that grew normally at the permissive temperature of 22 C but that arrested cell division at the restrictive temperatures of 36 C. (Additional mutations and genes have been identified since the initial studies.) The genes were termed **c**ell ***d***ivision **c**ycle defective, abbreviated *cdc1*, *cdc2*, etc. Recall from Chapter 3 that yeast can be grown as either a haploid or a diploid; Hartwell isolated these mutants in haploids and cultured them at the permissive temperature so that they grew normally.

The mutations had discrete terminal arrest phenotypes, recognized by their failure to execute one of the landmark events. For example, the mutation diagrammed in Figure 6.16 results in cells that arrest with buds forming but with no DNA replication. These landmark events are analogous to Hadorn's effective lethal phase. The first mutant defect that could be observed in a given mutant was termed its **diagnostic landmark**, which is comparable to Hadorn's concept of the phenocritical phase.

As shown in Figure 6.17, when grown at the restrictive temperature, a single cell progresses through the cell cycle to its next time of arrest. When a culture of growing yeast cells was shifted to the restrictive temperature, the cells arrested at the same point in the cell division cycle, even though the cells were at different stages in the cell cycle when shifted. With temperature shifts at different times

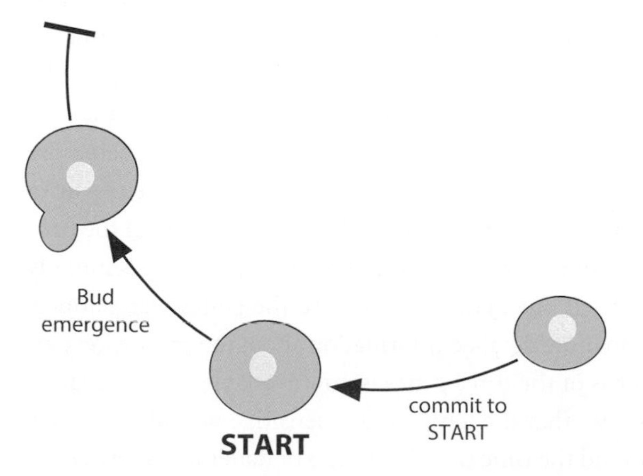

Figure 6.16 The phenotype of a *cell division cycle* (*cdc*) mutant. The drawing shows a mutant that initiates the cell division cycle and forms a bud but cannot initiation DNA replication at the restrictive temperature. The cells arrest at this stage, until they are shifted to the permissive temperature.

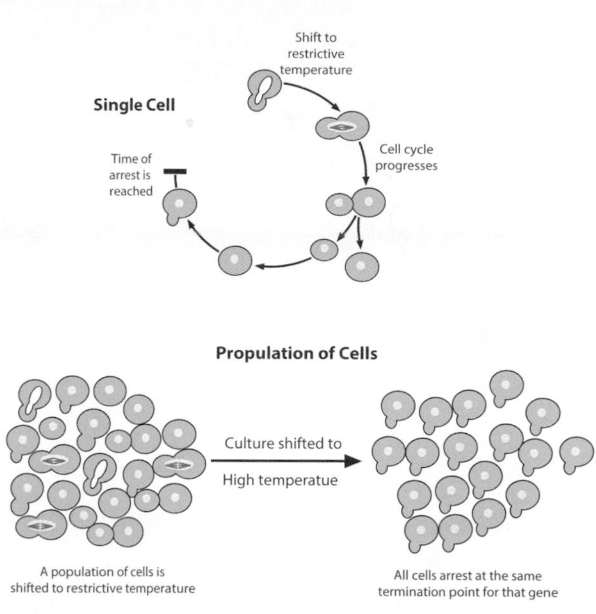

Figure 6.17 Temperature-shift experiments and the times of mutant arrest. Temperature-sensitive mutations in *cdc* genes were found to have characteristic times of arrest or termination points, which could be determined by a temperature shift experiment. When a cell was shifted to the restrictive temperature, it could complete some stages of the cell cycle but would arrest at the time in the next cycle, when the activity of the gene was needed. The conclusion about this example is that the function of this gene is needed after bud emergence but is not needed for the other events of the cell cycle. When a population of cells, originally at different stages of the cell cycle, is shifted to the restrictive temperature, all the cells arrest at the same cellular event. This implies that this cellular event requires the wild-type function of this gene.

in the cell cycle, a temperature-sensitive period for each mutation could be established; if a cell was shifted after that time, it completed its current cell cycle and did not arrest until the next cell cycle. Hartwell referred to this temperature-sensitive period as the **execution point,** the time in the cell cycle at which the defective gene activity is normally performed—or, alternatively, the time when the cell is no longer sensitive to shift-up experiments. He emphasizes that the execution point (and, by analogy, the temperature-sensitive period) could be allele-specific rather than characteristic for a gene. This is because the alleles could have differing amounts of residual gene activity after the shift to the restrictive temperature. As a result, a mutant allele with more residual activity after the shift up will be able to carry out more of the cell cycle stages than one with little residual activity.

Because the yeast cell cycle has an observable sequence of morphological events, it was possible to use the diagnostic landmarks to place different genes in the order of gene activity, as shown in Figure 6.18. For example,

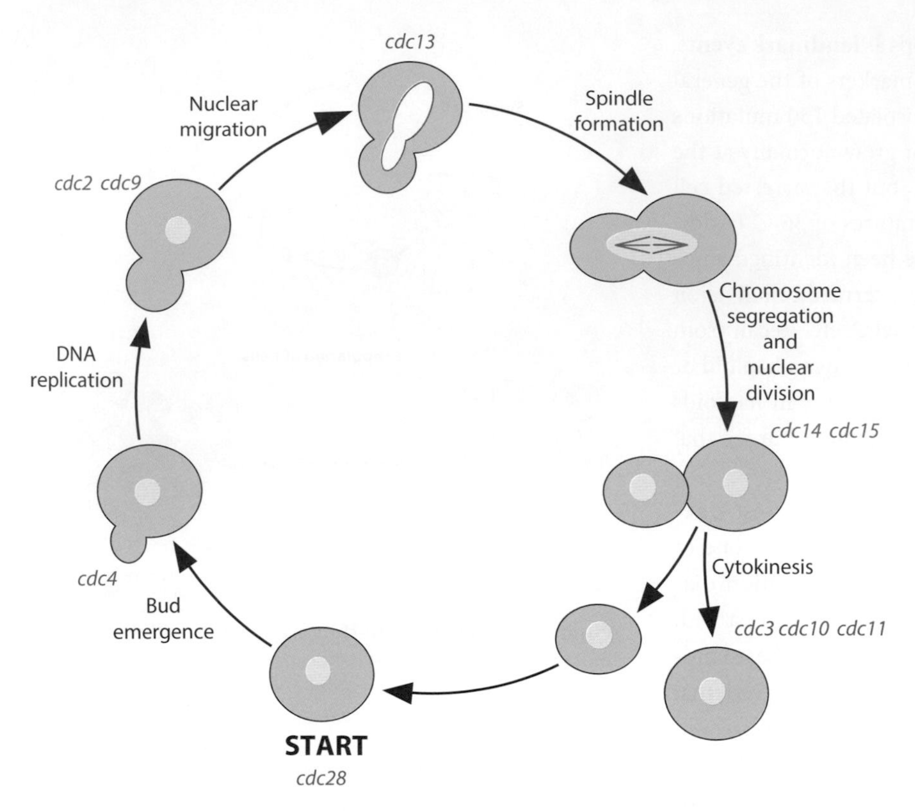

Figure 6.18 Using *cdc* mutations to characterize the cell cycle. Some of the *cdc* mutations used to define events in the eukaryotic cell cycle are shown, with the times when the function of the wild-type gene is needed. For example, the gene *cdc28* defines START, the commitment of a cell to enter the cell division cycle. Most of the *cdc* genes have orthologous genes in all other eukaryotes.

since spindle pole body emergence occurs before the chitin ring forms to separate the budding cells, mutants whose diagnostic landmark is spindle pole body emergence must affect an earlier stage in the cell division cycle than mutants whose diagnostic landmark is chitin ring formation.

It was also possible to use the temperature shifts to learn which events in the cell cycle occur independently; that is, a mutant that fails to carry out one landmark event may still be able to carry out other landmark events normally, because the events are regulated separately. For example, bud emergence can still occur even if chitin ring formation is blocked and vice versa, indicating that these processes are independently regulated. Therefore, although yeast cell division appears to be a cycle, the cycle branches into two or more separable pathways at certain points. This topic will be discussed again in Chapter 11, when the construction and interpretation of genetic pathways are described.

Before cloning a gene and analysing its expression became relatively routine experiments, temperature-sensitive periods were a common and usually easy strategy to infer when the activity of a gene is needed. The investigator simply had to move the cultures from one incubator to another and allow the organisms to grow. It is now possible to re-examine some of the results involving temperature-sensitive mutants and to compare the TSPs for gene activity with the times at which the corresponding genes are now known to be expressed. Case Study 6.1 describes the analysis of temperature-sensitive periods for three genes in *C. elegans* that affect the process of dauer larvae formation; the analysis was initially carried out many years before the genes were cloned, and more detailed information about the expression patterns of the three genes is now available. The case study shows that the temperature-sensitive period accurately found the time that the *process* of dauer larva formation occurs, which was somewhat different from the time at which some of the individual genes are expressed as part of this process. But it also demonstrates how accurately the temperature-sensitive period was able to capture that process.

6.4 Gene activity, gene expression levels, and biological noise

Any consideration of gene activity also touches on gene expression, since the two are clearly related to each other. We noted earlier that gene activity is usually based on inferences from a mutant phenotype. Gene expression, on the other hand, is usually measured directly from the cloned gene, without any necessary consideration of the mutant phenotype. If the gene is not expressed at all in the mutant, which would be a null mutant phenotype, gene expression and gene activity overlap each other, as we have described. But many changes in gene expression, particularly among natural populations, are not ones that completely shut off the expression of the gene like a null mutation. Instead, they are ones that affect the level at which the gene is being expressed. How do quantitative changes in gene expression affect gene activity and phenotypes?

The effect of quantitative changes in gene expression is a somewhat poorly charted territory in the landscape of genotypes and phenotypes. At first glance, there are two contrasting (and seemingly contradictory) observations for the effects that changing gene expression levels may have on phenotypes. On the one hand, many studies have shown that *overexpression* of the normal function for most genes has little or no effect on the overall phenotype of the organism. For instance, additional wild-type copies of genes are routinely inserted into the genomes of model organisms via microinjection, plasmid, electroporation, and transposable elements; while high copy suppressors in yeast are a well-defined phenomenon (as discussed in Tool Box 10.1) and investigators are always aware of the possible effects of overexpressing a gene, most of these effects are uncommon and relatively minor. Furthermore, the observation that most mutations are recessive to wild type shows that most genes are not sensitive to a twofold *reduction* in the level of gene expression either. These results suggest that neither overexpression nor underexpression of the wild-type function of a gene at physiologically relevant levels usually makes a significant change in the phenotype.

On the other hand, there are genes for which overexpression or underexpression clearly does produce a mutant phenotype, and thus has a major effect on gene activity; we discuss a few examples here. These quantitative questions become important, since very small changes in gene expression can be detected and correlated to changes in the phenotype, including the disease state, when the sensitive assays of microarrays, quantitative PCR, and RNA sequencing are used. It is widely thought that some or many of the variants discovered by genome-wide association studies could be changing the expression level of the genes, so-called qSNPs or eSNPs (for quantitative or expression single nucleotide polymorphisms), rather than the functions of the protein encoded by the gene. It is clear that changing the level of expression for a gene can have significant effects on its activity and on the phenotype it produces. These considerations lead us to ask about the effect of this variation in the expression levels of genes.

How much change in gene expression is biologically significant?

One of the underlying assumptions about the interpretation of gene expression data is that a change in the expression pattern of a gene or group of genes is correlated with a change in the phenotype of the organism, tissue, or cell. This may seem so obvious that it needs no further exploration, but hidden behind this assumption is a challenging question: how much change in gene expression (RNA or protein) is biologically meaningful? Current gene expression techniques, particularly those based on transcription, can detect very small changes in the expression level of a gene, as slight as a third more or fewer transcripts from a gene. Genomic techniques can use statistical tests to produce long lists of genes whose transcription levels are significantly different under different conditions. However, what is often missing from these studies is a sense that the *statistically* significant change in expression levels is also a *biologically* significant change in the phenotype.

Literature Link. Bar-Even et al. 2006, *Nature Genetics* 38: 636–43

As we noted above, there are many examples in which small expression changes have major biological relevance. Our example in Chapters 1 and 3 involves sex determination in flies, a process that depends on twofold differences in the copy number of certain key genes on the X chromosome; and these differences in turn lead to a twofold difference in the expression of these genes. Even changes smaller than twofold in gene expression levels would be found for some of the phenotypic differences between a Down syndrome child with trisomy 21 and an unaffected disomic 21; trisomics have a recognizable change in the phenotype. The analysis of trisomics has indicated that some phenotypes are sensitive to gene expression differences in the range of two versus three copies, a difference of 1.5-fold. In Chapter 4 we also noted that the expression of several genes important for human cancers, such as p53 and Rb, is tightly regulated and the genes probably must be present in two copies. Hence small changes in gene expression levels, both overexpression and underexpression, are not merely a theoretical question. The difference could have significant biological and clinical implications.

So the first answer to the question that opens this section is: "It depends on the gene." Many genes are relatively insensitive to changes in expression level, while others are very sensitive even to small changes. When working with a single gene, standard experiments can be performed to determine the sensitivity of the gene in question; but, for a study involving genome-wide changes in gene expression, this answer is unsatisfactory. If the goal of genome-wide expression experiment is to determine a transcription profile rather than to focus on the individual genes that control a process, this lack of a more precise answer is somewhat more acceptable. For example, if the goal is to determine the gene expression patterns that distinguish a metastatic tumor cell from a non-cancerous cell, it makes no difference which (if any) of the individual genes is dose-sensitive in its effects; what matters is that the profile is distinct. If the goal, however, is to find a therapy that prevents the cell from progressing into a tumor, it can be vitally important to determine which genes are driving the process.

The level of biological noise

A more complete answer to our question about how much change in gene expression is phenotypically significant comes from studies that combine microarrays, quantitative RT-PCR, and reporter gene assays to measure biological noise that affects RNA and protein expression. We describe one such study here, then return to this topic in Chapter 12. We are using the term **biological noise** here to refer to the stochastic variation in the level of gene expression that occurs under natural conditions. That is, if one could precisely measure the level of expression from the same gene under the same conditions in two identical cells, how much would the measurements fluctuate for biological reasons rather than for technical ones?

Biological noise arises from a complex and unknown combination of variations in all of the factors that affect gene expression: transcriptional initiation, the rate of mRNA synthesis and decay, initiation and rate of translation, protein turnover, and so on. In one study in eukaryotes, more than 4,100 yeast strains expressing green fluorescent protein (GFP) translational fusions to different individual proteins were monitored using flow cytometry to measure the protein expression level, as summarized in Figure 6.19.

Literature Link. Newman et al. 2006, *Nature* 441: 840–6

The reporter genes were inserted at their usual site in the genome by homologous recombination, and thus were under the control of the same regulatory region as the wild-type gene at that site. The assay allowed fluorescence measurements of single cells from cultures with more than 50,000 cells grown to mid-log phase in either rich or minimal media. More than 2,500 genes comprising nearly 40 percent of the yeast genome could be detected and scored in the assay. These assays were followed by microarray and quantitative RT-PCR experiments to determine RNA expression patterns, and the RNA and protein expression levels were compared.

A few general conclusions emerge from the study and have been generally supported by other studies. The first is that most of the variation in protein levels is also seen in mRNA levels. Thus, most of the regulation of gene expression occurs during transcription. On the other hand, twofold changes in mRNA levels were not as frequently seen as corresponding changes in protein levels. This indicates that most biological noise arises from the rate of mRNA synthesis and turnover, but that not all of these transcriptional fluctuations are reflected in protein levels. Additional regulation occurs post-transcriptionally, and it seems likely that mechanisms have evolved to buffer the level of protein expression from random fluctuations in the level of transcription. We discuss this idea in Section 12.7.

Second, the variation in protein expression may be different for different biological processes. For example, proteins involved in chromatin remodeling, ATP synthesis, the Krebs cycle, environmental sensing, and stress

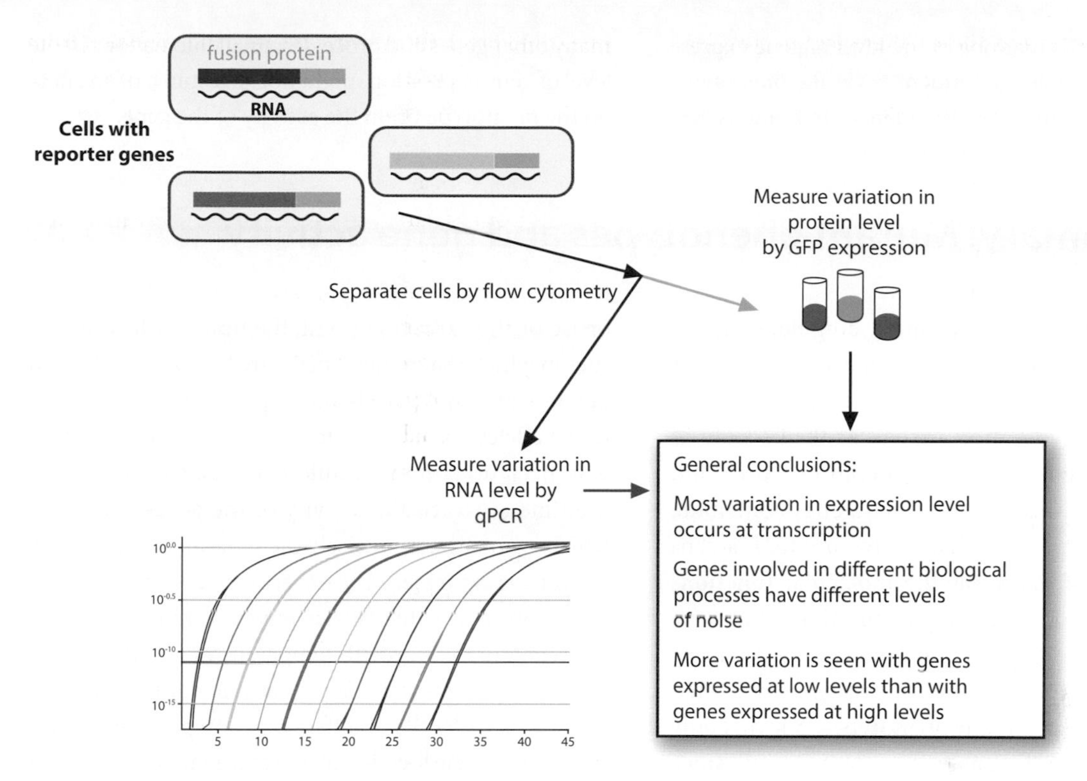

Figure 6.19 Measuring biological noise. Biological noise can be considered the naturally occurring fluctuation in the level of expression of a normal gene. In order to measure biological noise, yeast cells were constructed with GFP fusions to known genes, as represented here by the boxes with different types of cross-hatching. The reporter genes were integrated into the genome and expressed under their normal regulation. After the cells were separated by flow cytometry, the variation in the protein level was measured by the GFP fluorescence while variations in the mRNA level were measured by qPCR. The variations between mRNA and protein levels were then compared for each gene.

response have higher than average noise, as if the cell can tolerate variation in these processes. On the other hand, proteins involved in translation and protein degradation exhibit lower than average noise, as if variations in these processes are less easily tolerated by the cell. In general, the amount of protein noise is inversely proportional to protein abundance, so low abundance proteins have the greatest amount of variation in expression level. The same is true for mRNAs; mRNAs expressed at low levels in microarray studies have a much higher variance than those of greater abundance.

This inverse correlation between the amount of noise and mRNA abundance may reflect that biological noise arises from inherent imprecision in the machinery of gene expression. The number of transcripts produced from a gene may vary somewhat, because all biological processes have variability. For a gene expressed at high levels, a small change in the number of transcripts will not have much of an effect. On the other hand, for a gene expressed at low levels, a small change in the number of transcripts could change the expression twofold or threefold, either up or down. That is, while the *amount* of noise might arise from imprecision in gene expression, which should affect different transcripts similarly, the *impact* of the noise depends on the abundance of the normal transcript, and thus on the individual gene.

The phenotypic consequences of biological noise

We return to our discussion of small changes in gene expression and its effects on phenotypes in Chapter 12. But we can offer a summary here to help us think more carefully about gene activity and mutant phenotypes. From carefully controlled quantitative expression studies in some model organisms (both bacteria and eukaryotes) and in humans, it is clear that biological noise is prevalent. But what is not so clear is how much biological noise matters in the phenotype. For now, it seems that our first answer to the question is still the most accurate one—it depends on the gene. For some genes and some processes, small fluctuations in the level of gene expression can have notable consequences on the phenotype, even clinically significant consequences in humans. For many other genes and processes, small fluctuations in the level of gene expression might not have much of an effect on the phenotype or on the activity of the gene.

6.5 Summary: Mutant phenotypes and gene activity

Genetic analysis has undergone a complete change from the ways in which it was once done to the ways in which it is done now. Our ability to identify mutant or variant phenotypes is as ancient as the keeping of records by human populations, and our interpretation of how the activities of individual genes contribute to those phenotypes goes back for more than a century. For most of that time, the molecular functions of the gene were a matter of speculation. This is no longer true in the contemporary era of genetic analysis, when the molecular functions of genes are often known before their mutant phenotypes. But this only serves to make genetic analysis more powerful, since mutant phenotypes and molecular functions provide different kinds of insights into the activity of a gene in a biological process.

Mutant alleles themselves are helpful indicators of when and where a gene is active. The phenocritical phase or the execution point, the time at which or tissues in which the mutant effects are first observed, often demonstrates which cells and stages are affected. Conditional alleles, particularly temperature-sensitive alleles, have been used in many different organisms to analyse the time at which the activity of the genes is needed, which is not always precisely the same as the time when the gene is expressed. The experimental technique with temperature-sensitive alleles involves shifting mutants between different growth temperatures and thus could not be simpler. Mosaic analysis, carried out by producing organisms with cells of different genotypes, can be used to determine the location at which a gene is active. Most of these methods have been available for many years, and genetic analysis has used all of them successfully. All of them have been employed, to paraphrase McClintock, to make understandable the one kernel or cell or organism that is different.

Chapter Capsule

The activity of a gene is often understood by a careful analysis of its mutant phenotype. This careful analysis often depends on detailed observations of normal individuals and on comparing them to what happens in the mutant. Among the insights that are helpful in interpreting mutant phenotypes are the earliest times when effects are seen, which cells and tissues are affected, and at what time during the life cycle such effects are seen. Mosaic organisms, which have both wild-type and mutant cells, can be used to determine the cellular focus of gene activity. Temperature-sensitive mutants can often be used to determine the time at which the normal function of the gene is required. With contemporary methods, small changes in the level of gene expression can frequently be assayed, but the effect of these small changes on phenotypes is not easily predicted.

Case Study 6.1

Temperature-sensitive periods and gene expression

Temperature-sensitive (*ts*) mutations have been used for decades to estimate the time of gene activity. In this case study we discuss one example of a biological process in which *ts* mutations were an important part of the analysis. The genes involved in this study have subsequently been cloned and the biological process has been analysed in molecular detail. This allows us the opportunity to compare what was learned from genetic analysis before the genes were cloned to what has been learned using the cloned genes. The biological process that we will discuss is the formation of dauer larvae during the life cycle of *C. elegans*.

Thinking through the Experiment

1. Temperature-sensitive mutants are widely used in yeast, flies, and worms, as well as in bacteria and many other organisms, but are not widely used in mice or other mammals. Why not? (Don't overlook an obvious explanation.)

The dauer larva is an alternative stage in the nematode life cycle

Grown in the laboratory, *C. elegans* grows through four larval stages, termed L1 through L4, as summarized in Figure CS6.1. Each of these stages is punctuated by a molt in which the cuticle is shed and replaced. Under laboratory conditions at room temperature, the entire life cycle takes about four days. However, when the worm is starved, crowded, or otherwise stressed, it can enter a larval stage known as "dauer larva," as an alternative to the L3 stage; when this occurs, the worm molts directly from L2 into the non-feeding dauer larva, with a distinct cuticle and morphology. These are shown in Figure CS6.2. When dauer larvae (also known as dauers) are moved to fresh food on uncrowded plates, they molt to become L4 larvae and resume the life cycle. Because dauer larvae replace the L3 stage, the worm must detect the presence or absence of food at an earlier time in order to produce the appropriate cuticle and otherwise alter its metabolism. Temperature-shift experiments were used to determine when this signal detection about becoming either L3 or dauer occurs during the life cycle.

In order to understand the molecular and genetic basis of dauer signaling and formation, mutants that affected dauer formation were identified and sorted into a number of genes known as the *daf* (for ***da****uer* ***f****ormation*) genes. Null mutant alleles in some genes failed to form dauers even under stress conditions; these are known as the dauer-defective genes and will not be discussed further in this case study. Null mutations in other genes always formed dauers even under non-stress conditions; these are known as the dauer-constitutive (abbreviated Daf-C) genes, and a few of these genes will be the subject of this case study. We return to the analysis of the dauer formation pathway in the case studies for Chapters 11 and 12, where we discuss both the dauer-constitutive and dauer-defective mutations.

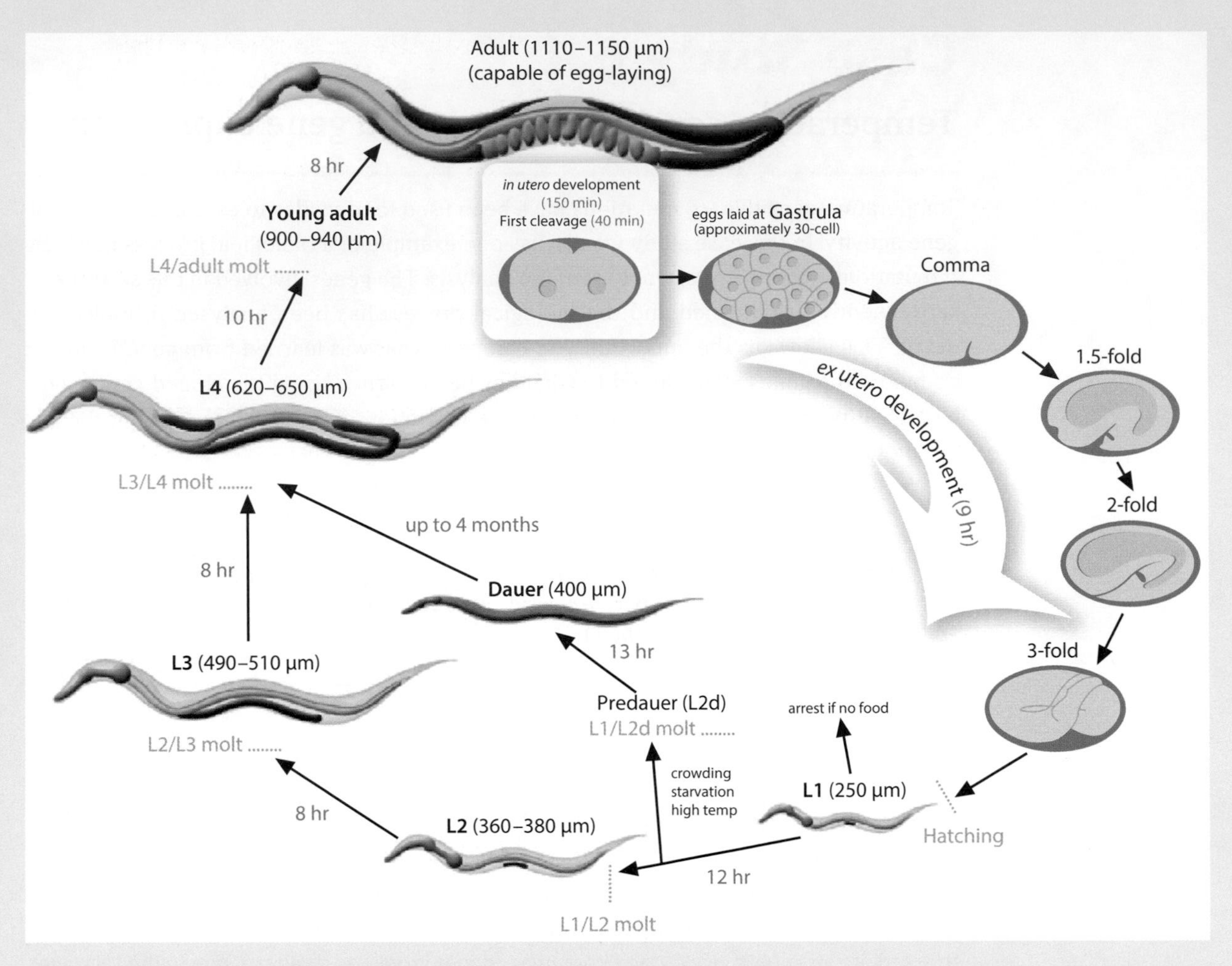

Figure CS6.1 The life cycle of *C. elegans*, showing the dauer larva stage. The times are the approximate duration of each stage at 20°. Under conditions of crowding, starvation, or high temperature during the late L1 stage, the worm will form a distinctive dauer larva rather than the L3 stage. The dauer larva molts to become an L4 stage larva, bypassing the L3 stage altogether. The time of commitment to dauer larva formation was determined by temperature-shift experiments. Reproduced from the Wormatlas.org, and used by permission.

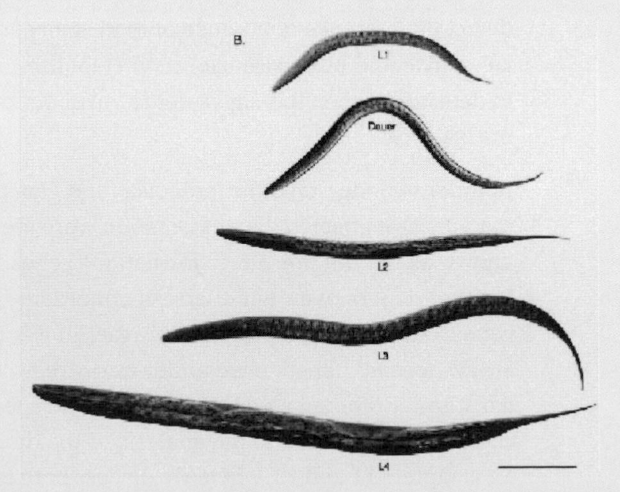

Figure CS6.2 The morphology of a dauer larva. Compared to an L3 larval worm, which is the equivalent age, the dauer larva is thinner and shorter. It is also a darker color and resistant to detergents, which does not show up in the figure. Reproduced from the Wormatlas.org, and used by permission.

Shift-up experiments suggest that the activities of the dauer-constitutive genes are needed well before dauer formation occurs

The dauer constitutive mutations lock the worms into the dauer larval stage before sexual maturation; so these mutations are effectively sterile, albeit with a distinctive phenotype. Many of Daf-C genes were defined by temperature-sensitive mutations so that the strains could be maintained as homozygote, a strategy for maintaining mutant strains introduced in Chapter 4 for *Drosophila* and in this chapter for the yeast cell cycle genes. A typical example is a temperature-sensitive allele of *daf-7*. When grown as a homozygote at the permissive temperature of 15°, the *daf-7* homozygote grows and reproduces normally. When grown at the restrictive temperature of 25°, the *daf-7* homozygote forms dauer larvae even when food is present. A similar phenotype is observed for mutations in several dauer constitutive genes. Some of these genes, including *daf-7*, were subjected to a temperature-shift experiment designed to determine when the activity of each gene is required. As a side note, dauer constitutive mutations imply a curious logic to dauer formation. The mutations always form dauers, so the wild-type activities for these genes must be involved in preventing dauer formation under non-stress conditions.

Thinking through the Experiment

2. Explain how a conditional mutation is used for strain maintenance. What might be an advantage of using a conditional mutation to maintain a strain rather than a balanced heterozygote, as described in Chapter 4? Are there disadvantages to relying on conditional mutations for strain maintenance?

The first complication in the temperature-shift experiments is that the life cycle of the worm is itself dependent on temperature, as is true for many organisms. That is, wild-type worms raised at the restrictive (warmer) temperature mature faster than worms raised at the permissive (cooler) temperature. As a result, simple chronological time—twenty-four hours after hatching for example—could not be used for the temperature shifts. For a culture of worms at 15°, the molt from L1 to L2 occurs about twenty-four hours after hatching; for a culture at 25°, this molt occurs about thirteen hours after hatching, and the worm is already in the middle of the L3 stage by 24 hours. Thus, in order to standardize the shift-up and shift-down experiments, the timing of the temperature shifts was related to developmental events such as hatching and the molts themselves, and worms were shifted up or down on the basis of their larval stage rather than chronological time.

In the temperature-shift experiments, populations of larval worms that hatched at the same time were maintained at one growth temperature until a particular molt had occurred, then shifted to the other temperature and assayed by counting the percent of dauer larvae among the total population. The slope of the transition between mostly dauer larvae and mostly wild-type larvae was characteristic for each mutant surveyed, as seen for mutations in *daf-4* and *daf-7* in Figure CS6.3. For example, for *daf-7*, the transition from mostly wild-type to mostly dauers in the shift-up experiments occurred over approximately six hours; for *daf-4*, the transition occurred much more gradually, over a period of about twenty-five hours. The physiological reason for this difference is not clear.

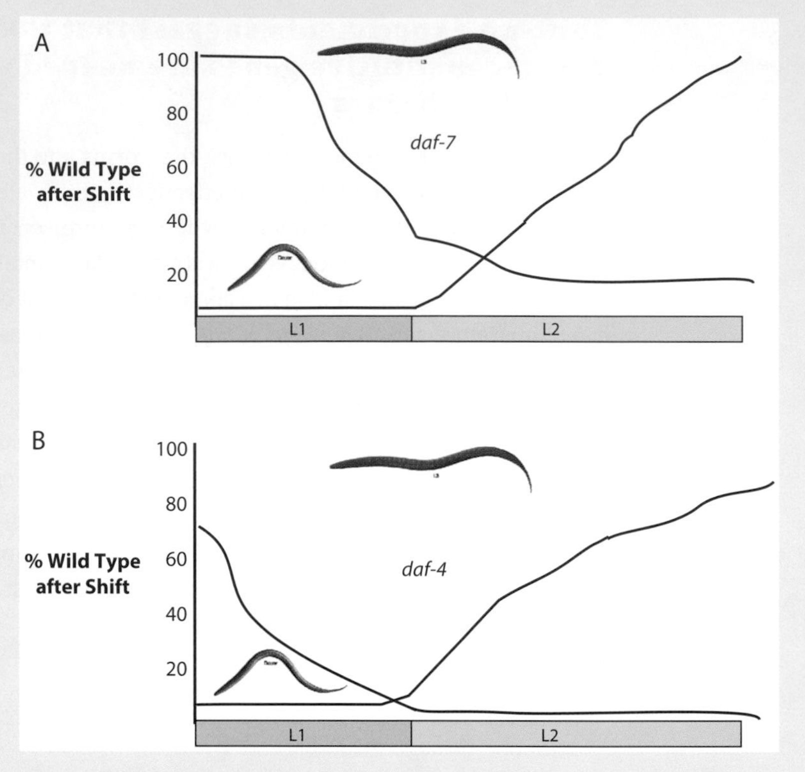

Figure CS6.3 Temperature shift experiments for *daf-7* (in panel A) and *daf-4* (in panel B). The X axis depicts the developmental stage of the worm, corrected for the temperature dependence of the life cycle. As described in the text, a population of worms is grown at one temperature and then shifted to the other temperature at particular stages. The percentage of worms that developed as wild-type worms is plotted on the Y axis. Shift-down experiments are plotted in blue, shift-up experiments are plotted in red. For example, if *daf-7* mutant worms are shifted up as newly hatched L1s, only a few percent develop as wild-type worms and most become dauers. Conversely, if *daf-7* mutant worms are shifted down as newly molted L2s, about 30 percent develop as wild-type and about 70 percent are dauers. The results with *daf-4* are similar. Note that, even when grown at the restrictive temperature for their entire lives, not all *daf-4* worms become dauer larvae; this is discussed more fully in Chapter 11.

Thinking through the Experiment

3. What might be some (speculative) explanations for the differences that are seen between different genes in the slope of these transitions? Although the explanation is not known, it may be helpful to consider what Hartwell concluded about the temperature-sensitive period and different alleles or genes.

Nonetheless, by taking the mid-point of each curve—the time at which half the population becomes dauers—it was possible to construct a temperature-sensitive period (TSP) for each mutant. Consider the data for *daf-7*, as shown in Figure CS6.3, beginning with the shift-up experiments plotted in red. If worms are shifted up before the L1-L2 molt, they become dauer larvae; if they are shifted up after the L1-L2 molt, they do not form dauer larvae and develop as wild type. Thus, in a simple interpretation, the temperature-sensitive period for *daf-7* activity begins in the middle of the first larval stage before the L1-L2 molt, well before the dauer larva itself actually forms after the L2 stage. Similar results were observed for several other mutants—shift-up experiments after the L1-L2 molt did not result in dauer formation, although the worm is homozygous for the dauer-constitutive mutation. The notable exception was the *daf-14* gene, in which the TSP is much earlier; shift-up experiments conducted even a few hours after hatching were largely ineffective in producing dauer larvae, which suggests that the activity of the gene is needed much earlier than that of the other dauer-constitutive genes; *daf-14* will not be considered in the remainder of this discussion but is included in the final figure.

One way to interpret the shift-up experiments is that the activity of most of these genes begins to be needed during the first larval stage. The appearance of the mutant phenotype following a shift up indicates that the activity of the gene is needed after the time of the shift. By the same reasoning, the shift-down experiments should define the times when the activity of the genes ends. A shift down *before* this time results in a wild-type phenotype, whereas a shift down *after* this time gives a mutant phenotype—the wild-type gene activity is being supplied too late to "rescue" the mutant defect.

This interpretation fits well with the results for some of the genes. Consider the shift-down experiments with *daf-7* again, as shown in Figure CS6.3, plotted in blue. The worms are raised at the restrictive temperature and then shifted down to the permissive temperature. If the period of gene activity has already occurred while the worm is at the restrictive temperature, the temperature shift down will result in only mutant worms; this begins to be observed for *daf-7* at approximately the time of the L1-L2 molt. Thus the temperature-sensitive period, as indicated by the appearance of mutant phenotypes in temperature shift-up and shift-down experiments, is slightly before or around the L1-L2 molt.

Not all temperature-sensitive mutations exhibit TSPs that can be interpreted as easily as the one for *daf-7*, but the best interpretation is that these shifts define the time when the gene activity is needed for the particular biological process. To appropriate the term used by Hartwell, the TSP is an execution point, a time when the function of the gene is needed. According to this interpretation, the functions of these *daf-C* genes are needed around the time of the molt from the L1 stage to the L2 stage.

Temperature-shift analysis has some inherent limitations

Because temperature-shift experiments are easy experiments to perform—they simply involve moving cultures from one temperature to another—they provide an attractive way to approximate the time of activity for a gene. There is an extensive, albeit older, literature dedicated to the detailed analysis of temperature-shift experiments that includes the use of temperature pulses, attempts to distinguish mutations that affect the assembly macromolecular complexes from mutations in the synthesis of the components, and so on. As seen with the TSPs in the preceding section, temperature-shift experiments have some inherent limitations that affect these interpretations.

First, the life cycle itself must be adjusted for temperature dependence, as was done carefully in the analysis of dauer formation and yeast cell cycle. It is much easier to use chronological time rather than developmental or diagnostic landmarks, but the appearance of certain developmental or morphological events provides the best timing method.

Second, as noted by Hartwell but often overlooked, the TSP is best regarded as characteristic for an *allele* rather than for a *gene*. In Chapter 4 it was noted that different mutations can have different phenotypes that are based on the amount of residual gene activity in the mutant. The same understanding should be applied to temperature-sensitive mutations. A shift up for some alleles may greatly reduce or eliminate the activity of the gene, whereas a shift up for other alleles may only slightly reduce the activity of gene. Such a difference, which might be unknown to the investigators as the temperature shifts are being done, could give rise to different TSPs for the same gene.

Third, since the life cycle itself is temperature-dependent, many other developmental processes are likely to be temperature-dependent too. If a given process is temperature-sensitive, *all* mutant alleles of a gene required for that process will be temperature-sensitive regardless of their molecular lesions. The TSP for such genes may be extremely broad, reflecting all the times when the function of the gene is needed.

Thinking through the Experiment

4. The limitation arising because the life cycle itself may be temperature-dependent can be addressed by relying on morphological or other events for timing rather than on the chronological time. What are some of the experiments or observations that could be done to address the possible limitation arising from the differing activities of each allele?
5. What are some of the experiments or observations that could address (or support) the possible limitation that the biological process itself could be temperature-dependent?
6. All experimental methods have limitations, but we usually learn to devise methods or controls to address them. If not, the method might be so flawed that the results are misleading. Do any of these limitations fundamentally alter the interpretation of temperature-shift experiments, or are they limitations that can be addressed with appropriate controls?

Because many temperature-shift experiments were carried out before assays for gene expression were readily available, little follow-up analysis has been undertaken in order to compare the conclusions from temperature-shift analysis with other types of data. Although this analysis is presented as representing the activity of individual genes, it can also be thought of as the time when the biological process affected by this gene is regulated.

Temperature-shift analysis and molecular analysis

The dauer formation pathway has become one of the best studied developmental pathways in worms, in part because many of the *daf* genes affect other processes, including aging. This subsequent analysis makes it possible to compare the thorough temperature-shift analysis done in 1981 with the contemporary knowledge of these genes derived from molecular biology and genomics.

Several of the dauer-constitutive mutations proved to be alleles of genes in a TGF-β signaling pathway. This is one of the standard signal transduction pathways in animal cells. For TGF-β pathways, the ligand binds to the type II receptor subunit, which phosphorylates the type I receptor subunit, and this in turn phosphorylates the downstream receptor Smad (R-Smad) proteins to regulate transcription. *daf-4* and *daf-1* encode the subunits of the TGF-β receptor, *daf-4* being a type II receptor that interacts with the ligand directly and *daf-1* being a type I receptor that is phosphorylated by *daf-4*. Reporter gene constructs with these genes showed that expression for these genes begins during the middle of embryogenesis, which is much earlier than the TSP for any of the mutants in the dauer formation pathway. Once expressed, the receptors are expressed and present throughout the life cycle, including at the larval stages. The receptors are expressed in many neurons in the head, including those involved in dauer formation, so they are present before any of the ligands are expressed.

daf-7 encodes a member of the TGF-β superfamily of proteins and is the ligand that binds to the receptor encoded by *daf-4* and *daf-1*. This is diagrammed in Figure CS6.4, with the protein expression data shown below the cellular pathway. The reporter gene analysis of *daf-7* expression is particularly illuminating. *daf-7* expression begins at about the middle of the L1 stage and continues throughout the subsequent larval stages. Importantly, *daf-7* is only expressed in the presence of abundant food, when dauers are not expected to form; when food is absent during the L1 stage, *daf-7* is not expressed. In addition, the expression of *daf-7* is limited to those neurons that are known from cell ablation experiments to be involved in dauer formation.

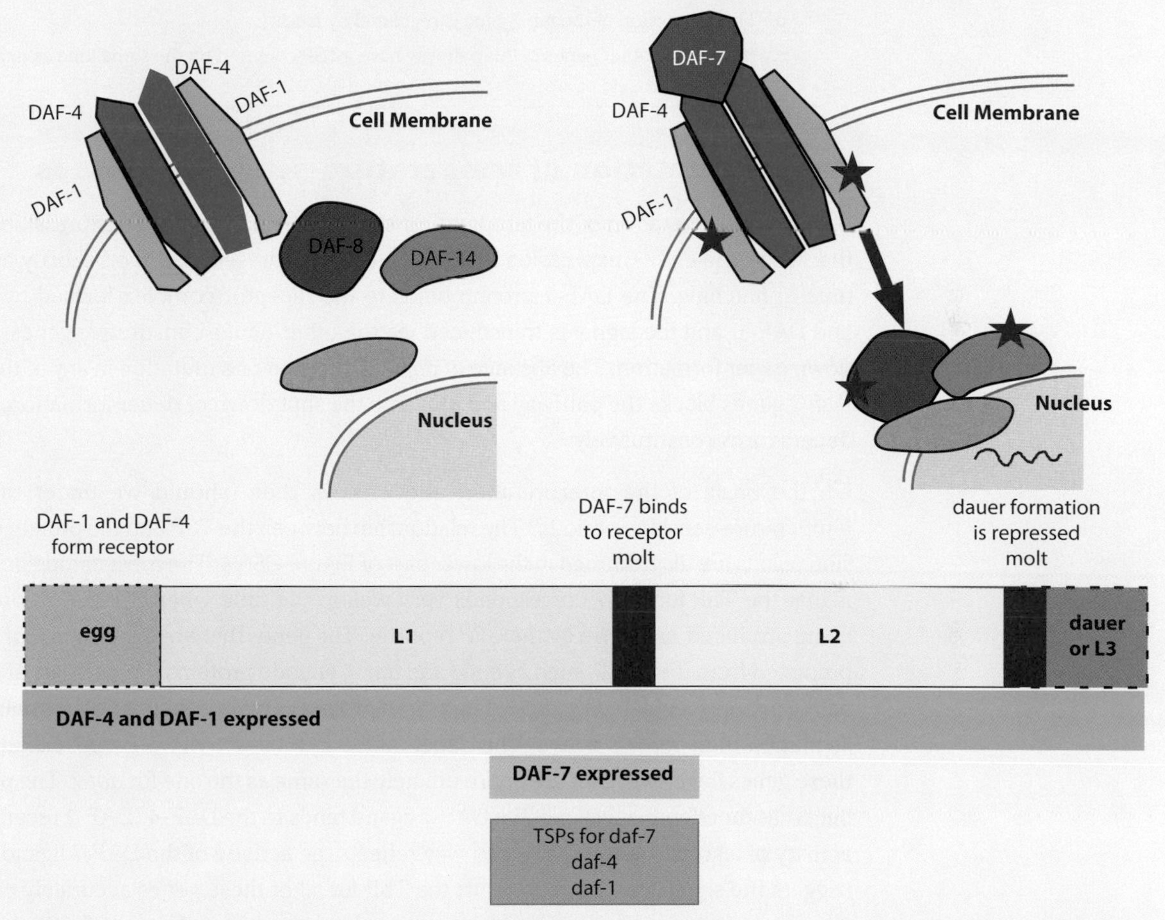

Figure CS6.4 Integration of the TSP with protein expression. The DAF-1, DAF-4, and DAF-4 proteins, as well as the proteins encoded by some of the other dauer formation genes such as *daf-14* and *daf-8*, are the components of a TGF-beta signaling pathway that shuts off dauer larva formation in the presence of food; the pathway is summarized here. DAF-1 and DAF-4 are the subunits of the receptor, DAF-4 being the type II subunit to which the DAF-7 ligand binds. *daf-8* and *daf-14* encode receptor Smad (R-Smad) proteins that rely the signal from the receptor to the nucleus to regulate transcription via a Co-Smad protein; the unlabeled gray shape is the Co-Smad protein, in this case DAF-3, which is a dauer-defective gene and is not discussed. When the DAF-7 ligand binds to the DAF-4 receptor subunit, DAF-4 phosphorylates DAF-1, as indicated by the red star. This signals the downstream R-Smad proteins, which are also phosphorylated, and transcription is initiated. The worm's life cycle, the protein expression results, and the TSP are shown below the cellular pathway; the protein expression data are summarized by green boxes, while the TSP is shown as a blue box. The *daf-1* and *daf-4* genes are expressed before hatching where their protein products form the receptor. At the end of the L1 stage, the presence of food triggers the expression of the *daf-7* gene and the DAF-7 ligand is made, which binds to the receptor to trigger the signaling pathway to repress dauer formation. The TSP for the *daf-7* gene corresponds to the time when the protein is made. The TSPs for *daf-1* and *daf-4* are also at this time, since the process depends on the activity of DAF-7.

Several of the other dauer formation genes encode proteins corresponding to known or suspected signaling molecules in the TGF-β pathway downstream of the receptor, as summarized in Figure CS6.4. These proteins are present in many other cells, in addition to neurons involved in dauer formation—possibly because this pathway regulates other processes in the worm. We will return to this pathway in the case studies for Chapters 11 and 12.

Thinking through the Experiment

7. On the basis of this information and before reading the subsequent section, develop a model for dauer formation in *C. elegans* that accounts for all of the following results.
   a. The *daf-1* and *daf-4* genes are expressed well before their TSP.
   b. The expression of the *daf-7* gene is regulated by food.
   c. Most of the other genes in this pathway have a TSP centered on the same time as *daf-1*, *daf-4*, and *daf-7*.

An interpretation of temperature-sensitive periods

A simple interpretation of the temperature-shift, molecular, and gene expression data is that wild-type *daf-7*+ expression is stimulated by the presence of food shortly after the time of hatching. The DAF-7 protein binds to the receptor complex formed by DAF-4 and DAF-1, and the signal is transduced via the other dauer-constitutive genes, to shut down dauer formation. The absence of *daf-7* expression or a mutation in any of the other Daf-C genes blocks the pathway and prevents the shut down of dauer formation, so that dauers form constitutively.

On the basis of the interpretation above, how, then, should we understand the temperature-sensitive periods? The relationship between the TSP and the protein expression results are diagrammed in the lower part of Figure CS6.4. The most compelling result is that the TSP for *daf-7* corresponds very well to the time when the DAF-7 protein is being produced, as shown by the GFP reporter. The genes that are expressed earlier than predicted from their TSP, such as *daf-1* and *daf-4*, encode proteins that are part of the signaling process regulated by *daf-7*. Thus these proteins are present but not active in dauer formation until *daf-7* is active. The dependence of their activity on *daf-7* explains why these genes have TSPs that are approximately the same as the one for *daf-7*. The pathway becomes functional only when the DAF-7 ligand binds to the DAF-4/DAF-1 receptor; the activity of all of the genes in the pathway reflects the activity of the DAF-7 ligand, which triggers the signal transduction. Thus the TSP for all of these genes accurately captures the time of gene activity for dauer formation—that is, the time of activity for the *biological process*, which corresponds to the activity of its key regulatory gene *daf-7*. Information on individual genes can be found on the website Wormbase at wormbase.org.

References

Gumienny, T. L., and C. Savage-Dunn. 2013. TBG-β signaling in *C. elegans*. *Wormbook*, July, doi 10.1895/wormbook.1.22.1. http://www.wormbook.org.

Hu, P. J. 2007. Dauer. *Wormbook*, August 8, doi 10.1895/wormbook.1.144.1. http://www.wormbook.org.

Swanson, M. M., and D. R. Riddle. Critical periods in the development of the *Caenorhabditis elegans dauer larva*. 1981. *Developmental Biology* 84: 27–40.

7

CHAPTER 7

Reverse genetics
Editing genes in yeast and mice

TOPIC SUMMARY

The process of genetic analysis often begins with a mutant phenotype, as discussed in Chapter 4, and moves forward from there to identify the corresponding DNA sequence in the genome, that is, the cloned gene, as described in Chapter 5. However, it is also possible to begin the analysis with the cloned gene and use that to identify a mutant phenotype, in effect to "work backwards" from the DNA sequence to the mutant phenotype to infer the gene's biological function. Approaches beginning from a cloned gene are sometimes called **reverse genetics**. This chapter introduces how reverse genetic analysis has been done for budding yeast and mice—organisms in which such approaches have long been employed. The next chapter extends this presentation to other organisms in which reverse genetic approaches, known as **genome editing**, are now feasible.

Two different experimental designs can be achieved by reverse genetics: **gene disruptions**, in which a mutation is made to disrupt or reduce the activity of the gene; and **gene replacements** and insertions, in which the genomic copy is replaced or supplemented by exogenous DNA sequences. For both designs, the changes are made in a single cell or in a small number of cells from which an entire organism is then produced.

INTRODUCTION

As discussed in the previous chapters, genetic analysis has two essential components: the phenotype arising from the mutated gene and the DNA sequence of the gene. While each component can contribute useful information by itself, the combination tells us much more about the biological process than either one alone can provide. Think again about the *white* gene in *Drosophila*, as summarized in Figure 7.1. As noted in Chapter 1, the history of *Drosophila* genetics began with white-eyed flies, and white-eyed flies are familiar to every geneticist who has worked on *Drosophila* during the past century. Decades of detailed information were obtained through the genetic analysis of the *w* mutant phenotype—information touching on many topics considered in other chapters, for example mapping, complementation, gene interactions, pleiotropy, allelic series, effects of chromosome structure, and gene dosage effects.

Just as significant as the function of the *w* gene itself, however, were the myriad experiments that used the *w* mutant phenotype as an experimental tool, as part of the analysis of other biological processes. From the white-eyed phenotype, we knew that the wild-type function of the *w+* gene was needed for proper pigmentation in the eye and in other tissues, but we did not know how the product of the *w* gene carried out these processes. Once the DNA sequence of the *w* gene was found—that is, when the *w* gene was cloned from wild-type and mutant flies—the mutant phenotypes arising from its many different alleles became more easily explained. The gene encodes a transporter protein of the ABC (ATP-binding cassette) superfamily, a family that also includes important genes for human genetics such as the CFTR gene. The protein made from the *w* gene uses ATP hydrolysis to provide the energy necessary to move pigment granules into the relevant cells in the fly, including the facets of the eye. Knowing the molecular function of ABC transporter proteins helped explain the mutant phenotypes of the *w* gene, and knowing the mutant phenotypes of the *w* gene helped explain the activity of other ABC transporter proteins.

For many decades, a gene's mutant phenotype, such as white eyes, was easier to obtain than its DNA sequence, and thus formed the traditional starting point for experimental genetics; the term "genetic analysis" was often used to refer only to experimental approaches that began with a mutant phenotype, the process described in Chapter 4. From that starting point, the gene could be cloned, as described in Chapter 5. But the world of genetic analysis has changed, and many genes are now cloned before a mutant phenotype has been found. In fact, for most of the 20,000 or so protein-coding genes in the human genome, no mutant phenotype is known; the situation is similar for many other eukaryotic organisms. We have the DNA sequence of the gene because the genome has been sequenced, which allows us to make very good predictions about the molecular functions of the protein product because of its sequence similarity to other known proteins; but we do not know the mutant phenotype arising from the gene. That is, we do not know how the functions of this particular protein affect the biological processes of this particular organism. Thus we need effective techniques to produce a mutant phenotype for genes recognized initially by their DNA sequence. In fact, as will be discussed in Chapter 9, an alteration of the DNA

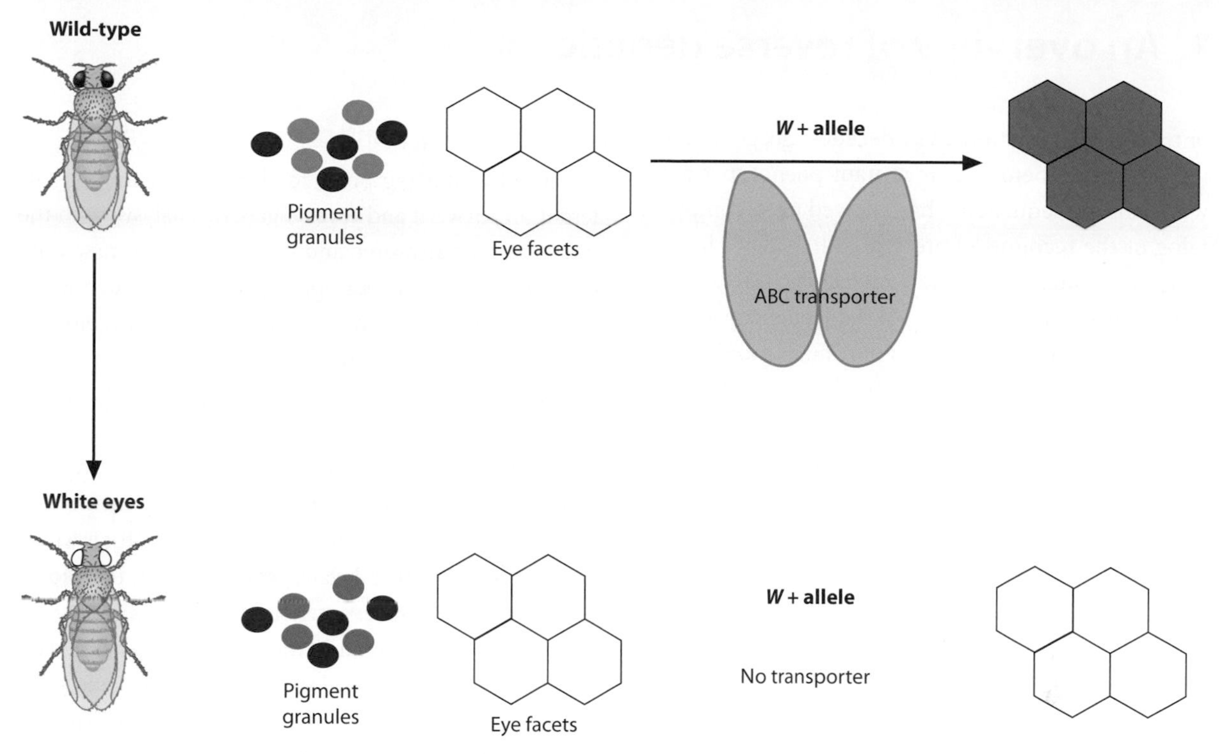

Figure 7.1 Genetic analysis using mutants. One of the first and most important mutations to be identified in eukaryotes involved the *w* gene in *Drosophila*. In wild-type flies, pigment granules (shown in scarlet and brown) are made. The activity of the wild-type w^+ gene is to transport these pigment granules into eye facet cells; in *w* mutants, the pigments are made but transport does not occur. Once the w^+ gene was cloned, it was predicted to encode an ABC transporter protein, which explains the phenotype in wild-type and mutant flies.

sequence appears not to be produce a mutant phenotype for many genes, so these genes would not have been found by the traditional type of genetic analysis.

In this chapter and the next, we describe how the genetic analysis of a biological process can begin by taking the DNA sequence of a gene (that is, the cloned gene) as its starting point. Because these approaches begin with the cloned gene and "work backwards" (in the traditional sense of genetic analysis) to find the mutant phenotype, they have been referred to as reverse genetics. The process of beginning with the mutant phenotype was sometimes referred to, then, as forward genetic analysis. But the distinction between traditional "forward" genetic analysis and "reverse" genetic analysis has become increasingly blurred, since each component is used to understand the other, and it is probably not very important to maintain this distinction. We use it here because it provides an intellectual and descriptive structure, but investigators would rarely think of these as being separate approaches.

In this chapter we describe reverse genetics for two model organisms: the budding yeast *Saccharomyces cerevisiae* and the mouse *Mus musculus*. In Chapter 8 we extend this description to how this type of analysis is being done in many other organisms; the process is commonly referred to as "genome editing" rather than reverse genetics but is conceptually similar. The techniques for finding a mutant phenotype for a cloned DNA sequence have been very well worked out in budding yeast and mice, so this chapter provides a foundation for thinking about the methods being used in other organisms. Whether beginning with the wild-type organism or the wild-type DNA sequence, the goal of genetic analysis is the same: find a mutant.

7.1 An overview of reverse genetics

In contrast to the situation a few decades ago, genes are now often cloned before their mutant phenotypes are known. Genome sequencing has allowed us to compile a catalog of the recognized protein-coding genes in an organism, and often to ascribe a molecular function to the protein product. On the basis of the sequence, we can say that the human genome has a particular number of C2H2 zinc finger proteins which bind to nucleic acids (often as transcription factors), the sea horse genome has a particular number of G-protein coupled receptors, and so on. Because we know the functions of these types of proteins in other organisms and can recognize orthologs based on sequence similarity, we are confident of the molecular functions of such gene products even in the absence of additional biochemical assays.

We also have effective and versatile methods to suggest some of the cellular roles of these genes and gene products. Using RNA-seq or microarrays, it is possible to determine what cells or tissues transcribe these genes (and thus probably use the protein product), and when these genes are being transcribed. We can find other genes with related expression patterns and biochemical activities and construct a realistic model for a particular gene interaction pathway; we may even be able to predict with some confidence which target genes are directly regulated by a specific C2H2 zinc finger transcription factor, although additional experiments are needed to confirm such predictions. Two characteristics of a gene that are central to genetic analysis and were once quite challenging to determine experimentally—the molecular function and the expression pattern—can now be learned quite readily, often using only data readily provided by genome projects with no further experimentation by an individual investigator.

Although the molecular function and the cellular expression pattern are important pieces of information for us to have about a gene, these do not tell us how the organism uses the gene in its biological processes. For that, finding a mutant remains the best approach. But, to find a mutant for the gene, we do not need to begin with the wild-type organism when we have the cloned gene in hand: we can begin with the cloned gene itself and learn how this particular gene is used. We summarize reverse genetic analysis in Figure 7.2.

Gene disruptions and gene replacements

Because the starting point for the investigation is different for forward and reverse genetic analysis, both the question to be answered and the scope of the answer are different. Reverse genetic approaches will not identify all of the genes that are involved in a particular biological process. Instead, the analysis is limited by the information that the geneticist has in advance. On the other hand, the only genes that are investigated are ones that are likely to be important to the biological process. Thus, reverse genetic methods do not identify all the genes involved in a process, but they do ensure that the investigator is working with at least one relevant gene. Unlike in traditional genetic analysis, in reverse genetic analysis it is usually not necessary to sort through all the mutants in genes that do not affect the process in order to find a relevant gene, since the change has been targeted to one specific gene. The two processes are combined when strain collections are made in which every recognized gene is altered, without regard to its predicted biological or molecular function; these genome-wide mutant screens are the topic of Chapter 9.

Generally speaking, two somewhat different but overlapping experimental strategies are used for reverse genetics and genome editing. One strategy is simply to disrupt the gene—that is, to make a targeted molecular change directly in the gene that impairs or alters its function. Gene disruptions are probably the most common and general strategy for reverse genetic analysis, and the ones that most closely parallel what is done when beginning with a wild-type organism and a mutagen. It might not matter very much to the investigator what type of gene disruption is made, although the most common ones are insertions and deletions as diagrammed in Figure 7.3; both of these alter the reading frame during translation and thus prevent a functional protein from being made. In terms of what we discussed about mutants in Chapter 4, both insertions and deletions are very likely to produce a null mutant.

On the other hand, the investigator *may want* to define the specific type of mutation in the gene, and thus replace part of the sequence of the gene with an altered sequence. For example, in addition to implementing a gene disruption designed to produce a null allele, it is often desirable to change a specific codon to a different

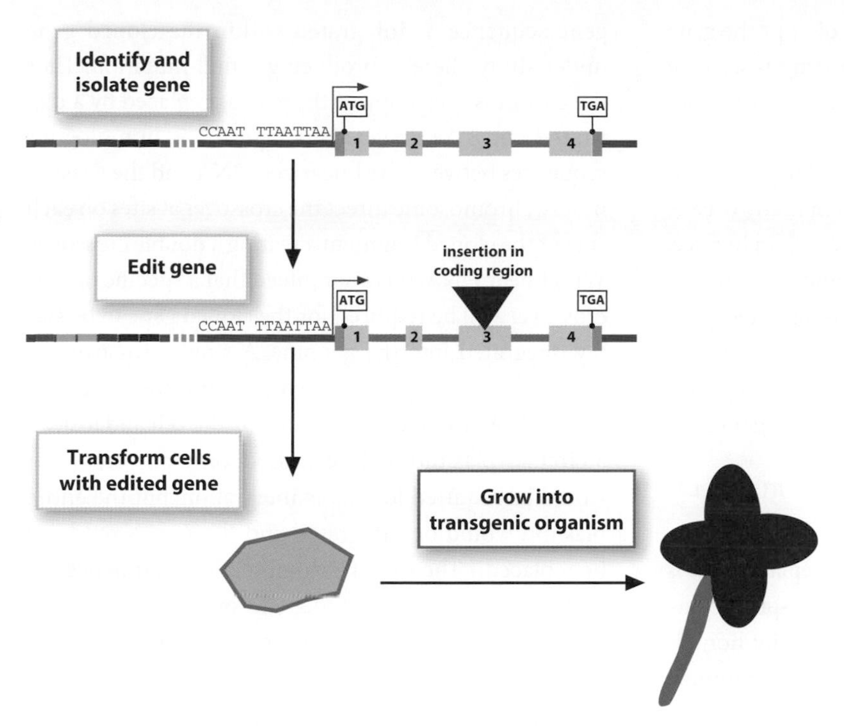

Figure 7.2 Reverse genetic analysis. Often the DNA sequence corresponding to a gene is isolated before the biological role of the gene is known. In reverse genetic analysis, the DNA sequence is altered or edited in vitro, shown here as an insertion into exon 3 that disrupts the reading frame for the coding region. The edited gene is transformed into one or a few cells, and a transgenic organism is grown from these transformed cells.

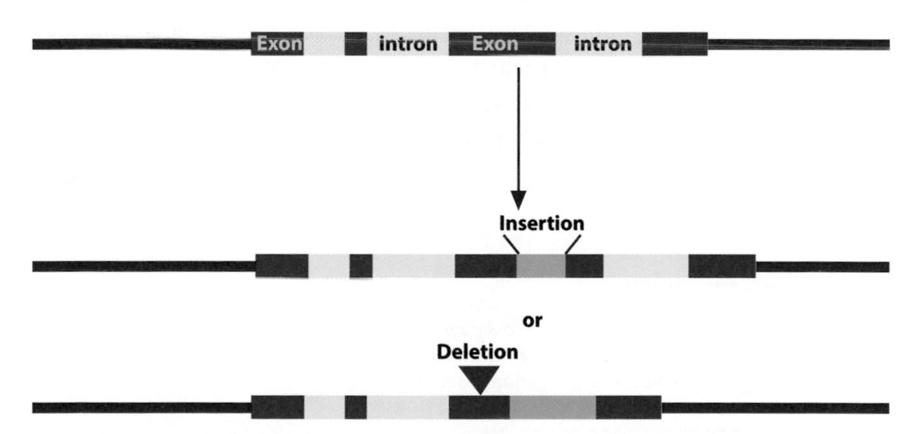

Figure 7.3 Gene disruptions. One of the most common strategies in reverse genetics is to make a gene disruption. Disruptions are often insertions or deletions that change the reading frame and produce a null mutation. The insertion is represented by a green box, while the deletion is a red triangle, both affecting exon 3. Such insertions and deletions may be as small as few bases or much larger.

codon so as to make a hypomorphic mutation, to target the change to one specific splice junction, or to alter the regulatory region of the gene, so that the gene is expressed ectopically (a neomorph) or at unusually high levels (a hypermorph). These examples require a strategy of targeted gene replacement. In this strategy, the cloned gene is altered in vitro, cells are transformed with the altered genes, and cells with the edited gene are introduced back into the organism. The DNA molecule with the specific change in the gene then replaces the copy in the genome. The two strategies of gene disruptions and gene replacements often overlap, since one common use of targeted gene replacements is to disrupt the gene by inserting a selectable nutritional or drug resistance marker, or by replacing it with a reporter gene construct of the type described in Chapter 1.

Reverse genetic experiments, particularly targeted gene disruptions, have been conducted in many organisms, but dominate genetic analysis in yeast and in mice, where targeted gene replacement is a standard or nearly standard procedure. In these organisms, targeted gene replacement has bestowed upon geneticists an unprecedented ability to manipulate and explore the functions of genes and other functional elements in the genome.

In gene replacement, a cloned gene is inserted at a specific site in the genome

The first requirement for efficient gene replacement is a method for integrating a cloned gene at a specific targeted site in the genome. This can be done by homology directed repair and recombination, as diagrammed in Figure 7.4. A cloned fragment of DNA that is to be integrated is introduced into the cell, typically as linear DNA, since linear molecules recombine more readily than circular ones. Recombination between homologous sequences on the cloned DNA fragment and the chromosome integrates the DNA fragment precisely into the chromosome. Often the site of integration will be the locus of the normal chromosomal gene, so that the chromosomal copy of the gene is replaced by the cloned copy.

In one of the most common procedures, a specific piece of DNA, such as a selectable marker or a reporter gene sequence, is integrated within the cloned gene under study, thereby producing a null mutation. Thus the chromosomal copy of the gene is replaced by a disrupted copy. As depicted in Figure 7.5, homologous sequences between the linearized DNA and the yeast or mouse chromosome direct the crossover at sites on each side of the cloned fragment, creating a double crossover. With linear DNA, it is guaranteed that a specific double crossover will be required for the cloned gene to be stably integrated into the genome. A single crossover on only one side will produce two broken DNA molecules that will be unstable or degraded by the cell and lost. (If a circular plasmid had been used, only one crossover would be required for stable integration, but the entire plasmid would be integrated and the gene would not be replaced.) The site of recombination, and hence the region that is integrated, depend on the homologous sequences chosen, so a specific location can be targeted. Crossovers and integration of the DNA fragment can occur precisely, without leaving gaps in the sequence.

All eukaryotic organisms can carry out homology-directed repair in this manner. As will be described in more detail in Chapter 8, homologous recombination is initiated by a double-stranded break (DSB) in the DNA sequence. Therefore, in principle, if the DSB can be specifically targeted to a defined location in the genome, any cell for which DNA fragments can be introduced into the nucleus could be used for this kind of reverse genetic approach. DSBs occur very commonly in genomes and, because DSBs cause genome instability if allowed to

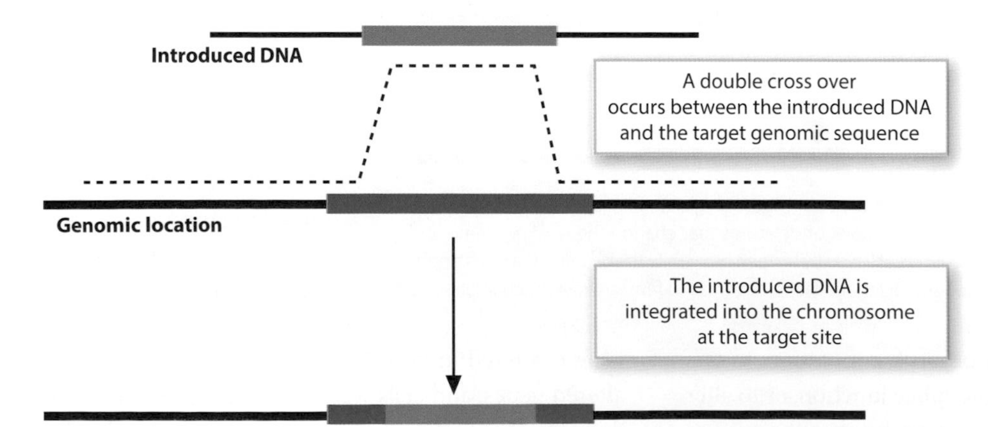

Figure 7.4 Gene replacement. Sequences in the genome can be replaced with introduced exogenous DNA whenever a double crossover occurs between the introduced DNA and the genomic target site. This usually depends on the homology directed recombination and repair system in cells. The introduced DNA sequence can then replace the copy in the genome.

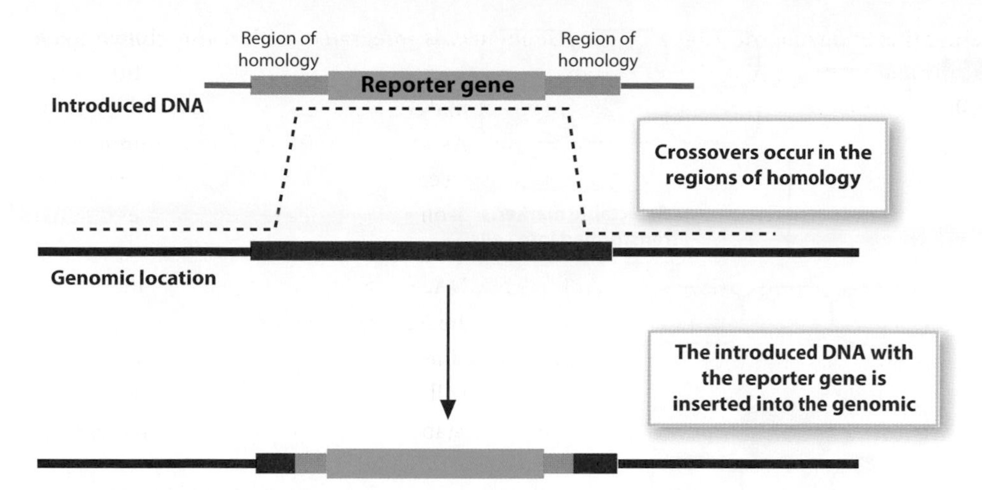

Figure 7.5 Inserting a reporter gene. A reporter gene (in green) is flanked by regions of sequence homology with the target site in the genome; the necessary sizes of these homologous regions vary among organisms. Crossovers between the regions of homology on the introduced DNA with the reporter gene and the target site will insert the reporter gene into the genome.

persist, cells have very effective repair mechanisms that prevent most of the damage.

As discussed in Chapter 8, two different types of repair mechanisms are found in cells, a more error-prone process known as **non-homologous end joining** (NHEJ) and a more precise process known as **homology-directed repair**. Homology-directed repair is the one that underlies homologous recombination and gene replacements of the type illustrated in Figures 7.4 and 7.5. While most organisms rely on NHEJ to repair the majority of DSBs, *Saccharomyces cerevisiae* primarily uses homology directed repair; thus, targeted integration by homologous recombination has been a routine procedure in yeast for decades. The ability to carry out gene replacements has been one of tools that makes yeast such a powerful model organism.

Literature Link. Haber 2000, *Trends in Genetics* 16: 259–64

Targeted gene replacement can be used in organisms with a lower rate of homology-directed repair, such as mammals, but usually an appropriate selectable marker has to be integrated during the recombination. That is, even if the rate of homologous recombination is very low, cells with the appropriate disruption or insertion can be selected using drug resistance markers and grown in culture. This approach has been used effectively to perform reverse genetics in mice. Without going into the molecular details about how the processes occur, we will describe gene replacement methods in first yeast and then in mice.

Literature Link. Karan 2000, *Current Opinion in Genetics and Development* 10: 144–50

While the emphasis in this chapter and in Chapter 8 will be placed on the strategy for making the edited gene, an altered gene is only one part of the process of reverse genetics or genome editing. The gene or genome is disrupted or replaced in one or a few cells in vitro, as summarized in Figure 7.6. Those cells are then identified (by using a selectable marker, for example), the molecular nature of the change is confirmed by DNA analysis, and the phenotypes of these single cells can be analysed. But it is often not enough to study the phenotype in a single mutant cell; the cell with the edited gene or genome must then be used to produce a mutant organism. Since yeast colonies are grown from a single cell, this requirement for a mutant organism is not very demanding. However, for multicellular organisms, the cell with the edited or mutated gene must somehow be grown into an organism. For some animals, this can be done by making the edits in stem cells; the process of producing a mutant organism from an edited stem cell is described for mice in Section 7.3. For genome editing in organisms other than yeast or mice, the choice of the cells to transform in vitro with the edited gene is an important consideration.

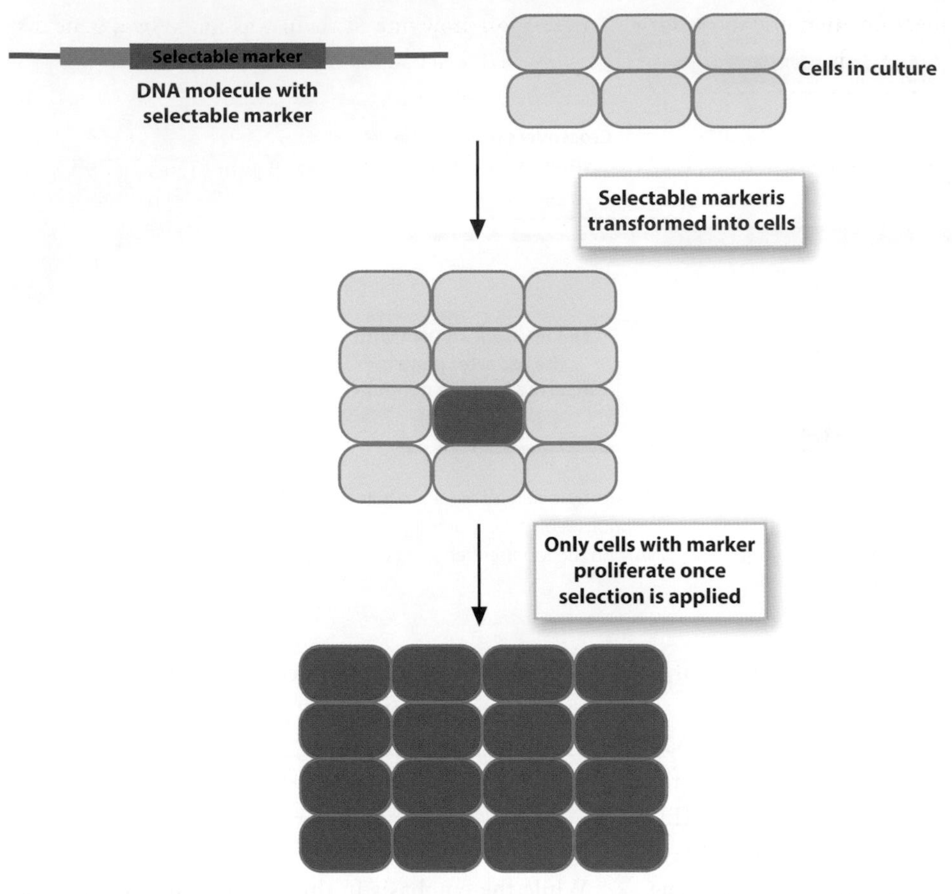

Figure 7.6 Selecting appropriately transformed cells. A DNA construct with a drug or a nutritional gene that can be used in a selection (shown in purple) is made in vitro and transformed into cells in culture. Even if only a few cells are transformed, those will be the only ones to proliferate once selection is applied. These cells can be used to produce a transgenic organism or tissue.

7.2 Genome editing in yeast

Gene disruptions and replacements have been done in *Saccharomyces cerevisiae* for more than thirty-five years, so the techniques for integration have been well documented. Furthermore, the applications for targeted insertions are very diverse, seemingly limited only by the investigators' goals and ingenuity. It is not an exaggeration to say that the linkage arrangements and gene locations in the yeast genome have been completely reconstructed for the convenience of the investigator. Cloned DNA is readily introduced by transformation into single cells, many nutritional markers are available for selection, and only about 100 base pairs of sequence homology on either side of the DNA are needed to be integrated for recombination to occur between the introduced cloned gene and the targeted chromosomal gene.

 ***Literature Link.*** Struhl, 1983, *Gene* 26: 231–41

Standard experimental strategies involve placing a gene downstream of different regulatory regions, to control its transcription, and replacing the wild-type chromosomal copy of a gene with an engineered version containing specific molecular changes. Yeast offers the geneticist more control over the type of mutation studied than any other eukaryotic organism. Many of the experiments that we describe in other chapters in this book depend on the ability to target yeast genes to specific locations, to disrupt specific genes, and to introduce specific genes. To cite one example that we will develop in more detail in Chapter 9, targeted insertions have made it possible for yeast geneticists to engineer deletion mutations for each individual gene, and thus to study each mutation in both single mutants and multiple mutants.

A simple example of a gene disruption is provided by the tubulin family of genes. We will encounter these tubulin mutants again in Chapter 10, so it is worth describing how one of them was made by a gene disruption. Recall that tubulin is a dimeric protein, with an α-tubulin subunit and a β-tubulin subunit that assemble precisely, as shown in Figure 7.7 (A). In yeast there are three genes, two α-tubulin genes named TUB1 and TUB3 and one β-tubulin gene named TUB2. Mutations

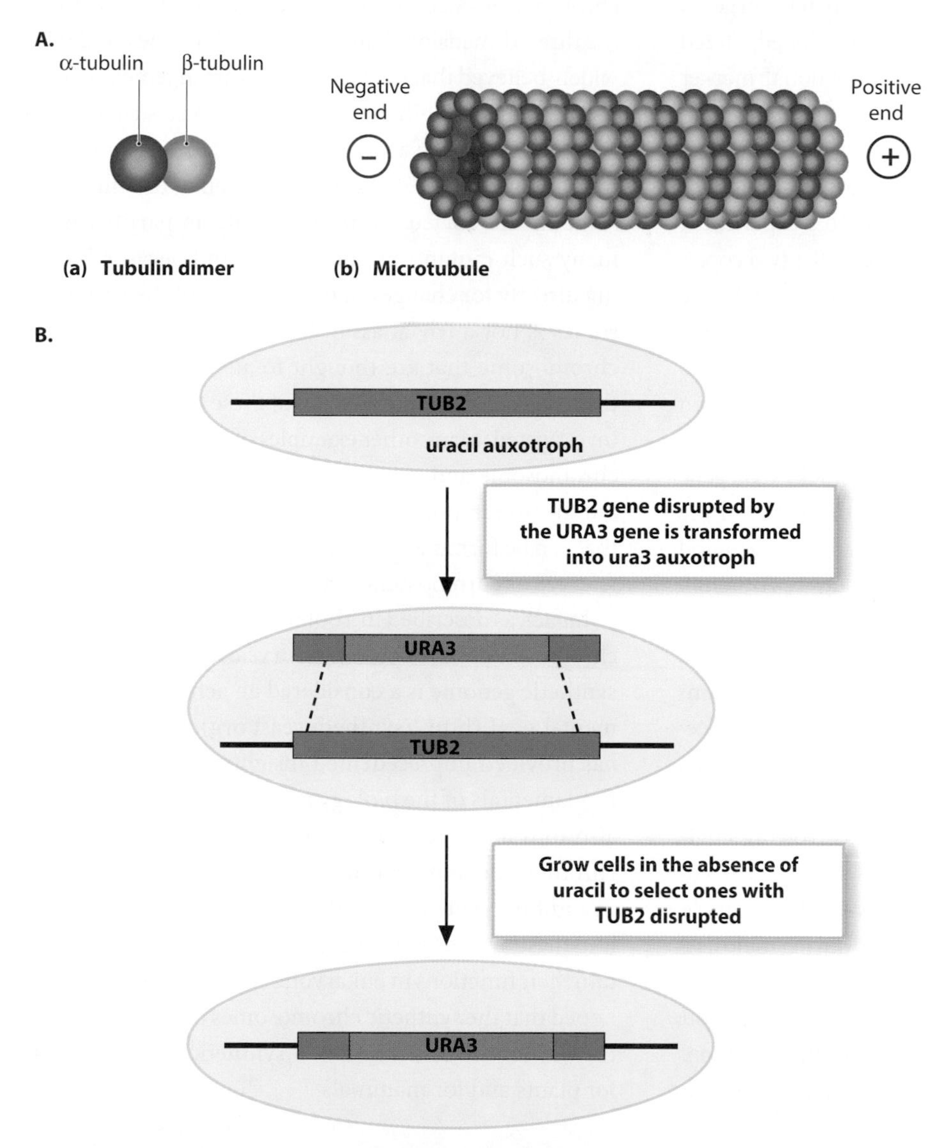

Figure 7.7 The URA3 gene inserted into the TUB2 gene. The tubulin protofilament shown in Part A consists of α-tubulin and β-tubulin. In yeast, α-tubulin is encoded by the TUB1 and TUB3 genes, while β-tubulin is encoded by the TUB2 gene. In part B, the wild-type URA3 gene has been cloned into the TUB2 gene with flanking regions of TUB2 on each side. This construct is transformed into a *ura3* auxtrophic mutant that cannot grow in the absence of uracil. A double crossover between the chromosomal TUB2 gene and the disrupted TUB2 gene, indicated by the dashed lines, replaces the chromosomal TUB2 gene with the disrupted gene. This disruption can be selected by the ability to grow in the absence of uracil. Since TUB2 is an essential gene, the disruption would involve only one of the two copies in a diploid cell; only one copy is shown in the figure, since the other copy of the gene is not disrupted or replaced. Panel A is reproduced from Meneely et al., *Genetics: Genes, Genomes and Evolution* (Oxford University Press, 2017), Figure 6.4.

in both TUB1 and TUB2 were initially identified in genetic screens by their increased sensitivity to the drug benomyl, which binds to microtubules.

Literature Link. Neff et al. 1983, *Cell* 33: 211–19

On the basis of the sequence homology with tubulin genes in other organisms, the β-tubulin gene was cloned, altered in vitro by inserting the selectable nutritional marker URA3 in frame into the coding region of the TUB2 gene, and the altered version was targeted to replace one copy of the chromosomal gene in vivo, as shown in Figure 7.7 (B). (Since TUB2 is an essential gene, this experiment was done in diploid cells and only one of the two copies of the TUB2 gene was altered.) The cloned TUB2 gene with the inserted dominant allele of the URA3 gene was transformed as linear DNA into *ura3* mutant yeast, so that homologous recombination could occur between the chromosomal TUB2 gene and the introduced gene. A double crossover replaced the chromosomal TUB2 gene with the disrupted gene, identified by the ability of the yeast strain to grow in the absence in added uracil and confirmed by subsequent molecular analysis. Although the frequency of the double crossover is perhaps one cell in a few thousand, the strong selection for the ability to grow in the absence of added uracil allows the insertions to be found even if they are infrequent. The general procedures and strategy outlined in this experiment have been employed many times for standard yeast genetics.

In addition to experimental questions involving single genes, broader questions about the functions and evolution of the genome can be addressed in yeast. For example, the number and positions of crossovers during meiosis is highly regulated in all eukaryotes, but the coordinating processes for crossover control are not well understood. Crossovers in yeast can be relocated to other positions by moving the centromere or blocks of heterochromatin, by moving genes from one location to another, and by changing the sizes of chromosomes, which has allowed the analysis of how each of these chromosomal characteristics might affect crossover control. Analogous experiments have been done to study the positions, roles, and coordination of origins of replication.

It is also known, in many eukaryotes, that particular chromosomes occupy certain relative positions within the three-dimensional nucleus at specific times, and it is widely believed that the location and arrangement of the chromosomes within the three-dimensional structure of the nucleus affect various functions. However, finding random mutants that affect the location of the chromosome in the nucleus is not so simple, in part because many such mutants are pleiotropic and because looking directly for changes in the positions of the chromosomes is not a trivial assay. By altering features of the chromosome that are thought to affect its position in the nuclear environment, some of these functions can be investigated. Many other examples of broad questions of chromosome and genome organization could be cited, but the overarching lesson is that nearly any hypotheses that can be formulated about the functions of structural features of chromosomes can be tested in yeast.

In fact, as described in Tool Box 7.1, entirely synthetic chromosomes have been made in yeast, and a completely synthetic genome is a considered an achievable experimental goal (http://syntheticyeast.org). Such an effort has provided unprecedented insights not only into the fundamentals of the biology and evolution of yeast, but also into more general biological questions involving chromosome and genome structure and organization. In addition, genes from other organisms can be included into these synthetic yeast chromosomes to offer insights into their functions in eukaryotic cells. Furthermore, it is hoped that the synthetic chromosomes of yeast can pave the way for the construction of synthetic chromosomes for plants and for mammals.

7.3 Targeted gene disruptions and replacements in the mouse

In thinking about how genome editing might be done in a mammal, we can make a comparison to yeast. Yeast can be grown as single cells in culture, and colonies arise from those single cells. Therefore the stable genetic modification of only one cell can result in the genetic alteration of a large group of yeast cells through asexual reproduction. Since a population of cells can arise from a single altered cell, it is possible to perform selections

TOOL BOX 7.1 The synthetic yeast genome

Budding yeast *Saccharomyces cerevisiae* offers geneticists the ability to manipulate and edit a genome that is unmatched by any other eukaryote. While genome-editing tools are rapidly being developed for other organisms, as described in Chapter 8, similar tools have been used in yeast for more than thirty years, so their use is quite mature. From the beginning of their research careers, yeast geneticists have been able to think about sophisticated biological questions without also having to wonder if the experimental techniques are feasible. Many of the techniques for genome editing, if not quite routine, are in common use.

The haploid genome of yeast consists of 12.1 Mb, distributed over sixteen chromosomes and encoding approximately 6,200 protein-coding genes; it is standard practice to refer to the haploid genome size in eukaryotes, but this is particularly relevant in yeast, since yeast can grow as either a haploid or a diploid cell. The chromosomes range in length from 230 kb to 1.5 Mb. Yeast chromosomes are structurally simpler than the chromosomes of most other eukaryotes, since their centromeres are less than 150 bp in length and have well-defined sequences; centromeres in other eukaryotes typically are comprised of a complex set of sequence repeats spanning more than 1Mb in length. The sequence simplicity of their centromeres allowed the construction of yeast artificial chromosomes (known as YACs), which have been key reagents in many genome projects. Stable transmission of a YAC through mitosis and meiosis in yeast requires only a centromere, telomeres, an origin of replication, and roughly 200 kb of other DNA, which can come from any organism. YACs have also been important in understanding the necessary structural elements of chromosomes and, more recently, in the development and assembly of synthetic yeast chromosomes in which all sequence elements have been defined by the investigator. This has led to a project that to design and construct an entire yeast genome with investigator-designed sequences throughout, a project known as Synthetic Yeast 2.0 or Sc2.0; this project and its progress is described at http://syntheticyeast.org/sc2-0.

The rationale for the Sc2.0 project, and for the field of synthetic biology generally, is found in the experiences of engineering and synthetic chemistry. Like chemists before them, biologists analyse the structures of macromolecules on the basis of their organization of subunits or components; in biology, these macromolecules are the ones found in nature as the outcome of evolution. The expectation for synthetic biology is that even greater insight will be gained from attempts to construct or synthesis the macromolecules, as has occurred with synthetic chemistry. The genome for any species is the product of evolution and not from design principles, but engineering a genome from scratch can help define key principles and constraints that existed during evolution, and thus can allow greater control of biological processes in the lab. Other synthetic biology projects have been building bacterial genomes, including the *E. coli* genome. One of the unifying goals of all of these synthetic genome projects is that the software, strategy, techniques, and so on—the design platform—will be generally applicable and will have many different experimental applications, only some of which might be envisioned now.

For the project to be considered successful, the yeast cell with the synthetic chromosome (and eventually the synthetic genome) must have retained all the biological activities needed for wild-type growth, division, and other functions under laboratory conditions; and its genome must be stable, without rearrangements or other mutations; in other words it must behave like a normal yeast cell. In addition, the strains should have design flexibility, so that other investigators may be able to manipulate the genome with relatively few additional modifications of the platform. The project currently involves ten or more different laboratories in the United States, Europe, and China, as well as several industry partners. Each group is assigned a different chromosome and uses a shared strategy for its synthesis and assembly. The first synthetic chromosome to be constructed was chromosome III, published in 2014; a series of papers published in March 2017 reported that six-and-a-half synthetic chromosomes have been made (the half being the right arm of chromosome IX). Substantial portions of every other chromosome have been constructed as well, and the project is on schedule to have synthetic versions of all chromosomes soon. A genetic strategy has been developed that will incorporate all these synthetic chromosomes into the same yeast strain, and a strain has been made in which two and a half of the synthetic chromosomes have replaced their natural homologues.

The synthetic yeast chromosomes (and genome) differ from the natural chromosomes in several ways. First, repeated sequences have been reduced or eliminated altogether, and the tRNA genes have been moved to a separate neochromosome. Many of the introns (of which yeast has relatively few by comparison to most eukaryotes) have been eliminated as well. As a result of these changes, the synthetic genome is expected to be about 94 percent of the size of genome of common lab strains; for example, the synthetic chromosome III is made up of 272.2 kb while the natural chromosome III is 316.6 kb . Second, the TAG stop codons are replaced with TAA stop codons; since some eukaryotes have a derived genetic codon with a single stop codon, this change is not expected to affect the fitness of the cell. The value of this change is that it frees up one of the codons in the genetic code, so that it can be reassigned to a different, even non-native amino acid; for example, in the synthetic genome projects in bacteria, TAG has been reassigned to selenocysteine, an amino acid that is found at a low frequency in nature (and not encoded by the genetic code), and has spectral properties that are advantageous for protein structural studies. Third, *lox* sites are incorporated into many locations throughout the genome to allow directed recombination at these sites; the Cre-*lox* system for site-specific recombination is widely used for genome editing in mice, as described in the chapter, and a similar strategy could be used here.

The synthetic sequences are made as "chunks" of less than 10 kb and then combined into longer blocks, which are referred to as "megachunks" of 30–60 kb. Short sequences to be used as PCR tags or bar codes (described in Chapter 9) are also included and used during the assembly process, to distinguish synthetic sequences from the natural sequences. Base substitutions in open reading frames that remove or add recognition sites for enzymes are also made; they facilitate the assembly of the smaller blocks of synthetic sequences into a longer chunk or megachunk. The synthetic chromosomes are built in modules from these megachunks, replacing a region of the natural chromosome with the corresponding synthetic megachunk. The strain with the synthetic replacement is then tested for wild-type fitness and growth, and any defects arising from that synthetic segment can be fixed and retested. The process follows a "design-build-assemble-test-learn" cycle meant to build synthetic chromosomes piecemeal—or, as the investigators put it, from the bottom up. (A top-down strategy involves deleting regions from a naturally occurring genome and testing the results, as has been used to define the minimal gene set needed for the replication and metabolism of the bacterium *Mycoplasma genitalium*.)

Many more details of the project and of the platform are provided in the papers and on the Sc2.0 website. This is an ambitious project, so it is difficult for anyone, including the members of the Sc2.0 consortium, to predict what insights and future experimental questions could arise from it. But it may be helpful to make a comparison to computer science, machine language, and software engineering. The genome consists of information encoded in a sequence; each position has four "bits" or states rather than the two bits found in binary systems, but the logic is the same, albeit more complicated. Just as we have done with computer software, we can write instructions using that sequence of bits and determine if the program can be executed properly, in this case if the cell can carry out its normal functions. It seems likely that none of the pioneers of binary computing imagined all the ways in which this technology has been used in the ensuing decades since it was designed. Similarly, it seems likely that we cannot yet envision all the applications that might arise from building and rewriting the genome of a eukaryotic cell.

Literature

Richardson, S. M. et al. 2017. Design of a synthetic yeast genome. *Science* 355: 1040–4.

for molecular events even if they occur rarely. A similar principle is true for mice; but, to understand the technique for reverse genetics in mice, we first need to briefly describe mouse embryogenesis, illustrated in Figure 7.8.

Embryonic stem cells can give rise to mouse embryos

In mice as in all placental mammals, an ovum is released from the ovary and fertilized as it moves down the oviduct, where it begins a series of mitotic divisions. At about the 8-cell stage, the cells rearrange themselves in a process known as compaction, with a corresponding change in cell contacts. Prior to compaction, individual cells termed blastomeres can be separated from one another. Shortly after compaction, a cavity known as the blastocoel forms in the embryo, a characteristic of the blastocyst stage. Implantation into the uterine wall occurs shortly thereafter.

The embryo itself develops from the inner cell mass arrayed in a layer along the top of the blastocoel at the blastocyst stage, as shown in the lower panel in Figure 7.8. In a mouse, the inner cell mass forms at about three-and-a-half to four-and-a-half days post coitum. The inner cell mass can be isolated from blastocysts and grown in culture indefinitely. The process is diagrammed in Figure 7.9. Such cells are termed **embryonic stem (ES) cells**. Cells in the inner cell mass are totipotent, that is, each of them is capable of giving rise to any cell types and tissues in the adult.

ES cells injected into the blastocoel at this stage can also become incorporated into the inner cell mass, and thus can also contribute to all the tissues of the mouse that

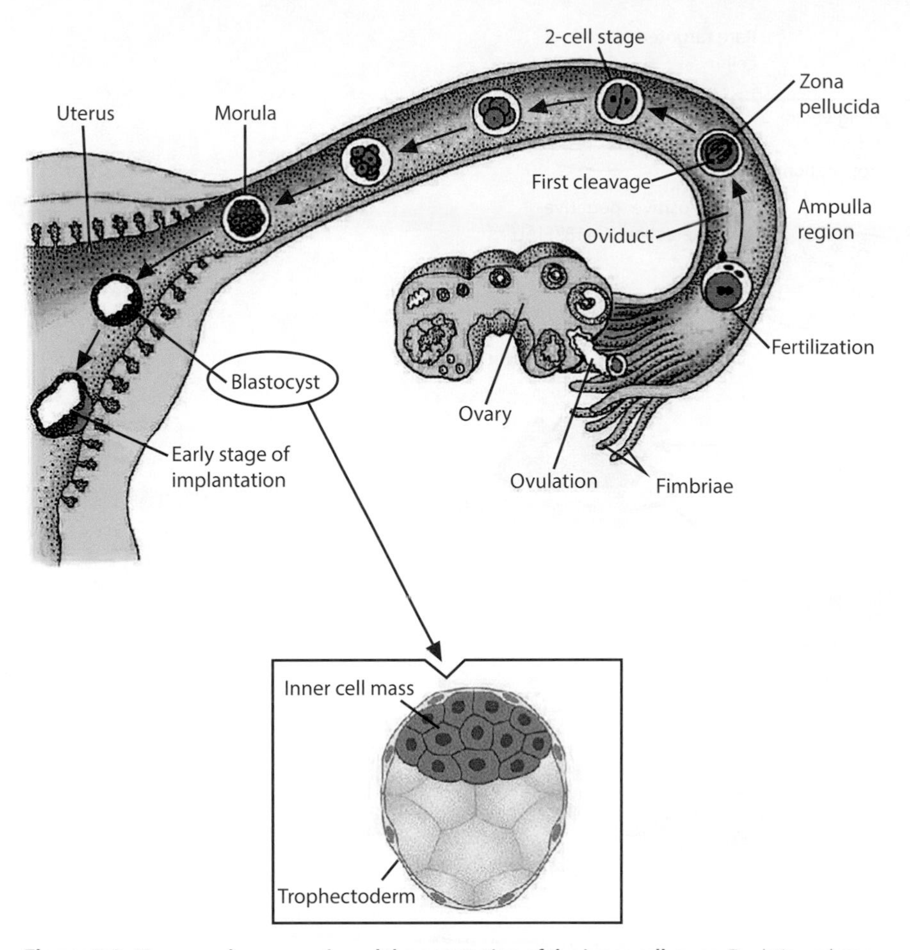

Figure 7.8 Mouse embryogenesis and the generation of the inner cell mass. Ovulation releases an oocyte that moves down the oviduct where it can be fertilized. After fertilization and the completion of meiosis, the embryo begins to divide mitotically. The dividing embryo reaches the blastocyst stage as it enters the uterus, approximately 3–4 days after fertilization, and implantation in the uterine wall follows. An isolated blastocyst is shown at the bottom as a diagram. A large cavity called the blastocoel forms, with the inner cell mass at one side. The cells of the inner cell mass will give rise to all the cells of the embryo and adult mouse. Redrawn with permission from Gilbert and Barresi, *Developmental Biology* (Oxford University Press, 2017).

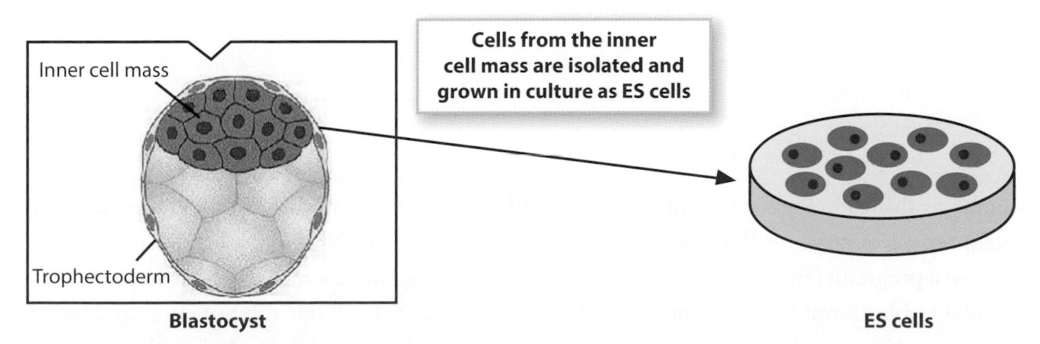

Figure 7.9 Inner cell mass and ES cells. The cells in the inner cell mass of the blastocyst will give rise to all cells and tissues in the adult. Cells of the inner cell mass can be isolated from blastocysts and grown indefinitely in culture, where they are referred to as embryonic stem or ES cells.

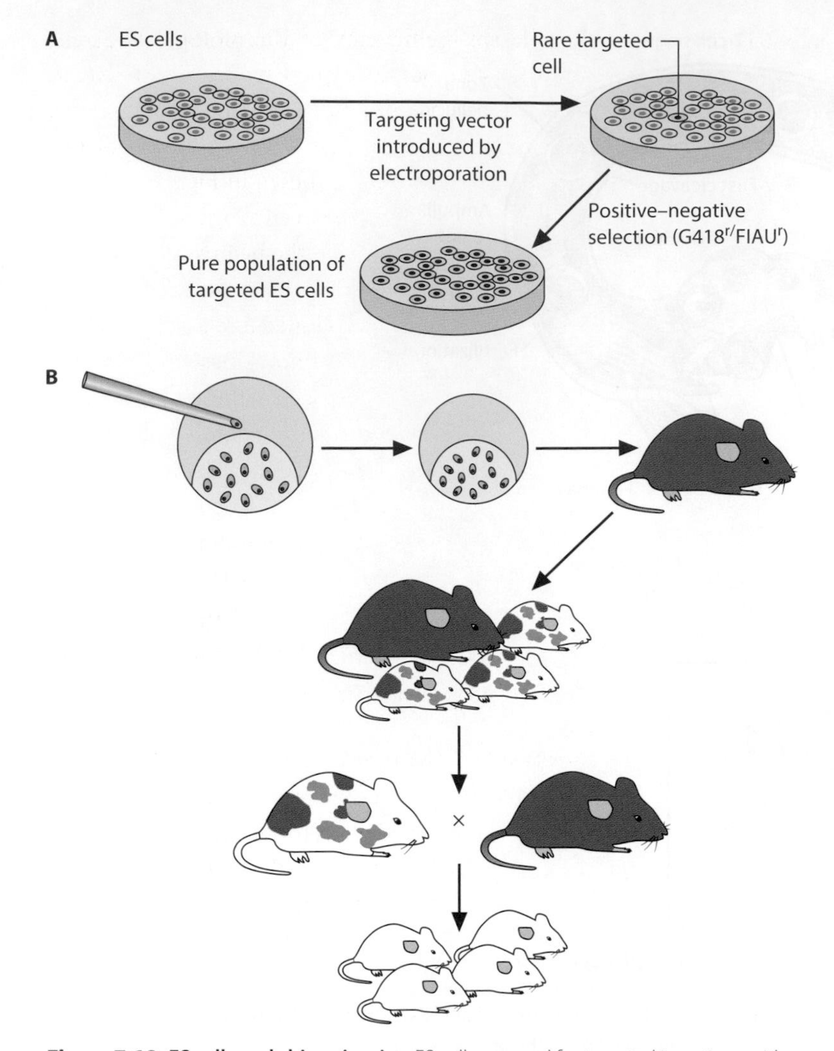

Figure 7.10 ES cells and chimeric mice. ES cells are used for targeted insertions with a selection for particular genetic events. The selected ES cells can be assayed molecularly, to ensure that the recombination event of interest has occurred. Those cells with the proper insert are injected into the blastocoel of an isolated blastocyst, and the injected blastocyst is implanted into a pregnant female host. In this example, the ES cells were derived from a white mouse and the recipient blastocyst was derived from a brown mouse, but other coat color combinations are often used. The pregnant female gives birth to a litter with some mice from her own embryos and others from the chimeric embryos. Mice derived from the chimeric embryos are recognized by the variegated coat color and are presumed to be chimeric in all cell types. The chimeric mice are mated to produce a mouse that derives entirely from the ES cells with the targeted insertion. Here those mice will be white. From Cappechi, 2005, Gene targeting in mice: functional analysis of the mammalian genome for the twenty-first century. *Nature Review Genetics* 6: 507, used with permission.

develop from this blastocyst, as shown in Figure 7.10. In order to produce transgenic mice, embryos at the blastocyst stage are isolated from a pregnant female, ES cells are introduced into an isolated blastocyst by injection, and the blastocyst is then inserted into another pregnant female or a hormonally treated pseudo-pregnant female. The blastocyst will implant into the uterine wall and will eventually give rise to a mouse that includes some of the engineered injected cells. Because there are very few cells in the inner cell mass and cells at this stage are totipotent, an introduced cell has the potential to develop into any structure in the mouse, including the germline. Thus the mouse developing from such an embryo will be a **chimera** of cells of more than one genotype.

In summary, the key cellular and developmental principles that make transgenic mice possible are as follows.

1. Cells of the inner cell mass can be isolated and grown in culture as ES cells.
2. ES cells can be reintroduced into a developing embryo at the blastocyst stage and will become incorporated into the inner cell mass.
3. The blastocyst can be reintroduced into a female and a chimeric mouse will develop with cells of both types.

Thus, when a gene is stably inserted into the genome of an ES cell grown in culture, a mouse may be produced with that gene insertion. This meets one of the criterion for reverse genetics: that an organism can be grown from a single altered cell or a small number of altered cells.

ES cells can be targeted for gene insertion using both positive and negative selections

A second criterion for efficient gene disruptions and replacements is that the introduced sequences can be targeted to specific sites in the genome. Cloned genes are typically introduced into ES cells by electroporation of linear DNA because large populations of cells can be treated simply and simultaneously. Again, just as in targeted gene replacements in yeast, integration depends on homology directed recombination between the introduced donor DNA fragment and the chromosomal sequence. Unlike in yeast, however, in ES cells extensive sequence homology is required for recombination to occur; most homologous recombination constructs have at least 8 kb of homology with the target sequence on either side of the DNA to be integrated. Homology-directed repair and recombination can occur at any location within that 8 kb region. Furthermore, unlike in yeast, the rate of homologous recombination in ES cells is very low, so a selection scheme is required to identify the ES cells with the proper insertion.

The relative inefficiency of the biological recombination process has prompted mouse geneticists to introduce both a positive and a negative selection, also called **counter-selection**. The targeting vector for this counter-selection process is shown in Figure 7.11. The gene under study is disrupted in vitro by the insertion of a drug selection marker, often the neomycin-resistance gene (abbreviated *neo*^{r}) that has been placed in an exon. The disrupted target gene is cloned into a targeting vector, which includes the thymidine kinase (abbreviated *tk*) gene from a Herpes virus. Each of these genes confers a specific drug response: the *neo*^{r} confers *resistance* to neomycin and its more stable analog drug G418; and the *tk* gene confers *sensitivity* to a drug known as gancyclovir. Note that these drug responses are opposite to each other and make it possible to select both for and against the cloned donor gene. The targeting vector is introduced into ES cells that are diploid for a normal copy of the target gene and that are naturally resistant to gancyclovir.

The introduced targeting vector will encounter one of three fates in ES cells, as summarized in Figure 7.12. In most cells, the targeting vector will not become integrated into the chromosome at all. The ES cells with no insertion are sensitive to G418, and so will not grow if G418 is included in the culture medium. Thus there is a positive selection *for* integration by using the *neo*^{r} gene and G418, which allows even rare integrations of donor sequences to be detected.

Literature Link. Capecchi 2008, *ChemBioChem* 9: 1530–43

On the other hand, the targeting vector can be inserted to produce G418 resistance by either of two mechanisms. Insertion can occur either by random insertion at any DSB location or by the homologous recombination at the desired location. Random insertion results in the

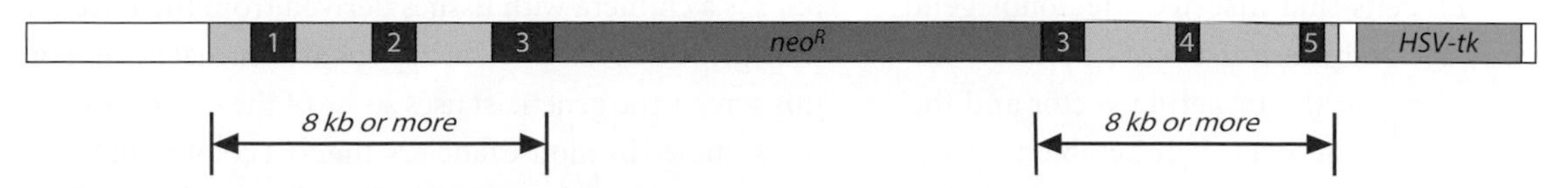

Figure 7.11 A targeting vector used for a gene disruption. The mouse gene is shown in blue, with light blue representing introns and the dark blue exons numbered. The *neo*^{R} gene has been inserted into the third exon. The vector also includes the thymidine kinase gene Herpes simplex virus (*HSV-tk*). The *HSV-tk* gene confers sensitivity to gancyclovir, whereas the *neo*^{R} gene confers resistance to neomycin or to its more stable analogue G418. At least 8 kb of homology are needed for homologous recombination between the targeting vector sequence and the chromosomal gene.

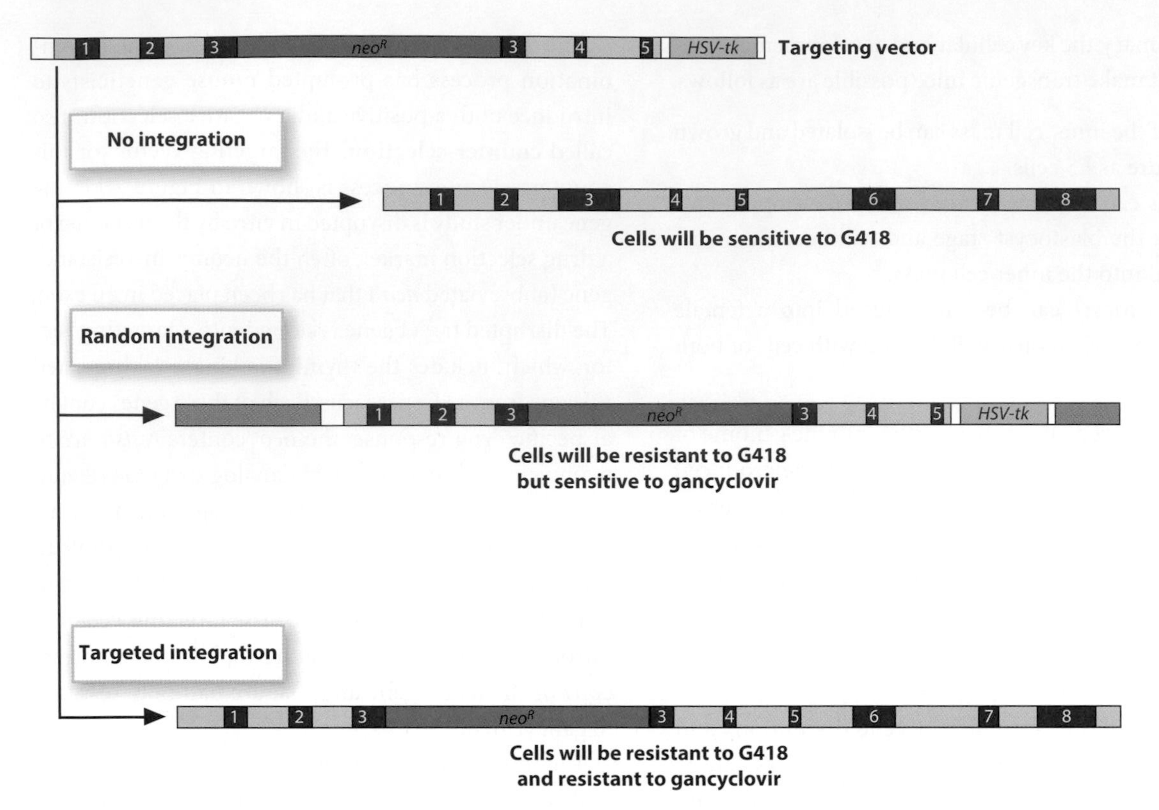

Figure 7.12 Selection and counter-selection for targeted integrations. Once the targeting vector has been introduced into ES cells, one of three events can occur, as shown here. If the vector fails to integrate into the chromosome in the ES cells, the cells will not grow in the presence of the neomycin analogue G418. G418-resistant cells can arise either from integration of the targeting vector at random sites (suggested by the light purple) or from targeted integration via homology-directed repair with the chromosomal copy of the gene of interest. These ES cells can be distinguished by a selection that is based on gancyclovir. If the vector inserts at random locations without relying on homologous recombination, the HSV-*tk* gene, which confers sensitivity to gancyclovir, will also be integrated. If the vector inserts by homologous recombination with the target gene, the HSV-*tk* gene is excluded (since it lies outside the region of homology) and the cells are resistant to ganciclovir.

integration of the entire targeting vector, and thus will confer not only resistance to G418 but also sensitivity to gancyclovir caused by the *tk* gene included on the vector. These untargeted events, which may be as much as 99 percent of the G418-resistant cells, can be selected *against* by including gancyclovir in the culture medium, so that no insert of the *tk* gene can grow.

The counter-selection with these two drugs ensures that only those ES cells that inserted the donor gene by homologous recombination events between the disrupted target gene on the targeting vector and the chromosomal copy of the gene will be able to grow and divide. These are the very cells that are needed to construct a chimeric mouse that has the engineered gene replacement. The ES cells that grow and divide in culture when subjected to this selection are isolated and introduced into a blastocyst embryo. As described above, these ES cells will become part of a new mouse.

There is yet one more screen in the gene replacement process, one that occurs among the mice born in the same litter. When the chimeric blastocyst is inserted into the surrogate mother, it is not the only developing embryo in her uterus. Embryos of her own genotype are also developing. How can one tell which of the newborn mouse pups is a chimera with tissues derived from the injected transgenic ES cells and which pups are normal mice? For this screen, the geneticist uses some of the earliest mutations studied by mouse fanciers: the coat color mutations.

As shown in Figure 7.13, the ES cells are often derived from a brown (or agouti) mouse and introduced into blastocysts that have been isolated from a black female. The chimeric mice will have a black and brown coat color,

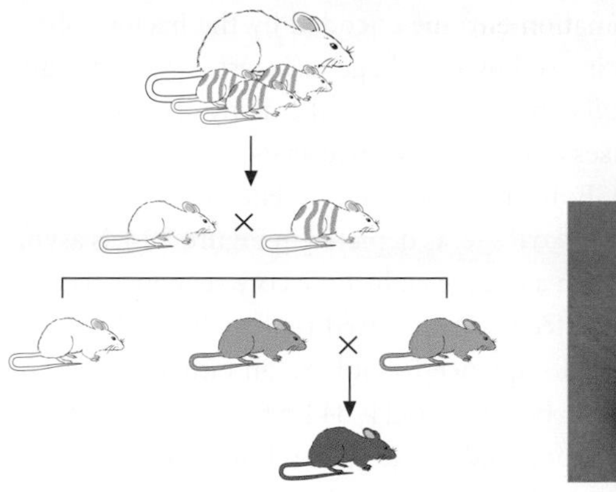
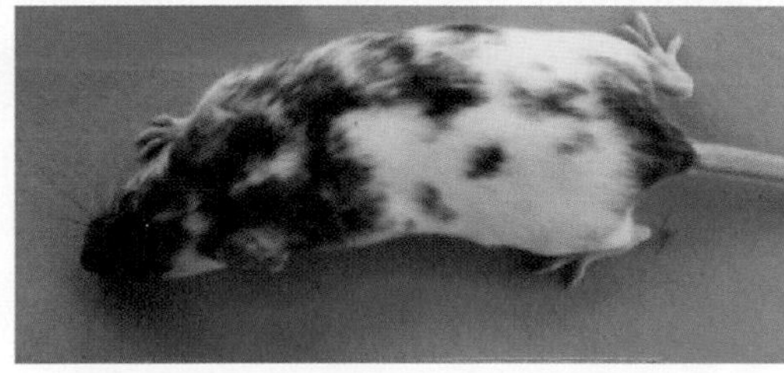

Figure 7.13 Coat colors indicate chimeric mice. Many ES cell lines are derived from the inner cell mass from brown mice; the injected blastocyst may be either black or white, as shown here. A mouse that arose from a chimeric inner cell mass will have a variegated coat color with both brown and white, in different patterns. These chimeric mice can be bred to produce offspring that are derived entirely from the engineered ES cells, as indicated by the brown coat. A PCR assay on cells from the tail is often used to confirm that the mouse is homozygous for the insert.

whereas those arising from a non-chimeric wild-type inner cell mass will give rise to entirely black mice. The exact pattern of black and brown in the mouse is not important, but the extent of brown fur can be helpful for recognizing the success of the process. A mouse that has only a few very small patches of brown fur may consist primarily of black non-engineered cells, whereas one that has more or larger patches of brown fur is likely to have more cells arising from the engineered ES cells. The coat color acts mainly as a marker of chimerism. These chimeric mice are also likely to be chimeric in their germline, and thus will give rise to two different types of gametes: gametes derived from ES cells with the disrupted gene and gametes derived from ES cells with the normal gene.

The chimeric mice are bred, and a homozygous strain is derived in which all the offspring are brown—meaning that all these cells arose from the engineered ES cells in a previous generation. The knockout can be confirmed by performing PCR or a similar assay on a sample of transgenic cells obtained from a snip from the tail, for example.

Literature Link. Capecchi 2005, *Nature Reviews Genetics* 6: 507–12

The ability to target gene knockouts has changed mouse genetics

Until gene knockouts were available, mouse geneticists relied on the types of mutant screens described in Chapter 4. These worked well, and many important and influential experiments in mouse genetics begin with random mutagenesis. However, the ability to target mutations to specific genes has added a new dimension to mouse genetics, and this general procedure has been used to create thousands of different **knockout mutations** in mice. The case study in this chapter describes how the mouse orthologue of the *patched* gene, introduced in the case study in Chapter 4, was targeted for disruption.

Once the mutation has been generated, characterization of the mutant phenotype is done as for other mutations. Many knockout mutations result in a lethal phenotype, so the mutation must be maintained as a heterozygote. Conditional mutations in which the gene is knocked out only in certain tissues or at certain times are described below and provide another method to investigate genes with widespread lethal effects.

In contrast to these severe effects of lethal mutations, many knockouts result in no obvious mutant phenotype, as has also been observed for mutations in other organisms. In fact this is one of the major benefits of reverse genetics and genome-wide mutant screen, since mutations in these genes would not have been identified in a traditional screen, which was based on mutant phenotypes. As will be described in Chapters 10 and 12, it is often the case that only the double mutant with knockouts in two separate genes shows a mutant phenotype, owing to issues of functional redundancy. Knockout mutations are expected to be null alleles with a complete loss of gene function, because the sequence of the targeted gene has been disrupted by the *neo*ʳ gene.

As noted in Chapter 4, null alleles are important for recognizing the earliest time at which a gene exerts its effect, but it is often helpful to have other types of alleles as well and to generate an allelic series of mutations. Most knockout mutations are recessive, but some are dominant loss-of-function alleles, including dominant negative mutations and mutations in haplo-insufficient genes. Modifications in how the gene is cloned into the targeting vector have allowed mouse geneticists to make different types of alleles. Some of these modifications are described in the next section.

Genes can also be targeted by Cre-*lox* or other recombination systems

Targeted gene replacements in ES cells can be detected because of the strong positive and negative selections that are imposed, even though they occur at very low frequency. A higher frequency of recombinants requiring much less sequence homology between the exogenous DNA and the target gene can be produced using **site-specific recombination** systems. With site-specific recombination, the DSB that is required to initiate the repair process is directed to a precise location by the combination of a specialized recombination enzyme (the recombinase) and a short sequence that is recognized by the recombinase. These site-specific recombination systems have allowed geneticists to replace the chromosomal gene with an altered version and to make reporter gene constructs. Targeting DSBs to precise sequences and locations in the genome is the core strategy for genome editing, and some other techniques to accomplish it are described in Chapter 8.

A common application of site-specific recombination in mice involves the Cre-*lox* system. Cre is a recombination enzyme encoded by the bacteriophage P1, and is used by the phage to insert its genome into the *E. coli* chromosome. Cre, like many other recombinases, uses a specific sequence as its substrate for making a DSB; in this case, the sequence is known as the *loxP* site. A *loxP* site, as depicted in Figure 7.14, is asymmetric, with a unique eight-base core sequence flanked by two thirteen-base inverted repeats. Thus the extent of required sequence homology on either side of the sequence to be integrated is 34 bp for Cre-*lox* mediated recombination, and Cre-*lox* recombination has been the standard method for producing targeted gene disruptions and replacements in mice.

Literature Link. Sauer and Henderson 1988, *Proceedings of the National Academies of Sciences USA* 85: 5166–70; tutorials from www.jax.org

The asymmetry of the eight-base *loxP* core sequence specifies the orientation in which the gene is inserted. The targeting vector used for Cre-*lox* techniques includes the *loxP* sites, but the procedure for inserting the altered gene is essentially the same as the one used for a standard gene knockout once the DSB is made by Cre. Two limitations with this method are that the *loxP* sites must be located in the target sequences and that Cre recombinase must be expressed in the genome of the host mouse. However, *loxP* sites can be inserted anywhere in the genome using the recombination and repair processes described above.

Once a mouse has been made with *loxP* sites at a certain location, it can be bred and the *loxP* site can be treated like any other genetic marker in generating other strains. Thus the first limitation has become less of an issue, as mouse strains with *lox* sites at the desired locations have been created. In addition, mouse strains with

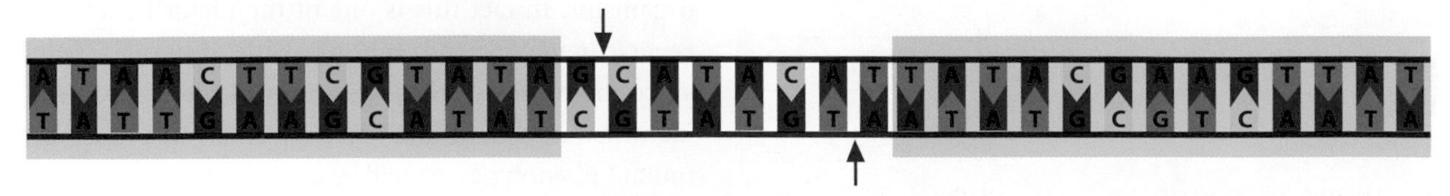

Figure 7.14 The Cre-*lox* system for site-specific recombination. Cre recombinase recognizes the specific sequence of the *loxP* site shown and cleaves it at the sites indicated by the arrow heads. Note that the *lox* site has palindromic arms, shown by the green boxes, flanking a unique sequence core. The center core sequence is asymmetric and confers directionality to a *loxP* site. In subsequent figures, this directionality is represented by a red arrow head.

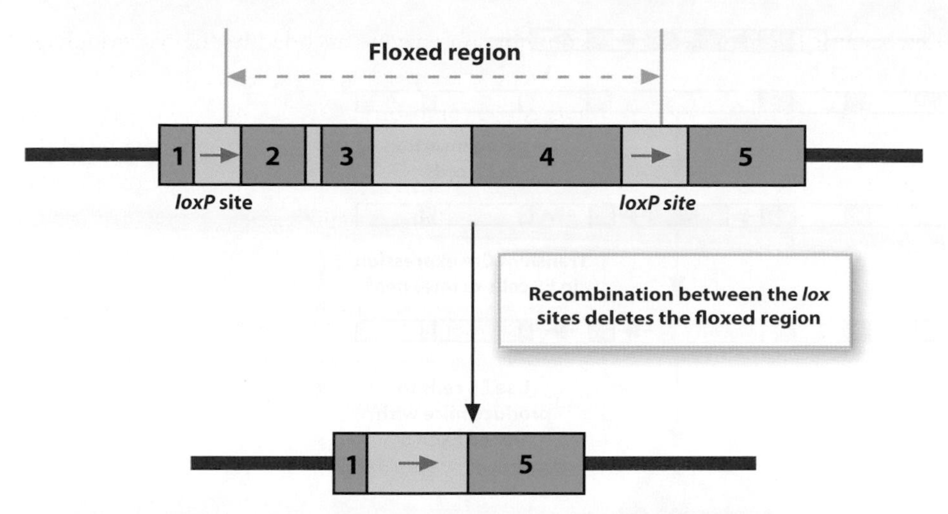

Figure 7.15 Floxed sites and deletions. Two *loxP* sites (orange arrow heads) in the same orientation are in introns and flank exons 2, 3, and 4; the sequence between the *loxP* sites is said to be floxed. The DSBs made by Cre at these *lox* sites allow recombination to occur, which deletes the sequence between them. Since both molecules retain a *loxP* site, Cre-mediated insertion can also occur in the reverse reaction. Flanking *lox* sites in other orientations can be used to produce other types of rearrangement of floxed regions.

the gene for Cre recombinase integrated into the genome have now been developed, so this is also not often a problem. More significant than these limitations is the advantage that the presence of the *lox* sites in the targeted gene allows for a number of additional manipulations.

Consider the situation when the *lox* sites are placed in direct orientation with respect to each other, as shown in Figure 7.15. When Cre makes a DSB at each *lox* site that initiates a crossover between the two sites, the sequence between the *lox* sites is deleted. A sequence that is flanked by *lox* sites is sometimes said to be "**floxed**." Thus, by assiduous placement of the *lox* sites, particular and limited portions of a gene can be targeted for specific deletion. Deletions of this type generate a true null allele—an advantage over gene disruptions using exogenous DNA such as the *neo*^{r} gene to interrupt the gene, since no extra DNA has been inserted.

Tissue-specific mutations can be made by regulating Cre expression

We discussed mosaic analysis in Chapter 6 as a method to determine which cells require the activity of the gene. The two strategies were either to introduce the wild-type copy of the gene in some cells of a mutant organism, or to remove the wild-type copy of the gene in some cells of a heterozygous organism. The specificity of the Cre-*lox* interaction provides a method to remove the wild-type allele only in particular cells, since a targeted deletion will only occur when Cre recombinase is being expressed; when Cre is absent, no DSB at the *lox* sites will occur and the target gene will not be affected. The specificity of Cre expression therefore allows an additional level of manipulation when making deletions.

In particular, by controlling the expression of Cre, it is possible to delete the target gene only in particular cells rather than throughout the entire mouse. This is an important advantage when working with genes that have widespread lethal effects. Mouse strains have been made in which the *Cre* gene has been integrated under the control of regulated or tissue-specific regulatory modules. The combination of the floxed gene with the specific expression of Cre provides a way to study gene function in a much more defined manner.

 ***Literature Link.*** Tsien 1996, *Cell* 87: 1317–26

For example, a gene such as *PTEN*, which is essential in early embryonic development, may have specific and additional functions in adult tissues, such as the pancreas. In a gene knockout made in the ES cells, the mice will die at the earliest developmental stage that

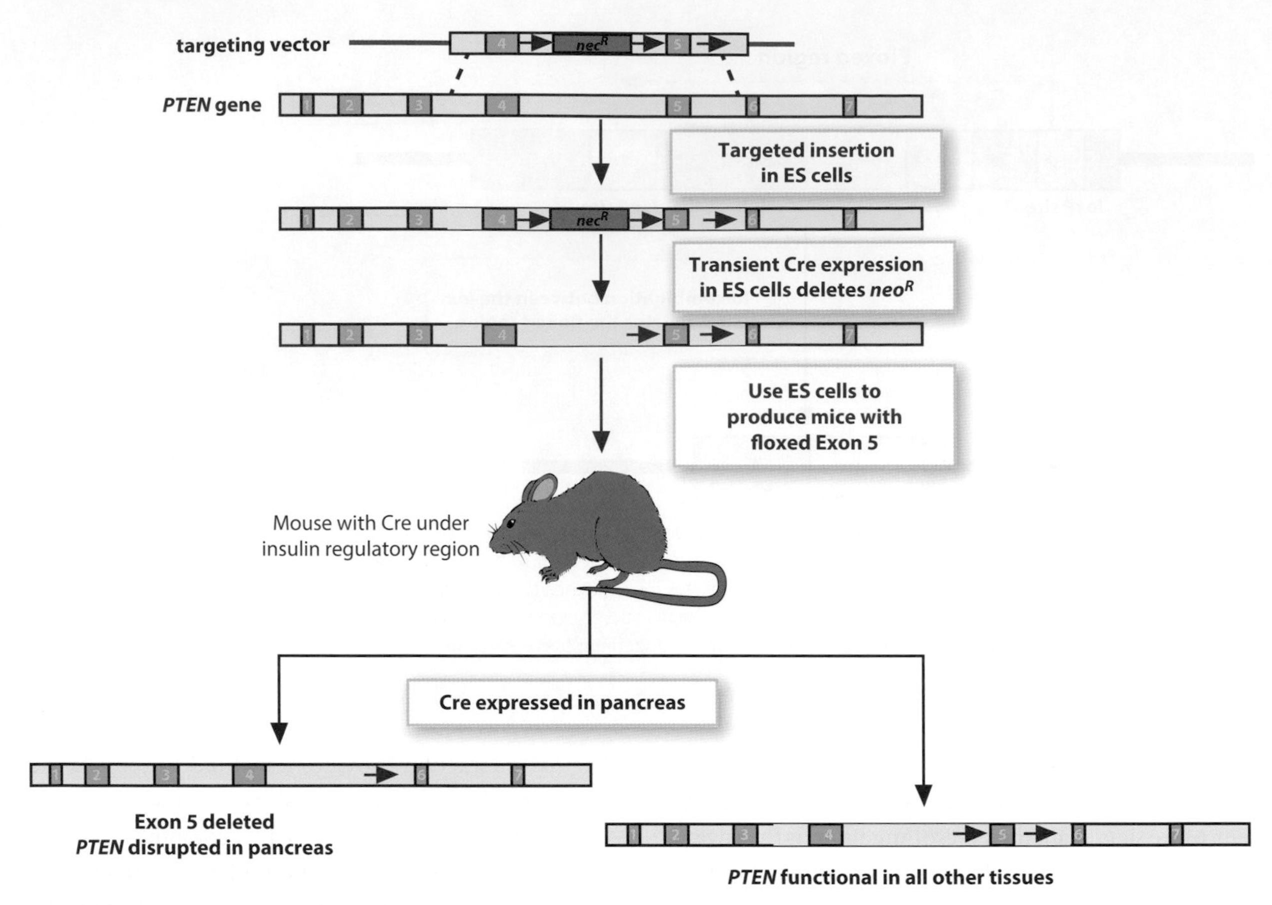

Figure 7.16 **Cre-*lox* deletion of *PTEN* in the pancreas.** A targeting vector is inserted in the *PTEN* gene by homologous recombination. In the targeting vector both the *neo* gene in the intron and exon 5 of the *PTEN* gene are flanked by *loxP* sites. Although recombination can occur at many locations, the recombinants with crossovers at the locations of the dashed lines are screened for; these have replaced the chromosomal *PTEN* gene with the edited gene. The actual *PTEN* gene has nine exons, of which seven are shown here. With transient Cre expression in the ES cells, the *neo* gene is deleted. These events can be detected in ES cells by appropriate molecular probes. The ES cells are used to make a transgenic mouse strain that also has the *Cre* gene expressed under the regulatory region from an insulin gene. Under the control of this regulatory region, Cre recombinase is expressed only in the pancreas, and exon 5 is deleted in those cells of the pancreas to produce a tissue-specific gene deletion that is to be analysed for the role of *PTEN* in pancreatic cells. In other cells Cre is not expressed, so the mouse can survive.

the gene is necessary, so any later functions cannot be recognized studied. However, these effects can be studied using a conditional knockout in which the gene is deleted only in certain tissues. An example is shown in Figure 7.16.

In order to study the effect of the *PTEN* gene specifically in the pancreas, the gene was replaced in ES cells with a version in which exon 5 is floxed; these ES cells were used to make a transgenic mouse. Another strain of mouse was made in which the Cre enzyme was placed under the control of the regulatory region from the insulin gene, so that *Cre* was transcribed only in specific cells in the pancreas. Both strains of mice are viable and fertile even as homozygotes, since neither the presence of the *lox* sites nor the expression of the Cre protein is detrimental by itself. The two mice were mated to produce offspring with the *Cre* gene expressed in cells in the pancreas and the *lox* sites flanking the region of interest in the *PTEN* gene. The expression on Cre then disrupted *PTEN* specifically and exclusively in the pancreas, allowing the investigator to determine the effects of the mutation in one cell type or organ, without the compounding effects of *PTEN* deficiency in other parts of the mouse at other times in development.

Literature Link. Zender et al. 2007, *Cell* 129: 838–9 and references therein

Knock-in mutations replace the coding region with an edited sequence and altered function

Another important use of the Cre-*lox* system occurs when the normal coding region of a gene is replaced by an altered gene. These have been nicknamed "**knock-in**" experiments because the altered function is *introduced* rather than knocked out. In these cases the gene is not deleted or disrupted, but instead is replaced by a specific and defined molecular defect. This is analogous to what occurs in homology-directed recombination and repair in yeast.

As noted in Chapter 5, some genetic diseases and syndromes are not due to a deletion or a gene disruption, and different mutations within the same gene can have different symptoms and phenotypes. A knock-in mutation that replaces the wild-type gene with an edited sequence is often the best method to produce an appropriate model and to test the functional importance of particular portions of a gene.

An example of one type of knock-in experiment is provided by the mouse orthologue of the human *CFTR* gene, and illustrated in Figure 7.17. As described in the case study in Chapter 5, the common mutation giving rise to cystic fibrosis in humans is a three nucleotide deletion in the *CFTR* gene, a mutation called delta F508 because a single phenylalanine residue is deleted. A gene disruption that deleted other portions of the gene may not provide a realistic mouse model for the most common mutation that occurs in this gene and the symptoms that arise from it; in fact CF patients who do not have the common delta F508 mutation frequently have somewhat different symptoms.

In order to create a mouse model that mimicked the most common form of cystic fibrosis, the normal coding region of the gene was replaced by a version with this three-nucleotide deletion made in vitro. Several different procedures have been used to create this knock-in mutation. One method was to make the desired alteration in the gene in vitro with flanking *lox* sites. Introduced into ES cells expressing Cre to make the DSB, this engineered allele can replace the normal allele of the gene in order to produce a mouse model with the same mutation found in human patients. Similar procedures have been used for many genetic diseases for which a common disease mutation is known.

Mouse genetics refers to gene disruptions and gene replacements as knockout, knock-in, and knockdown mutations

Procedures similar to those with *CFTR* can be used to make other types of hypomorphic alleles of a gene. This

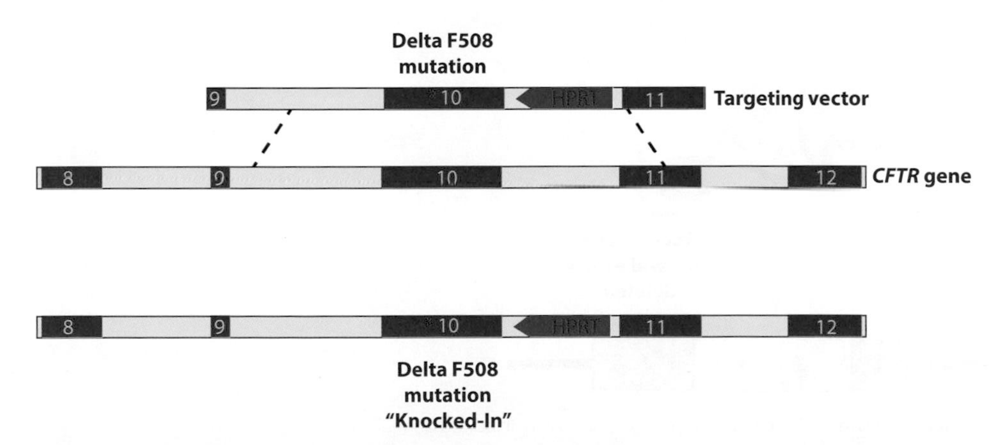

Figure 7.17 A knock-in of the *CFTR* deltaF508 allele. Most cases of cystic fibrosis among people of European descent are due to a specific three-base deletion in exon 10 of the *CFTR* gene, referred to as delta F508. This mutation was made in vitro in the mouse *CFTR* gene, as shown in the dark blue exon with the red star that indicates the mutation. The mutant gene was cloned by conventional molecular methods into a targeting vector, as shown at the top. In this particular case, the selection marker was the *HPRT* gene, which can be selected in HAT media, rather than the neo^R gene described in the chapter, but the rationale is the same. The *HPRT* gene was cloned into an intron in reverse orientation, so its expression would not interfere with normal CFTR expression in lung cells. Homologous recombination inserted this construct into the chromosomal *CFTR* gene in ES cells, as shown. These ES cells were used to make chimeric mice that carried the mutant *CFTR* gene.

is particularly useful for genes in which the null allele created by a gene knockout is lethal, but for which a hypomorphic allele, sometimes called a "knockdown" mutation (in keeping with the other nicknames), is not. Genes can also be engineered so that the resulting protein will have modifications in specific residues. For example, many proteins comprise several different domains or protein motifs such as a kinase domain, a DNA-binding domain, or a phosphorylation site. The effects of these motifs can be studied separately by using mutations that affect only one of them. As discussed in Chapter 4, an allelic series of hypomorphic and null alleles is useful to see the range of effects that arise from one gene rather than simply the null effects. Dominant alleles, such as constitutively active hypermorphic mutations and dominant negative mutations, can also be created in this way. Knockout and knock-in mutations are analogous to gene disruptions and gene replacements respectively, although using different terminology.

Knock-in experiments are also used to make reporter gene constructs in which *lacZ* or another reporter coding region replaces the coding region of the normal gene. With a reporter gene of this type, the normal expression

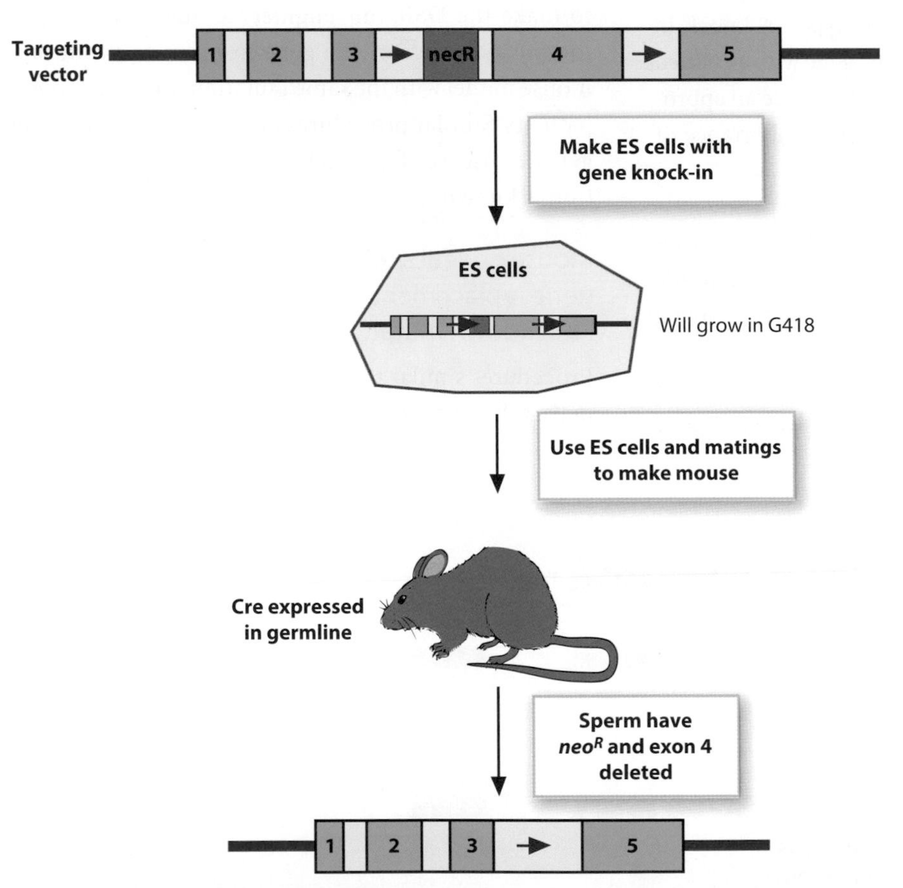

Figure 7.18 Germline knockout of *neo*^{R}. An investigator can produce a mouse with an edited gene but without the addition of *neo*^{R} in the genome of the animal. A targeting vector is made with the *neo*^{R} gene inserted and floxed; in this example, exon 4 is also floxed, so that the target gene will be disrupted. ES cells with this gene replacement or knock-in can be selected using G418 in the standard procedure. Once the ES cells are used to make a transgenic mouse, the mouse can be mated to produce a line that has both the edited gene and the Cre recombinase expressed in the germ line; in this case, Cre is expressed during spermatogenesis. When this mouse makes sperm, both the *neo*^{R} gene and exon 4 of the target gene will be deleted, so that the offspring will have the gene disruption but no *neo*^{R} gene.

can be easily monitored. Furthermore, since the reporter gene is expressed under the control of the regulatory region of the gene it has replaced, the normal expression pattern of the gene is revealed even if the pattern of expression is complex. Since the mutation made by the reporter construct is likely to be recessive with regard to gene expression and function, a heterozygote will have a normal phenotype so the reporter gene expression can be monitored in phenotypically wild-type mice. Although the procedure did not use a knock-in strategy, the *patched* gene disruption described in the case study also included the *lacZ* reporter gene, whose expression was important for understanding the complex roles of *patched* in mouse embryogenesis.

While the *neo*^{r} gene provides a powerful selection for identifying ES cells with the edited gene, the investigator may not want to have the inserted *neo*^{r} gene in the mouse once the chimeric embryo has been made. After G418 selection has been applied to find the ES cells, the *neo*^{r} gene is no longer needed and could be deleted. Thus, another powerful use of the Cre-*lox* technology places the Cre protein under the control of a germline specific promoter; when the chimeric mouse begins make gametes, the expression of the Cre in the germ cells will delete the *neo*^{r} selectable marker, and its offspring will not have the selectable marker but will continue to have the other edited genes.

An example of this use is shown in Figure 7.18. In this technique, the gene is altered in vitro with the *neo*^{r} gene flanked with *lox* sites introduced into ES cells by standard procedures, and the ES cells are injected into blastocysts from a mouse that has the Cre gene expressed only in the germline, for example under the control of a regulatory region from a gene expressed specifically during spermatogenesis. Since the *neo*^{r} gene is deleted in the sperm once the male mouse reaches sexual maturity and expresses Cre, the F_1 and succeeding generations will have the altered or the floxed allele without complications arising from the exogenous *neo*^{r} gene.

Literature Link. van der Weyden et al. 2011, *Genome Biology* 12: 224

7.4 Summary: Reverse genetics and genome editing allow specific types of mutations to be made and analysed

The "awesome power of yeast genetics" introduced in Chapter 3 owes much of its truth to the ability to manipulate genes in vitro and to target their reintroduction into the genome. It sometimes seems that it is possible to perform nearly any experiment imagined using yeast genetics, and recent work has allowed yeast geneticists to create synthetic chromosomes with any sequences of interest. Instead of relying on mutations occurring at random sites in the genome, genetic experiments in yeast can produce almost any desired mutation and determine its effect. The power of targeting the mutated gene is that the investigator can *replace* the genomic copy rather than having the genomic copy still present in the background. Thus the only version of the gene is the one that the geneticist has altered and introduced.

Our ability to manipulate genes and the genome in the mouse will probably never reach the level of sophistication in yeast, in part because the mouse is a more complex organism than yeast. Nonetheless, reverse genetic approaches have dominated mouse genetics. Since the mouse is the most thoroughly researched model organism for understanding human genetic diseases, flexible and efficient gene-targeting methods are regularly being developed. High-throughput genome-wide screens for gene disruptions in the mouse are underway, and it is certain that this technology will continue to evolve and improve. With the advent of CRISPR technology for genome editing, as described in the next chapter, it appears that we may be able to domesticate not only organisms such as yeast and mice but their genomes as well.

Chapter Capsule

Genetic analysis relies on having a mutant phenotype, but mutant phenotypes can be created by beginning with a cloned gene rather than an organism. In reverse genetic analysis, a cloned gene is used to identify a mutant phenotype for the gene and thereby to infer the function of the gene. These methods have been particularly well established in yeast and mice. Two different strategies have been widely employed. Gene disruptions or knockouts eliminate the functions of genes, usually by making deletions or insertions in the coding region. Gene replacements or knock-ins rely on homologous recombination to replace the chromosomal copy of a gene with the version created and edited in vitro.

Additional Reading

Richardson, S. M. et al. 2017. Design of a synthetic yeast genome. *Science* 355: 1040–4.

Capecchi, M. R. 2005. Gene targeting in mice: Functional analysis of the mammalian genome for the twenty-first century. *Nature Review Genetics* 6: 507–12.

Segal, D. J., and J. F. Meckler. 2013. Genome engineering at the dawn of the golden age. ***Annual Review of Genomics and Human Genetics*** 14: 135–58.

Case Study 7.1

Patched knockout mutations in mice

The *patched* gene in *Drosophila* was introduced in Chapter 4 as one of the segment polarity genes that affect the orientation of each segment in the embryo. Subsequent analysis in *Drosophila*, some of which is described in Chapter 11, showed that the *patched* gene is involved in a wide range of important development events through its interactions with *hedgehog* and *smoothened*; this is referred to as the *hedgehog* signaling pathway. The *hedgehog* pathway and orthologues of *patched* are found in most other animals, including mammals. Tissues that express *hedgehog* are also likely to express *patched*, so cDNA was prepared from *hedgehog*-expressing tissues. The mouse *patched* gene was cloned by finding cDNAs from these tissues that cross-hybridized to *patched* genes from other vertebrates. These experiments suggested that the mammalian *patched* gene encoded an important signaling protein during development and played a role in some human cancers; but they were not designed to reveal the full range of *patched* effects. For that, the investigators needed to make a gene knockout. We should note that subsequent genomic analysis has revealed that mice have two *patched* genes, known as *Ptch1* and *Ptch2*, with somewhat different expression patterns; the experiments to be described here were done with *Ptch1*.

The mouse *Ptch1* gene has fifteen exons, and a complete genomic clone was obtained by screening a mouse library. A portion of the gene was cloned into a standard targeting vector of the type diagrammed in Figure 7.11. On the targeting vector, a portion of the first exon was fused in frame with a *lacZ* gene and *neo*^{R} gene; this replaced part of the first exon, all of the first intron and second exon, and part of the second intron, as shown in Figure CS7.1. The HSV *tk* gene was also included on the targeting vector, downstream of the second

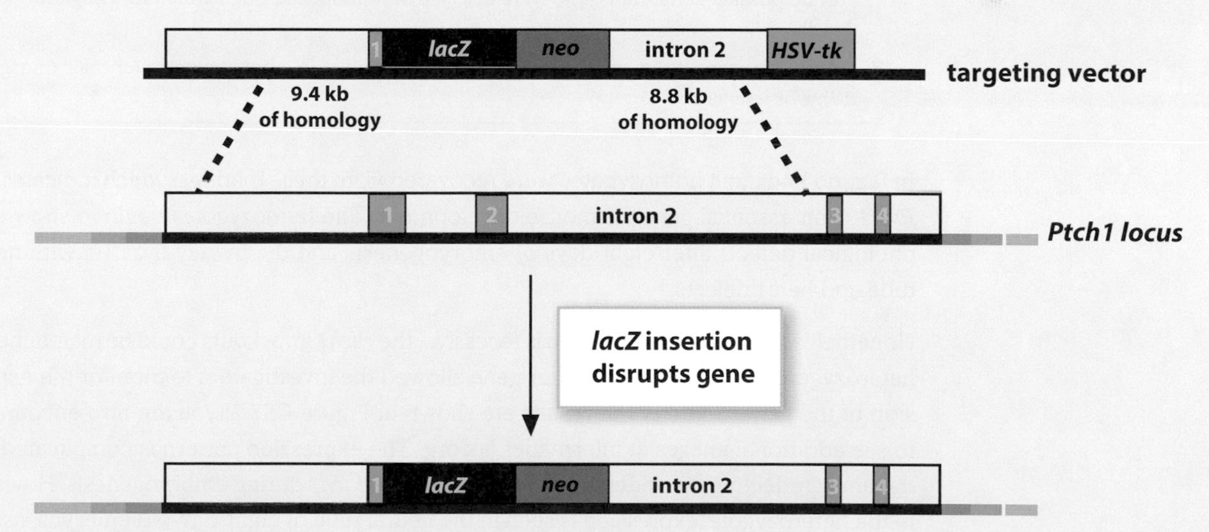

Figure CS7.1 Disruption of the mouse *Ptch1* gene. The targeting vector replaces most of exon 1, all of intron 1 and exon 2, and some of intron 2 of the *Ptch1* gene with the *lacZ* gene and *neo*^{R} gene, leaving 9.4 kb of the mouse *Ptch1* gene on the left side and 8.8 kb on the right side to allow homologous recombination between the targeting vector and the *Ptch1* locus on the chromosome. Locations are crossovers between the targeting vector and the chromosomal locus and are indicated by the dashed lines. The vector also includes the *tk* gene to select for targeted insertions. The *Ptch1* gene has fifteen exons and the drawing is not to scale.

intron. The targeting vector included 9.4 kb of sequence homology upstream of the first exon and 8.8 kb of sequence homology downstream of the neo^R gene.

Thinking through the Experiment

1. The *Ptch1* gene in the genome consists of fifteen exons, but the targeting vector had only a portion of the gene. Why was the entire *Ptch1* not needed on the targeting vector?
2. Describe how the neo^R and the *tk* genes will be used to identify ES cells with the desired insertion.
3. The *lacZ* gene plays two different roles in this experiment. What are these two roles?

This linear targeting plasmid was electroporated into ES cells derived from a brown mouse. The ES cells were grown in culture in the presence of G418 (the positive selection for integration) and gancyclovir (the negative selection against random integration). Three independent clones were expanded into colonies, and these were injected separately into blastocysts isolated from a black mouse. The injected blastocysts were introduced into black females and brought to term as chimeric brown and black mice. The chimeric mice were then bred among themselves to make heterozygotes, and the heterozygotes were mated, in the hope of finding a strain homozygous for the *patched* knockout.

Thinking through the Experiment

4. Outline how the coat colors in this experiment where used to identify the mice with the desired transgene. Did the coat color of the host recipient female matter?
5. Thinking carefully about mammalian developmental cell biology and chimerism, what are some of the limitations of using coat colors to find the mice with the transgene?
6. The investigators selected for targeted gene disruptions of the *Ptch1* gene rather than for random insertions at other DSB break sites. How would they have changed their selection strategy to find random insertions (which are much more common than targeted insertions) in ES cells? Which of the two roles played by *lacZ* in these experiments (see Question 3) would probably not be possible if the transgene were inserted at random sites, and which role might be still be feasible?
7. Would you expect the investigators to be able to find viable homozygotes for their transgene? Why, or why not?

In fact no knockout homozygotes were recovered from these matings, which indicates that *Ptch1* is an essential gene for mouse development. The homozygotes begin to show morphological defects after eight days of embryogenesis and die by day 9 or 10, with neural tube and heart defects.

Nonetheless, because the mutation is recessive, the *Ptch1* knockouts could be maintained as heterozygotes, and the *lacZ* reporter gene allowed the investigators to monitor the expression of the gene. Some of the results are shown in Figure CS7.2; you are also encouraged to see additional images at informatics.jax.org. The expression pattern is complicated and dynamic, reflecting the widespread signaling role of *Ptch1* during embryogenesis. However, in the heterozygote, expression is seen in the neural tube of eight-day-old embryos, as well as in other tissues. This expression pattern is consistent with the mutant defects seen in the homozygote.

Although heterozygous embryos developed normally, defects were seen in some heterozygous adults. A few *Ptch1* heterozygotes had extra digits (polydactyly) or fused digits

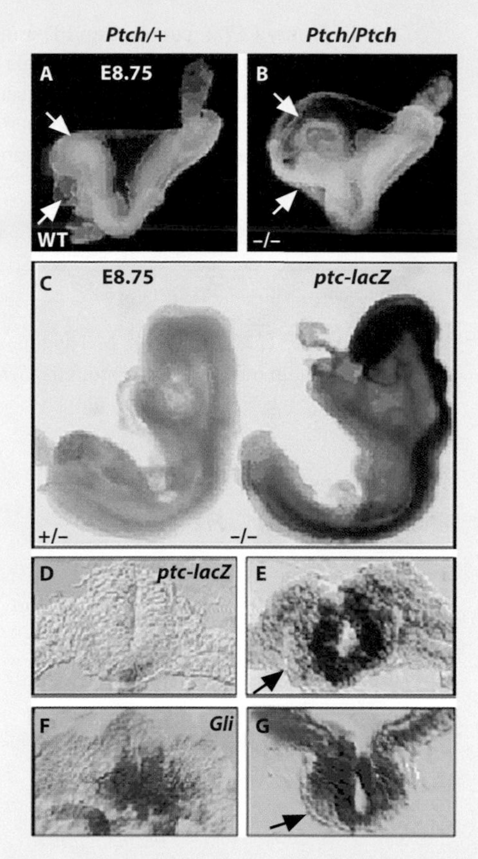

Figure CS7.2 Expression of the *Ptch 1*gene, in wild-type and mutant embryos. The gene disruption diagrammed in Figure CS7.1 included the *lacZ* reporter gene, whose expression is detected by staining for β-galactosidase, as shown here. Panels A, C, and D are heterozygous for the *Ptch1* mutant and show the expression of the gene in wild-type mice. Panels B and E and the right side of panel C show expression in homozygous mutant *Ptch1* embryos. Notice that Ptch1 expression expands over a region of the embryo that is wider in the homozygous mutant than in the heterozygote, which indicates that Ptch1 is a negative regulator of its own expression. Redrawn from Goodrich, Milenkovic, Higgins, and Scott, 1997, Altered neural cell fates and medulloblastoma in mouse *patched* mutants, *Science* 277: 1109–13, with permission.

(syndactyly), which indicated a broad and dosage-dependent role for *Ptch1* in limb development. More significantly, about 7 percent of the heterozygotes died prematurely, and most of those that were autopsied had distinctive cerebellar or other tumors. Although such tumors are rare in humans, mutations in the human *PTCH1* gene have been implicated in their formation, and the tumors in the mouse heterozygotes are histologically similar to the human tumors.

Thinking through the Experiment

8. The expression of the *Ptch* gene occurs before mutant phenotypes are observed. In addition, some mutant phenotypes are observed in heterozygous adults but not in heterozygous embryos. Discuss these results, remembering Chapter 6 on how to interpret mutant phenotypes.
9. Although there are two *Ptch* genes in the mouse (which was not known when these experiments were done), a lethal phenotype is seen in *Ptch1* homozygotes. What does this lethality suggest about the functional relationships between *Ptch1* and *Ptch2*? This question anticipates a topic taken up in Chapter 10.

10. In Figure CS7.2, panels A and D and the left side of panel C show the expression pattern in *Ptch1*/+ heterozygotes, while panels B, E, and the right side of panel C show the expression pattern in the same tissues at the same stages in *Ptch1*/*Ptch1* homozygotes. Discuss the differences between the expression patterns in *Ptch1* homozygotes and heterozygotes, and what this might tell us about the regulation of *Ptch1* gene activity.

Reference

Goodrich, L. V., L. Milenkovic, K. M. Higgins, and M. P. Scott. 1997. Altered neural cell fates and medulloblastoma in mouse *patched* mutants. *Science* 277: 1109–13.

8

CHAPTER 8

Genome editing

TOPIC SUMMARY

Genome editing, that is, making directed changes at specific sites in a genome, has long been done for yeast and for mice. Genome editing is now feasible for most other organisms because of the widespread adoption of a technique known informally as CRISPR, or more properly as CRISPR-Cas9. Genome editing by CRISPR or other techniques involves the following steps:

- searching the genome for the location of the desired edit,
- interrupting the DNA sequence at the specific location—such as between base pairs—by making a double-stranded break in the DNA, and
- editing the gene or site using the double-stranded break repair mechanisms native to eukaryotic cells.

For CRISPR-Cas9, the small guide RNA provides the search function, while the double-stranded break is made by Cas9 nuclease. Repair by non-homologous end joining often makes small changes or indels at the site to disrupt the coding region of the gene, while repair using homologous sequences can replace a DNA sequence or insert a larger sequence within a gene.

INTRODUCTION

The mantra of genetic analysis has long been "find a mutant." In traditional mutagenesis screens as described in Chapter 4, finding a mutant involved having stacks of petri dishes, racks of bottles, trays of seedlings, or something similar sitting next to the investigator waiting to be examined. When carrying out genetic analysis by using natural variation, finding a mutant means searching the planet for whatever random genetic changes arose and survived. Such screens are depicted in Figure 8.1 Panels A and B. With the genome-wide mutant screens to be described in Chapter 9, finding a mutant means obtaining a copy of the library of molecular clones in which every gene has been affected and patiently looking at the effects of each one, as shown in Figure 8.1 Panel C. Each of these methods brings its own thrill of discovery, with the anticipation that the next one to be examined will be the most informative mutant of them all. Through patience, perseverance, and perspicacity, thousands of mutants have been found and analysed, and many biological processes have been studied using these methods.

But each of mutants recovered from such a search usually has only a single type of mutation or molecular change in the gene—for example, a null allele or a hypomorphic allele. The mutation that is found may not have been the most desired one, so that other mutations in the same gene have to be found. It is always possible to return to the original search strategy or to use that first mutant to find additional alleles of the gene with other effects. On the other hand, it may not be feasible or

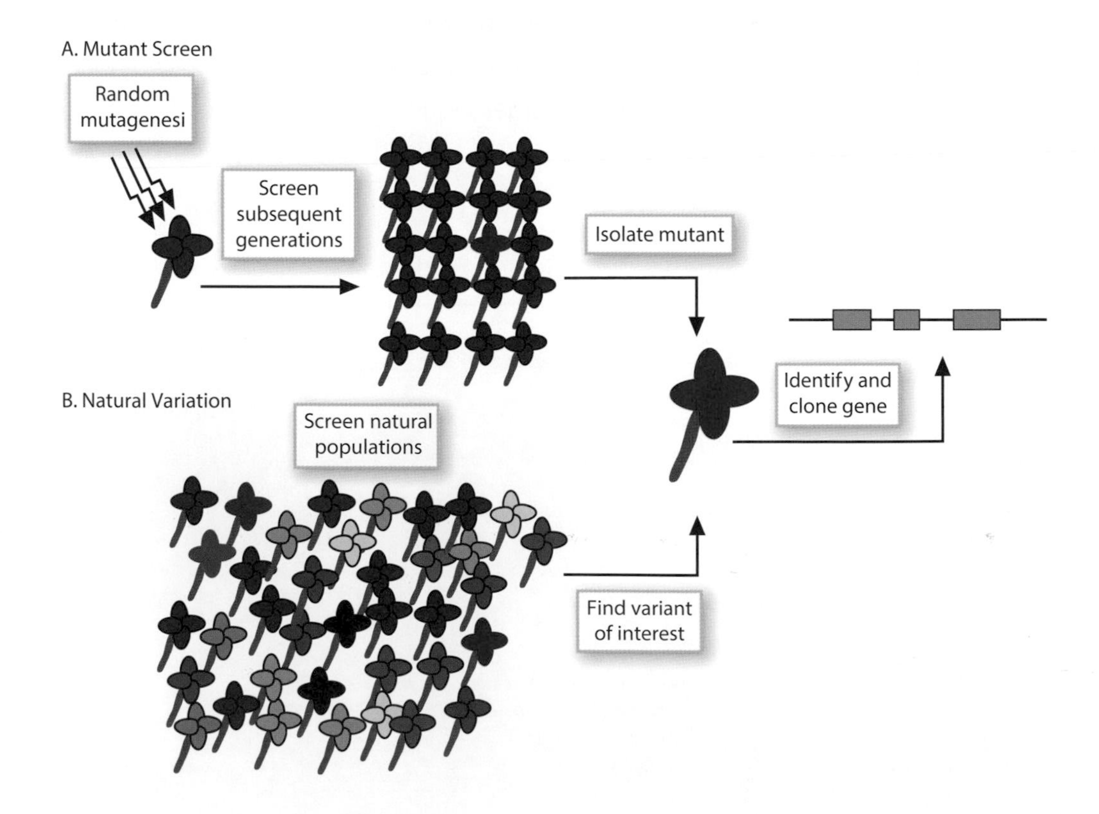

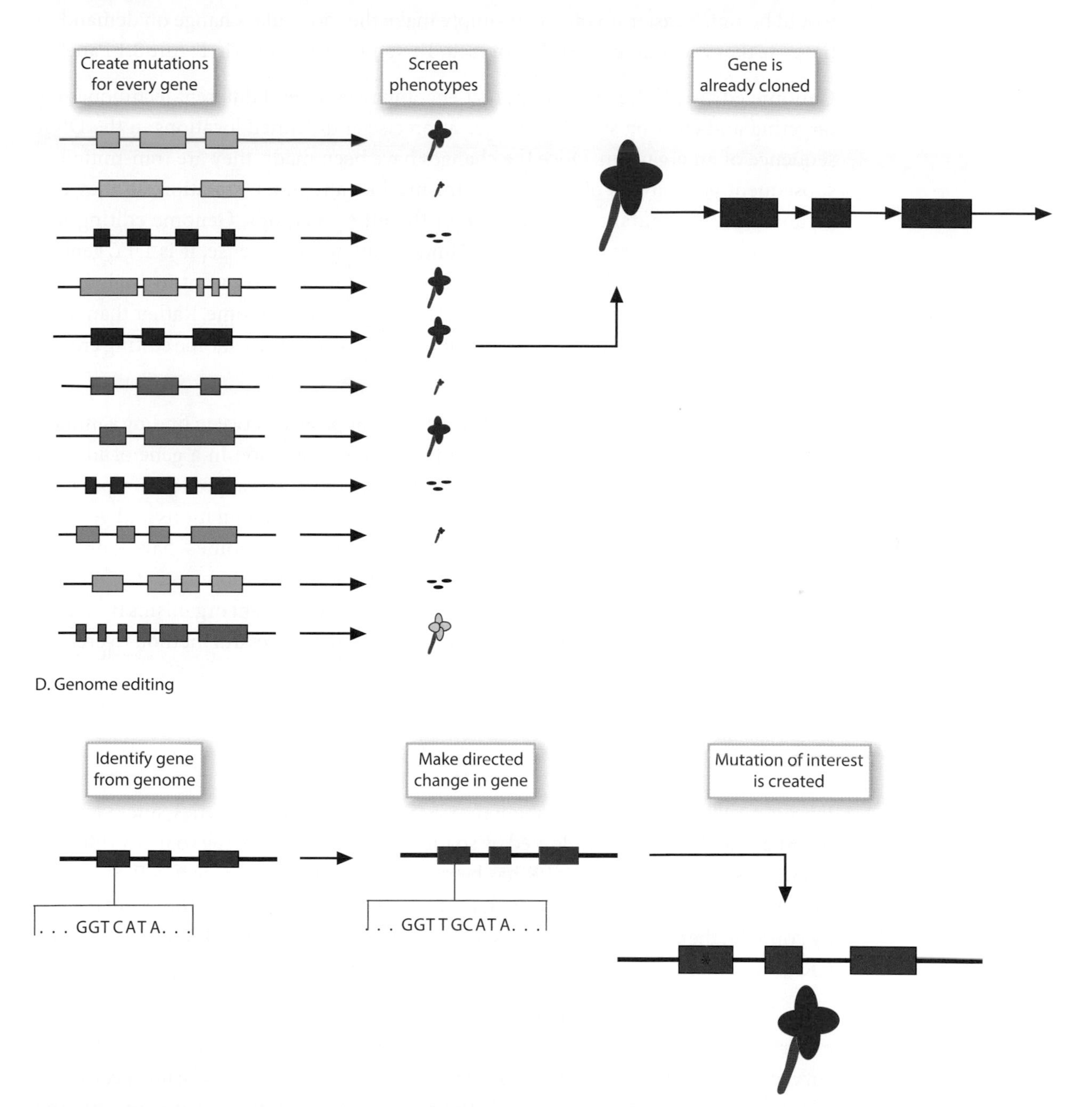

Figure 8.1 Strategies for finding a mutation in a gene of interest. Genetic analysis relies on having a mutant for the gene to be studied, so a mutant and its corresponding DNA sequence can be found by different methods. In this hypothetical example, a gene responsible for a blue pigment in purple flowers is under study, so the mutant has only the red pigment. A. In traditional mutagenesis screens, a strain that is of the wild type for the gene is treated with a mutagen in order to induce mutations randomly throughout the genome. Other than the mutations that are induced, the offspring are genetically uniform, which is shown here by the flowers with uniform purple colors. After several generations, a mutant with a phenotype of interest is found and isolated, as described in Chapter 4. From this, the gene is then cloned as described in Chapter 5. B. With naturally occurring variants, much more diversity is present in the original population, which is shown here in different shades of purple. The individual specimen with the phenotype of interest is found and isolated so as to identify the corresponding DNA sequence, similarly to what is done in traditional mutagenesis screens. C. In genome-wide mutant screens, to be described in Chapter 9, every gene in the genome is mutated first, and the mutants are stored as a library, regardless of their phenotypes. The library of gene disruptions is screened for those with the phenotype of interest. As seen here, many different phenotypes will be found and disruptions for some genes will result in no mutant phenotype. The clone encoding the gene of interest is identified by the phenotype. D. With genome editing, a mutation is created by a directed change in the DNA sequence of the gene of interest. In this example, the insertion of two bases changes the reading frame and disrupts the gene, so that no functional protein is likely to be made. The phenotype arising from the edited gene is then analysed.

productive to wait that long or to work that hard if it is not necessary. Finding a mutant would be much easier if you could simply make the molecular change on demand in the gene that you want to study by yourself. This strategy is shown in Figure 8.1 Panel D.

Genome editing is the broad term that encompasses several different techniques for targeting and creating specific changes at precise and defined locations in the DNA sequence of an organism. Once the changes have been made, they are transmitted to subsequent generations of cells or organisms, like any other mutation, so they can be analysed as a mutant found by one of the other methods. Genome editing fulfills many of the desires expressed by "find a mutant." In a sense, it is DIY genetic analysis—identify the gene and the specific mutation that you want to analyse and make the change yourself at the appropriate location in the genome. Rather than finding a mutant among the possible ones that may have occurred, genome editing creates at the outset the mutant that you want to find.

With genome sequences known, the distinction in approach between finding a mutant with an intriguing phenotype and creating a specific mutation in a gene of interest is not always clear-cut, nor is it logical to think about these approaches separately. As described in Chapter 7, the methods for creating specific mutations and genome changes in the laboratory—that is, for editing an organism's genome—have long been possible for the yeast *Saccharomyces cerevisiae* and for mice. More recently, genome editing has become feasible for nearly any gene in many different organisms by a technique known CRISPR, the latest and currently the most successful method to create a specific mutation at defined site in the genome.

Literature Link. Hsu et al. 2014, *Cell* 157: 1262–78

It is not an exaggeration to talk about the CRISPR revolution in genetics, since virtually no other method has been adopted by so many researchers studying so many different organisms as quickly as CRISPR has been: prior to 2012, references to CRISPR were found solely in the microbiology literature, and there were perhaps a thousand papers; by contrast, there were tens of thousands of papers published in 2017 alone that used CRISPR in many different organisms, and the number is increasing rapidly each year.

Since having a versatile method to carry out genome editing offers such a powerful experimental tool, the current techniques and reagents that are used for CRISPR and surveyed in this chapter might not be the same ones used in a few years; however, many of the underlying principles are likely to be the same even if the specific reagents and methods are different. Innovations in the technology of CRISPR and other types of genome editing should be expected. The relics of some of the previous genetic revolutions that made the CRISPR revolution possible—gel boxes and power supplies once used for DNA sequencing, antiquated thermal cyclers and bottles of mineral oil once used for PCR, and racks of restriction enzymes past their expiration dates—can be found in the storerooms, cabinets, and freezers of many genetics labs. Within a few years, the current reagents used for CRISPR might sit alongside them in tribute to the heroes that made the next revolution possible. While it is hard to predict the outcome of a revolution, it is exciting to watch, participate, and imagine what might happen next.

8.1 Genome editing: An overview

The genome is often compared to a manuscript in which the order of the letters of the DNA alphabet (A, T, G, and C) encodes information in the genome analogously to how the twenty-six letters of the English alphabet encode information in a written document. We can use the analogy with a manuscript to describe the process of genome editing. Imagine that you were tasked with using a word-processing program to make a specific edit at one location in a massive manuscript. (The King James or Authorized Version of the Bible consists of about 3.5 $\times$ 10^6 letters, while the complete works of Shakespeare are about 9 $\times$ 10^5 letters, so the human genome of 3 $\times$ 10^9 base pairs would be equivalent in length to about 800 books as long as the King James Bible or 3,500 books as long as the complete works of Shakespeare.) How will you go about making the specific edit? By thinking systematically about how manuscripts are edited, we can think also about how genome editing is done.

Genome and manuscript editing

Let's summarize the sequence of steps you would use to edit a manuscript, and compare these to some methods used to edit genes and genomes, ass summarized in Table 8.1. and depicted in Figure 8.2.

1. Search for and find the specific string of letters or characters at the location you want to edit. With most word processors, this search is done using the Find command, and then typing in a word or a phrase that you want the program to search for and locate. In genome editing, the search string is typically the nucleotide sequence surrounding the site where the edit will be made. The key consideration for this function is the length and exact sequence of the search string.

If the character string is short or inexact (that is, if wild-card characters or mismatches are allowed), multiple sites in addition to the one that you want to edit may be found, or the edit might be made at the wrong location. A sequence string that is too short or not specific enough results in **off-target effects**, in which the edits are made at locations in the sequence other than, or in addition to, the desired one.

On the other hand, a search string that is overly long might not be desirable either, at least in biological situations. In manuscript editing, there is very little cost or difficulty associated with making the search string longer. But when the search string is an oligonucleotide with a specific base sequence, there may be an additional cost per base in its synthesis and some added challenges associated with introducing a longer sequence than necessary into the cell or organism.

The first two editing methods in Table 8.1 require the insertion of specific recognition sequence into the gene or genome. The use of *loxP* sites for this purpose in mice was discussed in Chapter 7. The *FRT* sites (derived originally for certain plasmids in yeast) perform a similar function for a different editing method, and have been used in some organisms. This is analogous to how an editor or author might

Table 8.1 Editing methods

Editing Method	Search Function (sequence recognition)	Interrupt Function (nuclease making the DSB)	Comments
Cre/*Lox*	*LoxP*	Cre recombinase	Widely used in mice
Flp/FRT	*FRT*	Flp recombinase	Encoded by yeast plasmids, used in several organisms including *Drosophila*
ZFNs	Zinc finger recognition	*Fok*I nuclease	Requires different enzyme for each sequence to be edited
TALENs	TALE recognition	*Fok*I nuclease	Requires different enzyme for each sequence to be edited
CRISPR	sgRNA	Cas9 or related nuclease	Versatile since sequence recognition and DSB are provided by different molecules

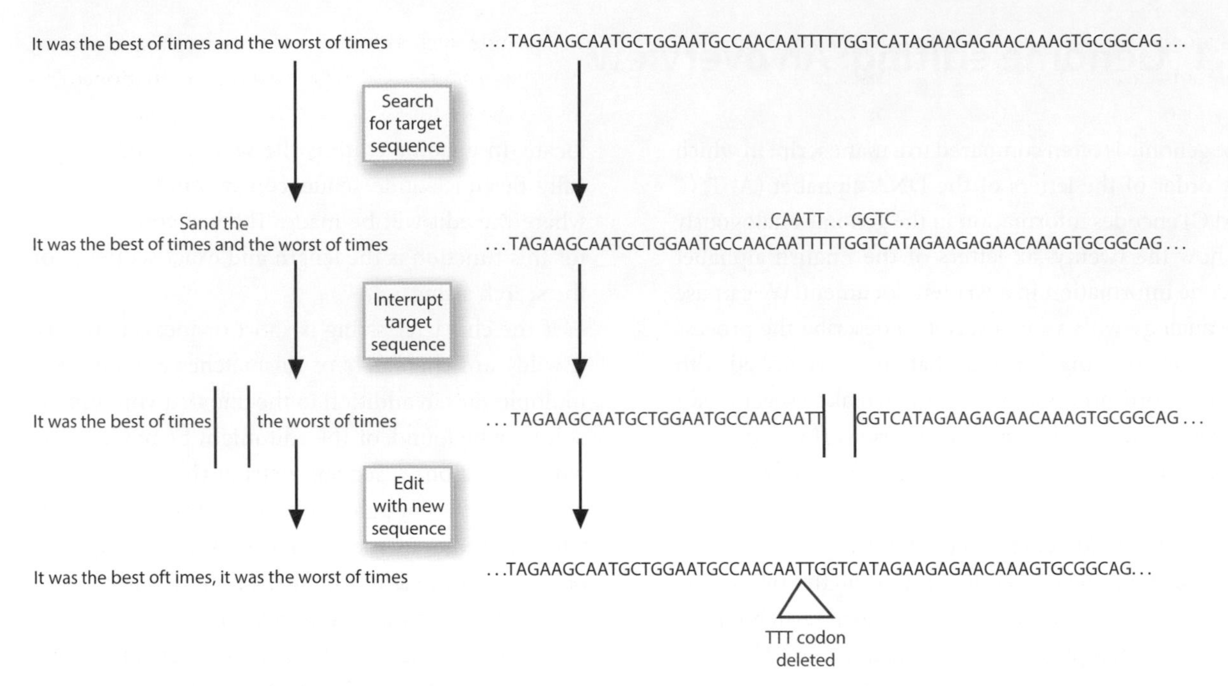

Figure 8.2 Editing manuscripts and genomes. Both a manuscript and a genome consist of long strings of characters. The editing steps in each case are analogous, as shown here for the opening sentence of of *A Tale of Two Cities* on the left and a sequence from the *CFTR* gene on the right. First, the location to be edited is searched for and found by matching a string of characters, either alphabetical letters or a nucleotide sequence. Second, the sequence is interrupted at the specific site where the edit is to be made. In manuscripts, this interruption is made with the mouse click or the space bar. In genomes, this interruption is a double-stranded break in the DNA sequence. Finally, the directed edit is made according to the wishes of the editor. For the manuscript, the characters in the original were directly replaced by other characters. In this example with the DNA sequence, certain base pairs are deleted.

insert a specific sequence of characters not found in the original manuscript to act as a bookmark when working on a long manuscript. For genome editing, the challenge is to insert the sequence bookmark (*loxP* or *FRT*) at the most appropriate location for the edit. The other editing methods rely on the sequence recognition capacities of enzymes—capacities that, as will be discussed, constitute one of the primary distinctions among these methods.

2. Interrupt the document to indicate the site of the edit as precisely as possible, including between individual characters. When editing a manuscript on a computer, this is the function provided by a click of the mouse. With a computer mouse that has been well calibrated, the click (the interruption in the sequence) can be located precisely between individual characters or letters. In genome editing, as we will see, this interruption involves creating a **double-stranded break (DSB)** in the DNA sequence to be edited. The endonuclease enzyme that makes this DSB is the other major distinction among these methods.

   The sequence interruption function in genome editing is often less precise than suggested by our mouse click analogy. Even when the location within the genome is appropriately and specifically found after Step 1, the endonuclease might not cut precisely between the designated bases at the site. Thus, while clicking a mouse at the appropriate location is nearly an unconscious action for most editors and writers, interrupting the sequence at the precise site to make the edit can be more variable and challenging in genome editing.

3. Edit the document at the site of the interruption. This third step in manuscript editing is the one that requires the most skill and experience. The editor can insert, delete, or rearrange small amounts of text, change individual letters, or move larger sections to a new location. Much of the skill required to be a good editor comes from the ability to make the proper change in the manuscript once the location and precise site have been identified. Likewise, genome editing can insert or delete characters, rearrange their

sequence, move larger blocks of DNA to other locations, and so on. As with manuscript editing, this is also the step in genome editing that requires the most skill and individual creativity, so it is the one that is the most difficult to describe exactly. While some commonly used techniques are used to search for the location and make the DSB at the precise site, there are different ways to edit the sequence once the DSB has been made.

The rationale for genome editing

Several different rationales underlie the desire for genome editing. First, as noted above, a rapid and versatile technique for genome editing provides a more efficient method than screening randomly generated mutants or large libraries of mutated genes. The investigator does not have to sort through all the mutations affecting other genes to find the one that affects the gene of interest. In addition, when relying on traditional mutagenesis screens or naturally occurring variation, mutations and variants at other locations in the genome of the organism can complicate the interpretation from change at the gene of interest. Thus, precise genome editing can also eliminate the need for outcrossing the mutant strain (described in Chapter 4) in order to replace any other mutations and variants that may affect the phenotype with the wild-type sequence.

Second, genome editing permits the investigation of a much wider range of biological questions, and it is not necessary to have a mutant phenotype in order to recognize the effect of the alteration. (This benefit is similar to the genome-wide mutant screens we will discuss in Chapter 9.) These biological questions can also be much more specific and focused than with other methods, as depicted in Figure 8.3. Thus genome editing can be used to ask whether a particular amino acid in a protein-coding region affects the function of the gene and its protein; it can be used to ask about the effects of specific bases in the splice site or in the promoter region on the expression of the gene; and it can be used to assign functions to specific enhancer sequences in regulatory regions. Genome editing can and has been used to ask about the functions of microRNAs and long non-coding RNAs without presumptions about what the phenotype might be, for example. In yeast, genome editing has been used to address broad but specific questions about genome organization and functions, such as the effects of relocating a centromere or heterochromatic region on recombination, or of changing the location or orientation of a replication origin; in fact genome editing in yeast has been used to engineer synthetic versions of entire chromosomes and large regions of the genome to determine the effects.

Literature Link. Richardson et al. 2017, *Science* 355: 1040–4

At the moment, the type of editing required to relocate or reorganize large parts of the genome has not yet been achieved in multicellular organisms, but it is not difficult to foresee that this will be possible in the near future. Many geneticists who work on other model organisms have suffered from a form of "yeast envy," because the ability to address specific and detailed biological questions in yeast seems to depend more on the creativity of the researcher than on the technical feasibility of the experiment. A general and versatile method for genome editing helps to narrow the gap between what can be done in yeast compared to other organisms and to allow an entirely new set of biological questions to be addressed.

Third, genome editing can be used to produce more representative models for specific mutations that give rise to particular diseases. We noted in Chapter 5 that different mutations in the same gene sometimes have different clinical symptoms, and thus may even be diagnosed as different diseases. The differences in phenotypes arising from mutation in the same gene could affect recommended therapies or the prognosis for the genetic disorder. From what we know about mutant phenotypes in model organisms, as described in Chapter 4, this is not a surprising result. It has long been recognized that amorphic and hypomorphic mutations, different hypomorphic alleles, and recessive and dominant mutations in the same gene can all have different phenotypes. On the other hand, it is also recognized that many cases of a genetic disease in a particular human population have the same molecular variant. A familiar example is cystic fibrosis, in which more than 70 percent of the cases among people of European ancestry are due to the same delta-F508 mutation in the CFTR gene. The ideal mouse model for cystic fibrosis should have this molecular variant in the mouse ortholog of CFTR, particularly since other mutations in the CFTR gene sometimes result in

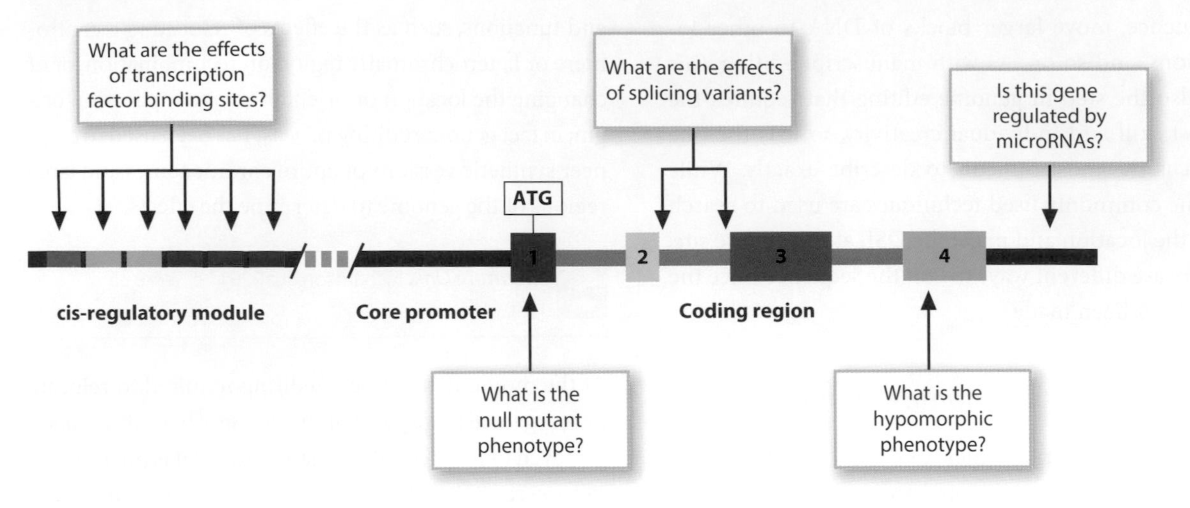

Figure 8.3 Some questions addressed by gene and genome editing. Gene and genome editing allows for very specific questions to be addressed. This schematic of a eukaryotic gene with four exons shows some of its functional components and the locations of edits that could be made. Null and hypomorphic alleles can be made for any part of the protein-coding region of the gene, so these locations are representative. In plant genes, the target sequence for microRNA regulation can be found at many locations, whereas for animal genes the target sequence is typically found in the 3′UTR, as suggested in this diagram.

different clinical symptoms; delta-F508 is the sequence example that was depicted in Figure 8.2. Researchers can thus make the exact molecular duplicate of a human mutation in a mouse gene.

To mark the site and initiate the editing processes, all the genome-editing techniques, including the ones for yeast and mice discussed in Chapter 7, rely on a DSB in the DNA sequence. As noted in Table 8.1, one of the principal ways in which genome-editing methods differ from one another is in the nuclease used to make the DSB. Hence, in order to understand the molecular basis for making edits in the genome using CRISPR or some other method, we first need to understand the cellular response to DSBs.

8.2 Double-stranded break response and repair

Double-stranded DNA breaks (DSBs) are very common in all living cells. Indeed, they are an essential part of natural molecular processes such as replication and meiotic recombination, during which their number is relatively small and their locations are highly regulated. DSBs arise even more commonly as errors during DNA replication and as damage caused by many different environmental agents, including ionizing radiation, chemical treatments that block topoisomerases (such as some chemotherapy drugs), and agents that generate reactive oxygen species.

Literature Link. Shrivastav et al. 2008, *Cell Research* 18: 134–47

In these cases, their numbers and locations are not regulated. Estimates of the number of DSBs experienced daily by mammalian cells vary widely, from tens to thousands, but they are definitely not rare; since DSBs affect the integrity of the genome, normal (that is, non-cancerous) cells rapidly repair more than 99.9 percent of them without further damage to the cell. In fact, cells constantly experience so many potentially damaging DSBs that the molecular mechanisms for repairing them are very well studied and are evolutionarily similar in bacteria and eukaryotes. DSB break repair is an essential process for genome stability and is often one of the processes that break down in cancer cells, resulting in the extensive genome rearrangements characteristically found in advanced tumors.

There are two mechanisms by which eukaryotic cells respond to DSBs. These two methods are **homology-directed repair** and **non-homologous end joining (NHEJ)**. Both are summarized in Figure 8.4. In most eukaryotes, DSBs are repaired by NHEJ much more commonly than by homology-directed repair. Much more detail could be provided about the molecular mechanism of each of these processes, but that level of detail is not needed to understand how their repair is important for genome editing, so we will limit ourselves to providing a general perspective.

Homology-directed repair

Homology-directed repair uses a highly similar sequence, present elsewhere in the genome as a template to repair the break. This was encountered previously in Chapter 7, when we described gene replacement techniques in yeast and mice. The length of homologous sequence that is needed for homology-directed repair

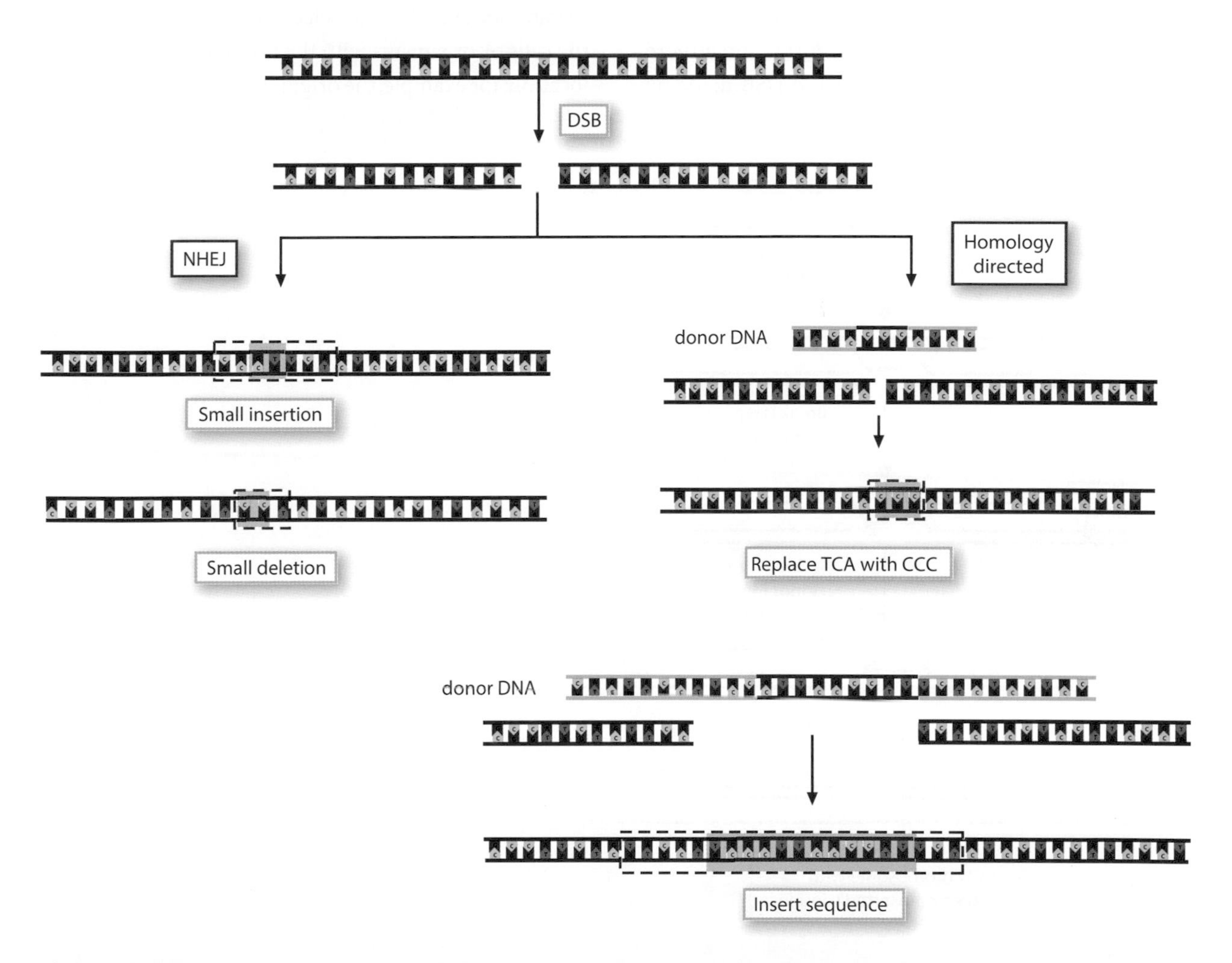

Figure 8.4 Pathways to repair double-stranded breaks. When a double-stranded break (DSB) in a DNA sequence occurs, two different pathways are used to repair it. In the absence of a homologous sequence to serve as a donor template, DSBs are repaired by NHEJ, as shown on the left. NHEJ often results in small insertions or deletions at the break site, seen by the red lines in the backbone and the shaded boxes. When a sequence with sufficient homology is present to serve as the donor, homology-directed repair can occur, as shown in the two examples on the right. The donor template (represented by the gray line in the backbone) need not match the original sequence precisely. The locations of possible crossover are indicated by the dashed lines. In one example, the donor has the sequence CCC, while the original sequence is TCA; the sequence from the donor can replace the sequence in the original. In the other example, the donor has a novel donor sequence (such as a reporter gene) flanked by sequences that match the original genome sequences. By homologous recombination, this donor sequence can be inserted into the original. The altered sequence is shown with the red backbone and shaded boxes.

varies among organisms, but commonly is several kilobases in the multicellular organisms for which this feature has been studied. Homology-directed repair can also occur between sister chromatids, paralogous members of a gene family, or other repeated sequences such as transposable elements, for example.

In homology-directed repair, a DSB is made in one chromosome and, through a process of strand invasion and the formation of an intermediate that involves a **Holliday junction**, one DNA molecule is joined to another with a highly similar but not necessarily identical sequence. This is summarized in Figure 8.5. A similar process occurs during meiotic crossing over, in which the DSB is made by the nuclease SPO11. Since there is no DNA sequence specificity for the DSB made by SPO11, homologous recombination can occur anywhere in the genome, but the likelihood that it will occur at one specific location chosen by the investigator is very low. (The quasi-random locations and low number of DSBs in eukaryotic germ cells provide the mechanistic basis for genetic maps.)

But DSBs are usually not repaired by homology-directed repair. It is estimated that the frequency of homology-directed repair of DSBs in mammalian cells is lower than 10^{-6} and may be as low as one cell in 10^{-9}. As a result of this very low frequency, genome editing applications based on homology-directed repair usually employ some additional selection or screen to identify the cells or organisms with the break at the appropriate location; for example, the original method for producing

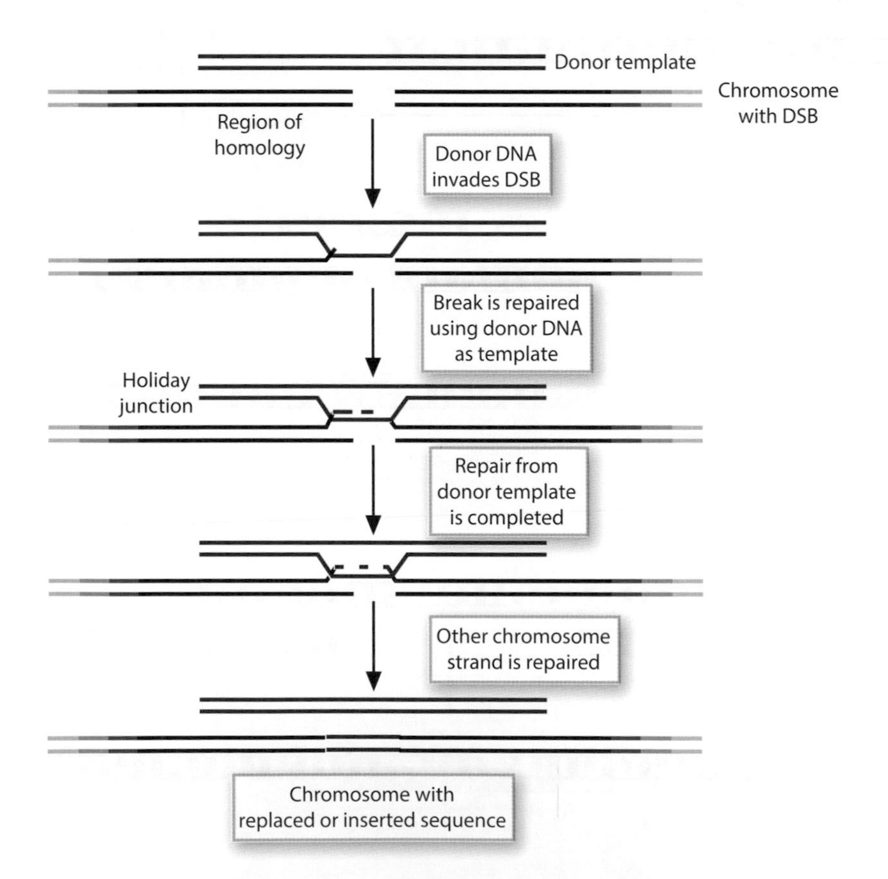

Figure 8.5 Homology-directed repair. Homology-directed repair of a DSB requires a template DNA molecule with a highly similar sequence to the chromosome; the donor template is shown here in red, with the regions of homology between the donor template and the chromosome flanking the DSB on each side. Because it relies on sequence homology between the two DNA molecules, homology-directed repair is usually considered to be precise. Donor template DNA first "invades" between the strands of the chromosomal DNA at the region of the DSB, where it provides the template sequence for the repair of the DSB; the repair DNA synthesis is illustrated by a dashed red line. Once this repair from the donor template is completed, the chromosomal DNA will have replaced its own sequence or inserted the sequence from the template.

transgenic mice, as described in Chapter 7 and illustrated in Figure 7.12, involved two different drug selections designed to identify ES cells with homology-directed repair of DSBs. Although the frequency of homology-directed repair is quite low in ES cells from mice, it is much lower in ES cells from other mammals, so the techniques described for mice have not been used for other mammals. In fact, even for mice, the use of Cre/*lox* for gene editing provides a much higher frequency of ES cells with the desired edit.

One of the primary reasons why genome editing has been so successful in yeast is that yeast cells resolve or repair DSBs principally through homology-directed repair rather than through NHEJ; in fact genes that encode some of the proteins necessary for NHEJ in other eukaryotes are absent from the yeast genome. In addition, yeast cells require only a few hundred base pairs of sequence homology for homology-directed repair to occur, less than a tenth of the amount of homology that is necessary in most other eukaryotes. Thus, genome edits by homology-directed repair can be efficiently carried out for nearly any location in the yeast genome. Nonetheless, most genome-editing experiments in yeast have also employed drug selections or nutritional auxotrophy to identify the cells with the appropriate edit.

Homology-directed repair can only occur when two DNA molecules with highly similar sequences are present in the same nucleus. Because homology-directed repair involves a template DNA molecule highly similar in sequence to the one with the DSB, it is generally considered to be very accurate and relatively error-free: the nucleotide sequence after the break is repaired will be almost identical to what was present before the DSB occurred. However, this might not be the desired outcome for some types of genome editing. On the other hand, crossovers can occur in the regions of homology regardless of what sequences are found in the non-homologous regions, as shown in Figure 8.6. Thus, in genome-editing applications, homology-directed repair can insert or delete specific sequences or genes, for example insert

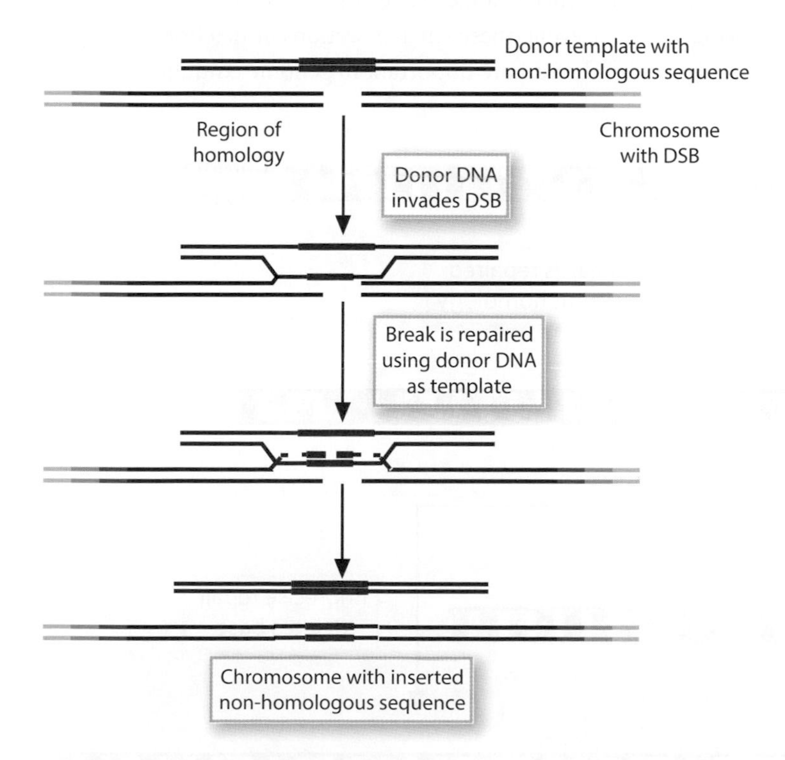

Figure 8.6 Inserting a non-homologous sequence using homology-directed repair. Since the regions of homology where the crossover occurs flank the DSB, sequences from the donor template that lie between the homology regions but are not similar to the chromosome can also be inserted; the non-homologous region, such as a reporter gene or some other construct, is shown in blue. When the donor DNA molecule is used as a template, the non-homologous region is also copied and is inserted into the chromosomal DNA, along with flanking sequences from the donor.

a reporter gene, or engineer a specific mutant allele by replacing the wild-type sequence with the mutant sequence. Homology-directed repair is likely to be the preferred pathway for genome editing that involves rearrangements of large regions of the genome as well, since it can be used to replace one sequence with another.

Non-homologous end joining (NHEJ)

For most eukaryotes other than yeast, DSB repair is most commonly by NHEJ, and homology-directed repair occurs much less often. For many applications in genome editing, the repair pathway involving NHEJ is sufficient to make an appropriate edit. Nonetheless, nearly all genome-editing applications still require a screen or a selection to determine which cells or organisms have the desired edited sequence.

In NHEJ, DNA molecules without extensive sequence homology are ligated together to repair the DSB on each molecule, as shown in Figure 8.7. Since the repair process brings together unrelated DNA sequences, NHEJ is the only mechanism available when no homologous sequence is available in the same nucleus as the molecule with the DSB. (The ligated molecules sometimes share microhomologies—that is, similar sequences of less than twenty base pairs at their ends—which may be generated during the repair process.)

The processes and the enzymes involved in binding, bridging, and ligating the ends of different DNA molecules are widely conserved among eukaryotes; at least a dozen evolutionarily conserved proteins have some direct role in NHEJ. Some of these proteins also play roles in homology-directed repair and are even present at the DSB site prior to repair, indicating that the two processes are in some form of regulated competition with each other.

NHEJ is usually described as being error-prone, since some bases are lost or added at the site of the repair. These changes arise because the DNA polymerase active during NHEJ lacks a proofreading function and because the joining process itself is imprecise Close analysis of NHEJ events indicates that at least a quarter of the repair events (and possibly more) are precise and involve no sequence changes, so "error-prone" is a relative term. In any event, these small insertions or deletions at the site of the edit are important in genome editing.

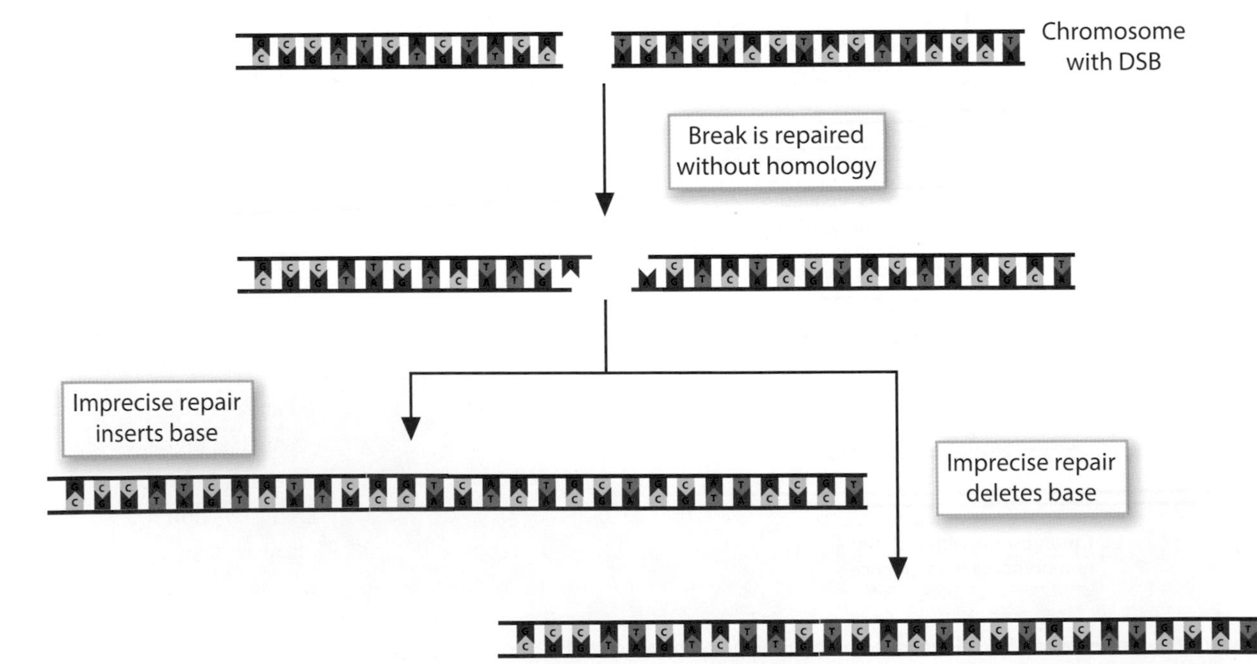

Figure 8.7 Non-homologous end joining. In most eukaryotes, the more common type of DSB repair is NHEJ. With no homologous sequence to use as a template, the DSB is repaired by joining the ends of two different DNA molecules. While the joining may be precise such that no changes occur, it is more often the case that the process is imprecise and one or a few bases are either inserted or deleted as a result. If the DSB is made within the protein-coding region of gene, the small indels arising from the imprecision of NHEJ produce a frameshift mutation and disrupt the function of the gene.

Suppose that the editing application is to make a null mutation in the protein-coding region of the gene; while the precise site of the mutation might matter, the exact nature of the change at that site may not be important, such that any indel that changes the translational reading frame is acceptable. Gene disruption like this is one of the most common applications of genome-editing techniques, so errors arising during NHEJ often create the mutations of interest without relying on homology-directed repair.

8.3 Before the revolution: Genome editing prior to CRISPR

In light of the power of genome-editing methods in yeast, it was inevitable that many investigators attempted to develop analogous methods in other model organisms. Of these, the one most widely used by far has been that of the site-specific recombination enzyme Cre and its target *lox* sites in mice, as described in Chapter 7, although the conceptually similar method of Flp recombinanse and *FRT* sites is also used. In our manuscript-editing analogy, the "search and find" function is provided by the *lox* or the *FRT* sites inserted at appropriate location in the genome, while the "interrupt" DSB is created by Cre or Flp at these sites.

As discussed in Chapter 7, genome editing is done in a single cell or in a small number of cells, which then are used to produce an altered colony or organism; it is important to consider what type of cell is used for the editing process. Yeast is unicellular, of course, so any edited cell can generate a colony or a strain with the appropriate change. This could also work well for many plants, which can be grown vegetatively from small numbers of cells. For mice, the edits are made in ES cells growing in culture, which are then reintroduced into a mouse embryo and used (through appropriate matings) to produce a strain of mice with the altered genome. By regulating the time and location of Cre expression, as discussed in Section 7.3, we can either edit the genome in all the cells or edit it only in particular cells of interest, while preserving an intact genome in other cells. In other organisms the genome might be edited in germ cells that could give rise to the mutant strain or line in a subsequent generation, or in stem cells that grow in culture and can be introduced into the organism at a particular time. If stem cells for somatic tissues are altered—for example, neuroblasts give rise to parts of a vertebrate's nervous system—the edited cells can differentiate into the tissue but the change is not propagated to the next generation of animals.

Before the use of the CRISPR-Cas9 method for genome editing, several techniques were developed, and two methods were more widely used than the others. These methods are known as zinc finger nucleases (ZFNs) and TALENs, which is based on the type of protein domain involved in sequence-specific recognition and binding. Each method was tried in many different cell types in different organisms, but neither was as versatile as CRISPR-Cas9; it is unlikely that many investigators will continue to use these techniques rather than CRISPR, but mutants made by these methods are still widely available. Each involved an engineered version of naturally occurring enzymes used to make a "programmable nuclease," that is, a nuclease that could be specifically targeted to a location in the genome in order to make the edit. Comparisons to other editing methods and to each other are found in Table 8.1.

Literature Link. Kim and Kim 2014, *Nature Review Genetics* 15: 321–34

Zinc finger nucleases (ZFNs)

ZFNs are fusion proteins made by attaching DNA sequences encoding C2H2 zinc finger domains to sequences encoding the DNA cleavage domain of the endonuclease *Fok*I in vitro, as shown in Figure 8.8. *FokI* is a modular enzyme, with its endonuclease domain separable from its own DNA binding domain. In the ZFN fusion proteins, the zinc finger domains provide the sequence-specific DNA binding to target the break to a particular location, while the endonuclease domain of *FokI* produces the double-stranded break that marks the genome-editing site. *FokI* functions only as a dimer, so two different monomer ZFNs need to be introduced to produce a functional nuclease; the binding sites for the monomers are 5–7 base pairs apart.

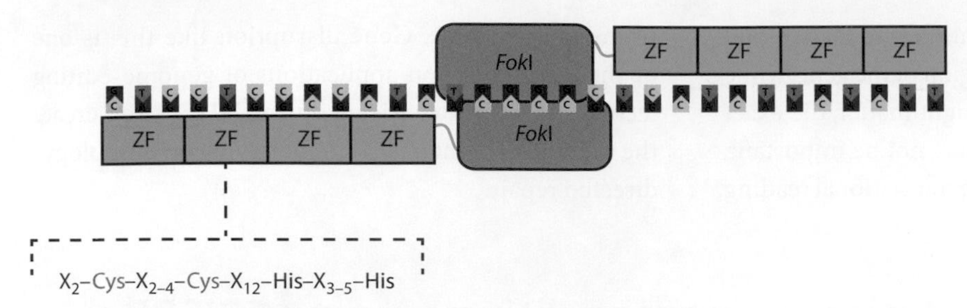

Figure 8.8 The structure of a ZFN. A zinc finger (ZF) nuclease has three or four zinc finger domains made as a fusion protein with the *FokI* nuclease; the diagram shows four zinc finger domains in each fusion protein and its binding site of approximately thirty base pairs. Since *FokI* acts a dimer, two fusion proteins are needed and the zinc finger domains need not be identical. The specific amino acid sequence in the zinc finger domain determines which three base sequence in the binding site is recognized, although the exact rules governing the specificity are not fully defined.

The zinc finger proteins fall into the broad category known as C2H2 zinc fingers with the structure shown in Figure 8.9, in which each zinc finger domain interacts with a different three-base target sequence, depending on the other amino acids in the domain; Figure 8.9 highlights in red those residues that are the most important in determining the binding specificity. As a result, locations in the genome can be searched for by using ZFNs with different amino acid sequences and different numbers and combinations of zinc finger domains. Once the location is found using the specificities of the zinc finger domains, and the editing site is marked by the DSB made by *FokI* endonuclease, repair occurs by NHEJ or (less often) by homology-directed repair, as with other applications.

Targeting ZFNs to known genes

An example of the use of ZFNs to produce mutations used two genes in zebrafish, the *golden* gene and the *no-tail* gene; as expected from the gene names, the *golden* gene affects pigmentation while the *no-tail* gene affects tail morphology, and both have easily observed mutant phenotypes.

Literature Link. Doyon, Y. et al. 2008, *Nature Biotechnology* 26: 702–8

Two different ZFNs, each with four zinc fingers, were used for each gene. To disrupt the gene, a mixture of two different mRNAs with the coding region for the endonuclease domain in-frame and downstream of the zinc fingers was injected into the fertilized fish egg, as shown in Figure 8.10. The egg translated the mRNA, producing fusion proteins that bound as a dimer to their preferred sites within the target genes in the genome. Once bound, the dimeric protein cleaved the target DNA, producing a substrate for repair by NHEJ, and thereby generating a mutant.

In this particular study, mutants were observed at a high rate, as many as a third of the injected eggs producing a mutation in the targeted gene. Most but not all the mutations were the small insertions or deletions arising from NHEJ, which indicates that the ZFN fusion protein

X_2–Cys–X_{2-4}–Cys–X_{12}–His–X_{3-5}–His

Figure 8.9 The amino acid sequence of a C2H2 zinc finger domain. A zinc finger protein usually has several zinc finger domains, whose precise sequence varies; the structure diagrammed in Figure 8.8 shows four zinc finger domains on each fusion protein, each of which interacts with a three-base sequence in the binding site. The overall sequence of a C2H2 zinc finger domain is shown at the top; X represents any amino acid, although the specific residue may affect the function of the protein. The amino acid sequence of the first zinc finger domain from the well-studied protein EGR1 is given as a specific example, with the positions of the cysteines (Cys or C) and histidines (His or H) highlighted. The amino acid sequence of the section between the two histidines affects the binding specificity, the most important positions being shown in red; changes in the amino acid sequence at these positions are likely to alter the binding specificity of the protein.

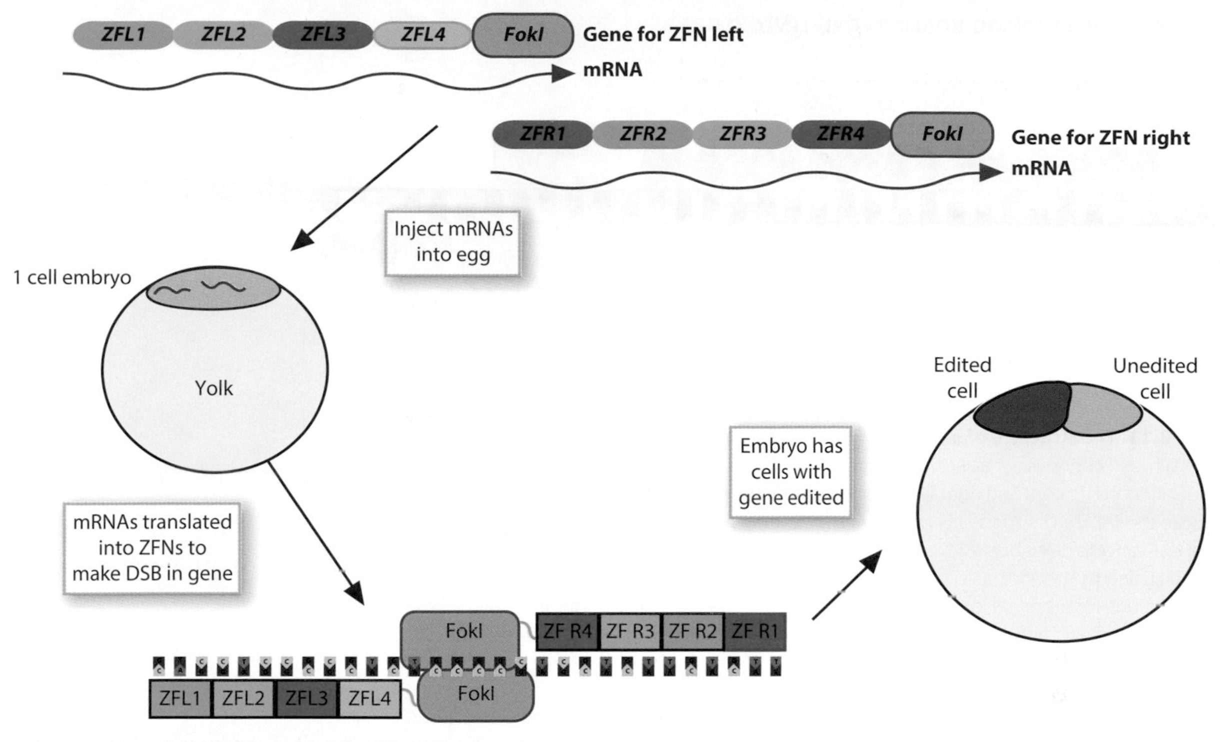

Figure 8.10 Gene editing example in zebrafish. As an example of gene editing using ZFNs in zebrafish, two DNA constructs were made. Each had the coding regions for four zinc fingers (labeled *ZFL1* through *ZFL4*, and *ZFR1* through *ZFR4*) of slightly different sequences, represented here by the different shades of green; the coding region for the ZF domains are fused to the coding region for *FokI*. The constructs were translated into mRNAs, and the mRNAs are injected into the one-cell embryo, as shown. The embryonic cells translate the two mRNAs into the dimeric ZFN proteins, which bind to the targeted gene, as shown. The developing embryo will be a chimera with some cells that have the edited gene (in pink) and others that do not have the edited gene.

worked as expected. Success rates vary, and other experiments found a much lower rate of success.

Because the mRNA is injected into the one-cell embryo as it is undergoing development, the gene disruption might not affect all of the cells of the adult fish. That is, the fish arising from the injected embryos could be mosaics, as shown at the bottom of Figure 8.10. (This is very similar to the situation with mouse ES cells, as described in Chapter 7.) Breeding schemes are carried out to produce homozygotes for the newly induced mutations, which appear to be faithfully transmitted in the germline. One intriguing modification was to inject the mRNA into eggs that are heterozygous mutant for the *golden* gene, so that homozygous *golden/golden* mutants could be recovered easily. This is a type of non-complementation screen, as described in Chapter 4, and works well for mutants like *golden* that are not lethal to the organism. In this case, the failure of the ZFN-induced mutation to complement a known mutation in *golden* provided the screen to ensure that the mutation arose in the desired gene.

The key to any genome-editing method is to identify an appropriate target site within the gene of interest. For ZFNs, this involves not only finding a gene-specific sequence but also constructing zinc finger domains that will bind to that sequence and not to off-target sequences that are highly similar. As a result, the sites need to be relatively large—about 30 bp for a dimer of two ZFNs each with four zinc fingers. This is a significant constraint, since a different pair of ZFNs is necessary for every gene being targeted, and the preferred binding site for most zinc fingers is not yet known. Thus not every gene has a site that can be targeted for editing by one of the studied zinc finger proteins, and different ZFNs need to be made for every site to be edited.

Transcription activator-like effector nucleases (TALENs)

A related strategy based also on fusion proteins with the *FokI* endonuclease domain uses different transcription activator-like effector proteins (TALEs) to determine

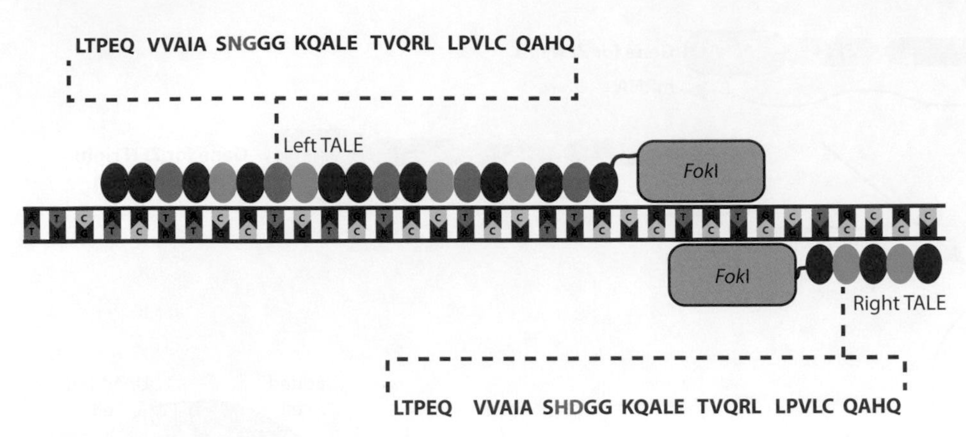

Figure 8.11 The structure of a TALEN. A TALEN is composed of a series of 33–5 amino acid TALE domains, each of which recognizes a specific base pair in DNA; each TALE domain is represented in the drawing by an oval of a specific color. The left and right TALE proteins are made as fusion proteins to the *FokI* nuclease as shown to compose the TALEN. The specificity of the TALE for its base pair arises from the two amino acids referred to as the repeat variable di-residue (RVD), highlighted in green in the expanded amino acid sequences; the colors of the TALE and its specific base pair correspond in the diagram. Since *FokI* functions as a dimer, two TALENs are needed but, for simplicity, only the left TALEN is diagrammed in full. The right TALEN also consists of about twenty TALE domains, although only the first five domains are shown.

the binding site. TALEs are derived from plant pathogenic bacteria of the genus *Xanethomonas.* TALE proteins have tandem arrays of 33–35 amino acid repeats, as shown in Figure 8.11, each one of which recognizes a single base pair in the DNA sequence. The specificity of binding comes from the residues at positions 12 and 13 in the amino acid repeat, which are known as the repeat variable di-residues or RVD; Asn-Asn or Asn-Lys recognize guanine, Asn-Ile recognizes adenine, His-Asp recognizes cytosine, and Asn-Gly recognizes thymine. Thus, once a specific site for the interruption is found, a sequence for the corresponding TALE domain can made in vitro and fused to the *FokI* endonuclease domain to produce a **TALEN** (TALE + Nuclease) for the site.

As with ZFNs, a plasmid or mRNA encoding the appropriate TALEN is introduced into the cell, which then translates it into the fusion protein to make the double-stranded break. Since *FokI* functions as a dimer, two different TALENs have to be introduced, although only one of the two TALENs is shown in full in Figure 8.11. As before, the double-stranded break is typically repaired by NHEJ, which results in small indels.

TALENs have the significant advantage over ZFNs that a binding protein can be constructed for any target sequence since the sequence requirements for the specificity of the binding protein are known. Although the size of the binding site is about 30 bp for both ZFNs and TALENs, appropriate sites can be found more readily for TALENs because the binding specificity of the fusion protein can be synthesized to match the DNA sequence at the site. However, the in vitro construction of the TALEN is somewhat cumbersome and, as with ZFNs, a different fusion protein is needed for each target site. In addition, the repeat structure of the TALE binding domain sometimes lends itself to inter- or intramolecular recombination, which can alter the structure of the resulting fusion protein and therefore its effectiveness.

Both TALENs and ZFNs were used effectively by some labs—*Nature Methods* named ZFNs Method of the Year in 2011—but neither technique achieved the general popularity or the ease of use offered by CRISPR-Cas9 when it was introduced as a genome-editing method in 2012. Each method had the limitation that a different programmable nuclease was needed for every site to be edited. CRISPR-Cas9 avoids this limitation, because the search function and the interruption functions are encoded by two different molecules, and the same "interruption function" nuclease could be used for any site. The specificity of searching with CRISPR is found in the base sequence of an RNA molecule rather than in the protein itself, and RNA molecules of different sequences are much easier to construct than polypeptides.

8.4 CRISPR

An overview of CRISPR-Cas9

As we will discuss shortly, the biological origins of CRISPR-Cas9 lie in the responses of some bacteria and archaea to invading nucleic acids such as viral genomes and plasmids. However, it may be helpful to describe the mechanism of action for CRISPR-Cas9 for genome editing before discussing the natural functions of these molecules.

Literature Link. Komor, Badran, and Liu 2017, *Cell* 168: 20–36

The location to be edited—that is, the "search function" from Table 8.1—is provided by an RNA molecule that is complementary in sequence to one DNA strand at the target location of the desired edit, and identical (except for uracil replacing thymine) in sequence to the other DNA strand, as shown in Figure 8.12.

The **Cas9** endonuclease makes the DSB needed for the edit; many CRISPR applications use the *Cas9* gene from the bacterium *Streptococcus pyogenes*. Cas9 is an RNA guided nuclease, which recognizes RNA:DNA hybrids and cuts the DNA sequence on each strand; thus the location of the DSB depends on the specificity of base pairing between RNA and DNA molecules rather than on the presence of a genomic sequence that can be recognized by the nuclease, as shown in Figure 8.13. The length of RNA:DNA complementarity recognized by Cas9 is about 20 bp. The role of the RNA molecule is apparently to act as a scaffold around which the domains of Cas9 fold themselves to provide the nuclease activity.

Since a complementary RNA sequence can be made for any DNA sequence, virtually any location in the genome can be targeted with reagents that are easy to make. This is a significant advantage over other methods. That is, RNA molecules of a defined length and sequence are relatively easy to make—by in vitro transcription or by commercial synthesis, for example—and the same nuclease is used for all edits. In addition, several different RNA guide molecules corresponding to different locations in the genome can be introduced in order to allow simultaneous editing of multiple sites.

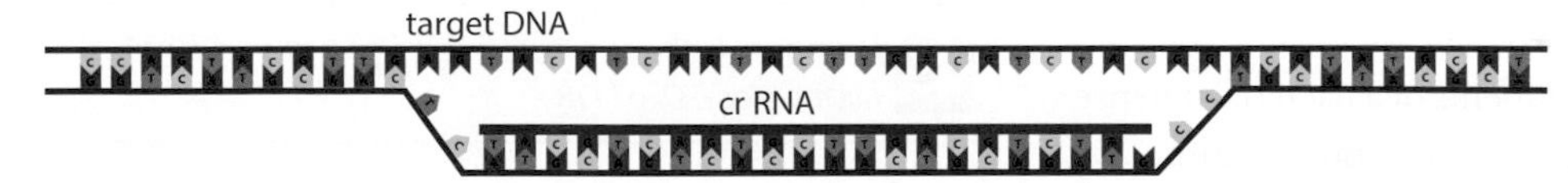

Figure 8.12 Using CRISPR to target a sequence for editing. With CRISPR, the location to be edited is targeted by an RNA sequence (labeled as the crRNA and shown with the purple phosphate backbone) that can base-pair with one strand of the DNA sequence, as shown here. The crRNA itself is, typically, twenty bases or more. The diagram shows that each base is paired between the crRNA and the DNA target, but the precise rules for targeting are not known and some mismatches do occur.

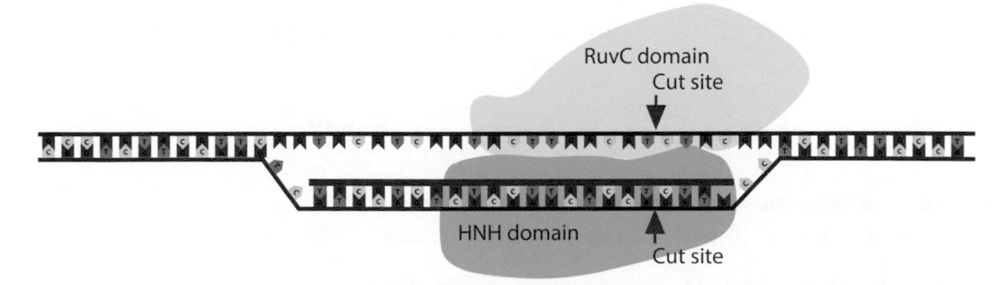

Figure 8.13 The domains of Cas9 nuclease. The Cas9 nuclease from *S. pyogenes* is the most commonly used enzyme to make the double-stranded break in the target DNA sequence. Cas9 has two separate domains that target the individual DNA strands and make the single-stranded cuts at approximately the locations indicated by the red arrow heads. Note the CGG sequence on the target DNA just downstream of the cut site by the RuvC domain; Cas9 from *S. pyogenes* depends on the presence of NGG as its PAM sequence to distinguish foreign invading DNA from sequences in the bacterial genome.

Under the correct reaction conditions, Cas9 can self-assemble and produce a double-stranded break in the DNA on its own with no additional proteins, which also simplifies the editing process. (Its need for an additional RNA molecule is discussed below.) Furthermore, Cas9 has the useful property that the protein has two separable nuclease domains, seen in Figure 8.13. One domain (referred to as the HNH domain) nicks the strand that is complementary to the sequence of the RNA, while the other domain (referred to as the RuvC domain) nicks the strand with the same sequence as the RNA.

Although we have focused our attention to the role of DSBs in genome editing, cells repair single-stranded nicks by base excision repair rather than by NHEJ, and the specificity of the nick repair is reported to be higher than that of DSB. Both DSBs and single-stranded nicks are used in various applications; broadly speaking, DSBs are more appropriate for inserting new sequences by homologous recombination, while single-stranded nicks are more appropriate for altering a sequence so as to correct or generate a specific mutant sequence.

With that background overview in mind, we will now discuss the underlying biology of CRISPR.

CRISPR is part of a natural bacterial process

CRISPR and *Cas9* are components of a bacterial defense mechanism against invading viruses and plasmids.

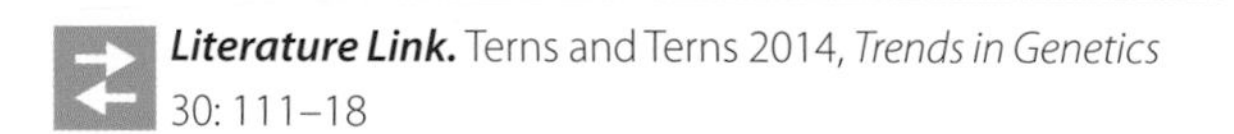

Literature Link. Terns and Terns 2014, *Trends in Genetics* 30: 111–18

When an increasing number of genomes from non-laboratory isolates of bacteria were sequenced, it was recognized that many such genomes had a cluster of **palindromic** repeats of about 28–40 bp with spacer sequences of about 35 bp between the repeats; such a cluster is summarized in Figure 8.14. As many as fifty repeats and spacers can be found in such an array, although most have fewer than that. To date, roughly 40 percent of bacteria and 90 percent of archaeae have such clustered repeat arrays. While the sequences of the palindromic repeats were highly similar for different isolates of the same bacterial species, the sequences of the spacers were unrelated to each other so each isolate had its own unique set of spacer sequences. This array of repeats and spacers was referred to as **CRISPR**: clustered regularly interspersed short palindromic repeats.

While the sequences at the CRISPR varied with the isolate, the location of the CRISPR array in the genome was similar for different isolates of the same bacterial species. CRISPRs are invariably located in the genome close to genes that encode functions related to nucleic acids such as nucleases, helicases, and DNA-binding proteins. These genes are referred to as the **CRISPR-associated (*Cas*) genes.**

Bacterial CRISPR systems have been classified into two classes (Classes 1 and 2) on the basis of the different combinations of *Cas* genes; many subtypes are also known.

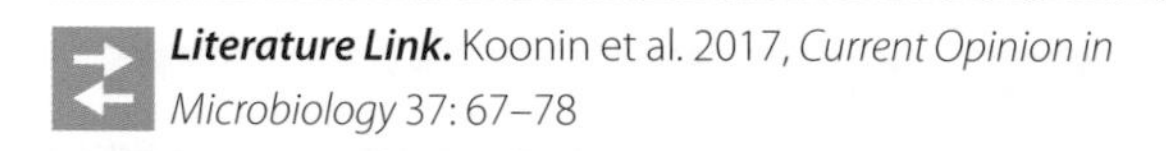

Literature Link. Koonin et al. 2017, *Current Opinion in Microbiology* 37: 67–78

The presence of a *Cas9* gene ortholog or related gene is the defining feature of Class 2 systems. The drawing in Figure 8.14 shows a Class 2 example with three *Cas* genes.

While it is the palindromic repeats that were recognized originally and gave CRISPR its name, the spacers are more significant for the specific functions of the

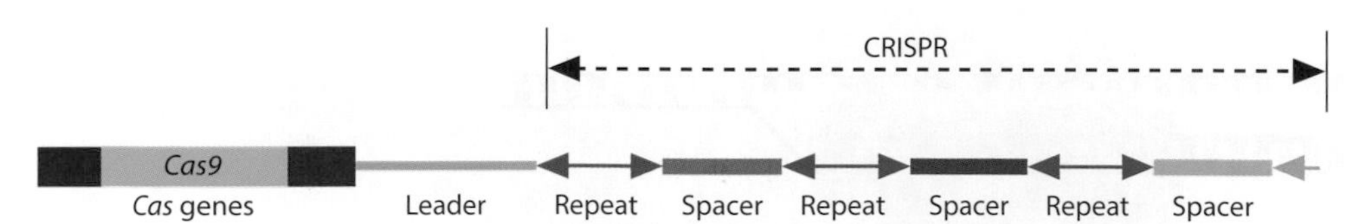

Figure 8.14 The structure of a Class 2 CRISPR system in bacteria. While the term "CRISPR" is often used broadly to encompass both, bacterial genomes have two components. The CRISPR array has palindromic repeats of about 30–40 bp (represented by the double-headed blue arrows) that are nearly identical in sequence to each other. These repeats are separated by spacer sequences of about 35 bp in length that differ in sequence from each other (represented by the different colors). Adjacent to the CRISPR array is a set of *Cas* genes that encode proteins involved in various functions related to nucleic acids; the defining feature of a Class 2system is the presence of a *Cas9* nuclease among the *Cas* genes, as shown here.

CRISPR-Cas defense system. Because the sequences of the spacers in the CRISPR array are specific to each isolate, CRISPR has microbiological applications such as identifying the presence of contaminating bacteria in food microbiology and tracking specific strains in epidemiology. But a key for developing CRISPR into a genome-editing tool lay in recognizing the origins and the biological role of these spacer sequences. While many of the spacer sequences were of unknown origin, those whose source could be identified correspond to a region of some viral (known as bacteriophage or **phage**) or plasmid genome; presumably those spacers of unknown origin also come from the genomes of uncharacterized phage.

The process that produces a CRISPR array is shown in Figure 8.15. When potentially lethal foreign DNA enters a bacterial cell (from the genome of a phage for example, as shown in the figure), the phage replicates and destroys most of the host bacteria. However, in bacteria that survive the infection, some of the Cas proteins fragment the foreign DNA into sequences of about 30 bp; these short fragments are known as **protospacers**.

Although many fragments of the phage genome are present, protospacers are not random sections of phage DNA. Sequences that become protospacers have a short conserved 3–7 bp motif, known as the **protospacer-adjacent motif (PAM)**, at their 3′ ends and thus are derived from regions of the invading phage genome immediately adjacent to these PAM sequences. The Cas9 protein relies on this PAM sequence for its function, so the PAM sequence is different for different Cas9 orthologs. For example, the PAM sequence for Cas9 protein from *Streptococcus pyogenes* is 5′-NGG-3′, where N can be any nucleotide. If the PAM region is mutated or absent, no response to the invading DNA occurs.

Protospacers without the adjacent PAM nucleotides are incorporated as a spacer at the 5′ end of the CRISPR array. As a result of this incorporation, the spacer sequences within CRISPRs act as databases of foreign DNAs that have been encountered previously by the ancestors of the current bacterial cell, a type of genomic record of past invaders, the spacers closest to the 5′ end of the cluster being derived from the most recent invaders.

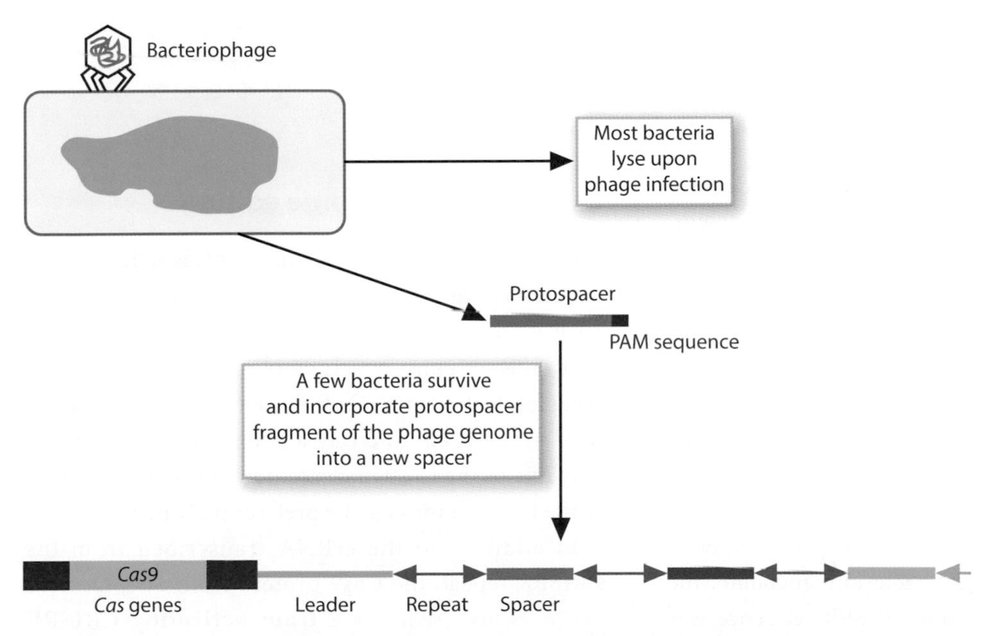

Figure 8.15 The origins of CRISPR spacers. The spacers in the CRISPR array are sequences from foreign genomes that have recently invaded the bacteria. In this example, the foreign genome comes from bacteriophage, shown infecting the bacterium at the top. The phage genome is light brown. Most of the bacteria infected by the phage are lysed but a few cells survive the infection. These bacteria acquire sequences from the phage genome immediately upstream of the PAM sequence (in red) characteristic for those bacteria; for *S. pyogenes,* the PAM sequence is NGG. The fragment of the phage genome with its PAM sequence is referred to as the protospacer. When these protospacers are incorporated into the CRISPR array as spacers, the PAM sequence is not included. Sequences from the most recent infections are incorporated near the beginning of the CRISPR arrays, as shown here.

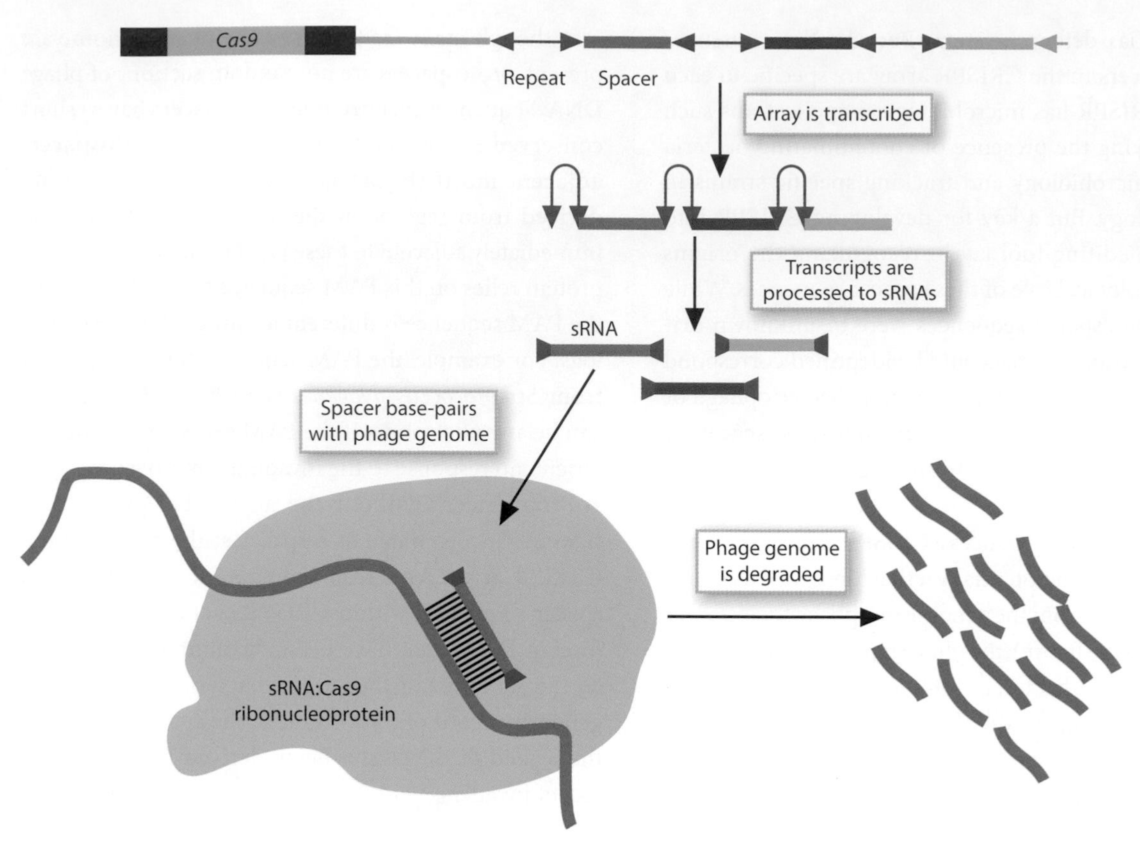

Figure 8.16 Spacers and subsequent infections. The spacers in the CRISPR array are part of a defense mechanism against subsequent infections. The array is transcribed, and the palindromic repeats form internal hairpins, as shown. The transcript is cleaved at these hairpins, producing small RNAs (sRNA) with the sequence of the spacer and part of the repeat at each end. The spacer RNA associates with the Cas proteins such as Cas9 nuclease and base-pairs with the complementary sequence in the phage genome; the phage genome has the PAM sequence (shown in red). Cas9 then degrades the phage genome.

The CRISPR array uses this spacer database to degrade foreign DNA from the same source when it is encountered again. Figure 8.16 illustrates how this occurs. When the CRISPR array is transcribed into a **CRISPR RNA (crRNA)**, the palindromic sequence within the repeat on the transcript folds back on itself to form a hairpin structure; the resulting RNA is then cut into a series of **small RNAs (sRNAs)** by one of the Cas proteins. These sRNAs can interact with Cas proteins such as Cas9, to form ribonucleoprotein complexes that carry out the different enzymatic functions encoded by the *Cas* genes. When the descendants of the bacterium encounter the foreign genome from which a CRISPR sequence was previously acquired, the variable spacer region of the crRNA guides the Cas nuclease to its complementary sequence in the foreign DNA, as shown in Figure 8.16. Because the PAM region is present on the foreign DNA but not on the crRNA, the Cas proteins are prevented from cleaving their own CRISPR array.

Cas9 is used in genome editing

While different bacteria use different *Cas* genes to carry out the cleavage reactions on the foreign DNA, the Cas9 nuclease from *Streptococcus pyogenes*, which has a Class 2 CRISPR system, is widely employed for CRISPR technologies. If the entire Cas9 protein is used for editing, each strand is cut to create a DSB with a blunt end. If only one of the Cas9 nuclease domains is active, either of the DNA strands can be preferentially nicked.

In addition to the crRNA transcribed from the CRISPR repeat, the Cas9 protein from *Streptococcus pyogenes* also requires a **trans-activating CRISRR RNA (tracrRNA)**; not all CRISPR systems require a tracrRNA. (The tracrRNA was originally called "tracRNA," but "tracrRNA" has become more widespread and can be pronounced "tracer RNA") The tracrRNA can form intrastrand base pairs to make a hairpin that helps to guide Cas9 to the appropriate location. In editing applications in the laboratory, the tracrRNA and the appropriate crRNA needed to target Cas9 are often combined into one RNA molecule known as the **single guide RNA (sgRNA)**, shown in Figure 8.17. Included in the sgRNA are approximately twenty nucleotides of the sequence to be targeted (that is, the crRNA) and the hairpin repeat formed by the tracrRNA, so the sgRNA is forty to fifty nucleotides in length. Various modifications to this RNA structure are being tested to improve the target specificity for Cas9 cleavage and reduce the possibility of off-target effects, and it seems likely that many variations on the sgRNA sequence and composition will be made as more is learned about the mechanism of action. At some positions in the sgRNA, exact base pairs are essential for targeting and cleavage, while mismatched base pairs can apparently be tolerated at other positions; this information will be important in the more specific and effective design of sgRNAs.

To carry out editing, the gene for *Cas9* and the sgRNA are introduced on a plasmid into cells by microinjection, viral transfection, or electroporation. Some applications have the Cas9 gene and the sgRNA encoded separately on the same plasmid, other applications put the two genes on different plasmids, while still others inject purified Cas9 protein directly along with the sgRNA or a plasmid with the sgRNA sequence. Among the eukaryotic species and cell types for which CRISPR-Cas9 editing has been successful are human cells, monkeys, mice, rats, zebrafish, *Drosophila*, *C. elegans*, yeast, *Arabidopsis*, and many crop plants; the wide range of successful applications suggests that the genomes of most organisms can be edited using CRISPR. Not all cell types are equally affected, however: experiments using dividing cells have typically been somewhat more successful than ones with non-dividing cells.

CRISPR-Cas9 has been used for so many different types of experiments and genome edits that no list of possible applications and ideas is complete, but finding an up-to-date review article is a good place to start.

Literature Link. Komor et al. 2017, *Cell* 168: 20–36; Turcotte et al. 2016, *Genetics* 204: 883–91

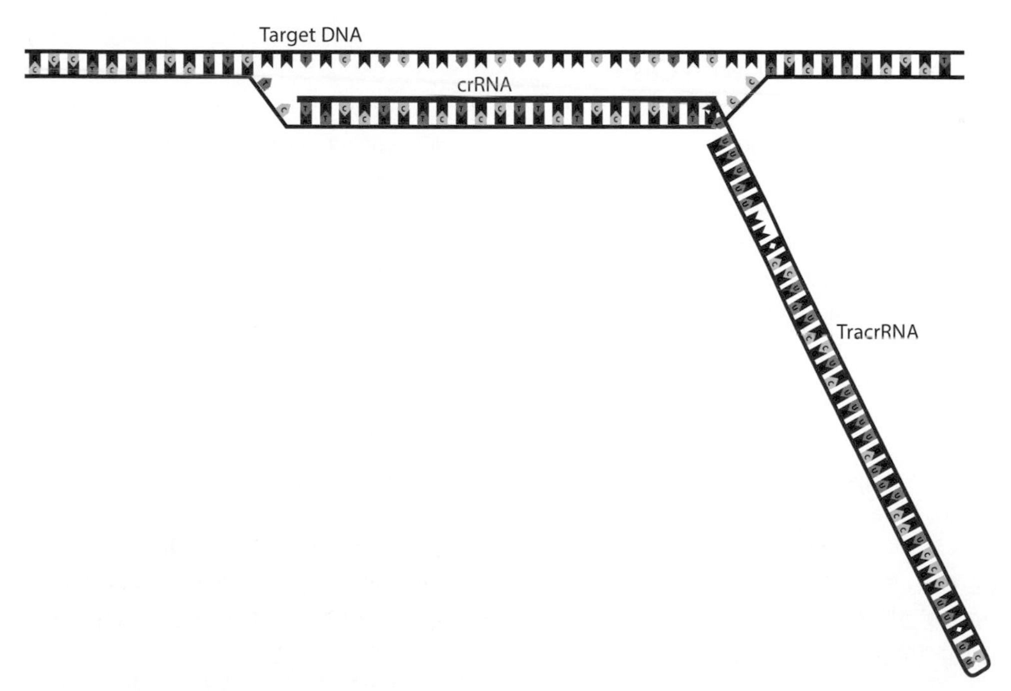

Figure 8.17 The structure of an sgRNA. The function of Cas9 from *S pyogenes* requires an additional RNA molecule known as the tracrRNA, as well as the crRNA that locates the position to be edited. For genome editing, these two RNA molecules are often combined into a single long transcript known as the single guide or sgRNA; the structure of such an sgRNA is shown with the phosphate backbone in purple. The sequences of tracrRNA vary, but the diagram uses a sequence common in *S pyogenes*. The tracrRNA forms a hairpin by intrastrand base pairing; some mismatches occur in the hairpin.

The most common applications, particularly in basic or agricultural research, are to create specific knockouts in genes based on the indels arising from NHEJ. These might be genes related to viral sensitivity such that eliminating the receptor protein on the cell surface could prevent the virus from binding and thus could confer resistance, knocking out a gene to create a sterile mosquito and thus reduce the spread of a mosquito-borne pathogen, for example. It is possible to use multiple sgRNA to alter members of a gene family or several genes simultaneously. In addition, specific sequences within a gene can be replaced or another gene (such as a reporter gene) can be inserted by providing a template for homology-directed repair. By using sgRNAs corresponding to sites on different chromosomes, one can create or potentially reverse a chromosomal translocation similar to those that occur in many cancers. The case study describes an application, namely editing the gene *MYBPC3* in human fibroblasts, embryos growing in culture, and sperm cells; mutations in *MYBPC3* are associated with hypertrophic cardiomyopathy and with an increased risk of heart failure leading to death.

All these applications are based on targeting the location of the DSBs in the genome and on the availability of additional templates, but the expression of Cas9 itself can also be regulated. Similar to what was described for the Cre recombinase in Section 7.3, the expression of Cas9 can be restricted to particular cells or times; *Cas9* genes whose transcription is regulated by light or by small inducer molecules have been constructed so that a gene may be edited in a few cells but not in all cells.

Another set of applications relies on "dead Cas9" or dCas9, an enzyme whose targeting function is intact but whose endonuclease function is eliminated. The dCas9 targeting domain can be coupled with the enzymatic module or domain from a different protein to direct the binding of that protein to a particular location. For example, dCas9 coupled with a methyltransferase has been used to regulate the epigenetic state and expression of the target gene. dCas9 applications are described in more detail in Tool Box 8.1. As with genome editing in yeast and mice, the applications may soon only be limited by the ingenuity of the investigators rather than by the availability of techniques.

TOOL BOX 8.1 Other applications with Cas9

The chapter describes how CRISPR-Cas9 can be used to knock out genes with indels generated by NHEJ or to replace and edit genes by homology-directed repair. These are currently the most common applications of CRISPR-Cas9, but they are not the only ones. Because the CRISPR-Cas9 system for targeting works well and can be done relatively quickly and inexpensively, it is being adapted for other purposes. The Cas9 applications described in this box are in common use, but many others are also being developed.

The Cas9 protein carries out two biochemically separable functions. First, it uses the complementarity between the sequences of crRNA and DNA of interest to direct its binding to a specific site in the genome. Second, it relies on its RuvC and HNH domains to function as an endonuclease and create DSBs or single-stranded nicks. Genome editing is based on the nuclease function, but the DNA binding function is also important. As described in the chapter, an RNA sequence can readily be made for any location in the genome, so Cas9 can be targeted to bind to nearly any site. This targeted DNA binding alone makes Cas9 a very useful experimental tool. Mutations have been made in the nuclease domains to create a "nuclease-dead" version of Cas9, referred to as dCas9. dCas9 retains the RNA-directed DNA binding ability but allows investigators to construct and target fusion proteins using domains with other functional activities.

Activating and repressing gene expression

Many experiments want to modify the level of expression of the gene without also altering its function. In fact, as discussed briefly in Chapter 5, changes in the level of gene expression have been implicated in many human diseases and have played important roles in evolution. Recall that transcription is initiated by the binding of RNA polymerase at the core promoter, so this initiation step forms the basis for several experimental techniques. An sgRNA can be created that will direct the binding of dCas9 corresponds to sequences at the core promoter, so that molecular events that occur at the core promoter can be altered or tweaked. Two of those are summarized in Figure B8.1.

Techniques generally known as CRISPRi (i.e. CRISPR interference, analogous to RNA interference or RNAi) are used to repress the initiation of transcription and thus shut down a gene. These were developed first in *E. coli*, in which dCas9 can be targeted by an sgRNA to bind at the core promoter. The binding of dCas9

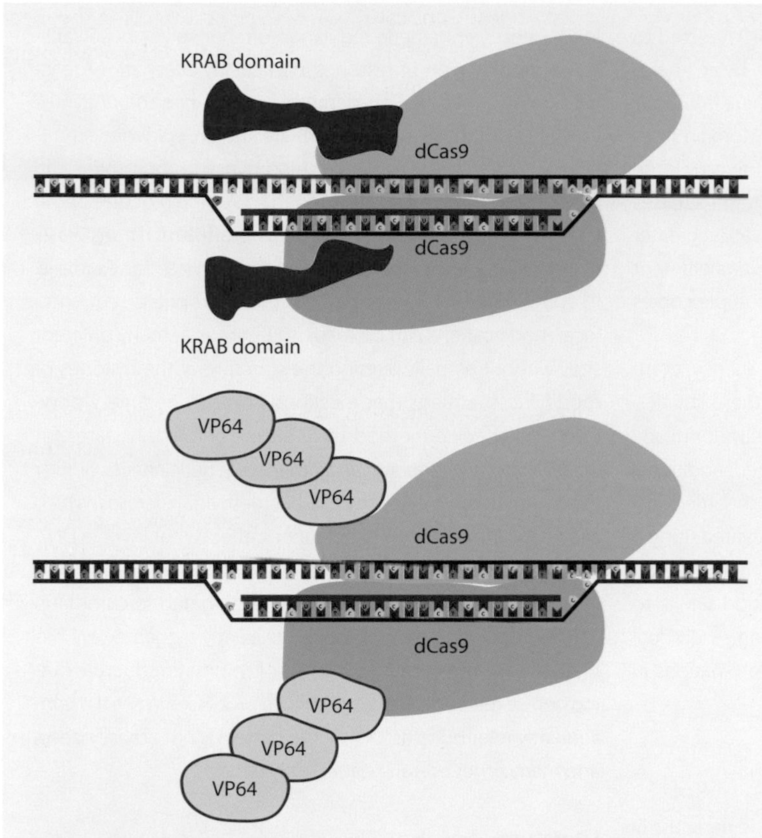

Figure B8.1 Modifications of Cas9. The ability of the Cas9 protein to bind to the crRNA allows it to be targeted to any sequence in the genome for modifications. If the nuclease domains of Cas9 are mutated to make a "dead Cas9" or dCas9, the RNA:DNA hybrid molecule will not be cut. Other enzymatic domains can be attached to dCas9 to make targeted proteins with specific activities. Part A. The KREB domain is a transcriptional repressor. A fusion protein between dCas9 and KREB represses transcription by ~50X, for a technique known as CRISPRi. Part B. The V64 domain from herpes simplex virus is a transcriptional activator. A fusion protein between dCas9 and one or more copies of the V64 domain can activate transcription up to ~50X, for a technique known as CRISRPa.

at the promoter apparently acts as a steric block to RNA polymerase, so dCas9 can repress transcription as much as 1,000-fold. Repression by dCas9 alone does not work this well in yeast or mammalian cells, possibly because dCas9 occupation alone does not fully prevent the binding of RNA polymerase to the promoter. However, a fusion protein can be made with the binding activity from dCas9 and a KRAB repression domain, as shown in Figure B8.1 Part A. KRAB domains are structures of 50–75 amino acids that are strongly associated with transcriptional repression in vertebrates. In the human genome there are more than 250 different transcriptional repressors with a KRAB domain, the largest category of transcription repressors; the domain from the human gene *Knf10* is often used in association with dCas9 for carrying out CRISPRi in mammalian cells. The fusion protein with dCas9:KRAB is 50–100 times more effective as a transcriptional repressor than dCas9 alone, and even more repression can be achieved with a combination of different sgRNAs to guide the proteins to different parts of the core promoter.

CRISPRi can be used as an alternative to RNAi for reducing the expression of a gene and may have some advantages over RNAi. Most importantly, RNAi blocks translation (as well as targeting mRNAs for degradation), so it works best for protein-coding genes. The vast majority of transcripts in vertebrate genomes are non-coding RNAs, however. Since non-coding RNAs by definition are not destined for translation, CRISPRi, which works by blocking transcriptional initiation, is often more effective at knocking them out or down. Thus CRISPRi can be used for many genes, but might be the method of choice for investigating the function of genes encoding non-coding RNAs such as long ncRNAs and microRNAs.

Just as repression can be achieved using dCas9 directed to the core promoter, gene activation can be done by an analogous method. This is referred to as CRISPRa, where "a" stands for activation. In this case, the VP64 activation domain from Herpes Simplex virus can be fused to dCas9, as shown in Figure B8.1 Part B; in mammalian cells, this fusion protein increases transcription twice or three times. Levels of activation greater than fiftyfold can be accomplished by using several different sgRNAs and dCas9:VP64 fusion proteins with multiple copies of VP64.

Libraries for carrying out CRISPRi and CRISPRa are now commercially available for most human genes. With these libraries, the effects of repressing or activating transcription for nearly every gene in the human genome can be observed. Additional refinements are allowing different genes within the same pathway or in related pathways to be repressed or activated using a combination of dCas9 fusion proteins and sgRNAs. If we envision genes as switches in a network of gene expression similar to an electrical circuit, these modifications are allowing individual switches to be flipped on or off in different combinations in order to determine the effects and interactions.

Epigenetic editing and more

While regulating the initiation of transcription is particularly useful for turning genes on and off, natural variation of gene expression in eukaryotes usually happens at the *pre-initiation* of transcription. That is, for every gene, the chromatin is modified by epigenetic effects to modulate transcription. As discussed in Chapter 3, epigenetic regulation often involves post-translation modification of histones, particularly acetylation/deactylation and methylation/demethylation. Since the enzymes that carry out these modifications are known, epigenetic regulation can also be investigated using dCas9 fusion proteins. In these applications, dCas9 is fused to the domain of one of the enzymes known to modify histones, a histone acetyl transferase, deacetylase, methyl transferase, or demethylase for example. In combination with an sgRNA of the appropriate sequence, local modifications can be made to histones at many different sites without actually altering the structure of the histones or modifying them at other locations. This can be a very powerful and specific method to fine-tune chromatin states for studying gene expression and other epigenetic effects. Similar experiments can be done with DNA methyltransferases, which allows the investigation of epigenetic effects that involve DNA methylation.

Many other experiments with dCas9 targeted to chromatin can be imagined. For instance, fluorescently tagged dCas9 has been used to monitor the dynamics of chromatin structure during gene expression and other cellular processes. While it is conveniently referred to as "dCas9," the protein is not actually dead, and many other useful lives certainly remain.

Literature

Gilbert, L. A. et al. 2014.Genome-scale CRISPR-mediated control of gene repression and activation. *Cell* 159: 647–61.

La Russa, M. F., and L. S. Qi. 2015. The new state of the art: Cas9 for gene activation and repression. *Molecular and Cellular Biology* 35: 3800–9.

Cas9 from *S. pyogenes* is only one of the *Cas* genes that have been applied to genome editing. Another method is to try the *Cas9* orthologs from other bacteria. The *Cas9* ortholog from *Staphylococcus aureus* is much smaller than the gene from *S. pyogenes*, and has been reported to be easier to work with.

One alternative to Cas9 is the related Cpf1 protein from *Francisella novicida* and related bacteria. While the overall method of action is similar between Cas9 and Cpf1, Cpf1 does not require a separate tracrRNA, and the protein is smaller than Cas9, which may make it easier to work with. In addition, Cpf1 cuts the two strands about four–five bases apart, generating a sticky end that may be repaired differently than the blunt end produced by Cas9 or that might be used in the same way as restriction endonucleases. The two enzymes also rely on different PAM sequences, so Cas9 and Cpf1 could be used to target different sites. It seems very likely that other alternatives to Cas9 and adaptations to the methods outlined here will be developed, in both academic and commercial laboratories.

CRISPR is proving to be versatile and simple enough to allow extensive editing of any genome, including the germline genomes of humans. This possibility can be seen to hold great promise for gene therapy and for the treatment of genetic diseases, but it also raises many ethical concerns. As often happens, the power of the technology has increased much more quickly than the discussion about moral and ethical questions arising from its use. This important topic is introduced in Human Angle Box 8.2.

HUMAN ANGLE BOX 8.1 Ethical considerations for human genome editing

The applications of CRISPR as a genome-editing tool are only being glimpsed. It appears that it will allow experimenters to edit the genome of any organism at will, which can affect not only that organism's survival and fitness but also its interactions with other organisms and with the environment. All of us can imagine benefits, from possible therapies for diseases to environmental remediation to improvements in our food supply. But it is also sobering and important to consider some potential risks, particularly when genome editing is done with humans. Virtually every powerful new technology raises ethical considerations, and CRISPR is certainly no exception. In fact, since the original paper showing that editing a gene could be done in sperm cells was published in 2017, many additional reports have appeared, including some that reported have resulted in babies with an edited genome.

The ethics of editing the genomes of humans can be considered in light of previous and ongoing efforts with human gene therapy. Most approaches to gene therapy involve adding a copy of a functional gene in certain somatic cell types; genome-editing applications in somatic cells are similar, except potentially more powerful. Rather than adding a copy of a functional gene, which might be mitotically unstable or change the genome in other ways, genome editing in somatic cells could allow investigators to replace the non-functional allele with a functional allele of the gene at the same locus in the genome. This could itself be an ethically fraught decision—after all, it involves deciding what constitutes a "functional" copy of a human gene when people vary widely in their genotypes—but it generally lies within the considerations that bioethicists and others have wrestled with for the past two decades or more. It offers the prospect of treating or even curing previously intractable genetic diseases; this prospect has to be weighed against the possibilities of undesirable (and perhaps unknown) side effects. In many countries, extensive regulatory oversight is already in place for human gene therapy trials, and it seems likely that genome-editing trials involving somatic cells could follow this same type of ethical framework and regulatory oversight.

But genome editing in animals can also be done in the germline, that is, in the sperm and ova themselves, or in the early embryo before the cells of the future germline have been set aside. This raises additional ethical considerations, since the genomic changes will be passed on to future generations, who will almost certainly be living under different conditions, with different advantages and challenges. For humans, parents already make decisions on behalf of their children routinely, but it somehow seems different to be making decisions about their own genetic constitution and the possible genetic makeup of their children as well.

No one can view this topic fully dispassionately and objectively; perhaps a fully dispassionate analysis is not what is required for decisions that are will ultimately be deeply personal for individuals, couples, and families. However, a very helpful and informative position statement on genome editing in the human germline was issued by the American Society for Human Genetics (ASHG) in August 2017. The working group that issued the position statement included members from several other organizations, and the statement itself has the endorsement of additional societies and organizations in the United States, Canada, the United Kingdom, Australia, Asia and Asia Pacific, and Southern Africa; while the list of signatory organizations is broad, there are serious concerns that some countries with active research programs in human genetics and genome editing do not have adequate ethics committee oversight or strong institutional review boards in place. A link to the full document is provided on the web site associated with the book.

This position statement appeared at about the same time as a report of an apparently successful effort to edit the *MYBPC3* gene in human sperm. This effort resulted in embryos in which the mutation had been correctly repaired, and the report in question is the basis for the case study in this chapter. Since that paper appeared, other reports of successful editing in the human germline have circulated. None of the published experiments was done with the intent of implanting the embryos with the editing genome, but clearly some of the apparent technical barriers for germline editing in embryonic cells are being solved. In fact it has been widely reported that babies with an edited genome have been born, although much of the information needed to verify these claims has not been made available. Thus the position statement from the ASHG is both timely and prescient. The position statement again notes that genome editing has the potential to treat otherwise intractable genetic diseases; in fact it is reasonable to argue that a treatment that can alleviate suffering for a patient or a family should always be considered as an option. On the other hand, there are concerns about possible undesirable side effects that a regulatory oversight committee needs to weigh.

The position statement is clear on its own, but some summary and reflection may be appropriate. The ethical considerations arise from many different sources, but the statement helpfully divides them into two groups: ethical considerations that arise from failure in genome editing and ethical considerations that arise from successes in genome editing. We will follow that same structure.

Ethical considerations from failures in the editing process

Ethical considerations from failures in the genome-editing procedure come from two different sources, both of which have been considered elsewhere in the book and both of which were addressed in the case study. First, as noted in Chapter 7 about genome editing in mice, all editing is done with the genome of one or a few cells. Thus the organism (in this case, the human embryo) may well consist of both "edited" and "unedited" cells for the same gene; in other words it could be a mosaic. This is definitely the case with transgenic mice, and we discussed in Chapter 7 how the investigator performs a series of crosses to create strains of mice in which all cells have the desired edit. Clearly this is not an option with humans. Thus there need to be considerations in place to address the likelihood that only some of the cells in the embryo will have the edited genome, or that only the cells with the edited genome will give rise to the embryo.

Second, the editing process may be subject to off-target effects, that is, genomic sites in addition to, or instead of, the desired location may also be edited. As discussed in the chapter, off-target effects have been a consideration for every method that relies on base pairing to target a location; at various times, it has been an important consideration in PCR primer design, and in siRNA and microRNA experiments. In these cases, off-target effects could be minimized as more information was gained about the experimental conditions, for example by recognizing which base pairs are essential for recognition and which ones (if any) can be variable, by making sequences that were longer or that had specific bases at certain positions, and so on. There was also a need to have greater understanding of the genome sequence itself, so that non-sequenced portions of the genome (which exist for the genomes of nearly all multicellular organisms) do not contribute to the off-target effects. This history of addressing the concern in previous techniques leads to some optimism that off-target effects might be reduced or eliminated with CRISPR-based genome editing as well, but this difficulty should not be brushed aside.

But there is another possible source of off-target effects in human genome editing that may prove to be even more problematic. Worms, flies, yeast, *Arabidopsis*, mice and so on are laboratory models organisms and, as noted in Chapter 3, the laboratory "wild-type" strain was derived from a very small original founding population that was then highly inbred. Thus, for these model organisms, there is a standard wild-type reference genome that is identical at most positions in the DNA sequence, and the laboratory variants differ from each other primarily or exclusively at defined locations. While there is a reference "human genome" that can be used for comparison, it is actually a composite or consensus genome; no one has this actual genome sequence, since all of us have polymorphisms throughout the genome. Thus, even if we know the "rules" for predicting off-target effects from base pairing, natural variation in the human genome will make it difficult to identify potential off-target locations in the patient or family. There may be off-target sites in one person's genome but not in another person's genome simply because of the number of polymorphisms. Databases of human genomic variation that can be used for this are both extensive and expanding, but we may always want to know the sequence of a single person's genome for predicting possible off-target effects. This seems like a potentially formidable complication for making site-specific edits.

Ethical considerations arising from successes in the editing process

Even if the genome editing of the human germline is successful, so that the appropriate edit is created at the specific location and at no other site, and applied to all the cells of an embryo, ethical questions still need to be considered. Many of these ethical questions have been discussed in relation to gene therapy, embryo research, and various reproductive technologies such as pre-implantation genetic diagnosis; but they are also relevant for germline editing. For individuals, the person (that is, the infant with an edited gene) has been "operated on" without her knowledge and without providing consent, in fact before consent is even possible. Parents are often the surrogate decision makers for their children in such situations, but there are consistent reports in which the maturing or adult child would have made a different decision from the one the parents made years before. An additional consideration that makes such decisions complicated is that the social structure of families is much more fluid than it was even a generation ago, so that the number of the people involved who may have different opinions is also less predictable.

There are also psychological and social pressures that enter into this decision-making. Every family is aware of pressures from peers about decisions regarding their children, and the difficult challenges of trying to "do no harm" to the child; this is the ethical principle known as non-maleficence. Just as there can be undue (and even unintended) pressures on families to do "what is right" for the children with regard to behaviors, school choices, career paths, hobbies and activities, and many other topics, the availability of genome editing may increase the pressure to conform to a particular social standard of acceptability; more generally, such social pressures may decrease our ability to accept differences or perceived imperfections in others. Clearly these depend on social norms that differ dramatically among humans, but it could affect our ability to accept, appreciate, and benefit from some of the differences and quirks that make us individuals. Many individual quirks are simply idiosyncrasies and personal differences, but some may be related to creative genius.

In addition, genome editing could very well have broader impacts on society. Human genetics may never escape the specter of eugenics, which again may provide an appropriate social caution as we think about germline genome editing. The history and the social forces that led to the eugenics movements a century ago are complicated and diverse, but one common element is that some social groups were in authority positions where they could make decisions that affected members of other social groups, who were often excluded from the decision-making processes. These societal differences in decision-making authority may be inevitable regardless of the social structures, but they need to be recognized and acknowledged.

Genetic analysis in model organisms can also provide insights and possible cautions here, as we consider genome editing in humans. As we discussed in Chapter 6, the connections between genotypes (which is what is being edited) and phenotypes are not always obvious. As geneticists, we have sometimes had more confidence than is warranted in our ability to predict the phenotypic effects of a genetic change. If this is true of model organisms living under laboratory conditions, it is likely to be even more true of humans.

In light of these and some other considerations, the ASHG position statement concludes with three propositions.

1. At this time, given the nature and number of unanswered scientific, ethical, and policy questions, it is inappropriate to perform germline gene editing that culminates in human pregnancy.
2. Currently there is no reason to prohibit in vitro germline editing in human embryos and gametes with appropriate oversight and consent from donors, to facilitate research on the possible future clinical applications of gene editing. There should be no prohibition on making public funds available to support this research.
3. Future clinical applications of human germline genome editing should not proceed unless, at a minimum, there is
    a. a compelling medical rationale;
    b. an evidence that supports its clinical use;
    c. an ethical justification; and
    d. a transparent public process to solicit and incorporate stakeholder input.

The issue of pleiotropy

The AHG statement provides an excellent introduction to the complex ethical issues associated with gene editing in humans. But perhaps another issue, familiar to geneticists and discussed in several places in this book (particularly in Chapter 12), should be considered as well. For many genes, we simply do not know their full biological functions. Most genes have pleiotropic effects on multiple biological processes, as discussed in Chapters 6 and 12; these effects can be unexpected. To carry out its functions, nearly every gene (or gene product) interacts with other genes (and gene products). These interactions are crucial, but extraordinarily difficult to predict, even in the yeast cell (as discussed in Chapter 12). From what we know in model organisms, we should almost expect that editing a gene to change one of its activities may also affect other activities and genetic interactions.

8.5 Summary: Genome editing and the many mechanisms to make mutants

The fundamental strategy for genetic analysis is to make a mutant and to characterize its phenotype. Nearly a century of experimentation has been based on mutants and mutant phenotypes, and much of the creativity of geneticists has been displayed finding, creating, classifying, and understanding mutant organisms. While CRISPR itself, and genome editing more generally, are a new and very powerful approach, the underlying rationale and logic are not very different from what has been done before. Genome editing revolutionized genetic mutant analysis in yeast and in mice and is very likely to do the same for many other eukaryotes, now that CRISPR provides a feasible mechanism.

With new tools, the specific mechanisms to find and create mutations have changed and will likely continue to change, but that has long been the case. Mutants were found in nature before they were created at high frequencies in the laboratory using X-rays, chemicals, and transposable elements. Nonetheless, these were random approaches; variants in every gene could potentially be identified and studied, but finding the mutants usually required persistence and sometimes a bit of luck. Genome-wide mutant screens avoid the randomness of the mutation process since every gene is altered; but it still requires considerable persistence to sort through all of the genes to find phenotypes of interest. Genome editing may offer a short-cut. With the sequence of the genome, we have the text; with CRISPR and other reagents, we appear to have the tools to edit the text so as to make it read whatever we want.

Chapter Capsule

Genome editing refers to the ability to make targeted and precise changes in the DNA sequence of a genome. The cell or organism that arises from the edited genome is then the mutant that can be used for further study. Genome editing has long been possible in yeast and mice and has been very powerful, and some experimental approaches have worked in other organisms as well, although with technical limitations. The recent development of techniques based on CRISPR-Cas9 now makes genome editing feasible in most (if not all) eukaryotes.

Genome editing depends on three steps—a search and find function, an interruption function to make a double-stranded break, and an editing function. For CRISPR-Cas9, the CRISPR array provides the search and find function, while Cas9 nuclease provides the interrupt and double-stranded break function. Editing the targeted double-stranded break depends on the two processes by which cells repair other double-stranded breaks: homology-directed repair and NHEJ. These can be used to disrupt the gene generally, to make directed changes to specific locations in the gene or genome, to mimic clinically important mutations, and potentially to reorganize the genome itself. In principle, any type of mutation can be created at any site in the genome.

Additional Reading

Chang, H. H. Y. et al. 2017. Non-homologous DNA end joining and alternative pathways to double-strand break repair. *Nature Reviews Molecular Cell Biology* 18: 495–506.

Farboud, B., A. F. Severson, and B. J. Meyer. 2019. Strategies for efficient genome editing using CRISPR-Cas9. *Genetics* 211: 431–57.

Foley, J. E. et al. 2009. Targeted mutagenesis in zebrafish using customized zinc-finger nucleases. *Nature Protocols* 4: 1855.

Hsu, P. D., E. S. Lander, and F. Zhang. 2014. Development and applications of CRISPR-Cas9 for genome engineering. *Cell* 157: 1262–78.

Kim, H., and J.-S. Kim. 2014. A guide to genome engineering with programmable nuclease. *Nature Reviews Genetics* 15: 321–34.

Komor, A. C., A. H. Badran, and D. R. Liu. 2017. CRISPR-based technologies for the manipulation of eukaryotic genomes. *Cell* 168: 20–36.

Richardson, S. M. et al. 2017. Design of a synthetic yeast genome. *Science* 355: 1040–4.

Shrivastav, M., L. P. De Haro, and J. A. Nickoloff. 2008. Regulation of DNA double-strand break repair pathway choice. *Cell Research* 18: 134–47.

Terns, R. M., and M. P. Terns. 2014. CRISPR-based technologies: Prokaryotic defense weapons repurposed. *Trends in Genetics* 30: 111–18.

Case Study 8.1

Genome editing in the human germline

As discussed in Chapter 7, there are two requirements for making a transgenic organism: the ability to introduce or edit the gene in selected cells, and the ability to produce an entire organism from the cells with the altered genome. CRISPR offers that possibility that the genome can be edited in nearly any cell type in any eukaryote; while the technical aspects of the editing process may be slightly different for different cells, the overall procedure is largely similar regardless of the cell type. What, then, is the optimal cell type in which to make the edits, so that the entire organism is composed of cells only with the edited genome?

For humans and other animals, there are a few good candidate cell types. On the basis of the procedures described in Chapter 7, using embryonic stem cells to make transgenic mice, the early embryo is one possibility. However, ES cells from humans and most other mammals have a much lower efficiency for homology-directed repair than mouse ES cells; this is one reason why the gene-editing techniques described for mice in Chapter 7 have not been widely used in other mammals. In addition, recall that mice arising in the first generation from edited ES cells are mosaics containing cells that have the altered genome and cells that have the original genome. These mosaic mice are bred to yield subsequent generations in which all the cells in the mouse have the altered genome. Clearly this is not acceptable for humans. Another possibility is to edit the genome in the zygote, the one-cell embryo arising immediately after fertilization, from which all the cells in the adult organism will ultimately arise. A third possibility is to edit the genome in the germline itself or in germ cells (sperm and ova), so that the fertilized egg and the zygote have an edited genome from the beginning.

The first successful attempt in the United States to edit the genomes in zygotes and in sperm cells was published in August 2017 and involved the gene *MYBPC3*. The paper, Ma et al. (2017), is the subject of this case study. Editing in the germline for humans is highly controversial, as discussed in the Human Angle Box 8.1. Thus it should be noted that this study did not intend to bring to term the embryos with the edited genome. However, the paper does address many of the ideas, controls, procedures, and technical challenges associated with human germline editing. Being the first such work published in a high-profile journal indicates that this is likely to become a new field; reports since then have certainly suggested that genome editing in humans is being actively pursued in laboratories around the world. While our case study is detailed, this level of detail should be helpful when subsequent studies with human gene editing are encountered.

Mammalian oogenesis and embryogenesis

Since the authors of this paper use cells from different stages of development for their experiments, it is helpful to review early embryogenesis in mammals and recall some of the specific terms. A human primary oocyte begins meiosis I with two copies of each homologous chromosome and two sister chromatids for each chromosome. Oogenesis in the primary oocyte begins and progresses before birth to the pachytene stage of prophase I before coming to the dictyate arrest, where the primary oocyte remains for many years (in humans) until ovulation occurs; there has been no net change in chromosome number or composition at the dictyate arrest, so each cell still has two copies of each homologue consisting of two sister chromatids. When ovulation occurs, the primary oocyte completes meiosis I to generate a secondary oocyte and the first polar body; the secondary oocyte has one copy of each homologue that consists of two sister chromatids, so there are two copies of each gene. The secondary oocyte begins meiosis II before arresting again (in humans) at metaphase II. Thus fertilization occurs between a sperm cell (which has one homologue comprised of one chromatid or one copy of a gene) and a secondary oocyte with one homologue and two sister chromatids (two copies of a gene) that begin to separate. The sperm pronucleus and the pronucleus of the secondary oocyte remain separate, but in the same cytoplasm. The oocyte pronucleus then completes meiosis II, producing an ovum and a second polar body, each with one homologue consisting of one chromatid, before the sperm and ovum pronuclei fuse to produce a single-celled zygote. The zygote begins the mitotic divisions of embryogenesis.

Thinking through the Experiment

1. Diagram the steps in human oogenesis and fertilization, noting what DNA molecules are available to be edited or to be used as a template for homology-directed repair and how many potential templates are present.

MYBPC3 and hypertrophic cardiomyopathy

Myosin binding protein C (MYBPC3), as encoded by the *MYBPC3* gene in humans, is found in the sarcomeres of the cardiac muscle, where it binds to the myosin heavy chain protein and interacts with actin to fine-tune the contraction of cardiac muscles. Mutations in *MYBPC3* are associated with a condition known as familial hypertrophic cardiomyopathy (HCM), which includes the thickening of the left ventricle as well as disorganization and stiffness of cardiac muscle; an estimated 15 percent of all cases of HCM (and perhaps 40 percent of cases of familial HCM) are due to mutations in *MYBPC3*. No single mutant allele of *MYBPC3* is especially common, and OMIM lists thirty different mutations in the gene, both recessive and dominant. Different mutant alleles appear to have different ages of onset for HCM, from infancy to adulthood, and different degrees of severity. The overall prevalence of HCM is about 1 in 500 adults. Many people live with HCM, but HCM is also the most common cause of sudden heart failure in otherwise healthy young athletes. The man who acted as the sperm and tissue donor for this study is a heterozygote for a dominant mutation in *MYBPC3*, a four bp deletion in exon 16 that produces a frameshift mutation; he has a history of HCM, managed by medications and an implanted defibrillator.

Thinking through the Experiment

2. What would be the advantages of editing a gene like *MYBPC3* with a dominant mutant allele and for a disorder like HCM, in which many people survive with proper therapy?

Editing in somatic cells

In order to test the effectiveness of different components used in CRISPR, the first experiments edited the gene in skin cells grown in culture. Induced pluripotent stem cells (iPSCs) were isolated and grown in culture from the man's skin fibroblasts; these would be heterozygous for the mutation. The authors electroporated the *Cas9* gene on a plasmid into the cells, along with one of two different sgRNAs and one of two different single-stranded oligodeoxynucleotides (ssODNs); these were designed to serve as the template for the repair.

Homology-directed repair using the ssODNs as the template could be distinguished from repair using the wild-type allele (which is also present), because the ssODNs have single base changes that do not affect the coding capacity but would be detectable when used as a repair template. Transfected cells were grown into colonies, and the editing results for *MYBPC3* determined by DNA sequencing. The goal of the experiment was to improve the reagents and the conditions, so the numbers are informative for thinking about other CRISPR experiments. Of sixty-one cell clones arising from the more successful of the two sgRNAs, seventeen had edits in *MYBPC3*; of these seventeen edited genes, seven were repaired by homology-directed repair while ten had indels arising from NHEJ, so seven of sixty-one cells had the desired edit. The other sgRNA had a lower rate of targeting efficiency (23 of 175 cells) and of homology-directed repair (3 of 23), one of which used the homologous allele as the template rather than the ssODN, since it did not have the single base change characteristic of the template.

From DNA sequencing of the *MYBPC3* gene, it was determined that the wild-type allele of *MYBPC3* was intact and not edited. In addition, in controls in which the better of the two sgRNAs and other components were introduced into homozygous wild-type cells, no editing of the wild-type alleles was found, indicating that the sgRNA was not producing so-called "on-target" editing effects which caused changes in the wild-type allele.

Thinking through the Experiment

3. Discuss how an on-target effect, in which the wild-type allele is edited to make a mutant allele, could arise by homology-directed repair.

Editing in early embryos and zygotes

Having identified an effective sgRNA that worked well in somatic cells, the next experiments attempted to edit the *MYBPC3* gene in newly fertilized oocytes in which the sperm came from the heterozygous man. Rather than inject *Cas9* encoded in a plasmid and require that the oocyte transcribe and translate the gene, the experiments used purified Cas9 protein with the sgRNA, to direct the edit and the ssODN so as to serve as the potential template. In control experiments with none of the CRISPR components added, about half of the fertilized oocytes were homozygous wild-type (9/19) and about half were heterozygous for the wild-type and deletion alleles (10/19), as expected in sperm from the heterozygous male.

When the CRISPR components were injected into newly fertilized oocytes at the pronuclear stage, the results were more variable. The injected oocytes were allowed to develop for three days, and both complete embryos and individual blastomeres were examined through DNA sequencing of the gene. Two thirds of the embryos (36/54) were homozygous and wild type in all their blastomeres; since half are predicted to be wild type homozygous through fertilization from a heterozygous male, the excess of wild-type homozygotes suggests that some of these may be the result of successful gene editing of the deletion allele in the one-cell embryo. The absence of mosaicism among the blastomeres is encouraging as well.

Consistently with the hypothesis that gene editing was successful, only five embryos were heterozygous for the two alleles rather than the twenty-seven (half of the fifty-four) expected from using sperm from a heterozygous male, and all five embryos were heterozygous in all blastomeres; presumably these are the ones in which no editing occurred, but the encouraging result is that the observed number is smaller than the expected number of heterozygotes. The remaining thirteen of the fifty-four embryos were mosaic, with at least one blastomere that was heterozygous for the wild-type and deletion allele, and/or a blastomere with the wild-type allele and the deletion allele with additional mutations. Several of these thirteen mosaic embryos were made up primarily of blastomeres with the homozygous wild-type allele, which indicates that the mutant allele was being successfully edited but that the editing occurred in the cell divisions after the one-cell stage.

Recall that the ssODN that was injected in order to be used as template had single base changes from the wild-type sequence. These changes allowed the investigators to determine which template DNA was being used to make the edits. Remarkably, the wild-type blastomeres in these mosaic embryos did not have the sequence change present in the ssODN, which indicates that it was not being used as the template for homology-directed repair. Rather it appeared that the wild-type maternal allele present in the oocyte pronucleus itself was being used as the template. In fact, when the ssODN was omitted from the injection, embryos in which the deletion allele was repaired in some blastomeres were found, showing that the maternal genome could serve as a template for the repair; most of these blastomeres had additional deletions, which suggests that both homology-directed repair and NHEJ were occurring, possibly at different embryo cell divisions.

The results discussed so far demonstrate that these CRISPR reagents and conditions can be used to repair the deletion allele of the *MYBPC3* gene; that the edit can occur by homology-directed repair, but then NHEJ also occurs; and that editing could apparently occur after the one-cell stage to produce mosaic embryos. Significantly, the ssODN introduced as a template for homology-directed repair is apparently not necessary, since most of the edits used the maternal genome as a template. However, there was significant mosaicism among the resulting embryos. Any level of mosaicism among human embryos would be unacceptable clinically, so the next protocols focused on solutions for that.

Thinking through the Experiment

4. Put together a chart or a table that compare the targeting efficiency of the experiment using iPSCs with the one using the pronuclear stage zygotes. What are some possible explanations for this difference in targeting efficiency?
5. In your chart or table, also compare the frequency of homology-directed repair to that of repair by NHEJ. Is the relative rate of homology-directed repair also different between the cell types? If so, what might be the significance of this?

The mosaic embryos arising from injection at the pronuclear stage principally had blastomeres of two different genotypes: either homozygous for the wild-type allele (one of which was the edited version) and heterozygous for the wild-type allele and an unedited deletion allele; or homozygous for the wild-type allele and heterozygous for the wild-type allele and deletion allele with additional mutations. This type of mosaicism indicates that there were two different sperm alleles in the developing zygote, either because the zygote had completed DNA replication at the time of the edit or because the editing processes continued during the subsequent mitotic divisions. In order to address these possible issues, the CRISPR-Cas9 reagents were co-injected with the sperm into a secondary oocyte at metaphase II, that is, at an earlier stage than in the previous experiments, in which the sperm had already fertilized the secondary oocyte. Since only one-sperm and one-oocyte pronuclei are present at this time, editing has to occur before DNA replication. In addition, the earlier time of injection makes it more likely that the CRISPR-Cas9 components will break down before the subsequent mitotic divisions.

Thinking through the Experiment

6. Diagram the differences in the experiment using cells at the pronuclear fusion stage, as done previously, and these experiments with the sperm and CRISPR components being co-injected. How many copies of the gene are present and available to be either edited or used as a template?

As before, the injected embryos were allowed to develop for three days after injection and the blastomeres of fifty-eight injected embryos were analysed by DNA sequencing. The best possible outcome is that the embryos would be homozygous for the wild-type allele, with no mosaicism among the blastomeres. An acceptable but less preferred outcome is that the embryos would be heterozygous for the wild-type allele and for an edited allele that has additional indels arising from NHEJ; such embryos would not be used for transfer or further development. The least desirable outcome is that the embryos continue to exhibit mosaicism, with blastomeres of more than one genotype.

All fifty-eight embryos had edited alleles, indicating that, in this experiment at least, the CRISPR-Cas9 method had 100 percent efficiency. In addition, none of the fifty-eight embryos was a mosaic and all consisted of blastomeres of the same genotype; this lack of mosaicism is the desired outcome. (One was homozygous for the wild-type allele but had blastomeres in which the edited wild-type allele used the maternal allele as the template and blastomeres in which the ssODN was used as the template; this is an acceptable outcome, since both alleles restore the wild-type *MYBPC3* sequence regardless of the template being used.) Of the fifty-eight embryos, forty-two were homozygous for the wild-type alleles in all blastomeres, showing that homology-directed repair had occurred, while sixteen were heterozygous for the wild-type allele and various indels, indicating that NHEJ had occurred. In other words, co-injecting sperm and CRISPR-Cas9 reagents into metaphase II oocytes was highly efficient in generating embryos with the edited allele and in avoiding mosaicism in the subsequent embryo. Some of these embryos were allowed to continue to develop in culture, to make sure that they were intact; almost three quarters of the embryos developed normally up to the 8-cell stage and half developed normally until the blastocyst stage, just like control embryos under the same conditions. Thus the gene was being successfully edited, all the blastomeres in the embryo had the genotype, and the embryo with the edited gene (and subjected to these experimental manipulations) showed normal development insofar as it was tested.

Thinking through the Experiment

7. In your chart or box comparing somatic cells with cells, add the pronuclear stage and include the numbers arising from co-injection of sperm and CRISPR reagents and the relative frequencies of homology-directed repair and NHEJ.

Off-target effects

As noted in the chapter, two of the major concerns with genome editing in the germline are mosaicism in the resultant embryo and off-target effects in which other locations in the genome are unintentionally edited. The experiments described so far addressed mosaicism; off-target effects were addressed by whole-genome sequencing. The most likely sequences for off-target effects, that is, regions of the genome with recognizable sequence identity to the sgRNA used for *MYBPC3* editing, were identified and analysed; sixteen sites were considered highly similar to the sgRNA sequence, while another seven sites had enough sequence identity to be considered as possible off-target locations. No indels were found at any of the twenty-three locations in the twenty-eight blastomeres screened. The whole genome was examined in a subset of these blastomeres, with the aim of finding off-target edits that might have been overlooked among the candidate sites; and some potential off-target indels were found at other sites. However, all these possible indels were located in repeated sequences and could be explained by sequencing errors rather than by off-target edits by CRISPR-Cas9. Thus it seems likely that no off-target edits occurred, at least none that was readily recognized by these filtering methods.

Thinking through the Experiment

8. What are the advantages and possible disadvantages of looking primarily at sequences with known homology for identifying off-target effects? What types of changes might have been observed, and what types might have been missed in these experiments?
9. Describe some of the follow-up experiments that could be done, or that have been done, since this original paper was published in 2017.

Summary

CRISPR-Cas9 is such a widely used method for genome editing that thousands of different examples could have been chosen for this case study. These experiments were the first to use CRISPR-Cas9 to edit the genome in germline cells in humans, so in many ways the example is not a typical one; most uses are simpler than this one. On the other hand, it is an important example to examine in detail.

Genome editing in the human germline raises ethical questions as well as technical ones. While this chapter does not address the ethical issues directly, since the embryos were not brought to term, it shows that many of the technical questions can be answered. Editing in the somatic cells derived from the patient was highly efficient, but genome editing in the germline was even more efficient. In addition, by appropriate (and apparently relatively modest) changes in the protocol, mosaicism among the cells of the edited embryo could be largely avoided. While there is no evidence of mosaicism arising in these embryos from co-injection of sperm and the CRISPR-Cas9 reagents, the numbers of embryos and blastomeres tested are still relatively small. Any mosaicism in the embryo is clinically unacceptable, so it is important that more embryos, more blastomeres, and other genes be tested. Nonetheless, it is highly encouraging that the mosaicism could apparently be eliminated. Subsequent investigators will have a successful protocol as their starting point.

It is also important to note that, for this gene at least, no off-target effects were observed. Certainly the elimination of off-target effects will also have to be developed separately for every sgRNA and gene sequence tested, but, as the sequence rules for sgRNAs are elucidated, these off-target assays could become a less formidable test. As happened with PCR primers and with siRNAs, the cases in which off-targets occurred proved to be informative by telling us which positions in the sequence were most important for avoiding off-target effects in the future and for anticipating what off-target effects might arise.

There are (as yet) unpublished but rumored reports of many other experiments that involve genome editing in the human germline. The transparency of this paper is significant. The authors present detailed information about their methods and results, which allows a more critical assessment of the experiments. In the end, it seems clear that genome editing in the human germline is technically feasible and that this needs to be taken seriously, both clinically and ethically.

Reference

Ma, H. et al. 2017. Correction of a pathogenic gene mutation in human embryos. *Nature* 548: 413–19. [Many commentaries on this paper have been written. Two helpful one are in *Nature* 548: 13–14 and 398–400.]

9

CHAPTER 9

Genome-wide mutant screens

TOPIC SUMMARY

Genome-wide mutant screens attempt to determine the mutant phenotype for every gene in the genome, regardless of its possible biological function. Two methods have been widely used for such screens: deletion of genes by targeted homologous recombination, used primarily in yeas; and disruption of a gene's expression by RNA interference, used in many multicellluar organisms. More recently, CRISPR has been applied to genome-wide mutant screens as well. Since the phenotype is not used as the basis for finding such mutants, genome-wide screens have identified unsuspected functions for many additional genes, even for biological processes that have been thoroughly studied by traditional gene-by-gene approaches. Genome-wide screens are rapidly becoming the method of choice for finding mutant phenotypes for many biological processes in many different organisms.

INTRODUCTION

For decades, geneticists have searched for mutants. The preceding chapters have described how mutations are induced and how the mutants are examined for phenotypic consequences. One by one, the genes are cloned. The cloned genes themselves can be used to generate gene disruptions and gene replacements. Little by little, gene by gene, genomes yielded their functional information. The pace of this type of genetic analysis has sometimes been slow and the path is often meandering, but usually the scenery has made the route worthwhile.

But, as more genomes have been sequenced and annotated, that type of mutant analysis has also changed. When the DNA sequence of a genome has been determined, all the genetic information that underlies every biological process has been revealed. The implications of this are enormous. Do we want to study complex developmental programs, simple or sophisticated behaviors, responses to pathogens, or other complicated biological questions? All the genetic information is available, waiting for us to use it. With thousands of sequenced genomes, genetics has begun to move from *gathering* information to *processing* information.

Despite these changes, the established tools of genetic analysis—such as mutant phenotypes, expression assays, and interaction experiments—remain tools through which we make sense of the information that genomics provides. The power of genome sequences has turbo-charged genetic analysis and has allowed us to examine many biological processes in ways that were not possible before. But the fundamental principles of analysis do not change.

In this chapter we introduce the concept of **genome-wide genetic analysis**, or **genome-wide** mutant **screens**. Genome-wide genetic analysis tests every gene in a sequenced genome for its effect on a biological process, and retains the mutants regardless of whether they have a recognizable mutant phenotype.

While genome-wide analysis for mutant screens have been used for relatively few years by comparison to traditional genetic screens, the results have already been remarkable and many new insights have been gained that were not recognized with traditional genetic screens. In this chapter we will describe some of the techniques that are used to perform genome-wide mutant screens and some of the results that have arisen from these mutant screens. The methods can be used for any biological process, and many new experimental options are now open to us. Genome-wide mutant analysis presented in this chapter can identify every gene in the genome using targeted mutations. The task is to identify the genes involved in a biological process; the tool is still mutations in the gene, the same as we used in Chapters 4 and 6. The technology has changed but the principle has not: using the genome sequence as the starting point, *find a mutant*.

9.1 Genome-wide mutant screens: An overview

Identifying mutations in every gene in a genome is a big task. The yeast *Saccharomyces cerevisiae* has an estimated 6,200 protein-coding genes, the fruit fly *Drosophila melanogaster* about 14,000, the nematode *C. elegans* about 21,000, *Arabidopsis* about 27,000, and a mouse about 21,000 genes. Even a single mutant allele in each of these genes would require substantial effort, and geneticists typically want multiple mutant alleles of the gene. Generations of geneticists have been identifying mutations in these organisms, but only a fraction of the genes in any of these organisms had been mutated—at least, until genome-wide screens were developed. The development of genome-wide mutant screens has identified new genes involved in well-known biological processes and has made it possible to consider biological questions that were unapproachable about a decade ago. Genome-wide mutant screens are rapidly becoming the standard approach for all mutant hunts in model organisms.

Before beginning to describe how genome-wide mutant screens are carried out and what we have learned from them so far, we need to define what they are. The term "**genome-wide**" is commonly used to describe these procedures, but is not very well defined. Our definition is as follows: *a genome-wide mutant screen is one in which the investigators use genome sequence information to design specific molecular tools to disrupt or alter the functions of all or nearly all of the identified genes.* Our definition lays out the basic requirements of a genome-wide screen: a sequenced genome for which most of the genes have been identified, molecular probes that are specific to each gene, and a method to disrupt the function of the gene.

Genome-wide mutant screens have two significant advantages when compared to standard genetic screens that use randomly induced mutations. The first is that the gene has already been cloned before the mutant phenotype has been identified. The process of connecting a mutant phenotype to a DNA sequence, as we described in Chapter 5, now happens immediately, in fact before the mutations have been created. The second advantage is that *all* the genes are being tested at the same frequency, without regard to a suspected physiological function or phenotype. This is in contrast to what happens in traditional genetic screens or reverse genetic analysis. Neither of these other methods has the capability to test all of the genes.

A comparison among traditional forward genetic screens, reverse genetic screens, and genome-wide screens is shown in Figure 9.1. A forward genetic screen begins with a single mutant organism, as shown at the top of the figure; a reverse genetic analysis begins with a single mutated gene, as shown in the middle. A genome-wide screen, as shown as the bottom, begins with mutants for every cloned gene.

In Section 4.5, and more fully in Text Box 4.2, we introduced the concept of **saturation**—identifying all of the genes that can be found by a particular mutant screen. As discussed there, in a traditional genetic screen based on randomly induced mutations, the frequency at which the same gene is mutated allows the investigator to infer how many genes could be found and, perhaps more significantly for our purposes here, how many genes have *not yet* been found. We illustrate this in Figure 9.2 with a hypothetical process affected by the four genes shown. Among the first thousand mutant samples tested, one of the genes (the blue one) is mutated. In the second thousand samples, a second gene (the brown one) is mutated, and so on for 5,000 samples. But again, the estimate of the number of genes is based on statistical methods and assumptions about the underlying probability distribution, so certain genes will almost surely be missed. In our Figure 9.2, one gene is found three times, one gene is found twice, and one gene is found once; but one gene (the one in purple) is not mutated at all, so its role in the biological process would not be detected.

A genome-wide screen tests every gene but does not rely on statistical sampling to estimate how many genes are left unexamined, since it mutates every gene on the basis of knowledge of its DNA sequence. Unlike in a traditional genetic screen, all genes will be mutated, but no gene will be mutated more than once in a genome-wide screen (unless that is the investigator's intent). This is shown in in the lower half of Figure 9.2. Each gene is mutated and the mutants are divided into pools, which are then tested for their mutant phenotypes.

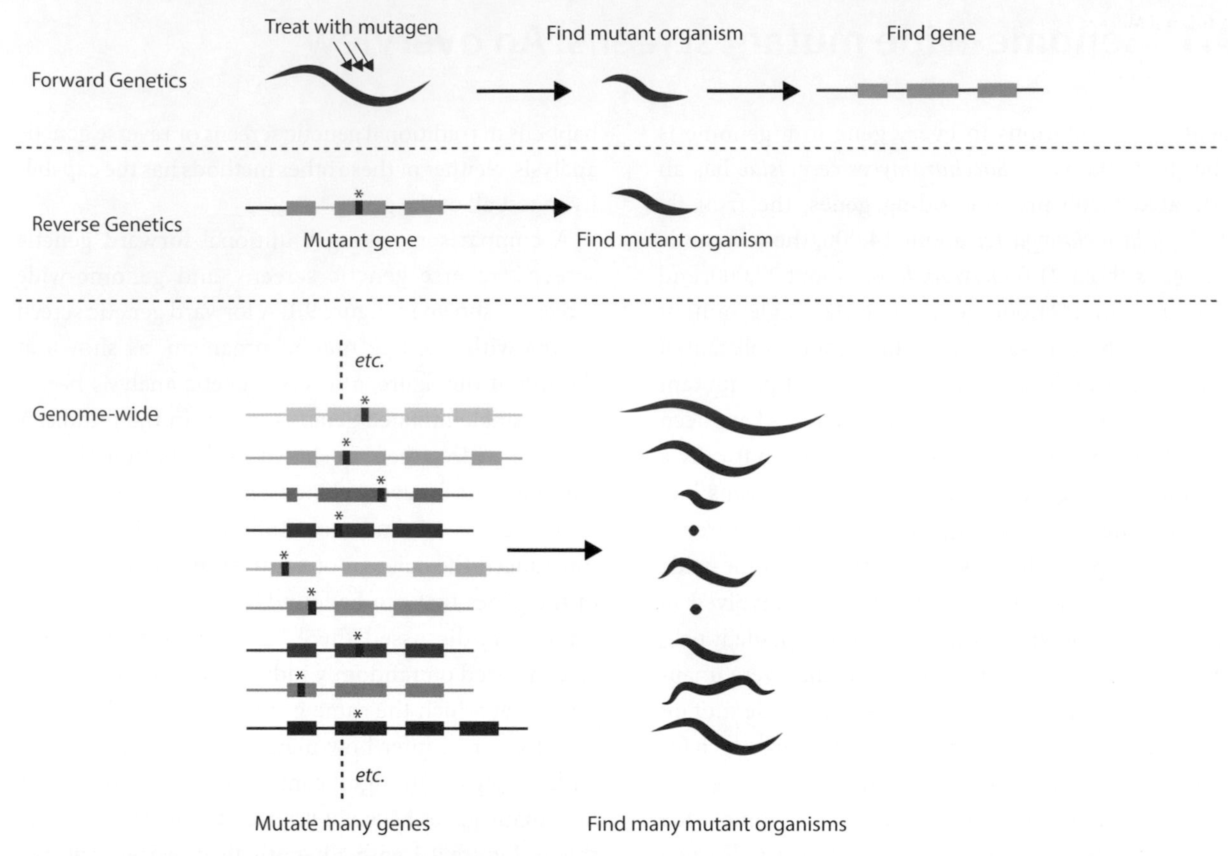

Figure 9.1 A comparison of forward and reverse genetic analysis with genome-wide mutant screens. In forward genetics, as described in Chapter 4, a wild-type organism is treated with a mutagen, mutant organisms are found, and the gene is cloned. The mutant phenotype is found before the gene is identified. In reverse genetics, as described in Chapter 7 for yeast and mice, the gene is identified first and is mutated in vitro. The mutant gene is introduced back into the organism and the phenotype is scored. In genome-wide screens, all genes are altered in vitro and reintroduced individually into the organism or cell without regard to their possible phenotypes or functions. The phenotypes are then scored.

Genome-wide screens identify new genes that affect well-described biological processes

The assurance that every gene has been mutated in a genome-wide screen presents an enormous advantage, which may be best illustrated with an example. *Saccharomyces cerevisiae*, like all organisms, repairs damage to its DNA induced by environmental agents. The process of DNA repair has been studied very thoroughly and cleverly by traditional mutant analysis, and many yeast mutants that cannot repair some type of DNA damage are known. All the genes identified by traditional genetic analysis have been cloned, and orthologues of these genes have been identified in many other organisms. Many of the orthologues have themselves been tested for a role in DNA repair, either by mutant analysis or by biochemical assays. DNA repair is a biological process thoroughly studied, at least through the traditional methods of genetics.

However, in a genome-wide screen, *every* gene in the yeast genome was individually mutated and *every* mutant was tested for its effect on DNA repair. Most of the mutants that were found to affect DNA repair proved to be alleles of genes that had already been identified by traditional gene-by-gene analysis. (It is reassuring to know that the genome-wide screen also finds the genes that have been so thoroughly studied, since it verifies the validity of the screen.) But new genes involved in DNA repair were also found. Specifically, mutants in three genes that had never been identified by traditional analysis were found to have observable defects in DNA repair. Furthermore, each of these genes has orthologues in other species, including humans, which suggests that they are fundamental to the process. This is an exciting example of the power of genome-wide screens. By being able to test mutants in all possible genes, we learned that we did not have all

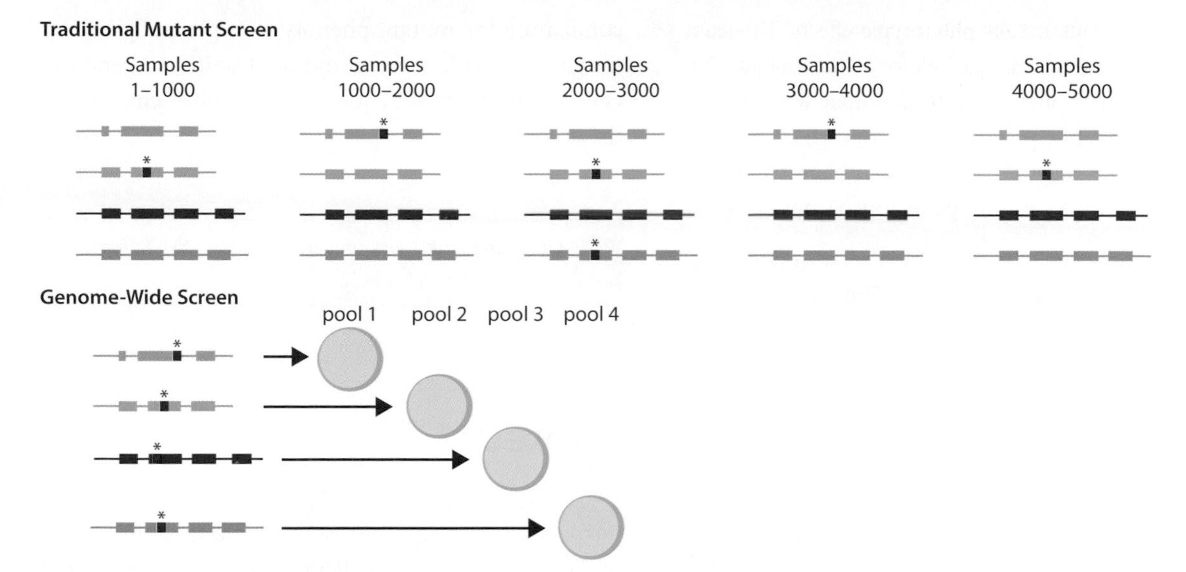

Figure 9.2 Saturation in traditional genetic screens and genome-wide mutant screens. Four different genes are involved in this hypothetical process, as indicated by the four different colors in the gene structures diagrammed on the left. In the traditional mutant screen, the first 1,000 samples yield a mutant in the blue gene, the second 1,000 samples yield one in the brown gene, and so on. After five 5,000 samples, the blue gene has been mutated three times, the brown gene twice, the green gene once, and the purple gene has not been mutated at all. Thus the role of the purple gene in the process has not been recognized and the gene remains undetected. In a genome-wide screen, all genes are mutated by directed methods first, without regard to what process they may affect. Each gene is tested as part of a pool with other genes in the genome. Because all genes are mutated, the role of the purple gene in this process would be detected. Furthermore, no gene needs to be mutated more than once.

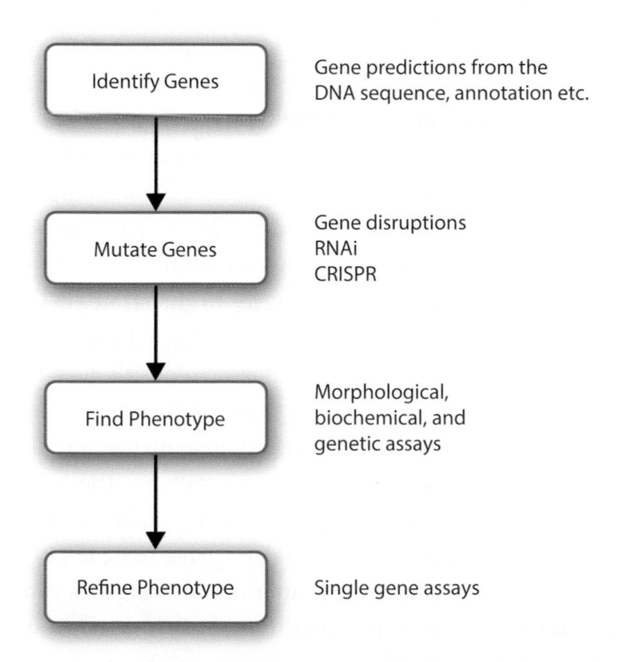

Figure 9.3 A flow chart of a typical genome-wide mutant screen. The first step is to identify all genes in the genomes using gene prediction methods such as similarity to other genes, the presence of certain recognizable features, for example start and stop codons, and transcript annotated. Then genes are perturbed or mutated by gene disruptions, RNAi, or CRISPR. Third, phenotypes are found by a range of different assays. Finally, the role of single genes involved in the process is confirmed and refined, usually by methods used for other mutants.

the parts of a process that has been thoroughly analysed in many different species. Other examples will be mentioned throughout the chapter.

Although the steps of a genome-wide screen differ in each organism and in each screen, genome-wide analysis can be summarized as follows, and is depicted in flowchart form in Figure 9.3.

1. **Identify** all the genes to be mutated. For the largest scale of genome-wide screens, this set includes all the genes in the genome. However, this large goal is often subdivided into more manageable smaller goals. For example, the set of genes to be mutated could contain all the ones expressed in a particular pattern, such as all the genes expressed in lymphomas, or all the genes of a particular functional class, such as G-protein coupled receptors. The same techniques and strategies used to test every gene in the genome can be adapted to testing every gene that fulfills particular criteria.
2. **Develop** methods for disrupting or perturbing genes. Most of the chapter will be spent describing these methods, which include targeted gene deletions in yeast, RNAi screens in many organisms, and, more recently, the use of CRISPR.

3. **Examine** the mutants for phenotypic effects. The team of investigators may want to look for all the mutants that have a particular phenotype; or they may want to look for as many phenotypes as possible for each mutant.
4. **Confirm** and follow up on the mutants that are found, often with gene-by-gene approaches.

We will consider each of these in turn, most of our attention being on the first three steps. The fourth step, of confirming the mutant phenotype, is generally similar for genome-wide screens and for traditional gene-by-gene methods, so it is addressed in other chapters in the book.

Literature Link. Carpenter and Sabatini 2004, *Nature Reviews Genetics* 5: 11–22

9.2 Identifying the genes to be mutated

It may seem obvious, but the first requirement for a genome-wide mutant screen is to have a genome that has been sequenced. Our definition requires going beyond a sequenced genome, however: the genes themselves must also be identified. In the jargon of genomics, as discussed in Chapter 2, the genome must be **well annotated**. This means that the coding regions of genes must be identified and the corresponding protein sequences, if any, must be inferred. These protein sequences must have been compared to other sequences (through BLAST or a related tool) in order for the researcher to infer functions on the basis of similarity. Furthermore, the locations of exon-intron boundaries must be identified, or at least predicted by some reliable method. Pseudogenes and non-coding RNA genes should be identified.

Annotation of a genome eventually goes well beyond these basic parameters to include the location of repeat elements, inferences about regulatory regions, comparisons to the linkage arrangement in related species, and so on. However, the annotation needed for a genome-wide mutant screen is more fundamental. Where in the genome are the genes? What is their likely structure and what can be inferred about their function? Because the structures of predicted protein-coding genes are continually being refined, particularly for additional splice variants, a genome-wide screen from a few years ago might not be a genome-wide screen now. Furthermore, annotation of non-coding RNA genes (which can be tested by CRISPR, as we discuss in Section 9.5) is constantly changing.

Once the investigator is confident enough that the gene structures have been well annotated for his or her purposes, it becomes necessary to think about which of the annotated genes will be tested. The ultimate goal may be to test all genes in the genome, without regard to location, expression, and hypothesized functions. Such a project will certainly provide the most complete information, although often with a lot of work. But, for a variety of reasons, it is often necessary or helpful to test a subset of the genes initially; indeed, it is unlikely now that a single laboratory, or even a small group of laboratories will want to test every gene for every conceivable phenotype, as was done originally for genome-wide mutant screens in yeast and *C. elegans*. The subset to be tested can be defined by either arbitrary or functional means. For example, some genome-wide screens in *C. elegans* have tested all genes on each chromosome without regard to their postulated functions, analogously to the strategy for subdividing the genome in traditional genetic screens as described in Section 4.4. Since map position—and not functional criteria—is the basis for inclusion in the screen, the mutant could have any imaginable mutant phenotype, or even a wild-type phenotype. The map location simply becomes one arbitrary criterion for dividing a large number of objects (genes to be tested) into a more manageable group until all the genes can be tested, but it is not one that most investigators would use nowadays.

Most genome-wide mutant screens have used some functional criteria to divide the large number of genes into more manageable subsets. For instance, microarray or RNAseq data provide information on gene expression that comprises a set of functional criteria for a genome-wide screen: the genes to be tested may be ones that are expressed in embryonic stem cells, in the germline during sexual maturation, or in particular cancer cell types. Many other examples of expression-based subsets have also been used, including genes whose expression is induced or repressed by drugs, hormones, or an environmental stimulus. The interplay between gene expression and mutation, which was described for traditional gene-by-gene cloning in Chapter 5, becomes all the more important in genome-wide approaches, since the expression results confirm and extend the results of the mutant screen, and vice versa.

Another characteristic that has been used to test subsets of genes is the postulated function of the genes that is based on sequence comparisons. As genomes are sequenced or annotated, the sequences are searched for similarity with all other DNA sequences; for protein-coding regions, these tests are based on inferred amino acid sequences. Such similarity searches are an important part of the annotation process designed to refine predictions about the functions of genes. One outcome is to provide a complete set of proteins in one family, such as all the G-protein coupled receptors or all the serine proteases. Some gene families in multicellular organisms have dozens or hundreds of members, so the process of testing each paralog uses the same overall methods and strategy as genome-wide screens. The standards for inclusion among the genes to be tested can be even more loosely defined or combined with expression data to produce the subset of genes to be tested, for example all coiled-coil proteins expressed in the germline.

Although some of these procedures may not seem like genome-wide screens in that not all genes are being tested, it is often reasonable to limit the scope of the screen. This is not so different from searching for mislaid car keys: it makes sense to focus on the search on the locations where they were most likely to have been left. One important difference is that a search for lost keys has one very specific goal, and the search will stop when that goal is achieved. Genome-wide screens, whether limited to certain subsets or not, have much wider goals, closer perhaps to producing an inventory of all objects—whether lost or not, whether keys or not. All genome-wide screens begin somewhere, but have the same ultimate destination: to test mutations in all of the genes in the genome.

9.3 Disrupting and perturbing genes

Once the genes have been identified and annotated, the second necessary part of a genome-wide mutant screen is to develop a method or methods for producing mutations in every gene, regardless of its predicted phenotype. As noted in our definition of genome-wide mutant screens, these methods rely on molecular tools that are developed from sequence information for the genes. Two different approaches to producing genome-wide mutant phenotypes have been widely used: gene disruptions and **RNA interference (RNAi)**. In *Saccharomyces cerevisiae*, genes can be specifically targeted for disruption or deletion by homologous recombination, as described in Chapters 3 and 7; yeast strains with targeted deletions have been constructed and screened for nearly all the 6,130 genes in the yeast genome. The second approach does not affect the structure of the gene itself, but instead disrupts or perturbs its expression. This method, known as RNAi, has been widely applied to many organisms and has very broad applications and significance; in fact the significance of RNAi was recognized in 2006, when the Nobel Prize for Physiology and Medicine went to Andrew Fire and Craig Mello. More recently, genome-wide mutant screens based on CRISPR have begun to be used in some organisms and are likely to become much more widely used in the next few years. We begin by describing the gene disruption methods used in yeast.

Genome-wide mutant screens in yeast have been performed by targeted gene disruption

More than 96 percent of the 6,130 predicted genes in yeast have been disrupted by targeting insertions into the genome by homologous recombination. These strains are maintained as diploids, either as homozygotes or as heterozygotes; because 1,159 genes are essential for growth, these are maintained as heterozygous diploids, whereas haploid deletion strains have been made for each of the 4,757 non-essential genes. The deletion strain collections are commercially available and are widely used in yeast genetics.

Literature Link. Giaver and Nislow 2014, *Genetics* 197: 451–65

The original collection of gene disruptions was made by inserting a kanamycin-resistance gene into each target gene, disrupting the coding region of the target gene with an easily selected marker. A different disruption cassette was made for each of the 6,100 targeted genes. The process of targeted integration is detailed in Figures 9.4 and 9.5. Figure 9.4 shows the PCR strategy to amplify the *kan*^{R} gene with specially designed primers to produce an insertion cassette, which is the first step in the process. The primers to amplify the *kan*^{R} gene were approximately seventy-four base pairs long and included both gene-specific and gene non-specific components. This can be illustrated through a careful examination of the right (or downstream) primer in Figure 9.4; the left or

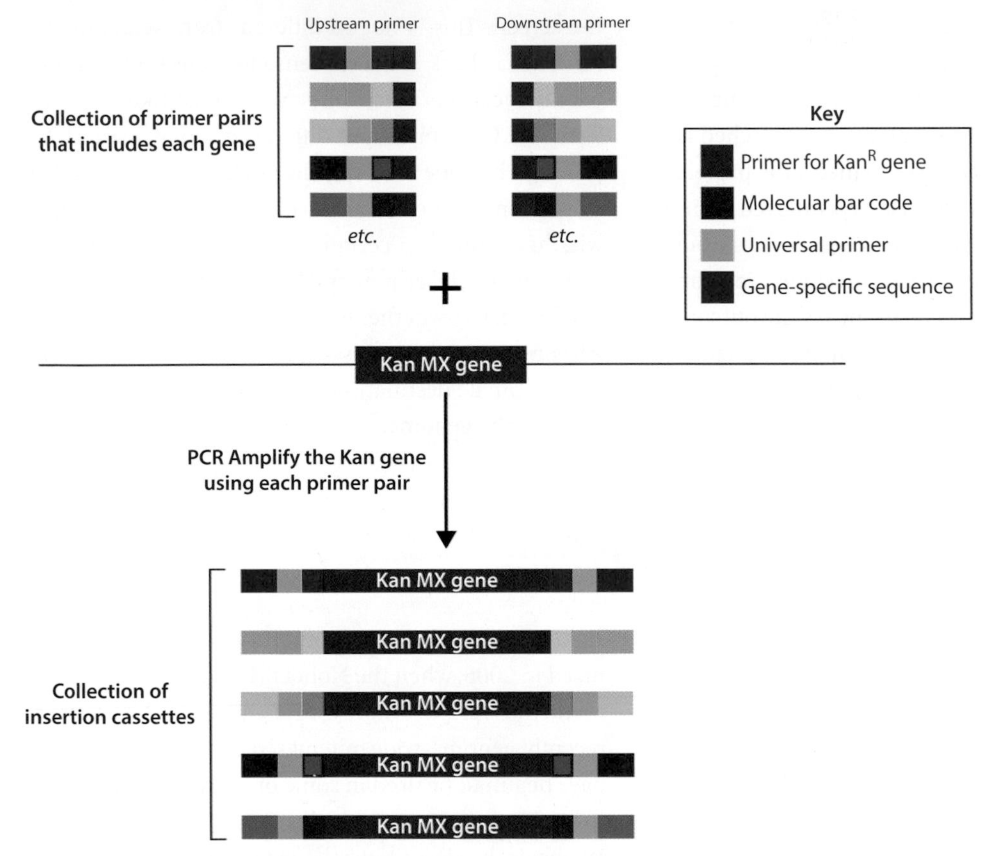

Figure 9.4 A schematic diagram of the PCR primers and the insertion cassettes used in genome-wide gene disruption in yeast. A collection of PCR primer pairs is made, one upstream and downstream primer pair for each gene in the genome. Five pairs of primers are shown at the top. Each primer is seventy bases long and has four sequence components, two of which are specific to that gene and two of which are common to all primers. The common components are shown in the same color in each primer, whereas the gene-specific components are shown as different shades of the same color. The color key for the primer components is shown at the top right. Each primer pair has sequences corresponding to the target gene in the genome, here shown in shades of blue. The gene-specific sequence is indexed with a specifically defined sequence known as molecular bar code, shown in shades of red. Note that each gene-specific sequence in blue is associated with one specific bar code in red. By knowing the bar code sequence, the investigator could determine which gene has been disrupted. The components that are common to all primer pairs are the sequences to amplify the *kanomycin-resistance* gene (in purple) and the sequence used to amplify the molecular bar code (in green). By using an adjacent universal primer, all bar codes could be amplified without regard to their own sequence. Each primer pair is used to amplify the *kanomycin-resistance* gene, which results in a collection of insertion cassettes specific to each gene in the genome. Although five cassettes are shown here, the actual genome-wide screen had nearly 6,000 different insertion cassettes, one for each gene in the yeast genome.

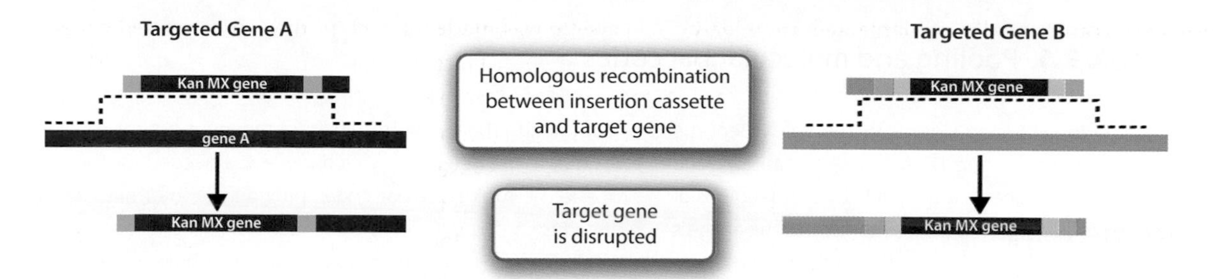

Figure 9.5 The insertion cassettes are used to disrupt the genes. Two of the insertion cassettes from Figure 9.5 are shown here; they correspond to different genes in different yeast strains. The insertion cassettes are transformed at random into yeast cells. Homologous recombination between the ends of the insertion cassette and the target gene inserts the *kanomycin-resistance* gene and the specific molecular bar code into each gene. Kanomycin is used to select for yeast cells that have an insertion, and the bar code will be used in Figure 9.6 to determine which gene has been disrupted.

upstream primer has a similar structure in reverse. The first part of the primer sequence corresponds to eighteen nucleotides from the end of the *kan*^{R} gene and is used to amplify the *kan*^{R} gene. The next part is a sequence of twenty nucleotides referred to as the **molecular bar code**. This sequence is different for each gene knockout. We will describe the molecular bar code in detail later on, so for now we will focus on the general structure of the primer sequences.

Immediately downstream of the molecular bar code is a gene non-specific sequence, the same for each of the 6,100 genes. This sequence is not found in the yeast genome. This universal sequence allows a single primer pair to be used to amplify the molecular bar code regardless of the gene. With more contemporary methods using high-throughput sequencing rather than PCR amplification, this sequence would probably not be necessary. Immediately downstream of the universal primer is a gene-specific sequence of eighteen nucleotides corresponding to a sequence in the gene to be disrupted. This sequence serves to target the insertion to a specific gene and is different for each insertion cassette; in Figure 9.4, these sequences for targeting recombination into specific genes are shown in various shades of blue. Note, again, that the same structure is shown in reverse in the upstream primer sequence: the region of specific yeast homology comes first, followed, in order, by a gene non-specific sequence, then by the upstream molecular bar code sequence, and by a portion of the *kan*^{R} gene. These two primers, used together, will amplify the *kan*^{R} gene with a long sequence tail at each end as an insertion cassette, parts of which are specific to each gene in the yeast genome and parts of which are general to all 6,100 insertion cassettes.

Molecular bar codes allow identification of individual genes from among a pool

Almost all genome-wide mutant screens involve some method for **pooling** the collection of strains or cells with the disruption. The strategy of pooling and its relationship to the molecular bar code, probably the most unusual and clever aspect of this collection of insertion cassettes, are discussed in Tool Box 9.1. In yeast gene disruptions, the molecular bar code is a sequence of approximately twenty nucleotides at each end that is not found in the yeast genome; the upstream and downstream sequences are different, and the bar codes are different for each yeast gene. (That is, there are 6,100 different bar codes.) The bar codes provide the investigator with information about which gene has been disrupted in each strain: this code is a unique identifier for each gene, in the same way in which a bar code is a unique identifier for an item in a retail store or for a personnel badge and card key in a corporation.

It is instructive to consider what would happen had these molecular bar codes been omitted. A typical insertion cassette for a single yeast gene would include the *kan*^{R} gene flanked by sequences corresponding to the yeast gene targeted for disruption. In our drawing in Figure 9.4, these are the blue elements. If this standard primer sequence were used, the *kan*^{R} gene would insert into the targeted gene by homologous recombination, as described for a typical yeast gene in Chapter 7. While this procedure works for a single gene, it has a serious

TOOL BOX 9.1 Pooling and molecular bar codes

The project to disrupt every gene in the yeast genome required the use of a unique DNA sequence tag for each gene. Each gene is disrupted individually by using regions of homology to integrate the selectable marker, but the analysis of these disruption mutants is most efficiently done by looking at pools of mutant strains. The unique sequence tag, also known as a molecular bar code or DNA signature tag, allows the investigator to find the particular gene among the pool of mutants.

The pooling process can be illustrated by a thought experiment. Imagine a biological process in yeast that requires the products of twenty genes—a metabolic pathway, for example. Yeast has a genome of about 6,100 genes, so the twenty genes that affect this pathway will be less than 0.5 percent of the gene disruptions—99.5 percent of the disruptions will not affect this pathway. Rather than individually screening through thousands of mutants that don't affect the pathway, we can pool the mutants into groups of, say, 100 mutants. Now the process of screening the genome involves screening sixty pools of mutants rather than 6,000 individual mutant strains. Even with 100 mutants in a pool, most of the pools will not have a mutation that affects our pathway, so all those pools with those mutants can be set aside. The number of pools that have a mutant of interest will be no greater than twenty, since there are twenty genes that affect our pathway (a number we don't know at the outset); it could be smaller than twenty if two or more mutants that affect the process end up, by chance in the same pool.

Now those pools that have a mutant of interest can be themselves subdivided and pooled, and each pool tested. By a regular cycle of testing and subdividing pools of mutants, the individual mutants can be identified with less effort than would be needed to test each mutant separately. But, even with this method of pooling mutants, several rounds of growth and plating are needed, and the process could take several weeks.

The bar codes provide an even more effective method to identify individual genes within a pool that includes a mutant of interest. The original pool of mutants can be thought of as the *input* group. This entire group is then subjected to the phenotypic assay—growth under particular nutritional conditions, for example—and the pool of strains that *passed* the test assay are collected. These strains define the output group, genes that *do not* affect the process of interest. For example, these are the mutants that were able to grow normally despite the lack of a nutrient; whatever gene is affected in each mutant, it is *not* needed for the process. By comparing the input group of mutants with the output group (in effect, by subtracting the output results from the input group), the investigator can determine which genes are needed for the process. The molecular bar codes provide the information required for a comparison to be made. Each gene is tagged with a bar code during the mutation process, so the investigator compares the bar codes rather than trying to isolate individual mutants. The bar codes that are missing among the output group identify the genes of interest. Because the mutants of interest are the ones that are missing, this is also known as a *dropout screen*.

Molecular bar codes or DNA signature tags were first used for high throughput analysis with pathogenic bacteria. Pathogenesis is often a difficult process to study in the lab because so many hosts must be tested. An early use of bar codes involved *Salmonella enterica* virulence in mice. Bar codes of about forty nucleotides were incorporated into transposable elements, which were then inserted randomly throughout the *Salmonella* genome. The mutant bacteria were pooled and tested in mice; bacteria were collected from mice that became ill. These *Salmonella* mutants had retained the ability to become virulent. The virulent output pools were compared to the input pools using radioactive hybridization or a restriction digest in the universal sequence flanking each bar code to identify the genes that were *not needed* for virulence.

Detecting bar codes

A variety of methods have been used to detect the bar codes in the output pool. The key element is the sensitivity to detect the absence of individual bar codes in a large pool. Some investigators have used PCR amplification or restriction digests based on the universal sequence flanking each bar code. As techniques have developed, the bar codes have usually been detected by hybridization to a microarray, as was done in yeast. All the bar codes from the input group are placed in a defined array pattern; bar codes in the output group are amplified by PCR and hybridization to the array. Those bar codes that do not show a hybridization signal identify the genes that are needed for the process. Rather than hybridization, the signals could also be detected by direct sequencing, but this was not feasible for the original yeast collection.

Although they were first used in gene disruption screens, molecular bar codes have recently been applied in genome-wide screens in mammalian using RNAi and siRNA. The strategy is essentially similar to what is done with the gene disruption screens: first, one creates a pool of "mutagenic agents" tagged with a DNA signature bar code (in this case, a set of siRNA clones); second, one uses these clones as an input group to produce mutant phenotypes; and, third, one uses microarrays or sequencing to compare the output group with the input group in order to detect genes with no effect and to infer which mutants are needed for the process.

Literature

Mazurkiewicz, P. et al. 2006. Signature-tagged mutagenesis: Barcoding mutants for genome-wide screens. *Nature Review Genetics* 7: 929–39.

limitation when applied to a genome-wide screen in which 6,100 different genes are disrupted simultaneously. How will the investigator know which gene has been disrupted in each yeast strain?

It is necessary to have a gene-specific sequence, that is, a bar code, for each gene in the genome. The investigators designed the bar codes knowing which unique sequence has been assigned to and associated with each gene-specific homology region. When the gene homology regions target recombination to a specific gene, the recombination inserts not only the selectable *kan*^{R} gene but also the bar code that allows the specific gene to be tagged and identified in a large scale assay. The bar codes were particularly important before it was feasible to sequence many yeast colonies rapidly.

Gene disruptions are targeted by recombination

The disruption procedure is similar to other yeast transformation experiments discussed in Chapter 7, and is shown in Figure 9.5. A library of insertion cassettes is created by PCR, then transformed into yeast cells under conditions in which most yeast cells will receive at most a single cassette. It is worth thinking for a bit about transformation efficiency, since this is often a step that can be overlooked. In the typical molecular genetics experiment in which the goal is to get a plasmid into the bacterial cell, a high concentration of the plasmid is used relatively to the number of cells; this high concentration helps to ensure that most cells are transformed with the plasmid. However, if the concentration of the plasmid is too high and if the cell can tolerate more than one plasmid, some cells will have two plasmids. It becomes difficult to know which plasmid or mutant is the one producing the effect when more than one are present in the same cell type. Thus, in most genome-wide experiments of this type, the investigators will use lower transformation efficiency, typically a concentration such that only about half the cells acquire the plasmid, or even fewer. Since the plasmid (or the virus in transfection experiments to be described for RNAi and CRISPR screens) includes a selectable marker such as drug resistance, transformed cells can be found among the population of cells in which most have not been transformed. The lack of transformants, that is, of cells with no plasmid, is much less of an experimental concern than the presence of more than one plasmid.

In the cell, homologous recombination between the insertion cassette and the target gene inserts the *kan*^{R} gene and the molecular bar code. As a result, each gene is individually disrupted and tagged in a different yeast cell. The cells are grown in the presence of kanamycin to select for transfomants that inserted the *kan*^{R} gene. The collection of transformed cells can then be tested under a wide range of growth conditions and subjected to other phenotype assays. For example, the cells can be grown in the absence of uracil, to identify the genes needed for uracil biosynthesis. Such an example is shown in Figure 9.6. The mutant yeast cells are pooled and grown in absence of uracil. The cells that do not need additional uracil—that is, uracil prototrophs—will grow, whereas those that require uracil—uracil auxotrophs—will not grow.

The investigator now uses the molecular bar codes to determine which gene has been disrupted in each of the cultures, as shown in Figure 9.6. The bar code is amplified for each gene using the gene non-specific primer and the results are displayed as DNA microarrays. The amplified bar codes from each of the two cultures are hybridized to these defined arrays and the results are noted, as shown in Figure 9.6 for our five disruption mutants. Each spot corresponds to the bar code for a gene disruption, so the absence of hybridization indicates a failure of that cell to grow, that is, it indicates which mutants have "dropped out" during culturing. The disrupted gene is inferred to be needed for growth in the absence of uracil. The microarray display of the bar codes allows the investigator to view the complete collection of 6,000 gene disruption strains at one time without having to sequence any of the mutants to find out which gene has been disrupted.

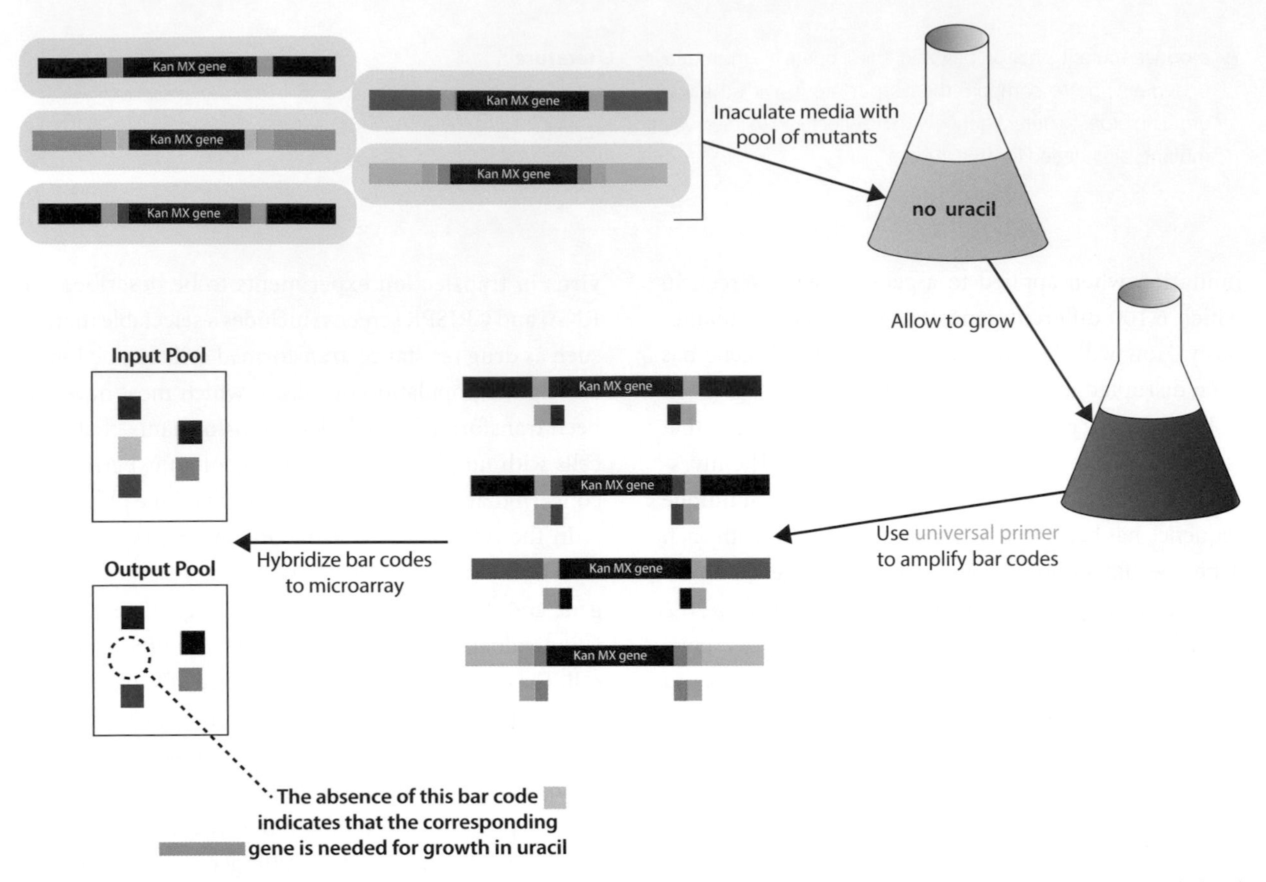

Figure 9.6 The use of the molecular bar code. Once disruptions are made for each gene, as shown in Figure 9.5, the mutants are pooled and tested for a mutant phenotype. In our drawing, each yeast cell at the top left has a single gene disruption, and the pool of mutants is grown in media that lack uracil. Once the culture has grown, the cells are removed and assayed by PCR. The universal primer will amplify each bar code in the culture, as shown. The collection of bar codes is hybridized to a microarray that has all of the input bar codes in a defined pattern, here the same one as shown in the pool of mutants. The bar code for any gene that is needed for growth in the absence of uracil will not hybridize in the output array. This is often referred to as a dropout screen, since the desired mutants are the ones that failed to grow and are not present in the final population.

Our simple example with uracil biosynthesis greatly understates the power of this approach. A better demonstration is provided by some numbers. In the ten years after it was made, the yeast deletion collection was cited in the scientific literature more than 1,000 times, with more than 100 different assay conditions. Some of these are described below, in discussion of phenotypic assays. More than 5,000 of the roughly 6,000 genes have been found to have a mutant phenotype by at least one assay, and many additional genes have been found even for processes such as sporulation and DNA repair, which have been thoroughly studied by other procedures. All these numbers have continued to increase. Although the process seems completely removed from an individual investigator carefully streaking yeast cells on different culture plates, the intellectual basis and the fundamental question remains the same: *find a mutant*.

9.4 RNAi and large-scale mutant analysis

One of the most powerful methods of mutant analysis for genome-wide screens involves the use of double-stranded RNA (dsRNA) to interfere with gene expression, in the technique called RNAi. An overview of the process of RNAi is depicted in Figure 9.7. In a typical RNAi experiment, double-stranded RNA corresponding

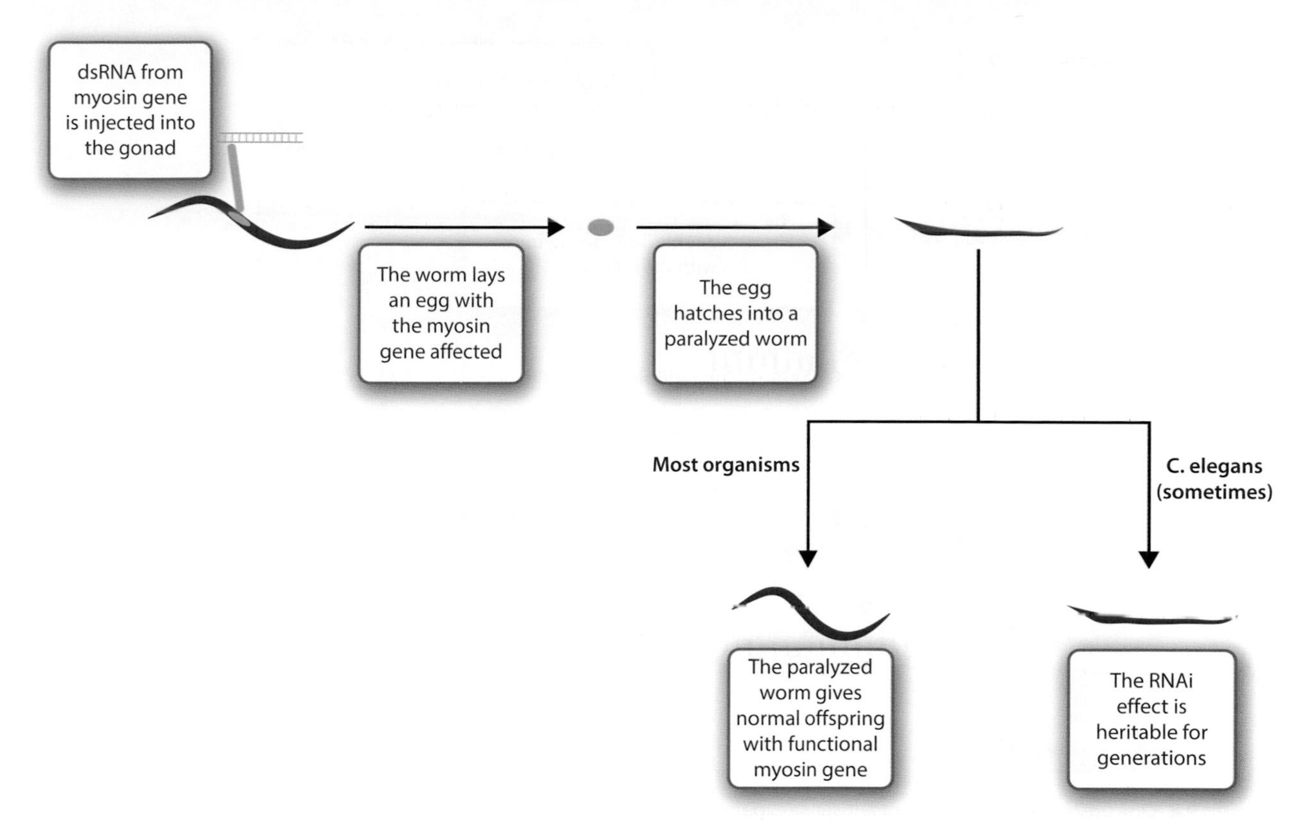

Figure 9.7 An overview of RNAi. The example here uses a muscle myosin gene from *C. elegans.* dsRNA is made in vitro, corresponding to a portion of the coding region of the myosin gene. This dsRNA is injected directly into the worm gonad. The injected worm lays eggs, some of which have gotten the injected dsRNA. The affected egg hatches and the resulting worm is paralysed, as if it had a mutation in the myosin gene. The gene itself is not structurally altered in the paralysed worm, so in most cells and organisms its offspring will not have the dsRNA and will not show an RNAi phenotype. *C. elegans* is unusual among animals, since the RNAi phenotype can be heritable and some offspring will continue to be affected.

to a portion of the transcript from a gene is introduced into the cell or organism. The dsRNA specifically blocks or reduces expression of that gene, thereby producing a mutant phenotype.

Let's consider an example from one of the early RNAi experiments in *C. elegans*. When dsRNA that corresponds in sequence to some of the exons encoding a myosin heavy chain gene is introduced into *C. elegans*, the offspring of the injected worm are paralysed with a phenotype that resembles a structural change in the myosin heavy chain gene. In most organisms, the effect is not heritable. (*C. elegans* is unusual in that the RNAi effect can be heritable and propagated for some generations.) That is, the paralysed worms have an intact and unmutated gene for the myosin heavy chain and will thus produce offspring with normal movement. In classical genetic terms, a non-heritable mutant phenotype induced by environmental agents is referred to as a **phenocopy**, and the phenotypes arising from RNAi are probably most accurately called phenocopies rather than mutants. However, the use of RNAi has become quickly widespread, its practitioners calling these "mutants," so it seems pedantic to insist on calling them phenocopies.

The antecedents of RNAi lie in other experiments

Although RNAi as an experimental method was first attempted on animals about twenty-five years ago, it is rooted in an older process, known as **antisense technology**. Antisense techniques involved producing a single-stranded RNA or another modified nucleic acid that is complementary to the mRNA or sense strand. The single-stranded RNA was predicted to form a double-stranded hybrid with the mRNA and thereby block its translation. This hypothetical mode of action for antisense experiments is shown in Figure 9.8.

Many antisense experiments were successful, but many more were unsuccessful and not reported (except

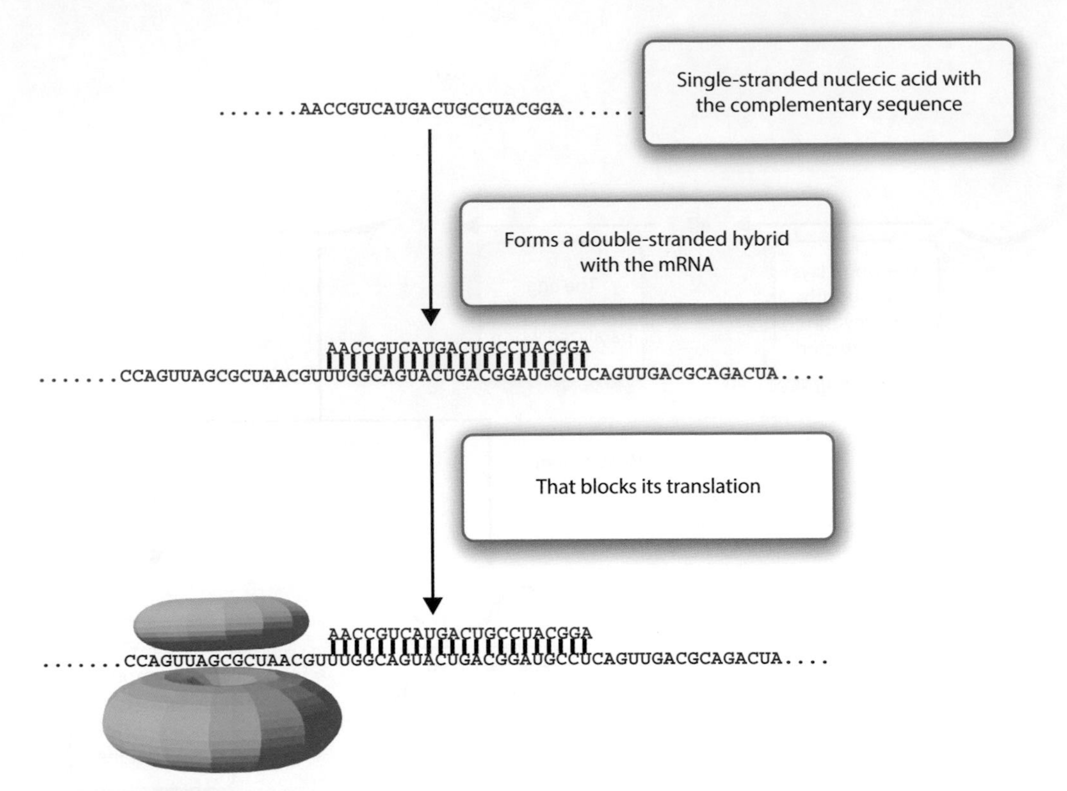

Figure 9.8 One postulated mechanism of antisense experiments. A single-stranded nucleic acid, either RNA or DNA or a synthetic molecule, is made that is complementary in sequence to the target mRNA. The single-stranded molecule is shown in red; the target mRNA is shown in blue. This single-stranded molecule is introduced into the cell, where it is postulated to form a double-stranded hybrid with the mRNA. This double-stranded hybrid is then postulated to be blocked in translation, as shown at the bottom of the figure. Although antisense experiments of this type do block gene expression, the actual mechanism is rarely explored. It seems likely that many of these antisense experiments work by triggering RNA degradation via the RNAi pathway.

by word of mouth), mostly without investigation of the actual method by which the RNA molecule was exerting its effect. An important control for understanding the mechanism of antisense effects was the introduction of a single-stranded *sense* RNA. If the *antisense* strand were blocking translation by forming a double-stranded hybrid with the mRNA, the sense strand should not produce an effect, being unable to form this hybrid. Although this control worked in some antisense experiments, it failed in others. Careful experiments in *C. elegans* with purified single-stranded and double-stranded RNA revealed that the dsRNA hybrid formed in vitro had the greatest effect. This suggests that silencing in some of the antisense experiments arose from a small amount of dsRNA that contaminated the sense or antisense RNA preparations. RNAi as an experimental technique arose from these carefully done controls on antisense experiments.

Similar effects had first been observed in plants, particularly for flower color in petunias. Figure 9.9 is a diagram of the key experiment, done some years earlier than the work in *C. elegans*. The petunia variety had light purple flowers, the result of a hypomorphic mutation in the pigmentation gene. Recall from Section 4.6 in our discussion of classifying mutations that additional copies of a hypomorphic allele are expected to produce a phenotype more similar to the wild type, in this case more deeply purple flowers. When multiple copies of the mutant pigmentation gene were introduced as a transgene, the expression of the normal gene was unexpectedly blocked rather than increased, and white flowers were produced instead. Further investigation of this effect revealed that it was mediated by RNA, and was also seen with RNA viruses that replicate through a dsRNA intermediate. Apparently the plant has mechanisms for synthesizing a double-stranded RNA from a sense

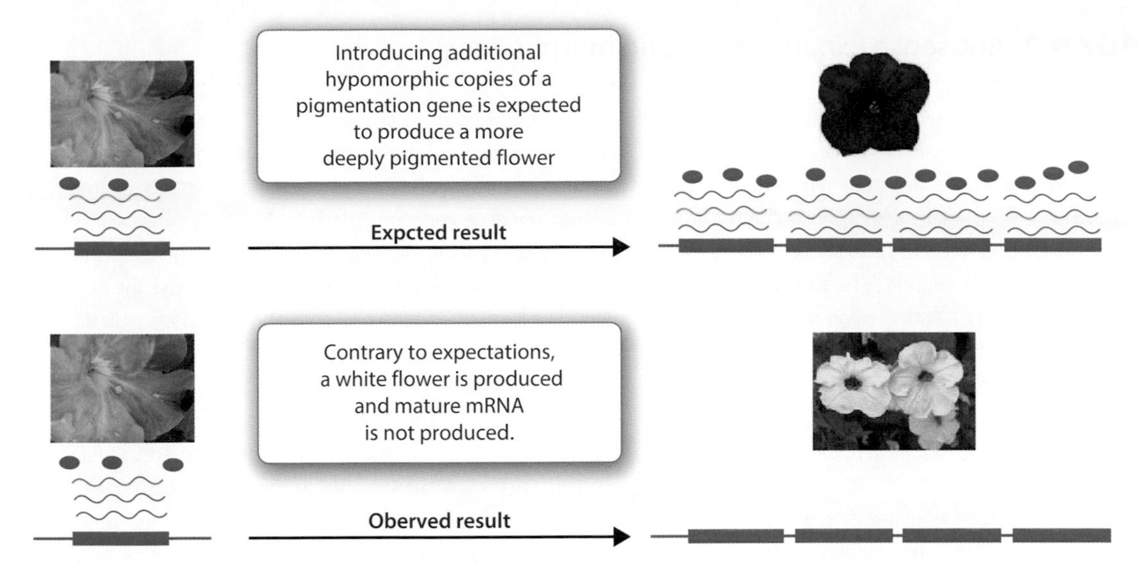

Figure 9.9 Post-transcriptional gene silencing in petunias. A petunia with light purple flowers is shown at the left. The flower is lightly pigmented because it has a hypomorphic mutation in the blue pigmentation gene shown beneath the flower; the gene is transcribed and protein is produced, but in reduced amounts by comparison to darkly pigmented petunias. Additional copies of the pigmentation gene were introduced into the plant. The top half of the figure summarizes the expected result of introducing additional copies of the gene, while the bottom half of the figure shows the actual result. Increasing the copy number of a hypomorphic mutation is expected to result in more pigmentation, a more nearly wild-type phenotype, as described for an allelic series in Chapter 4. Contrary to expectations, gene expression was decreased rather than increased and a white flower resulted. Further investigation showed that the effect occurred post-transcriptionally, through RNA degradation. Post-transcriptional gene silencing in petunias was one of the first demonstrations of the RNAi pathway.

transcript. In plants, the effect is called **post-transcriptional gene silencing** (PTGS), but the general mechanism for gene silencing by dsRNA appears to be the same in plants and animals. Thus, although the experimental history of RNAi is recent, the evolutionary history of the response to foreign dsRNA pre-dates the divergence of plants and animals. The RNAi response appears to be an ancient mechanism to protect eukaryotic organisms from RNA viruses and the effects of transposable elements, both of which involve dsRNA molecules.

Although in this chapter we will focus on RNAi and its role in genome-wide mutant screens, antisense experiments are still the preferred method for reducing the expression and function of the gene for some organisms. For example, in vertebrates such as the clawed frog Xenopus and the zebrafish Danio, antisense experiments are widely used as a means to knock down individual genes. The antisense molecules are not, however, the naturally occurring RNAs, but a synthetic type of molecule referred to as a **morpholino.** Tool Box 9.2 describes morpholinos and their use in these organisms.

As noted in Chapters 1 and 2, small non-coding RNAs are important naturally occurring regulators of gene expression. Such **microRNA**s have been found in both unicellular and multicellular eukaryotes, and the majority of genes in multicellular organisms are regulated by microRNAs. It was realized relatively quickly that the mechanisms by which RNAi occurs experimentally are largely the same as the normal cellular response involving microRNAs. Curiously, RNAi is able to exert an organism-wide effect on both plants and worms, which suggests that the dsRNA or the gene silencing effect is somehow passed between cells in these organisms. In other organisms, most of the RNAi screens have been done on cells in culture. Whether performed on organisms or cells, RNAi has had a profound impact nonetheless.

RNAi has proved to work in all multicellular plants and animals that have been thoroughly tested, although not equally well for all genes or in all tissues. The case study in Chapter 3 used an RNAi screen to identify genes involved in regeneration in planaria, an organism for which no mutations are known. Because the technique has such promise for altering gene expression, which could provide useful therapeutic agents, many experiments have been directed at understanding the mechanism of RNAi

TOOL BOX 9.2 Antisense experiments using morpholinos

The strategy of using antisense nucleotides to block gene expression has taken many different forms. One of the best uses is the application of a synthetic structure similar to an oligonucleotide known as a Morpholino™. (Morpholino™ is the trade name given by Gene Tools and AVI BioPharma to these molecules, which are not found in nature.) The structure of a Morpholino is shown in Figure B9.1, with the comparison to an RNA sequence on the other strand. Note that, on the Morpholino, the sugar ring of the nucleotide (deoxribose or ribose) is replaced by a chemically similar morpholine ring. The morpholine rings are connected by very stable phosphorodiamidate linkages rather than by the familiar phosphodiester backbone of nucleic acids. The nucleoside bases adenine, guanine, cytosine, and thymine are the same as those in DNA, which allows the possibility of forming a complementary base-paired structure between the Morpholino and a nucleic acid.

Morpholinos are typically synthesized to have about twenty-five bases, enough to provide a transcript-specific interaction with the complementary sequence in mRNA. Unlike RNAi and microRNA interactions, the interaction between mRNA and its complementary Morpholino to form a double-stranded structure does not trigger degradation of the mRNA, but instead appears to block translation of the message. They are not charged, and thus do not have ionic interactions with proteins. Because the rings are not nucleotides and the backbone is not a phosphodiester, the Morpholino is not targeted for degradation by endonucleases and does not trigger an innate immunity response. Thus Morpholinos work by the antisense mechanism that many assumed lay behind the initial antisense nucleic acid experiments. Careful controls have shown that Morpholinos block protein expression extremely well for a wide range of different transcripts, usually with few effects other than those on the target sequence. However, it has been recommended that an independent assay be used to monitor their effectiveness, for instance using an antibody to the protein product of the gene of interest.

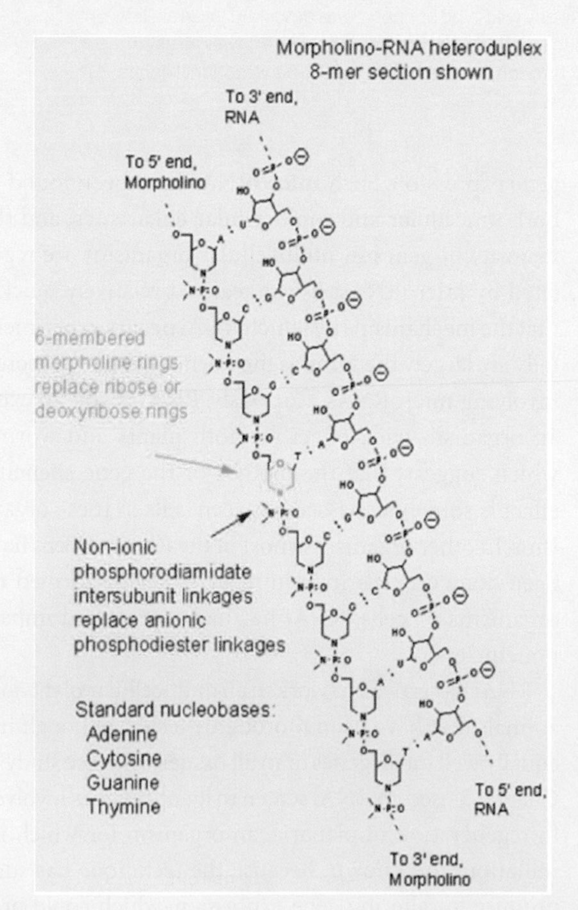

Figure B9.1 A comparison of the Morpholino structure with RNA. Note that Morpholinos have the same bases as DNA but a morpholine ring rather than a deoxyribose sugar, and a different, non-ionic backbone. Like nucleic acids, Morpholinos have a 5′ and 3′ direction.

Morpholinos have been very widely used in vertebrates with large, easily injected eggs and embryos, such as frogs and fish, and in some invertebrates such as sea urchins. They have also been used in cultured cells with electroporation as the means for delivery. They have not been widely adopted in other organisms, in part because techniques involving RNAi have been so well developed in many other organisms. A Morpholino sequence complementary to the 5′ end of the target mRNA is synthesized commercially. This is then injected into the egg or embryo, which blocks the binding and progression of the ribosomal initiation complex, and thus prevents translation. Other applications have used Morpholinos to block RNA splicing of a specific exon by binding at the splice donor or acceptor sites, which resulted in exclusion of an exon or in alternative and modified splicing. The effect of the Morpholino lasts for as many as five to seven days, long enough for an injection into a fertilized frog or fish egg to block expression until after most of organogenesis has occurred.

As with any technique that relies on base complementarity, off-target effects are a concern. One common control is to co-inject the Morpholino with a modified mRNA for the target gene. The modified mRNA codes for the same amino acid sequence as the original target gene, but with a different nucleotide sequence. Thus the co-injected mRNA is not blocked by the Morpholino, and the wild-type function is restored. Such a control demonstrates that the Morpholino effect is specific to its target and not some other effect.

Literature

Eisen, J. S. and J. C. Smith. 2008. Controlling morpholino experiments: Don't stop making antisense. *Development* 135: 1735–43.

Stainier, D. R. Y. et al. 2017. Guidelines for morpholino use in zebrafish. *PLoS Genetics* 13: e1007000.

and improving its effects. Many features that are common to RNAi effects for most genes and most organisms have been identified, but differences in the details have also been found. Procedures that work well for many genes in one organism may not work for all genes in that organism, and may not work very well at all in another organism. In addition, although we write of the mechanism by which RNAi works, it is likely that there are different mechanisms with some steps in common.

The effectiveness of RNAi for any particular gene is also likely to depend on the normal transcription pattern for that gene, which is often not known or not easily correlated with a specific effect. This is particularly true in the case of describing genome-wide screens in which the goal is to test hundreds, thousands, or even tens of thousands of genes as rapidly as possible. The procedures are sometimes presented as if one common protocol fitted all genes, but this is clearly not true. With those cautions, we will describe some of the protocols that are used and the mechanism by which RNAi is thought to work. We will describe the procedure and mechanism generally, then note some of the specific differences that are appropriate in dealing with different organisms.

Literature Links. Ahringer 2006, WormBook, doi/10.1895/wormbook.1.47.1, http://www.wormbook.org; Heiger et al. 2018, *Genetics* 208: 853–74

The method to introduce dsRNA depends on the cells and organisms

The first requirement for conducting an RNAi screen is to be able to introduce the dsRNA into the cell or the organism. Three widely used methods are shown in Figures 9.10, 9.11, and 9.12. The dsRNA can be transfected for cells grown in culture, as for mammalian cell lines or *Drosophila* cells (Figure 9.10).

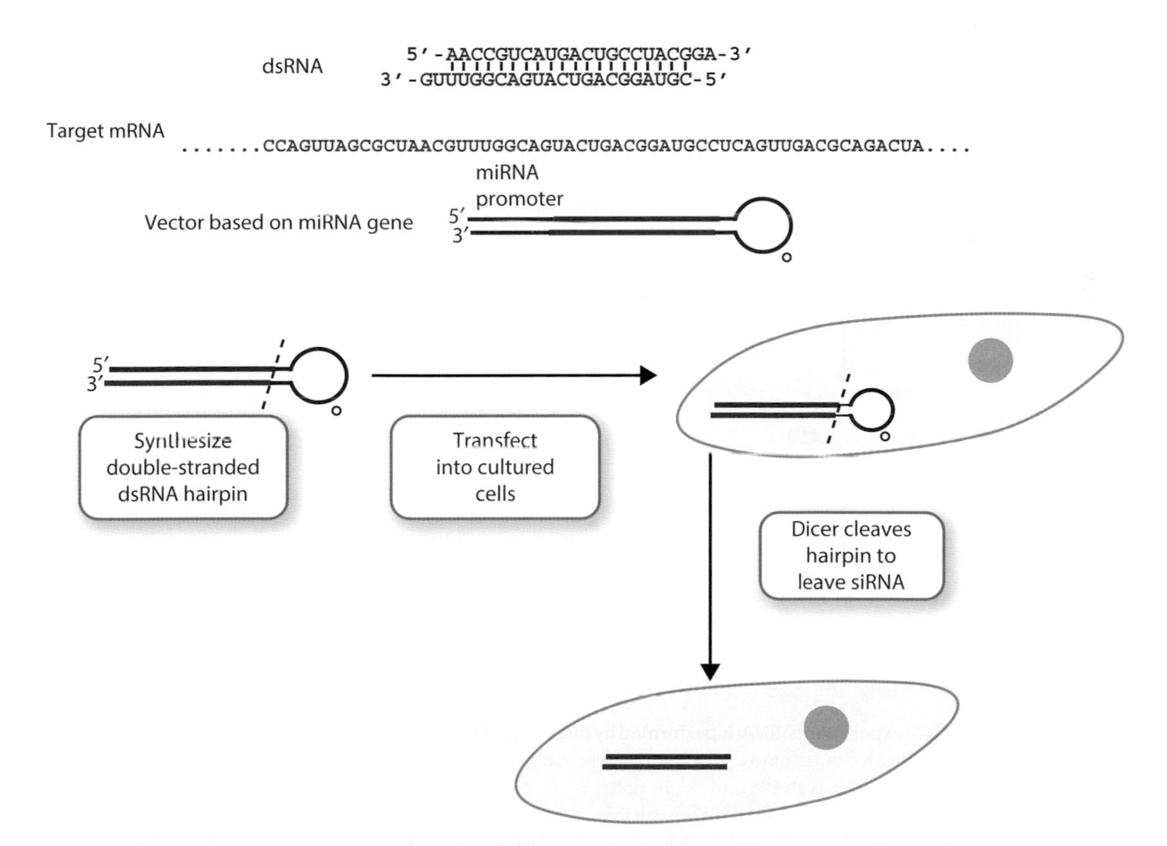

Figure 9.10 Introducing dsRNA for RNAi experiments: dsRNA is transfected into mammalian cells as a hairpin. The target mRNA is shown in blue at the top of the figure; the dsRNA to create RNAi is shown with the complementary strands in red and blue. This is cloned, as DNA, into a vector whose backbone structure is based on a microRNA gene. A RNA hairpin is synthesized in vitro, with a stem of twenty-twy nucleotides, including a two base overhang at the 3′ end. The loop is usually about ten nucleotides long. The dashed line indicates the site of future cleavage by Dicer. The hairpin RNA is transfected into mammalian cells in culture, where it is cleaved by Dicer to make an siRNA, which triggers RNAi. The RNAi effect in the cultured cells is transient.

The vector used for transfection is often a modified microRNA gene. The dsRNA is cleaved and induces a transient effect in the cultured cells. Alternatively, for worms and also for flies, the dsRNA can be injected directly into the germline, either that of the adult mother (in worms) or that of the developing embryo (in flies), as summarized in Figure 9.11. The effect of the dsRNA is then seen in their offspring. Injection is a laborious process, and only a few experimenters who work with worms have used this method for a genome-wide screen. More commonly, injection is used to test individual genes that have been identified by other means.

The most common method in *C. elegans* is to introduce dsRNA by feeding the worms on an appropriate bacterial strain, as shown in Figure 9.12. As noted in Chapter 3, worms in the laboratory eat *E. coli*, so the source of the dsRNA is the worms' food. Especially designed RNAi plasmids are available in which a cDNA insert is flanked by T7 promoters, as shown at the top of Figure 9.13. These plasmids are transformed into an *E. coli* strain in which T7 RNA polymerase has been inserted into the chromosome under the control of a *lac* promoter region, so expression of the polymerase is induced by the addition of IPTG. Thus, upon IPTG induction of the bacteria—that is, the worms' food—T7 RNA polymerase

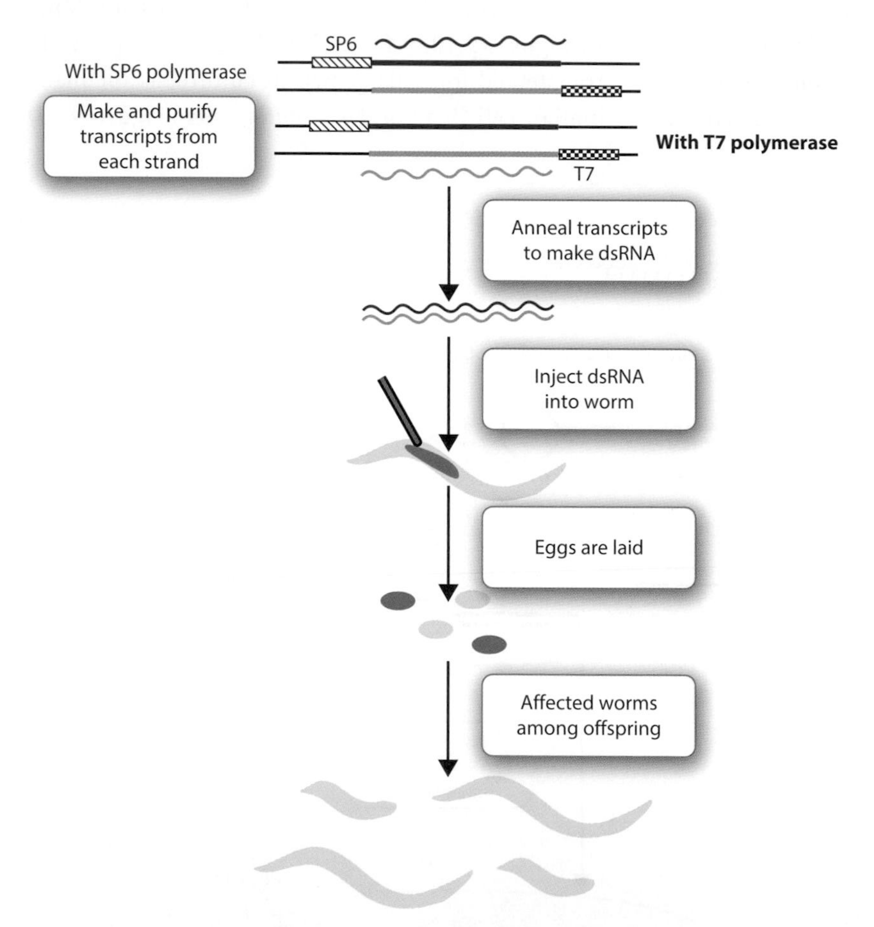

Figure 9.11 Introducing dsRNA for RNAi experiments: RNAi is performed by direct injection of the dsRNA. A cDNA from the gene of interest is cloned into a vector with different transcriptional promoters on each side of the cloning site. A vector with SP6 and T7 promoters is commercially available, but other promoters can be used. Two reactions are done in vitro in separate vials, one with SP6 polymerase added and the other with T7 polymerase added. Each reaction produces one single stranded RNA molecule, which are annealed in vitro to make dsRNA. As in Figure 9.7, the dsRNA is injected into the animal, which lays eggs with the dsRNA. The eggs hatch, and some of the offspring show the RNAi phenotype, here a dumpy worm. Similar procedures are performed by injecting directly into eggs of *Drosophila*.

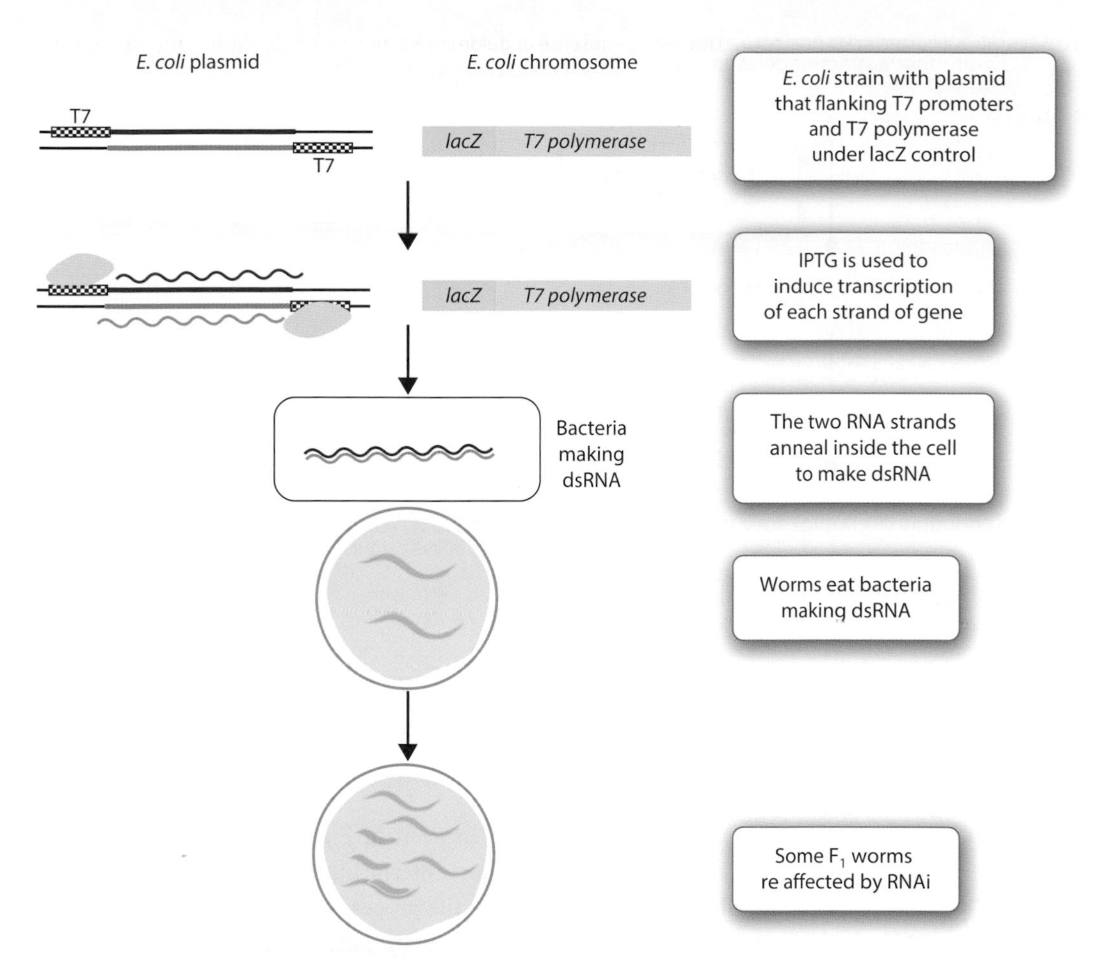

Figure 9.12 Introducing dsRNA for RNAi experiments: RNAi is performed by feeding. It will be recalled from Chapter 3 that the worm *C. elegans* eats *E. coli*, and molecules introduced into the gut have a systemic effect throughout the worm. The cDNA for the target gene is first cloned into a plasmid that has T7 promoters on each side. This plasmid is transformed into an *E. coli* strain that has a T7 polymerase gene under the control of the *lacZ* promoter integrated into the chromosome. The relevant genes on the plasmid (on the left) and the chromosome (on the right) are shown. Both plasmid and chromosome have other selectable markers, which have been omitted for simplicity. The bacteria with the plasmid are induced by IPTG, which results in the production of T7 polymerase. T7 polymerase works on the plasmid promoters and transcribes both strands of the cDNA for the target gene. These two complementary RNA strands can anneal with each other inside the bacterial cell, to make a dsRNA molecule. The bacteria making the dsRNA are spread onto plates, and worms are placed on the spread plates. As the worms eat the bacteria, they ingest the dsRNA, which then induces RNAi in the target gene. The worms reproduce, and some of their offspring exhibit the RNAi phenotype, here shown as a dumpy worm. A library of *E. coli* strains with each worm gene cloned into the plasmid vector is commercially available, and genome-wide screens in *C. elegans* are usually done using this feeding method.

is expressed, which in turn results in the transcription of the plasmid insert. Since the T7 promoters flank the insert, each strand is transcribed and a dsRNA hybrid is made by the bacteria. The worms eat the bacteria, and the dsRNA spreads from the gut throughout the organism.

Although the feeding procedure may seem strange, it works well for *C. elegans* (although not for related nematodes) and is extremely easy to do. Libraries with cDNA inserts from nearly all the predicted worm genes are available, which makes a genome-wide screen no more technically demanding than growing bacteria and feeding them to worms. Many genome-wide screens have been done in *C. elegans* using RNAi by feeding. In direct comparisons involving the same genes in the same genetic strains, RNAi induced by injection is considered more reliable than is feeding, but feeding works very well as an initial screen to find genes of potential interest.

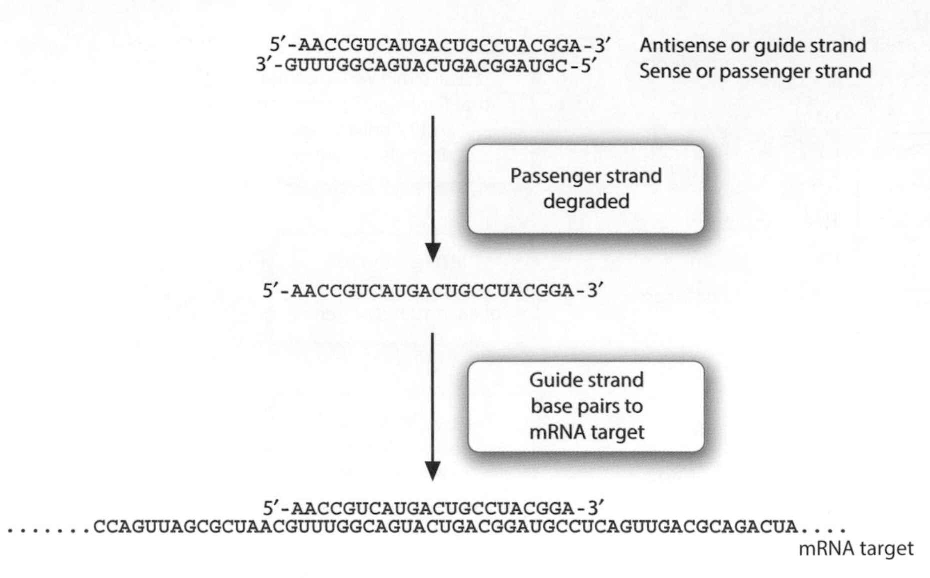

Figure 9.13 The structure of the siRNA. The active molecule for RNAi is double-stranded siRNA with a two-base overhang at each end, as shown here. The passenger strand is degraded, and the guide strand, which is complementary to the target mRNA remains to make the dsRNA. It seems likely that which strand in an individual siRNA is degraded is random, but that only the ones capable of forming a double-stranded hybrid are recognized and protected. In a subsequent step in the RISC, one strand is degraded to leave the guide strand, which targets RISC to the mRNA by base pairing.

The mechanism of RNAi relies on normal cellular functions

How, then, does the dsRNA produce its silencing effect? It may be helpful to compare RNAi with what investigators believed was happening in antisense experiments. The experimentally generated antisense strand that base-pairs with its target mRNA is the critical step, and gene silencing by RNAi probably includes both a block to translation and mRNA degradation. This overall mechanism of RNAi is shown in Figure 9.13. In *C. elegans* or in flies, dsRNA that is 300 or more nucleotides long and corresponding to the gene is introduced; with such a length, many different parts of the transcript are represented. One strand is identical in sequence to the mRNA (the sense or passenger strand), whereas the other is the complementary sequence (the antisense or guide strand). The long dsRNA is then cleaved into fragments of twenty-two nucleotides by the enzyme Dicer. Orthologues of Dicer have been found in all multicellular organisms for which genomic information is available. This is consistent with an evolutionarily conserved mechanism.

The RNA fragments produced by Dicer have a characteristic structure—a central dsRNA region of about nineteen nucleotides with a two- or three-base overhang at each end, as seen in Figure 9.13. These fragments are called short interfering RNAs (siRNAs), and their appearance in the process is common to all organisms. The sense strand of the siRNA is then degraded, leaving the antisense strand. In flies and mammals, the antisense strand becomes incorporated into a protein complex referred to as RISC (RNA-induced silencing complex), where the siRNA targets this complex to the specific, complementary mRNA. RISC degrades the corresponding mRNA without degrading the antisense strand of the siRNA, so the same complex can target and degrade many copies of an mRNA. Because the same RISC is used repeatedly, the mRNA from a gene can be completely degraded and no protein is produced. Not all the mRNA population is degraded, so RNAi is best considered to be knocking down gene expression rather than completely knocking it out.

For worms and plants, an additional or slightly different step has been described, which helps to explain the inheritance of RNAi in these organisms. Once the sense strand of the siRNA is degraded, the antisense strand forms a double-stranded hybrid, with the mRNA as shown in Figure 9.13. Such a double-stranded hybrid, with its overhangs, can serve as a primer for an RNA-dependent RNA polymerase, which makes multiple copies of long dsRNAs. These are cleaved to make more siRNAs, which target the RISC to the complementary mRNA. The amplification step apparently does not occur in flies or mammals, since no RNA-dependent RNA polymerase has been found. The evolutionary advantage, if any, of RNA-dependent RNA polymerases is not obvious; but there is an experimental advantage, namely that many different parts of the mRNA become targeted for degradation and little protein is produced from the target gene.

The procedure for carrying out RNAi in mammalian cells is somewhat different because we have a better immune system. In particular, a long dsRNA of this type triggers an anti-viral interferon response, and interferons degrade dsRNAs non-specifically. Therefore a long dsRNA does not yield siRNA, which makes it inappropriate for RNAi. A key insight in the development of RNAi for knocking down gene expression in mammalian cells came from the recognition that RNAi require siRNAs with a particular structure, a molecule that can be synthesized in vitro or in vivo. Such a synthetic siRNA is too short to trigger the interferon response. The synthetic siRNAs are incorporated into the RISC, bypassing the need for Dicer, and gene expression can be blocked. One limitation of this procedure is that the short synthetic RNAs target a very specific region on the mRNA; different siRNAs corresponding to different parts of the same mRNA sometimes have different effects, some working very well and others not working at all. Therefore most investigators working with mammalian cells design multiple siRNAs for different regions of the same transcript.

Literature Link. Moffat and Sabatini 2006, *Nature Reviews Molecular Cell Biology* 7: 177–87

Despite the variability in mechanism, RNAi is proving to be a powerful method to knock down gene expression in mammalian cells, and different procedures for introducing siRNA have been tried. In addition to being transfected as a synthetic molecule, the siRNAs can also be made from plasmid or lentiviral vectors that are introduced in the cells. One such vector is diagrammed in Figure 9.14. Libraries representing tens of thousands of siRNA sequences have been constructed and are in use, from both academic and commercial laboratories. Libraries are also constructed with the insert as a longer

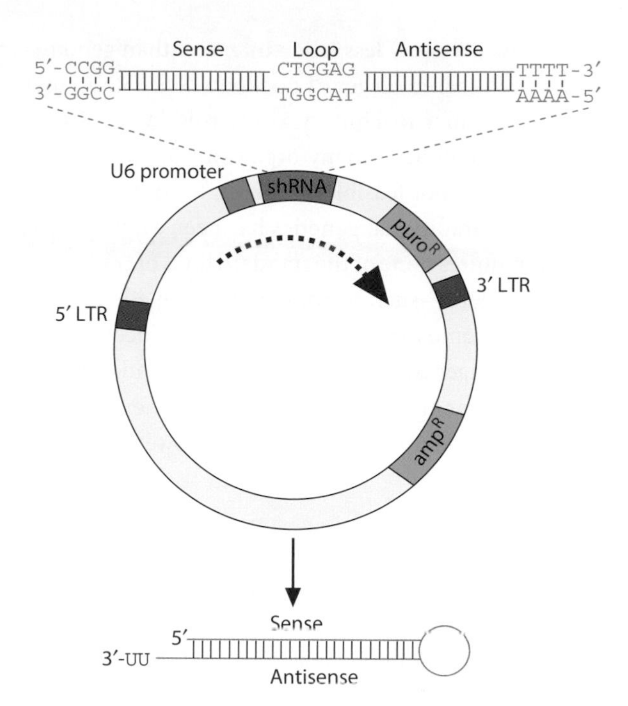

Figure 9.14 One plasmid used for shRNA libraries in mammalian cells. The sequence corresponding to the short interfering RNA (siRNA) is shown at the top; the siRNA sequence corresponds to the two arms flanking the loop. The loop sequence will be cleaved and is not important. The sequence corresponding to the short hairpin RNA (shRNA) is cloned into the expression vector plasmid in the middle of the drawing. In this particular library, the shRNA sequence is until the control of the promoter from the U6 gene, which encodes an RNA involved in splicing, so the shRNA will be constitutively transcribed. Other promoters can also be used. Drug markers are included on the plasmid for growth in bacteria (ampicillin) and in mammalian cells (puromycin). The dashed arrow indicates the direction of transcription, and the viral LTRs are involved in transcriptional regulation and termination. When the sequence corresponding to the shRNA is transcribed, the transcript forms the short hairpin shown at the bottom of the figure by intrastrand base pairing. Cleavage of the loop produces the siRNA involved in RNA interference.

sequence, which has an inverted repeat structure. When this insert is transcribed, the presence of the repeats results in the formation a hairpin or a stem-loop RNA, providing the necessary dsRNA for effective RNAi. Many of these libraries place the siRNA sequence under the control of an inducible promoter, allowing the investigator to trigger RNAi with an environmental signal.

Literature Link. Pei and Tuschl 2006, *Nature Methods* 3: 670–6

Genome-wide screens based on RNAi are increasingly common in the genetics literature. RNAi is an inexpensive and relatively simple method to reduce gene expression, certainly less labor-intensive than genome-wide screens based on gene knockouts. Since RNAi uses the same cellular machinery as microRNA regulation, it can also be used in many organisms for which gene knockouts are not feasible, and even for organisms for which little traditional genetics has been available. The process requires none of the standard tools of traditional genetic screens—such as mutations, a genetic map, or balancer chromosomes. Even if RNAi produces no new therapeutic agents, it has provided an easy and powerful means to silence gene expression in diverse experimental systems, opening them up to the power of genetic analysis.

RNAi has some known limitations

Despite its attractiveness for knocking down gene expression, RNAi has some limitations. First, the effects are variable for different genes and different organisms, for reasons that are largely unknown. In any method that attempts to perform a uniform assay on tens of thousands of diverse genes, some genes will prove to need individual attention. This variability is not too surprising to anyone who has done traditional genetic screens and realizes that even traditional mutants have variable expressivity, as described in Chapter 6.

Second, RNAi-induced phenotypes are probably best regarded as similar to genetic hypomorphs, as defined in Chapter 4, rather than as null mutations. In fact most investigators refer to "RNAi knockdowns," recognizing that the RNAi phenotype might not be the same as the null phenotype. With genome-wide screens involving thousands of genes, the construction of an allelic series for each gene is not feasible. However, as more is learned about the effectiveness of different siRNA sequences, it is not inconceivable that the equivalent of an allelic series will eventually be created by RNAi. In fact, with heat-inducible promoters for the dsRNA transcription, the equivalent of temperature-sensitive mutations has been made in *Drosophila*. This may be the first step toward producing hypomorphic mutations intentionally.

The problem of cross-reactivity

The third limitation is that non-specific or cross-reactive effects between a target gene and another gene have also been seen, as shown in Figure 9.15. Cross-reactivity or non-specific effects can arise because the siRNA is quite short—it has only twenty-two nucleotides—and mismatches are tolerated at some positions in the sequences. In Figure 9.15, two mismatches are shown between the siRNA and the cross-reactive sequence, but the molecules will base-pair with each other despite the mismatches. Thus, since absolute identity is not to be required for all positions in the siRNA, the region of complementarity is actually shorter than twenty-two nucleotides. Although these dsRNA molecules have a mismatch by comparison to their naturally occurring target sequence, gene expression is still blocked. The tolerance for mismatches increases the probability that the same or a closely related sequence will be found at other sites in the genome, either another member of the same gene family or a functionally unrelated gene. Thus, in complex genomes, transcripts with unrelated functions that happen to have short regions of sequence identity may be inadvertently targeted by the same siRNA. This cross-reactive effect is analogous to the off-target effects of CRISRPR—which also tolerates some mismatches between the target site

Target Gene: Exact Match to mRNA

3′ –AACCGUCAUGACUGCCUACGGA– 5′
. . . . 5′ –**CCAGUUAGCGCUAACGUUUGGCAGUACUGACGGAUGCCUCAGUUGACGCAGACUA**– 3′. . . .

Cross-reactive Gene: Near Match to mRNA

3′ –**AACC - GUCAUGA*CU*GCCUACGGA**– 5′
5′ –**UAUGCAUAAGCUAUUGGGCAGUACUGAAGGAUGCCUCAGUGACUUCGGA**– 3′

Figure 9.15 The cross-reactive effects of siRNA. The siRNA molecule is shown in red, with the target mRNA shown in blue. The siRNA has been synthesized to correspond in sequence to a portion of the target mRNA, and will form an exact match to that sequence. Cross-reactive effects are seen when other, unrelated mRNA molecules (such as this one shown in black) have a sequence that matches or nearly matches the siRNA sequence. In this hypothetical case, the cross-reactive sequence has one additional base and one mismatched base with the siRNA, as indicated by the red stars.

and the sgRNA. (RNAi appears to be more subject to mismatches than CRISPR, possibly because the core sequence is shorter for RNAi.) Certainly experiments in which the specific effect on the target gene has not been verified should be viewed with some caution.

In mammals, and perhaps in other organisms, non-specific effects also can arise from physiological sources, in addition to this unexpected sequence similarity. Some evidence indicates that even such short sequences as found in siRNA can sometime induce in mammals an interferon response that degrades RNA generally. In addition, it should be recalled that microRNA molecules similar to siRNAs are involved in a large number of physiological processes. Since RNAi takes advantage of these normal cellular processes, the machinery might be overwhelmed or shut down if too much dsRNA is used. As with many other aspects of RNAi experiments, these off-target effects will become easier to avoid as more is learned about the normal cellular response and better designs are incorporated.

9.5 CRISPR in genome-wide mutant screens

As discussed in Chapter 8, genome editing using CRISPR–Cas9 has quickly and dramatically altered the processes designed to make mutations in individual genes. It is probably not surprising, then, that genome-wide mutant screens based on CRISPR are being developed for many organisms and in use for mammalian cell types. Unlike RNAi, CRISPR-based screens generate true mutants, which are heritable. The lessons learned from RNAi screens have greatly informed the strategies and approaches for CRISPR-based screens.

Literature Link. Hart et al. 2017, *G3* 7: 2719–27; Doench 2018, *Nature Reviews Genetics* 19: 67–80

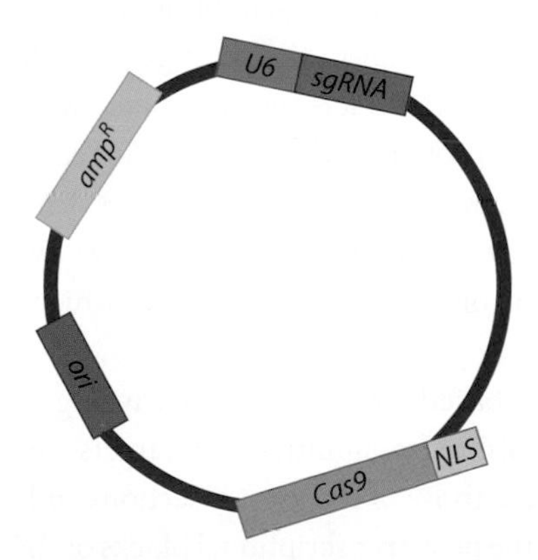

Figure 9.16 A lentiviral CRISPR vector. Many vectors for genome-wide CRISPR screens are based on a lentiviral backbone, as simplified here. The key elements are the origin of replication, drug resistance markers (typically more than one is used), and the sgRNA. In many cases, the sgRNA is inserted downstream of the promoter for the human U6 small RNA gene. This diagram also includes the Cas9 gene with a nuclear localization signal (NLS) in the vector, but separate Cas9 vectors are also widely used. While these are some of the important elements, most commercial vectors are more complicated than this.

A CRISPR-based screen requires that two components are introduced into cells; first, the RNA directed nuclease Cas9 (for genome-wide screens, often as the gene that encodes Cas9 rather than the enzyme itself); and, second, one or more sgRNA specific to each gene. For screens in mammalian cells, these are often transfected together on a single lentiviral vector that also encodes a drug selectable marker. An example is diagrammed in Figure 9.16. Such vectors are transfected into populations of cells at low efficiency, so that, as discussed in Section 9.3 in the case of yeast disruption plasmids, cells do not have more than a single sgRNA. The methods are still being developed, and some approaches introduce (and select for) the *Cas9* gene first, and then transfect the cells with sgRNAs. In any case, Cas9 will use the sgRNA to target the gene to be edited, and the cells with the edited gene are identified.

Many of the same questions that have been introduced with the other types of genome-wide mutant screens are also relevant here. Because genome editing using CRISPR-Cas9 is such a recent and powerful experimental tool, and because libraries with all the components are commercially available for use in individual laboratories, some of these considerations, such as the importance of using a low transfection efficiency, the effectiveness of pooled samples, or the possibility of off-target effects, are the same as for other methods. Other considerations are a bit different for CRISPR-based screens, so we will focus on them.

The choice of sgRNA and Cas9

With RNAi, any effective siRNA for the gene targeted the same mRNA to block the expression of the gene, so it was fairly uncommon to see different phenotypes arising from different siRNAs. Investigators used multiple siRNAs for the gene to reduce the impact of off-target effects (since different siRNAs would have different cross-reactivity with other possible targets) and because not all siRNAs work equally well. These same considerations hold for the libraries of sgRNAs, and screens typically use several (between 2 and 10) different sgRNAs per gene. However, a significant difference and potential advantage of using different sgRNAs is that different mutations can be made in the same gene. As discussed in Chapters 4 and 5, different mutations in the same gene often produce related but distinct phenotypes. Thus, by a judicious choice of sgRNAs, an investigator can create an allelic series for the gene. Of course, if different phenotypes are seen for different mutations in the same gene, the possibility of off-target effects should be considered, just as geneticists using traditional screens need to be certain that only a single mutation was present if different phenotypes were observed.

A standard set of sgRNAs for human or mouse genes are commercially available. Most of these are expected to produce indels in the gene by NHEJ, as discussed in Chapter 8. These will be null alleles of the gene, but an investigator can then design sgRNAs for other target sites and other mutations.

Additional flexibility is provided by the versatility of Cas9. As noted in Chapter 8, Cas9 is an RNA-directed nuclease, but the nuclease function is separable from its targeting function. As a first choice, the investigator can use the native form of Cas9, which generally produces null alleles or gene knockouts. But Cas9a and Cas9i can be used to activate or inhibit the expression of the gene by using an sgRNA directed at the transcriptional start site, as discussed in Chapter 8. Proteins with the Cas9 targeting domain coupled to other enzymatic domains can also be introduced. The combination of different sgRNAs to target specific locations in the gene and different Cas9 derivatives to produce different effects at those locations offers exceptional power. Having found a gene from its null allele phenotype in the genome-wide screen, the investigator can then explore many more aspects of its function and activity than in the case of other types of mutant screens.

The choice of cell types

As discussed in Chapter 7, mutations are typically produced in one or a few cells that may be grown into transgenic organisms. For CRISRP-based genome-wide mutant screens in humans, the phenotypes are usually scored in cells of a single type, and then inferences are made about the effects on the organism as a whole. Thus, the best choice of cell types becomes an important consideration. This is also true, but not discussed, for RNAi-based screens in mammals; if the cell type does not express the gene in question, no mutant phenotype will be seen, regardless of other factors. Often the cell type chosen reflects the phenotype that is expected. For example, RNAi or CRISPR screens for cancer genes are done in cells affected by the cancer.

One advantage of CRISPR is that the mutations generated by Cas9 are usually repaired by NHEJ, as discussed in Chapter 8. The process of NHEJ is largely, if not entirely, the same in different cell types. Thus, once mutations are found in a gene with a particular sgRNA in a genome-wide screen with one cell type, it may not be difficult to use the same sgRNA to test the same mutation in other cells types. Of course, the challenge of connecting the phenotype seen in individual cells in culture with the phenotype that might arise in the organism is still present.

The choice of genes and genomic elements

Recall that the first consideration in genome-wide mutant screens, as outlined in Figure 9.3, is the choice as to which genes to perturb. This choice is dependent on both the annotation of the genome and the investigator's own interests. Except in the yeast gene disruption collection and in RNAi screens, the choice was limited to the exons of protein-coding genes. The yeast insertions are large changes targeted to the exons. RNAi screens work at post-transcriptional steps, either by transcript degradation or by translational blocks, which also target exons.

CRISPR-based screens can similarly target exons; but they offer more possibilities. The targets can be much smaller than those of the yeast insertions and more specific than the post-transcriptional blocks of RNAi, which is an advantage. But, because the target for CRISPR is the DNA sequence itself, there are even greater advantages. CRISPR can be used to disrupt genes for non-coding RNAs that, since they are not translated, are not good targets for RNAi screens. CRISPR-generated mutations may be the best strategy for identifying the functions of such genes.

Furthermore, neither gene disruptions or RNAi screens can target regulatory regions, which CRISPR can do. One (hypothetical) CRISPR screen, for instance, might use an sgRNA that corresponds to the enhancer sequence or binding site for some transcription factor. Such a screen would examine the effects of a transcription factor not by altering the protein itself but by changing its binding sites. The sequence does not need to be transcribed to be a target, so other targets could include replication origins, chromatin conformation regions, centromeric repeats, and many more features.

9.6 Screening for mutant phenotypes

While most of the attention in genome-wide screens is given to the techniques involved in creating the mutants, screening the newly created mutants is equally important. As with a traditional genetic mutant hunt, the phenotypes that are recognized depend on the eyes and assays of the investigators. With genome-wide screens that test thousands or tens of thousands of genes, this becomes an important large-scale problem. Many possible phenotypes could be scored and, in many cases, the genome-wide screen is done with one or a few specific phenotypes in mind. For example, the genome-wide screen may be done by one set of investigators looking for genes that affect chromosome behavior in meiosis. Any mutant phenotype that does not affect meiosis may be noted but is not pursued. This is similar in logic to traditional gene-by-gene genetic screens, but boosted by the power of the genome-wide screen.

Some of the earliest genome-wide screens attempt to catalogue as many possible phenotypes as can be observed. Such screens are similar in principle (if not in detail) to the screens for the first mutants in any model organisms. In these cases, the assays often test for very broad phenotypes, such as growth and viability or fertility, and often use a form of dropout screens, as described. The initial genome-wide screen of the yeast deletions tested growth under eight different conditions, and subsequent tests were done for many other effects. The RNAi screens of the *C. elegans* genes recorded mutant phenotypes that can most readily be observed in worms—viability, fertility, general morphology, and movement. Subsequent experiments tested for many other effects, for example lifespan and behavior.

In many ways, characterizing the phenotype of mutations generated by genome-wide screens is similar to characterizing the phenotype of mutations generated by gene-by-gene screens; but there is one key difference. Since all the strains with gene disruptions or perturbations are maintained, they are available to be tested by another investigator with another idea for a screen. Thus the same strains can be tested repeatedly. This is only slightly different from what has been done with traditional genetic screens, but the difference is very important, as illustrated in Figure 9.17. In both traditional genetic screens and genome-wide screens, mutants are found and characterized; these strains are saved, so that they can be available for other investigators. This is the gene in blue shown at the top in Figure 9.16. The difference between the two methods lies in the strains that *do not* show a mutant phenotype, the genes shown in green or in purple in Figure 9.16. In traditional genetic screens, these strains are not recognized at all and are discarded. They may have phenotypes that are interesting or important to another investigator, but, because the original investigator was not looking for those specific phenotypes, the strains are not saved. Other investigators will need to produce and identify the mutants themselves. This is inevitable, since no investigator can look for every possible phenotype; and not every strain can be maintained if it is not known to have a mutation.

In a genome-wide screen, by contrasts, all the gene disruptions and perturbations are saved, even if the original investigator sees no mutant phenotype. Retention for the strain collection depends *not on a mutant phenotype*, but rather *on a mutant gene*. Thus, other investigators need only to screen the original collection of genes and not to produce their own collection of mutations each time. This is a very important advantage of genome-wide screens, and it is one of the reasons why many more phenotypes can be tested by genome-wide

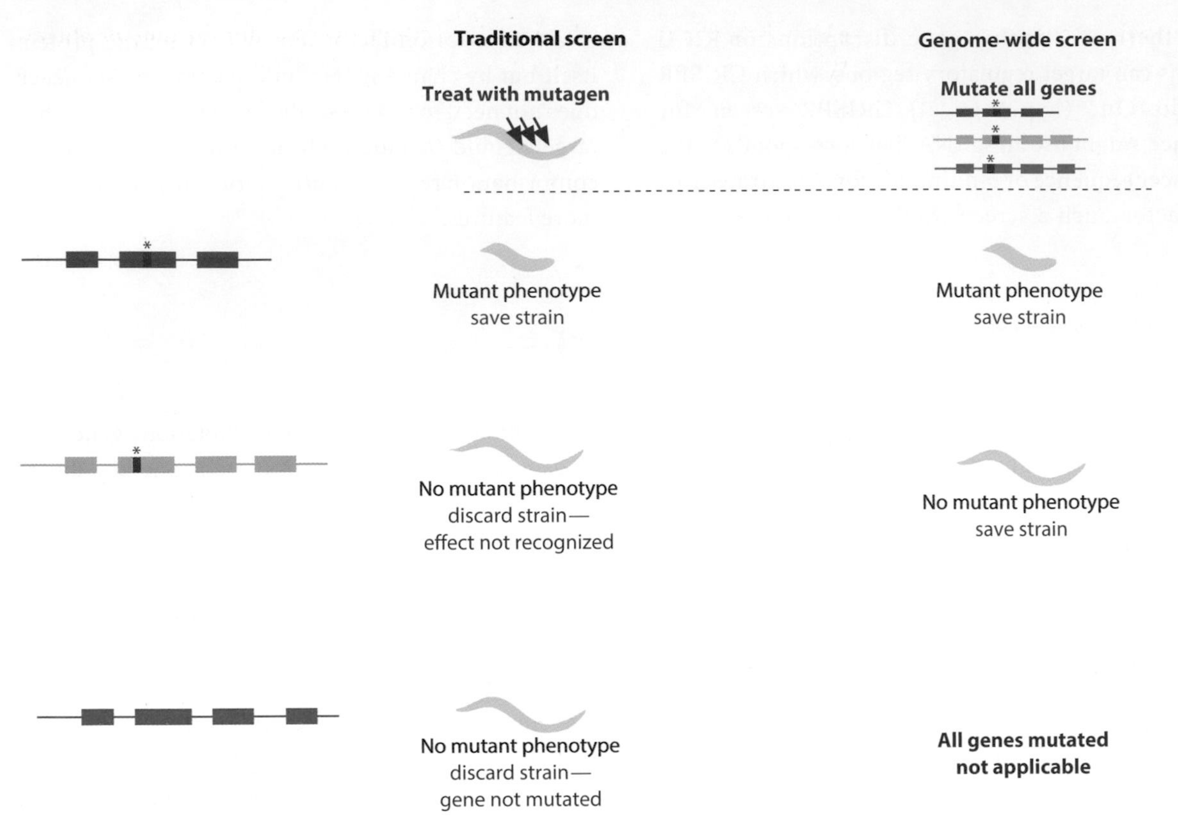

Figure 9.17 A comparison summary of traditional genetic screens with genome-wide screens. Three different genes are shown in the diagrammatic structures on the left. In a traditional screen, genes are mutated at random and mutant phenotypes are found. In a genome-wide screen, all genes are mutated by directed methods before looking for phenotypes. If the mutation causes a mutant phenotype, as shown for the gene in blue, the two approaches are the same. But if the mutated gene does not give a mutant phenotype, as shown for the green gene and the purple gene, traditional methods miss the effect that genome-wide screens find.

screens. Of course, for technical reasons, investigators may repeat the genome-wide screen; perhaps they do this because they are working with a more refined set of gene predictions or with a better method to perform the perturbation, whether RNAi or CRISPR. Nonetheless, the collection of affected genes is saved and available to other investigators with other assays.

This may also be a difference between the yeast disruption collection, the RNAi screens in model organisms, and CRISPR-based screens, although in this case the difference may be related to historical or social factors in the research. Both the yeast disruption screen and most of the genome-wide RNAi screens were done in one or a few collaborating laboratories, which then made the collection available to others. The collection included both those genes that had a mutant phenotype and those that did not. Because CRISPR-based screens are being done in many different laboratories and are rarely "genome-wide" in the sense that all genes are tested and included, those genes that do not have a mutant phenotype for one laboratory might not be so readily distributed to be tested by other laboratories. On the other hand, perhaps the procedures will be simple enough and the types of mutations sought individualistic enough for this difference with genes that show no effect might not to matter much.

Because genome-wide screens involve examining thousands or tens of thousands of mutants, attempts are being made to score phenotypes by automated or high-throughput methods. For example, the phenotypic assay might involve expression of a reporter gene that can be detected in cell sorting or through other automated methods. Mutants that affected expression of the reporter could be detected in a high-throughput assay.

Automated methods to perform these high throughput assays are being developed as well. With the number of mutant strains that are generated by genome-wide screens, high-throughput methods are very attractive and new assays will regularly be developed. But, whatever the method, the assay for the mutant phenotype will only be as refined as the observer, and many phenotypes will be missed in the initial screens. Because all strains exist for subsequent analysis, this is not usually a problem.

9.7 Confirming the effects

Reading the results of a genome-wide screen is roughly similar to reading a comprehensive encyclopedia: abundant amounts of information are available but little of it can be assimilated at one time. Most of us will go to the part that interests us the most, extract the information that we want, and ignore the rest. The outcomes of a genome-wide screen are in effect a database, and the precise roles of the individual genes usually need to be pursued separately, through the traditional methods of genetic analysis described elsewhere throughout the book.

One method that is available in some model organisms is to compare the mutant phenotype from the genome-wide screen to that observed in a traditional single gene screen, as previously described in Chapter 4. As with a traditional mutant phenotype, it may be possible to determine whether the mutant is rescued or complemented by a wild-type version of the gene or whether a related (possibly opposite) mutant phenotype is observed when the candidate gene is overexpressed. Most of the other methods for analysing mutant phenotypes that are described in the book can also be used for phenotypes arising from a genome-wide mutant screen.

Genome-wide mutant screens, like traditional mutant screens, provide the material for analysing nearly any biological process. Compared to traditional genetic screens, genome-wide screens have the significant advantages that all genes can be tested in a statistically unbiased manner and that any candidate gene has already been cloned. Although the field of genetics has not quite moved to this situation, it is not hard to envision a time when all genetic screens for any biological process will be done as genome-wide screens. In fact, for some model organisms, that day is nearly upon us.

9.8 Lessons from genome-wide screens

Every genome-wide mutant screen yields its own lessons; the principal result may be the library of mutant strains itself. These screens can also be viewed as a whole rather than as a collection of individual effects, which may provide additional insights into global aspects of physiology, evolution, and genetics that are not readily available by other methods. For example, genome-wide screens recognized that few X-linked genes in *C. elegans* are involved in gametogenesis, a result that could not have been readily shown by traditional genetic screens.

It may seem foolish to try to draw even more general conclusions or cross-species comparisons from the genome-wide mutant screens that have been done so far. After all, different methods have been used in each model organism, and various levels of "genome-wide" analysis have been achieved. At a fundamental level, the genome-wide screens are not comparable. Among eukaryotes, only for *Saccharomyces cerevisiae* have deletions been made for nearly every predicted gene. In other organisms such as worms and flies, the phenotypes are primarily the result of RNAi screens. Thus, in comparing the results of these screens, we may be comparing null mutations with hypomorphic mutations. Despite these cautions about the nature of the mutant phenotypes and the number of genes, we will attempt to draw a few lessons from the results of genome-wide mutant screens in yeast, flies, and worms.

Lesson 1: New genes are found even for well-studied processes

The first and perhaps most important general lesson is that genome-wide screens work very well and find genes that have been missed or might have been missed by traditional genetic screens. At the beginning of the chapter we cited one example from yeast and new genes involved in DNA repair. The genome-wide screen identified three genes that had never before been implicated in DNA repair. Another example from yeast involves galactose metabolism, a pathway that has been thoroughly studied by traditional genetic means. From the genome-wide selection in which each deletion was tested for its ability to grow in the absence of galactose, ten genes were found that had escaped detection in traditional gene-by-gene mutant selections. Finding new genes that are involved in such familiar cellular processes may be sufficient justification for genome-wide screens. There is great satisfaction in knowing that all possible genes have been tested for an effect and that no one critical gene remains to be found.

But this is not the only justification: new phenotypes are also found for known genes. In the yeast genome-wide screens, additional mutant phenotypes were also found for genes that had been studied previously; for example, two genes known to affect the calcium-regulated phosphatase calcineurin were also found to affect the ability of yeast cells to grow under ionic stress. Similar examples can be found from nearly every study that uses genome-wide screens to investigate a biological process; new genes are found, and previously known genes are found to have unexpected effects.

Lesson 2: Many genes do not have a mutant phenotype

The second general lesson is that many genes do not appear to have a mutant phenotype, at least in strains with a single mutated gene. For geneticists who make their living from mutant phenotypes, this result is a bit humbling. For example, among the yeast deletion collection, only 18.7 percent of the deletions affected the ability of the yeast cells to grow in rich media. For RNAi experiments in worms, about half the genes have recognizable mutant phenotypes; the remaining genes have no mutant phenotype under the RNAi conditions tested.

These results can be understood in a number of ways, as summarized in Figure 9.18. A trivial explanation is that the procedure itself failed to affect that particular

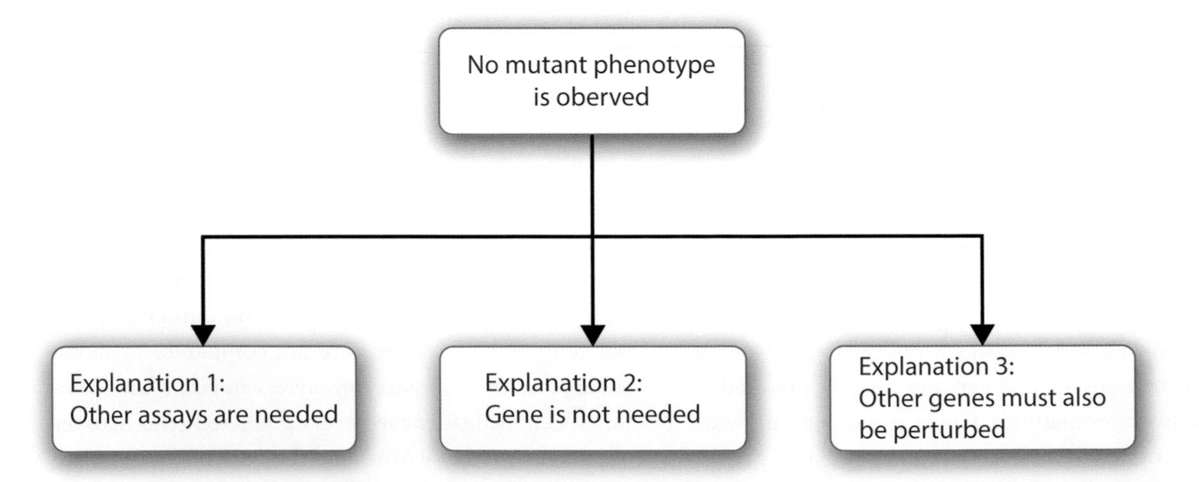

Figure 9.18 Three possible reasons why no mutant phenotype is observed for some genes in genome-wide screens. The first and least interesting explanation is that different assay conditions are needed to see the mutant phenotype. The second and third explanations are that the gene is truly not needed under any current conditions, or that the gene produces a mutant phenotype only when another gene is mutated as well. These are discussed in more detail in subsequent chapters.

gene. This is especially relevant for the screens based on RNAi and the gene in question has escaped the effects of RNAi. This possibility can be addressed by additional tests more specific to that gene, or by comparing the RNAi results with the effects of deletions, when available. Experiments do not always work, and the reasons might not be interesting.

The more intriguing possibilities arise when the gene was perturbed or mutated but no mutant phenotype is seen. The first explanation is probably the easiest. Only a limited number of phenotypes and environmental conditions are tested in any one screen, and when thousands of mutants are being scored, subtle defects can be easily missed. Thus, under one set of assay conditions, the strain looks normal, whereas under another set, the strain has a mutant phenotype. This explanation requires that more assay conditions be used. This certainly explains some of the result with yeast disruptions and RNAi screens, since new phenotypes for known genes are often observed when different conditions are tested. It is probably impossible to know whether this explains all or even most of the genes for which not mutant phenotype has been observed. After all, there are always other conditions to test and other phenotypes to assay. In fact, with repeated screens under a wide range of genetic conditions, a phenotype has been found for about 5,000 of the 6,130 genes in yeast. As mentioned previously, one of the strengths of genome-wide screens is that the mutant strains are retained for further tests regardless of their original phenotype.

The second and third explanations are biologically the most intriguing. Alterations or even deletions in some genes might simply have no effect on the organism that can be detected even in the most careful and thorough assay. Genomes and genes are the outcome of millennia of evolutionary pressures, and some genes may be functional but not needed at all. A logical conclusion is that, at some point in the evolutionary future of the organism, this gene will become a pseudogene, but that has not happened yet. But, for some other genes, the mutant effects may be apparent only in combination with other mutations in the genome. That is, gene products interact with one another and compensate for each other such that the loss of one gene shows an effect only if other genes are also altered. This will be discussed in Chapters 10, 11, and 12.

Literature Link. Walhout 2018, *Science* 360: 269–70

9.9 Summary: Genome-wide mutant screens attempt to identify phenotypes for every gene in the genome

The genome sequence and the list of annotated genes are sometimes compared to a parts list for a complicated machine. The parts list is certainly important as a first step to ensure that everything that will be needed is present. We know that we will not have to run to the store at the last minute to buy one more piece that we had overlooked. But knowing how much sheet metal and how many screws, bolts, and nuts hold it together does not allow us to build an automobile. Likewise, knowing how many putative transcription factors, potential receptors, and so on will be needed does not tell us how a complex biological process works. However, with genome-wide mutant screens, we have the opportunity to begin to understand even the most complex biological questions. We have all the parts, and we have tested them for their role in the biological question of interest. Genome-wide mutant screens combine the information from genome projects with the power of traditional genetic analysis, in an attempt to infer a function for every gene in a genome. By itself, a genome-wide mutant screen provides new information even about well-studied biological questions. But, even more importantly, genome-wide mutant screens provide the tools that allow us to answer more complicated questions than we yet know how to ask.

Chapter Capsule

Genome-wide screens quickly become the standard method in those organisms whose genomes have been sequenced. The steps in a genome-wide mutant screen are:

- identifying the genes to be mutated
- disrupting or perturbing genes
- examining the mutant strains for phenotypic effects
- confirming the mutants that are found, usually by gene-specific methods.

Every gene has been disrupted in yeast *Saccharomyces cerevisiae* by homologous recombination. For multicellular organisms, RNAi has become the best method for perturbing genes, and CRISPR-based screens are becoming more common.

Additional Reading

Beronja, S. et al. 2013. RNAi screens in mice identify physiological regulators of oncogenic growth. *Nature* 501: 185–90.

Boutros, M. et al. 2006. Analysis of cell-based RNAi screens. *Genome Biology* 7: R66.

Carpenter, A. E., and D. M. Sabatini. 2004. Systematic genome-wide screens of gene function. *Nature Reviews Genetics* 5: 11–21.

Doench, J. G. 2018. Am I ready for CRISPR? A user's guide to genetic screens. *Nature Reviews Genetics* 19: 67–80.

Farboud, B., A. F. Severson, and B. J. Meyer. 2019. Strategies for efficient genome editing using CRISPR-Cas9. *Genetics* 211: 431–57.

Giaever, G., and C. Nislow. 2014. The yeast deletion collection: A decade of functional genomics. *Genetics* 197: 451–65.

Grunenfelder, B., and E. A. Winzeler. 2002. Treasures and traps in genome-wide data sets: Case examples from yeast. *Nature Reviews Genetics* 3: 653–61.

Moffat, J., and D. M. Sabatini. 2006. Building mammalian signalling pathways with RNAi screens. *Nature Reviews Molecular Cell Biology* 7: 177–87.

Ngo, V. N. et al. 2006. A loss-of-function RNA interference for molecular targets in cancer. *Nature* 441: 106–10.

White, J. K. et al. 2013. Genome-wide generation and systematic phenotyping of knockout mice reveals new roles for many genes. *Cell* 154: 452–64.

Unit III

Interactions, Pathways, and Networks

CHAPTER 10

10

Gene interactions Suppressors and synthetic enhancers

TOPIC SUMMARY

Nearly every biological process is the outcome of genes interacting with each other. Investigators often want to find functionally related genes on the basis of their interactions; interactions can be detected using either the mutant phenotypes or the cloned gene products.

If the investigator begins with a mutation in a particular gene, interacting genes involved in the same cellular or physiological process can be identified on the basis of the phenotypes of double mutants. These genetic interactions include:

- **suppression**—mutations in a second gene that overcome or compensate for the defect in the original mutant. Such mutations include high copy suppression (see **high copy suppressor**), that is, overexpression of the wild-type function of a second gene that overcomes the defect in the original mutant—a strategy widely used in yeast;
- **synthetic enhancement**—mutations in a second gene that exacerbate the defect of the original mutant. A related approach is **non-allelic non-complementation**, in which recessive mutations in two different genes fail to complement each other.

If the investigator begins with a cloned wild-type gene, proteins that interact directly with the original gene product can be identified by

- yeast two-hybrid selection;
- co-immunoprecipitation of protein complexes.

INTRODUCTION

Finding a gene with an intriguing phenotype is the early step in the genetic analysis of the biological process, as discussed in Chapter 4. However, one gene and its phenotype are rarely the end point of the analysis, and the investigator usually wants to find as many other genes as possible that affect the same process. Naturally, the investigator can continue to repeat the same mutant screen and find more genes (as discussed for embryonic segmentation mutants in *Drosophila* in Chapter 4) or use genome-wide methods (as described in Chapter 9) to find genes with similar phenotypes. But in this chapter we describe how any gene can be used as a starting point to find other genes that affect the same biological process or pathway. The methods for doing this use two contrasting but complementary approaches. Genetic approaches start from the mutant phenotype of the gene, while molecular approaches begin with the cloned gene or the protein it produces. The genetic and the molecular approaches are both powerful; they often identify the same genes, but each also identifies the genes the other does not.

We will first describe the methods that rely on the mutant phenotype; these are suppression, synthetic enhancement, and non-allelic non-complementation. We will then briefly discuss two applications that begin with the cloned gene: the yeast two-hybrid screen; and co-immunoprecipitation. Because these approaches use different starting material and may or may not find the same genes, the results from a combined approach are usually much more complete than from either alone.

Suppressor analysis—that is, finding mutations in a second gene that overcome the effects of mutations in the first gene—has a long history in genetic analysis in both bacteria and eukaryotes. There is an extensive literature gong back several decades on characterizing the effects of suppressor mutations. Synthetic enhancers—that is, mutations in a second gene that result in a more severely mutant phenotype (see **synthetic enhancement**)—had not been so frequently applied in traditional genetic approaches. However, as described in Section 9.8, genome-wide mutant screens revealed that many genes were undetected in traditional approaches, in part because many genes appear not to have a mutant phenotype on their own. Synthetic enhancement has become the tool that allows geneticists to identify the biological process affected by these genes; mutations in two different genes, neither of which has a mutant phenotype, can together have an informative mutant phenotype. In fact, as will be discussed more fully in Chapter 12, synthetic enhancement is probably much more common than suppression when gene interactions are thoroughly characterized.

Genetic approaches like suppression and synthetic enhancement can identify genes that interact in the same biological process or that affect the same function, but they might not tell us precisely how the genes or the gene products interact with each other. Genetic interactions can be indirect, and the gene products might not have any direct physical connection to one another. Thus, molecular approaches that find interacting molecules, particularly proteins, yield a different type of information, so that a more complete set of gene interactions can be characterized for a process. Yet another type of gene interactions, namely epistatic interactions, allows the investigator to construct a pathway by which the genes interact with each other to affect a biological process; epistastic pathways are described in Chapter 11.

10.1 Gene interactions affecting a biological process: Overview

So far in this book we have focused on single genes. We have discussed ways of finding a mutation that affects a biological process, strategies to connect that mutant phenotype to a DNA sequence, and guidelines for interpreting mutant phenotypes. We have discussed approaches that allow investigators to edit the gene or genome directly, and to perturb the functions of every gene in the genome by targeted methods. Our premise has been that every mutant gene tells a story about its role in that particular biological process, so our mantra has been to find a mutant.

But few biological processes are the outcome of a single gene acting in isolation. In fact the examples of single gene phenotypes familiar from introductory genetics class, such as blood types and rare genetic diseases in humans, are somewhat atypical. Most phenotypes are affected by multiple genes that interact among themselves in various ways, so that determining the number of genes and their interactions are important, if occasionally difficult, aspects of genetic analysis. How, then, can an investigator find these additional genes and characterize their interactions?

One obvious way to find functionally related genes is simply to repeat the mutagenesis procedure employed previously, searching for more mutants that have similar phenotypes, as we discussed for *Drosophila* segmentation in Chapter 4. We might also look for genes whose DNA sequence or expression pattern is similar to that of our original gene. These could be combined and expanded using genome-wide method, so that many candidate genes affecting the same process can be identified. These approaches have been very successful in many different organisms and for many different processes. But the functional interactions among these genes could remain unknown.

This might be illustrated with an analogy. Imagine that the biological process that we want to understand is analogous to some machine or item of furniture comprised of multiple component parts; we will think of it as a large desk with drawers. The approaches we have described so far have allowed us to catalog the different parts of a desk, and even to recognize (on the basis of sequence comparisons and mutant phenotypes, for instance) the function of many individual parts. For our desk, we can distinguish the bolts from the nuts from the screws, the top surface from the legs, and so on. But these parts don't give us direct experimental information about how they are assembled in the completed desk. We can make inferences, often very accurate ones, about which part fits together with which other part, and maybe even about what substructures in the desk could be assembled independently of other substructures. We could make a good guess that the sides, back, bottom, and front of each drawer go together and could probably be assembled independently of the legs, for instance. But, if we want to understand how the desk was assembled or how a biological process occurs, we would like to have more complete information about how these parts (genes and gene products, in the case of biological processes) interact with each other and fit together.

This chapter and the next two describe approaches that can provide some of the "assembly instructions" for a biological process, that is, they teach us how to recognize, characterize, and interpret gene interactions. Rather than focusing on the activity of single genes, we will focus on two genes acting together. As a point of explanation, which reflects the history in which such experiments were done, we will often refer to the "original mutation" (or gene) and the "second mutation" (or gene). Often an investigator has found one gene, either from a mutation or from a DNA sequence, which is then used as a starting point for finding additional mutations or genes. Attention is given to the effects that the second gene has on the first gene, and so to their interaction. This becomes a strategy for identifying genes that affect the same process, even if we do not initially recognize their functions or how they might be related to each other. Functionally related genes are found because they interact in some way with the first gene. The nature of that interaction then enables us to understand the relationship of the individual genes to the biological process that all of them affect.

In practice, there is a close relationship between the tasks of identifying and characterizing functionally related genes and of ordering these genes into a pathway or network. However, to make the explanation clearer, we will describe the techniques for finding functionally related genes in this chapter and the techniques used to place the genes and their products into logical pathways in Chapter 11. We will also focus on a few processes in model organisms to illustrate the strategy rather than attempting a more thorough discussion of gene interactions.

10.2 Genetic interactions: Suppressors

We begin this chapter by describing the genetic interactions of suppression, synthetic enhancement, and non-allelic non-complementation. These approaches start with a well-characterized mutant phenotype in a single gene. We will see how these methods have been used to investigate examples such as muscle formation and movement, vulva development, and neurotransmitter function and release in worms and axonal guidance in *Drosophila*; many other examples could also be used. As in Chapters 4 and 5, where we also outline general strategies for genetic analysis and for connecting mutant phenotypes with a DNA sequence, this chapter's case study, on spindle morphogenesis in budding yeast, can be read as a parallel chapter that describes a specific applications of these genetic interaction principles.

Suppressors and enhancers modify the phenotype of another mutation

Once a mutation has been found that affects the biological process of interest, a common and powerful strategy to find functionally related genes is to search for mutations that modify the original mutant phenotype. The approach can be summarized as follows, the terminology and outcomes being diagrammed in Figure 10.1. The mutation *a* is found as the original mutation. A homozygous mutant designated *a*/*a* is treated with a mutagen, and the offspring in a subsequent generation are examined. These will include a wide variety of double mutants, all of which are still homozygous for the original *a*/*a* mutation. Among these are double mutants that carry a second mutation that modify the *phenotype* of the *a*/*a* mutant strain. Because they affect the same mutant phenotype, these modifying mutations identify candidate genes that affect the same biological process as the original mutant gene. Different modifying mutations can affect the mutant phenotype in different ways, so the double mutants will also often have different phenotypes from each other. We are using homozygous diploids in this example, but the analysis of haploid organisms is similar.

We will note in passing that many mutations will be found in this screen in which the phenotype of *a*/*a* is observed but not modified. Thus the phenotypes of the original mutation and any second mutations can both be observed. An example can be used from Mendel himself. Mendel found that the round versus wrinkled peas phenotype was inherited independently of the green versus yellow phenotype, as discussed in Chapter 1. In fact, since the round versus wrinkled phenotype could be observed independently of the green versus yellow phenotype, these two genes *act* independently as well, and the function of one is apparently unrelated to the function of the other. This provides our first and simplest general rule of gene interactions, which we will consider more fully in Chapter 11; if the double mutant shows the phenotypes of each of the two genes, the genes act independently of each other and probably affect separate biological processes.

But let's return to those situations in which the phenotype of the original mutation has been modified by a mutation in another gene. These modifying mutations fall into two major categories (see Figure 10.1). The first category comprises those that make the mutant phenotype less severe and more like wild type. These mutations are called **suppressors**. Our figure shows that the original mutation (*a*/*a*) results in a worm that has a dumpy morphology. A suppressor mutation (*b*) in a second gene overcomes that defect, so that the *a*/*a; b*/*b* double mutant worm has a morphology that is more like that of the wild type. In the notation of genotypes, the *a*/*a* strain has a mutant phenotype, whereas the *a*/*a; b*/*b* double mutant strain has a more nearly normal phenotype; so the *b* mutation is a considered a suppressor of the *a* mutation. The ability to restore the wild-type function in a double mutant, as suggested by the phenotype, is the defining characteristic of a suppressor

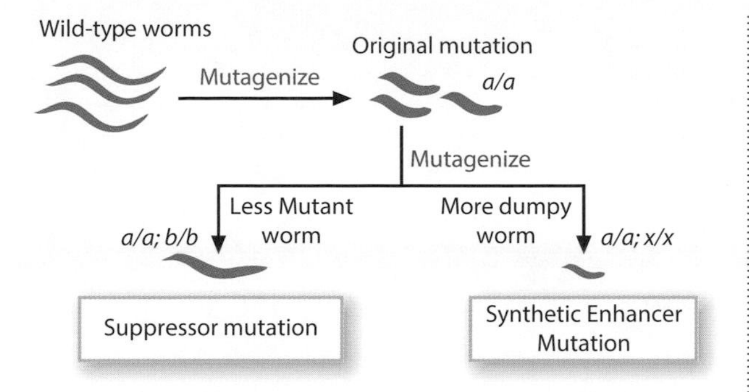

Figure 10.1 Suppressors and synthetic enhancers. Genetic interactions are found when the original mutant (here shown as Dumpy worms) are mutagenized, and additional mutants are found. A second mutation that makes the first mutant more closely resemble the wild type is a suppressor. A second mutation that makes the first mutant more severely mutant is a synthetic enhancer.

mutation. It should be noted that the suppressor mutation makes the double mutant phenotype more similar to the wild-type phenotype, although it still may not be completely like the wild-type phenotype in all the characteristics.

The second category of modifying mutations comprises those that make the mutant phenotype more severe. These mutations are generically called enhancers. It is an unfortunate confusion that the name "enhancer" is used both for these modifying mutations and for regulatory sequences that specifically stimulate the expression of a gene. To avoid confusion, we will refer to these modifying mutations as synthetic enhancers (see **synthetic enhancement**, **synthetic enhancer mutation**), although, as we note in Chapter 12, other terms might be used as well. Again, in our example in Figure 10.1 the original mutation produces a worm with a dumpy morphology. A mutation in a different gene results in a double mutant that has an even more severely dumpy phenotype than the *a/a* single mutant. In the notation of genotypes, the *a/a* strain has a mutant phenotype, but the *a/a; x/x* double mutant strain has a more severely mutant phenotype, and the *x* mutation is said to be a synthetic enhancer of the *a* mutation. This more severe mutant phenotype is the defining characteristic of a synthetic enhancer.

Suppressors and synthetic enhancers are two of the most useful tools in the genetic analysis toolbox. They have been used for decades in genetic analysis to find genes with related functions and to determine the nature of the relationships between different genes.

Literature Link. Botstein and Mauer 1982, *Annual Review of Genetics* 16: 61–83

Until recently, suppressors were more widely used than synthetic enhancers. This is in part because suppressor mutations have proved to be very successful at identifying functionally related genes, whereas enhancer mutations were thought to be less likely to identify a specific functional relationship. To use an analogy familiar to homeowners with plumbing problems, many different actions can make a problem worse (enhancers), but relatively few are able to fix it (suppressors). Fixing the plumbing problem requires that the nature of the problem be recognized and addressed. Turning a leaky pipe into a flooded basement might not require such a specific interaction or insight.

In the past few years, as genome-wide screens for mutations have become more widely used (this is discussed in Chapter 9), synthetic enhancers have been found to be extremely useful. It is true that, as in the analogy with the leaky pipe and the flooded basement, many mutations can result in a more severe mutant phenotype when combined with another mutation, and can do so even if the mutations have unrelated functions. However, the key to a successful exploitation of synthetic enhancers lies in identifying those modifying mutations that are specific to the function of the original mutation rather than simply making some general physiological problem worse. Our greater knowledge of genomes and a greater ability to manipulate genes allows us to determine these specific interactions more readily, so the real power of synthetic enhancers has emerged.

In fact synthetic enhancers gained special attention when the genome-wide mutant screens found that many mutations produced no mutant phenotype, as discussed in Section 9.8. Our example with the dumpy worm becoming more severely mutant might not be the most interesting or informative case; perhaps many mutations that affect growth rate rather than the effects of the dumpy mutation could result in a more severely mutant worm. However, as summarized in Figure 10.2, the most informative and significant type of synthetic enhancement occurs when the original mutation has no mutant phenotype, and only the double mutant shows a mutant or lethal phenotype. We will return to these in Section 10.3.

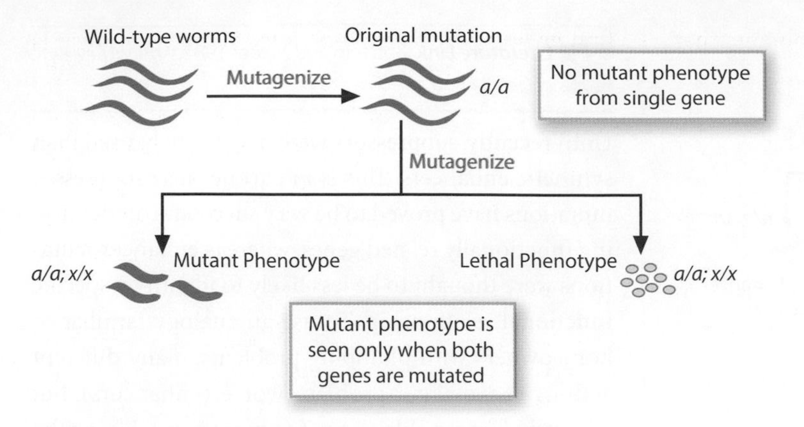

Figure 10.2 Synthetic enhancement when no visible mutant phenotype is observed. In addition to synthetic enhancement in which the mutant phenotype becomes more severe, as in Figure 10.1, synthetic enhancement also occurs when the original mutation has no mutant phenotype on its own. A mutant phenotype is observed only when a second gene, which has no mutant phenotype on its own, is also mutated. This phenotype could be visible, such as a Dumpy worm, or lethal; synthetic enhancement in which the double mutant is lethal is called synthetic lethality.

Suppressor and enhancer gene nomenclature can be extremely confusing

Gene names and genetic nomenclature can be confusing, as discussed in Section 4.5. This often becomes even more pronounced with suppressors and synthetic enhancers, in which the nomenclature varies between different model organisms and can be confusing even to experienced geneticists. The important thing to keep in mind is that, in general, genes are named after their *mutant* phenotype rather than after their wild-type function. For example, the *white* gene in *Drosophila* is so named because the mutant phenotype is white eyes. The wild-type *w+* gene produces red eyes. A similar general rule of nomenclature applies to naming suppressors and enhancers. The mutant phenotype of a suppressor gene is, by definition, suppression of another mutation. Therefore, given a suppressor gene name such as *sup-5+* in *C. elegans* or *SUP45* in budding yeast, it will be the *mutant* version—*sup-5* or *sup45*—that acts as the suppressor. The function of the wild-type *sup-5+* or *SUP45* gene may have nothing to do with suppression.

But do the names *sup-5* or *sup45* tell us anything about which mutation is being suppressed, or about the wild-type function of these genes? In the case of yeast and worms, the answer is usually no, although, as in most cases of genetic nomenclature, exceptions exist. Suppressor gene names in *Drosophila* usually follow different rules, and often do include the name of the mutation being suppressed. From the gene name *su(Hw)* in *Drosophila*, we can infer that a *mutation* in the gene *su(Hw)* acts to suppress the defect caused by a *mutation* in the gene *Hw* (*Hairy wing*). The *wild-type* function of the *su(Hw)* gene is not to suppress either the mutant or the wild-type version of *Hairy wing*. In fact, in this example, it turns out that the wild-type function *su(Hw)* has nothing to do with the wild-type function of *Hw*, and the suppressor affects mutations in genes in addition to *Hairy wings*.

Suppressor mutations like these were found and named because they suppressed another mutation. Fewer mutant screens were done specifically to find enhancer mutations (although the example with **position effect variegation** (PEV) was given in Section 2.5), and so fewer genes are named according to their ability to enhance other mutations, but a similar convention is used for naming synthetic enhancers. The genes are, again, named after their mutant phenotype, and the name of the mutant phenotype being enhanced is often included in the gene's name. This is probably to show that the interaction is specific to that mutant phenotype or gene and that the enhancer mutation is not simply making many mutations more severe by a non-specific interaction. For example, from the gene name *E(spl)*, we infer that a mutation in the gene *Enhancer of split* acts as a synthetic enhancer of the *split* mutation in *Drosophila*. The wild-type function of the *E(spl)* is not to enhance the wild-type function of *split*; it is the mutants that show the interaction, and the genes are named after the mutant phenotype.

This type of nomenclature can be difficult to follow. Many suppressor or synthetic enhancer genes were first identified and named because they have mutant phenotypes of their own, and their effect on other genes was only subsequently recognized. To try to reduce the confusion, the different genes that will be discussed in this chapter are listed in Table 10.1, along with the type of interactions that they exhibit.

The history and use of suppressor mutations

As with many other types of genetic analysis, the use of suppressors began with *Drosophila*, when A. H. Sturtevant, in 1920, recognized that suppressor analysis could find genes with related functions. Suppressor analysis was subsequently more fully developed as a powerful technique in bacterial and phage genetics. The reason for this is that suppressors are often rare, so that large populations of double mutants need to be examined. Thus suppressors are more widely used in yeast than in worms or flies, and more widely used in worms and flies than in mice, which reflects the ease with which thousands or tens of thousands of double mutant organisms can be examined.

In yeast, worms, and flies, suppressor analysis has proven to be an extremely powerful method to find genes that modify or interact with the original gene. As we shall see in our examples, suppressor mutations often find genes that might not have been so easy to find by simply doing a more extensive search for the

Table 10.1 The genes discussed in this chapter

Organism	Wild-Type Gene	Homozygous Mutant Phenotype	What Does This Example Illustrate?
Saccharomyces cerevisiae	*TUB2*	Cold-sensitive arrest of cell division	Used throughout the case study
C. elegans	*unc-54+*	Mutant is paralysed	Beginning mutation for many of the suppressor screens
C. elegans	*sup-5+*	Suppresses nonsense mutations in many genes	Informational suppressors
C. elegans	*myo-3+*	Suppresses *unc-54* alleles	Bypass suppressors
Arabidopsis	*CO+*	Delayed flowering	Epistatic suppression
Arabidopsis	*FT+*	Flowering time is delayed by suppression of *CO*	Epistatic suppressor
C. elegans	*ace-1+* *ace-2+* *ace-3+*	Reduced levels of acetycholine esterase. A mutant phenotype is seen only when more than one gene is mutant.	Synthetic enhancement by paralogous genes
C. elegans	*lin-8+* *lin-9+*	No mutant phenotype alone No mutant phenotype alone Multiple vulvae in the double mutant homozygote	Synthetic enhancement by functional redundancy
Drosophila	*slit+*	Axonal guidance defective	Non-allelic non-complementation with *robo* mutations
Drosophila	*robo+*	Axonal guidance defective	Non-allelic non-complementation with *slit* mutations
C. elegans	*unc-13+*	Paralysed	Non-allelic non-complementation with *unc-64* mutations
C. elegans	*unc-64+*	Paralysed	Non-allelic non-complementation with *unc-13* mutations
Arabidopsis	*BRI1*	Dwarf plant phenotype	Protein interactions by Y2H
Arabidopsis	*R, RIN*	Pathogen response	Protein interactions by Y2H and co-IP

more mutations of the same type as the original one. Recall that the defining characteristic of a suppressor mutation is to modify the phenotype of another mutation. Therefore, even if the suppressor mutation has a different phenotype from the original mutation or has no other mutant phenotype on its own, it can be found by virtue of its interaction with the original gene. With suppressors, we do not need to know what phenotype to look for among the double mutants—if it suppresses the original mutation, the suppressor probably defines an interacting gene that is worth studying. The suppression phenotype reflects some aspects of the underlying biology that may not be apparent to the investigator by other means.

Suppressors are particularly easy to find if the original mutation has a pronounced and severe phenotype so that correction of that defect is more obvious. Notable biological examples that have given up their secrets to suppressor analysis include many different cells or organisms that cannot grow or reproduce under certain growth conditions; worms that cannot move; flies that do not fly toward a light source; yeast cells that cannot repair particular kinds of DNA damage; plants that cannot flower; and many more. In fact, when the original mutation has an easily recognized phenotype and it is possible to screen lots of mutagenized offspring, suppressor analysis has proven to be so useful that it is nearly compulsory if one wants to do a thorough investigation of the phenotype and the biological process.

Figure 10.3 shows an example of a mutant used to find suppressors in yeast. Yeast cells with the cold-sensitive *tub2-406* mutation in the β-tubulin gene do not complete nuclear division, so no growth is seen at the restrictive temperature (the low temperatures in this case). Thus the phenotype of *tub2-406* is easy to score in populations of cells; the cells do not divide at low temperature, which made it relatively easy to find suppressors. These mutant yeast cells were treated with a mutagen, replated, and their progeny screened for cells that could grow at the restrictive temperature. Several different suppressor mutations were found and form the basis for the case study.

Another example of the identification of suppressors is provided by *C. elegans*. Recessive mutations in the gene *unc-54* result in paralysis. The paralysed *unc-54/unc-54* worms were mutagenized and suppressors were identified by searching subsequent generations for worms capable of movement. These will be double mutants of the genotype *unc-54/unc-54; sup/sup* in which an additional mutation—the suppressor abbreviated *sup*—has reversed the phenotypic effects of the *unc-54* mutation and allowed the worm to move. Although we have shown the suppressor as a homozygote, the suppressor mutation may be dominant or recessive, just like any other mutation. The assumption of this screen is that the second mutation is identifying another gene that affects movement. The combination of the mutation in *unc-54* and this second mutation *somehow* restores movement to paralysed *unc-54/unc-54* worms. This "somehow" in the preceding sentence is the crux of suppressor analysis. That is, once a suppressor is found, we still want to know how it acts to restore the function to the original mutation.

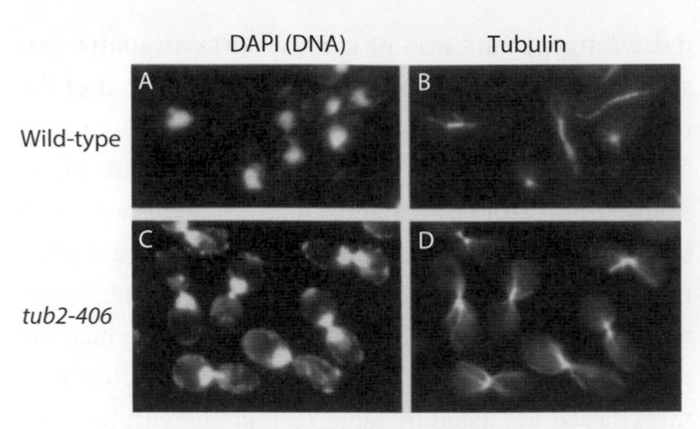

Figure 10.3 The phenotype of the yeast β-tubulin mutation *tub2-406*. Wild-type budding is shown in the top two photos, with the *tub2-406* mutant phenotype at the bottom. The DAPI stain in the left panels indicates the effect on chromosome segregation whereas the α-tubulin antibody staining in the right panels shows the mitotic spindle. Notice from the DAPI stained cells that the *tub2* mutant arrests with its nuclear DNA in the neck of the budding cell and does not complete nuclear division, as the wild type does. In addition, microtubules in the *tub2* mutant do not assemble into normal length spindles, as shown in the α-tubulin stained cells. Because the *tub2-406* mutation prevents nuclear division at the restrictive temperature, suppressors of this phenotype could be found by looking for normal division. This mutant is discussed extensively in the case study. From Pasqualone and Huffaker 1994, *STU1*, a suppressor of a β-tubulin mutation, encodes a novel and essential component of the yeast mitotic spindle, *Journal of Cell Biology* 127: 1973–84, with permission.

Literature Link. Hodgkin et al. 1987, *Trends in Genetics* 3: 325–9

Although finding suppressors seems uncomplicated, particularly when the original mutant phenotype is easy to observe, the follow-up genetic characterization is usually more demanding. The main complication is that the suppressor's phenotype is initially defined only in relation to its effect on another mutant. Therefore all strains used for the initial outcrossing, mapping, and maintenance of the suppressor mutation, as discussed in Chapter 4, must also carry the original mutation. Thus, one of the first steps in working with a suppressor is to determine whether it has a mutant phenotype of its own, independent of its suppressor property. If it does, this separate mutant phenotype can be used instead of the suppression phenotype and simplifies the routine crosses of genetics.

Suppressor mutations can be either intragenic or extragenic

The principal motivation for most suppressor screens is to find additional genes that affect the same biological process as the original gene. Thus we want to determine quickly whether the suppressor does in fact define a different gene or whether it is another mutation in the original gene, such as a reversion. This is usually done by determining whether the suppressor maps to a different genetic location, which can be done concurrently with determining whether the suppressor has a mutant phenotype of its own.

Suppressors that map onto the same gene as the original mutation are called **intragenic suppressors**. They include true revertants, which affect the base or amino acid affected by the original mutation and pseudorevertants, which affect a different base or amino acid in the same gene. In addition, many dominant hypermorphic or antimorphic alleles of a gene that overexpress or alter the function of the gene can be suppressed by recessive null alleles of the same gene that eliminate the gene function. While important, particularly in the intensive study of a single gene, these intragenic events are not of interest for our present purpose, since they do not define a new gene. We are interested in the genes known as **extragenic suppressors**.

Extragenic suppressors fall into three main functional classes

Once it has been determined that a mutation maps to a different location and acts as an extragenic suppressor, the first goal of a suppressor screen has been accomplished: we have found a new gene. The analysis of suppression is only beginning, however. We now want to know *how* the suppressor mutation restores the original function. The answer will help determine whether the suppressor mutation does in fact define a gene with a function related to that of the original gene. Many decades of genetics can be summarized in the flowchart of suppressor analysis presented in Figure 10.4, so we will refer to this figure often.

 ***Literature Link.*** Prelich 1999, *Trends in Genetics* 15: 261–6

In order to understand an extragenic suppressor's mechanism of action, two fundamental questions have to be answered.

- Does the suppressor mutation also affect other mutant alleles of the original gene? In other words, is the suppressor *allele-specific*? The suppressor may interact with the original *mutation* but might not be interacting with the function of the original gene.
- Does the suppressor mutation also affect other genes in addition to the original gene? In other words, is the suppressor *gene-specific*? The suppressor may be interacting with the function or activity of the gene rather than the specific mutation in the original gene.

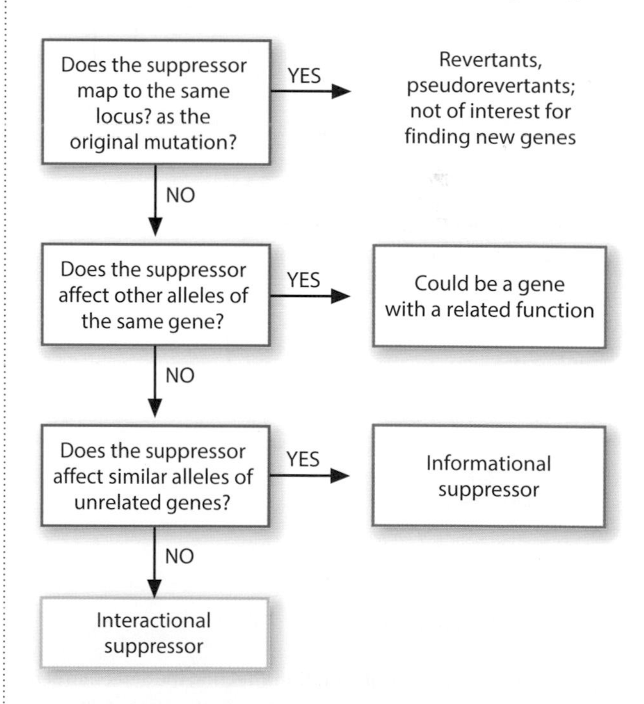

Figure 10.4 A flowchart illustrating the strategy for the characterization of suppressors. The arrows indicate logical questions and may not indicate the order in which the experiments are done. The suppressor is mapped, tested with other alleles of the same affected gene, and tested against similar alleles (such as null alleles, splicing defects, and so on) of genes with unrelated functions.

In answering these questions, it is important to realize that the suppressor was identified because it modified the phenotype of a particular mutant allele of a single gene, and it is this specific mutant allele that is being suppressed. This mutant allele could represent any one of many different molecular lesions that might be found for that gene. A suppressor mutation could therefore be modifying the effect of a particular mutant *allele* by correcting its specific molecular defect, or the suppressor could be modifying the general effect of the mutated *gene* by compensating for the overall loss of its function.

We will illustrate this with a hypothetical situation, drawn in Figure 10.5, which is similar to our example of genetic analysis in Section 1.6. (Although our example is hypothetical, a screen similar to this one found genes involved in signal transduction in *Drosophila*.) *Drosophila*, like other insects, will fly toward a light source. In order to understand this process, we find a mutant fly that, unlike wild-type fruit flies, does not move in the direction of blue light. The mutant can still move and it is still attracted to white light (for example), so this is not some general inability to fly or to detect light. We infer from this phenotype

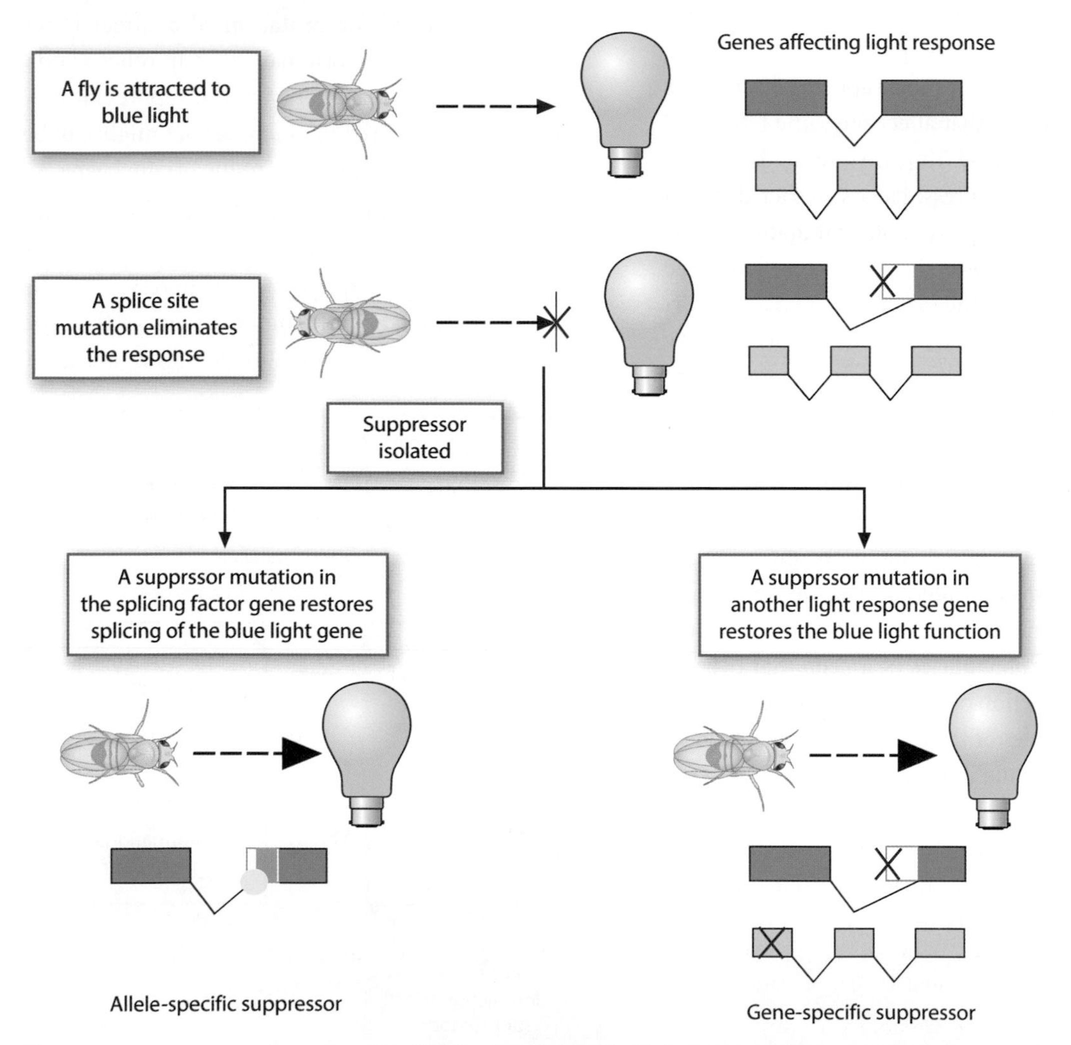

Figure 10.5 Suppression of mutation affecting the detection of blue light in *Drosophila*. The wild-type fly is attracted to blue light, as shown at the top. Two genes that affect this response are shown as gene diagrams, in blue and green. A mutant that alters the splicing of a blue light detection gene (as shown by the red X in the blue gene diagram in the second drawing) eliminates the response to blue light. By finding flies with a restored response, suppressor mutations are identified. An allele-specific suppressor, on the left, restores the splicing defect of the blue light sensing gene. This suppressor will not suppress another mutation in the gene but could suppress splicing defects in functionally unrelated genes. A gene-specific suppressor, on the right, affects another gene, indicated as the yellow gene. This type of suppressor will affect many types of alleles of the blue-light sensing gene.

that the mutant gene impairs the ability to detect blue light specifically; and we want to use the mutant fly to find other genes involved in light detection. The mutant flies are themselves mutagenized, and their offspring are exposed to blue light. A double mutant is found that can move toward blue light, and the suppressor mutation maps to a different location, so it is an extragenic suppressor. In other words, a new gene that may affect the ability to detect light has been identified. How do we characterize it further and understand its method of suppression?

Suppose that the mutation in our original mutant fly is defect in the splicing of the mRNA from a blue-light detector gene, that is, the molecular lesion is a splicing defect. Our extragenic suppressor could then represent a mutation in a gene involved in the splicing process itself, thereby fixing the specific defect (splicing) rather than the function of the gene (blue light detection). What should we expect when we analyse this type of suppressor? Think again about the flowchart in Figure 10.4 and ask yourself what else this suppressor would affect. If the suppressor is truly a change in splicing itself, this suppressor may be able to modify and suppress similar splicing-defect mutations in other genes regardless of the biological functions of those genes. On the other hand, it will have no effect on mutant alleles of the blue-light gene that involve molecular lesions other than splicing, such as nonsense mutations, missense mutations, or deletions. This type of suppressor is thus *allele-specific* for the defect in the blue-light detector gene, but it is *gene-nonspecific*, as it can also suppress splicing mutations in other genes. This type of suppressor is referred to as an **informational suppressor**. The suppressor does not identify a gene that affects light detection at all, but instead finds a gene involved in a fundamental informational transfer process in the cell. Splicing and translation are two fundamental cellular processes for which informational suppressors have been the most informative. Unless we are studying this particular cellular process, these are probably not the types of suppressor mutations of the greatest interest to us.

However, this is only one possibility for our suppressor mutation. Suppose that our suppressor mutation could correct the blue-light defect by affecting another gene involved in light detection; in our figure, this would be the gene in yellow. It might, for example, modify such a gene so that the modified gene product, which previously responded to light in a different spectrum, now reacts to blue light. Exactly how this suppressor works is unimportant for now: the important point is that it restores the biological function lost in the original mutant. Furthermore, because it affects the light-sensing function, this suppressor could also modify the phenotype of many different mutant alleles of the blue-light detecting gene, in addition to the splicing defect in the original mutation; and this would include missense alleles, nonsense alleles, and deletions. Such a suppressor is therefore *allele-nonspecific* in its ability to modify the phenotype of many different molecular lesions in this gene. But the suppressor is unlikely to affect any mutations in genes that are not involved in light detection. Thus it is *gene-specific* to the light-detecting pathways.

Although the terms "allele-specific" and "gene-specific" are widely used in suppressor analysis, they could be defined in a more informative way: "allele-specific" suppressors modify the molecular lesion that caused the missing function; "gene-specific" suppressors affect the function itself. Although the term "gene-specific" is often used to distinguish these two types of suppressors, *function-specific* might be a more accurate description of the effect. A gene-specific suppressor will often suppress mutations in several different genes if all the genes have related functions. The key to distinguishing allele-specific effects from gene-specific or function-specific effects is to determine (1) whether many different alleles of the same gene are suppressed, and then (2) whether alleles of functionally unrelated and functionally related genes are also suppressed. These function-specific suppressors define the genes that of the greatest interest to us when it comes to finding genes that affect the same biological process.

Finally, a suppressor mutation can be both allele-specific and gene-specific. These are the most specific suppressors in that they suppress one mutant allele of only one gene, and thus affect *both* the molecular lesion of the allele and the function of the gene. This type of suppressor is known as an **interactional suppressor** (Figure 10.4). We discuss it more fully in what follows.

The two tests involved in suppressor analysis are, then, to determine whether the suppressor modifies the phenotype of other mutant alleles of the *same gene* and whether it modifies the phenotype of mutant alleles of *functionally unrelated* genes. These two tests will determine the type of suppressor even when the molecular lesion of the original mutation is unknown. In general, if suppression is being used to identify additional genes with related functions, interactional suppressors and gene-specific suppressors are the most useful, because they affect the function of the gene, whereas informational suppressors will be the least informative as they simply affect the defect in the mutation. We will consider each possibility in turn, beginning with the interactional suppressors and working up the flowchart until we conclude with the function-specific suppressors.

Interactional suppressors are specific to both the gene and the allele

Many biological processes have been studied by suppressor analysis, so many types of suppressor mutations are known. A very few suppressors affect only the original mutation that is used and only the original gene that is tested; that is, the suppressor is *both* allele-specific to that particular mutational lesion *and* gene-specific to that particular genetic function. Although rare, these suppressors are among the most informative of all classes of suppressors. Nearly all, if not all, instances of such highly specific suppression have involved a direct physical interaction between the product of the original gene and the product of the suppressor; this is true whether the products of the two genes are RNA or proteins. This is why these mutations are called interactional suppressors.

To appreciate how interactional suppressors work and why they are both allele-specific and gene-specific, we will use an example of a ligand binding to its receptor. Although our example is hypothetical, similar examples are known both in bacteria and in yeast. A schematic diagram of a ligand binding to its receptor is shown in Figure 10.6, and this binding triggers some downstream process. In the original mutant, the process fails to occur because a mutation in the ligand changes its conformation, so the mutant ligand can no longer bind to the receptor. The suppressor mutation, this time in the receptor, restores triggering of the downstream process by altering the conformation of the receptor at the binding site, which allows the mutant ligand to bind. The specificity between the two mutations is the key to recognizing interactional suppressors. Other mutations that alter the function or conformation of the ligand will not be suppressed since these other mutant ligands will still be unable to bind to either the wild-type receptor or the mutated receptor. This suppressor is thus allele-specific.

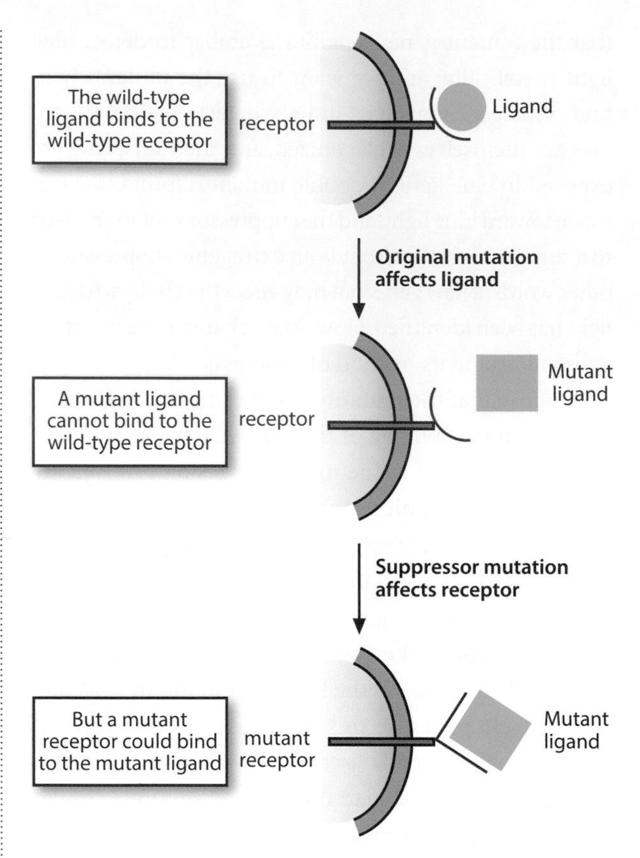

Figure 10.6 Interactional suppressors are specific to both the mutant allele and the gene. A ligand (shown in blue) binds to a receptor (in pink) on the surface of the cell. A missense mutation in the gene encoding the ligand results in an alteration of the shape of the protein, which reduces or eliminates the function of the ligand because it can no longer interact with its receptor. A suppressor mutation in the gene that encodes the receptor may be able to correct the defect and restore the interaction with the mutated ligand. The suppression is allele-specific and will not extend to mutations in the ligand gene with other defects.

In addition, the suppressor will not suppress mutations in genes other than that for the ligand; thus the suppression is gene-specific.

While it is convenient and appropriate to think of interactional suppressors in terms of ligand receptor interactions and other protein–protein interactions, they have also been used to analyse interactions between DNA and binding proteins, DNA and RNA, RNA and proteins, and RNA and RNA. For instance, in the structural analysis of tRNA with rRNA at the ribosome during translation, a mutation in the tRNA sequence that could no longer base-pair with rRNA, could be compensated for—that is, suppressed—by a change in rRNA that restored base pairing. Interactional suppressors are an important tool in the analysis of a biological process, because they not only identify genes with a related function but also help elucidate the type of relationship between the genes. Interactional suppressors provide information about direct physical interactions and the sites or residues that are involved in that interaction—information that other types of suppressors cannot provide. They are thus highly sought after, and suppressor screens can be structured to make their recovery more likely. Because they are specific to both the function of the gene and the defect in the mutant allele, interactional suppressors are often rare; in order to find and use them, it must be possible to screen large number of mutants easily. For this, it is essential to use as the original mutation a missense allele that is known or thought to be at the interaction site of the gene product, since only this type of allele is expected to be suppressed by an interactional suppressor. The case study in this chapter illustrates how interactional suppressors affecting a gene involved in yeast spindle morphogenesis were sought and discovered.

Informational suppressors affect the molecular lesion in the specific allele but not the function of the gene

Some suppressors affect only particular mutant alleles, but they suppress such alleles for several different genes, even when the genes have unrelated functions. Suppressors of this type are known as informational suppressors. In our flowchart in Figure 10.4, informational suppressors are allele-specific suppressors that affect only the molecular defect of the allele but have nothing to do with the function of the gene itself. They are called informational suppressors because they often affect the general transcription and translational machinery, and thus the information flow from gene to protein. When we want to find genes with functions related to the original gene, informational suppressors are less helpful than interactional and gene-specific suppressors.

Informational suppressors were first identified in bacteria and were subsequently found in yeast, *Arabidopsis*, worms, and flies. A eukaryotic example was introduced earlier, as part of our general description of the process of finding suppressors.

Literature Link. Hodgkin et al. 1987, *Trends in Genetics* 3: 325–9

The *sup-5* gene in *C. elegans* was identified by its ability to suppress a mutant allele of the muscle-specific gene *unc-54*. Subsequent work showed that *sup-5* affected only a few particular null alleles of *unc-54*; this suggests that *sup-5* mutations are suppressing some specific molecular defect in the *unc-54* null allele rather than the muscle defect of *unc-54* mutations in general. More significantly, *sup-5* also suppresses some mutant null alleles of many other genes, including ones that have nothing to do with muscle function. The inference is that, since the genes being suppressed have no common function, suppression must be related to the molecular nature of the alleles.

Once the genes and the mutations were sequenced, it became evident that the mutant alleles suppressed by *sup-5* are nonsense mutations resulting in a premature UAG stop codon; the effects are shown in Figure 10.7. You may recall that UAG is a stop codon because no naturally occurring tRNA molecule has the complementary CUA anti-codon. *sup-5* is a mutation in a tRNA gene that results in the amino acid tryptophan being inserted at the UAG amber stop codon and preventing the premature termination of translation, as diagrammed in Figure 10.7. Thus *sup-5* is an example of a common type of informational suppressors: mutations in tRNA genes. It may be helpful here to recall how the genes were named. The normal function of the *sup-5+* gene is to encode a tRNA that inserts tryptophan at UGG codons; it has the complementary CCA as its anti-codon. The mutated gene with the CUA anti-codon now inserts tryptophan at UAG stop codons.

Mutations in tRNA genes are probably the best-studied group of informational suppressors, particularly in bacteria where they were instrumental in understanding the mechanism of translation. However, informational suppressors have also been identified as mutations in the genes for ribosomal proteins and in rRNAs, translation elongation factors, components of mRNA processing, splicing, and many other genes involved in transcription or translation. Again, an example was used in our earlier discussion of suppressor nomenclature. The original *Hairy wing (Hw)* mutant allele in *Drosophila* described earlier arose from the insertion of a *gypsy* transposable element into the *Hw+* gene; inactivation of the gene by the *gypsy* insertion results in the Hairy wing phenotype. The suppressor mutation *su(Hw)* suppresses the defect by affecting *gypsy* transposable elements, and

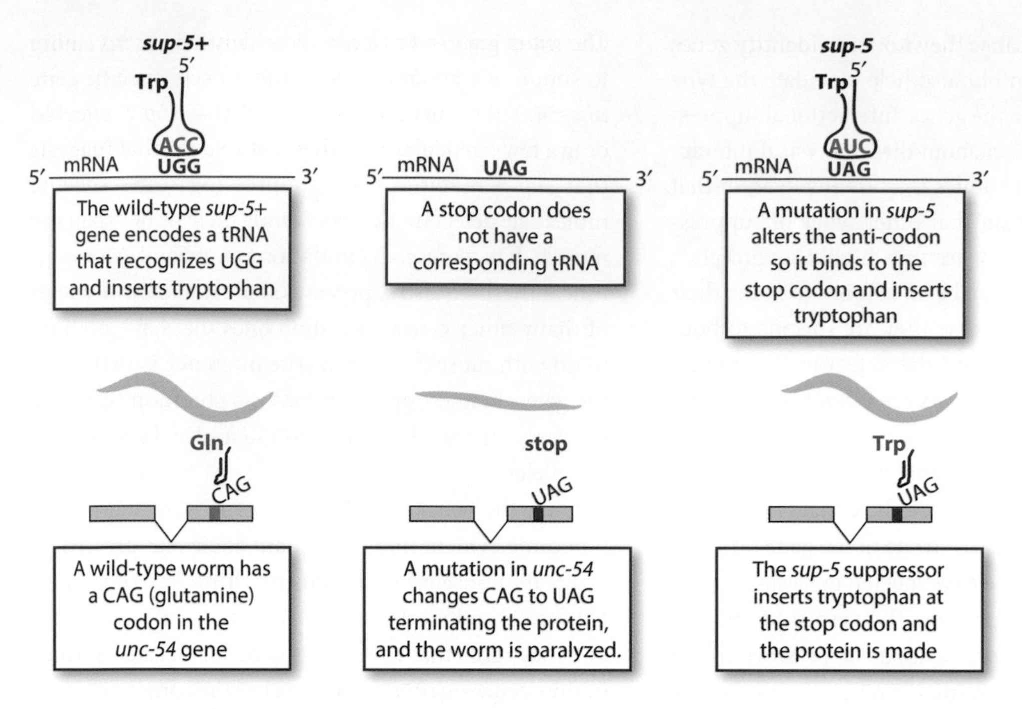

Figure 10.7 Informational suppression involving tRNA genes. The tRNA gene *sup-5* from *C. elegans* is illustrated, but suppressor tRNAs similar to this have been studied in many bacteria and yeast. The top half of the diagram illustrates suppression in terms of the tRNA gene. The wild-type *sup-5+* tRNA recognizes UGG codons and inserts a tryptophan. If the UGG codon is mutated to UAG, the mRNA cannot be translated since no tRNA gene naturally recognizes UAG. This generates a stop codon. However, a compensatory mutation in the *sup-5* tRNA can suppress this stop codon if it recognizes UAG and inserts an amino acid. Although this is shown as inserting the same amino acid as in the wild-type protein, this may not be the case. The lower half of the diagram shows suppression as observed from the phenotype of the worm and the alleles of *unc-54*. In the wild type, the gene has a CAG codon that is recognized by a tRNA. A mutation replaces the CAG with UAG. As a result, the protein is not made and the worm is paralysed. The suppressor *sup-5* mutant tRNA inserts a tryptophan at this stop codon, allowing the muscle protein to be made and movement to be restored. The protein in the suppressed worm has a missense replacement of glutamine with tryptophan, so movement may not be completely normal. Suppressor tRNAs will also act on the naturally occurring stop codons, generating read-through at the C terminus for some proteins. As a result of reading through natural stop codons, many suppressor mutant strains grow poorly.

thus acts as a suppressor for many other mutations that involve *gypsy*. It has nothing to do with the function of the *Hairy wings* gene itself, and a wide variety of other genes with other functions have mutant alleles that can be suppressed by *su(Hw)*.

Informational suppressors have been important to our understanding of the fidelity of transcription and translation but, because they suppress the molecular lesion in a particular allele and not the functional defect in the gene, they are not very useful for finding additional genes that affect the same biological process. This is not a failure of the suppressor screens, as they are in fact picking up interacting genes. It is simply that suppressors screens in the past worked so well that we now understand the overall information flow in the cell.

Gene-specific suppressors affect many different alleles of a gene, but usually only a few other genes

We turn now to the third class of suppressor mutation in Figure 10.4: those that suppress many mutant alleles of the original gene but are specific to this gene in their effect. These suppressors are the class found most frequently in a suppressor screen. They usually identify other genes that affect the same biological process—exactly the result we are looking for. In contrast to interactional suppressors, however, **gene-specific suppressors** often identify functionally related genes whose gene products do not interact directly.

Broadly speaking, there are two ways in which a suppressor of this type can work. These two ways are illustrated in Figures 10.8 and 10.9, but we caution the reader that these diagrams are suggesting somewhat specific modes of action for concepts that are often more abstract. The underlying notion for suppression is that the cell or organism may have multiple genes and pathways that carry out related functions. As shown in Figures 10.8 and 10.9, the original mutation affects a gene in one of the two pathways. A suppressor can arise with a mutation in a gene in a parallel or related pathway or in the same pathway.

For example, as shown in Figure 10.8, a mutation that overexpresses or ectopically expresses an alternative pathway may be able to restore wild-type function, because it bypasses the need for the original gene that was mutated; hence these are sometimes termed **bypass suppressors.** Bypass suppressors that work by affecting a different pathway are known in many different organisms and for many different processes. More examples will be encountered in Chapters 11 and 12, so we will introduce just one example here. As noted before, the *C elegans unc-54* gene encodes a myosin heavy-chain isoform, so mutations in the gene cause paralysis. One suppressor that restores movement to *unc-54* worms is a dominant mutation in another myosin heavy-chain gene known as *myo-3*; the *myo-3* mutation suppressed many different *unc-54* alleles, which suggests that its wild-type function is related to the function of *unc-54*. The suppressor mutations are dominant hypermophic mutant alleles that cause a stable overexpression of a myosin heavy chain, which is usually a minor component of the body-wall myosin. The mechanism of suppression was discovered when *myo-3* was cloned and analysed. In the context of Figure 10.8, the *unc-54* mutation knocked out the top pathway, so the worm is paralysed. The dominant mutation in *myo-3* is represented by the overexpression of the alternative pathway, which produces enough myosin to restore movement.

Bypass suppressors were among the early evidence that genomes contain redundant information. In the case of *myo-3* and *unc-54*, the bypass occurs because

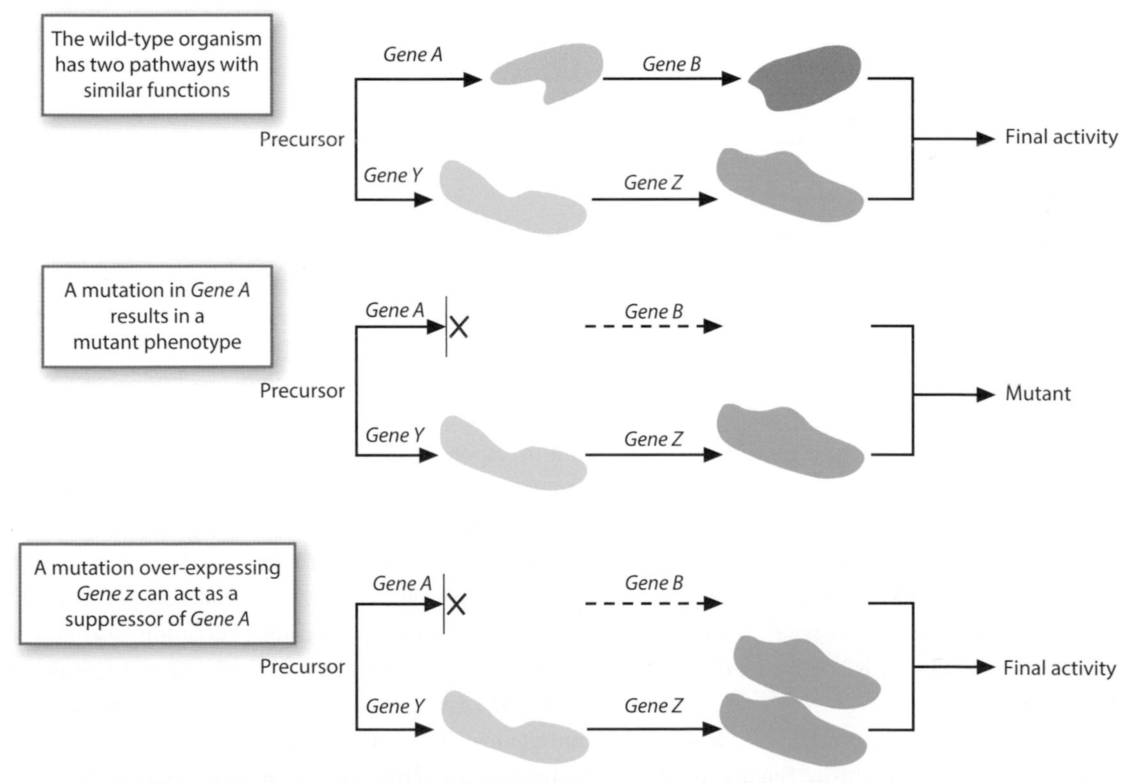

Figure 10.8 Bypass suppression. The organism has two pathways or process that are both necessary for the final outcome. A mutation in the *a* gene results in a mutant phenotype. This mutant phenotype of *a* can be suppressed, for example, by a mutation that overexpresses the other pathway or process, which is here suggested by a hypermorphic mutation in the *z* gene. Many other types of bypass suppression are known, so this diagram is illustrative.

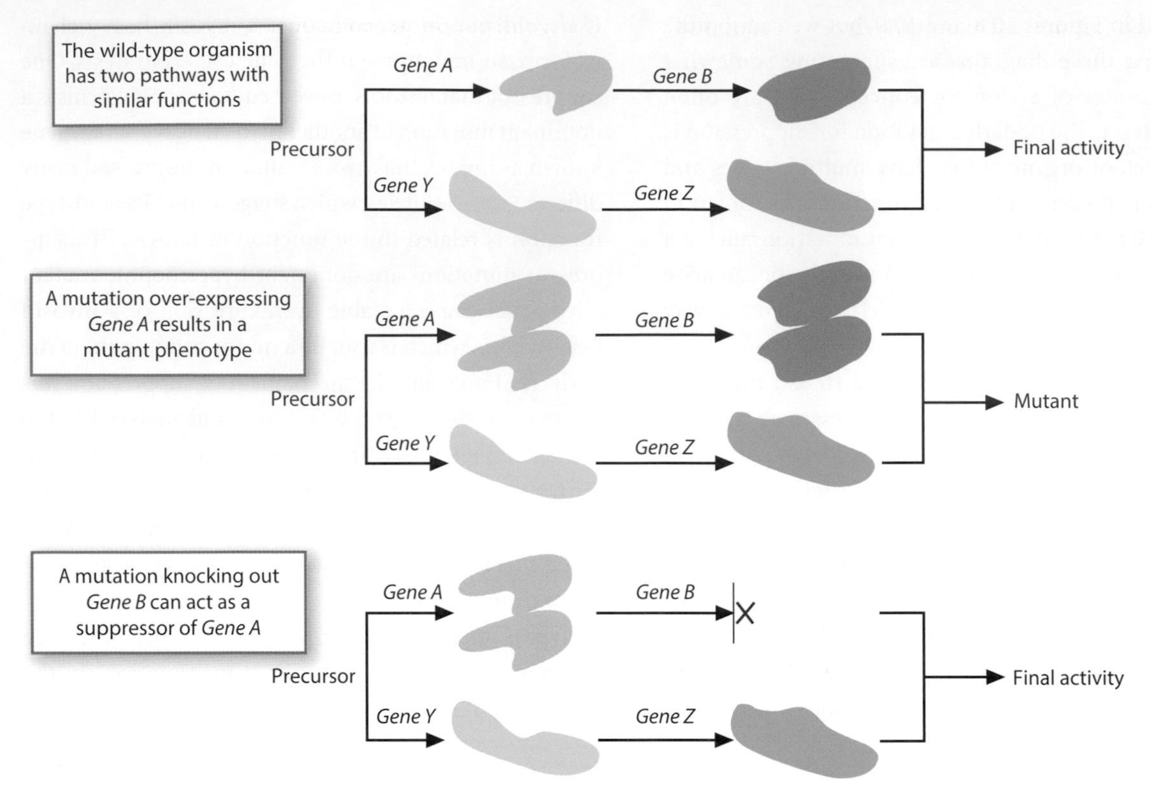

Figure 10.9 Epistatic suppression. The organism has two pathways or process that are, both, necessary for the final outcome. A hypermorphic mutation in the *a* gene results in a mutant phenotype. This mutant phenotype of *a* can be suppressed, for example, by a mutation that knocks out the function of another, downstream gene in the same pathway or process. Many other types of epistasis are known, so this diagram is illustrative.

the worm has paralogous copies of myosin genes with partially redundant functions. We now know from the many eukaryotic genomes that have been sequenced that paralogs and gene families such as the myosin genes are a common phenomenon, as discussed in Chapter 1. Given the abundance of gene families, bypass suppression that activates another member of a gene family is expected to be rather common.

However, not all bypass suppression is due to duplicate genes. Occasionally bypass suppression indicates that the organism has redundant information or redundant pathways to a downstream phenotype. This is not due to similar genes but to alternative pathways that are partially redundant.

The second way in which a suppressor can arise is with another gene in the same pathway (Figure 10.9). For example, the original mutation could be a dominant hypermorphic mutation, so that its mutant phenotype arises from the overproduction or unregulated activity of that pathway. A mutation affecting a gene downstream in the pathway could act as a suppressor, since it eliminates the effects of overproduction.

For example, a gene in *Arabidopsis* known as *CONSTANS* (*CO*) plays a major role in regulating flowering time in response to day length. The expression of the wild-type *CO* is induced in response to long-day conditions; correspondingly, a *CO* null mutation significantly delays flowering under long-day conditions. Transgenic plants that constitutively overexpress *CO*, which is effectively the same as a *CO* hypermorphic mutation, also flower much earlier than wild-type plants. Thus turning on the *CONSTANS* gene appears to turn on flowering. The transgenic plant with constitutive expression of *CO* was mutagenized and mutations that delayed flowering—that is, suppressors of *CO*—were found. These defined additional genes involved in the flowering response under long-day conditions, including the genes *FT* and *SOC*, which appear to be direct targets of regulation by *CO*.

Literature Link. Onouchi et al. 2000, *Plant Cell* 12: 885–900

Suppressors such as these, which affect the same pathway, are sometimes known as epistatic suppressors; since **epistasis** will be described in detail in Chapter 11, we will defer a more detailed discussion of this type of suppression until then. Often the only way to determine whether the suppressor affects another pathway or another step in the same pathway is by a more extensive analysis, and the distinction between bypass and epistasis is somewhat more historical than practical. The example of functional suppression in the case study involves spindle morphology in yeast, although the mechanism of suppression here is not unmistakably due to either bypass or epistasis. Note that, in all of our examples of suppression, the suppressor mutation itself could have a mutant phenotype that may be more apparent than the suppression phenotype.

Suppression can occur through many different mechanisms, depending on the activities of the genes involved. We have used hypermorphic or overproducing mutations in our examples because these are among the easiest ones to imagine and because this is a relatively common situation. Suppression that involves overexpressing the wild-type function of one gene and a loss-of-function mutation in another gene forms the strategy for high copy suppression, as discussed in Tool Box 10.1. For example, recall from Chapter 3 that yeast contains naturally occurring plasmids that can be at high copy number in the cell. The high copy number of the plasmid with the wild-type copy of the gene generally implies that the wild-type function of the gene is being over expressed; that is, the high copy number plasmid is the equivalent of a hypermorphic mutation in the gene.

TOOL BOX 10.1 High copy suppression

We have discussed suppression in terms of how a *mutation* in one gene alleviates the defect of a *mutation* in another gene. But another type of suppression arises when overexpression of the *wild-type* allele of one gene ameliorates the effect of a mutation in a different gene. In this case, the wild-type allele of a gene is acting as an extragenic suppressor. This technique is called **high copy suppression** and has been applied particularly in yeast, where it is easy to overproduce a wild-type gene product by using a high copy plasmid, as discussed in Chapter 3. High copy suppression can be considered analogous to the ability of a hypermorphic mutation to suppress by the knockout of another gene, in either the same or a related pathway. The suppressor gene itself is not mutated, simply overexpressed.

Although high copy number yeast plasmids are widely used, other methods to overproduce a wild-type gene product can also be used, in other organisms as well as in yeast. For example, the wild-type copy of a gene can be cloned in vitro, so that its transcription is under the control of a constitutive or an inducible promoter and regulatory region. Upon induction, the gene is expressed at high levels, higher than might be found physiologically. These are diagrammed in Figure B10.1.

No matter by which mechanism the strains are made, the conceptual strategy underlying a high copy suppression screen is similar to that of other types of suppression. For example, suppose that the original mutation is a null allele that prevents yeast cell division. A library of wild-type genes on high copy number plasmids is transformed into the mutant strain such that each cell has the high copy plasmid for one

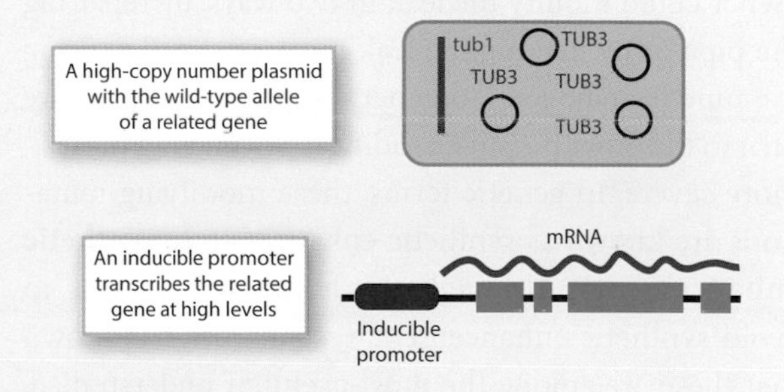

Figure B10.1 High Copy Suppression. High copy suppression in yeast occurs when the mutant phenotype of one gene is overcome by overexpressing a wild-type copy of a separate gene. In the top example, the mutant phenotype of *tub1* is suppressed when the wild-type gene *TUB3*, another member of the gene family, is present on a high copy plasmid and thus overexpressed. Overexpression can also be accomplished by placing the wild-type copy of a gene (marked in blue) under the control of an inducible promoter (marked in red) that allows high levels of transcription, as illustrated in the lower figure.

gene. Colonies that grow and divide normally are selected for and thereby suppress the mutant phenotype. The plasmid from these suppressor colonies is isolated and the gene on the plasmid is analysed.

High copy suppression has advantages over conventional suppressor studies, including that the genes being screened for suppression have already been cloned. This avoids mapping and cloning the suppressor gene, which are among the most challenging steps in conventional suppressor analysis in other organisms. The function of the gene on the plasmid is also recognized, because it acts as a suppressor regardless of the phenotype it might have. The plasmid with the cloned gene can also be readily transformed into other genetic backgrounds, simplifying the manipulations to test other alleles and other genes for suppression. By comparison to the methods of suppression already described, high copy suppression is relatively easy to perform in eukaryotes, especially yeast, and is more directly connected to the molecular analysis of the genes involved.

An example of high copy suppression involving two α-tubulin genes in yeast is described in the case study. As in the *C. elegans* example with the hypermorphic allele of *myo-3* suppressing the null allele of *unc-54*, the nature of the suppression in the yeast case is clear from the molecular analysis of the genes involved and from the recognition that the suppressor gene encodes a similar product, which can directly substitute for the mutant product. It is important to realize, however, that it is the *overexpression* of the related gene that confers suppression. The endogenous wild-type copy or copies of the suppressor gene that are present in the wild-type genome have not been sufficient to cause suppression on their own, possibly because they may not be expressed in the relevant cell type or at the required time.

10.3 Genetic interactions: Synthetic enhancers

We compared a cell or an organism with a mutation to a homeowner with a leaky pipe. In our analogy, the homeowner could modify the leak in two ways: by repairing the pipe or by making the leak more severe. Repairing the pipe is analogous to genetic suppression. Now we turn to the other possible modification, making the leak more severe. In genetic terms, these modifying mutations are known as synthetic enhancers (see **synthetic enhancement**). Although the homeowner wants to avoid synthetic enhancement, recent work has shown that these are among the most useful, if understudied, types of modifying mutations when it comes to understanding a biological process.

Synthetic enhancers are mutations that exacerbate the effect of the original mutation

Many different modifying mutations can occur that make a mutant phenotype more severe. Although such mutations can simply be called enhancers, for the reasons described at the beginning of the chapter we will call them synthetic enhancers. We call them "synthetic" because the effect of the mutant enhancer is seen only when it occurs in combination with a mutation in a different gene; the word "synthesis" means the combination of two or more things into a new thing. Historically, synthetic enhancement has not received the attention that suppression has, and has not been as widely used as a genetic tool.

Literature Link. Guarente 1993, *Trends in Genetics* 9: 362–6

In part, this arises from the point made at the beginning of the chapter: apparent enhancement may arise because two mutations, each of which is mildly deleterious to the organism, combine in a non-specific fashion to make the organism even less viable, or even dead. If I am already in a bad mood because I have a cold, the lack of a parking space close to my building can put me in a really foul temper. This type of non-specific enhancement is rarely interesting or informative.

Synthetic enhancement includes two different but related effects on mutant phenotypes. The first effect is that a mutant phenotype arising from one gene is made more severe by a mutation in a second gene. For example, imagine that a mutation results in a phenotype that is 60 percent penetrant; in other words, among a population of genotypically mutant organism, 60 percent of them display a mutant phenotype and 40 percent have a wild-type phenotype. A synthetic enhancer of this mutant phenotype will result in a higher penetrance, so that perhaps 100 percent of the genotypically mutant organisms are phenotypically mutant. We will use this effect in describing synthetic enhancement of the dauer formation phenotype in *C. elegans* in the case study in Chapter 11, so a more complete discussion is found there.

Another, potentially more interesting, situation arises when the mutation in one gene results in no obvious mutant phenotype. The mutant phenotype is only observed when a second specific gene is also mutated. These are also known as the synthetic enhancers; the mutant phenotype is not just made more severe by the second mutation, it only is observed when both genes are mutant. We will use this effect in what follows, in describing genes that affect vulva formation in *C. elegans*. In many cases the mutation in the first gene results in no mutant phenotype, whereas the mutation in the second gene results in a lethal phenotype. This is known as **synthetic lethality**; the concept is the same as in other examples of synthetic enhancement, but the phenotype is lethality. In all these manifestations, the significance of synthetic enhancement lies in the *specificity* of the interaction. In terms of genotypes, the *a/a* mutant has a wild-type phenotype and the *b/b* mutant has a wild-type phenotype. However, the double mutant *a/a; b/b* has a mutant phenotype. In this case, the *b* mutation is a synthetic enhancer of the *a* mutation (and the *a* mutation is a synthetic enhancer of the *b* mutation). Enhancement does not arise from the combination of two different and quite unrelated deleterious effects, since each mutant alone has a wild-type phenotype and the interaction is specific to these two genes (and possibly to a few others with related functions).

As discussed in connection with suppressors, synthetic enhancement is evidence that genomes have redundant information. It can be used as a tool to understand the function of a gene whose activity is partially or completely redundant with another gene or pathway. The ability to carry out genome-wide mutant screens and to make targeted mutations has sparked new interest in synthetic interactions, which are currently in wide use. As will be discussed in Chapter 12, synthetic enhancement is probably more common than suppression.

Synthetic enhancement can involve duplicate or paralogous genes

Our first example illustrates how synthetic enhancers can indicate the presence of functionally redundant paralogs, even if such genes have not been previously detected. The genes involved in this example encode the enzyme acetylcholine esterase in *C. elegans*, and the experiments were done many years before any of the genes or the genome was sequenced. The situation is shown in Figure 10.10. Acetylcholine esterase breaks down the neurotransmitter acetylcholine and can be detected by a relatively simple colorimetric assay in worms, shown in Figure 10.10 as "ace activity." Mutations were identified by finding worms with reduced histochemical staining for the enzyme. Worms homozygous for a null allele of the gene for acetylcholine esterase (*ace-1/ace-1*) have reduced levels of the enzyme by colorimetric assays but have no obvious defect in either movement or behavior, as summarized in the phenotypes on the right of graph in Figure 10.10. Given the widespread importance of acetylcholine as a neurotransmitter, the lack of a mutant phenotype was initially puzzling.

To address this, the *ace-1* mutant strain was itself mutagenized; among the offspring were some that had a distinctive uncoordinated movement. Standard mapping experiments showed that the uncoordinated mutant phenotype was due to a second unlinked recessive mutation in a hitherto unknown gene, which was then called *ace-2*. As summarized in Figure 10.10, the uncoordinated worm was a double mutant, homozygous for the *ace-1* mutation and an *ace-2* mutation. Colorimetric assays for the enzyme indicated a further reduction of acetylcholine esterase in these uncoordinated

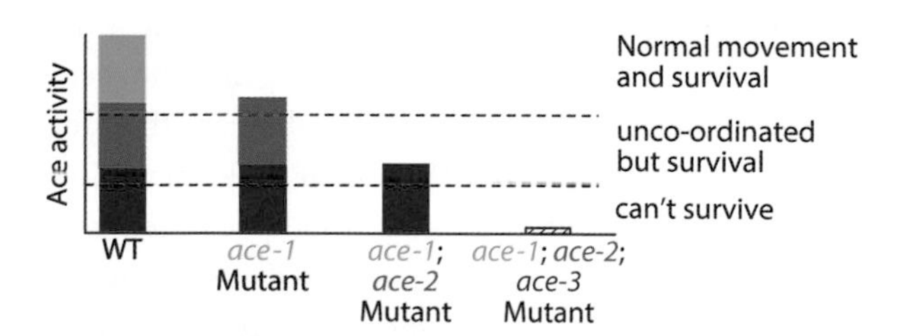

Figure 10.10 Synthetic enhancement, as illustrated by the acetylcholine esterase (*ace*) genes in *C. elegans*. Each of the three genes, *ace-1*, *ace-2*, and *ace-3* makes a different isoform of the enzyme, as shown in the different colored bars. The total Ace activity is based on histochemical stains. Threshold levels for different phenotypes are inferred from the results with the mutants. In worms homozygous mutant for either *ace-1* or *ace-2*, the amount of acetylcholine esterase is reduced. However, the residual amount produced from the two remaining genes is enough to ensure normal movement. A double mutant *ace-1; ace-2* reduces the amount of the enzyme further, leaving only the *ace-3* gene active, and has impaired movement but is still alive. The triple mutant that completely lacks the activity of the enzyme does not survive. The presence of three genes was inferred from the results with synthetic enhancement.

ace-1/ace-1; ace-2/ace-2 double mutants. Both *ace-1/ace-1* and *ace-2/ace-2* single mutants have a wild type phenotype, but the double mutant has the uncoordinated mutant phenotype.

Although it was not certain until both genes were cloned, the reduction of acetylcholine esterase in each of the *ace* gene mutations suggests that *ace-2* also encodes acetylcholine esterase. The interpretation of this interaction is that normal movement requires a particular threshold level of acetylcholine esterase, and wild-type worms make levels of the enzyme well above this threshold. Even if one of the *ace* genes is knocked out, the level of the enzyme is above this threshold, and the worm moves normally. Only when both *ace* genes are knocked out does the enzyme level fall below the movement threshold, so that the double mutant worm is uncoordinated.

However, even the double mutant has some residual acetylcholine esterase by colorimetric assays. This suggested the presence of yet a third *ace* gene. The *ace-1; ace-2* double mutant was mutagenized to search for this third gene. This time, synthetic lethal mutations were found; these are mutations that resulted in lethality only when both *ace-1* and *ace-2* were also homozygous mutant. These studies identified the third acetylcholine esterase paralog *ace-3*. As shown in Figure 10.10, this suggested the presence of another mutant phenotype threshold, this time one needed for survival.

Mutants that knock out only one of *ace-1*, *ace-2*, or *ace-3* have a wild-type phenotype. The inference is that, even if the isoform encoded by one *ace* gene is not being made, enough of the enzyme is made from the other two wild-type genes that the worm is normal. However, each knockout of an additional *ace* gene results in a more severe mutant phenotype. If two genes are knocked out, the worm is uncoordinated. If all three genes are knocked out, the worm cannot survive (likely because it cannot flex enough to break through the egg shell). The number of acetylcholine esterase genes was not known at the time the genetic analysis was done. The presence of three genes with redundant functions was inferred from the synthetic enhancement; and it was confirmed when the genome was sequenced many years later. By beginning with one mutation that had no obvious mutant phenotype, it was possible by synthetic enhancement to find the other genes that had related functions.

Synthetic enhancement and bypass suppression can be indications of the same effect

Genomes have paralogous genes for many gene families; with *unc-54* and *myo-3* and with the *ace* genes, we have seen two examples in which genetic evidence showed this well before the genome had been sequenced. But *myo-3* and *unc-54* are an example of suppression, while the *ace* genes are an example of synthetic enhancement. How, then, can the existence of genes with duplicate or highly similar functions account for two apparently opposite genetic effects?

The key to thinking about this phenomenon lies in the nature of the mutations for each gene and in the threshold of gene activity at which a mutant phenotype is observed. For synthetic enhancement to occur, as in Figure 10.10, a knockout of the first gene has no effect on the phenotype, because the product of the second gene is sufficient for activity above the critical threshold. Thus a mutant phenotype will only be observed if the second gene is also knocked out.

In the example of bypass suppression in Figure 10.8, the knockout of the first gene results in a *mutant* phenotype. However, this mutant phenotype can be suppressed by overexpression of the second gene. The investigator will observe this as bypass suppression. In each example, the two mutated genes have related functions. The differences between these two situations lie in two characteristics: the critical threshold of gene activity needed to observe a mutant phenotype; and the nature of the mutation in the second gene. In each case we have been able to use a mutation in one gene as our tool for finding other genes with related functions.

Synthetic enhancement can involve parallel biological pathways

Another example from *C. elegans* illustrates that synthetic enhancement need not be limited to one gene. In fact entire pathways of genes with overlapping functions can be found by synthetic enhancement screens. In Chapter 3 we described genes that affect vulva formation in worms. Some of the mutations resulted in a phenotype referred to as Multivulva or Muv, in which the mutant worms have multiple partial vulvae. This phenotype is shown in Figure 10.11. When these Muv mutants were mapped and analysed, one Muv mutant strain proved to be a double mutant homozygous for

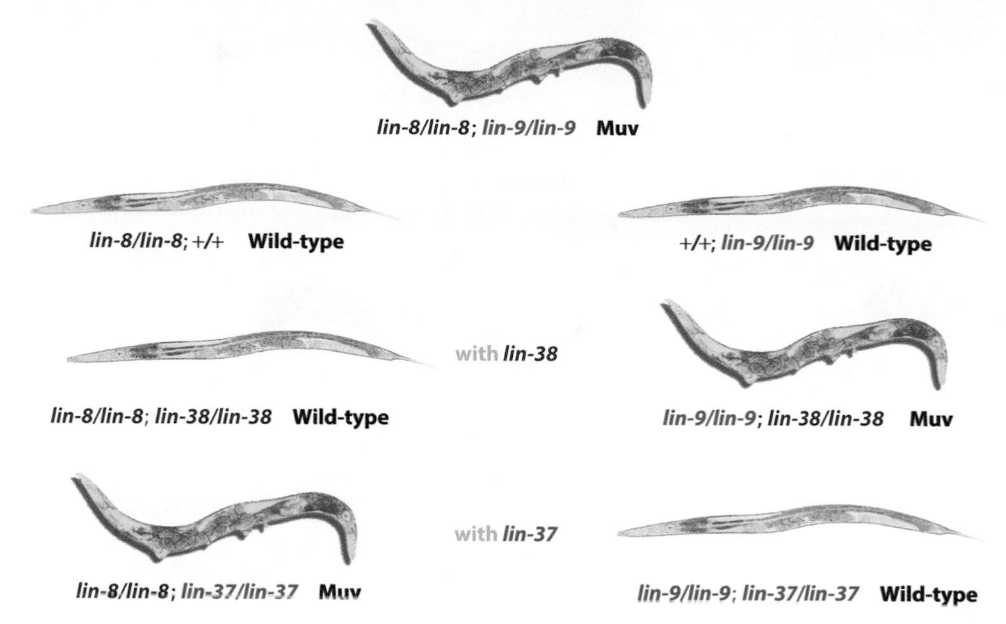

Figure 10.11 Synthetic enhancement, as illustrated by genes affecting vulva lineages in *C. elegans*. A double mutant that is homozygous mutant for *lin-8* and *lin-9* exhibits a multivulva (Muv) phenotype. A worm that is homozygous mutant for either *lin-8* or *lin-9* alone exhibits a normal phenotype. The genes function in redundant pathways, and both pathways must be mutant before a mutant phenotype is seen. This is illustrated by mutations in two other genes. The *lin-38* gene works in the same pathway as *lin-8*, so a *lin-8; in-38* double mutant is of the wild type. However, *lin-38* and *lin-9* are synthetic enhancers of each other, so a *lin-9; lin-38* double mutant has a Muv phenotype. Similarly, *lin-37* works in the same pathway as *lin-9*, so it does not enhance *lin-9* mutations but does act as synthetic enhancer of *lin-8*, and is also a synthetic enhancer of *lin-38*.

two unlinked recessive mutations in two different genes. These two genes were named *lin-8* and *lin-9*. Note from Figure 10.11 that the *lin-/lin-8* and *lin-9/lin-9* single mutant strains each have a wild-type phenotype, so it was fortuitous that this double mutant was observed. (In fact it is possible that each gene had been mutated in previous mutant screens but, because the mutant phenotype is only seen in double mutants, the mutations were not recognized.) Since both mutations needed to be homozygous in order to observe the Muv phenotype, *lin-8* and *lin-9* are synthetic enhancers of each other, which suggests that they affect duplicate or overlapping functions.

If this were the only effect of the two genes, the mechanism of synthetic enhancement between *lin-8* and *lin-9* could be similar to that observed in the preceding example of the *ace* genes. A difference became apparent, however, when each of the single mutant strains was used in a direct screen for additional synthetic enhancers, as shown in Figure 10.12. That is, *lin-8/lin-8* worms, which have a wild-type phenotype, were mutagenized and Muv offspring were recovered. Likewise, *lin-9/lin-9* worms were mutagenized and Muv offspring were recovered. These Muv offspring are double mutants, with the phenotype arising from synthetic enhancement. If *lin-8* and *lin-9* were simply functionally redundant paralogs, the synthetic enhancers of *lin-8* would include alleles of *lin-9* and possibly a few other functionally related genes. More significantly, if *lin-8* and *lin-9* were paralogs, the genes that were synthetic enhancers of *lin-8* would also be expected to be synthetic enhancers of *lin-9*. This is what was observed in the functionally equivalent *ace* genes; each mutation acts as a synthetic enhancer of the others.

However, this is not what was observed for *lin-8* and *lin-9*. When *lin-9/lin-9* worms were mutagenized and synthetic enhancers were recovered, several genes in addition to *lin-8* acted as synthetic enhancers of *lin-9*. Besides, several genes in addition to *lin-9* acted as synthetic enhancers of *lin-8* in the parallel experiment. Most significantly, the mutants that enhanced *lin-8* did not enhance *lin-9* and did not enhance each other; likewise, genes that were synthetic enhancers of *lin-9* did not enhance *lin-8* and did not enhance each other. Two examples are shown in Figure 10.11, with the genes

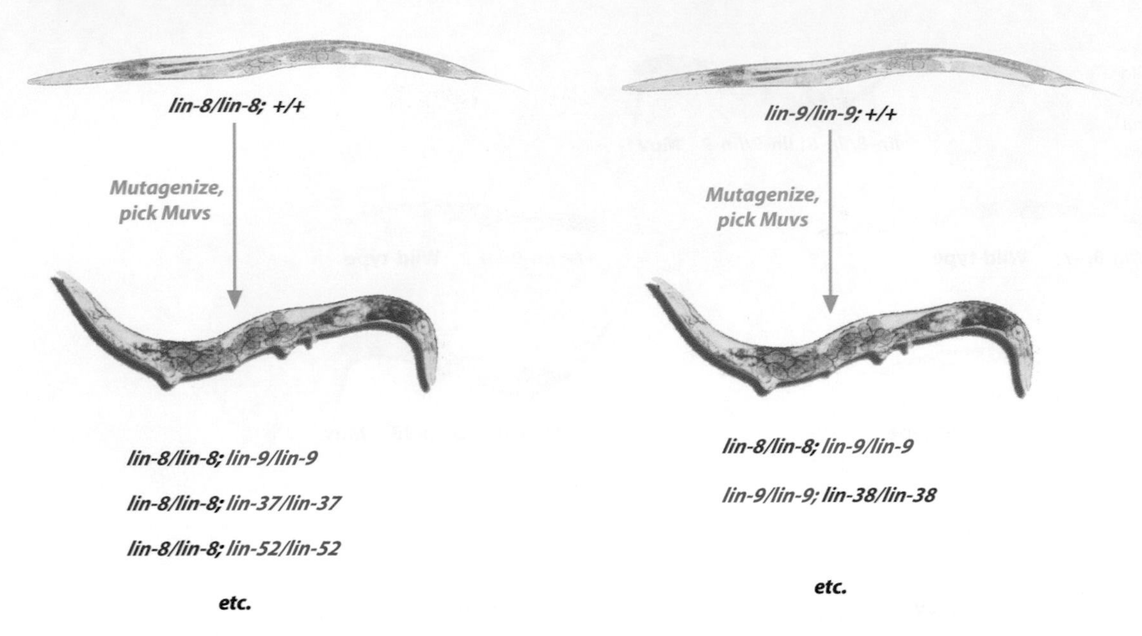

Figure 10.12 Screening for additional synthetic enhancers. The Muv phenotype in these mutants only arises when a gene from class A (such as *lin-8*) and a gene from class B (such as *lin-9*) are simultaneously mutant. In order to find additional class B mutations, a *lin-8* worm, which has a wild-type phenotype, is mutagenized and Muv mutants are found. These are class B mutants, including *lin-9*. A similar screen for additional class A mutant can be done by beginning the mutagenesis with a *lin-9* mutant strain.

lin-38 and *lin-37*. Mutations in *lin-38* act as synthetic enhancers of *lin-9* but not as synthetic enhancers of *lin-8*. Likewise, mutations in *lin-37* act as synthetic enhancers of *lin-8* but not as synthetic enhancers of *lin-9*. As might be predicted, *lin-37* and *lin-38* are also synthetic enhancers of each other. The synthetic enhancers fell into two distinct classes—the *lin-8*-related genes and the *lin-9*-related genes.

The newly identified genes were not paralogs of either *lin-8* or *lin-9*, and they were not paralogs of one another. Instead the two sets of synthetic enhancers were defining two functionally redundant pathways, both involved in vulval development. The mutant Muv phenotype from these pathways is observed only when both pathways are mutated; if either pathway is functional, the worm is wild type. The two classes of genes, called class A and class B, now comprise at least eight different genes. A Muv mutant phenotype is observed for these genes only when the worm is doubly mutant for both a class A mutation and a class B mutation. *lin-8* and *lin-38* are members of the A class, whereas *lin-9* and *lin-37* are members of the B class.

Literature Link. Ferguson and Horvitz 1989, *Genetics* 123: 109–21

Molecular analysis of these genes has shown that, unlike the cases above, the two classes do not encode duplicate genes. In fact genes in class B appear to affect cell-cycle progression in G1 and include the *C. elegans* ortholog of the human *Rb* gene, a cell cycle-related gene that is mutated in human retinoblastoma. Given the relevance of the *Rb* gene to human cancer, its role has been extremely thoroughly analysed. But none of the class A genes encodes a protein that is known to interact with Rb. Clearly the two pathways have related functions. However, the functional redundancy of the pathways, inferred from the synthetic enhancement, remains to be solved.

As more genomes are being sequenced and analysed, there has been renewed interest in synthetic enhancers and in functionally redundant pathways. Figure 10.13 depicts the general logic for synthetic enhancers and functionally redundant pathways, a concept that we consider in much more detail in Chapter 12. We noted in Chapter 9 that genome-wide mutant screens have shown that many genes have no mutant phenotype alone. For instance, only about 20–25 percent of genes are essential for growth under lab conditions, the remaining 75–80 percent of the knockouts giving no mutant phenotype. However, most of the genes that appear to be wild type as

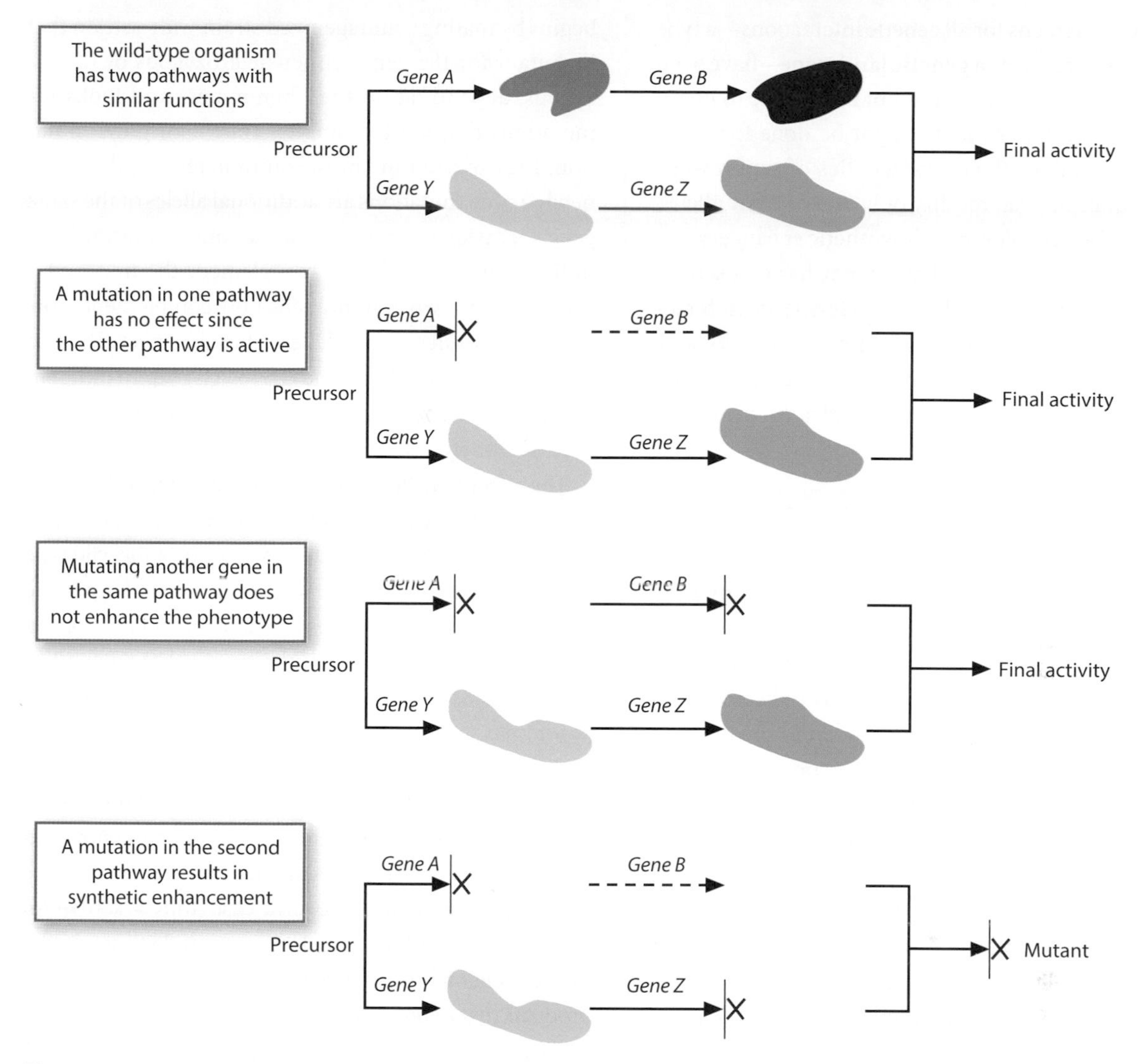

Figure 10.13 Synthetic enhancement and two pathways. Two hypothetical pathways with partially redundant functions are active in the wild type. Either of the pathways is sufficient for wild-type function, so a mutation in gene *a* does not result in a mutant phenotype because the other pathway is still functional. A second mutation that knocks out a gene in the other pathway (gene z) will result in a mutant phenotype by synthetic enhancement, while a second mutation that knocks a gene in the same pathway (gene *b*) will have no additional effect.

single mutant strains have a synthetic lethal phenotype in combination with another mutation.

It is somewhat humbling for a geneticist to realize that much of what we have learned from Mendelian genetics has come from a minority of the genes in the genome, namely those that have an easily recognized mutant phenotype. Potential synthetic enhancer mutations are probably found in every mutagenesis screen but, because there is no mutant phenotype, the mutation has been overlooked. Synthetic enhancers are easier to find when the first gene has a mutant phenotype and the second mutation makes it more severe. On the other hand, this enhancement is more likely to be non-specific or to involve genes whose functions are not closely related. The synthetic enhancers in which no mutant phenotype is observed from the first gene are much more difficult to find but are probably much more likely to identify genes with related functions.

The findings described in these experiments fulfill the goal of a synthetic enhancement screen: to use mutations in one gene in order to find functionally related genes. Thus, synthetic enhancement screens can be a very useful method to find other genes affecting the same biological process. As we will discuss in Chapter 12, genome-wide screens for all genetic interactions—a type of process referred to as a **genetic landscape**—have been done in yeast; it is daunting but not impossible to imagine that similar experiments might be done for other organisms. Many of the characteristics of suppressors, such as mutations that modify only one or a few alleles, have not yet been explored with synthetic enhancers. For example, most synthetic enhancement has focused on null alleles, but the possibility of allele-specific interactions for one or both genes should not be overlooked. It seems very likely that we have only begun to explore the potential information that can be gleaned from synthetic enhancers.

Non-allelic non-complementation occurs when both mutations are heterozygous

Synthetic enhancement, as described so far, involves organisms that are mutated for two different genes, neither of which has a mutant phenotype on its own. In the cases we have described, each gene has to be homozygous mutant for an effect to be seen and no wild-type alleles are present in either strain. A possibly related phenomenon occurs in *heterozygotes* and is referred to by the somewhat unwieldy name of non-allelic non-complementation—or, particularly in yeast, as unlinked non-complementation. This is most often found unexpectedly during a mutagenesis scheme, and its general relevance for gene interactions is not yet clear.

In Section 4.5 we described the use of a non-complementation screen or an F_1 screen to identify additional mutant alleles in a gene of interest; such a screen is summarized in Figure 10.14. In it, the investigator begins by mating a mutagenized strain with a strain that is mutant for the gene (either heterozygous or homozygous, depending on the phenotypes) and looks for mutations that fail to complement the original mutation. The expected and most common result is that these newly arisen mutations are additional alleles of the same gene. Occasionally, however, mutations in a second and unlinked gene will fail to complement the first mutation; this is known as non-allelic non-complementation. The name, while awkward, describes the effect. The two mutations are not alleles and may even map to different chromosomes, but they still fail to complement to provide the function that the other mutation lacks.

The effect may be understood by looking at the genotypes. In non-allelic non-complementation, both the *a*/+ heterozygote and the *b*/+ heterozygote are of the wild type because each mutant allele is recessive to wild type. The *a*/*a* and the *b*/*b* homozygotes display a mutant phenotype, often a very similar mutant phenotype. Although all of the mutations are recessive, the *a*/+; *b*/+ double heterozygote has a mutant phenotype. (For comparison and review, with synthetic enhancement the *a*/*a* genotype has a wild-type phenotype, whereas the *a*/*a*; *b*/*b* genotype has a mutant phenotype.) This is not a rare phenomenon, and examples have been reported in many different organisms for many different processes. The case study describes an example with spindle formation in yeast. Frequently the non-complementing genes have been shown to encode products that interact physically, which suggests that non-allelic non-complementation might be particularly helpful in analysing protein complexes. But other examples involve genes whose products do not appear to interact

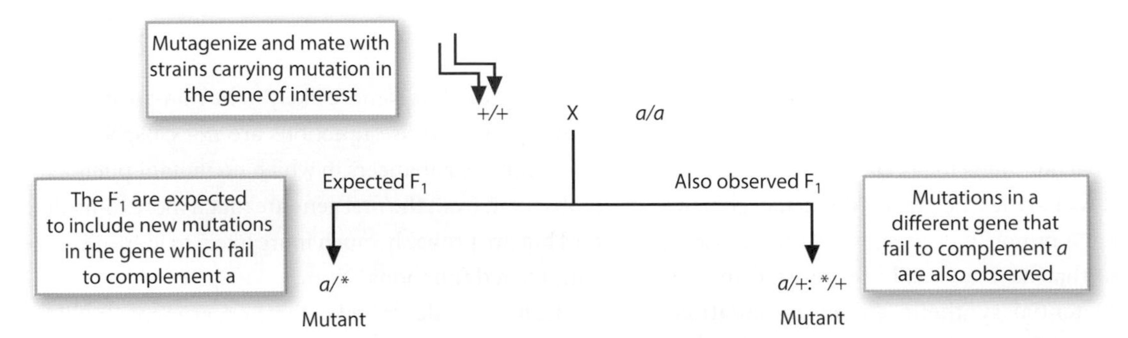

Figure 10.14 Non-allelic non-complementation. Non-allelic non-complementation is often found in F_1 mutant screens, as shown here. The goal of an F_1 screen is to find additional alleles for a gene of interest. In a F_1 screen, mutagenized individuals are crossed with ones with the recessive mutant allele *a* in the gene of interest, and F_1 individuals with a mutant phenotype are identified. Most of these will be additional alleles in the gene of interest that fail to complement the initial mutation *a*. However, a recessive mutation in another gene may also be found. The new mutation fails to complement *a* but lies in a different gene, so it is not an allele of the original gene of interest.

physically, so the inference of a physical interaction should be drawn with caution. Even if a physical interaction does not occur, non-allelic non-complementation can be an effective, and possibly underused, approach to finding genes that affect the same biological process.

Two general models have been proposed to explain non-allelic non-complementation, each of which can be illustrated with an example. First, the underlying biological process may be sensitive to *dosage* of the two products, such that a simultaneous reduction in both proteins produces a mutant phenotype. An example of this type of genetic interaction is seen between two genes involved in neuronal migration in *Drosophila*, as depicted in Figure 10.15. During neuronal migration, the ligand encoded by the *slit* gene binds to the receptor encoded by the *robo* gene. Mutations in either *slit* or *robo* are recessive, so each heterozygote is normal. Mutant homozygotes for either gene have axons that cross the ventral midline inappropriately, resulting in a mutant phenotype; as expected for an allelic series, the null allele of *slit* has a more severe mutant phenotype than a hypomorphic allele. Note, however, that flies heterozygous for both *slit* and *robo* have a mutant phenotype, with axons that cross the ventral midline; these are indicated by the arrows in the picture of the lower right of Figure 10.15. The interaction is not allele-specific, as many different alleles of each gene fail to complement each other.

The hypothesis of a dosage effect was explored using *slit* hypomorphs and null alleles.

Literature Link. Kidd et al. 1999, *Cell* 96: 785–94

Many alleles of *slit* have been found, and it is possible to rank the strength of the alleles on the basis of the penetrance of the *slit* mutant defect. Different alleles in the series were tested for non-complementation with *robo*. The most severe loss-of-function mutants in *slit* homozygotes are also the most severe mutations for failing to complement *robo* in double heterozygotes. In fact the allele order derived from the phenotype in *slit* homozygotes is similar to the allele order derived from the phenotype in *slit*/+ *robo*/+ double heterozygotes. This is consistent with non-allelic non-complementation arising from a dosage effect of interacting proteins.

A second example of non-allelic non-complementation has postulated that the two mutations each result in a

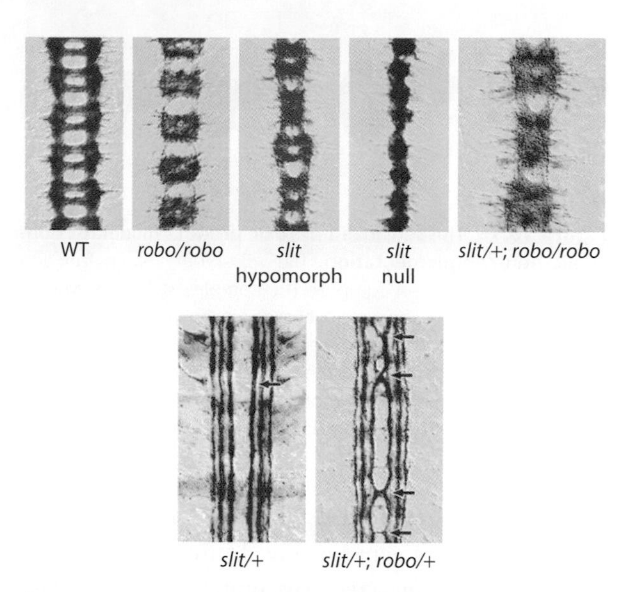

Figure 10.15 Non-allelic non-complementation between *robo* and *slit* in the *Drosophila* central nervous system. The embryonic nervous system is stained with a monoclonal antibody that recognizes all commissural and longitudinal axons. Top row. In wild-type embryos, axons cross the midline in a regular pattern. In *robo* mutant embryos, the longitudinal axons are reduced and the commissural axons cross and re-cross the midline. A similar phenotype is seen in *slit* hypomorphic alleles and null allele; in the *slit* null allele, the longitudinal axons do not form, and the axon scaffold is greatly compressed. A similar compression is seen in a *robo* homozygous and *slit* heterozygote; compare the width of the scaffold in the second picture and the fifth picture. This is a synthetic enhancement of *robo* by *slit* in *slit/+; robo/robo*. Bottom row. *A slit/+* heterozygote is normal, as is a *robo/+* heterozygote (not shown). However, a *slit/+; robo/+* double heterozygote has axons that cross the midline abnormally. The two genes exhibit non-allelic non-complementation. From Kidd, Bland and Goodman, 1999, Slit is the midline repellent for the robo receptor in *Drosophila*, *Cell* 96: 785–94. The order of the original figures has been rearranged.

sensitized product in a macromolecular complex, as diagrammed in Figure 10.16; this type is sometimes referred to as a poison model. In this model, an altered product of the first gene has no visible phenotype in a heterozygote, but it renders the entire complex sensitive to a second defect; that is, it "poisons" the structure of the complex in some way. The second defect could come either from a second mutation in the same gene (that is, a homozygote) or from a mutation in an interacting gene (that is, two heterozygotes, and hence non-allelic non-complementation). An interaction with the sensitized product is not necessarily expected to follow an allelic dosage series but might depend on the particular defect in each allele.

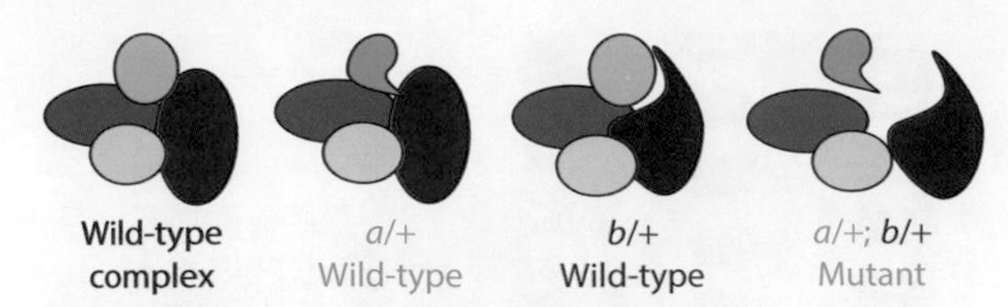

Figure 10.16 The sensitized or toxic product model for non-allelic non-complementation. The wild-type proteins function as part of multi-protein complex. The complex still forms when a recessive mutation in one gene occurs; this is shown here as a change in the blue subunit. Similarly, a recessive mutation in a second gene, shown here as an alteration in the purple component, also has a wild-type phenotype, and the protein complex forms. In the double heterozygote *a*/+ *b*+ the complex cannot form normally and a mutant phenotype arises.

This model has been used to explain the interaction between two genes involved in neurotransmitter release. In *C. elegans*, a recessive mutation in *unc-64* fails to complement a recessive mutation in the unlinked gene *unc-13*. The proteins encoded by both *unc-64* and *unc-13* are involved in neurotransmitter release at the synapse. They are part of a larger complex of interacting proteins: a mutation in one gene, while having no mutant phenotype on its own, makes the complex more sensitive to mutations to other proteins in the complex. In contrast to *robo* and *slit*, some alleles of *unc-13* and *unc-64* do not show non-allelic non-complementation, and the interactions do not follow an allelic series. This result suggests that the interaction between *unc-13* and *unc-64* is not simply a dosage effect. Instead, it has been inferred that the mutation in one gene (*unc-13*) alters the complex enough to make it more sensitive to mutations in other genes (for example, *unc-64*) that affect the same process.

Like synthetic enhancement, non-allelic non-complementation has usually been observed between genes involved in the same or closely related processes. Although it is difficult to know how commonly this type of interaction occurs, non-allelic non-complementation has proved to be a useful and informative property of the two genes when it has been encountered. The case study describes examples of synthetic enhancement and non-allelic non-complementation in yeast mitotic spindle morphogenesis.

10.4 Genetic interactions: Summary

Many different concepts, with many years of thoughtful genetic analysis supporting them, have been discussed in this chapter so far. For most readers, it is unlikely that these concepts were covered in an introductory genetics class, so the ideas might be quite new even though geneticists have used them for a long time. Thus, before moving to our next topic, it seems appropriate to make a very brief summary.

The preceding sections have shown how interactions between two genes can be found on the basis of their effects on the mutant phenotype in the double mutant. By comparison to the original mutant, the double mutant might resemble the wild type more closely (that is, be an example of suppression) or might be more severely mutant (that is, be an example of synthetic enhancement). The interactions are usually observed in homozygotes (although suppressors can be either recessive or dominant), but non-allelic non-complementation occurs between heterozygotes.

These genetic interactions, that is, those based on mutant phenotypes, may indicate that the relevant gene products interact with each other directly; but a direct interaction between gene products is not required. This is both a strength and a possible weakness. Because the interactions are based on gene activity, which may not be easily defined in molecular terms, many significant but unexpected *functional* interactions are found. On the other hand, some of the more important *molecular* questions—such as about the types of molecules, about their actual contacts with each other, or about their stoichiometry and expression levels—are not answered. The genetic methods based on mutant phenotypes can often be used to provide a broad-brush picture of a biological process. The details of the process require methods that are based on cloned genes. We address these methods next.

10.5 Molecular interactions between proteins

Protein–protein interactions are central to every cellular process. In fact it is hard to think of a protein that does not carry out its function by interacting with other proteins, or of cellular processes that do not depend on protein–protein interaction. In this section we briefly describe two widely used methods to look directly at protein–protein interactions. These methods rely on having cloned one or more of the genes of interest; they require some additional molecular tools; and they have proved to be adaptable to many organisms. The methods will be described here as supplemental to the genetic methods, and we will not attempt an exhaustive review; that would be more appropriate to a biochemistry course. The first method to be described, the yeast two-hybrid assay, uses a transcriptional activation procedure designed to detect binary protein interactions. The second method, co-immunoprecipitation, relies on the specificity of antibody:antigen interactions to precipitate a macromolecular complex. Many other biochemical methods have been developed to analyse protein complexes that are beyond the scope of this book.

Yeast two-hybrid assays use a genetic approach to discover protein–protein interactions

Protein interactions in macromolecular complexes have been analysed for decades, by biochemical methods such as cross-linking and complex purification. The technical challenges of these biochemical experiments led to the development of simpler alternatives to examine protein–protein interactions. Of these, the most widely used is the **yeast two-hybrid (Y2H) assay**, a genetic method that uses transcriptional activation as the assay for protein interactions. The method uses the selection power of a yeast genetic screen to examine protein interactions. Because it is fundamentally a genetic screen, the yeast two-hybrid assay allows the screening and detection of large libraries of cloned genes, and can be scaled up to examine the entire genome. Yeast two-hybrid selections are also useful for detecting transient interactions or other interactions that might not be stable during a purification scheme. Because the method is widely used, there are many variations and modifications for specific purposes. Our description will cover only the most general version.

The basis of the yeast two-hybrid screen (abbreviated Y2H) is shown in Figure 10.17. Y2H depends on the biochemical properties of the transcriptional activator protein GAL4. GAL4, like many transcription factors, has two separate domains that are capable of acting independently of each other. The DNA binding domain (abbreviated DB) is needed for sequence-specific binding. The activation domain (abbreviated AD) interacts with the general transcription machinery to activate transcription. Each of these domains is cloned separately into a yeast plasmid that is maintained at low copy number in the yeast cell. The two plasmids also include antibiotic resistance markers that allow them to be grown in bacteria and a nutritional marker that allows their analysis and maintenance in yeast. (This type of centromere plasmid is described in Chapter 3.) In Figure 10.17, the plasmid with the DNA binding domain includes the *LEU2* gene, whereas the plasmid with the activation domain includes the *TRP1* gene as its nutritional marker. These are the two plasmids of the two-hybrid assay.

Suppose that the investigator has a gene of interest, *Gene A*, and wants to find proteins that interact directly with the protein product of *Gene A*. *Gene A* is cloned in frame to the sequence for the GAL4 DB domain, so that, when this plasmid is transcribed and translated, a fusion protein will be made that has the GAL4 DB fused to Protein A. This is called a **bait plasmid**. A second plasmid has the transcriptional activation domain of GAL4 fused in frame with the gene to be tested. This is called a **prey plasmid**. A more powerful version has a library of prey plasmids in which an individual gene is cloned with the AD domain, so that many different genes are represented as prey in the library.

Literature Link. Yu et al. 2008, *Science* 322:104–10

The two plasmids are transformed into a yeast cell that has a GAL4-binding site upstream of some easily selected marker, *HIS3* in Figure 10.17; alternatively, haploid cells with each plasmid are mated to produce a diploid cell that has both plasmids. The *HIS3* gene is only transcribed, so

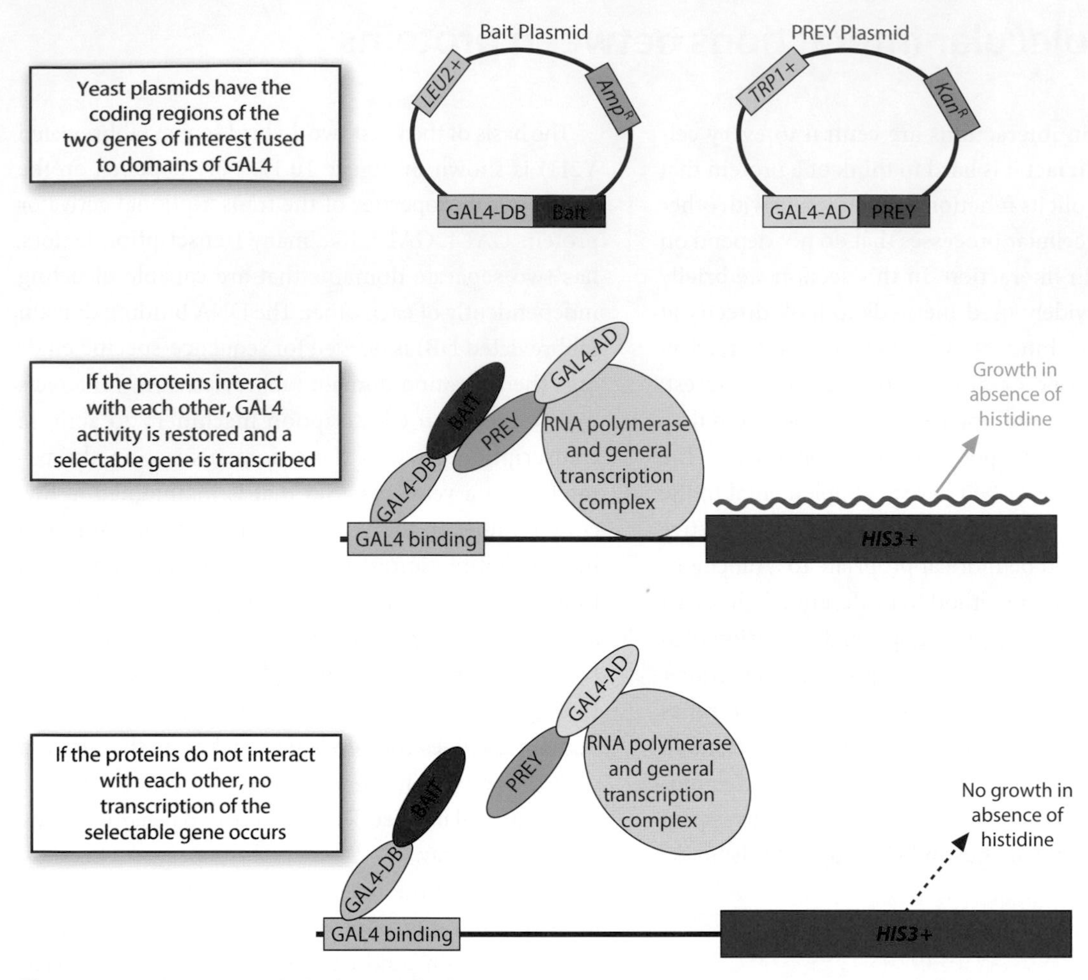

Figure 10.17 The yeast two-hybrid system for detecting protein–protein interactions. The yeast two-hybrid screen relies on the domain structure of the transcriptional activator in yeast GAL4. The GAL4 DNA binding domain (DB) binds a specific nucleotide sequence. The GAL4 transcriptional activation domain (AD) interacts with the transcriptional machinery. The gene-specific transcriptional activator is found when the GAL4 DB is associated with the GAL4 AD. Each domain is cloned into a yeast centromere plasmid with a selectable drug marker (ampicillin- or kanamycin-resistance in the figure) that allows selection in *E. coli* and a nutritional marker (*LEU2* or *TRP1* in the figure) that allows selection in yeast. The gene of interest, known as the bait, is cloned in frame with the GAL DB. Genes to be tested, known as the prey, are cloned in frame with the GAL AD. Each prey plasmid has a different gene to be tested, so a library of prey genes is constructed. The two plasmids are transformed together into host yeast cells that have the *GAL4* binding site inserted upstream of one or more reporter genes, here the *HIS3* gene. The host yeast cell will both express the DB-bait fusion protein and the AD-prey fusion protein. If the bait and prey proteins interact with each other, the two domains of the GAL4 protein are brought together and transcription of the reporter gene is activated. Thus, in the figure, the yeast cells would be plated on media that lack tryptophan (to select for the prey plasmid), leucine (to select for the bait plasmid) and histidine (to select for the interaction). If the bait and the prey proteins do not interact with each other, the cells will not grow in the absence of histidine.

the cell is only able to grow in the absence of histidine, if the GAL4 binding site is activated. Each plasmid produces its bait or prey fusion protein. The bait protein can bind to the GAL4 binding site, but lacks the ability to interact with the transcriptional machinery and to activate transcription. On the other hand, the prey protein can activate transcription but cannot bind to the GAL4 binding site. However, if the bait and the prey interact physically with each other, the two different domains of the GAL4 protein are brought together, and transcription of the *HIS3* gene under GAL4 control occurs. If the bait and the prey interact with each other, the cell can grow in the absence of added histidine; if the bait and prey do not interact with each other, the cell cannot grow in the absence of added histidine. Thus there is a powerful and easy selection for protein–protein interactions.

In this simple form, Y2H provides a relatively easy method to test postulated interactions between two proteins. It is also possible to map the portion of each protein that is needed for the interaction to occur. This is described in the case study for the interactions between proteins in the spindle in yeast. Y2H is extremely powerful, and a well-designed prey cDNA library allows thousands of possible interactions to be screened at once.

The method does have certain limitations, both in that it may fail to detect some interactions (false negatives) and in that it may find interactions that are not physiologically meaning (false positives). Failure to detect an interaction that does occur naturally could happen because the fusion proteins do not adopt the same conformation found in the cell, or because only binary interactions can be detected, or because one or both of the proteins have to be covalently modified (e.g. phosphorylated) for the interaction to occur.

Modifications of the basic Y2H assay have reduced these potential shortcomings and allowed more interactions to be detected. In general, Y2H finds most of the interactions that occur within the cell, even if these are transient.

A more serious caveat to the Y2H method comes from the potential for false positives; that is, proteins interact when expressed in this form in a yeast cell that do not interact physiologically. For example, two proteins that are capable of interactions in Y2H might be expressed at the different times or in the different cells under physiological conditions, so that Y2H is detecting an interaction that does not occur within the cell. One way to avoid some of these false positives is to compare interaction results with expression data, so as to determine whether the interaction is physiologically feasible. The suspected interactions can also be tested by other methods, such as co-IP, as described in what follows.

Despite their limitations, Y2H screens are widely used and extremely informative. Part of their utility comes from their relative simplicity, particularly as yeast strains with multiple reporter genes under GAL4 control have been developed. Furthermore, no protein purification or molecular reagents are required other than cDNA clones for the genes of interest and a good cDNA library. Unlike in purification procedures, one protocol can generally be established for many different interactions and used for genes from many different organisms, which is an enormous advantage. The same procedure is used for detecting interactions whether the proteins are derived from mammals, higher plants, or invertebrates. Another significant advantage is that further investigations of the interaction, such as mapping important amino acids or regions of the proteins, can be done through routine manipulations of the cDNA clones rather than the proteins.

The greatest advantages of Y2H screens arise when using a cDNA library as the prey. In this case, a library is made in which each cDNA is fused to the activation domain. If the cDNA library is sufficiently complete, essentially all the genes in a genome can be screened for their ability to interact with a bait protein using only a few plates of yeast growth medium in a few days. Depending on the quality of the cDNA library and the number of colonies scored, one can be confident that even interactions among minor components or components with low affinity have been found. Just as significantly, the gene responsible for each of these interactions is already cloned. For these reasons, the yeast two-hybrid selection has become the most widely used method to look for interacting proteins.

A Y2H assay was used to examine brassinosteroid signaling in *Arabidopsis*

To illustrate their effectiveness, yeast two-hybrid screens were used to identify components of the brassinosteroid signal transduction pathway in *Arabidopsis*. Remember from Section 3.5 that brassinosteroids are plant hormones with a variety of developmental roles, many due to their positive effects on cell elongation. The primary brassinosteroid receptor is BRI1, a receptor-like kinase. Null mutants of *BRI1* have a dwarf phenotype and are insensitive to exogenous brassinosteroids. In an attempt to better understand the function of BRI1, Y2H screens for proteins interacting with the intracellular kinase domain of BRI1 have been done. These Y2H screens have identified both positive and negative regulators of BR signaling, resulting in the model summarized in Figure 10.18.

The interacting protein that acts as a positive regulator of BR signaling was another receptor named BAK1. BAK1 does not itself act as a brassinosteroid receptor, but its activity is required for normal brassinosteroid signaling. BAK1 interacts with the BRI1 receptor both in Y2H assays and in vivo, and the interaction is stronger after BRI1 binds brassinosteroids. Null mutants in *BAK1*

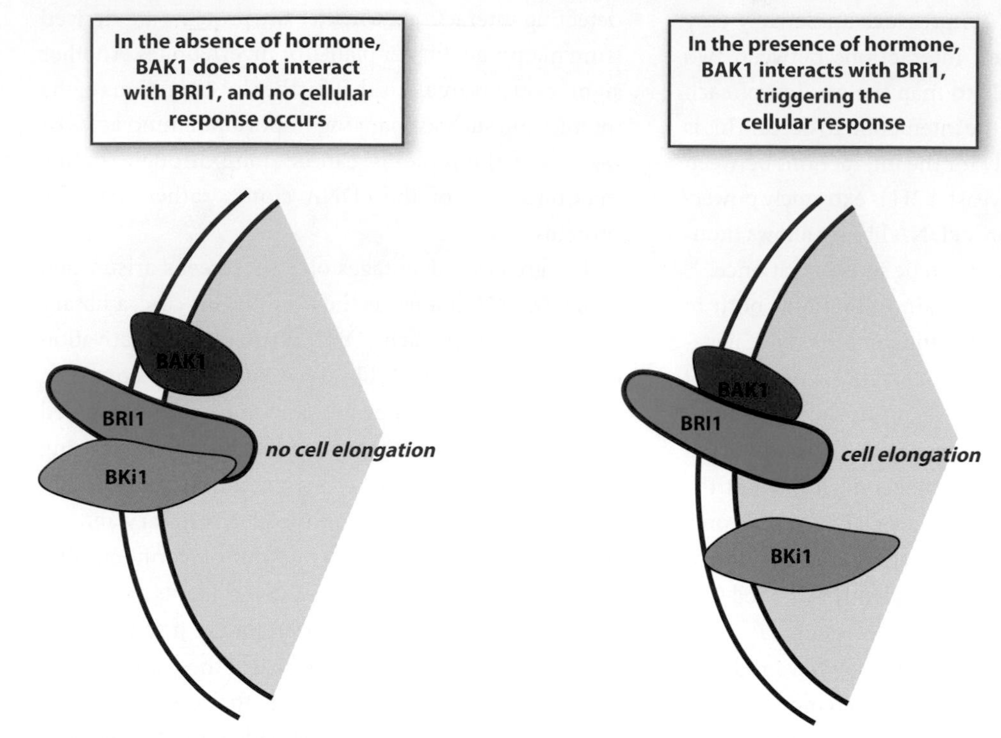

Figure 10.18 BR11 and brassinosteroid signaling. BRI1 is a receptor for brassinosteroid growth hormones that produce cell elongation that is regulated by BAK1 and BKI1, as shown. In the absence of the growth hormone, BKI1 is associated with BRI1 in the plasma membrane, and no cell elongation occurs. When the hormone is present, BKI1 dissociates from BRI1, allowing BRI1 to associate with BAK1 and to trigger cell growth and elongation.

have a very weak brassinosteroid insensitive phenotype, and a kinase inactive version of BAK1 acts as a dominant negative mutation, interacting with the BRI1 receptor, but unable to transduce the signal to downstream genes. Interestingly, the *BAK1* gene was also identified as a suppressor of *bri1* mutations, demonstrating the utility of having approaches that identify both physical and genetic interactions.

One of the proteins identified as a negative regulator of brassinosteroid signaling, BKI1, also clarified how BAK1 acts as a positive regulator. BKI1 is localized to the plasma membrane, like BRI1 and BAK1. In contrast to the results with BAK1, overexpression of *BKI1* in plants results in dwarfism and greatly reduced sensitivity to brassinosteroid. In the absence of brassinosteroids, BKI1 binds to the kinase domain of BRI1, blocking the interaction between BRI1 and BAK1. Upon brassinosteroid treatment, BKI1 dissociates from the plasma membrane, allowing BAK1 to interact with BRI1 and elicit downstream signaling events. Thus BKI1 acts as a brassinosteroid signaling inhibitor by preventing the interaction of BRI1 and BAK1 when the hormone is absent.

Co-immunoprecipitation is a physiological standard for protein–protein interactions

Our emphasis on gene-based tools to detect interacting proteins should not be interpreted as an underestimation of the value of biochemical methods that are based on the proteins themselves. Many sophisticated biochemical methods to look at protein–protein interactions have been developed. Here, we will briefly describe one widely used method, namely **co-immunoprecipitation (co-IP)**. The basis of the procedure is that one antibody directed against one protein in a macromolecular complex will co-precipitate the other proteins in that complex, as shown schematically in Figure 10.19.

Co-IP was introduced in Chapter 2 as a method to analyse chromatin complexes involving protein interactions with DNA; it is referred to as chromatin immunoprecipitation or ChIP, but immunoprecipitation was first applied to protein–protein interactions. In order to perform co-IP, the protein of interest is purified and used as an antigen to produce an antibody. Cells or tissues that express the protein of interest are lysed and the macromolecules are extracted under gentle conditions

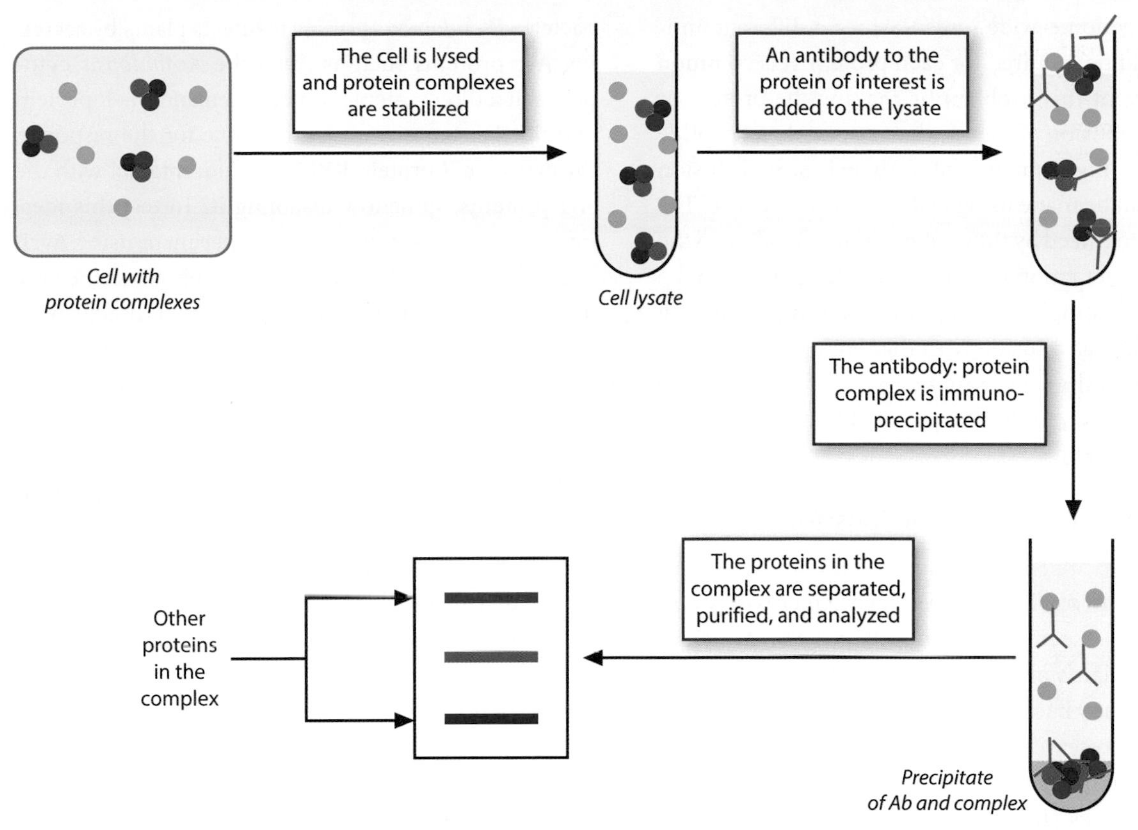

Figure 10.19 Co-immunoprecipitation to detect protein–protein interactions. Proteins within the cell are represented by small colored balls. The blue protein is found in a multimeric complex with the red and the purple proteins, but not with the green protein. The cell is lysed by a procedure that preserves the macromolecular complexes; alternatively, the complexes can be reversibly cross-linked. An antibody that is specific to the blue protein is added to the lysate, and the antigen–antibody complex is precipitated. The components of the precipitate are then separated on a denaturing gel. Although the antibody was directed only against the blue protein, the purple and the red proteins are also seen on the gel because of their interaction with the blue protein. The green protein, which is not part of the stable complex, is not found.

that preserve the macromolecular interactions. In some protocols, the proteins in the cells are reversibly cross-linked prior to extraction. The cell extract is incubated with the antibody against the protein of interest, which then precipitates not only the original protein but also other proteins with which it has stable interactions. The precipitated complex is separated on a denaturing gel, and the individual proteins are analysed.

Co-IP in the form just described has its advantages and limitations. One advantage is that co-IP experiments can find previously unknown interactions, involving even minor components, and many polypeptides. For example, co-IP experiments have shown that at least fifty-five proteins are needed for accurate transcriptional initiation and elongation in eukaryotes, which probably would not have been found by genetic methods alone. Co-IP also provides a good approximation of the stoichiometry of the molecular components. Its most fundamental strength is that it approximates physiological conditions and gives few false positives—proteins that can be co-precipitated probably do form a complex in vivo. For this reason, co-IP experiments are often used as the standard for confirming protein–protein interactions initially detected by other methods such as interactional suppressors or yeast two-hybrid assays. For example, the Y2H interactions between BRI1 and BAK1 and between BRI1 and BKI1 described above were then also demonstrated by co-IP to occur in vivo.

One limitation in the method described here is that, when compared with the other methods that we have described, it is more technically challenging and depends heavily on the ability of the antibody to precipitate the complex. Thus it is not easily scaled up to look at interactions for many different proteins (as would be done in a genome-wide screen), since a different antibody would be required for each protein. It is common to circumvent this problem by making the protein of interest be a fusion protein with some unrelated peptide for which a high-quality antibody exists. Such fusion proteins can be made using cDNA libraries in vitro. The peptide recognized by the antibody is its epitope, and the protein of interaction is then epitope-tagged within the cell and the complex (see **epitope-tagging**). Thus any protein made as a fusion with that epitope can be precipitated with the same antibody and roughly the same protocol. This allows for the protocol to be scaled up to genome-wide scale.

The power of protein interactions: An example from plant pathology

A combination of Y2H and co-IP experiments in *Arabidopsis* led to a recent paradigm shift in the field of plant pathology. Plants are subject to infection and disease caused by many bacteria, fungi, viruses, and nematodes. For decades, research in many different crop species had noted a "gene-for-gene relationship" between so-called resistance genes (*R* genes) in the host and avirulence genes (*avr* genes) in their pathogens. For example, *Arabiposis* uses the R protein RPM1 to defend against the AvrB protein of *Pseudomonas syringae*. A standard model proposed that the protein product of a specific *R* gene recognizes a single Avr protein from one pathogen via direct protein–protein interactions, which are similar to antigen–antibody interactions in mammals.

Genetic experiments in *Arabidopsis* and various crop species supported such a model. A pathogen with a recessive mutation in an *avr* gene could avoid the resistance conferred by the corresponding *R* gene. Likewise, a plant with loss-of-function mutations in an *R* gene can lose resistance to a specific pathogen.

The simplicity of this model and its intellectual parallel with the mammalian immune system made it very appealing. It makes the testable prediction that direct protein–protein interactions should occur between the specific R proteins and specific Avr proteins, as is true in the immune system. In fact, in contrast to the immune system, very few examples of direct protein–protein interactions between R/Avr protein pairs could be demonstrated. Thus a different model had to be considered.

Mutants in *Arabidopsis* and the protein interactions assays described here allowed a direct test of one such model, as summarized in Figure 11.20. The pathogenic bacteria *Pseudomonas syringae* infects plants by secreting Avr proteins such as AvrB the *Arabidopsis* cytoplasm. If a direct interaction between and the R protein and the Avr protein was responsible for the response, the cytosolic R protein RPM1 shouldt interact with the Avr proteins, somehow disabling it. To test this idea, investigators performed a Y2H experiment using AvrB as the bait with an *Arabidopsis* cDNA library as the prey in order to identify the plant proteins that directly bind AvrB. The RPM1 was not found among the interacting proteins, which refuted a key prediction of this model. However, a protein called RIN4 was among the interactors with AvrB, and the interaction between RIN4 and AvrB was confirmed by co-IP with an epitope-tagged version of AvrB.

A similar co-IP experiment with an epitope-tagged version of RPM1 showed that RPM1 and RIN4 also interact in vivo in the absence of pathogens. Thus RIN4 interacts directly with both AvrB and RPM1, which do not appear to interact with each other. Further co-IP experiments showed that RIN4 also interacts with other Avr proteins and other R proteins. A revised model, termed the guard hypothesis and shown in Figure 10.20, has been proposed. In the absence of a pathogen, RIN4 is found in a protein complex with one of the R proteins. Upon

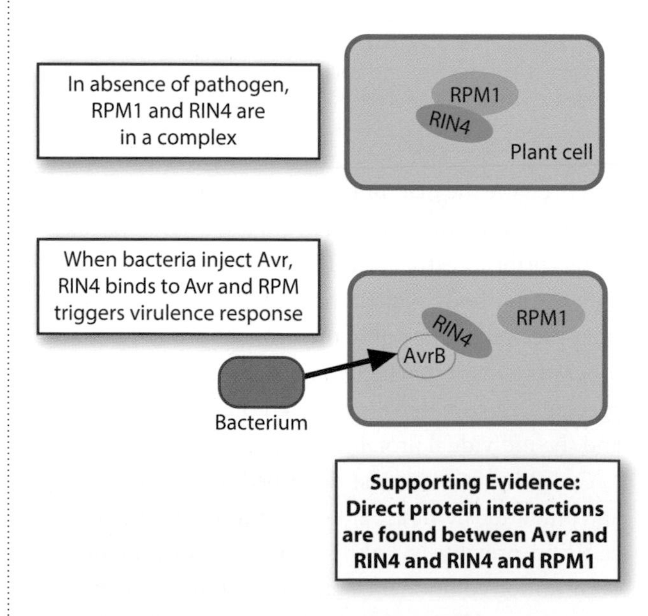

Figure 10.20 The guard hypothesis in *Arabidopsis*. Although there is a specific correspondence between the Avr protein of a bacteria pathogen and the plant R protein, direct protein interactions were not observed. Direct protein interactions were observed between the AvrB protein and the RIN4 protein and between the RIN4 and RPM1, as well as between RIN4 and other Avr and R proteins; this leads lto the guard hypothesis, as shown.

infection, the Avr proteins are expressed by the bacteria, which bind to and destabilize RIN4. The destabilization of RIN4 signals, in turn, the R proteins that an infection has occurred, and the plant defense pathways are engaged. Thus, rather than recognizing the pathogens themselves, R proteins detect the presence of an infection by guarding the proteins commonly targeted by the pathogen. Rather than preventing the destruction of these targets, these R genes recognize the damage and increase the defense response. Additional examples of this mechanism involving different guarded proteins have since been uncovered in *Arabidopsis*. This supports the validity of the model.

10.6 Summary: Genetic interactions, both genes and proteins

Almost every biological process is affected by multiple interacting genes and their protein products. In Chapter 4 we have described methods to find genes that affect a biological process, beginning with a wild-type organism. The analysis of phenotype in this gene and its product has been described in the subsequent chapters; but all these chapters focused on single genes. In the present chapter we have added new genetic tools that rely on this previously identified gene, either in the form of a mutant phenotype or a cloned gene, or both. Two of the methods, suppression and synthetic enhancement, require only that the investigator have a mutation in the gene of interest. The gene of interest does not need to be cloned for these methods to be used, although that is certainly an advantage. The other two methods, co-immunoprecipitation and yeast two-hybrid screens, require that the gene of interest be cloned and, in the case of co-IP, that an epitope-tagged protein or specific antibodies have been made against its protein. These latter two methods do not require that mutant alleles exist for the gene, although, again, that is certainly an advantage. In practice, an investigator will probably use several of these methods to find genes related to the gene of interest, because no single method provides all the information that is necessary, or even all the other related genes that might be identified.

The main strengths of the genetic methods using mutant phenotypes are that relatively little needs to be known in advance about the biological process and that many interactions can be found and characterized. But these strengths also point to the major limitation that the functional interactions might not be direct interactions between the gene products. Y2H and co-IP, which require a cloned gene but not a mutant phenotype for the gene, find direct interactions between the protein products, but might not tell us about the functions of those interactions. A full analysis of the functional and the direct interactions requires both types of experiments.

Think of the genetic and molecular analysis of a biological process in terms of solving a jigsaw puzzle, where each gene or protein is represented by a puzzle piece with its own shape and depicting part of a scene. Suppose it is a piece that depicts part of the roof of a barn. In this analogy, the molecular and biochemical methods correspond to finding an interlocking puzzle piece with the right shape, regardless of what scene the piece depicts. If the two pieces fit together in all particulars, we can have great confidence that they sit next to each other in the puzzle. We are defining the pieces by their physical interaction.

By contrast, the genetic methods look only at the scene that is depicted and not at the shape of the piece. The genetic methods are likely to yield far more genes than the molecular methods would yield—in our puzzle analogy, there are more pieces that depict parts of the farmer's barn than pieces that fit together with our individual piece. Some of those pieces are next to each other in the puzzle, but other pieces are far apart. Thus, in order to complete the puzzle, we have to use both the scene that is depicted and the shapes of individual pieces. A common strategy in putting together a jigsaw puzzle is first to sort the pieces by scene, and then to use the shapes; similarly, in order to analyse a biological process, the investigator will often use the genetic interactions first and then rely on the physical interactions.

Having used these methods, the investigator will have identified a substantial number of the genes that affect the process—in principle, every gene or protein that affects the process. In the course of finding these genes, the investigator has got glimpses of the underlying logic of the process itself. In the next two chapters we will describe how these methods can be used together, to yield an even more complete picture of the biological process.

Chapter Capsule

Many different genes and their polypeptide products work together to carry out a biological process, so an investigator will want to define as many of these as possible. Genetic approaches like suppression and synthetic enhancement find functionally related genes by examining changes to the mutant phenotypes that occur in double mutants. Genetic interactions find both direct and indirect interactions, so they are especially useful for obtaining a broad picture of the activities of the genes in the biological process; however, they usually do not provide many details about the specific roles. Information about the specific roles of the proteins is found using molecular approaches such as yeast two-hybrid screens or co-IP, the latter being considered the standard for proving that the proteins are part of the same complex in vivo.

Additional Reading

Botstein, D., and R. Mauer. 1982. Genetic approaches to the analysis of microbial development. *Annual Review of Genetics* 16: 61–83.

Brookfield, J. F. Y. 1997. Genetic redundancy: Screening for selection in yeast. *Current Biology* 7: R366–8.

Guarente, L. 1993. Synthetic enhancement in gene interactions: A genetic tool comes of age. *Trends in Genetics* 9: 362–6.

Hodgkin, J. et al. 1987. Suppression in the nematode *Caenorhabditis elegans. Trends in Genetics* 3: 325–9.

Prelich, G. 1999. Suppression mechanisms: Themes from variations *Trends in Genetics* 15: 261–6.

Thomas, J. H. 1993. Thinking about genetic redundancy. *Trends in Genetics* 9: 395–9.

Case Study 10.1

Genetic analysis of spindle morphogenesis in budding yeast

The use of suppressors and synthetic enhancers to study gene interactions has a rich history. In this case study we focus on one biological process, the morphogenesis of the mitotic spindle in budding yeast *S. cerevisiae*, starting with a mutation in a gene for β-tubulin. Several properties make this a good example. First, nearly all the tools described in this chapter were used in the experiments. Second, cytological experiments on spindle morphogenesis in yeast and other organisms provided an extensive background by which to interpret the genetic outcomes. Third, more detailed molecular and biochemical experiments confirmed and extended the conclusions from the genetic analysis through suppressors and enhancers. An overview of the yeast cell cycle and some of the genetic and molecular methods that were used are described in Section 3.2.

Thinking through the Experiment

1. Why is it so important to have cytological, molecular, and biochemical methods to confirm and extend the genetic analysis?
2. What advantages does the genetic analysis provide that cytological, molecular, and biochemical methods might not?

Budding and cell division: Overview

In the course of dividing *S. cerevisiae*, a cytoplasmic bud emerges early in the cell cycle and grows during cell-cycle progression. During S phase, the cell nucleus migrates to the neck region, which connects the mother cell to the bud. Mitosis partitions the chromosomes into the two cells (which are now nearly the same size), and cytokinesis completes the division (Figure CS10.1).

As in other eukaryotic cells, chromosome separation depends on the function of the mitotic spindle. The spindle is formed of microtubules, which are heterodimeric polymers of two paralogous proteins, α-tubulin and β-tubulin (Figure CS10.2). Associated with the microtubules in the spindle are many other different proteins, known generally as microtubule-associated proteins, which enable it to carry out its function of segregating sister chromatids during mitosis. Because the amino-acid sequences of α-tubulin and β-tubulin are highly conserved across evolution, vertebrate tubulin genes were used to carry out the molecular cloning of the tubulin genes from yeast. This molecular analysis showed that yeast has two structural genes for α-tubulin, *TUB1* and *TUB3*, and one structural gene for β-tubulin, *TUB2*.

A cold-sensitive mutation of *TUB2*

The roles of each tubulin gene were studied through genetic analysis. Gene disruptions revealed that both the α-tubulin gene *TUB1* and the β-tubulin gene *TUB2* are essential for cell growth and division under normal lab conditions, whereas *TUB3* is not. Because the deletion mutations are lethal, it was necessary to use conditional mutations of *TUB1* and *TUB2* in the further analysis of yeast cell division; using conditional mutations in essential genes for strain maintenance is described in Section 4.3.

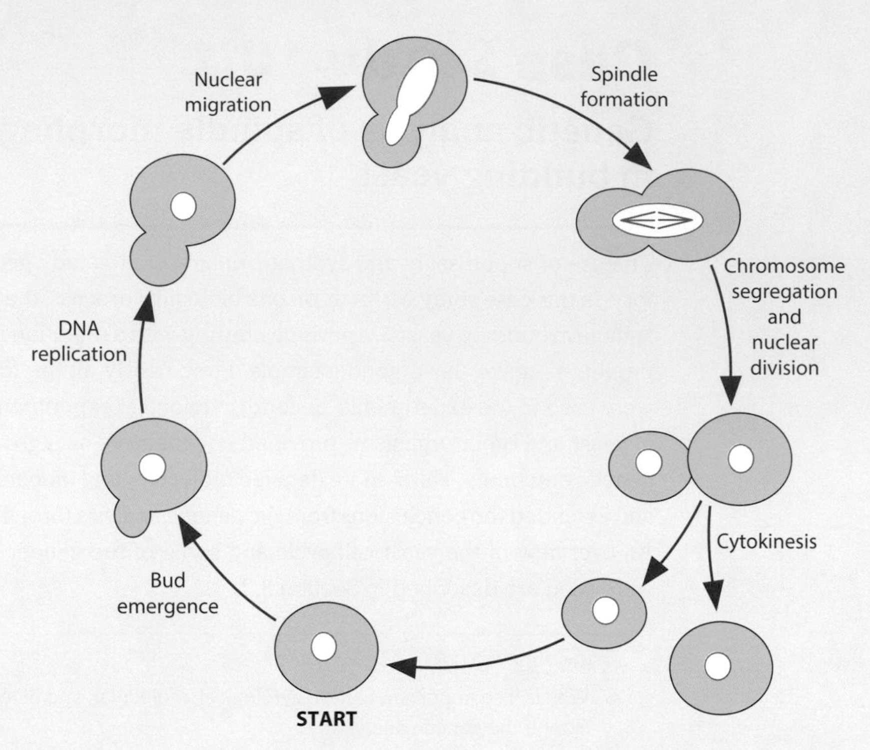

Figure CS10.1 The cell division cycle of the budding yeast *Saccharomyces cerevisiae* is diagrammed. A bud begins to form and the nucleus migrates to the neck of the mother cell and bud. The mitotic spindle is established and the chromosomes are segregated to the two cells.

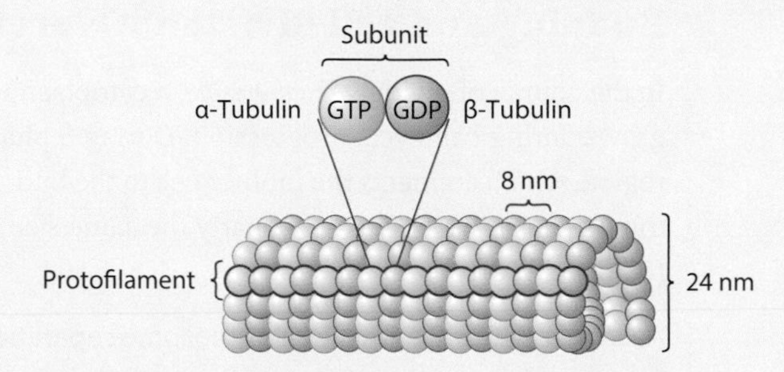

Figure CS10.2 A drawing of the microtubule. Note the alternating structure of the α-tubulin and β-tubulin subunits and the close physical contact between the two tubulin subunits.

Thinking through the Experiment

3. How are conditional mutations found in a genetic screen, and what advantages do they offer?

Of particular interest are cold-sensitive mutations in *TUB2*. At the permissive temperature of 28 C, these mutant cells grow and divide normally. At the restrictive (low) temperature of 14 C, the *tub2* mutants accumulate as cells with large buds with a single undivided nucleus and no mitotic spindle (Figure CS10.3). Not all the *tub2* cold-sensitive mutations have identical phenotypes, possibly because the mutations affect different amino acids in β-tubulin. One cold-sensitive allele of *TUB2*, *tub2-406*, arrests with large buds and undivided chromosomes at the region of the bud neck. The nucleus is found in the bud neck; this indicates that the cytoplasmic microtubules needed for movement of the nucleus during budding appear to function normally in this mutant. The presence of non-separated chromatids, however,

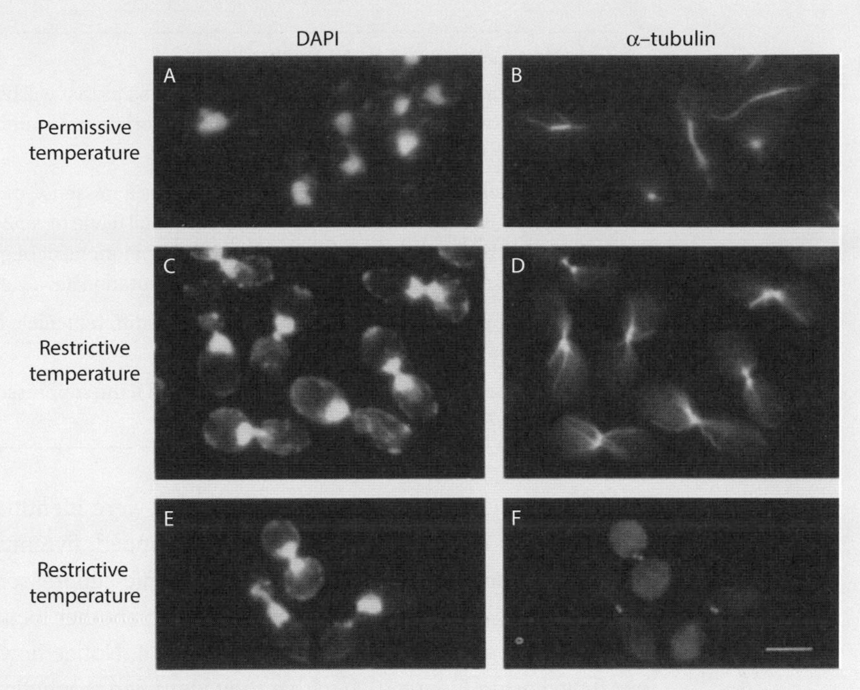

Figure CS10.3 The phenotype of the mutant *tub2-406* at the permissive (a and b) and restrictive temperature (c through f). Panels a, c, and e are DAPI stained cells that show the chromosome behavior. Panels b, d, and f are stained with an anti-α-tubulin antibody to show the mitotic spindle. Notice that the nucleus arrests at the neck of the bud at the restrictive temperature (c and e) and that the spindle does not assemble normally (d and f). From Pasqualone and Huffaker 1994, *STU1*, a suppressor of a β-tubulin mutation, encodes a novel and essential component of the yeast mitotic spindle, *Journal of Cell Biology* 127: 1973–84, with permission.

indicates that this mutant has a specific defect in spindle microtubules. Thus the *tub2-406* mutation separates the functions of β-tubulin in cell movements and spindle structure and affects only spindle structure. It was therefore used as the original mutation to search for suppressor mutations.

Thinking through the Experiment

4. What is the significance of the observation that *tub2-406* affects spindle formation but not other cellular functions provided by β-tubulin?

Suppressors of *tub2-406*: Interactional suppressors

Because *tub2-406* is cold sensitive and affects only one of the several phenotypes of a *tub2* null allele, it seems very likely that *tub2-406* represents a missense mutation. In fact subsequent molecular analysis revealed that *tub2-406* is a missense mutation in which methionine is substituted for valine at amino acid 100. This cold-sensitive growth defect was used to find genes that affect spindle morphology and function by looking for suppressors of *tub2-406*. The genetic selection involved mutagenizing *tub2-406* mutant cells and looking for yeast colonies that grew and divided normally at low temperature. Such suppressors could be gene-specific for many β-tubulin mutations and could thus identify genes that affect the functions of the β-tubulin protein. Alternatively, the suppressors might be allele-specific and affect only the specific defect on the *tub2-406* lesion.

Thinking through the Experiment

5. With Figure 10.4 as your guide, summarize the steps that will be done to characterize the *tub2-406* suppressors. Give as much detail as to the experimental methods, the possible results, and their interpretation as possible.
6. It was likely from its phenotype that *tub2-406* is a missense mutation (as it proved to be). In this case, a missense mutation was used, since a null allele of *tub2* does not allow cell division at all. But what are some advantages and possible limitations of beginning a suppressor analysis with a missense mutation? Would other missense mutation have yielded the same results as *tub2-406*?
7. What suppressors would have been found if a different allele of *tub-2* had been used and which suppressors would not have been found?
8. What type of suppressor is unlikely to be found if the suppressor analysis begins with a missense mutation?

Both categories of gene-specific suppressors were identified. Eleven different suppressor mutations of *tub2-406* were found and mapped: five mutations mapped to the *TUB2* locus itself and were presumed to be intragenic suppressors; two mutations mapped to the *TUB1* locus; and four mutations mapped to another locus, subsequently named *STU1*. These results are summarized in Figure CS10.4. Notice how closely these suppressors fit our flowchart in Figure 10.4, in that they identified α-tubulin—a protein known to interact

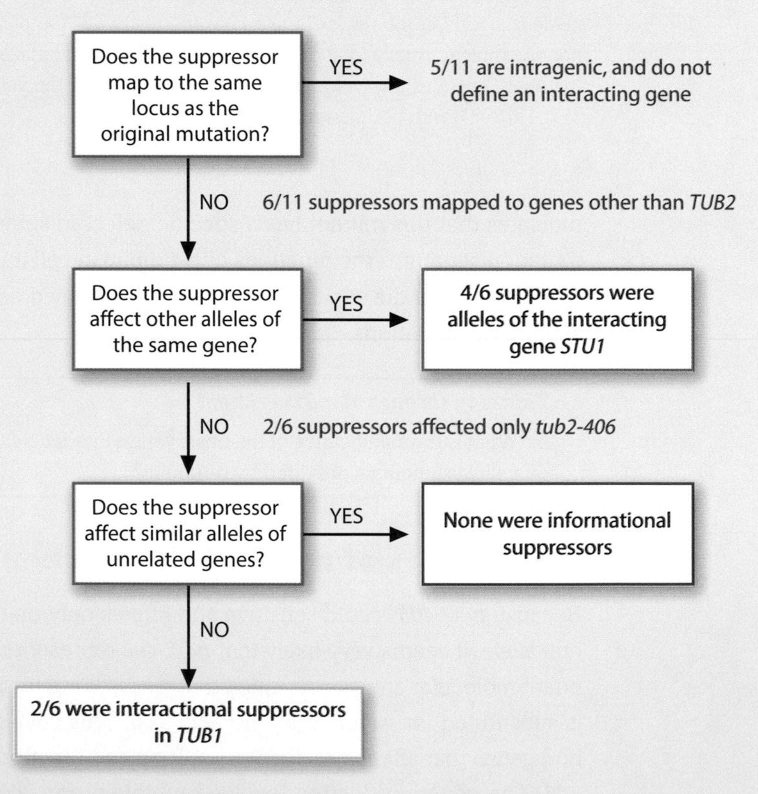

Figure CS10.4 A flowchart showing the results of the genetic screen for suppressors of the cold-sensitive mutation *tub2-406*. Eleven suppressors of *tub2-406* were identified. Five suppressors map to the same locus as *tub2* mutations and are classified as intragenic suppressors. Of the remaining six suppressors, two were specific to both the gene and the allele and four were specific to the gene but not the allele. (No informational suppressors were found.) The four allele-nonspecific suppressors identify a new gene known as *STU1*. The two interactional suppressors are alleles of the α-tubulin gene *TUB1*.

with β-tubulin—and another previously unknown gene that may have a function related to β-tubulin.

The next step in the flowchart is to ask whether the suppressors affect other alleles of *tub2*, that is, to determine whether they are allele-specific to *tub2-406*. The four *stu1* mutations did in fact suppress several alleles of *tub2* in addition to *tub2-406* and will be discussed in the next section. In contrast, the two *tub1* mutations isolated as suppressors of *tub2-406* (one of which, *tub1-108*, will be used in subsequent experiments) did not suppress other alleles of *tub2*. In fact *tub1* mutations other than these two suppressors do not suppress *tub2-406* or other *tub2* mutations. This immediately indicated that these *tub1* suppressor mutations represent allele-specific or interactional suppressors of *tub2-406*.

Thinking through the Experiment

9. Explain the rationale outlined here for the interpretation of these *tub1* mutations.

10. Do you expect these *tub1* mutations to have a mutant phenotype on their own, in addition to their suppression of *tub2-406*? If so, what cellular functions of tubulin do you expect are the ones most likely to be altered by these *tub1* mutations, and which ones might not be affected?

11. Table CS10.1 will be used to keep track of the different genetic interactions that are observed, each line representing a different experimental result. The *tub* mutation, the interacting gene, and the phenotype for the double mutant are given. For some experiments, the key result is also given, whereas in others you will need to supply it. For every experiment, you will need to supply the interpretation of the result. Complete Experiment 2.

The inference is that the *tub2-406* mutation disrupted a specific interaction with α-tubulin. The suppressors are postulated to be mutations on α-tubulin (*tub1*) that restore the interaction with the mutated β-tubulin; but it is also postulated that they cannot restore the function lost in other mutations in β-tubulin. In support of this inference for interactional suppression, the *tub1* alleles that suppressed *tub2-406* map to a specific location on α-tubulin, a region of α-tubulin that is apparently needed for its interaction with β-tubulin in spindle formation (Figure CS10.5). Thus allele-specific, gene-specific suppressors were used to map the region of interaction between α-tubulin and β-tubulin in microtubules; confirmation of this inference is described in the last section of this case study.

Suppressors of *tub2-406*: *stu1* and functional interactions

As summarized in Figure CS10.4, four suppressor mutations of *tub2-406* mapped to the same gene and failed to complement each other for suppression. This gene was named *STU1*, which stands for "suppressor of tubulin." The *STU1* gene was cloned by complementation: a plasmid with the wild-type *STU1* gene was identified by its ability to complement the suppression in *stu1 tub2-406* strains and thus to display the cold-sensitive arrest phenotype of *tub2-406*. Molecular analysis of the complementing plasmid showed that the *STU1* gene, as defined by these suppressor mutations, encoded a previously unknown protein. Although the function of STU1 could not be inferred from sequence similarity (i.e. from the similarity of its sequence to those of other known proteins), several lines of genetic evidence indicated that *STU1* is needed for spindle formation.

Thinking through the Experiment

12. Outline the experiment by which the *STU1* gene "was cloned by complementation."

13. What results might convince you that *STU1* is needed for normal spindle function?

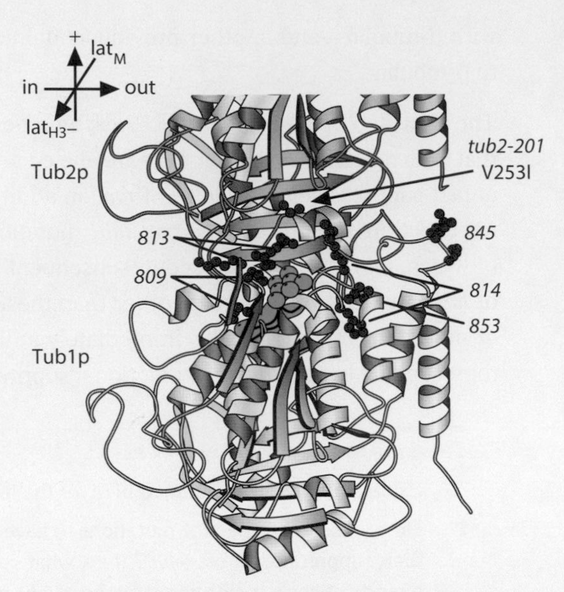

Figure CS10.5 The location of the interactional suppressors as projected onto the three-dimensional ribbon diagram of the tubulin subunit. Tub2p is shown at the top, Tub1p below. The residues altered in the intearctional suppressors are shown in pink and red; residues identifed by the suppressor analysis and alanine scanning mutations are located predominantly at the interface between the two protein. From Richards et al. 2000, Structure-function relationships in yeast tubulins, *Molecular Biology of the Cell* 11: 1887–903, with permission.

First, both of the tested *stu1* mutations suppressed not only *tub2-406* but also other alleles of *tub2* with similar spindle defects. This indicates that suppression of β-tubulin function is not allele-specific to *tub2-406* but is related to the function of β-tubulin in spindle morphogenesis. Even more convincingly, the *stu1* mutations did not suppress alleles of *tub2* with defects in cytoplasmic microtubules, indicating that *STU1* is not needed for these functions of β-tubulin. Thus, from the suppression analysis itself, it seemed likely that *STU1* defines a gene needed for spindle morphogenesis.

Second, *stu1* deletion mutations made by gene disruptions had spindle defects similar to those seen in *tub2-406*, like the *stu1* point mutations isolated as *tub2-406* suppressor alleles. That is, many of the *stu1* mutant cells arrested with large buds and an undivided nucleus located at the neck, as shown in Figure CS10.6. Thus the suppressor gene had a mutant phenotype on its own, which resembled the mutant phenotype of the mutation being suppressed. This provides additional evidence that the *STU1* is functionally related to the spindle morphogenesis function of *TUB2*.

Third and finally, an epitope-tagged version of STU1 was produced and, by using an antibody directed against the epitope tag, the protein was found to localize specifically to the spindle. This directly confirmed the inferences from genetic analysis that the STU1 protein is involved in spindle formation. Significantly, this epitope-tagged version of the *STU1* gene complements the *stu1* gene disruptions. In other words, the only functional STU1 protein in the cell is the epitope-tagged protein; and the cell divides normally.

Thinking through the Experiment

14. Which of the three results persuaded you that *STU1* is involved in spindle formation, and why? What does each of these results add to the interpretation of *STU1* function?

15. Complete Experiment 3 in Table CS10.1.

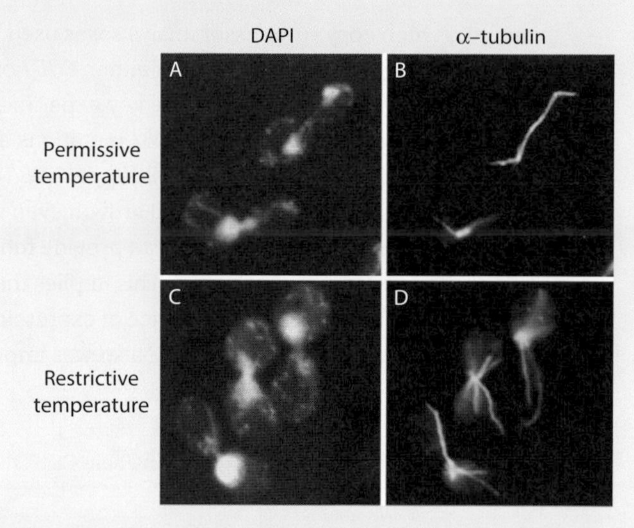

Figure CS10.6 The phenotype of *stu1* mutant cells. DNA is visualized by DAPI staining (panels A and C) and the spindle by an antibody directed against α-tubulin (panels B and D). The cells in A and B are grown at the permissive temperature and show a wild-type phenotype. The cells in C and D were grown at restrictive temperature and show the *stu1* mutant phenotype. Note the nuclear arrest at the bud neck (in panel C) and the abnormal spindle formation (in panel D). This phenotype suggests that STU1 has a function related to TUB2. From Pasqualone and Huffaker 1994, *STU1*, a suppressor of a β-tubulin mutation, encodes a novel and essential component of the yeast mitotic spindle, *Journal of Cell Biology* 127: 1973–84, with permission.

High copy suppression and the α-tubulin genes

Let's return to the role of the two α-tubulin genes, *TUB1* and *TUB3*. Sequence analysis of the two genes for α-tubulin—*TUB1* and *TUB3*—showed that they are paralogous but not identical. However, the two genes behave differently in genetic analysis. Yeast cells in which *TUB1* is disrupted cannot grow and divide, which indicates that the *TUB1* encodes an essential function. In contrast, *tub3* mutants grow and divide normally. Furthermore, a *tub1* mutant strain with a wild-type copy of the *TUB3* gene on the chromosome cannot grow even when *TUB3* is also on a low copy number. How, then, can two slightly different versions of the same protein, α-tubulin, have mutations with such different phenotypes?

Thinking through the Experiment

16. Two different possible explanations for the results with *TUB1* and *TUB3* will be proposed. Before reading on, how would you explain these results with *TUB1* and *TUB3*?

Two possible explanations for these results come to mind. First, it might be that the TUB3 isoform of α-tubulin is not involved in growth and cell division under normal lab conditions; that is, TUB3 is functionally distinct from TUB1 despite their sequence similarity. Second, it may be that the two proteins have very similar functions and the difference in their mutant phenotypes reflects a difference in the level of expression. Under the second hypothesis, mutations that knock out *TUB3* are wild type in phenotype, because *TUB1* provides enough α-tubulin to compensate, whereas mutations that knock out *TUB1* cannot be compensated by the normal and low level of expression from *TUB3*.

These two hypotheses can be tested by varying the dose of the genes. Under the first hypothesis, changing the dose will have no effect since the genes are functionally distinct. Under the second hypothesis, increasing the dose of *TUB3* should be able to suppress a *tub1* mutant. A high copy suppressor analysis was used to test these hypotheses. Plasmids were made carrying the wild-type copy of either *TUB1* or *TUB3*, and grown to high copy number in different yeast mutant strains. The key experiment involves the yeast cells with *TUB3* on a high copy plasmid. In this case, *tub1* mutations are suppressed and the cells grow. Thus it appears that overexpression of wild-type *TUB3* using a high copy number plasmid can provide enough α-tubulin for normal growth even when *tub1* is absent. This high copy suppression indicates that either gene can provide functional α-tubulin for normal cell growth if expressed at a high enough level. This implies that the difference seen in the mutant phenotype for the two genes is a difference of expression level. The significance of the different expression levels of *TUB1* and *TUB3* also was important in the next set of experiments.

Thinking through the Experiment

17. Complete experiments 4 and 5 in Table CS10.1.

Non-allelic non-complementation among *STU1*, *TUB2*, *TUB3*, and *TUB1*

Thus far we have shown how suppressor analysis has been used to identify potential regions of interaction between α-tubulin and β-tubulin and to find a new gene involved in spindle morphogenesis. We have also seen how a high copy suppressor experiment revealed that the two α-tubulin genes are functionally equivalent despite different mutant phenotypes. In this section we describe experiments that reveal a more complex set of interactions between these genes involved in spindle morphogenesis. The experiments involve varying the wild-type or mutant dosage of two genes simultaneously, either in double heterozygotes or in homozygous double mutants. The reader is encouraged to use Table CS10.1 to keep track of the gene interactions being tested.

Non-allelic non-complementation involving *stu1* and *tub1*

Recall that certain specific *tub1* mutations and all *stu1* mutations suppress the cold-sensitive *tub2-406* mutation in β-tubulin and restore spindle microtubule function in a *tub2-406* mutant at restrictive temperature. The *TUB1* gene encodes α-tubulin, and *STU1* encodes a novel protein that localizes to spindle microtubules and affects spindle function. This raises the question of how the *tub1* mutations and the *stu1* mutations affect each other.

Note, from lines 6 and 7 in Table CS10.1, that two of the *tub2-406* suppressors, *tub1-108* and various *stu1* alleles, are recessive so that the heterozygous suppressors do not grow and divide at low temperature. To give a more specific example, *tub1-108* is one of the allele-specific suppressors of *tub2-406*. A *TUB1/tub1-108*; *tub2-406/tub2-406* strain is cold-sensitive and shows the spindle morphology defect characteristic of *tub2-406* mutants. Likewise, a *STU1/stu1*; *tub2-406/tub2-406* strain also has the mutant phenotype of *tub2-406*. These results have been included in Table CS10.1.

In contrast to the results with heterozygotes in one of the suppressor genes, a *STU1/stu1*; *TUB1/tub1-108*; *tub2-406*; *tub2-406* strain grows and divides at low temperature. That is, a strain that is heterozygous for both suppressors simultaneously has the *tub2-406* mutation suppressed. To express this result somewhat differently, although *stu1* and *tub1* define distinct genes, mutations in *stu1* fail to complement the *tub1-108* in their ability to suppress *tub2*. Although each of the suppressors is recessive, a strain that is heterozygous for both of

Table CS10.1 Gene interactions in spindle morphogenesis

A. Non-allelic non-complementation

***tub* gene**	**Interacting gene**	**Phenotype**	**Interpretation**
tub2-406/ tub2-406	*STU1/stu1*	Cold-sensitive arrest	*tub2* mutant phenotype
tub2-406/ tub2-406	*stu1/stu1*	Growth at low temperature	Suppression of *tub2-406*
tub2-406/ tub2-406	*TUB1/tub1-108*	Cold-sensitive arrest	*tub2* mutant phenotype
tub2-406/ tub2-406	*tub1-108/tub1-108*	Growth at low temperature	Suppression of *tub2-406*
tub2-406/ tub2-406	*TUB1/tub1-108; STU1/ stu1*	Growth at low temperature	Non-allelic non-complementation
TUB2/tub2 (cs)	*TUB1/TUB1*	Normal growth	*tub2* mutation is recessive
TUB2/tub2 (cs)	*TUB1/tub1-1*	Cold-sensitive arrest	Non-allelic non-complementation

B. Synthetic enhancement

***tub* gene**	**Interacting gene**	**Phenotype**	**Interpretation**
TUB1/tub1	*TUB3/TUB3*	Normal growth	*tub1* mutation is recessive
TUB1/tub1	*TUB3/tub3*	Cold-sensitive arrest	Non-allelic non-complementation
tub1(cs)/tub1(cs)	*TUB3/tub3*	Cold-sensitive arrest	*tub1(cs)* mutant phenotype
tub1(cs)/tub1(cs)	*tub3/tub3*	Arrest at all temperatures	Synthetic enhancement
tub2(cs)/ tub2(cs)	*STU1/STU1*	Normal growth at 30 °C	Conditional *tub2* mutant phenotype
TUB2/TUB2	*stu1/stu1*	Normal growth at 30 °C	Conditional *stu1* mutant phenotype
tub2(cs)/ tub2(cs)	*stu1/stu1*	Cellular arrest at 30 °C	Synthetic enhancement

them suppresses *tub2-406*. In other words, the mutations *tub1-108* and *stu1* show non-allelic non-complementation for their ability to suppress *tub2-406*. (Yeast geneticists often refer to this as unlinked non-complementation rather than non-allelic non-complementation.)

Thinking through the Experiment

18. Fill in experiment 8 in Table CS10.1, paying particular attention to the interpretation of this result.

The non-allelic non-complementation suggests that *tub1* and *stu1* may interact genetically with each other, since both show genetic interactions with *tub2*. The interaction between *tub1* and *stu1* may or may not be a direct physical interaction between the two proteins. As described at the end of Section 10.3, this type of non-allelic non-complementation could be an example of a toxic product. From this model we infer that the mutation in *stu1*, although recessive, somehow affects the structure of the mitotic spindle. This altered structure, although functional, is then further compromised by the mutation in *tub1*, so that the spindle with both mutations cannot assemble properly. Either mutation, alone, can be tolerated as long as the wild-type protein is also present, but the combination of both mutations is defective.

Thinking through the Experiment

19. Two different models for non-allelic non-complementation were proposed in the chapter. Do these experiments support or contradict either of these two models? What type of experiment might be imagined, even if it could be technically difficult, to rule out or support one or both models?

Non-allelic non-complementation involving *tub1* and *tub2*

Non-allelic non-complementation is also observed between certain α-tubulin and β-tubulin mutations. A genetic screen was designed to find mutations that failed to complement cold-sensitive mutations in *tub2* but were not themselves alleles of *TUB2*. One of the mutations found in this screen was an allele of the α-tubulin gene *TUB1*, designated *tub1*-1. This is also an example of non-allelic non-complementation. Both the *tub1*-1 mutation and the *tub2* mutation are recessive (see Experiment 9 in Table CS10.1). However, a strain that is heterozygous for both *tub1*-1 and *tub2* has a cold-sensitive mutant phenotype similar to that seen in *tub2* (see Experiment 10 in Table CS10.1). In this example interactional suppressors, molecular assays, and cytological assays have all shown that these two proteins contact each other directly in the mitotic spindle. The non-allelic non-complementation between mutations in the two genes provides one more piece of evidence for this interaction. Just as significantly, the molecular and cytological evidence justifies our interpretation that non-allelic non-complementation often involves proteins with direct physical interactions.

Non-allelic non-complementation and synthetic enhancement involving the two α-tubulin genes

We have been discussing the genetic interactions between the two different tubulin genes *TUB1* and *TUB2* and the microtubule-associated protein encoded by *STU1*. Another interaction was also observed between the two different α-tubulin genes, *TUB1* and *TUB3*. As described above, mutations in *tub1* are recessive and arrest cell division while mutations in *tub3* have no mutant phenotype alone. Experiments using high copy suppressors have indicated that the difference between these two phenotypes arises from a difference in expression level of each protein.

Just as interactions with *overexpression* of the genes were described above, we can also test the effect of gene dosage by *lowering* the dose of each gene using heterozygotes. A strain that is heterozygous for both mutations, *TUB1/tub1*; *TUB3/tub3*, arrests cell division at low temperatures. Thus the two α-tubulin genes also exhibit non-allelic non-complementation. Given what we know about the effects of gene dosage from high copy suppressors, we infer from this that the non-allelic non-complementation is also indicative of a dosage effect.

The dosage effect is further illustrated by results with homozygous double mutant cells. A *tub1*; *tub3* double mutant strain arrests cell division under growth conditions when either *tub1* or *tub3* single mutants can grow. In other words, the two α-tubulin genes, *TUB1* and *TUB3*, display both dominant and recessive interactions; they display both non-allelic non-complementation and synthetic enhancement.

Thinking through the Experiment

20. Fill in experiments 12 and 14 in Table CS10.1.

It is worth thinking about these examples in more depth. As discussed in the main text, two different models have been postulated to explain non-allelic non-complementation. The interaction between *TUB1* and *TUB3* is a clear example of gene dosage effects with two closely related genes. *Over expressing* wild-type *TUB3* on a high copy plasmid *suppresses* the mutant defect in a *tub1* mutation. *Reducing expression* of *TUB3 enhances* the phenotype of *tub1* mutations. In heterozygotes for the two genes, this is seen as unlinked non-complementation. In homozygotes for the two genes, this dosage effect is seen as synthetic lethality. These effects are expressed graphically in Figure CS10.7.

The interactions between *STU1* and *TUB1* and between *TUB1* and *TUB2* cannot be so easily explained as a gene-dosage effect arising from differences in gene expression level. Several observations indicate that a different explanation is needed.

- First, the proteins encoded by these genes are not closely related in amino acid sequence, so it is unlikely that these proteins are functionally redundant.
- Second and more significantly, the effects are seen with specific mutant alleles of *tub1* and *tub2*, but not with gene knockouts. An allele-specific effect of this type is not consistent

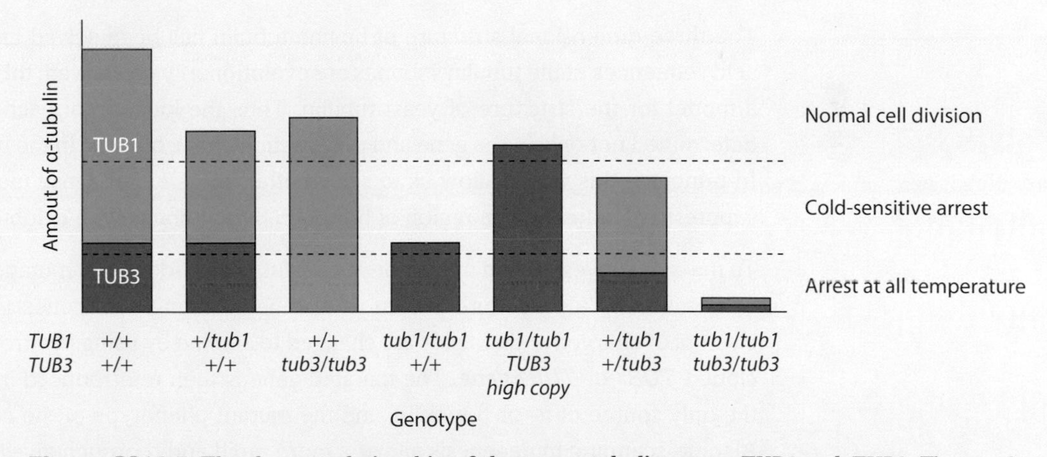

Figure CS10.7 The dosage relationship of the two α-tubulin genes *TUB1* and *TUB3*. The purple boxes represent the amount of α-tubulin produced from *TUB1*, and the green boxes represent the amount from *TUB3*. The genotype of each tubulin gene is shown below the boxes. Wild-type cells make more α-tubulin from the TUB1 gene than from the TUB3 gene. For simplicity, only two different mutant thresholds are illustrated. Cells with an amount of α-tubulin above the upper threshold have normal growth. This result is seen in wild-type cells, in heterozygous ***tub1*** strains that have normal *TUB3* function, or in ***tub3*** mutants that have normal *TUB1* function. A ***tub1*** homozygote arrests during mitosis; this particular allele shows a cold-sensitive arrest, but this varies with different alleles. This arrest is suppressed when *TUB3* is overexpressed on a high copy number plasmid. ***tub3*** mutations also show non-allelic non-complementation with ***tub1*** mutations and the strain heterozygous for both genes shows a cold-sensitive arrest. Furthermore, the ***tub1; tub3*** double mutant arrests cell divisions under conditions where the corresponding ***tub1*** mutant will grow.

with an effect of gene dosage when the effect should become more obvious as the gene function is reduced.

- Third, overexpression of the wild-type function of one of the genes does not act as a suppressor of mutations in the other gene. Thus a toxic interaction model, as described in the main text, more easily explains these effects.

Thinking through the Experiments

21. Which of the observations in the bulleted list above persuades most that a toxic product model is more likely than a gene dosage model? Is any of them consistent with either explanation for non-allelic non-complementation?

Confirmation of the genetic results by biochemical and molecular assays

All the genetic interactions discussed in this chapter were used in the analysis of the mitotic spindle in yeast. Microtubule assembly into the spindle has been studied by a variety of other methods in addition to traditional genetics, and this fact serves to justify our confidence in the genetic analysis.

Confirming interactional suppressors

Certain mutations in the α-tubulin gene *TUB1* are interactional suppressors of *tub2* missense mutations, and we implied that such suppressors define sites of interaction between the α- and β-tubulin proteins. Do these mutations in fact affect amino acids that lie at the sites of interaction between the two proteins?

The three-dimensional structure of bovine tubulin has been solved and, since the amino acid sequences of the tubulin subunits are evolutionarily conserved, this structure provides a model for the structure of yeast tubulin. Thus the location of each mutation could be determined not only in the gene and polypeptide chain, but also in the tubulin heterodimer. In principle, this would allow us to see whether the *tub2* missense mutations used in the suppressor studies lie in a region of β-tubulin that interacts with α-tubulin.

To make the investigation even more powerful, many additional mutations in *tub1* and *tub2* were made by a procedure known as alanine-scanning mutagenesis. In this procedure, clusters of charged amino acids are changed to alanine by using in vitro mutagenesis of the cloned *TUB1* or *TUB2* gene. The mutated gene is then reintroduced into the yeast cell as the only source of α- or β-tubulin, and the mutant phenotype of the cell is characterized. Alanine-scanning mutagenesis allows a more rapid and thorough investigation of different regions of a protein than single amino-acid substitutions. In particular, charged residues that are replaced with alanine are often on the surface of the protein and thus are more likely to affect interactions with other proteins. From the phenotypes of these mutations, different interactions could be tentatively assigned to regions of the proteins, which could then be compared to the expected location on the three-dimensional structure.

Thinking through the Experiments

22. On the basis of your knowledge of amino acid structures, why would the substitution of alanine for other amino acids be a particular good strategy for finding sites of interactions?

The results of these experiments for α- and β-tubulin provided striking confirmation of the value of the suppressor experiments, as shown in Figure CS10.5. The cold-sensitive *tub2* mutations suppressed by specific mutations in *tub1* do in fact alter amino acids that are predicted to lie on the surface of β-tubulin, either in the region of interface between the two proteins or on the lateral surfaces of β-tubulin. These are sites on the tubulin protein that other proteins involved in spindle assembly might contact. Therefore, by comparing what we know about the three-dimensional structure of the tubulin heterodimer with what we have inferred from the suppressor analysis, we find that the genetic suppressor analysis gives a good approximation of the regions of interaction. Because this comparison of molecular and biochemical analysis works so well in an example when both are known, we have greater confidence that interactional suppression between other proteins identifies good candidates for amino acids that lie at the regions of interaction, even when the structures of these proteins are not known.

The role of the *STU1* gene and protein

Having been identified in *Saccharomyces*, *STU1* orthologs have now been found in many other organisms, including humans. These proteins are also involved in spindle functions. In addition, a number of different *stu1* mutations have been studied in yeast, including several temperature-sensitive alleles. These *stu1* mutations have their own mutant phenotype, independently of their ability to suppress a *tub2* missense allele. The phenotype of these *stu1* mutations, like that of the original *stu1* mutation identified as a *tub2* suppressor, displayed clear defects in the mitotic spindle.

One of the temperature-sensitive alleles of *stu1* has been used for its own suppressor analysis. The suppressors of this *stu1* mutation included mutations of *tub2*. That is, *stu1* mutations act as suppressors of a cold-sensitive *tub2* mutation and, in turn, certain *tub2* mutations can suppress certain *stu1* mutations.

Further investigations of these suppressors suggest an even more complex picture. The *tub2-406* mutation affects a site on the lateral surface of the β-tubulin protein, a location that is exposed and might well interact with the Stu1p protein or with another microtubule-associated protein. However, the *tub2* mutations that acted as *stu1* suppressors did not affect residues in the same region of the β-tubulin protein, which suggests that these *tub2* mutations affect the folded structure of β-tubulin, and the altered structure of the protein changed its interaction with Stu1p.

The genetic analysis indicating that Stu1p is a component of microtubules that interacts specifically with β-tubulin was confirmed from two types of molecular experiments. First, in biochemical extraction experiment, Stu1p is found physically associated with microtubules. Truncated versions of Stu1p, made in vitro and reintroduced into the cell, identified the most likely portion of Stu1p needed for microtubule binding.

The association with microtubules by biochemical extraction assay did not indicate whether Stu1p interacts with α-tubulin, with β-tubulin, or with both. To determine what interacts directly with Stu1p, the region of Stu1p involved in microtubule association (as determined in the truncation experiments) was used in a yeast two-hybrid assay, as described in the main text. The microtubule binding region of the Stu1p was cloned into a yeast two-hybrid vector in which it is fused to the GAL4 activation domain. Plasmids with either *TUB1* or *TUB2* fused to the GAL4 DNA-binding domain were transformed into the cells, and a yeast two-hybrid assay was performed with each gene. The assay clearly showed that the microtubule-binding region of Stu1p interacts with β-tubulin and not with α-tubulin. Again, this is precisely what is expected from the suppressor analysis of the genes—suppressors of Stu1p were found in *TUB2* but not in *TUB1*.

But this was not the end of the experiments. The region of β-tubulin that interacts with Stu1p was identified by repeating the yeast two-hybrid assay with the *tub2* mutations induced by alanine-scanning mutagenesis. The assumption is that, if the interacting region of *tub2* has been mutated, no interaction will be seen with the binding region of Stu1p. Such mutated versions of *tub2* that no longer interacted with Stu1p were found. The interacting region of *tub2* as identified by this assay was compared with the three-dimensional structure of tubulin, as before. This revealed that the putative interface between β-tubulin and Stu1p is a patch of amino acids on the lateral surface of β-tubulin, on the same face of the β-tubulin protein as the *tub2-406* mutation used in the original suppression studies. Therefore the physical interaction detected by the yeast two-hybrid assay defined precisely the same part of the proteins as was defined by the suppressor analysis. Once again, the striking similarity of results from the yeast two-hybrid screen and the suppression analysis gives us confidence that the suppressor analysis is providing meaningful information about protein–protein interactions, even when the structures are not known.

Summary

The mitotic spindle in yeast allows the analysis of an unusually rich set of interactions by both genetic and molecular assays. Because the mutant phenotype is a cell division arrest that can be scored easily, nearly all the types of genetic interaction screens described in the chapter were exploited in this analysis. This included multiple types of suppressor analysis, synthetic enhancement, and non-allelic non-complementation. Furthermore, because tubulin is a protein whose biochemical properties are characterized and whose three-dimensional structure has been solved, it is possible to combine the results of the genetic analysis with biochemical analysis and molecular analysis, including yeast two-hybrid assays. The combined results provided a far more complete picture than either approach could have done alone.

References

Huffaker, T. C., M. A. Hoyt, and D. Botstein. 1987. Genetic analysis of the yeast cytoskeleton. *Annual Review of Genetics* 21: 259–85.

Pasqualone, D., and T. C. Huffaker, 1994. *STUI1*, a suppressor of a β-tubulin mutation, encodes a novel and essential component of the yeast mitotic spindle. *Journal of Cell Biology* 127: 1973–84.

Reijo, R. A. et al. 1994. Systematic mutation analysis of the yeast β-tubulin *gene. Molecular Biology of the Cell* 5: 29–43.

Richards, K. L. et al. 2000. Structure-function relationships in yeast tubulins. *Molecular Biology of the Cell* 11: 1887–903.

Schatz, P. J., F. Solomon, and D. Botstein. 1988. Isolation and characterization of conditional-lethal mutations in the *TUB1* α-tubulin gene of the yeast *Saccharomyces cerevisiae. Genetics* 120: 681–95.

Stearns, T., and D. Botstein. 1988. Unlinked noncomplementation: Isolation of new conditional-lethal mutations in each of the tubulin genes of *Saccharomyces cerevisiae. Genetics* 119: 249–60.

Yin, H. L. et al. 2002. Stu1p is physically associated with β-tubulin and is required for structural integrity of the mitotic spindle. *Molecular Biology of the Cell* 13: 1881–92.

11

CHAPTER 11

Epistasis and genetic pathways

TOPIC SUMMARY

Genes or their products modify the activity of other genes, with the result that a phenotype is the outcome of genes interacting with each other in a pathway or a network. Genetic methods such as suppression and synthetic enhancement can be used to find functionally related genes, while epistasis can often be used to infer a logical order in which genes function in a pathway. Each of these genetic methods compares the phenotype of an organism that is mutant in two different genes with the phenotype of single mutants. In epistasis, the double mutant has the mutant phenotype of one gene, while the phenotype of the other mutant is obscured; the gene whose phenotype is observed is said to be epistatic. The phenotype of the double mutant is then used to infer which gene is downstream of the other in the pathway.

INTRODUCTION

Genetics has prospered from the analysis of single genes with a simple mutant phenotype. But genetic analysis of a biological process will almost never stop with the analysis of only one gene. Many different genes contribute to each biological process, and the investigator will want to identify and characterize as many of them as possible. With genome-wide mutant screens, such an inventory of all the genes involved in a biological process is feasible.

But having an inventory of all of the individual genes is only part of the process of genetic analysis. A biological process can be thought of as a pathway where the genes and the proteins constitute individual components or steps. The pathway may be long or short, straight or branched, but genetics can be used to reveal its underlying logical structure. Genes, RNA, and proteins interact with other genes, RNA, and proteins. In fact it could be argued that every identified gene or gene product leads to a larger question involving other genes and gene products. If the protein is a kinase, we immediately wonder about its substrate; if it is a DNA-binding protein, we wonder about its regulatory targets. Even when we know all the parts of the pathway, we still need to know their interactions in order to understand a biological question.

In the previous chapter we described the use of suppressors and synthetic enhancers as genetic methods to identify functionally related genes. These approaches find interacting genes (and their products), but might not provide all of the information needed to interpret these interactions. In this chapter and the next, we will consider several different methods that expand genetics beyond the analysis of individual genes and proteins to the analysis of several genes interacting in a biological process. This chapter builds on genetic approaches, primarily using mutant phenotype in double mutants, to infer the pathways by which genes affect a biological process. The key concept is **epistasis**, a classical genetics term, which means that the phenotype of one gene masks the phenotype of a different gene. Epistasis is a relatively simple but powerful method to construct the logical pathway of the gene interactions that underlie a biological process.

11.1 Epistasis and genetic pathways: An overview

Genes and proteins work together with other genes and proteins to carry out a biological process. The preceding chapters of this book have described how to find and characterize individual genes, even how to find all the genes that affect a biological process. In Chapter 10 we described how one gene could be used as a starting point to find additional genes involved in the same process. In this chapter we will describe how to expand genetics from understanding the contributions of a single gene to interpreting how genes interact to produce a phenotype.

In learning how genes (or, more often, their protein products) interact with one another, we turn from the analysis of single mutants to the analysis of double or even triple mutant strains. That is, rather than examine the phenotype of a cell or organism that has a mutation in one gene, the investigator examines the phenotype of a cell or organism that has mutations in several different genes—most commonly, in two different genes. To simplify slightly, the double mutant may have one of four possible phenotypes with respect to each of the single

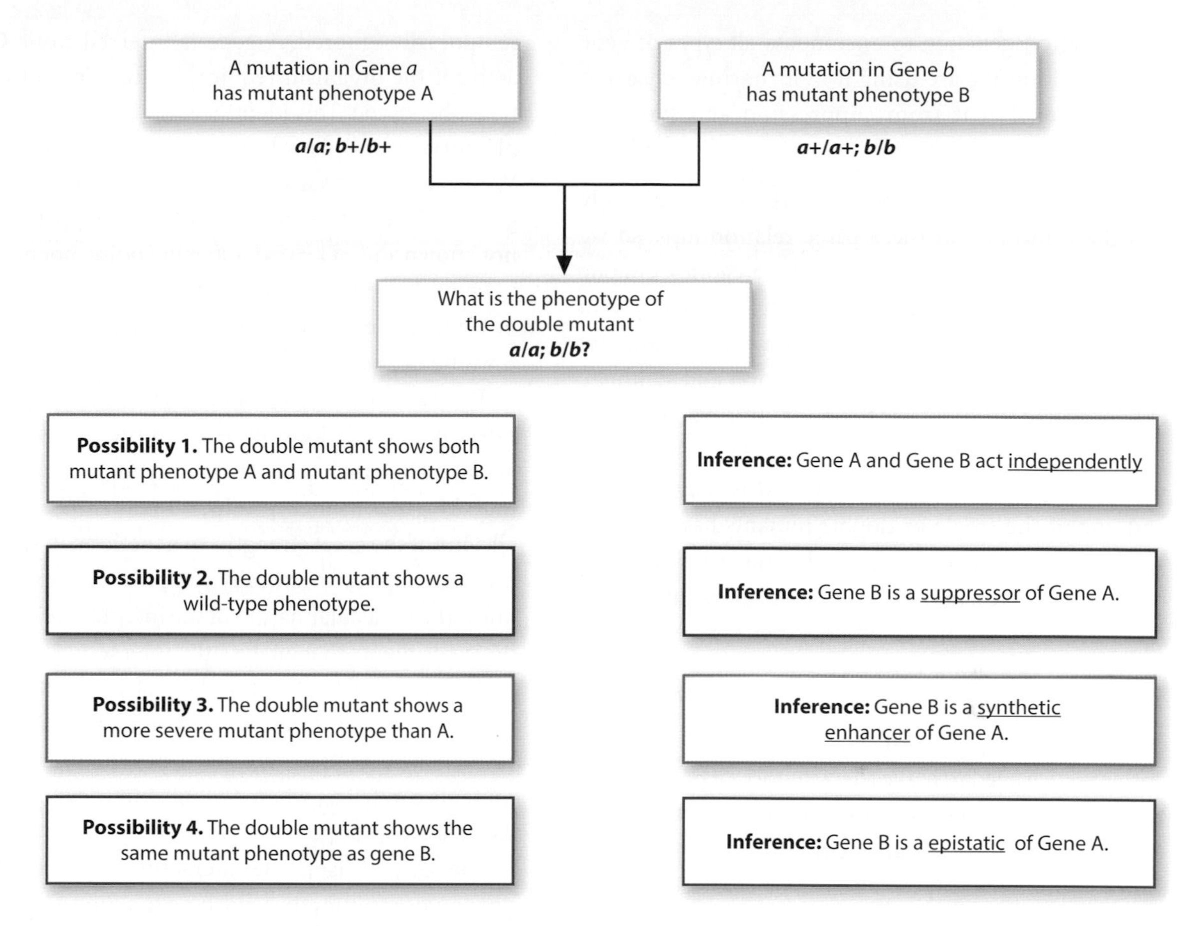

Figure 11.1 Gene interactions that could be observed in a double mutant. A double mutant between two genes, summarized here as Gene *a* and Gene *b*, can have any of these four phenotypes. If the mutant phenotypes of both *a* and *b* are observed, the genes probably have independent functions. If the double mutant shows a more wild-type phenotype than the single mutant *a*, suppression is occurring. If the double mutant has a more severe mutant phenotype, or if it has a mutant phenotype when neither single mutant does, synthetic enhancement is occurring. Suppressors and synthetic enhancers are discussed in Chapter 10. If the mutant phenotype of only one gene is observed and the other is masked, epistasis is occurring. The phenotype that is epistatic can be used to inferred the cellular pathway by which the genes interact.

mutants. We summarize these possibilities in Figure 11.1. First, the double mutant may exhibit both mutant phenotypes. Mendel encountered such an example in using peas that were doubly mutant for round versus wrinkled and for yellow versus green. We learned long ago that these traits are inherited independently of one another; this principle was taught as Mendel's law of independent segregation. But an overlooked lesson from that same experiment is that these genes *act* independently of one another: the color of the pea has no effect on the shape of the pea. So the phenotype of the double mutant is indicating that these genes have different functions.

What if the two genes do not function independently of one another, so that the phenotype of one mutant will affect the phenotype of the other? There are three other possible double mutant phenotypes:

- For example, the double mutant may exhibit a phenotype that is of the wild type or more like the wild type, showing neither of the phenotypes of the individual mutations. This type of interaction is **suppression**, which is discussed in Section 10.2.
- Alternatively, the double mutant may have a more severe mutant phenotype than either of the two single mutants, or may have a mutant phenotype when each single mutant shows a wild-type phenotype. This type of interaction is **synthetic enhancement**, which is discussed in Section 10.3.
- Finally, only one of the two mutant phenotypes is seen and the other one is obscured or masked. This type of interaction is known as epistasis, and it will be the main topic of this chapter. (We note that the term "epistasis" is sometimes used, particularly in

population genetics, to encompass all types of gene interactions. We are using it in its narrow sense and distinguishing it from suppression and synthetic enhancement.)

Some double mutant combinations cannot be neatly classified into any of these three relationships, so we also consider the limitations of these categories. But one way to think about the difference between the topics in Chapter 10 and this chapter is to compare mutant and wild-type phenotypes. With suppression and synthetic enhancement (particularly the most informative type of synthetic enhancement), either the single or the double mutant has a wild-type or nearly wild-type phenotype, and one of the single or double mutants has a mutant phenotype. With epistasis, both the single and the double mutants have a mutant phenotype, so the interpretation hinges on which mutant phenotype is seen.

The logic of epistasis requires close attention

Before describing epistasis and the analysis of genetic pathways, we need to make a few important points about the logic that will be used. First, recall that the normal functions of the genes are being inferred from those genes' *mutant* phenotypes. In nearly all cases, the alleles used for the genes under study are recessive null alleles, so the inference is based on the absence of function of a gene; however, exceptions that use other types of mutant alleles are important and will be noted.

Second, in working with mutant phenotypes, we are making inferences about *gene activity* rather than *gene expression*. This concept was discussed in Chapter 6. When a gene is described as being ON or OFF, the molecular biology of gene expression is being temporarily ignored. A gene is inferred to be ON or active if the organism requires its function for a wild-type phenotype. This identification is based on mutant phenotypes and usually not on a molecular analysis. When we write that gene *A* turns OFF or inactivates gene *B*, the mechanism by which this happens is often not known and, at least as a starting point, not necessary to know. Any mechanism by which the wild-type activity of Gene *A* is needed for the wild-type activity of Gene *B* would be included.

Third and most significantly, just because we are often inferring the relationship between the two genes from an analysis of their mutant phenotypes, we can't assume that the genes or their products interact molecularly at all. One of the great strengths of genetic analysis in general also applies specifically to epistasis: one need not know the molecular nature of the interaction in order to do experiments that analyse the interaction. Of course, as discussed in Chapter 10, if the molecular nature of the interaction is never investigated, some key insights will be missed. As with many aspects of genetic analysis, the mutants can tell us where to look, but they may not tell us what we will find.

The logic of using the interactions between mutant phenotypes to infer the wild-type functions of the genes can become intricate. As was discussed with single mutants in Chapter 6, one has to think carefully about what specific phenotype is being used as basis for the interpretation. A form of epistasis can also be done using a mutation in one gene and the wild-type phenotype for another gene. This analysis is usually a little easier to grasp and it sets the stage for the analysis using two mutant phenotypes. We will begin with that approach.

11.2 Combining mutants and molecular expression assays

Let us consider an analysis of the genetic interaction between two genes that uses one mutant phenotype and one wild-type phenotype. In this particular kind of experiment, the investigator has a mutation in one of the two genes and a molecular reagent such as an antibody or a reporter gene for the expression of another (potentially) interacting gene. While these experiments are often not considered to be classical epistasis in the same way that two mutants are, their interpretation helps us think about experiments that work with two mutant phenotypes in classical epistasis.

We introduced the *patched* gene in *Drosophila* in the Case Study in Chapter 4. Although much of our prior discussion of *patched* centered on its role in the segmentation of the embryo, the gene also has a later role in wing development, as we noted in Chapter 4 when discussing some hypomorphic alleles of *patched*. In this chapter we will use its phenotype in wing development as the basis for understanding its genetic pathway.

As noted in Chapter 3, the wings and other limbs in adult flies are derived from particular pockets of cells set aside during embryonic development and known as imaginal discs. The cells of the imaginal wing discs are subdivided into an anterior compartment and a posterior compartment; these compartments and the border between them are not evident morphologically but were revealed by various developmental genetic experiments. Mutations in *patched* (*ptc*) affect the development of the anterior compartment, with no effect on the posterior compartment. For example, mutations in *ptc* result in loss of specific structures in the anterior wing compartment and overgrowth of other anterior structures. Antibodies to the Ptc protein showed that *patched* is expressed throughout the anterior compartment, but most strongly in the cells close to the compartment boundary; it is not expressed in the posterior compartment.

A *ptc* mutation changes the expression pattern of other proteins expressed in the wing

Another gene involved in imaginal disc development is *dpp*, needed for outgrowth of the discs into normal size limbs; in *dpp* mutants the wings are quite small, or even absent. Antibodies to the Dpp protein showed that expression was confined to the cells on the anterior side of the compartment boundary. This is diagrammed in Figure 11.2. Thus both the Dpp and the Ptc proteins are present in some of the same cells at about the same time, although Ptc is expressed in a broader region than Dpp is. In light of the similar expression pattern, it is logical to ask whether the expression of one gene affects the expression of the other.

The combination of the antibodies designed to detect the presence of each protein and the mutations in each gene allowed investigators to answer a fairly simple question. Does the *patched* gene affect the expression of the Dpp protein? (The results of the analysis are here slightly simplified, for clarity.) To answer this question, the Dpp

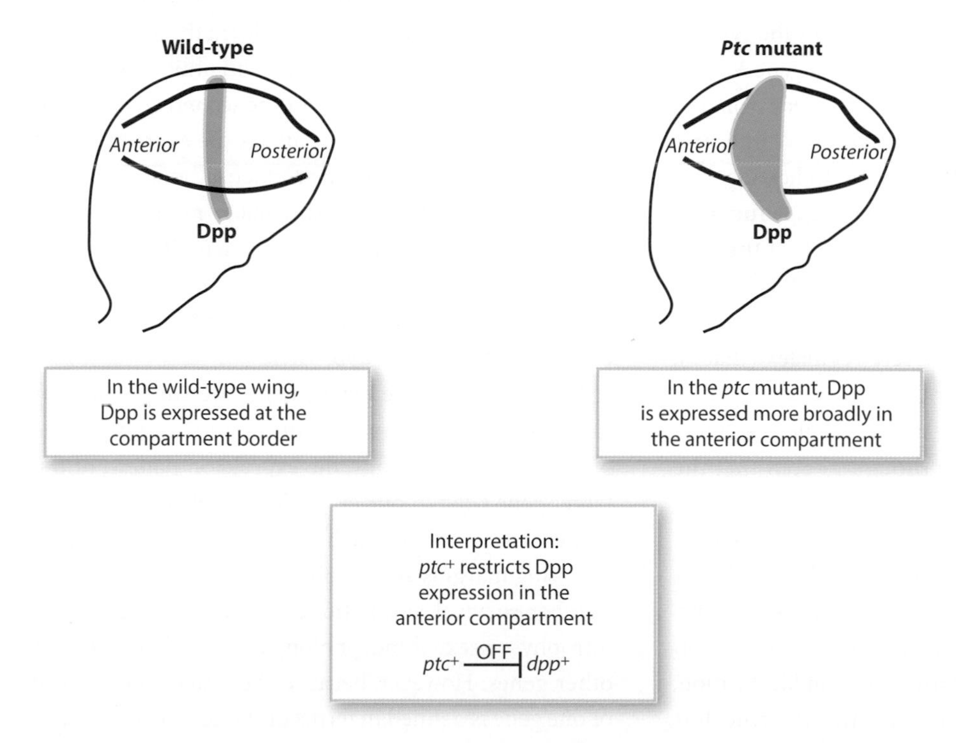

Figure 11.2 The effect of *ptc* mutations on Dpp protein localization in the wing disk. In the wing imaginal disk in wild-type *Drosophila*, Dpp protein is found in a band across the middle. In a *ptc* mutant, the localization of the Dpp protein expands into the anterior compartment of the wing disk. The inference is that the wild-type role of the ptc^+ gene is to repress or turn off Dpp expression in the anterior compartment, which is expressed in the pathway diagram at the bottom. This symbolism refers to the logical relationship without implying a direct interaction or a mechanism.

expression pattern was examined in a *ptc* mutant fly and compared to the wild-type pattern; the results are shown in Figure 11.2. In a *ptc* mutant, the expression of Dpp protein expands over a broader region of the anterior compartment than it does in a wild-type fly, and cells that *do not* express Dpp in a wild-type fly *do* express Dpp in a *ptc* mutant. The inference is that the expression of Dpp depends on *ptc* expression; in particular, *ptc+* turns off *dpp+*. In contrast, the expression pattern of Ptc, as detected by antibody staining, is the same in wild-type and in *dpp* mutant flies, so the localization of the Ptc protein does not depend on the activity of the *dpp* gene. By a similar analysis, *ptc+* was also found to turn off the gene *wingless* (*wg+*), which indicates that the effect of *ptc* applied to other genes expressed in the anterior compartment; we will not include *wg* in our discussion but it behaved like *dpp* in all experiments that we discuss.

Literature Link. Capdevila et al. 1994, *EMBO Journal* 13: 71–82

The logical pathway describing these interactions is also depicted in Figure 11.2. Note especially the symbol used for turning *off* the activity of the other gene. We are inferring the activity of the wild-type *ptc+* gene from the effects of the *ptc* loss of function mutation. We interpret the results diagrammed in Figure 11.2 as indicating that, in normal flies, the activities of the *dpp+* gene function *downstream* the *ptc+* gene, and that the role of the *ptc+* gene is to turn off the *dpp+* gene in cells in the anterior compartment. To express this relationship another way, the wild-type *ptc+* gene negatively regulates (that is, turns off) the expression of Dpp.

Although this looks like a simple result, the interpretation is worth a bit more thought. Recall that the mutant phenotype can be used to infer the gene function but does not provide details about that function. We have inferred that *ptc+* turns off Dpp, but we do not know from this experiment how that regulation occurs. The experiment does not tell us whether the effect is on transcription, splicing, translation, protein localization, or some other molecular process for the *dpp* gene. Just as significantly, the experiment does not tell us whether the interaction between *ptc* and *dpp* is direct, or whether other genes are involved. In fact many other genes could be involved in this process. The experiment simply describes a functional relationship between the normal activity of the *ptc+* gene and the localization of the Dpp protein. Put another way, the interaction described this way gives us the *logic* by which *ptc* and *dpp* interact, but does not tell us the *mechanism*.

Interpreting the results of this experiment

But the logic of many complicated processes is often enough to provide useful information, at least as a starting point for understanding the entire process. Consider an example from daily life. When I need to stop my car suddenly, it is sufficient for me to know that, if I push down hard on the pedal in center of the floor, the car will stop. I do not need to know the mechanism by which this pedal exerts its effect, only that it happens. While it might be helpful to understand hydraulics, brake pads, and rotors, this will not change the simple logic—step down here and the car stops. So it is with this simple example of gene interaction: we infer that *ptc+* turns off Dpp because Dpp is expressed in *more* cells in the *absence* of *ptc* activity.

The combination of genetic mutations with molecular reagents is a frequently used analytical method in genetics, with thousands of examples in the literature. By using mutant alleles of one gene and antibodies raised against the wild-type protein of the second gene, we can determine whether one gene affects the expression and localization of another gene product; although antibodies were used in these experiments to monitor expression, many molecular reagents can be used. Microarray experiments that compare the wild-type expression pattern with the expression profile in a mutant are another example of this logical approach.

In one sense, these experiments are simply obtaining a more detailed analysis of the mutant phenotype of *patched*. Conceptually, the expression pattern of the target gene (*dpp* in our example) is a phenotype of *ptc* in the same way in which its effect on segmentation and wing margins are the phenotypes of *ptc* and bristle number, survival, antibiotic sensitivity, amino acid prototrophy, or sexual morphology are the phenotypes of other genes. However, because the mutant phenotype of one gene is defined in terms of the activity of another gene or genes, this approach allows us to build a *pathway* through which the genes or their proteins work in a way that morphological phenotypes alone do not.

In our example, we placed *ptc* upstream of *dpp* (and *wg*) in a functional pathway and showed that *ptc* is a negative regulator of these two genes or their products. This approach can be extended to include many other genes and proteins to construct a longer or more detailed logical pathway. The strength of this method is that the investigator needs to have only two components—a mutation in one gene and a molecular reagent to monitor the expression of the other gene—and no prior knowledge of the functions of these genes or of their interaction.

A thoughtful consideration of this approach will also recognize its limitations. Although we can say that *ptc+* turns off the *dpp+* gene, we have no idea whether the interaction between *ptc+* and these downstream genes is direct or indirect; we don't know whether the mechanism by which ptc^{+} affects *dpp+* is similar or different from the way *ptc+* affects wg^{+} or other genes in the pathway. We have not investigated how the activities of *dpp+* and wg^{+} are related to one another. We have no idea how many other genes might be involved and what they may be doing, and we have no idea of the mechanism by which the upstream gene is affecting the activity of the downstream gene. Of course, this method also requires that molecular reagents are available to monitor the expression of at least one gene. Despite these limitations, we can begin to build a pathway or a network by which genes interact to determine a more complicated phenotype.

By understanding the rationale behind this experiment, we can now consider some other genetic methods that do not rely on molecular reagents or cloned genes, and thus may be even simpler to perform. These genetic methods use only the mutant phenotypes to infer the pathway of interactions among genes without knowledge of any of the molecules involved. This is one of their advantages: the interpretation of the interaction does not require knowledge of the molecules involved. All that is required is the ability to construct and analyse a double mutant strain.

11.3 Epistasis and genetic pathways

At the beginning of the chapter we discussed that a double mutant could have the mutant phenotype of only one of the two genes. Since gene activity can be inferred by observing mutant phenotypes, epistasis refers to the ability of mutations in one gene to mask the effects of mutations in a second gene. That is, in a double mutant, only one of the two mutant phenotypes is observed; the other is not seen.

An example used in many textbooks involves genes for coat color in mammals, as shown in Figure 11.3. In the textbook example, there are two genes represented by *C* and *B*. In order for color to be observed, the *C* allele must be present: *cc* animals are white. At the *B* locus, black (*B*) is dominant over brown (*b*), so that the genotypes *BB* and *Bb* are black while the genotype *bb* is brown. Double mutant animals that are genotypically *cc BB*, *cc Bb*, and *cc bb* are all phenotypically white, because the *C* allele is needed to express color. We say that a mutation in the *C* gene to *cc* is epistatic to the *B* gene: *C* masks the effects of *B*. To put this another way, if the *C* allele is not present, the genotype at the *B* locus has no effect on the phenotype—or we say that, in *cc* animals, the genotype at the *B* locus is irrelevant to the final phenotype.

Let us think about this result in light of our previous discussion, using expression pattern as our phenotype. The *C* allele is needed to turn ON expression of the *B* locus, because in *cc* mutants, the *B* locus is functionally OFF. As discussed previously, we do not know the molecular mechanism by which the *C* allele affects the expression or function of the *B* locus, or whether the interaction between the two loci is direct or indirect. But we have been able to infer a simple genetic logic: the dominant *C* allele turns ON the function of the *B* gene, as symbolized in the pathway in Figure 11.3. Note also that we have introduced a new symbol, the arrow indicating that the activity is turned on.

Epistasis is widely used to place the genes that affect the same biological process into a logical pathway or framework of interactions. Epistasis has also been used to help us understand the logic of some important biological pathways long before molecular or biochemical information was available. To cite two examples in addition to the ones we will use in this chapter, both the Ras/Raf signal transduction pathway and the apoptosis pathway were first analysed using epistasis with mutants in flies or worms.

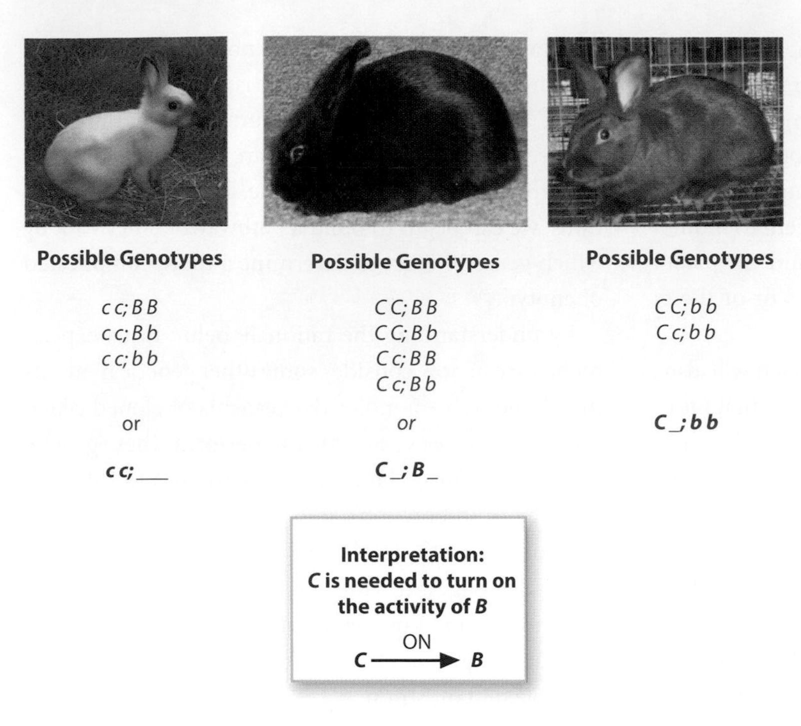

Figure 11.3 Epistasis involving coat colors. In this example of epistasis involving mammalian coat color, the wild-type *C* allele must be present for coloration, and the *B* gene determines black vs. brown. The possible coat color genotypes for each phenotype are listed below each picture, and the epistasis pathway inferred from these phenotypes is shown at the bottom of the figure. If the *C* allele is absent (that is, if the rabbit is *cc*), the fur is white regardless of the genotype of the *B* gene; thus we infer that the *C* allele is needed to turn on the function of the *B* gene.

But epistasis, like any powerful tool, needs to be used with caution. One of the first warnings is that the method describes a logical pathway, not a biochemical one. It is tempting to think of epistatic pathways in terms of metabolic pathways or biosynthetic pathways, but this is usually not the best way to approach epistasis analysis; in fact, to do so is to interpret the results before all of the relevant data have been collected. A significant strength of the genetic approach is that epistasis experiments do not require any prior molecular or biochemical information about the genes or the process they affect. For this reason, epistasis experiments are often an early step in the genetic analysis of a biological question. Epistasis provides a logical framework inferred from the phenotype of double mutants. It does not require having all the steps in the pathway or knowing what any of those functions are. But having the logical pathway then allows the investigator to design new experiments in order to learn the molecular and biochemical mechanisms by which the process occurs.

Although there are many published examples where epistasis has been used to infer a pathway of gene interactions, relatively few authors have provided detailed descriptions of the logic that they used to produce to produce the pathway. In the remainder of this chapter, we will take apart some of these experiments and show how epistasis was used in the construction of a few particularly well-studied genetic pathways, attempting to reconstruct the logic of the analysis.

Literature Links. Botstein and Mauer 1982, *Annual Review of Genetics* 16: 61–83; Avery and Wasserman 1992, *Trends in Genetics* 8: 312–16; Huang, and Sternberg 2005, *Methods in Cell Biology* 48: 97–122

General amino acid control in budding yeast

Budding yeast has an intriguing regulation of amino acid biosynthesis known as general amino acid control. When starved for one amino acid such as histidine, the yeast cell activates (or, more formally, de-represses) the transcription of genes that encode biosynthetic enzymes for not only for histidine but also for arginine, tryptophan, and lysine, as diagrammed in Figure 11.4. It is easy to

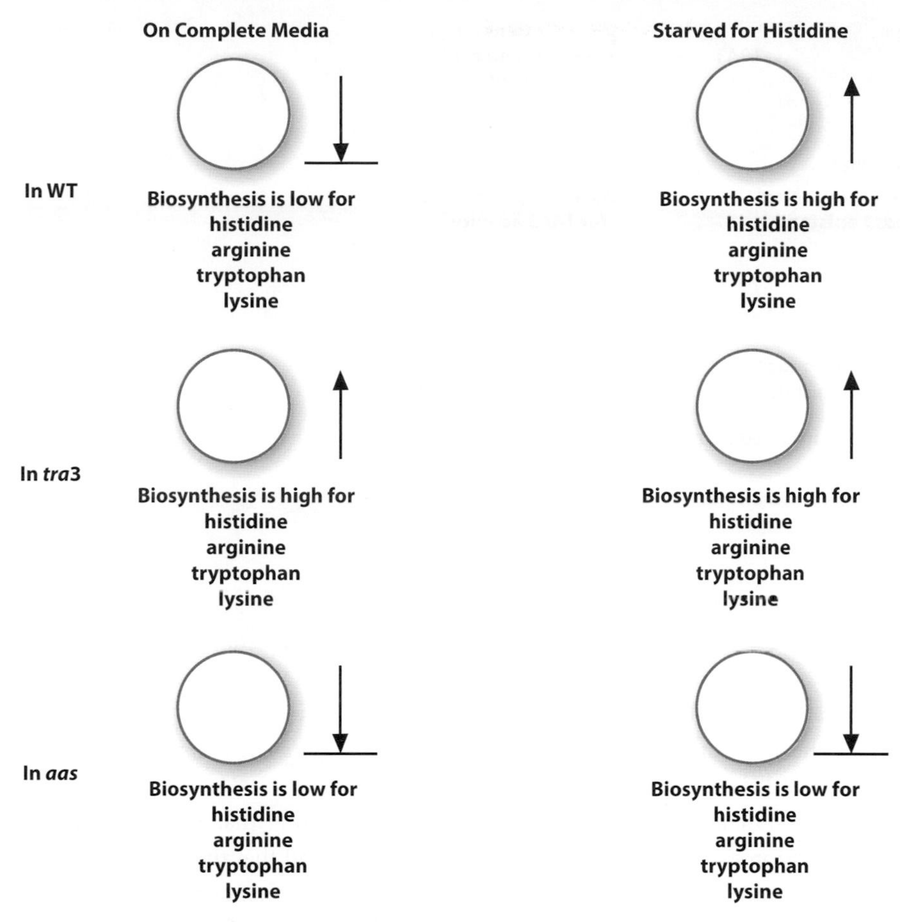

Figure 11.4 The general amino acid response in yeast. When wild-type yeast cells are grown in complete media, amino acid biosynthesis is low, as shown at the top left. When the cell is starved for one amino acid, biosynthesis of several different amino acids increases, as shown on the top right. Two different categories of mutant phenotypes are known to affect this response. In *tra3* mutants, general amino acid biosynthesis is elevated even when the cells are grown in complete media. In *aas* mutants, amino acid biosynthesis is not elevated even when the cell is starved. Three different *aas* genes are known, *aas1*, *aas2*, and *aas3*.

imagine how such a regulatory system may be advantageous in nature since yeast cells would rarely be starved for only one amino acid without also being starved for the other ones; in the absence of one amino acid, the cell begins to synthesize many amino acids for itself.

Literature Link. Hinnebusch and Fink 1983, *Proceedings of the National Academy of the Scinces* 80: 5374–8

Mutations that affect general amino acid control fall into two classes, summarized in Figure 11.4 as *tra3* and the *aas* genes. Recessive *tra3* mutants constitutively activate general amino acid control such that amino acid biosynthetic genes are transcribed regardless of the growth media. This mutant phenotype indicates that the wild-type *TRA3* gene acts as a negative regulator of the process, and its normal function is to turn off general amino acid control.

In contrast, mutations in any of the three *AAS* (*AAS1, AAS2,* and *AAS3*) genes cannot activate general amino acid control under starvation conditions. Hence the *AAS* genes are positive regulators of the general amino acid control and must somehow turn on the process. Thus the *TRA3* gene and the *AAS* genes have opposite mutant phenotypes and affect general amino acid control in opposite functional ways—the wild-type *AAS* genes turn the process on, and the *TRA3* gene turns the process off. This is summarized in Figure 11.5.

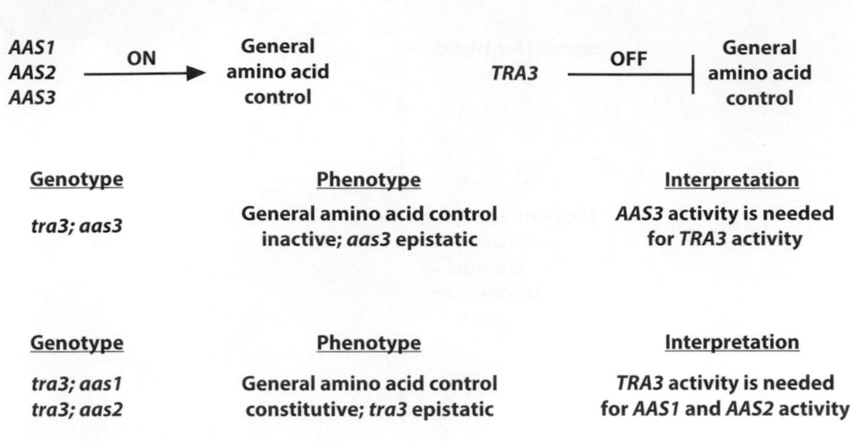

Figure 11.5 **The wild-type roles of the genes involved in the general amino acid response.** The three *AAS* genes are needed to turn on the general amino acid response, whereas the *TRA3* gene is needed to turn off the general amino acid response. These roles are inferred from the phenotypes of loss of function mutations in each gene, as shown in Figure 11.4.

In order to understand the pathway by which these genes regulate general amino acid process, double mutant strains were made that carried the *tra3* mutation and one of the *aas* mutations. The *tra3; aas3* double mutant shows the phenotype of *aas3* gene and cannot activate general amino acid control even under starvation conditions. In other words, the mutation in *aas3* is epistatic to the mutation in *tra3*: the mutation in *aas3* *masks* the effect of the mutation in *tra3*. We have summarized this in Figure 11.5.

Mutations in the two other *aas* genes, *aas1* and *aas2*, have a mutant phenotype similar to that of *aas3* and do not activate general amino acid control. However, the phenotype of a double mutant strain between one of these genes and *tra3* is different from the *tra3; aas3* double mutant. Both an *aas1; tra3* double mutant and an *aas2; tra3* double mutant have the phenotype of a *tra3* mutant and do not repress general amino acid control. Therefore a mutation in the *tra3* gene is epistatic to mutations in *aas1* or *aas2* genes. Thus, we cannot conclude that either the TRA phenotype or the AAS phenotype is inherently epistatic. Rather the phenotype depends upon which pair of genes is being tested.

Let us think about these results as a logical pathway of interactions, which are summarized in Figure 11.6. When starved of amino acids, the *AAS+* genes are turned ON and the *TRA3+* gene is turned OFF. Although we know that the genes have opposite mutant phenotypes and hence functions, we do not know which gene is the upstream regulator of the other. Epistasis tells us that the activity of *TRA3* depends on the activity of *AAS3* since, when *AAS3* is mutant, the genotype of *TRA3* is irrelevant; *aas3* mutants are epistatic to *tra3* mutants such that *aas3; TRA3* and *aas3; tra3* have the same phenotype. Figure 11.6 lists this as Result 1. This means that the activity of *AAS3* is *downstream* of the activity of *TRA3*.

With a similar logic, noting that *tra3* mutants are epistatic to *aas1* and *aas2* and that the *TRA3* gene and the *AAS* genes have opposite phenotypes, we can infer the activity of *TRA3* downstream of *AAS1* and *AAS2*, as shown, as Result 2 with its interpretation in Figure 11.6. In other words, the phenotype of *TRA3* masks the phenotype of *AAS1* or *AAS2*. We can then put these two pathways together, to construct a combined longer pathway with all three genes (not distinguishing between *AAS1* and *AAS2*).

Although inferred from the mutant phenotypes, the pathway is drawn with the wild-type gene functions. Recall that the *AAS* genes are needed for the general amino acid response and the *TRA3* gene turns off the general amino acid response. The function of the *AAS1* and *AAS2* genes is to turn off *TRA3* when the yeast cell is starved for amino acids. When either *AAS1* or *AAS2* is knocked out, *TRA3* is not turned off, and the yeast cell does not respond to amino acid starvation. The general amino acid response depends on turning off *TRA3*.

TRA3 turns off *AAS3* but, since *TRA3* is not active when the cells are starved, *AAS3* is not turned off, so it is active. On the other hand, when the yeast cells are grown in complete media with all amino acids, the *AAS1*

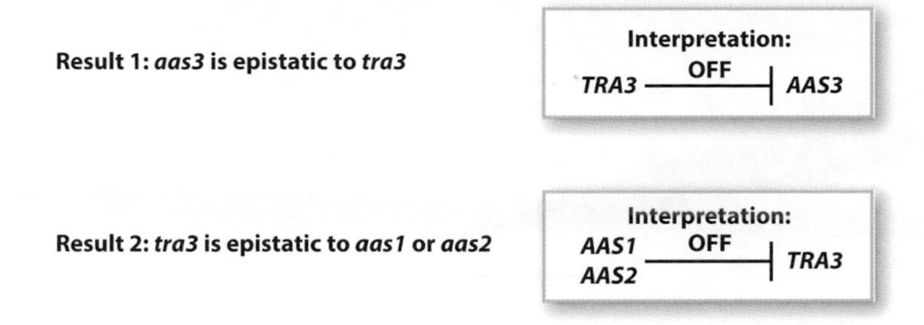

Inferred Combined Pathway

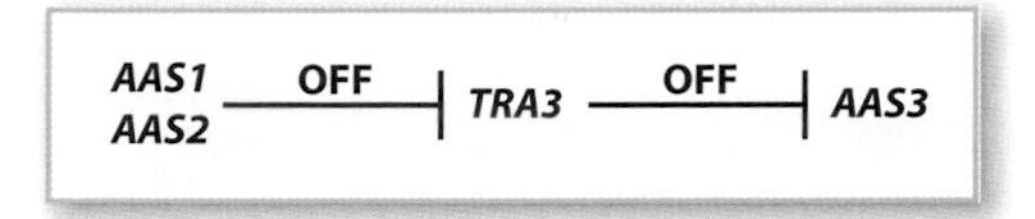

Figure 11.6 The inferred pathway for the general amino acid response. As discussed in the text, double mutants were made for each *aas* gene and *tra3*, and the phenotypes observed. The phenotypes showed clear epistasis for one of the two mutants, which allowed the construction of a logical pathway by which the genes interact. Details are described in the text.

and *AAS2* genes are not active. Since they are not active, *TRA3* is active, and *AAS3* is turned off so the cells do not overproduce amino acids that are not needed. The general amino acid (via *AAS3*) is turned off in complete media. The positive regulation of general amino acid control by *AAS1* and *AAS2* is actually the outcome of two negative or repression steps.

In this pathway we have not separated the roles of *AAS1* and *AAS2* or ordered them with respect to each other. A double mutant between these two genes would not be informative. Neither of these mutants can activate transcription of the biosynthetic genes, so the double mutant will also not activate transcription of the biosynthetic genes. Since this is the phenotype of each of the single mutants, we cannot determine which phenotype is being observed in the double mutant. Thus we have to leave the relationship between *aas1* and *aas2* unresolved, at least for now.

This pathway illustrates many of the important points about epistatic analysis in a relatively simple system of control. For example, although we have inferred the order in which each of the genes functions, nothing about the molecular mechanism has been revealed. Many other genes could be involved in this process, but we have not included them in the pathway. Some of the interpretations are subtle, so we will now reinforce many of these same points with another complex process, in an attempt to reach some general principles for constructing and interpreting gene interaction pathways.

Sex determination in *C. elegans*

Differences in sexual morphology are among the most familiar and fundamental biological phenomena. As discussed in Chapter 3, *C. elegans* has two sexes, males and hermaphrodites; pictures of hermaphrodites and males are shown in Figure 11.7. The genetic control of sex determination in worms is one of the best examples of the use of epistasis to analyse a complex biological question. The primary signal for sex determination in normal diploid *C. elegans* is the number of X chromosomes: males have one and hermaphrodites have two. The hermaphrodite is essentially a somatic female that makes some sperm, which it can use for internal self-fertilization. Our example will concentrate on the somatic tissues; the regulation of sex in the germline will be ignored.

The somatic differences between males and hermaphrodites are apparent from Figure 11.7. Males have an elaborate tail, used in mating, whereas the hermaphrodite tail is morphologically simple. Hermaphrodites have a vulva and associated structures for egg laying that males lack. There are also differences in size and behavior, and in the shape of the gonad. In fact, if one

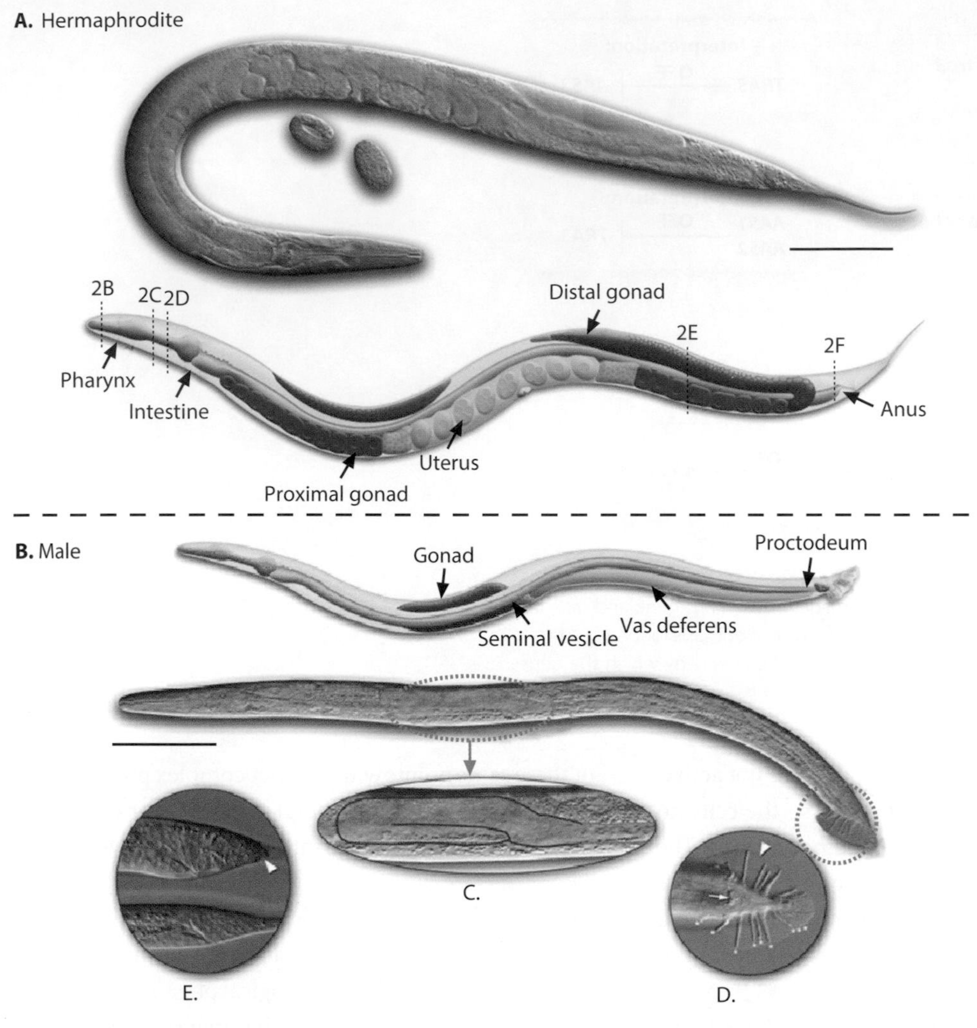

Figure 11.7 Sexual morphology in *C. elegans*. Hermaphrodite and male worms are shown; males have one X chromosome (1X), while hermaphrodites have two X chromosomes (2X). The sexual dimorphism in the somatic cells includes differences in size and in the tail, as well as in behavior. Since the discussion in the text only deals with somatic cells and not the germline, the 2X hermaphrodites are referred to as females. Images used by permission from the Wormatlas.org.

examines the cell lineages, approximately 40 percent of postembryonic cell lineages show evidence of sexual dimorphism. Thus males and hermaphrodites can be distinguished at the gross morphological level, as well as by examining individual cells or lineages.

Mutations in a number of different genes that affect somatic sex determination have been identified, and epistasis among these genes has established the genetic pathway of sex determination.

Literature Link. Hodgkin 1987, *Annual Review of Genetics* 21:133–54; Zarkower 2006, Wormbook.org

Our discussion will focus on three of the genes, but many more genes have been found using suppression and synthetic enhancement screens of the type discussed in Chapter 10. The three genes that we will discuss are, in order, *tra-1*, *her-1*, and *tra-2*. Their mutant phenotypes and their inferred roles in wild-type worms are summarized in Figure 11.8. All of these genes are defined by mutations that result in sex reversal of one sex with little effect on the somatic sex differentiation of the other sex. Recessive null mutant alleles of the *tra-1* gene result in both 1X and 2X animals developing as males. Thus, the wild-type function of *tra-1+* is apparently dispensable in 1X animals—they become males

	Mutant phenotype		Inferred role in Wild-Type
	1X	2X	
tra-1	Male	Male	Needed for female development. ON in 2X, OFF in 1X
her-1	Female	Female	Needed for male development. ON in 1X, OFF in 2X
tra-2	Male	Male	Needed for female development. ON in 2X, OFF in 1X

Figure 11.8 Mutations in genes affecting sex determination in *C. elegans*. In the wild type, worms with one X chromosome are male and worms with two X chromosomes are female in the somatic cells. The mutant phenotypes of three genes in 1X and 2X animals are summarized. In *tra-1* or *tra-2* mutants, both 1X animals and 2X animals develop as males. Thus these genes must be needed for normal female development in 2X worms, but not needed in 1X worms. In *her-1* mutants, both 1X animals and 2X animals develop as females. Thus, this gene must be needed for normal male development in 1X worms but not needed for female development in 2X worms. In a formal sense, when the *tra-1*+ and *tra-2*+ genes are active or ON, the *her-1*+ gene is inactive or OFF, and vice versa.

even the absence of *tra-1* function—but is needed for 2X animals to become hermaphrodites. We can say that, in an animal with one X chromosome, *tra-1*+ activity is OFF whereas in an animal with two X chromosomes, *tra-1*+ activity is ON.

Negative pathways involve mutations with opposite phenotypes

In contrast to the *tra-1* mutant phenotype, recessive mutations in the *her-1* gene result in both 1X and 2X animals developing as hermaphrodites. This implies that the wild-type function of *her-1*+ is needed *for male* development in 1X animals—but not hermaphrodite development in 2X animals. (The "*her*" gene name is based on the mutant phenotype in which hermaphrodite development occurs rather than on the wild-type function of the gene.) Because the hermaphrodite has somatic sexual characteristics similar to those of females in related species and we are ignoring the germline, we will refer to this phenotype as female. From our summary chart in Figure 11.8, in an animal with one X chromosome, *her-1*+ activity is ON, whereas in an animal with two X chromosomes *her-1*+ activity is OFF. Therefore the activity of the *tra-1*+ gene must be ON when the activity of *her-1*+ is OFF and, conversely, *tra-1*+ must be OFF when *her-1*+ is ON. In a logical sense, the role of one of the genes is to turn off the activity of the other. The *AAS* genes and the *TRA3* gene involved in general amino acid control in yeast were another example of this—that one gene must be inactive when the other is inactive. Note, again, that we are describing a functional role for the genes, and the molecular biology of this process is being set aside for now.

What is the order in which these two genes act? In broader terms, which gene turns off the activity of the other? This question can be answered by making a double mutant strain and examining the epistatic interaction between the two genes. We will scrutinize the logic of this procedure in detail. The step-by-step analysis is as follows.

Step 1. Determine whether the pathway involves positive or negative interactions. Genes can interact in either of two ways. First, when one gene is on, the other gene is off. Alternatively, when one gene is on (or off), the other genes is also on (or off). These interactions are interpreted differently, so it is imperative to infer the correct type of interaction pathway between the two genes. One useful way to determine whether the genes are in a positive or negative relationship is to compare their mutant phenotypes. In a positive relationship, when gene A is ON, gene B is also ON. We recognize this type of relationship because the mutants have similar phenotypes. In our example with general amino acid control described previously, the *aas* mutations are in a positive relationship with each other because the mutants have the same or similar phenotypes. By contrast, in a negative relationship, the genes have opposite and distinct mutant phenotypes. Thus, in this type of interaction, when gene A is ON, gene B is OFF and vice versa.

Let us return to the example in which we see that, in a normal male, *tra-1+* will be OFF whereas *her-1+* will be ON. In a normal hermaphrodite, *tra-1+* will be ON and *her-1+* will be OFF. Therefore *tra-1* and *her-1* must be in a negative relationship. This pattern of interaction is the easiest one to analyse by epistasis; it was the same result we observed with the *AAS* and *TRA3* genes.

Step 2. Determine the type of alleles. For most purposes, it is important that null mutations be used for each gene. Because a null allele eliminates the function of the gene, the double mutant represents the situation when each gene is OFF. Null alleles are the one type of mutation whose gene activity is unambiguous—there is none.

Step 3. Compare the phenotype of the double mutant to the phenotype of each of the single mutants. The third step is to make the double mutant and compare its phenotype to that of each single mutant. The simplicity of this experiment should not hide its subtlety. The interpretation of the phenotype of the double mutant is *the* crucial step in doing epistasis experiments, and the one that is the most frequently debated by collaborators and competitors. Because interpretation of the double mutant phenotype is the crucial and sometimes most subjective result, the genetic relationship inferred from epistasis needs to be tested by other means, such as additional genetic or molecular tests. It is important to examine as many different aspects of the phenotypes as possible to confirm this interpretation.

We said, when discussing *tra3* and *aas3*, that the double mutant had the phenotype of an *aas3* mutant. This result places the wild-type *AAS3* gene downstream of the wild-type *TRA3* gene; as we summarized it previously, *AAS3* is needed for *TRA3* activity. This might seem a bit counterintuitive; and, since interpretation of the double mutant phenotype is the crucial step, let us use a non-genetic analogy to explain the reasoning. On a car, the brake pedal and the tires are in a negative relationship: when the brake is active, the tires are not moving; and when the tires are moving, the brake is not active. Defective brakes result in a moving car, whereas flat tires result in a stationary car. Those are the "mutant phenotypes" of brakes and tires. In the formal sense we have been using, brakes and tires are in a negative relationship.

Now consider a "double mutant" car in which the brakes are defective and the tires are flat. The observed phenotype is that the double mutant car is stationary; in other words, it has the phenotype of the tire defect. To use the genetic term, mutant tires is epistatic to mutant brakes. To express this relationship another way, if the tires are flat, the activity of the brakes make no difference to the phenotype of the car. Because the mutant tire phenotype is epistatic to the mutant brake phenotype, we can infer that tires function downstream of the brakes, or that functional tires are needed for the activity of the brakes to be observed. Previously, the result that *aas3* mutants are epistatic to *tra3* mutants allowed us to place *AAS3* downstream of *TRA3* in that pathway.

Now let us use the same logic in our biological example. On the basis of the phenotypes of null mutations in the genes, *tra-1* and *her-1* are in a negative relationship. Thus the activity of one of the genes is to turn off the activity of the other. We summarize two possibilities in Figure 11.9. Either the activity of *tra-1+* turns OFF *her-1+*, or else the activity of *her-1+* turns OFF *tra-1+*. Suppose that the biological pathway is that *tra-1+* turns OFF *her-1+* (Possibility 1). In the double mutant, both genes are mutant and thus inactive. If the role of *tra-1+* is to turn OFF *her-1+*, the double mutant should look like the *her-1* single mutant—the mutation that knocks out the activity of *her-1* renders the activity of *tra-1+* irrelevant in the same way in which flat tires render the activity of the brakes irrelevant.

A similar logic can apply to Possibility 2 in Figure 11.9, where the proposed pathway is that *her-1+* turns OFF *tra-1+*. In this case, the double mutant with both genes inactive should look like a *tra-1* mutant worm. From this example, the one with general amino acid control, and the broken-down car, we can draw a general conclusion about genes in a negative pathway.

PRINCIPLE **In a negative pathway, the phenotype of the downstream gene is epistatic to the phenotype of the upstream gene.**

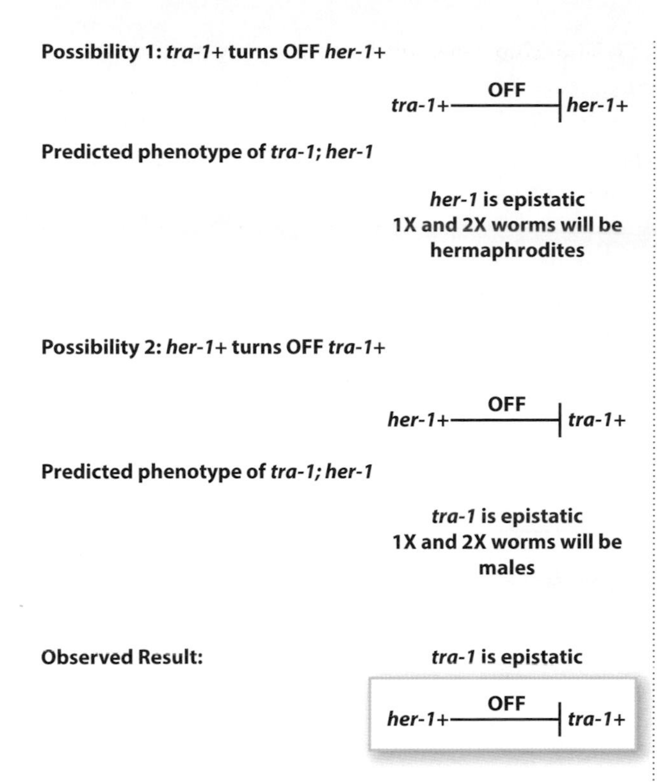

Figure 11.9 Epistasis between *tra-1* and *her-1*. The two genes are in a negative relationship, since one must be ON or active when the other one is OFF or inactive; this means that there are two possible pathways. In Possibility 1, *tra-1+* turns OFF *her-1+* in the wild type. In the double mutant, *her-1* is inactive because it is mutant so the activity of *tra-1* is irrelevant. Thus the double mutant will have a Her-1 mutant phenotype, and *her-1* is predicted to be epistatic. In Possibility 2, *her-1+* turns OFF *tra-1+*. The double mutant will have a Tra-1 mutant phenotype, and *tra-1* will be epistatic.

Having stated the analysis, what is the actual result with worms? The double mutant worm carries null alleles of both *tra-1* and *her-1*. In the double mutant, both 1X and 2X animals are males, exhibiting the *tra-1* mutant phenotype. Since the *tra-1; her-1* double mutant has the same phenotype as the *tra-1* single mutant, the *tra-1* mutant phenotype is epistatic. We infer that, in the normal pathway, *tra-1* is the downstream gene and that *her-1+* turns OFF *tra-1+* (Possibility 2 in Figure 11.9).

It is important to remind ourselves what this result does and does not mean. Most significantly, it does not mean that the *her-1+* gene and the *tra-1+* gene or their products *directly* affect one another. Epistasis describes a genetic interaction that results in a mutant phenotype. Many possible molecular relationships involving the two genes are consistent with the same result from epistasis. It is possible that the *her-1+* gene product is a direct repressor of *tra-1* transcription; it is also possible that the *her-1+* gene product directly interacts with the *tra-1+* protein, somehow blocking its function. However, it is also possible that the two genes are expressed at different times, in different tissues, or, in this case, in different sexes. As a result, there may be no direct molecular interaction between the genes or their products. In fact, any number of genes could act in the pathway between *her-1+* and *tra-1+*. Epistasis by itself cannot distinguish among these and many other possibilities.

On the other hand, since we are comparing phenotypes, epistasis does not require that we have a molecular or biochemical understanding of genes, their products, or the pathway they affect. If we want to control the movement of our car, we do not need to know how many mechanical steps occur between the brake pedal and the tires. In fact the epistatic pathways with sex determination in worms and general amino acid starvation in yeast (and many others) were published years before the molecular functions of any of the relevant genes were known.

Positive pathways involve mutations with similar phenotypes

We now have established a portion of the logical pathway for somatic sex determination, namely that *her-1+* turns OFF *tra-1+*. Epistatic interactions involving other genes have allowed investigators to infer the entire sex determination pathway. We'll limit ourselves here to considering only one such interaction, because it demonstrates an important extension of the use of epistasis.

The *tra-2* gene was identified with a mutant phenotype similar to *tra-1*: recessive null alleles of *tra-2* result in a male phenotype in both 1X and 2X animals, as summarized in the Figure 11.8. The null mutant phenotypes of *tra-1* and *tra-2* are not identical, a result whose significance will be apparent shortly. Thus *tra-2* must also be in negative relationship with *her-1*. Since our previous results showed that *her-1+* lies upstream of *tra-1+*, different pathways could explain the relationship among *tra-1*, *her-1* and *tra-2*, as shown in Figure 11.10. For example, *tra-2+* could turn OFF *her-1+*, which in turns OFF *tra-1+* (Possibility A). Alternatively, *her-1+* might turn OFF one of the *tra+*

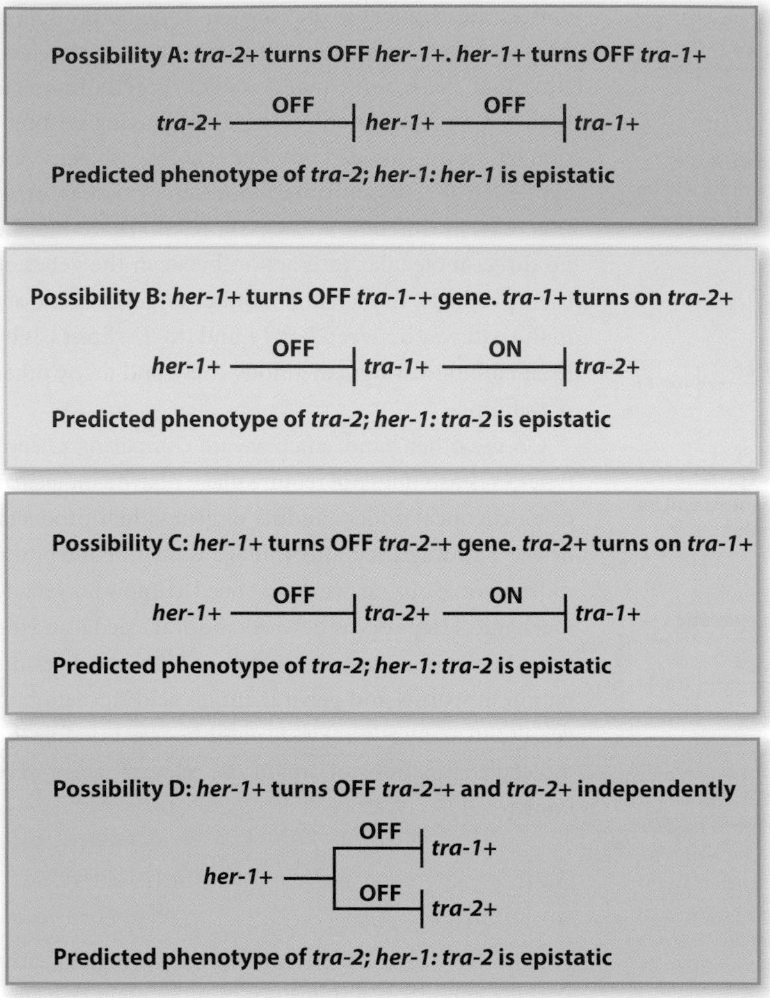

Possibility A: *tra-2+* turns OFF *her-1+*. *her-1+* turns OFF *tra-1+*

tra-2+ —OFF—┤ *her-1+* —OFF—┤ *tra-1+*

Predicted phenotype of *tra-2; her-1*: *her-1* is epistatic

Possibility B: *her-1+* turns OFF *tra-1-+* gene. *tra-1+* turns on *tra-2+*

her-1+ —OFF—┤ *tra-1+* —ON—┤ *tra-2+*

Predicted phenotype of *tra-2; her-1*: *tra-2* is epistatic

Possibility C: *her-1+* turns OFF *tra-2-+* gene. *tra-2+* turns on *tra-1+*

her-1+ —OFF—┤ *tra-2+* —ON—┤ *tra-1+*

Predicted phenotype of *tra-2; her-1*: *tra-2* is epistatic

Possibility D: *her-1+* turns OFF *tra-2-+* and *tra-2+* independently

her-1+ —OFF—┤ *tra-1+*

her-1+ —OFF—┤ *tra-2+*

Predicted phenotype of *tra-2; her-1*: *tra-2* is epistatic

Figure 11.10 The possible interactions among *her-1*, *tra-1*, and *tra-2*. The two *tra+* genes are in a positive or parallel relationship with each other and in a negative relationship with *her-1+*. From results in Figure 11.9, *her-1+* is upstream of *tra-1+*. This leaves four possibilities for *tra-2+*. In Possibility A, *tra-2+* turns OFF *her-1+*, and *her-1+* turns OFF *tra-1+*. In Possibilities B and C, *her-1+* turns OFF one of the *tra+* genes and the *tra+* genes turn ON each other. Either of the *tra+* genes could be upstream of the other. In Possibility D, *her-1+* turns OFF *tra-1+* and *tra-2+* in separate pathways, and the *tra* genes act independently. If *her-1* is epistatic to *tra-2* in a double mutant, Possibility A is the most likely pathway. The other three pathways predict that *tra-2* is epistatic to *her-1* in the double mutant. In fact *tra-2* is epistatic to *her-1*, so Possibility A is ruled out.

genes, which turns ON the other *tra+* gene. In this latter hypothesis, either of the two *tra* genes might be the more upstream gene, and either might be the most downstream gene; these are possibilities B and C in Figure 11.10. As yet another possibility, *her-1+* could turn OFF both *tra-1+* and *tra-2+*, but those genes are independent of each other (Possibility D). This last is the most subtle one to interpret, so we will reserve discussion for later.

Since *tra-2+* and *her-1+* are in a negative relationship, we can make a *tra-2; her-1* double mutant animal and interpret the results as we did previously for *her-1* and *tra-1*. If possibility A is correct and *her-1* is epistatic to *tra-2* (that is, if the double mutant has a *her-1* phenotype and both 1X and 2X worms develop as hermaphrodites), then our interpretation is relatively straightforward with the pathway shown; this is the outcome that occurred with the genes involved in general amino acid control. If, on the other hand, *tra-2* is epistatic to *her-1* (so the double mutant has a *tra-2* mutant phenotype), any of the other models could be correct. The double mutant has been made and the *tra-2; her-1* double mutant resembles a *tra-2* mutant, such that both the 1X and 2X animals show male somatic development. This epistasis places *tra-2+* downstream of *her-1+* in the pathway, thereby eliminating the first possibility.

How, then. can we distinguish among the other possibilities in Figure 11.10? We said earlier that epistasis works best when the mutations have opposite phenotypes, as in all of the preceding examples. The *tra-1* and *tra-2* mutant phenotypes are similar to each other in that 1X and 2X animals are males. As a result, the *tra-1*; *tra-2* double mutant will also be male, which is uninformative. The mutations do not have opposite phenotypes. That is, *tra-1+* and *tra-2+* are in a *positive* relationship so that, when one of genes is active, the other gene is also active, since the genes are responsible for female development in 2X animals. In previous examples we have not worked with a positive relationship and the *aas1* and *aas2* genes were not ordered with respect to each other using epistasis. How can such an experiment be done?

Two approaches were used to determine the order of *tra-1+* and *tra-2+* in the pathway, and can often be applied for determining epistasis between genes in a positive relationship. Both are subject to other interpretations, so it is helpful to describe both.

1. **Using subtle differences between the two phenotypes.** The phenotypes of the mutants in each gene may be similar, but not identical in all regards. If this is true, then the phenotype of the double mutant can be compared to each of the single mutants, particularly as it applies to these subtle aspects. It is especially important to work with null alleles in this case, since slight differences could also be due to hypomorphic alleles for one or both genes.
2. **Using a dominant mutant allele for one of the two genes.** It may be possible to use a hypermorphic mutation for one of the two genes. As noted in Chapter 4, dominant gain-of-function or hypermorphic mutations often have a mutant phenotype that is the opposite phenotype from the recessive null mutation for the gene. Such a difference could allow an unambiguous recognition of epistasis between the two genes. In this case, it is particularly important to understand the nature of the hypermorphic allele. If the dominant mutation introduces a novel phenotype or acts in some other way than a hypermorph, the results of the epistasis analysis could again be misleading.

A third approach, which was not available at the time for *tra-1+* and *tra-2+*, would have been to use molecular reagents, such as an antibody for the protein made by one or both genes. As discussed in Section 11.2, such experiments are common and would allow us to tell which gene's activity is dependent of the function of the other gene, or whether the genes' functions are independent. Suppose that we found that an antibody to TRA-2 is expressed in the same pattern in a *tra-1* mutant as in the wild type; this would tell us that the localization (and presumably the function) of TRA-2 does not depend on *tra-1+* activity, so Possibility B is ruled out but Possibilities C and D remain. We could then ask about the expression or localization of TRA-1 in a *tra-2* mutant compared to the wild type. If the expression of TRA-1 changes in a *tra-2* mutant, then we conclude that its activity depends on the activity of *tra-2+*, and Possibility C is the most likely pathway. If on the other hand the expression of TRA-1 does not change in a *tra-2* mutant, and the expression of TRA-2 does not change in a *tra-1* mutant, then the activities of the two genes are independent of each other and Possibility D is the most likely pathway. But notice that, to distinguish among these three possible pathways, we had to use a somewhat different phenotype of each gene—its expression pattern or protein localization rather than its morphological mutant phenotype.

Since molecular reagents for *tra-1* and *tra-2* were not available, but mutant alleles for each gene were available, let us consider in more detail the approaches that were used.

As noted before, *tra-1* and *tra-2* mutant do have some differences in their mutant phenotypes, so we could use those as our assay. Most noticeably, *tra-2* mutant males do not exhibit male mating behavior, and the male tail is not as fully formed as in *tra-1* mutant males or normal males. Thus we could make a *tra-2; tra-1* double mutant and look at those subtle differences. If it resembles *tra-1* mutants, then we hypothesize (as above) that the activity of *tra-2+* must depend on *tra-1+*, which favors Possibility B. If, on other hand, the double mutant has the phenotype of *tra-2* for those subtle differences, we hypothesize that its activity does not depend on the function of *tra-1+*, which rules out Possibility B and favors Possibilities C or D.

But this can be tricky. Here is a situation when it is important to know that one is working with null alleles. Some hypomorphic *tra-1* mutant alleles resemble a null *tra-2* allele and do not have the normal-looking male tail; an analysis with such a *tra-1* hypomorphic mutant would be difficult to interpret, since the double mutant will probably have the partial transformation of the tail. We have also made the assumption that all aspects of somatic sex determination are regulated by these genes, but this has not been verified (at the time the experiments were done). For example, the reason why *tra-1* mutant males exhibit normal male mating behavior but *tra-2* mutant males do not was attributed to a "subtle difference" in the phenotypes; it may be, however, that mating behavior is regulated by as a separate pathway from the rest of somatic sex determination (e.g. tail formation). Therefore it is helpful to have multiple phenotypic differences between *tra-1* and *tra-2* mutants to examine in order to confirm that the interaction between the two genes can be generalized to different tissues and aspects of the phenotypes.

When the double mutant between *tra-1* and *tra-2* was analysed, it had the phenotype of a *tra-2* mutation; the mutant males have the less complete transformation of the tail that is characteristic of Tra-2 mutants and do not exhibit male-like mating behavior. Thus, in this interaction, *tra-2+* is epistatic to *tra-1+*. But what does this tell us about which gene is upstream or downstream in the pathway? We will state the conclusion before developing the explanation. In positive pathways, the epistatic gene lies upstream.

PRINCIPLE **In a positive pathway, the phenotype of the upstream gene is epistatic to the phenotype of the downstream gene.**

Illustrating positive pathways using non-biological examples

Positive pathways are familiar to us, so we can use another non-biological analogy to explain how to interpret them. My desk lamp and the circuit breakers in the building are in a positive pathway, with the circuit breakers upstream of the lamp. If a circuit breaker is off, all electrical appliances on that circuit go out, including my desk lamp. But, if only my desk lamp goes out and other appliances on that circuit in the building are working, the problem lies in my lamp and not in the building circuit breaker. Knowing the pathway, I can interpret the "phenotypes."

Suppose that my desk lamp suddenly goes out. I can deduce the source of the problem (i.e. the source of the "mutation" in the pathway) by looking at "subtle differences" in the phenotype of the two defects. In this case, the subtle difference is other electrical appliances. If other electrical appliances in the building are still functional, then the darkness has the characteristics of a knockout of the desk lamp, the equivalent of the downstream gene. On the other hand, if other electrical appliances are also not working, then the darkness has the characteristics of a knockout of the circuit breaker, the equivalent of the upstream gene. (At the risk of overapplying a simple analogy, this also tell us which appliances are on the same circuit and which ones are on independent circuits, just as mutant phenotypes can tell us which processes are affected by the same gene.)

Suppose that I turn off my desk lamp—so it is analogous to a null mutant—and I can move the circuit breaker to the off position as well. This is the equivalent of making a double mutant in the pathway. What is the expected phenotype of this double knockout? In addition to the desk lamp, other electrical appliances will also be off. By using subtle differences in the phenotype of my darkened room, I can infer the pathway. The double mutant has the mutant phenotype of the *upstream* switch, the circuit breaker.

Returning to somatic sex determination

The epistasis experiments with *tra-1* and *tra-2* allowed the tentative construction of a genetic pathway involving these three genes, possibility C in Figure 11.10. The *her-1*; *tra-1* and *her-1*; *tra-2* double mutants showed the mutant phenotype of the Tra mutants and developed as males. This information placed the *tra-1+* and *tra-2+* genes downstream of *her-1+* in this negative step of the pathway. We previously noted that the *tra-1*; *tra-2* double mutant has the phenotype of *tra-2* mutants, placing *tra-1+* downstream of *tra-2+* in a positive step in the pathway. So the pathway can be summarized by saying that the wild-type function of *her-1+* is to turn OFF *tra-2+*. The wild-type function of *tra-2+* is then to turn ON *tra-1+*. This does not happen when *her-1+* has turned OFF *tra-2+*.

Using a dominant allele to confirm the results

Because subtle differences in the *tra-1* and *tra-2* mutant phenotypes could also arise from pleiotropic differences between the two genes, it is useful to have another method to infer the same information. As we noted above, Possibility D is not completely ruled out unless we have some additional evidence that *tra-1* and *tra-2* lie in the same pathway for the relevant phenotypes. We said previously that a second way to analyse the epistatic relationship between two genes in a positive pathway is to use a dominant gain-of-function or hypermorphic mutation for one of them. This principle was also introduced in Section 10.2 in discussing suppression by epistasis. Such a hypermorphic mutation has been used to show that *tra-1* and *tra-2* do work in the same pathway and to confirm the order of their activities.

A dominant allele of *tra-2* has the opposite phenotype from recessive loss of function alleles. *tra-2* dominant mutations, abbreviated *tra-2 (dom)*, do not affect somatic sexual development in 2X animals, but cause 1X animals to develop as females. It is possible to make a double mutant between *tra-2 (dom)* and *tra-1* and examine the phenotype. Since the phenotypes of *tra-2 (dom)* and *tra-1* are opposite, we can apply the principle for interpreting epistasis that we used for negative relationships. In this particular case, the phenotype of a *tra-2(dom); tra-1* double mutant resembles a *tra-1* mutant. Thus, the activity of *tra-2+* must depend on the function of *tra-1+*, which also favors Possibility C, and we can place *tra-1+* downstream of *tra-2+* in the same pathway.

Again, this epistatic relationship can be illustrated by our analogy with the building electrical circuit breakers and my desk lamp. Suppose that I force the circuit breaker into the ON position and use electrical tape to hold it in place. I have produced, in effect, a dominant gain of function mutation in that part of the pathway (and an electrical hazard that we will ignore). Now if the desk lamp goes out (the "loss of function" phenotype), the problem can only lie at that step in the pathway because we know that the upstream step is constitutively active.

This then explains wild-type sex determination, at least as it relates to these three genes, as diagrammed in Figure 11.11. Look at this figure and notice what we have been able to infer about the complex process of sex determination using only the careful analysis of double mutant phenotypes. In a 1X animal, *her-1+* is normally ON or active in some way. Because it is ON, it turns OFF *tra-2+*, which is needed for hermaphrodite development. Since *tra-2+* is OFF, *tra-1+* is also OFF, and male development ensues. In a 2X animal, *her-1+* is normally OFF or inactive. Because it is OFF, *tra-2+* is ON, which turns on *tra-1+*, and normal hermaphrodite or female development ensues.

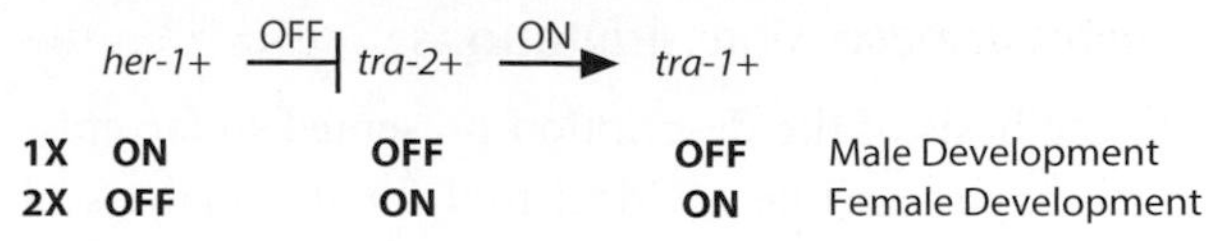

Figure 11.11 The relationship among *her-1+*, *tra-2+*, and *tra-1+*, as determined by epistasis. Double mutants placed *her-1+* upstream of both *tra-2+* and *tra-1+*. A double mutant between *tra-2* and *tra-1* showed the mutant phenotype of Tra-2 in subtle characteristics. Thus, if *tra-2* is inactivated, the activity of *tra-1* is irrelevant. This places *tra-2+* upstream of *tra-1+*. In 1X animals (males), *her-1+* is ON, which then turns OFF *tra-2+*; since *tra-2+* is not active, *tra-1+* is not active and male development ensues. In 2X animals (females), *her-1+* is OFF so *tra-2+* is ON. This activates *tra-1+*, and female development ensues.

The pathway inferred from epistasis informs further experiments to understand sex determination

This pathway has implications for additional experiments and their interpretation. For example, the *her-1* gene product and the *tra-1* gene product are unlikely to interact with each other directly, because they do not encode consecutive steps in the pathway. Further analysis has supported this inference. Each of these three genes was cloned, and molecular analysis of these genes suggested that *her-1* probably encodes a secreted protein that could function as a ligand, *tra-2* encodes a membrane-spanning protein, and *tra-1* encodes a transcription factor related to the GLI family. Epistasis suggests that the *her-1* and *tra-2* gene products could interact with each other, being consecutive steps in the pathway, so a direct physical interaction between a possible HER-1 ligand and a possible TRA-2 receptor is a reasonable hypothesis. In fact a direct physical interaction between HER-1 and TRA-2 has been demonstrated by a yeast two-hybrid assay and by a co-immunoprecipitation assay, methods described in Chapter 10.

Epistatic analysis has limitations

On the basis of the description presented so far, epistasis may seem like an ideal tool for the analysis of complex biological problems. First, the process is fairly easy to perform, since it only involves making a double mutant. Second, the interpretation of the double mutant phenotype also seems straightforward. There are two simple principles to make the analysis of the double mutant easy. In addition, the procedure does not require detailed knowledge of the physiology or biochemistry of the process or of the molecular biology of the genes or their products. Furthermore, epistasis has a history of success in solving complicated biological pathways. Most eukaryotic signaling pathways were deduced at least in part using epistasis in yeast, worms, or flies.

But epistasis simply presents a logical model of gene functions without providing information on the molecular biology of the process. In some situations, a faulty interpretation of epistasis can be misleading. It is worth considering how an investigator might be misled.

First and most commonly, the phenotype of one gene is not completely epistatic to the other gene. As a result, the double mutant shows some combination of both single mutant phenotypes. This presents complications for the investigator. Is it more informative to consider those aspects of the phenotype where one gene is clearly epistatic, or is it more important to look at the aspects of the phenotype that do not show obvious epistasis? The most conservative interpretation of this result is that the two genes under analysis act in parallel pathways that, together, contribute to the phenotype. An example of such an analysis involving dauer larvae formation in *C. elegans* is considered in the case study.

Second, the alleles may not be null alleles. This may occur because the investigator does not have true null alleles to work with, or because the investigator has chosen to work with non-null alleles (for reasons of viability and fertility, for instance.) In this case, the mutant phenotypes may be difficult to interpret. They might indicate genuine epistasis between two genes with slightly different roles (such as *tra-1* and *tra-2*), or they might be indicative of phenotypes arising from reductions in gene activity caused by non-null alleles.

Certainly working with dominant hypermorphic alleles to analyse a positive pathway needs to be approached with caution. Not all dominant alleles are hypermorphic with a simple overproduction phenotype. The overproduction may also result in expression at specific times or in tissues that do not usually express the gene. Overexpression can result in unpredictable relationships among gene products whose stoichiometry is sensitive to dosage. And, like hypomorphs, hypermorphs fall onto a continuum of gene activity, which is often difficult to predict.

In addition, it is important to analyse the phenotypes carefully and using multiple assays, including morphological, cellular, and molecular examination, if possible. Misleading or ambiguous interpretations may be detected when investigators attempt to use the proposed pathway to interpret the relationships of additional genes, only to find unpredicted results. Some of the results in the case study will illustrate this. The presence of internal inconsistencies is often a signal that some aspect of the epistatic pathway has been interpreted incorrectly. Epistasis is often a good "first approach" that will suggest numerous additional experiments. It is important to be thorough in these follow-up experiments and to make sure that the results are internally consistent before placing too much confidence in the inferred pathway. The examples just discussed made all the possible double mutants involving these genes, and analysed other alleles that we have not discussed to confirm the pathway. Helpful as they are, epistatic pathways should rarely be understood as the complete solution to the relationship between genes in a set. But, to summarize one more significant cellular pathway sorted out by epistasis, we turn again to a familiar set of genes in *Drosophila*.

11.4 The *ptc* pathway

The analysis of the *patched* (*ptc*) gene in *Drosophila* has been a recurring theme throughout the book, so we complete the picture now. In Section 11.2 we described the experiments that placed *ptc+* upstream of the gene *dpp+*, but we did not discuss other genes in this pathway. Several other genes were known or suspected from the original analysis to be involved in the same biological process as *ptc* and *dpp*. Like *ptc*, the genes *hedgehog* (*hh*) and *smoothened* (*smo*) are segment polarity genes, and all have proved to be evolutionarily conserved from flies to humans, with one ortholog of *smoothened* and several orthologs of *hedgehog* and *patched* in vertebrates. Epistasis was also used to infer the signaling pathway that includes these genes, a pathway that has been shown to have implications for developmental patterning in nearly all animals and for cancer in humans.

The effect of the mutations on downstream target genes

Each of the mutants has a distinctive morphological phenotype in *Drosophila* embryos and in adults that could have been used to perform an analysis of epistasis. The most informative phenotype is the effect of these genes on the expression pattern of *wingless* and *dpp* in the wing disk. As before, we will focus on the effect on *dpp*, but the effect on *wingless* is similar and was used as an additional phenotype to check the interpretation; some experiments were done initially with *wg* rather than with *dpp*. Recall from Figure 11.2 that, in a *ptc* mutant animal, the expression of the normal Dpp protein extends to a more anterior part of the wing than it does in flies that are wild-type for *ptc+*. Thus the wild-type function of *ptc+* includes restricting *dpp+* expression in the anterior part of the wing, and we can say that, in the wing disk of a normal fly, *ptc+* turns OFF *dpp+* (Figures 11.2 and 11.12). As in all such epistasis experiments, the molecular mechanism by which this regulation occurs is not known and not needed for an interpretation of the interaction pathway.

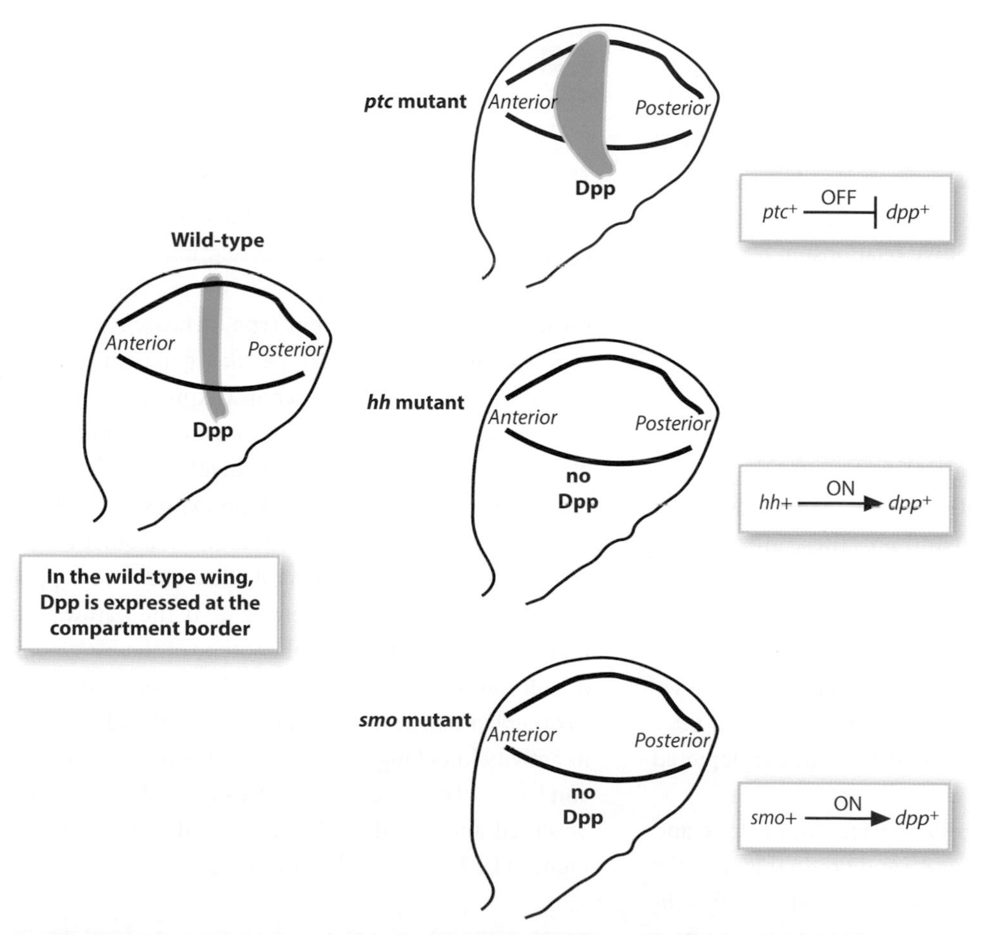

Figure 11.12 The normal role of *ptc+*, *hh+* and *smo+* on the expression of *dpp+*. This has been inferred from changes in expression pattern using antibodies and reporter genes for Dpp. The *ptc+* gene turns off *dpp+* (or prevents it from being expressed in the anterior part of the wing), while *hh+* and *smo+* are needed for this expression of *dpp+*.

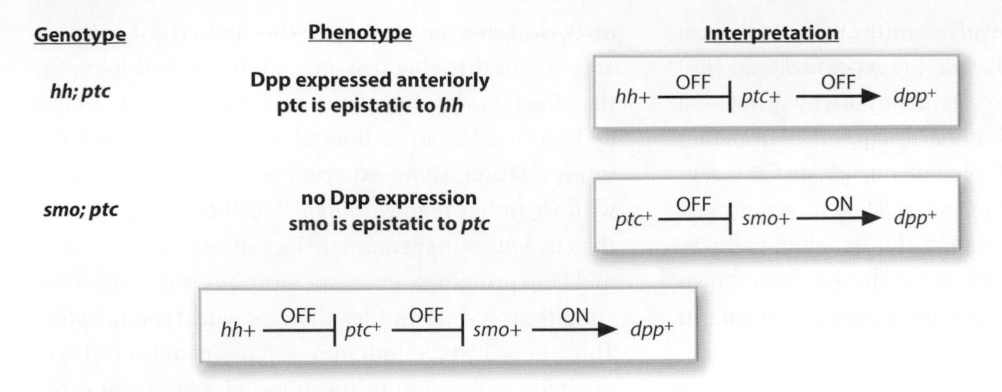

Figure 11.13 The mutant phenotype of the single and double mutants. The results of epistasis involving each pair of the three genes allowed the construction of the pathway among them, as shown.

Similar experiments were also done to determine the effect of *hedgehog* and *smoothened* mutations on Dpp protein expression and localization. In contrast to the results with *ptc* mutants, Dpp is not expressed in a *hh* mutant or in a *smo* mutant. Thus the wild-type functions of *hh+* and *smo+* must be required for the activity (or localization) of *dpp+*, as summarized in Figure 11.12. The expression of Dpp becomes a convenient phenotype for analysing the epistatic interaction among *ptc*, *hh*, and *smo* mutants. From this we can say that both *ptc+* and *hh+* and *ptc+* and *smo+* are negative relationships, while *hh+* and *smo+* are in a positive relationship.

Literature Link. Capdevila et al. 1994, *EMBO Journal* 13: 71–82; Alcedo et al. 1996, *Cell* 86: 221–32; Martinez Arias, Baker, and Ingham 1988, *Development* 103: 157–70

Epistasis involving negative pathways

Double mutant combinations were made between *hh* and *ptc* and between *smo* and *ptc*. These are summarized in Figure 11.13. In the *hh; ptc* double mutant, Dpp expression has the phenotype of a *ptc* single mutant, so *ptc* is epistatic to *hh*. On the basis of the rule for a negative relationship we derived earlier, we place the epistatic gene downstream, and we can infer that *hh+* turns OFF *ptc+* in wild-type flies, as depicted in Figure 11.13.

The experiment with *smo* and *ptc* can be done and interpreted just as it was for *hh* and *ptc*. In this case, the double mutant *smo*; *ptc* has the phenotype of *smoothened*, and Dpp is not expressed in the wing. Since *smo* is epistatic to *ptc*, we infer that *ptc+* turns OFF *smo+* using the same principle for negative pathways as before, as shown in Figure 11.13. Combining these results, the epistasis model for this pathway suggests that *hh+* turns OFF *ptc+* and *ptc+* turns OFF *smo+*. *smo+* then turns on Dpp (Figure 11.13).

Literature Link. van den Heuvel and Ingham, 1996, *Trends in Cell Biology* 6: 451–3; Ingham et al. 2000, *Currents in Biology* 19: 1315–18

The use of a *hh* hypermorphic mutation

An additional experiment helped to confirm this model by directly investigating the relationship between *smoothened* and *hedgehog*. Since these genes are in a positive pathway, a different type of *hedgehog* phenotype was used. The wild-type *hh+* gene was cloned under a constitutive promoter and expressed in flies by procedures similar to those described in Chapter 3. The overexpression resulted in the equivalent of a *hedgehog* hypermorphic mutation, and Dpp expression spreads anteriorly in this *hh constitutive* or hh^{const}. In other words, the hh^{const} resembles the *ptc* null allele and is the opposite of the *smo* mutant phenotype. If the pathway in Figure 11.13 is correct, the phenotype of the hh^{const}; *smo* double mutant should be like *smo*, the downstream gene. In other words, since *hh+* requires a functional *smo+* for its activity, knocking out *smo* makes the hh^{const} irrelevant and Dpp will not be expressed. This was in fact the result observed, and the inferred pathway of the three genes in Figure 11.13 has more data to support it.

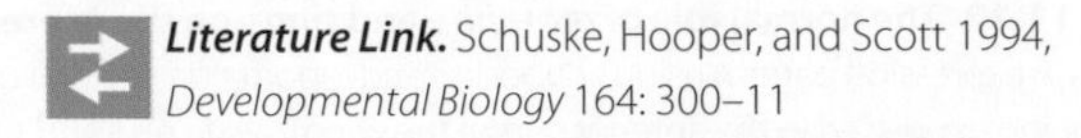

Literature Link. Schuske, Hooper, and Scott 1994, *Developmental Biology* 164: 300–11

The importance of the epistasis results became obvious when each of these three genes was cloned. *hh*+ encodes a secreted protein, which could function as a ligand; this role is consistent with its upstream position in the pathway. However, both *smo*+ and *ptc*+ encode trans-membrane proteins. Simply on the basis of the molecular analysis and in the absence of epistatic analysis, it would be difficult to predict which of these two genes is the receptor for Hh. However, because *ptc* acts in the pathway between *hh* and *smo*, the most reasonable genetic model is that *ptc*+ encodes the receptor for the Hh ligand.

Literature Link. Hooper and Scott 1989, *Cell* 59: 751–65; Nakano et al. 1989, *Nature* 341: 508–13; Chen and Struhl 1996, *Cell* 87: 553–63

Biochemical confirmation of the epistasis pathway

Biochemical analysis using co-IP with epitope tagged versions of the proteins (as described in Section 10.5) has confirmed this relationship among *hh*, *ptc*, and *smo*. Vertebrate orthologs of each of these three genes were made as fusion proteins with epitope tags and antibodies to the epitope tags were used to immunoprecipitate the proteins complexes from cells. The antibody to Hh co-precipitated Ptc but not Smo, which indicates that a direct physical interaction occurs between Ptc and Hh, but not between Hh and Smo. Similarly, the Smo antibody co-precipitates Ptc but not Hh, and the Ptc antibody co-precipitates both Hh and Smo. Thus Ptc, the middle gene of the three-gene pathway, physically interacts with both Hh and Smo; but Hh and Smo do not physically interact with each other. This suggests that Patched is in a complex with Smoothened in the membrane. The binding of the Hh ligand to the Ptc protein may then release Smoothened from the Ptc-Smo complex, and it is able to turn on the expression of the downstream genes Dpp. The molecular biology underlying the epistatic pathway is diagrammed in Figure 11.14.

Literature Link. Beachy et al. 1010, *Genes & Development* 24: 2001–12

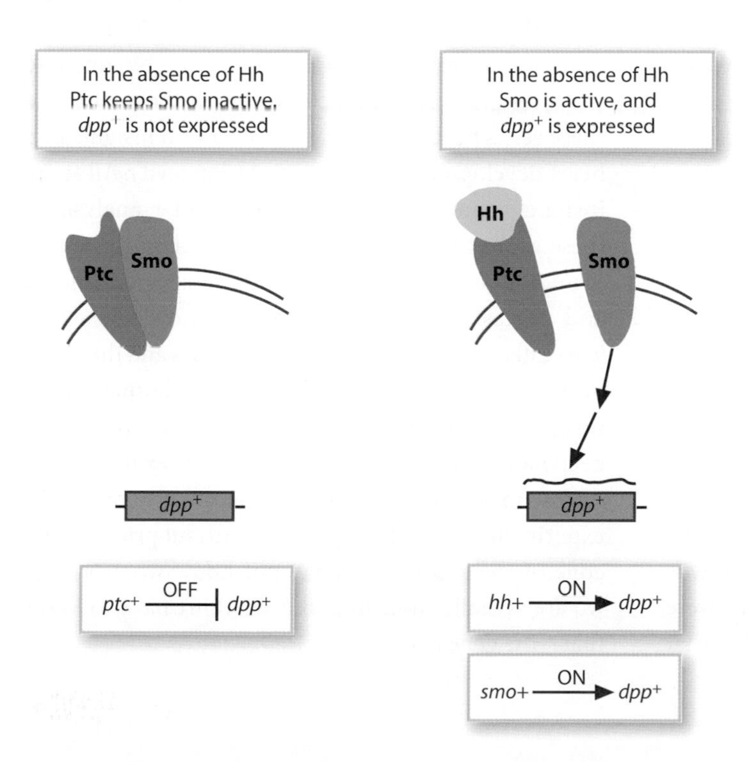

Figure 11.14 The Hh-signaling pathway in *Drosophila*. In the absence of Hh, Ptc and Smo are in the complex and the downstream genes (such as *dpp*) are not active. When Hh binds to Ptc, Smo is freed from this complex and activates the downstream genes.

The concordance between the epistasis pathway derived from genetic interactions in *Drosophila* and the physical interactions demonstrated with vertebrate proteins reinforces the power of each approach. The *hedgehog*, *patched*, and *smoothened* pathway in *Drosophila* is an evolutionarily conserved and crucial signaling pathway in the development of most if not all animals. The literature on these genes is vast. The ligand is encoded by a *hedgehog* single gene in *Drosophila*, but by a family of related genes in vertebrates. Of these, the best studied is the gene *Sonic hedgehog* (*SHH*), which has a very broad role in the development of the gut, the limbs and digits, the neural tube and the brain, the eye, the heart, the notochord, even left–right asymmetry. In light of this expression pattern, it is not too surprising that people with inherited mutations in *Sonic hedgehog* have severe developmental problems, for example holoprosencephaly, cleft lip and palate, and many more. Most such unfortunate individuals die in utero. Somatic mutations in *Shh*, *patched* (*PTCH*), and *smoothened* (*SMO*) have also been associated with basal cell carcinomas, medulloblastoma, rhabdomyosarcoma, and other human tumors. *SHH* and *SMO* are proto-oncogenes, while the *PTCH* gene family acts as tumor suppressors. Thus the epistatic pathway inferred from *Drosophila* wing development and embryo segmentation has given us fundamental insights into human development and into some of the most common forms of human cancer.

11.5 The pathways unveiled

The descriptions we have given of the genetic analysis of general amino acid control, worm sex determination, and the wing pattern development in flies have withheld molecular information until the pathways have been inferred from the mutants. The approach has been intentional, so that the strengths and weaknesses of purely genetic screens can be seen more easily. None of the experiments was done exactly as described here. Some of the details of the mutant phenotypes have been simplified, complications such as temperature sensitivity and viability have occasionally been ignored, and some genes with significant molecular roles have been omitted. In addition, the genetic analysis of these genes often occurred at the same time as their molecular cloning, so the genetic and molecular analyses were usually part of the same description, and the results from molecular biology experiments were important in interpreting some of the genetic results. In none of these cases were the mutants the only source of information about how the pathway affects a biological process, although the mutants were often the earliest evidence. It is also worth noting that these same principles were very widely used in bacterial and phage genetics many years before they were applied in eukaryotes.

But the underlying conclusion is that epistatic analysis works, and often leads to information with implications well beyond the original question. No one had the foresight or the audacity to begin the analysis of these genes because they expected to gain insights into heart development, skin cancer, or longevity. All those insights came later, when the molecular analysis of these genes found their orthologs in other organisms, including humans. The use of suppressors, enhancers, and epistasis gives us the expected knowledge about gene interaction in yeast, worms, and flies and the unexpected knowledge into key processes in humans. Such insights about these complex processes would have been hard to obtain without the ability to examine gene interactions in model organisms. Because we can do the experiments in model organisms without prior knowledge of the process, we may not know what process we are actually modeling. But we don't have to know that—the genes can tell us.

Chapter Capsule

Every biological process is the outcome of genes or gene products working together in pathways and networks. These pathways may be linear or branched, and may involve many genes or few genes. Gene interactions based on the phenotype of double mutants can be used to infer a logical order of gene function in the pathway.

- In a *negative* pathway in which the wild-type function of one gene is to turn OFF the other gene, mutations in the two genes have opposite phenotypes. The double mutant has the mutant phenotype of the *downstream* gene in the pathway.
- In a *positive* pathway in which the wild-type function of one gene is to turn ON the other gene, mutations in the two genes have similar phenotypes. The double mutant has the mutant phenotype of the *upstream* gene in the pathway.
- When there are *parallel* or functionally redundant pathways for the same biological process, a double mutant with mutations from two different pathways may have a mixed mutant phenotype that is a *composite* of the two mutants. Alternatively, the double mutant may show *synthetic enhancement* of one or both of the single mutations, producing a more severe mutant phenotype than either single mutant alone.

Additional Reading

Avery, L., and S. Wasserman. 1992. Ordering gene function: The interpretation of epistasis in regulatory hierarchies. *Trends in Genetics* 8: 312–16.

Botstein, D., and R. Mauer. 1982. Genetic approaches to the analysis of microbial development. *Annual Review of Genetics* 16: 61–83.

Huang, L. S., and P. W. Sternberg. 2006. Genetic dissection of developmental pathways. *WormBook*, June 14, 2006, http://www.wormbook.org.

Case Study 11.1

Dauer larva formation in *C. elegans*

Our case study for this chapter draws together the use of suppression, epistasis, and synthetic enhancement, the gene interactions described in this chapter and the previous one. The analysis of this pathway – or, more appropriately, these pathways – led to unexpected insights into a number of fundamental processes, including mammalian longevity.

Dauer larvae in worm biology

The process of dauer larvae formation was described in the case study for Chapter 6, in the context of analysing temperature-sensitive periods, and some of the mutants were introduced there. As a brief review, under stress conditions such as starvation or crowding, *C. elegans* will form a distinctive developmental stage called the dauer larva, as an alternative to the third larval L3 stage. Molecular and biochemical analysis has shown that the signal that triggers dauer larva formation is a pheromone secreted by all worms; high concentration of the pheromone indicates crowding, which triggers dauer formation. A worm can stay as a dauer larva for months, until conditions are favorable, at which time it molts to the L4 stage. Many wild isolates of *C. elegans* have been as dauer larvae.

Dauer formation has been studied as a behavioral response pathway, and recessive mutations in more than thirty genes have been identified that have some defect in dauer formation. All the mutations we will consider are recessive null alleles for these genes, although hypomorphic alleles also exist. The genes, called *daf* (dauer formation-defective), fall into two broad classes: Daf-d (defective) and Daf-c (constitutive). Daf-d (defective) mutants cannot form dauer larvae even under stress conditions. An example of such a gene is *daf-3*.

In contrast to the dauer-defective genes, Daf-c mutants *always* form dauers, regardless of conditions. The worms arrest at this stage and never mature to the next larval stage or adulthood. Examples of Daf-c mutants include the genes *daf-7* and *daf-11*. Both genes are defined by mutations that are temperature-sensitive, so that, at the restrictive temperature, these mutants form dauers regardless of the conditions.

Thinking through the Experiment

1. What are the wild-type roles of *daf-3+*, *daf-7+*, and *daf-11+*?
2. Based on the mutant phenotypes, which of these genes are in a negative relationship with each other and which ones are in a positive relationship with each other? Summarize them in a chart or a drawing.
3. The Daf-c mutants are all temperature-sensitive, but the genes are defined by null alleles. Explain how the null mutant alleles for *daf-7* and *daf-11* could be temperature-sensitive. What does this imply about the process of dauer larvae formation?

Suppression analysis was used to find additional genes affecting dauer formation

Recall from Chapter 10 that one way to find additional genes that affect the same biological process is to look for suppressor mutations. A suppressor is a mutation that suppresses or overcomes the effect of another mutation in another gene. The phenotype of the Daf-c mutants such as *daf-7* and *daf-11* easily lends itself to the identification of suppressors to find additional genes affecting dauer formation. A worm that arrests as a dauer larva (a Daf-c phenotype) does not mature and lay eggs at the restrictive temperature. One can mutagenize the Daf-c worms and look for suppressors that allow the Daf-c worms to mature and give offspring, as shown in Figure CS11.1.

Thinking through the Experiment

4. In Figure CS11.1, the inferred pathway is that the *daf-C* gene is turning off or inactivating the *daf-D* gene. How does this inferred pathway allow the investigator to find suppressors of the *daf-C* gene? It may be helpful to review Section 10.2 on gene-specific suppressors.
5. Suppose that the suppressors were isolated by mutagenizing an apparent null mutant in *daf-7*. What would be the next experiments to characterize this suppressor?
6. Some of the suppressors isolated by mutagenizing a *daf-C* gene such as *daf-7* and *daf-11* suppressed both *daf-7* and *daf-11*, while other affected only one of the two genes. How would you interpret this result and place these genes in a pathway?

We should insert a clarifying note here about suppression and epistasis in these strains. Mutations in the *daf-d* genes are found as suppressors of the *daf-c* mutations. Strictly speaking, these *daf-d* mutations are epistatic to the *daf-c* mutation. That is, the double mutant *daf-c; daf-d* have a Daf-d mutant phenotype and they cannot form dauers at all. They do not form dauers, which is the suppression of *daf-c*; but, under normal conditions in the lab, even

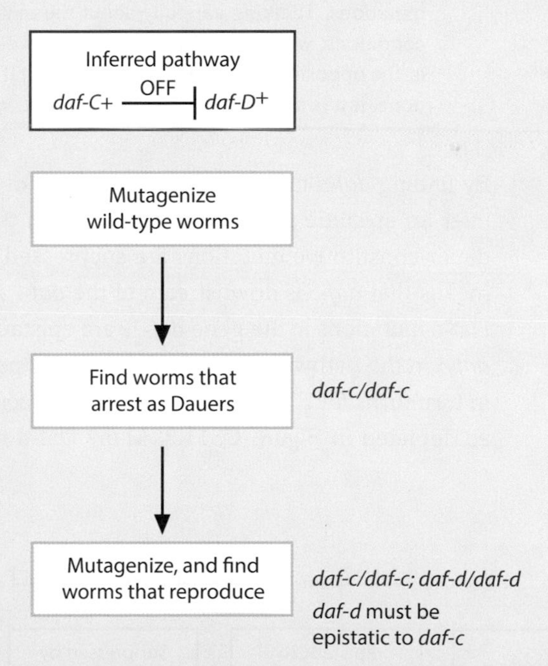

Figure CS11.1 Finding *daf-d* genes as suppressors of a *daf-c* gene. In a negative pathway such as the one shown at the top, the double mutant has the phenotype of the downstream gene. In this case, a mutation in a downstream *daf-d* gene will act as a suppressor of a *daf-c* gene, and the double mutant will not form dauers. A *daf-c* mutant is found by its constitutive dauer formation. This *daf-c* mutation is mutagenized and allowed to reproduce at the permissive temperature for two generations before the population is shifted to the restrictive temperature where the Daf-c phenotype is observed. If a mutation has been induced in a downstream *daf-d* gene, the double mutant will grow and reproduce, whereas the *daf-c* single mutant arrests development. Such a screen was used to find *daf-d* genes downstream of different *daf-c* mutations.

wild-type worms don't form dauers so a Daf-d mutant phenotype would not be observed. Thus the Daf-d mutants look like the wild type, and it is the suppression phenotype that is monitored and analysed.

Note, from our inferred pathway in Figure CS11.1, that the *daf-d* gene is placed downstream of the *daf-c* gene. One principle in the chapter is that mutations in genes acting downstream in a negative pathway (such as this one) are epistatic to mutations in upstream genes. Thus we would predict that different *daf-d* mutations will be found if we began the experiment with different *daf-c* genes. Consider the hypothetical pathway in Figure CS11.2 and the predicted outcomes. If we began our screen with a *daf-c1* mutant, we could recover both *daf-d1* and *daf-d2* mutations, since both genes are downstream. If we had begun our screen with a *daf-c3* mutant, we would not have recovered either of these *daf-d* genes as epistatic suppressors. When we began the screen, we did not know the order of the genes, so any *daf-d* gene could be used as a starting point. However, by knowing this principle of epistatic suppression and observing which genes are isolated as suppressors of which other genes (and which are not), we can put together a pathway by which these genes interact.

Thinking through the Experiment

7. A new *daf-d* mutation is found as a suppressor of *daf-c3* from Figure CS11.2. What is its predicted effect on *daf-c1* and *daf-c2* and its likely position in the pathway?
8. A new *daf-c* mutation is tested with these *daf-d* mutations. *daf-d2* is epistatic to it, but *daf-d1* is not. What is the most likely location of this new *daf-c* gene in the pathway?
9. This pathway was assembled by identifying *daf-d* genes that are epistatic suppressors of *daf-c* mutations. Thinking carefully about the mutant phenotypes, and possibly a bit about growth conditions, would it be feasible to find *daf-c* mutations as suppressors of *daf-d* mutants? (That is, the opposite of what was actually done.) If it is feasible, describe how such a screen could be done. If it is not feasible, describe why not.

By finding *daf-d* mutations as suppressors of different *daf-c* mutations, it was possible to infer an epistatic pathway among some of the genes. As shown in Figure CS11.3, *daf-7* dauer-constitutive mutations are suppressed by mutations in the Daf-d gene *daf-5*. This means that *daf-5* is downstream of the *daf-7* in the epistatic pathway. On the other hand, Daf-c mutations in the gene *daf-12* are epistatic to *daf-5*, so *daf-12* must by downstream of *daf-5* in the pathway. This gives the simple pathway that *daf-7+* turns OFF *daf-5+*, which in turn turns OFF *daf-12+*; when *daf-7+* is active, *daf-5+* is turned OFF, so *daf-12+* is ON, as depicted in Figure CS11.3. Many Daf-d mutations were found by these suppressor

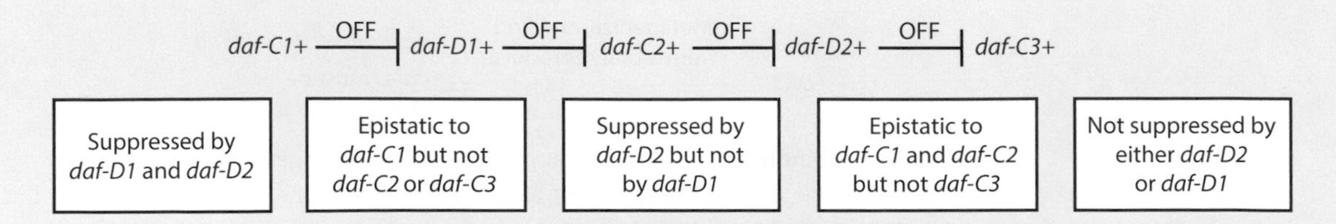

Figure CS11.2 Ordering the dauer-constitutive and dauer-defective genes into a pathway. In this and subsequent figures, the dauer-defective genes are shown in blue and the dauer-constitutive genes are shown in red. No gene has both types of alleles. The *daf-c* genes will be suppressed by mutations in the downstream *daf-d* genes; or, stated another way, the *daf-d* mutations will be epistatic to any upstream *daf-c* mutation. This allows the genes to be ordered into a pathway.

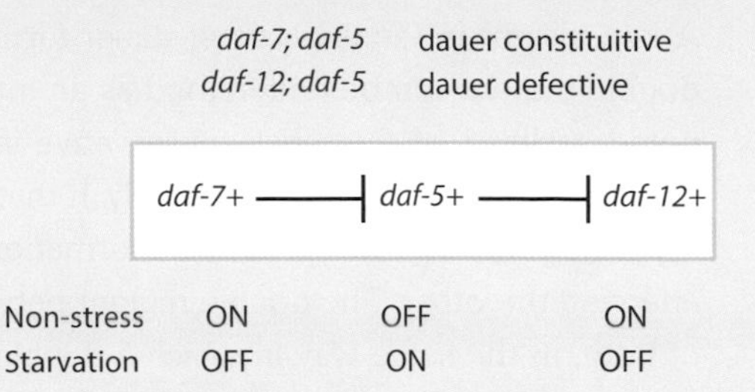

Figure CS11.3 Interactions among dauer genes. A simple pathway involving *daf-7*, *daf-12*, and *daf-5* is shown; *daf-7* and *daf-12* are *daf-c* genes, while *daf-7* is *daf-d*. Using the same principles of epistasis as discussed in the chapter, double mutant phenotypes allow these three genes to be placed into a pathway.

experiments because it is easier to assay the ability to suppress a Daf-c mutation than to assay the failure to form dauers.

Interactions among the dauer mutations are not always simple to interpret

As more mutations were identified and epistasis between the mutations was performed, some results were difficult to interpret. Some of the subtleties of epistasis analysis can be illustrated with three genes. These subtleties provide some helpful insights into the power and limitations of epistasis, suppressors, and enhancers.

One complicating feature is that the Daf-c mutants are not 100 percent penetrant, even in null alleles, so there are always some worms that are homozygous mutant in genotype but do not form dauers. This means that a quantitative assay based on a population of worms rather than on the phenotype of a single worm is essential for recognizing the phenotypes of each of the mutants. A Daf-c mutant may result in 70 percent dauer formation; suppression by a *daf-d* mutation may give only 10 percent dauers, under the same conditions, in the same mutant strain, so the phenotype is more like the wild type, but dauers are still forming. In other mutant combinations, epistasis works very clearly. A double mutant between *daf-7* (constitutive shown in red) and *daf-5* (defective shown in blue) is defective for dauer formation, the phenotype of *daf-5*, since almost no dauer larvae are formed by the double mutant. Since *daf-5* is completely epistatic to *daf-7*, the two genes are inferred to be in a linear pathway with *daf-5+* downstream of *daf-7+*. This part resembles the negative pathways we have considered previously.

By contrast, a double mutant between *daf-5* (Daf-d) and *daf-11* (Daf-c) is only partially dauer-defective. In a population of *daf-5; daf-11* double mutants, most of the worms were dauer-defective (the Daf-5 mutant phenotype) but some of the worms were capable of forming dauers (the Daf-11 mutant phenotype). This result presents a complication in thinking about how *daf-5* and *daf-11* are related to each other because the phenotype of the double mutant is not clearly either Daf-defective or Daf-constitutive.

Thinking through the Experiment

10. On the basis of a conservative interpretation of epistasis, how would you interpret the phenotype of the *daf-5; daf-11* double mutant?

A key insight in the analysis of dauer formation is that the phenotype of a *daf-5; daf-11* double mutant can be understood as an intermediate of the two mutant phenotypes. We noted earlier that the most conservative interpretation of an intermediate phenotype is that the two genes act independently. If that interpretation is true here, there could be two separate parallel pathways to dauer formation, *daf-5* affecting one of the pathways and *daf-11* affecting the other. The double mutant between the two pathways exhibits characteristics of both, in the same way in which a double mutant between any two independent genes shows two phenotypes. However, since clear epistasis is observed between *daf-5* and *daf-7*, they are hypothesized to be in the same pathway but different from the pathway with *daf-11*.

Synthetic enhancers support the presence of two pathways

The presence of parallel pathways affecting the same biological process was introduced in Chapter 10 in the discussion of the multivulvae mutations in *C. elegans* (Section 10.3). In that section, the idea was developed that the presence of two pathways with functional redundancy can be inferred from synthetic enhancement. Synthetic enhancement refers to the ability of one mutation to enhance or make more severe the mutant phenotype of another gene.

The presence of two pathways in dauer formation is supported by synthetic enhancer. experiments. For example, neither of the Daf-c mutants, *daf-11* or *daf-7*, is 100 percent penetrant, and some genotypically mutant worms are phenotypically of the wild type; in careful experiments, about 68 percent of *daf-11* mutants form dauers, and a similar percentage (65 percent) is observed for *daf-7* mutants. The exact percentages are probably not too significant, but the presence of non-dauer worms could be an indication of the functional redundancy of the two pathways; although one pathway is mutant, the other pathway is still functional and some worms form dauers.

If two pathways are present, the expectation is that knocking out both of them simultaneously will result in synthetic enhancement. The quantitative phenotypes of the Daf-C mutations can be used to detect such an interaction. In line with the hypothesis of parallel pathways, the *daf-11; daf-7* double mutant is completely penetrant, and all worms form dauers. Thus the two mutations are synthetic enhancers of each other, supporting the hypothesis that *daf-11* and *daf-7* are in two different pathways leading to dauer formation.

Thinking through the Experiment

11. The data for some of the single and double mutant combinations are summarized in Table CS11.1. Using this information, place these genes together into one or more pathways. While this looks tedious, it is not difficult once the basic principles are grasped.

These two pathways are shown in Figure CS11.4. The *daf-7* pathway includes *daf-3* and *daf-5* downstream of whereas the *daf-11* pathway does not include *daf-3* as a downstream gene. As noted above, more than thirty genes affect dauer formation, of which sixteen are shown in Figure CS11.4. We can make predictions about what should be observed when those other genes are tested in double mutant combinations with *daf-3*, *daf-7*, and *daf-11*, as well as among themselves. The genes in one pathway are expected to exhibit clear-cut suppression (by epistasis) with either *daf-3* or *daf-5*. They are not expected to be synthetic enhancers of the genes in the same pathway, in particular of *daf-7*. Genes in the other

Table CS11.1 Phenotypes of single and double mutants affecting dauer formation in *C. elegans*

Genotype	Phenotype
daf-3	Cannot form dauers
daf-6	Cannot form dauers
daf-10	Cannot form dauers
daf-12	Cannot form dauers
daf-2	90% dauers
daf-4	70% dauers
daf-8	70% dauers
daf-21	70% dauers
daf-3; daf-8	Cannot form dauers
daf-3; daf-21	70% dauers
daf-3; daf-2	90% dauers
daf-6; daf-8	70% dauers
daf-6; daf-21	70% dauers
daf-6; daf-2	90% dauers
daf-10; daf-8	70% dauers
daf-10; daf-21	Cannot form dauers
daf-10; daf-2	90% dauers
daf-12; daf-8	Cannot form dauers
daf-12; daf-21	Cannot form dauers
daf-12; daf-2	90% dauers
daf-2; daf-4	90% dauers
daf-4; daf-8	70% dauers
daf-4; daf-21	100% dauers
daf-4; daf-3	Cannot form dauers

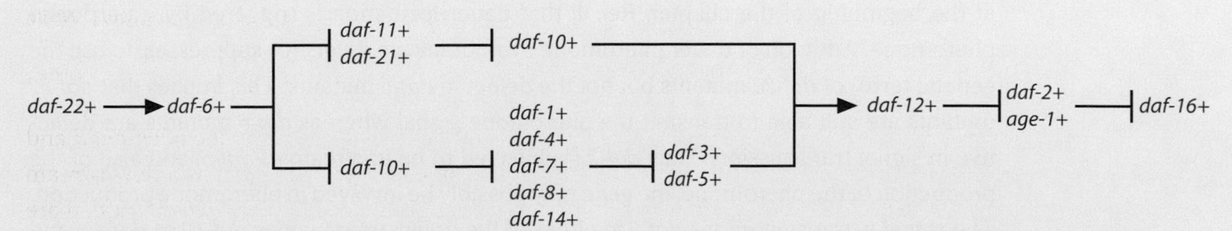

Figure CS11.4 The pathway for dauer formation in *C. elegans*, as inferred from genetic interactions involving double mutants. As before, genes in blue are needed for dauer formation so that mutations cannot form dauers. Genes in red are needed to prevent dauer formation, and mutations constitutively form dauers. Genes stacked at the same position cannot be ordered using gene interactions alone. The pathway begins and ends with common steps, but has parallel pathways in the middle. Genes in the same pathway exhibit epistasis and suppression, while genes in two different pathways show synthetic enhancement.

pathway are expected to exhibit clear-cut suppression by epistasis with *daf-11*. Genes in different pathways are expected to give a mixed phenotype when Daf-c and Daf-d mutations are combined, but Daf-constitutive mutations in two different pathways are expected to exhibit synthetic enhancement.

These predictions have been tested for many of the different genes, and the presence of two pathways in Figure CS11.4 leading to dauer formation is consistent with all of the experiments. For example, mutations in *daf-5* suppress the constitutive phenotype of *daf-7* and *daf-1*, but give a mixed phenotype when tested with the constitutive mutants *daf-11* and *daf-21*. This places *daf-5* downstream of *daf-1* and *daf-7*, but in the same pathway as these genes and *daf-3*. Mutations in *daf-21* act as synthetic enhancers of the constitutive mutations in *daf-1*, *daf-4*, and *daf-7* (among other genes), but not of mutations in *daf-11*. This places *daf-21* in a different pathway from *daf-1*, *daf-4*, and *daf-7*.

Thinking through the Experiment

12. Notice in Figure CS11.4 that some genes are shared by both pathways. Describe some of the epistasis or suppression experiments that are expected when one of these gene is tested.

13. The genes at the same step in a pathway, such as *daf-11* and *daf-21*, often have a positive relationship and have not been ordered with each other. Describe an experiment that might be able to order these two genes in a pathway, including what experimental tools would be needed.

14. *daf-10* is common to both pathways, but is not clearly upstream or downstream in either pathway. Describe a few results that would give rise to this placement of *daf-10*.

The two pathways share some common steps

The results with several of the genes are particularly informative for thinking about the presence of two pathways. Notably, some genetic steps are common to both of the pathways and others are unique to one pathway. Genes common to both pathways exhibit epistatic interactions with genes that do not show interactions with each other. For example, dauer-defective mutations *daf-22* and *daf-6* are suppressed by Daf-c mutations in either of the pathways. This indicates that these two genes are the most upstream in each pathway and that the two pathways must diverge after these two genes.

These two genes were themselves placed in order through a type of experiment described at the beginning of the chapter. Recall that dauer formation is triggered by a particular pheromone. Addition of dauer pheromone overcomes the defect (or suppresses, to use the genetic term) of *daf-22* mutants but not the defect in *daf-6* mutants. This implies that *daf-22* mutants are still able to transmit the pheromone signal whereas *daf-6* mutants are defective in signal transmission. Thus *daf-22* is inferred to be upstream of *daf-6*, and also of the production of the pheromone; the gene may possibly be involved in pheromone production. The signal is transmitted via *daf-6* to either of the downstream genes *daf-10* or *daf-11*, and then via either of the two pathways.

The pathways reconnect downstream, and the final signaling steps are shared between the two pathways. Mutations in the Daf-d gene *daf-12* mutations are epistatic to Daf-c mutants in both pathways. This suggests that *daf-12* acts downstream of both pathways, possibly to integrate the signal generated from each one. Also downstream of both pathways—in fact, downstream of *daf-12*—are the genes *daf-2, age-1,* and *daf-16.* Actually *daf-16+* encodes a transcription factor, so presumably its target genes are the same regardless of the branch of the dauer formation pathway. DAF-16 probably has many transcriptional targets, including some of the genes involved in the morphological changes between wild-type worms and dauer larvae. The inducing signal at the most upstream end of the branched pathway is the same, and the responding signal at the most downstream end is the same; the functional redundancy occurs between the branches in the middle of the pathway.

This careful analysis reveals the power and limitations of this type of genetic analysis. The power is that the functional roles and relationships of more than fifteen genes in complicated and interlocked signaling pathways have been described. In addition, some of the dauer mutations were used to find additional genes, some of which had been previously recognized for their role in other processes (such as thermotaxis or osmotic pressure avoidance), but not for their role in dauer formation. Because of the thoroughness of the screens to find the new genes, it is likely that most or all of the genes involved in this aspect of dauer formation have been found.

On the other hand, the role of some genes is not conclusive even when the epistasis results are clear; notice that *daf-10* is placed in both branches of the pathway but upstream in the *daf-7* and *daf-3* branch and downstream in the *daf-11* branch. (A molecular analysis of the *daf-10* gene reveals that it encodes a transport protein, so its role is common to both pathways but occurs at different points on each pathway.) It is also worth noting how carefully this analysis was done, and that some aspects were only possible because of the nature of the mutant phenotype and the versatility of working with *C. elegans*. The synthetic enhancement experiments were possible because even null alleles are not 100 percent penetrant, which could be recognized because large numbers of worms could be grown and examined. The phenotypes had to be carefully and, in this case, quantitatively analysed. In fact other aspects of the phenotypes of these mutants, such as cell morphology and dye uptake, have not been discussed here but were also used in determining the relationships. Finally, many different double mutant combinations have been used to confirm the resulting pathway, which can only be done in an easily manipulated genetic model organism.

Dauer larva formation in *C. elegans* might seem like a fairly arcane biological process to receive such a level of scrutiny. But the dauer pathway in *C. elegans* has interesting implications for human biology. The branch of the pathway that includes *daf-3* and *daf-7* is a TGF-β signaling pathway, one of the fundamental signaling pathways in eukaryotes. In fact *daf-7* encodes a ligand in the TGF-β family. Many of the other genes in this branch of the dauer formation pathway are related to known human oncogenes. On the other hand, the branch of the pathway that includes *daf-11* and *daf-21* regulates cyclic GMP signaling, another fundamental biochemical property in all eukaryotes; *daf-11* is expressed in the neurons in the snout that have been implicated in dauer formation.

Perhaps most interesting are the genes downstream of both signaling pathways that, by genetic logic, are responsible for integrating the different signals for dauer formation. *daf-2* encodes an insulin receptor, whereas *daf-16* encodes a transcription factor of the forkhead family, often found downstream in the insulin pathway. Orthologs of both genes are known in many other organisms, including humans. Mutations in the human ortholog of *daf-16*, FOX01A, are associated with the tumor rhabdomyosarcoma. Just as interestingly, each of these mutants in the insulin branch of the dauer formation pathway has an unusually long lifespan in *C. elegans*. Mutations in the insulin-signaling pathway are associated with longevity in flies and mice as well, and are being investigated for effects on aging and lifespan in humans. The mutants gave insight into questions that we did not have the insight to pose ourselves.

References

Albert, P. S., and D. L. Riddle. 1988. Mutants of *Caenorhabditis elegans* that form dauer-like larvae. *Developmental Biology* 126: 270–93.

Hu, P. J. 2007. Dauer. *WormBook*, August 8, 2007 http://www.wormbook.org.

Riddle, D. L., M. M. Swanson, and P. S. Albert. 1981. Interacting genes in nematode dauer formation. *Nature* 290: 668–71.

Thomas, J. H., D. A. Birnby, and J. J. Vowels. 1993. Evidence for parallel processing of sensory information controlling dauer formation in *Caenorhabditis elegans Genetics* 134: 1105–17.

Vowels, J. J., and J. H. Thomas. 1992.Genetic analysis of chemosensory control of dauer formation in *Caenorhabditis elegans*. *Genetics* 130: 105–23.

12

CHAPTER 12

Pathways, networks, and phenotypes

TOPIC SUMMARY

An organism can be thought of as an organized system made up of networks of interacting components. Using the sequences of the genome, the component parts can now be nearly fully catalogued and their network of physical interactions can be studied in detail. Examples of networks of physical interactions include transcription factor proteins and their DNA binding sequences, microRNAs and their target mRNAs, and protein–protein interactions. Attempts are being made to map each of these networks completely, at least in some organisms or in limited tissue types. These interaction networks are also being integrated into systems-level descriptions that include not only these physical interactions but also the genetic interactions. Genetic interactions are much harder to categorize and analyse, but this level of understanding is necessary if one tries to predict the phenotype from the underlying genotype. A nearly complete genetic interaction network for yeast cells has been done and serves as a template for thinking about genetic and physical interactions more generally. But even the most complete networks of gene interactions do not yet capture some of the dynamic features of cells and organisms.

INTRODUCTION

Genetics is about genotypes and phenotypes. Genomics has made it feasible, almost routine, to determine the DNA sequence of the genome for a species, which is to say, the full genotype for all its processes. As we noted in Chapter 2, although the DNA sequence of the genome may be thought of as a blueprint for the species, the sequence must be extensively annotated in order to be interpreted. Annotation will never be complete, since there will always be more information to be garnered from analysis of the DNA sequence, but many effective genome annotation methods are in place so genome annotation can be thorough if not comprehensive. Complete genotypes may become obtainable.

But we still have to connect genotypes with phenotypes. Traditional molecular genetic analysis—that is, the tools that were used before genomes could be fully sequenced—began with phenotypes and worked toward genotypes, as described in Chapters 4 and 5. The connection between phenotype and genotype can often be made for single genes, as discussed in Chapter 6. However, genes don't work in isolation from one another, so we expanded our genotype from one gene to two and asked in Chapters 10 and 11 how these, together, give rise to phenotypes. This brings us a bit closer to connecting phenotypes and genotypes.

But, again, genes don't often work singly, and they don't usually work only in pairs to determine the phenotype. The full phenotype is the outcome of many genes (plus the environment and the interactions between the genotype and the environment) working singly, in pairs, in pathways, and in many other interacting combinations. Can we really expect that we can predict that complete phenotype from the genotype? What additional ideas and methods are needed to make the connection between the sequence of the genome—that is, the complete genotype—and the complete phenotype of the organism?

In this chapter, we begin to explore the interaction networks among the genes in eukaryotic cells. For the most part, these networks are based on physical interactions, such as between transcription factors and their target sequences, between microRNAs and their target mRNAs, and between proteins. These physical interactions, as we noted in Chapters 10 and 11, are an essential aspect of understanding the actions of the genes and gene products in generating phenotypes but they do not provide the entire picture; this also requires the genetic interaction networks—suppression, synthetic enhancement and epistasis. Genetic interaction networks are more difficult to investigate on a genome-wide scale; so far such investigations have been attempted only in yeast. So it is fair to write that "we begin to explore" as we open the last chapter of this book. This chapter promises to be an introduction to what we have not yet learned, which brings us back where we started in Chapter 1—confident that we are correct so far, yet aware that our knowledge is still incomplete.

12.1 Pathways, networks, and phenotypes: An overview

The phenotypes that are most familiar to both geneticists and students of genetics arise from allelic differences in one gene or in a few genes, each gene having a significant impact on the phenotype. This is what we encounter in blood types, in many rare genetic diseases and syndromes, and in model organisms. But genes and gene products interact with each other, so most phenotypes do not arise from the activity of a single gene. The interactions that give rise to these phenotypes can be analysed by genetic and molecular tools, as described in Chapters 10 and 11, typically on a relatively small number of genes with a limited range of phenotypes and a specific biological question. Even with these relatively limited and specific phenotypes, the interactions and the pathways can be remarkably intricate.

As we expand our analysis from the activities of one or two genes to the activities of the thousands or tens of thousands of genes encoded in a eukaryotic genome, our tools, and even our thought processes, can be overwhelmed. The concept of determining the pathways in which the sequence of gene activities occurs is appealing, but the scale is very defined even for the most complicated pathways. Within the cell and the organism, these pathways connect to one another, and a gene can play a role in more than one pathway. The idea of linear pathways that we can break down into discrete steps quickly gives way to networks of pathways, for which it might not be feasible (or accurate) to break into sequential steps. When we talk about more than a few genes, and particularly when we are trying to discuss the actions and interactions of thousands of genes, we have to think about networks or systems. Some new approaches to genetic analysis have as their central tenet that the organism, cell or biological process under study is treated as an organized system; this field has come to be called **systems biology**.

Literature Link. Chuang et al. 2010, *Annual Review of Cell and Developmental Biology* 26: 721–44

The exact definition of "systems biology" is somewhat fluid among its practitioners. For our purposes, a system has a large number of components, each of which has its individual characteristics and functions; furthermore, these components interact among themselves, both directly and indirectly, to produce a functional entity. Many of the concepts discussed in this chapter have been considered in the preceding chapters, hence much of this chapter is a reflection and a synthesis of topics already introduced; in fact some geneticists think (at least privately) that all the approaches described previously work so well that a systems approach does not add much to our understanding of how phenotypes arise from genotypes. But it does seem helpful at least to think of the genes, their interactions, and the pathways as part of a larger system and to consider some ideas from systems analysis.

Systems biology is feasible thanks to genome sequencing and genome-wide screens, as described particularly in Chapter 9. In a well-annotated genome, the genes and gene products have been catalogued. Furthermore, many of the characteristics and functions of these individual components can be deduced or demonstrated using methods and strategies described in the preceding chapters. In a system, these individual components interact with one another, just as genes and gene products interact in the cell. We have described genetic and molecular methods to identify and analyse these interactions in Chapters 10 and 11. So, in principle, we have both elements needed for a systems approach: an inventory of components with known functions and a catalogue (at least a partial one) of the interactions between these components.

Some properties in a network emerge from the interactions of its components

In thinking about an organism as a biological system, one point of exploration is the extent to which the properties of the overall system can be inferred from the properties of the individual components; another is to establish which properties of the system are emergent rather than being inferred from the parts (see **emergent property**). One useful contrast between systems thinking and what has been done previously is to recall that the powerful experimental approaches of molecular biology are largely reductionist (see **reductionist approach**); that is, these approaches reduce a complex biological system to its parts, which can be analysed in detail. Systems biology is largely synthetic; in systems biology, the parts identified by the reductionist approaches of molecular biology are intellectually reassembled into a working system.

Frequently cited examples to illustrate reductionist and emergent approaches are supplied by questions around the human brain. To what extent can we explain the functions of the human brain—memory, learning, communication, information processing, consciousness, and so on—by knowing the actions and interactions of individual neurons and their molecules? Are there properties that cannot be explained by reductionist approaches and that only emerge when the system as a whole is understood? Since we don't yet know the answer, this question borders on personal opinion and philosophy. Reductionist approaches clearly work exceptionally well, and should be vigorously employed; but we don't yet know whether there are limits to this approach, and where these limits would lie.

Systems biology is frequently explained with the help of analogies with other familiar systems. For example, the traffic pattern in a particular locality is a system. The number and movement of individual cars comprise the component parts. These components interact with each other via a series of highways, pathways, or links. A direct interaction between the cars occurs when one car follows another immediately along the same highway or stops to give way to another car at an intersection. Each car functions separately from the other car, but the movement (or function) of one car also affects directly the movement of another car. An indirect interaction in this highway system arises when the number of cars on one highway is particularly heavy, so much so that some cars are diverted to another highway. This rerouting then also slows down the movement of cars on the second highway; although no car on one highway interacts directly with any car on the other highway, the overall function of the traffic pattern through the locality is affected. By knowing the number and movement of cars in one part of the traffic system, a traffic engineer can infer or predict the impact on another part of the system.

This analogy illustrates several features of systems biology. For an individual commuter, the properties of the entire system provide useful knowledge insofar as they affect single components—his car and his route to work, for example. Hearing that there is heavy traffic on one road, the savvy commuter chooses an alternative route. In such an example, the properties of the system as a whole can also be used to predict the effects on individual components. We know what happens if a car breaks down on that particular bridge and how that affects the traffic flow on our road. In addition, the properties of the highway system are dynamic rather than static; the direct and indirect interactions between cars are different at rush hour and in the middle of the night. A traffic engineer can predict the system-wide effects of closing a particular road at a given time or of building an additional connecting highway.

Systems biology is not yet as sophisticated as traffic engineering in being able to infer the properties of the system from individual components or to predict the effects on the whole system of changing one component. In particular, the analysis of indirect interactions and dynamic changes in biological systems is in its infancy. As we noted in Chapters 10 and 11, the great advantage of the genetic analysis of interactions is that both direct and indirect interactions are found; a weakness can be that direct and indirect interactions are not immediately distinguished from each other. For that reason, we will initially concentrate on the procedures used to analyse direct interactions among biological components, and then turn to indirect interactions and some dynamic changes later in the chapter. While many biological processes such as phosphorylation and metabolism have been studied as interaction networks, we will consider three types of direct interactions between biological components known to regulate the overall properties of the organism—that is, its phenotype—that are being studied on a genome-wide scale. These direct interactions are:

- interactions of transcription factor proteins with their target DNA binding sequences;
- interactions between microRNAs and their mRNA targets;
- interactions between proteins.

These interactions are diagrammed in Figure 12.1: on the left, a conventional diagram shows them; on the right, a network diagram captures them. Transcription factor regulation, microRNA regulation, and protein interactions are shown. In each case, we will discuss how the interactions are determined experimentally; and we will do so often using tools and approaches that have been developed in previous chapters. We will also consider

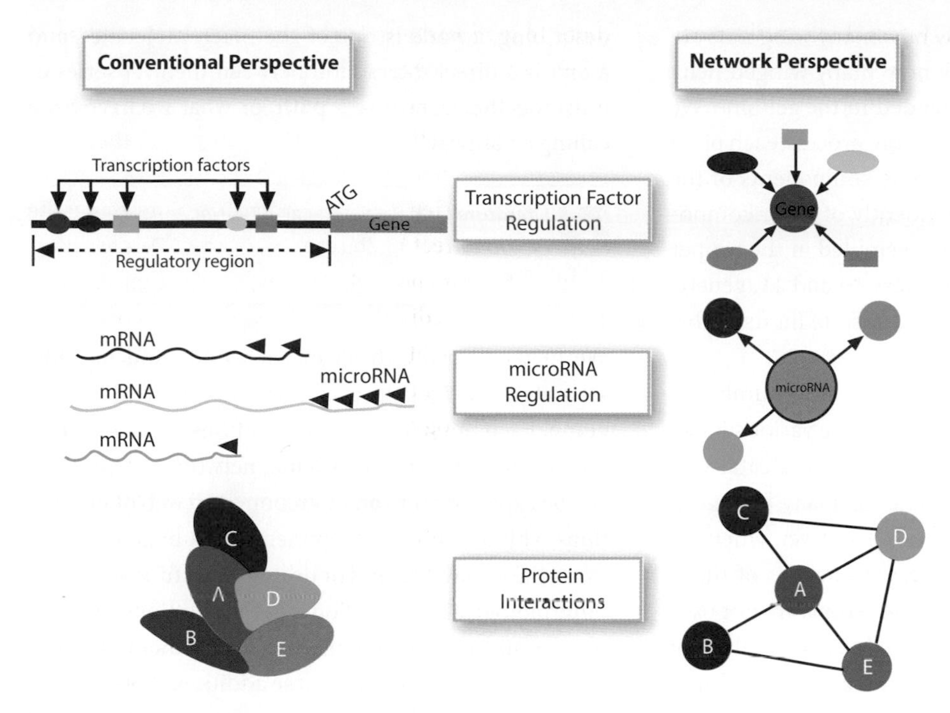

Figure 12.1 Three biological examples of interaction networks. On the left side of the figure, transcriptional regulation, microRNA regulation, and protein interactions are shown as they might be conventionally diagrammed. These interactions are plotted in network format on the right. Note that the connecting lines have arrows showing that one macromolecule is the target of regulation by another in transcription factors and microRNAs networks; no directionality is shown for protein interaction networks.

the methods assessed to test the functional significance of the interactions. The interactions, together, can be thought of a comprising a **network**. For our purpose, the system will have one or more networks that depict these interactions.

Two complementary questions lie at the core of a systems approach to genetic analysis.

- Can the properties of a biological system be inferred and understood from the functions and interactions of individual genes? To return to our earlier analogy, how well can the traffic flow in a locality be predicted on the basis of knowing the actions and movements of every individual car?
- Can the activity and functions of individual genes be inferred from the properties of the biological system? In our analogy, can we predict how a particular car will behave on the basis of knowing the overall traffic pattern?

Although we cannot yet answer these questions fully for biological networks, the framework for thinking about them promises some new insights into genetics and biological systems.

12.2 Networks and graphs

In Section 10.1, genetic analysis of a biological process was compared to assembling a desk; imagine that the desk is shipped in a flat box that itself looks nothing like the final product. If we are careful and thorough in our assembly process, we first take an inventory with the parts list and identify every piece and its apparent function. We try to understand not only what it does but when and where it is likely to be needed. To make sure that this analogy does not stray too far from genetics, this is similar to what is done in genome annotation described in Chapter 2—a process in which all genes, their likely function, and their expression patterns are characterized. Just as we know how many wing nuts are present in the desk, we know how many winged helix DNA binding proteins are encoded in the genome. We then begin the step-by-step assembly process, each piece being connected to the next. Some components of the desk can be assembled independently of other components, while others have to be assembled in the proper sequence. As discussed in Chapters 10 and 11, genetic and physical interactions form the basis of the assembly instructions.

If our desk assembly has a relatively small number of component parts, it is possible to list the various interactions that occur between components as a catalogue or as a list of steps; this is feasible for many biological processes and phenotypes as well. However, when we consider the genome as a whole, with its tens of thousands of component parts, the interactions are not usually depicted in a list or in a table. (Lists and tables of interactions for each individual gene can often be found in the databases for the organism described in Chapter 3 and referenced in Table 3.1.) Instead the interactions are often plotted as a graph.

The differences in how a table and a graph are used to show us the data will be familiar to travelers. We use a timetable—that is, a table of connections—to determine whether a certain train goes from Station A to Station B, or when the next train arrives. We use a route map—that is, a graph or a chart of connections—to work out the best plan for a more complex itinerary. In a graph, a station is depicted as a **node** or a vertex, and the travel between them as a connection or **link**, of the type shown in Figure 12.2. For biological networks of the type we are describing, a node is one of the macromolecules, and a link is a direct interaction between them. A series of links together comprise a **path**, or what we have been calling a pathway.

Literature Link. Barabasi et al. 2004, *Nature Reviews Genetics* 5: 101–13; Zhu et al. 2007, *Genes & Development* 21: 1010–24

We may originally think of each node and its links separately, but this approach does not capture the entire graph. Each node has its own set of links in addition to the one described in the original network, so the graph can be expanded into an interconnected web of interactions with additional components; in a biological network, these may be genes or proteins that function in the same biological process. Some of the additional interactions connect among themselves, while other links go to different components yet. These additional interactions are plotted in a similar way and a network of connected nodes arises, as shown in Figure 12.2.

By analogy, consider what happens when a new subscriber signs onto a social network site. The new subscriber finds friends and acquaintances who are also subscribers and invites them to be friends on the social network. A small network forms, composed initially of the relationships between the new subscriber and his or her friends. Each friend has his or her own group of network friends as well. Some of the new subscriber's friends will also be friends among themselves, so they are already connected in the social network. Others will be "friends of friends" but are not known to the new

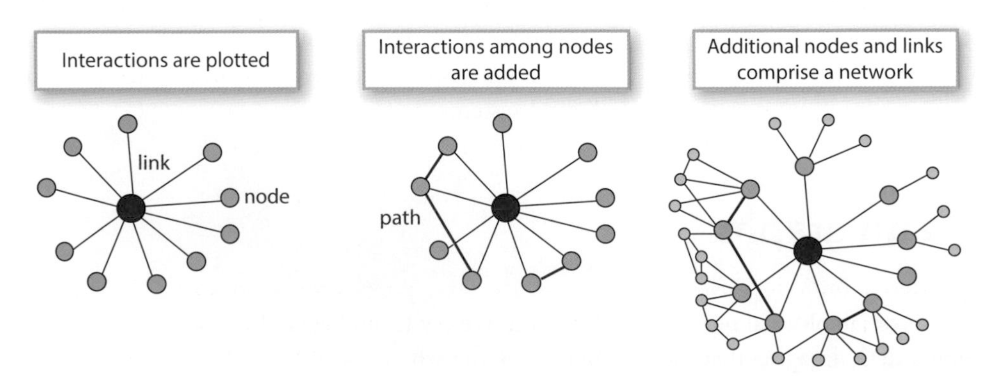

Figure 12.2 A simple network. A hypothetical protein interaction network is shown. Each protein is represented by a node and each interaction or link between the proteins is represented by a line connecting two nodes. The protein studied initially (marked in red) interacts with ten other proteins (marked in green). These proteins also have interactions with each other, which form paths through the network. Each protein interacts with additional proteins (in blue), too—so nodes and links are added as the network grows.

subscriber. Some of these people will invite the new subscriber to become their friend as well, so the subscriber's social network becomes more complex, as more links and nodes are added.

The network can be used to infer the functions of its components

In previous chapters we described different methods to infer the function of a gene and to determine how it contributes to a phenotype: mutations that knock out or change its function; sequence similarity to other genes or proteins whose function is known; and expression patterns. A fourth method to infer the function of an unknown gene is to analyse its interactions. Analysis of the interactions can be done by looking at an individual node in a network and noting all of its interactions. Most fundamentally, a network allows us to infer the function of genes and proteins whose function was not known from other methods. For example, imagine a protein of unknown function whose interactions include a number of proteins known to be involved in RNA processing, modification, and splicing, as shown in Figure 12.3. Because this protein is found in association with proteins related to RNA processing, a reasonable hypothesis is that the unknown protein is also involved in RNA processing.

In addition to being a valuable method of inferring the function of an individual gene or protein, the interactions can be analysed by studying the properties of the entire network. The plot of a biological interaction network comprises a graph whose properties can be studied by the field of **graph theory**. Graph theory studies the characteristics or the **topology** of the entire graph rather than the properties of individual nodes and links. Some

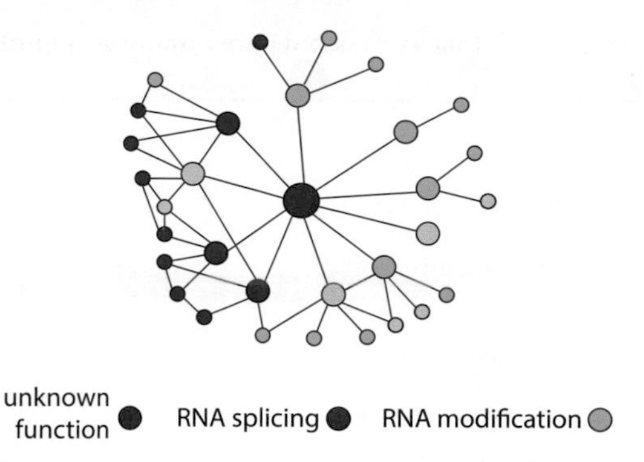

Figure 12.3 Interactions suggest biological functions. In this hypothetical graph of protein interactions that is based on interactions observed in yeast and other eukaryotic cells, the biological function of the protein marked in red is not known. The proteins marked in blue have been shown to be involved in splicing of RNA, while the proteins marked in green have functions in modifying RNA structure or sequence. The proteins marked in gray have unknown or unrelated functions. Since the red protein has numerous interactions with proteins involved in RNA splicing or modification, a reasonable hypothesis is that the red protein is also involved in RNA processing.

authors have referred to it as molecular cartography—a field whose objective is to draw a map of the macromolecular connections that occur within a system. It is worth recalling the two questions that framed our introduction to systems biology. The network graph can give insights into the roles of individual nodes—proteins or genes in our case. The topology of the entire network graph can help to synthesize the properties of these individual components into a functional system. It is not yet clear how much the principles of graph theory can be directly applied to interpreting biological interaction networks, or which of these principles are the most valuable. Thus we will focus on a few examples of biological networks and on the roles of individual genes and proteins.

12.3 The interactions between transcription factors and DNA sequences

The regulation of transcriptional initiation occurs through the interaction of particular proteins—transcription factors—with specific DNA sequences—their binding sites. An example is shown in Figure 12.4, where a conventional view of transcriptional regulation is replotted as a network, both in terms of the target gene and in terms of one of the transcription factors. As transcriptional regulation is considered in terms of a biological system or network, the components are the transcription factors together with their specific DNA target sequences.

The binding sites for transcription factors have been very thoroughly studied using reporter genes, **chromatin immunoprecipitation (ChIP)** assays, and other types of

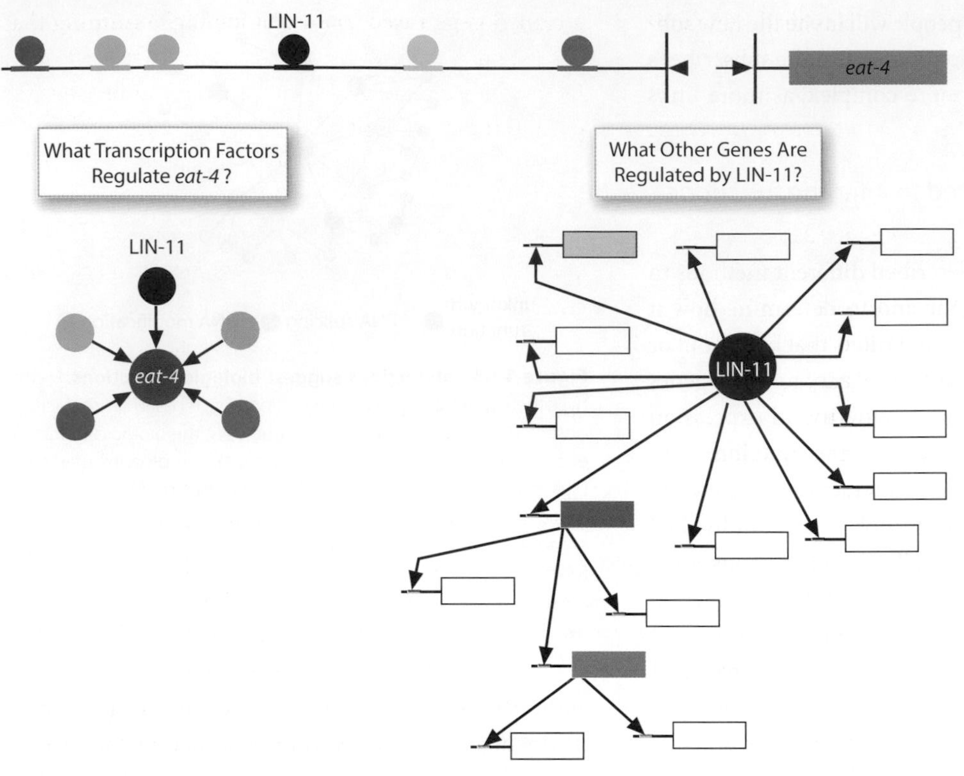

Figure 12.4 Depicting transcriptional regulatory networks. A portion of the cis regulatory region of the *C. elegans* gene *eat-4* is used as an example to show two questions rased in transcriptional regulation. In the conventional view, at the top there are six enhancer sites for five different transcription factors, including LIN-11. This can be asked in graph form as the number and identity of the transcription factors that regulate *eat-4*, our gene of interest. Alternatively, the other genes regulated by LIN-11 can be determined and plotted as a graph; as shown on the right, there are 11 target genes including *eat-4*. One of these target genes itself encodes a transcription factor (marked in pink) that regulates some additional genes, one of which also encodes a transcription factor (marked in brown), and a network of regulation is built up. Only some of the targets of the latter two transcription factors are shown.

experiments, some of which are described in Chapter 2. But can the properties of the individual components be reassembled into a network of interactions, and what does that tell us about the logic of transcriptional regulation?

In order for us to analyse this network on a genome-wide scale, the specific DNA sequences bound by individual transcription factors have to be identified and the association between the regulatory region and the gene being regulated has to be established. As suggested in Figure 12.4, two different approaches can be taken. First, the investigator can begin with an individual gene, identify its regulatory region, and then find the transcription factors that bind to this region. At its most basic level, this approach does not consider what other genes might be regulated by the same transcription factor. Second, the investigator can begin with a known transcription factor and ask what genes are regulated by this transcription factor. On a genome-wide scale, this is the more common approach.

The interactions between transcription factors and their binding sites

In order to identify the genes regulated by a transcription factor, which thus might be coordinately regulated, the investigator begins with a transcription factor and identifies its binding sites throughout the genome using a ChIP assay, as described in Chapter 2. The DNA sequence that co-precipitates transcription factor or the chromatin-associated protein is then analysed through hybridization to a microarray (ChIP-chip) or through direct sequencing (ChIP-seq). This begins to address the questions of how many different genes are regulated by one transcription factor, and how the expression patterns of these genes are coordinated.

Transcription factors are readily classified into families such as C2H2 zinc finger proteins or homeodomain proteins, on the basis of sequence similarities in their DNA binding domain. Proteins with closely related amino acid sequences in their DNA binding domains usually bind to similar nucleotide sequences in the genome. A quick search of the domain database known as **pfam** on the National Center for Biotechnology Information (NCBI) website identifies more than 500 different DNA binding domains used to classify proteins, most of which are also found in transcription factors. Thus it is comparatively easy to identify a gene product as a potential transcription factor; approximately 5–10 percent of proteins encoded in a eukaryotic genome are transcription factors.

 ***Literature Link.*** https://pfam.xfam.org

Thousands of ChIP experiments have been carried out, in many different organisms and under many different conditions. The ENCODE and modENCODE websites (Chapter 2) and the genome database websites for different model organisms (Chapter 3) catalog many of these experiments, and data from more transcription factors and more conditions are added regularly. These datasets serve as our best way to understand how many genes are regulated by a single transcription factor and how the expression of these different genes is coordinated. One of the limitations of ChIP as a genome-wide approach is the need to have a different antibody for each transcription factor protein. Thus, in many large projects, the transcription factors are synthesized in vitro, from a cloned gene fused to an epitope tag. Assuming that the fusion protein binds like the normal transcription factor, an antibody to the epitope tag can be used for ChIP rather than antibodies to the transcription factor. This obviates the need to have specific antibodies to each transcription factor.

Another limitation of ChIP experiments is that binding interactions vary widely depending on experimental conditions and on biological condition such as tissue type or developmental stage. In fact transcription factor binding is one the key properties that *determine* different biological conditions. Thus the assay must be repeated for many hundreds of different conditions, so that we may view all the interactions that regulate the transcription, and it will never be certain that all the possible interactions have been found. Yet another challenge for ChIP assays is that a transcription factor is often present at very low levels in only few cells and at specific times, particularly when considering the complex regulatory networks involved in metazoan development. Thus, many transient interactions that are critical to cellular differentiation could be missed. These limitations are worth noting, but they do not rule out the importance of the method.

While ChIP is by far the most commonly used method to analyse and catalog the interactions of transcription factors with their binding sites, it is not the only method. A method referred to a yeast one-hybrid (Y1H) assay begins with an individual gene and searches the catalog of known transcription factors to identify all the ones that regulate a specific gene; then it builds the larger network of interactions. Y1H is described in Tool Box 12.1.

TOOL BOX 12.1 Yeast one-hybrid assays

Chromatin immuno-precipitation is by far the most commonly used technique for connecting transcription factors with their target genes; by determining the genes regulated by the same transcription factor, regulatory networks can be worked out, as the chapter discusses. An alternative approach is to begin with the target gene and to find the transcription factors that regulate it. The protocol has been termed a "yeast one-hybrid (Y1H)" approach, as diagrammed in Figure B12.1. The strategy and set-up are similar to those in yeast two-hybrid assays, described in Section 10.5.

In a Y1H procedure, the bait vector has the predicted regulatory region of the gene of interest fused as a transcriptional reporter upstream of the coding region of a selectable marker or reporter gene such as HIS3 or *lacZ* genes. The bait vector with the DNA regulatory sequences is integrated into a chromosomal location in the yeast genome. The prey library consists of plasmids, each with a cDNA for a known or suspected transcription factor fused to the Gal4 activation domain. (Recall from the Y2H assay that the Gal4 activation domain interacts with RNA polymerase II and the general transcription factors.)

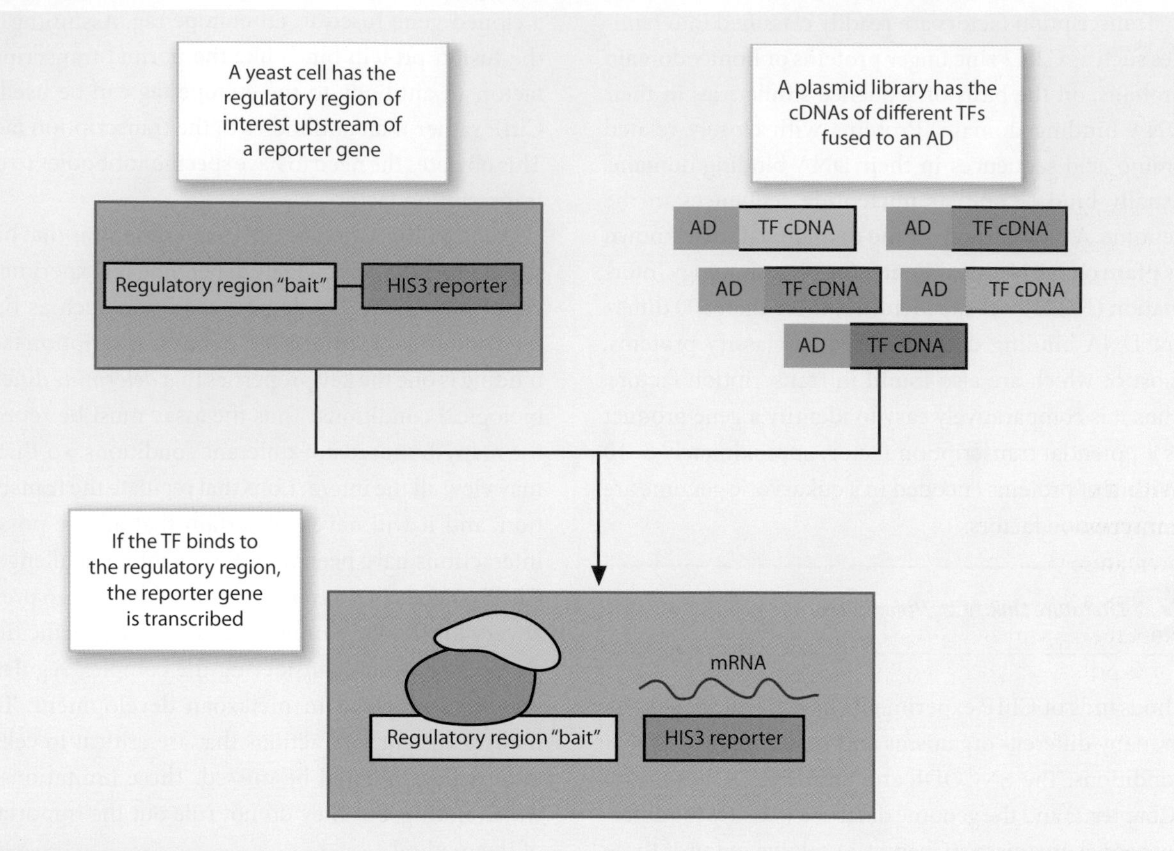

Figure B12.1 The yeast one-hybrid screen. The goal of a yeast one-hybrid screen is to identify the transcription factors that bind to the regulatory region of a gene of interest. The regulatory region is fused to a reporter gene such as HIS3, and this is integrated into the yeast chromosome. A cDNA library is made with known or suspected transcription factors (TF cDNA) fused to the activation domain of the GAL4 protein (shown in green). This library is transformed into yeast cells, with the regulatory region driving HIS3 expression. If the transcription factor binds to regulatory region, the activation domain can interact with RNA polymerase and the reporter gene is transcribed. In the example shown, the cell would grow in the absence of histidine.

Transcription factor prey libraries for *C. elegans*, *Drosophila*, *Arabidopsis*, and humans are commercially available. The prey plasmids are transformed into the bait strains and, as with the Y2H screen, interactions are detected using reporter gene expression. For example, in the diagram in Figure B12.1, an interaction between the bait sequence and a prey transcription factor results in the ability of the cell to grow in the absence of added histidine.

Yeast one-hybrid assays have so far been used for a somewhat limited subset of gene regulatory regions and transcription factors. One published screen examined the interactions between seventy-two different regulatory regions from genes expressed in the nematode endoderm used as baits and 117 transcription factors used as prey; a total of 283 interactions were detected and examined. The results were comparable to what was found for transcriptional regulatory networks using ChIP-on-chip assays. The number of transcription factors binding to a particular regulatory region ranged from one to fourteen, with a mode of four interacting proteins per regulatory region. On the other hand, a few transcription factors regulate many different targets—the maximum observed was twenty-seven genes for one transcription factor. Most transcription factors regulate very few genes, however. Seventy transcription factors regulated a single promoter and twenty-eight regulated two promoters, so ninety-eight (or 84 percent) of the 117 transcription factors tested have just one or two gene targets.

A system of interactions in which many transcription factors control only one or two targets will result in a highly compartmentalized and fragmented network; imagine what communication would be like in a work environment where 84 percent of the people spoke to only one or two other people. The network is held together by the few highly connected transcription factors that interact with many genes, just as a few highly interactive and social co-workers can hold an office environment together. In the set of interactions in the worm and in transcriptional regulatory networks in yeast, the transcriptional factors that interact with many targets are essential; elimination of these genes (by mutation or by RNAi) is approximately five times more likely to be lethal to the cell or to the worm than the elimination of a gene with few target sequences.

The Y1H assay in worms tested the interaction of perhaps 10 percent of the known or suspected transcription factors with a very small percentage of the regulatory regions, and in one highly defined differentiated tissue. This is clearly just a beginning, albeit a promising one. The number of different transcription factors and the number of different potential targets is very large, so testing all of them by direct experimental means will be a daunting task, but the tools are available.

Literature

Fuxman Bass, J. I. 2016. Gene-centered yeast one-hybrid assays. *Cold Spring Harbor Protocols* 12. doi: 10.1101/pdb.top077669.

Deplancke,B. et al. 2006. A gene-centered *C. elegans* protein-DNA interaction network. *Cell* 125: 1193–205.

Different types of transcriptional regulation are seen among the networks

With the vast amount of data generated from different types of experiments, it is challenging and perhaps premature to attempt to draw general and overarching lessons about systems of transcriptional regulation. Nonetheless, we can point to a few examples that suggest some principles of eukaryotic gene regulation.

Endogenous and exogenous regulation in yeast cells

More than 250 ChIP-chip experiments have been performed in yeast, and more than 10,000 interactions between transcription factors and binding sites have been detected. The first analysis used epitope-tagged versions of 106 known transcription factors, so that the same antibody was used to precipitate different transcription factors.

Literature Link. Luscombe et al. 2004, *Nature* 431: 308–12; Hu and Gerstein 2006, *Proceedings to the National Academy of Sciences, USA* 103: 14724–31

The strains were grown in three different media, and the interactions were tabulated; while the results from these early experiments have been refined and expanded since their initial publication, the overall conclusions have not changed very much. Out of the 6,270 yeast genes in the genome, 37 percent (2,343) were bound by one or more of these transcription factors, and a third were bound by two or more of the proteins. The focus on these assays is, however, on the behavior of the transcription factor proteins themselves rather than on the genes being regulated. On average, a transcription factor bound to thirty-eight different regulatory regions, but the range is significant: a few transcription factors bound to no regulatory regions under these growth conditions, while a few others bound to 181 different regulatory regions.

One potentially important distinction in the types of transcriptional regulation in yeast was termed endogenous and exogenous network components, as illustrated in Figures 12.5 and 12.6. **Endogenous** network components, as illustrated in Figure 12.5 are characteristic for genes whose expression is fundamental to the basic function of the cell and restricted to particular times. Genes involved in the cell cycle constitute an example of endogenous components. As a rule, these genes tend

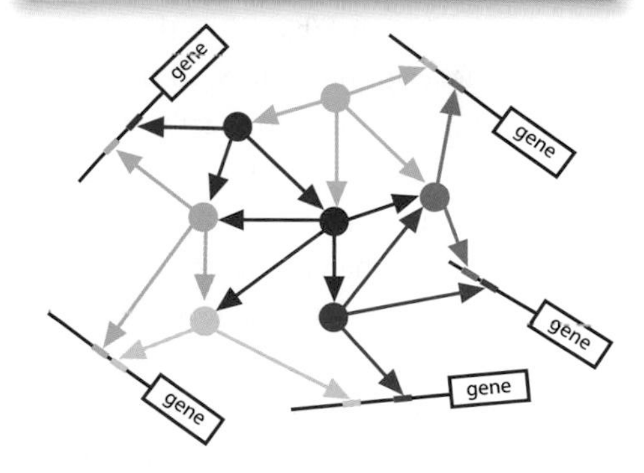

Figure 12.5 Endogenous transcription networks in yeast. The network diagrams in Figures 12.5 and 12.6 show the transcription factors as colored circles with arrows representing the binding site to regulatory regions for different genes; genes that do not encode transcription factors are depicted as box diagrams. These data are redrawn from yeast networks. In endogenous networks, most of the interactions occur between the transcription factors with relatively few other target genes; each target gene that is not a transcription factor is regulated by more than a single transcription factor. Note that the red gene encodes a transcription factor that has only other transcription factor genes for its targets. This type of network is expected to result in cellular events that are highly regulated and closely coordinated; this example has been redrawn from some transcription factors regulating cell division.

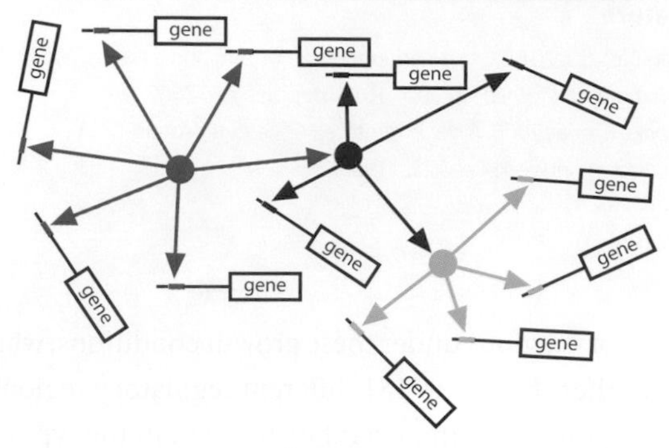

Figure 12.6 Exogenous transcription networks in yeast. The network diagrams in Figures 12.5 and 12.6 show the transcription factors as colored circles with arrows representing the binding site to regulatory regions for different genes; genes that do not encode transcription factors are depicted as box diagrams. These data are redrawn from yeast networks. In exogenous networks, each transcription factor has many targets that do not encode other transcription factors, and most target genes are regulated by a single transcription factor. Thus the regulatory network is not highly coordinated, but response could be rapid. This example has been redrawn from a part of the network that affects the stress response.

to be regulated by a fairly high number of transcription factors, each of which has only a few additional target genes that are not other transcription factors. Furthermore, these transcription factors are themselves regulated by other transcription factors, so that the average path (i.e. the steps required to connect two nodes) in an endogenous network is longer than the average path in the total yeast cell – in other words, the number of steps is greater in the former than in the latter. The regulatory network for endogenous components is therefore highly interconnected, very specific, and sharply defined—it is a set of genes expressed coordinately under specific conditions, with many cross-connections among them.

An **exogenous** network includes target genes that respond to environmental stimuli. Figure 12.6 summarizes the properties of this type of network. These target genes tended to be regulated by few transcription factors, but each transcription factor has many targets. In contrast to endogenous networks, the transcription factors have fewer connections to each other, so that the path between two nodes in an exogenous network is on average short. Such a network would seem to allow for a more rapid but less highly coordinated response, which involves genes with many different functions. One inference is that the genes needed for responses to unpredictable situations such as stress responses are described by an exogenous network when a few genes respond quickly and directly contact many other targets. On the other hand, genes needed for types of normal cellular organization such as cell division, when changes in transcription occur at predictable times, can be described as an endogenous network.

Hierarchical levels of regulation in flies and worms

One of the primary goals of genome annotation is to identify the binding sites throughout the genome for every transcription factor, using ChIP with epitope tag proteins as the principal assay. Such binding information can also be used to compile networks of regulatory information. Note that these graphs we will show represent the connections among *transcription factors*; connections to genes encoding other types of proteins are not shown but can usually be found on the modENCODE site. An example for the *prd* gene is shown in Figure 12.7 Panel A; a more complete example is drawn from the modENCODE data on *Drosophila melanogaster* in Figure 12.7 Panel B. The reader is strongly encouraged to log on to modENCODE.org and examine the set of interactions in order to gain a better appreciation for this description.

 ***Literature Link.*** http://www.modencode.org

We can summarize a few of the findings. In flies, a typical transcriptional factor gene is regulated by an average of twelve other transcription factors, but a few genes are regulated by more than fifty transcription factors. These numbers are illustrative rather than definitive, since many transcription factors have not been tested extensively and the number of transcription factors per regulated gene ranges widely. For example, as shown in Figure 12.7 Part A, *prd* is regulated by seventeen transcription factors and regulates thirty other transcription factors; at least six transcription factors (*eve*, *kn*, *run*, *Tri*, *Kr*, and *cad*) have two-headed arrows, which shows that these both regulate *prd* expression and are themselves regulated by Prd.

Information similar to what has been found in flies is also found in other organisms, so it is likely that this

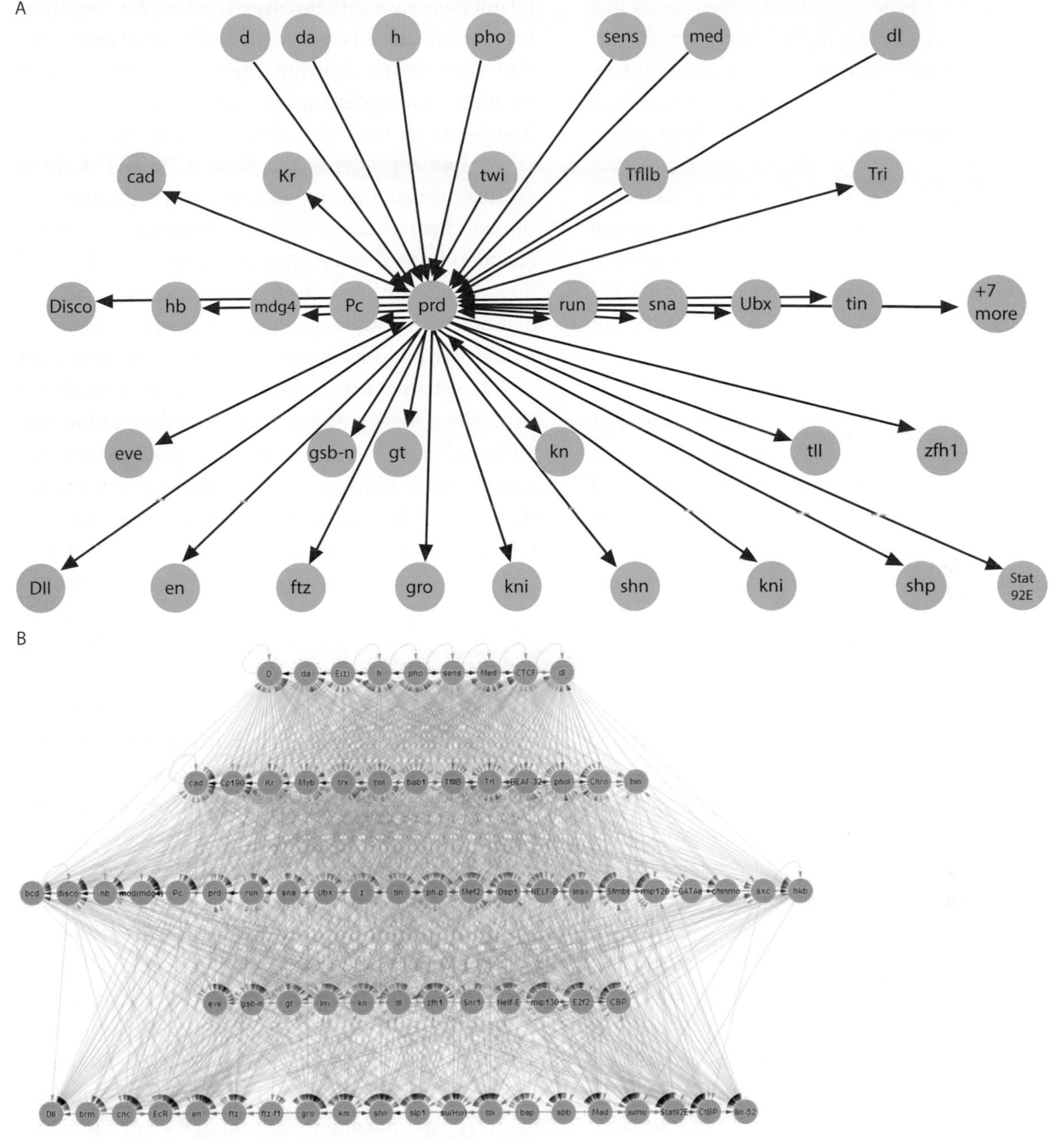

Figure 12.7 Transcription factor networks from *Drosophila*. As depicted on the modENCODE site, each green circle is a transcription factor gene, with the arrows showing the direction of regulation. Part A shows the example for the *paired* (*prd*) gene. Note that the regulatory network is hierarchical, with nearly all of the regulatory interactions to a lower tier. All the interactions involving *prd* itself are shown, but the interactions among the genes that regulate *prd* or that are regulated by *prd* are not shown. Part B shows the entire transcription factor network in *Drosophila*, as taken from the modENCODE website. Note that transcription factors form a hierarchical structure of regulation, regulation occurring primarily across and down the levels rather than to a higher level. Transcription factors in the lowest tier regulate many other genes, but do not regulate many other transcription factors except each other. The reader should explore the network at modENCODE.org to appreciate the complexities of the interactions.

reflects some general principles about the interactions among transcription factors in metazoan transcriptional regulation. Notice, first, that the graphs in Figure 12.7 show a clear hierarchy of regulatory relationships; the hierarchy arises from the direction of regulation as represented by the arrows, such that transcription factors encoded by genes at the top of the hierarchy are not themselves the targets of many other transcription factors. There are very few regulatory interactions that go up from a lower tier to a higher tier; in *Drosophila*, more than 92 percent of the regulatory interactions are with genes in a lower tier, and only 8 percent feed back to the transcription of a gene at the higher level. Target transcription factor genes at the bottom of the hierarchy such as *sbb* are regulated by most genes, show less tissue-specific expression, and are more likely to the essential when knocked out by mutation or RNAi. This pattern with master regulatory genes has also been inferred from genetic experiments with mutants, so the different types of analysis give a consistent picture.

The genes at the top or in the middle of the hierarchy include some of known master regulators of worm or fly development, such as the Hox genes (*Ubx*), the gap genes (*hb* and *Kr*), and some of the pair-rule genes (*prd* and *eve*) discussed in the case study for Chapter 4. These genes have few genes regulating them but have more interactions among themselves. It is tempting, and perhaps informative, to see a parallel with workplace organization charts in this network. The transcription factors at the top are the top "executives" in the organization; they have relatively few direct connections to the "worker" transcription factors at the bottom. Their control is exerted through their interactions with each other and through their effects on the "mid-level manager" transcription factors, which interact among themselves and directly regulate the transcription factors at the bottom of the hierarchy. Although these regulatory diagrams are so far based only on a subset of transcription factors and their target genes in very few organisms, the consistency with which these patterns are observed—and the correlations between these networks and observations with genetic interactions, as described below—suggest that this type of hierarchy is a basic strategy for transcriptional regulation in metazoans.

12.4 microRNA and mRNA interactions

The regulation of transcription found by examining transcription factors and their binding sites is only part of the complicated process of transcriptional regulation. The roles of microRNAs in gene regulatory networks have also been a fundamental goal of genome annotation projects. Recall that microRNAs (miRNAs) down-regulate their target genes by binding to complementary sequences in the corresponding mRNA of the target. This double-stranded RNA molecule either blocks translation of the mRNA or targets it for degradation. This is depicted as a network in Figure 12.8, which shows the interactions between microRNAs and some transcription factors in *C. elegans*. Since most (but certainly not all) transcription factors stimulate or promote the expression of their target genes, the role of transcription factors and that of microRNAs are often in opposition to each other: it is a role of up-regulation in the case of transcription factors, of down-regulation in the case of microRNAs.

Because these interactions rely on Watson–Crick base pairing, computational predictions for miRNA target sites in mRNAs are much easier than for transcription factor-binding sites or protein interaction sites. Many different algorithms are being used to predict target sequences for miRNAs from direct sequence alignments and comparisons. Because it is much simpler to predict a base pairing interaction computationally that it is to verify it functionally in vivo, the number of predicted interactions is usually much greater than the number of experimentally verified regulatory interactions.

As with transcription factors, the average number of target genes regulated by miRNAs vary widely. The first two miRNAs to be identified in *C. elegans* are illustrative. According to the database at miRBase.org, the *lin-4* miRNA apparently has nine verified target mRNAs (including *lin-14*, the original interaction identified genetically) and 135 predicted targets. The *let-7* miRNA has forty experimentally verified targets and more than 530 predicted targets.

Literature Link. http://www.mirbase.org

The transcriptional regulatory networks in the ENCODE-related databases, like the ones in Figures 12.8 and 12.9, typically show the predicted and verified miRNA regulatory interactions as well as the transcriptional factor interactions. Like transcription factors, the miRNAs also appear to fall into distinct hierarchical

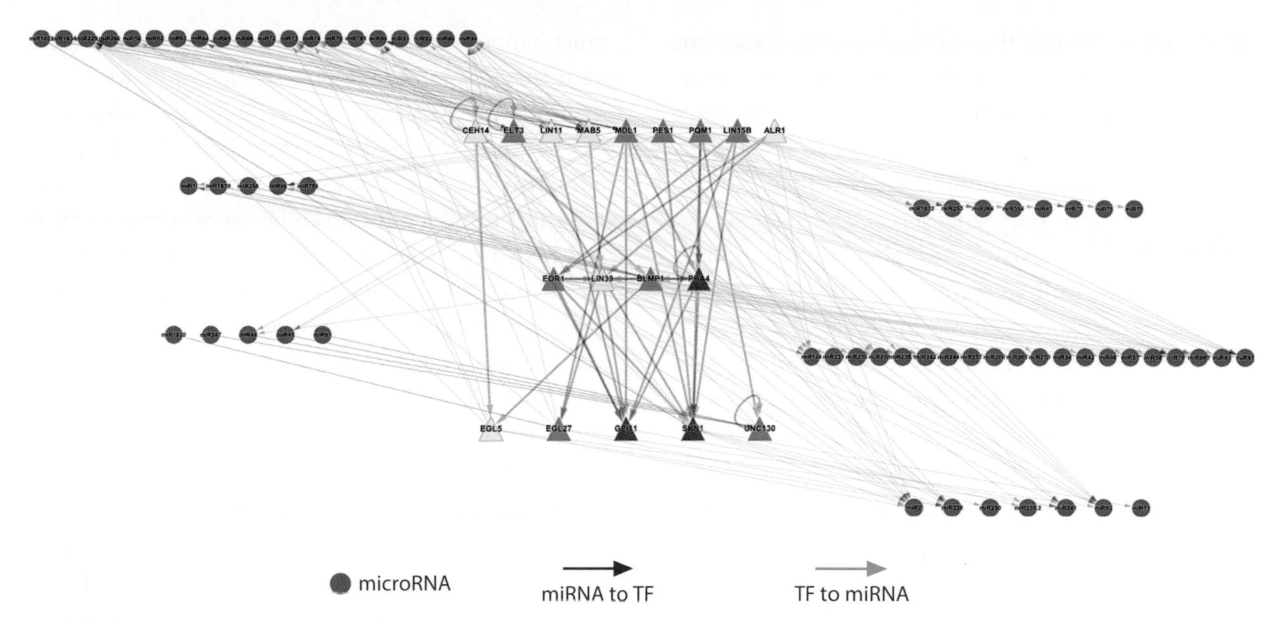

Figure 12.8 An integrated network of transcription factor and microRNA regulation of gene expression in *C. elegans*. Interactions among eighteen transcription factors and microRNA genes are shown. Different types of transcription factors are represented by different colored triangles in this figure, but that is not relevant to our discussion. The red arrows represent regulation of the TF gene by microRNA, whereas the green arrows represent regulation of the microRNA by the TF. Note that TFs at the bottom level regulate the transcription of microRNAs, which regulate TF genes at the higher levels, providing a feedback from the lower level to the higher level via microRNA regulation.

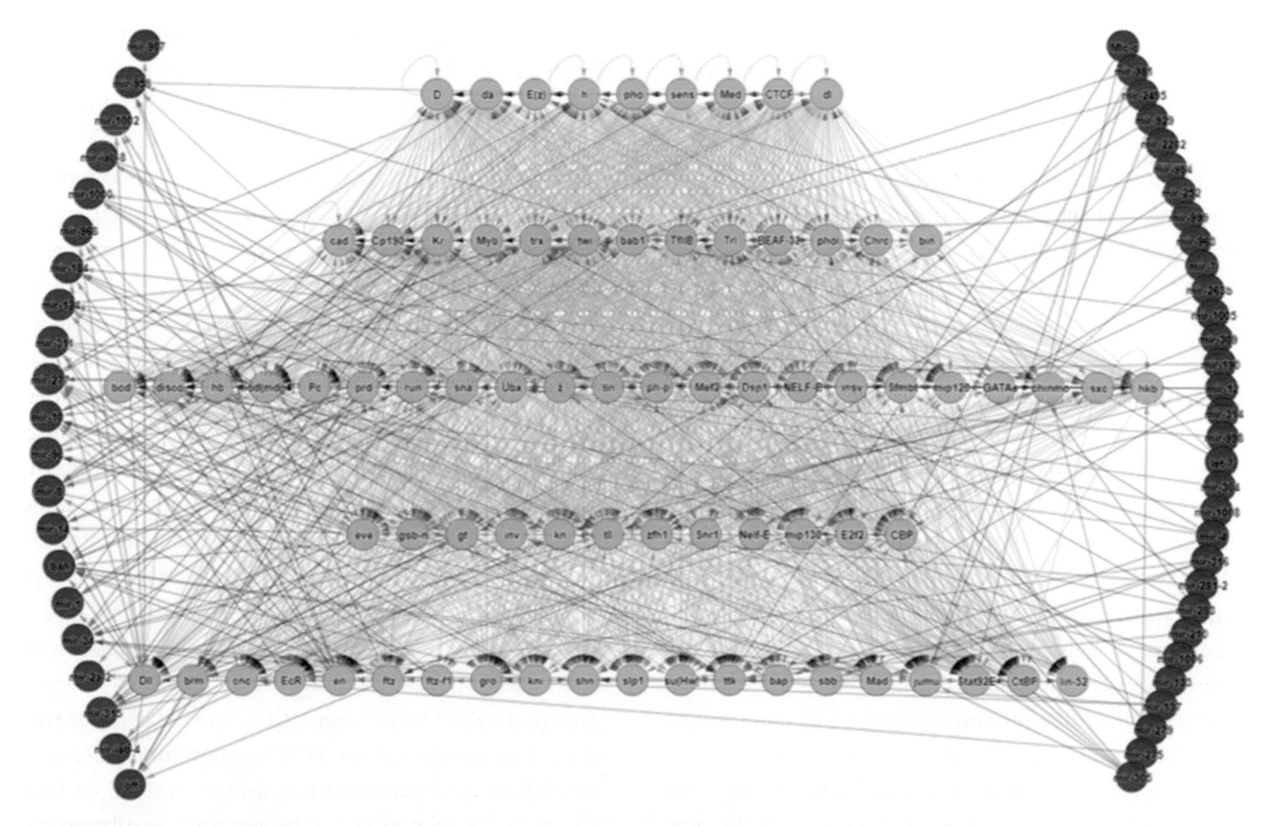

Figure 12.9 The integrated transcription factor and microRNA regulation in *Drosophila*. As in Figure 12.7, transcription factor genes are shown as green circles, and regulation by the transcription factor is shown as green lines. Genes encoding microRNAs are red circles with regulation by microRNA to transcription factor shown as red lines. Reproduced from the modENCODE site (modencode.com).

levels: some regulate the expression of transcription factors, others are primarily the targets of transcription factor regulation. Interestingly, transcription factors at the lowest level of the hierarchy regulate fewer other transcription factors but more miRNAs than do transcription factors at the higher levels. This suggests that feedback from the lower to the higher levels of the hierarchy occurs via miRNA regulation rather than through transcription. That is, as shown in Figure 12.9, *engrailed* (*en*)—a transcription factor at the bottom of the hierarchy—does not directly regulate the transcription of any transcription factors at the higher tiers, but regulates the expression of the miRNA gene *mir-958* and of eight other microRNAs, which in turn regulate the expression of genes at these higher levels.

Transcriptional regulation of miRNA might reflect timing of events

The evolutionary roles and strategy of microRNAs in transcriptional regulation are intriguing. (We recognize that a term such as "evolutionary strategy" borders on a teleological view of evolution and is used here only as a shorthand.) In some examples such as that of developmental timing in *C. elegans*, transcription of the microRNA genes is tightly regulated temporally. The microRNA gene is expressed at specific times, while transcription of the target mRNAs is less tightly regulated, and the mRNA already exists in the organism when transcription of the microRNA is initiated. Thus an existing mRNA is shut off through the temporal expression of the miRNA.

Many other microRNA and mRNA genes exhibit the opposite pattern of expression. In many cases the microRNA pre-exists, while the mRNA transcription is temporally regulated. That is, the microRNA is present when the transcription of the mRNA begins. It is intriguing to consider the consequences—or, more carefully expressed, the relative selective pressures—that might contribute to these different modes of regulation, as summarized in Figure 12.10.

One speculation is that this difference is related to the timing by which the process occurs on one hand, and to the tolerance for variation in expression on the other. The microRNA *lin-4* shuts down *lin-14* which regulates molting. Molting in *C. elegans* occurs quite rapidly, over the space of a few hours, so the signal to initiate molting must happen rapidly. The initial transcript of *lin-4* is 94-nucleotide long (a length similar to that of other microRNA genes), and it is processed to a siRNA of 23 nt. The pre-mRNA for its target *lin-14* has 1,939 nucleotides, which are then processed to an mRNA of about 1620 nt. Thus, assuming that other steps in transcription and RNA processing occur at similar rates, regulation of microRNA transcription is twenty times faster than regulation of its mRNA target would be, which can allow for the expression of a protein-coding gene to be quickly shut down, as shown in top half of Figure 12.10.

Conversely, as discussed in Chapters 1 and 2, we now know that a low level of transcription occurs throughout the genome at all times; that is, the genome is pervasively transcribed. The role of this low level of transcription is not yet known, but it is not difficult to imagine that it might be wasteful, if not deleterious, to have each of these transcripts present at a low level and potentially available for translation. Availability for translation might be set by microRNA regulation. The presence of microRNAs in a cell can effectively set a threshold for expression in order to reduce the effects of this background transcription, as suggested in the lower half of Figure 12.10. For a protein-coding gene to be expressed and carry out its function, the transcription of the gene must exceed the level set by its regulatory microRNAs. When only a few transcripts are present from a gene, the microRNAs block their translation or target them for degradation, so that the protein is not made. When the gene is highly transcribed, the number of transcripts is greater than the capacity of the microRNAs.

Although we have focused here on some of the broad organization of transcriptional networks in this discussion, these networks also reveal the transcriptional regulatory interactions for specific individual genes. While we have enormous amounts of interaction data already, these represent only a subset of the transcription factors, the microRNAs, and the biological conditions under which they operate. These are laying the groundwork for many additional experiments. For an individual investigator, it may not be necessary to isolate the components of the transcriptional regulatory system for a particular set of genes; the questions might focus instead on how this network of regulatory interactions produces the phenotypes we see.

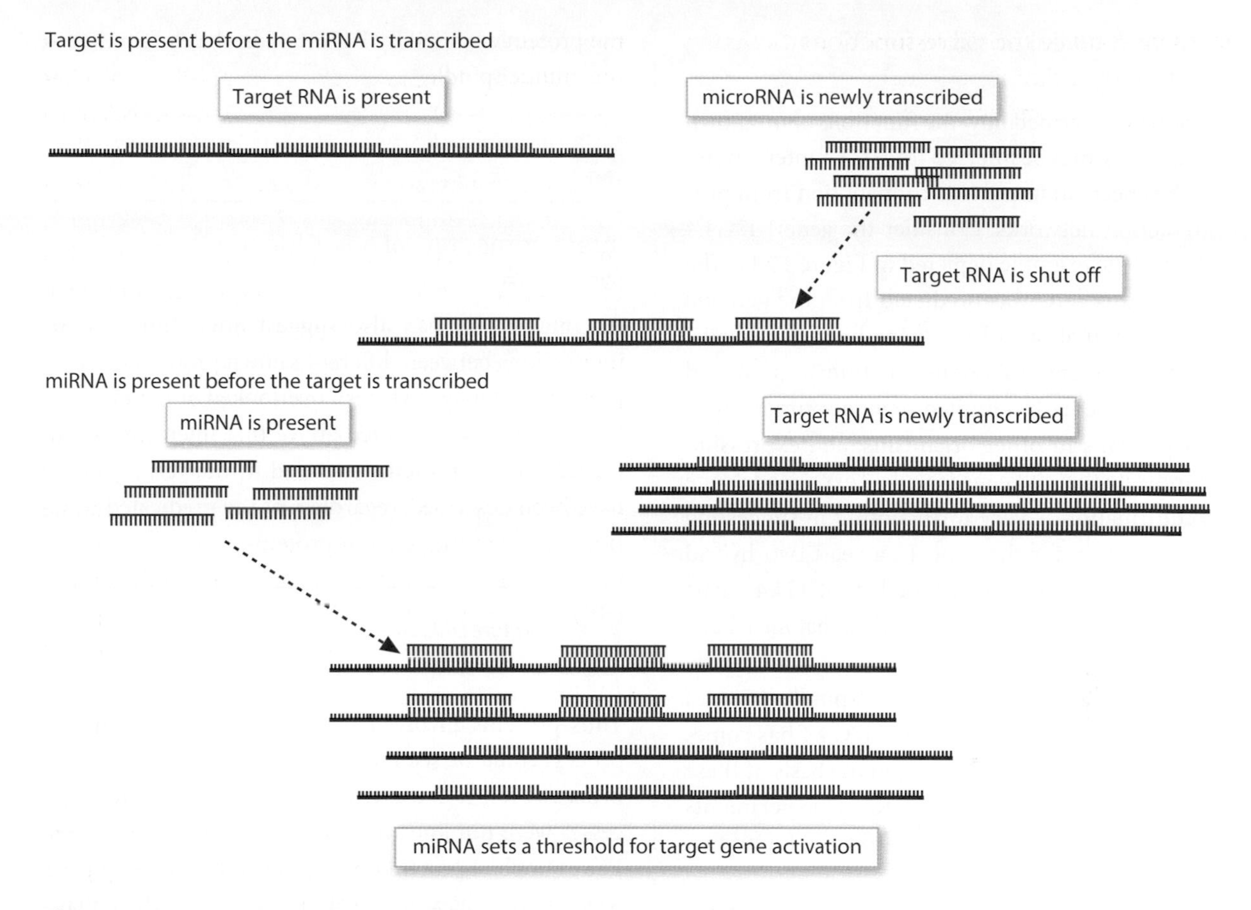

Figure 12.10 Two strategies of gene regulation involving microRNAs. Gene regulation involving microRNAs can occur by regulating the initiation of transcription of the microRNA or the initiation of transcription of the mRNA target. In some examples of regulation, the mRNA target is already present in the cell when transcription of the microRNA begins. The microRNA is newly transcribed to shut down the preexisting mRNA. This situation is seen in the regulation of developing stages in *C. elegans*, and may allow the organism to shut off a gene rapidly. In other examples, the microRNA is already present when the transcription of the target mRNA is initiated. This situation, which is common, may allow the organism to set a threshold for gene activation and thus to reduce the effects of "biological noise" in gene activation.

12.5 The interactions between proteins

Most, if not all, proteins in a cell carry out their functions in complexes with other proteins, so protein interaction networks have been extensively studied in model organisms. The complete set of protein–protein interactions that occur in an organism are referred to as an **interactome.** Protein interactions are studied on genome-wide scale using yeast two-hybrid (Y2H) screens, a technique described in Section 10.5. Interactions involving all the proteins encoded in the yeast genome will be discussed. In multicellular eukaryotes, a complete set of protein–protein interactions is not yet available, so the term "interactome" is also applied to the set of protein interactions detected within a particular cell type or under certain conditions; for example, in worms we can write about the "germline interactome."

As with other types of networks, the interactions between two proteins as detected by Y2H are plotted as a graph. From such a network it is possible to examine a single protein node and to catalog all its interacting proteins by following the links. We showed an example in Figure 12.3. This is one of the most generally useful results to arise from interactome experiments. However, analysis of other properties of these networks has revealed some novel characteristics of cells and organisms.

Interactomes reveal possible functions for unknown genes

We previously described how the functions of unknown genes can sometimes be inferred from their interactions, and this has been an important contribution from protein interaction networks. Consider the gene F47G4.4 in *C. elegans*, the example depicted in Figure 12.11. The existence of this gene was predicted from the genome sequence; no mutations of this gene had been identified, to allow inferences about its function; its predicted amino acid sequence has limited sequence similarity to those of proteins in other organisms; and expression studies indicate that the gene is widely expressed. None of this information is very forthcoming in suggesting a specific function for the gene. In a yeast two-hybrid screen, however, the protein encoded by F47G4.4 is part of an interaction network with proteins that *have* been found in worms and in other organisms, particularly proteins associated with the mitotic spindle. We can make a reasonable guess that gene F47G4.4 has something to do with spindle function. On the basis of these results, investigators performing RNAi experiments paid particular attention to cell division and found that the protein encoded by F47G4.4 also has a function in the mitotic spindle.

Literature Link. Srayko et al. 2000, *Genes & Development* 14: 1072–84

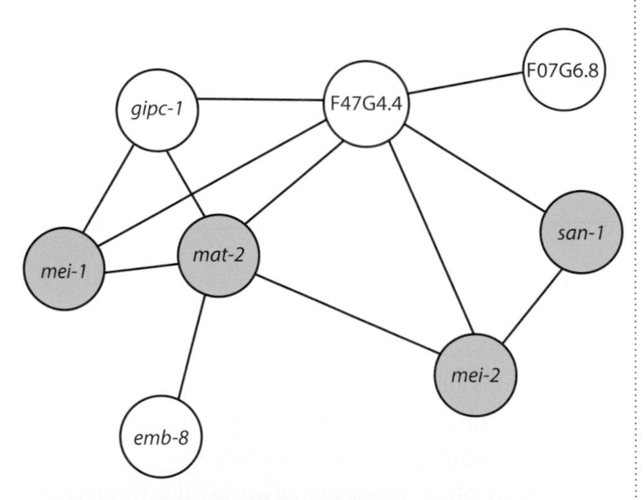

Figure 12.11 Inferring the function of a gene. The logic is similar to that of the general example in Figure 12.3, this example being drawn from experiments in *C. elegans*. The gene F47G4.4 from *C. elegans* is highlighted in red; its function was not known. It is part of the protein interaction network for the gene *mat-2*, whose interactions are shown. The proteins encoded by the genes shown in blue, including *mat-2*, are known to be involved in the function of the mitotic spindle. Since F47G4.4 interacts with spindle-associated genes, its role in spindle function was tested by RNAi and gene deletion experiments. The data are replotted from information on Wormbase. All the interactions for *mat-2* are shown, but only some of the interactions for the other genes and proteins are included.

Interactomes reveal novel functions for known genes

An interactome can also suggest novel functions or interactions between different known proteins or components that may have been overlooked or not detected by other assays. We noted above that interactions that include all the proteins encoded in the yeast genome have been described, regardless of the predicated functions of any of the genes or proteins.

Literature Link. Yu et al. 2008, *Science* 322: 104–10

Thus proteins involved in one cellular or molecular process could be seen to interact with those of another process even if these two cellular or molecular processes were not known to be connected to each other. For example, a protein involved in ribosome biogenesis was found to have physical interactions with proteins from the spliceosome. This suggested an unexpected connection between ribosome biogenesis and splicing. Furthermore, the same ribosomal biogenesis bait protein interacted with a protein involved in removing the 5′ cap from transcripts. Interestingly, the decapping protein also interacted with proteins from the spliceosome, which suggests that mRNA decapping, splicing, and ribosome biogenesis are physically and perhaps functionally linked to one another in a previously unknown manner.

While using associations like this is a very helpful method to suggest functions for proteins, some caveats should also be recognized. Most of these will be familiar to anyone who has friends in a social network, but it is worth being reminded of them. These cautions arise because the complexities of different kinds of associations are reduced to a single link, and all associations in the interactome are not in fact the same. Similarly, not all our friends on our social network have the same type of relationship with us. For example, some molecular interactions can be stable while others are quite transient, but the stability of the interaction is not distinguished in an interactome. The stability of the interaction does not necessarily indicate the importance of the association for function, however, and some physical interactions probably have little effect on the function of the gene. Again, many people whom I have known for a long time in my social network do not have much impact on my current behavior, whereas others, whom I have known for a much shorter time, have a significant impact. In addition, unless other information is also integrated into the network (as described below), the interactome depicts all the interactions as occurring at the same time, which is distinctly not the case biologically. This has an obvious parallel in social networks, which do not distinguish when and where the interaction was formed. In fact, while associations provide a helpful starting point for inferring functions, the functions typically need to be verified experimentally.

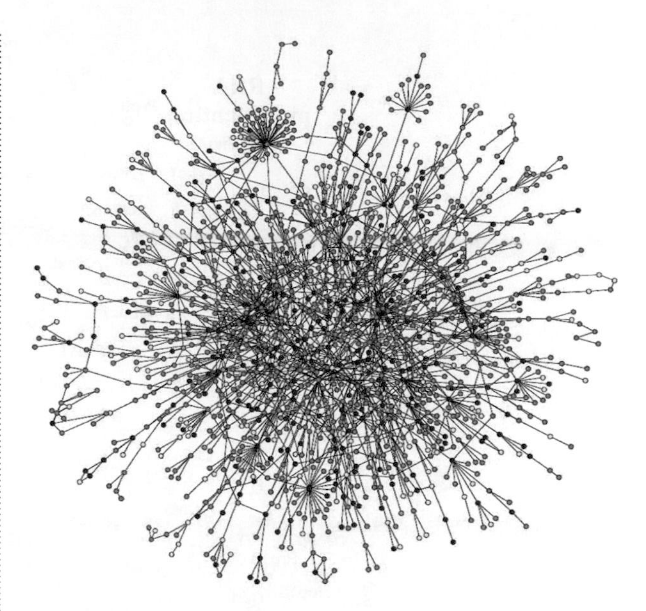

Figure 12.12 The yeast Interactome. The complete or nearly complete set of protein–protein interactions found among the 6,200 proteins in the yeast *Saccharomyces cerevisiae*. Note that some proteins interact with a large number of other proteins; these are known as hubs. Other proteins interact with only one or two other proteins. The median number of links is approximately 3.8 per gene, but this number does not capture the structure of the network.

The protein interactome is characterized by hubs

One of the most striking and visible features when protein interaction networks are graphed is that proteins vary widely in their number of interactions; in graph theory, this number of connections per node is referred to as **degree**, and the variance among different nodes is know as **degree distribution**. Without applying the statistical methods used to describe a degree distribution rigorously, we will use a more intuitive and observational approach, with the interactome among the complete set of yeast proteins in Figure 12.12. The median number of links per protein (or the average degree) is 3.8. Notice that the number of interactions per gene (that is, the degree distribution) varies widely: many proteins have more than ten interactions with other proteins, while some have twenty interactions or more. On the other hand, many proteins have only one or two interactions with other proteins. Clearly the mean number of connections is not very informative in describing the network graph. This observational conclusion is confirmed by more rigorous statistical tests that go beyond the scope of this book.

In other words, interactomes are characterized by a series of hubs; a **hub** can be defined informally, as a node with many connections, or formally, by statistical methods, as a node with many more connections than expected by chance. Airline flight patterns are also based on hubs, that is, airports with a particularly large number of connections, so the concept is probably familiar. What do these hubs tell us about biological interactions in the cell or in the organism? In some cases, as shown by the examples in Figure 12.13, the hubs correspond to known biological protein nanomachines, such as the polyadenylation complex or the proteasome. Other hubs correspond to other large protein complexes in the cell, such as the mitotic spindle or the secretory vesicle. In these cases, the interactome plot is re-creating what we already suspected from other types of analysis. It is reassuring that a novel approach such as network analysis is identifying proteins known to have many interactions. However, many other hubs do not correspond to protein complexes we knew about previously. In these cases the hub could be revealing some unsuspected property of the cellular organization or of its evolution.

Properties of networks with hubs

It is known from non-biological systems such as flight patterns that a network organized by hubs has important features that are relevant to biological systems. First and foremost, the deletion of a hub has profound consequences for the entire network. A knockout of a hub, whether by gene disruptions, RNAi, or mutation, is at

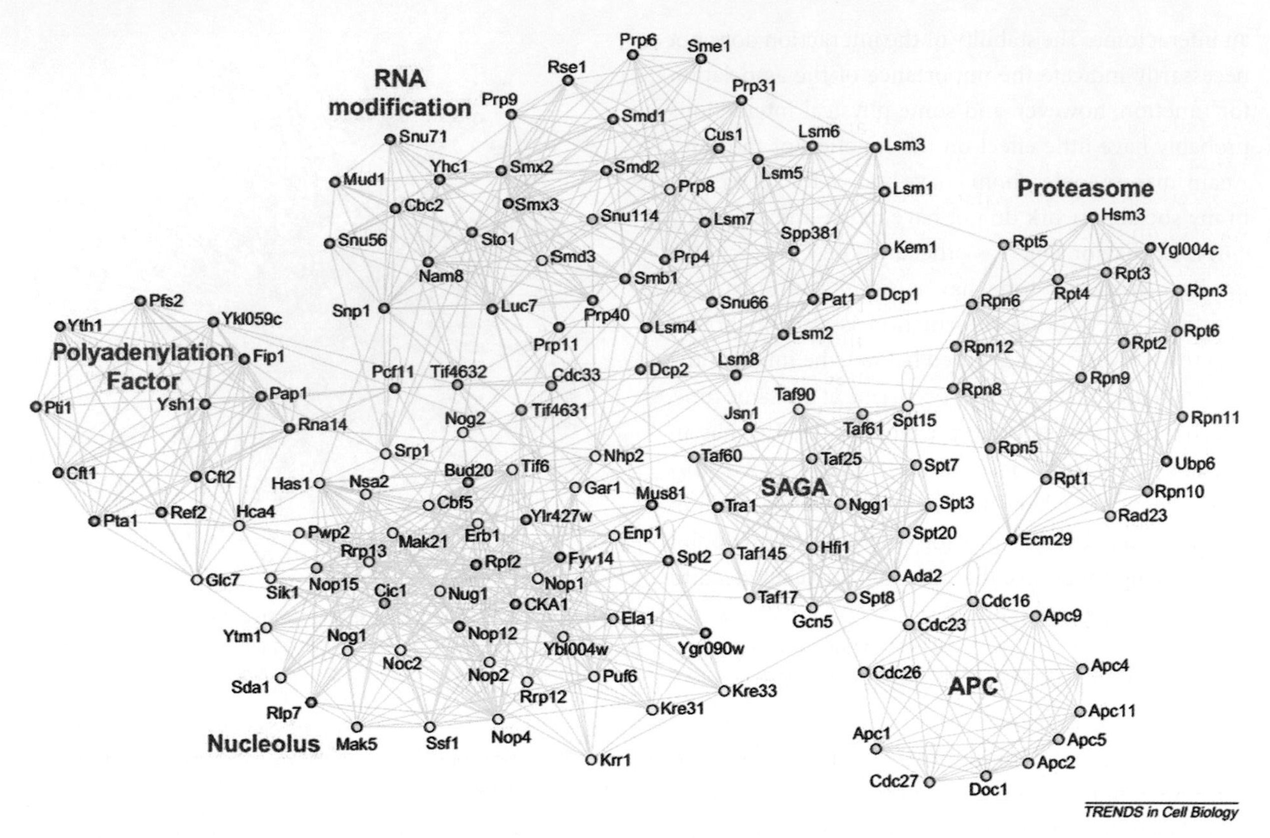

Figure 12.13 Some hubs from the yeast interactome. The yeast interactome shows that some proteins involved in common functions cluster together, having extensive interactions among themselves. Among the identifiable clusters are proteins involved in RNA modification, the proteasome for protein degradation, the anaphase promoting complex (APC), a histone acetyltransferase complex (SAGA), the nucleolus, and polyadenylation. Source: Figure 4 from Bader et al. 2003,Functional genomics and proteomics: charting a multidimensional map of the cell, *Trends in Cell Biology* 13: 344–56, used by permission.

least three times more likely to be lethal to the organism than knockouts of other genes. This property of hubs is referred to as **attack vulnerability**—targeted disruptions of a hub make the network or the system more likely to collapse.

There is a corresponding and opposite effect to attack vulnerability of hubs; and it, too, is characteristic of these types of networks. As can be seen in Figure 12.12 especially for proteins plotted nearly the periphery, many proteins have only one or two interactions. Deletion of such a node has very little effect on the structure and function of the network—or, in this case, on the growth of the yeast cell.

Another important property is that hub genes tend to correspond to more evolutionarily conserved genes, although there are numerous and important examples of non-hub nodes that are also highly conserved. This evolutionary conservation of hubs reveals another important feature of the interactome network: genes that are connected in one organism (such as yeast) are likely to interact in another organism (such as worms or humans). This increases both our understanding of the network in each organism and our insights into the organizational hierarchies of the cell.

The evolutionary conservation of a hub is partly a consequence of the number of its connections. Every protein interaction that is important for the survival of the cell imposes a constraint on selection. In other words, components in any complex system that interact with many other components are necessarily resistant to change. This becomes something of a tautology—hubs are defined by having many connections, a property that constrains their evolution. Hubs are more likely to be essential because they have many connections. As a result, hubs are more likely to be both essential and conserved.

Types of hubs can be distinguished by the times of their interactions

As noted previously and as is familiar from social networks, not all interactions with a hub occur simultaneously. Attempts have been made to use expression data, such as microarray information in yeast, in order to examine how interactions change over time and space. These attempts have distinguished several types of hubs, which have somewhat different properties. The original experiments referred to these hubs as "date hubs" and "party hubs," but we will call them "sequential hubs" and "simultaneous hubs." They are illustrated in Figure 12.14.

Literature Links. Han et al. 2004, *Nature* 430: 88–93; Chang et al. 2013, *Scientific Reports* 3: 1691

Sequential hubs (or date hubs) are ones in which the proteins are expressed at different times and often in different locations in the cell. Thus a sequential hub has one set of interactions first, then another. Simultaneous hubs (or party hubs) are ones in which the interacting proteins are expressed at the same time and place. These hubs often correspond to functional modules and protein complexes, including what is called nanomachines. Examples include proteins involved in yeast budding and cell polarity, protein folding, and DNA replication.

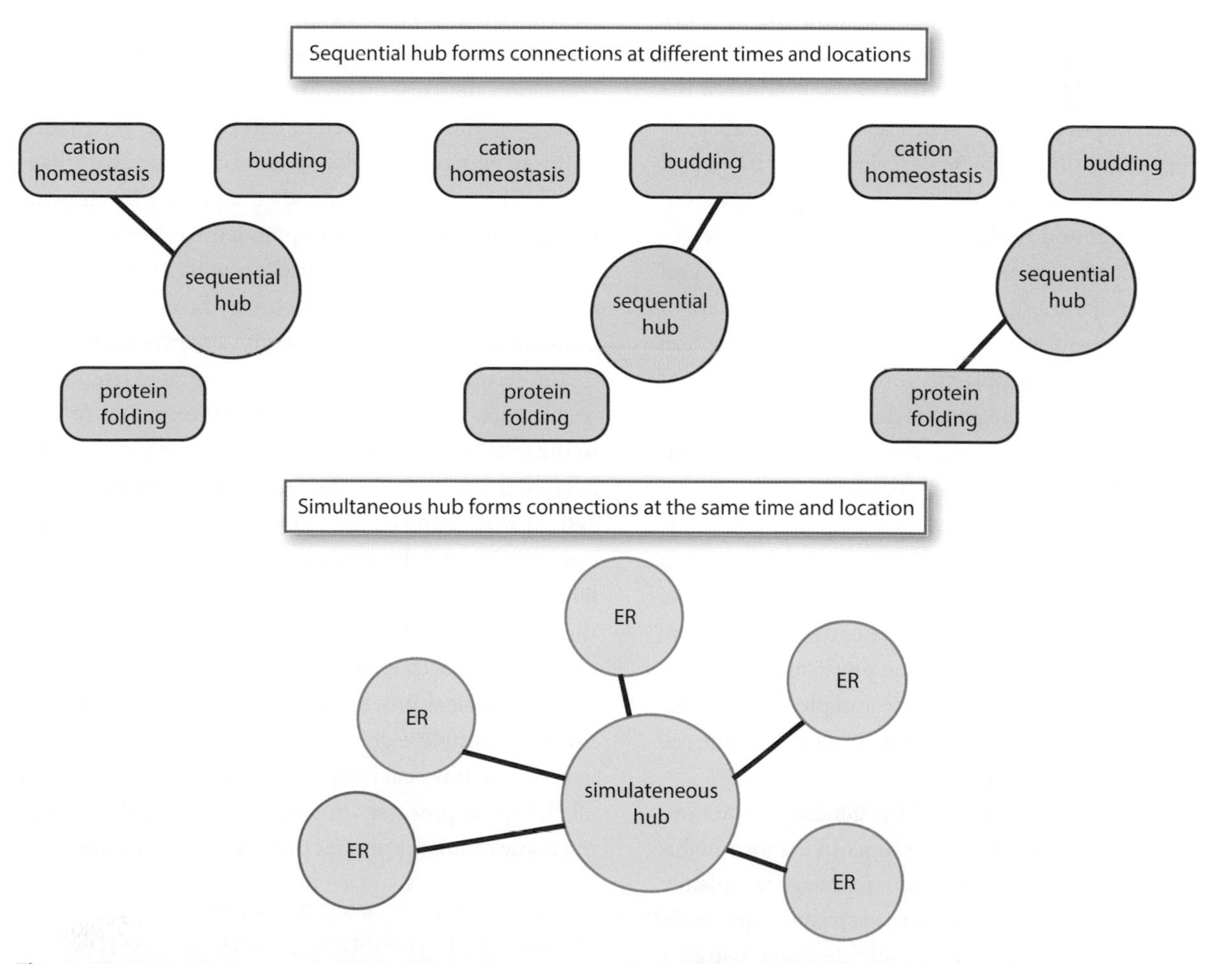

Figure 12.14 Simultaneous and sequential hubs. Using microarray data, different times of hubs can be distinguished. A sequential (or date) hub forms connections to one group of proteins, then to another group at a different time or location, and so on but does not interact with these proteins at the same time. A simultaneous (or party) hub forms all of its connections at the same time and location. These data are stylized from Han et al. 2004, Evidence for a dynamically organized modularity in the yeast protein-protein interaction network, *Nature* 430: 88–93, which shows Cmd1 as the sequential hub and Sec22 as the simultaneous hub to different modules of complexes associated with the endoplasmic reticulum.

Sequential hubs tend to make connections between simultaneous hubs. For example, calmodulin is an example of a sequential hub that connects a number of different simultaneous hubs at different times and locations, each of which involves calcium ions.

An intriguing result was observed when mutations that deleted one of the hubs were tested for genetic interactions. Sequential hubs were more than twice as likely to act as a synthetic enhancer with their interacting partners than were simultaneous hubs, while simultaneous hubs were no more likely than non-hubs to act as synthetic enhancers. This suggests, as implied in Chapter 10, that disrupting one simultaneous hub knocks out the module or pathway, so that further disruption of those partners has no additional effect. In contrast, disrupting a sequential hub tends to disrupt one pathway, and interactions are defining an alternative pathway involved in the same process.

12.6 Genetic interaction networks

Networks based on physical interactions, including transcription factor binding, microRNA and target mRNAs, and protein interactions, show the *presence* of an interaction without also showing the *importance* of the interaction. To a geneticist, a description of the functional importance of a gene includes its mutant phenotype, so that the complete description of a biological process needs to include the genetic interactions as well as the physical and molecular data. By genetic interactions we mean suppressors, synthetic enhancers, and epistasis. In one sense, these genetic interactions are of a different type altogether from the physical interaction data. As pointed out repeatedly during our discussion of genetic interactions, the genetic data do not necessarily imply a physical interaction, and physical interactions between the gene products are not necessary for observing a functionally important genetic interaction. Furthermore, while the presence of a physical interaction between two proteins is likely to indicate a functional relationship between them, the nature of that functional relationship is not revealed by the physical interactions along. The combination of genetic and physical interactions will comprise the assembly instructions for the cell or the process.

A complete analysis of all of the genetic interactions for every gene in the genome seems like a formidable task. It requires having a mutation in each gene—ideally, more than one type of mutation in each gene—and making and analysing double and multiple mutant strains. We will limit our ideal experiment to double mutant strains, since multiple mutants introduce a staggering number of additional interactions to consider and strains to construct.

Three examples will be used to illustrate how the analysis of genetic interactions on a wide scale is working so far. The case study for this chapter describes the integration of the protein interaction network and the genetic regulation network involving dauer larva formation in *C. elegans*. This is governed in part by a TGF-β signaling pathway, which was explored in the case study for Chapter 11. This case study extends the picture by showing how an integrated systems approach has been used for one well-defined signaling process with a limited number of components in a metazoan, so it connects the processes described in Chapter 11 with a network approach.

The second and third examples are drawn from studies in yeast. The second example in the chapter returns to the yeast secretory pathway, introduced in Chapter 3, Section 3.2 to describe the interactions that occur in a system with many components within a single eukaryotic cell. The cellular process is complicated and a fairly large number of genes are involved; on the other hand, the cellular process is well studied and the data can be readily summarized and tested for their applicability to other biological processes. The third example extends this to a complete genome-wide analysis of pair-wise genetic and physical interactions that represent most or all biological processes that occur in yeast cells, including many cellular processes that are not well understood.

The secretory pathway shows many basic principles of synthetic interactions

The two examples that we will describe for the yeast cell are built on the foundation of the targeted gene knockouts described in Chapter 9. Recall that targeted deletions for each of the yeast genes has shown us that

18.7 percent of genes are essential for growth under lab conditions. (Different laboratories using somewhat different versions of the knockout library have reported slightly different numbers. In their review in 2014, Giaever and Nislow report that 18.7 percent of the open reading frames are essential when tested in homozygous diploid strains, or 1,106 genes. We will use that number, although the fundamental point is the same.) Overall, the essential genes are more likely to be evolutionarily conserved than the non-essential genes. For example, 38 percent of the essential genes have a clear-cut human ortholog, whereas about 20 percent of the non-essential genes have human orthologs.

How can the prevalence of such non-essential genes be explained? We suggested some possible answers for this question in Section 9.8, but this time we will answer it more directly. One postulate is that some gene knockouts survive because the genome has a paralog with a duplicate function. The acetylcholinesterase genes in *C. elegans* were used in Chapter 10 to illustrate this possibility. While this certainly explains some of the yeast knockouts, it is not the explanation for most of the non-essential genes. Only about 8.5 percent of the non-essential genes have a duplicate copy predicted to be carrying out the same function elsewhere in the genome; double mutants knocking out both paralogs are frequently lethal. (Not surprisingly, less than 1 percent of the essential genes have an obvious paralog.) Thus nearly 75 percent of the genes in the yeast genome have functions that apparently are not needed for growth and division in the lab as determined by analysing single mutant phenotypes. In the context of this chapter, these mutations correspond to deleting random nodes, which have little or no effect on the overall structure and function of the network.

But, as we noted in Chapter 10 with the multivulva pathway in *C. elegans*, there may be duplicate functions without apparent sequence or biochemical similarity between the corresponding gene products. The genetic approach to studying functional redundancy is to use synthetic mutations, that is, to compare the phenotype of a double mutant with the phenotypes of each of the two single mutants. Our particular interest here is not that synthetic interactions will occur, but to consider which pairs of genes produce a genetic interaction and what that reveals about the underlying functional organization of the cell.

An extensive analysis of synthetic interactions has been performed with the secretory pathway in yeast, introduced in Chapter 3 and shown in Figure 12.15.

Literature Link. Tong et al. 2004, *Science* 303: 808–13; Schuldiner et al. 2005, *Cell* 123: 507–19

The secretory pathway is genetically complicated, involving 173 genes (in these papers) that act in ten distinguishable steps. Three of these ten steps are translocation of the secretory vesicle to the Golgi apparatus, vesicle budding, and fusion with the plasma membrane.

Many double mutants have been created that involve two genes with defects in some aspect of the secretory pathway, and a few general conclusions about synthetic enhancement emerged from this analysis. Synthetic lethal interactions are commonly found among genes that affect the same step in the pathway and are sometimes observed between genes that affect different steps. Synthetic lethal interactions are occasionally found between a secretory gene and a gene not involved in secretion, but these are less frequent than interactions between secretory genes. For example, thirty genes are involved in fusion of the secretory vesicle with the plasma membrane. There are forty-three synthetic lethal interactions among these genes, and thirteen additional synthetic lethal interactions with genes involved in other steps in the secretory process. Thus, 63 percent (forty-three out of sixty-eight) of the synthetic lethal interactions observed for these genes are found among genes that affect the same step, and an additional 19 percent (thirteen out of sixty-eight) of the synthetic lethal interactions involved genes in the same pathway but not in the same step. There were also twelve synthetic lethal interactions with genes not involved in secretion, some of which could reflect previously unknown relationships between processes with the cell.

These results with the yeast secretory pathway appear to be typical when expanded to other double mutant combinations and other cellular processes. They suggest that most, but certainly not all, synthetic lethal interactions will be observed among genes that have similar phenotypes or affect similar functions—either genes encoding proteins in the same complex or genes encoding proteins involved in other functionally related processes.

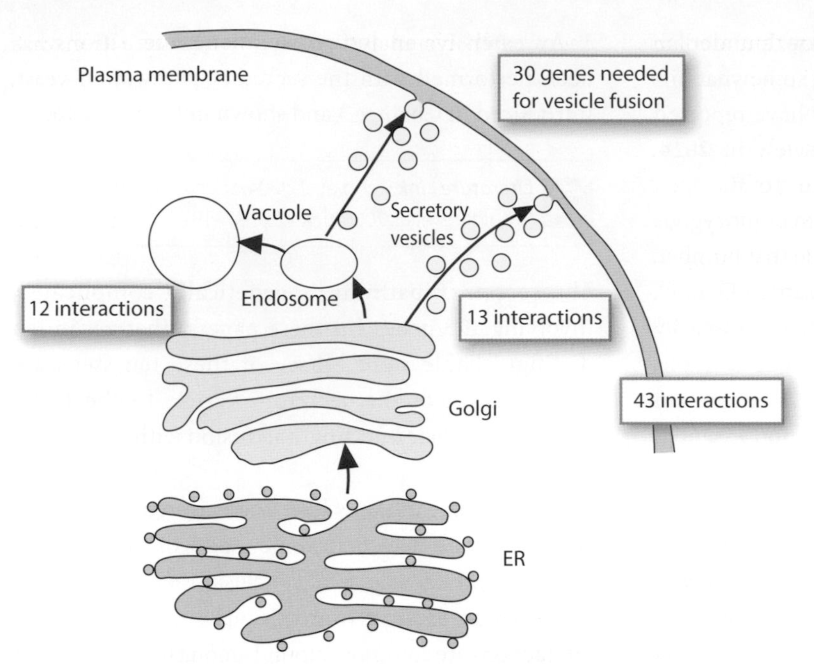

Figure 12.15 Genetic interactions in the yeast secretory pathway. Genetic analysis of the protein secretory pathway in yeast was introduced in Chapter 3. This figure summarizes an analysis of the synthetic interactions involving double mutants with thirty genes involved in the fusion of the secretory vesicles with the plasma membrane. The number of genetic interactions involving these thirty genes is highlighted for different steps in the secretory pathway. Most of the synthetic interactions (forty three of them) are among genes involved in membrane fusion, and many of the others (thirteen) are with genes involved in other steps with the secretory vesicles. Only twelve interactions were with genes not directly involved with protein secretion.

A nearly completely and unbiased set of gene interactions reveals the genetic landscape of the yeast cell

This network of genetic and physical interactions has been extended beyond a single biological process in yeast by examining pairwise genetic and physical interactions on a genome-wide scale. The work by Costanzo et al. is an ambitious effort to survey the entire set of functional interactions in yeast—what they termed "the genetic landscape of the cell" or "the complete cellular wiring diagram," and what we have considered to be the assembly instructions. It is likely that, for the foreseeable future, all other efforts to examine genetic interactions in eukaryotes will be compared to this seminal work, since no multicellular organism can yet be analysed in this way. While we will describe this work in some detail, the papers contain much more information than we can distill, and some of the most important lessons to be learned may turn out to be ones that we are currently overlooking.

Literature Link. Costanzo et al. 2010, *Science* 327: 425–31; 2016, *Science* 353: 1381; http://thecellmap.org

Defining the interactions

The project constructed and analysed the phenotypes of double mutants using the yeast targeted deletions for both essential and non-essential genes, as well as temperature-sensitive hypomorphic mutants for many essential genes grown at semi-permissive temperatures, so that a weak mutant phenotype was observed. The phenotype used was fitness or cell division and growth, which could be assayed quantitatively. Networks based on different combinations of mutants were constructed and had somewhat different outcomes—deletions of essential genes paired with other, deletions of non-essential genes paired with each other, and so on. The mutants and the processes they affected were not biased toward any particular biological function. The fitness was scored, the network diagrams graphed, after which the biological functions for the genes superimposed on the graph. Thus the overall lessons of the analysis are expected to be generally applicable.

The procedure used a method called **synthetic genetic array (SGA)**, which is shown in Figure 12.16. A strain of one mating type with targeted gene disruption or other mutation (referred to as the query gene) is crossed with the library of mutant strains of the opposite mating type, and the double mutant cells are individually plated and grown in an array. The growth (using colony size) of each of the single mutants and of the double mutant was plotted and measured quantitatively. If the two single mutants did not interact genetically, the growth of the double mutant should be the multiplicative product of the growth of the individual mutants; in other words, it should exhibit the combined phenotypes of both genes. Double mutants that grew worse (as determined statistically) than predicted on the

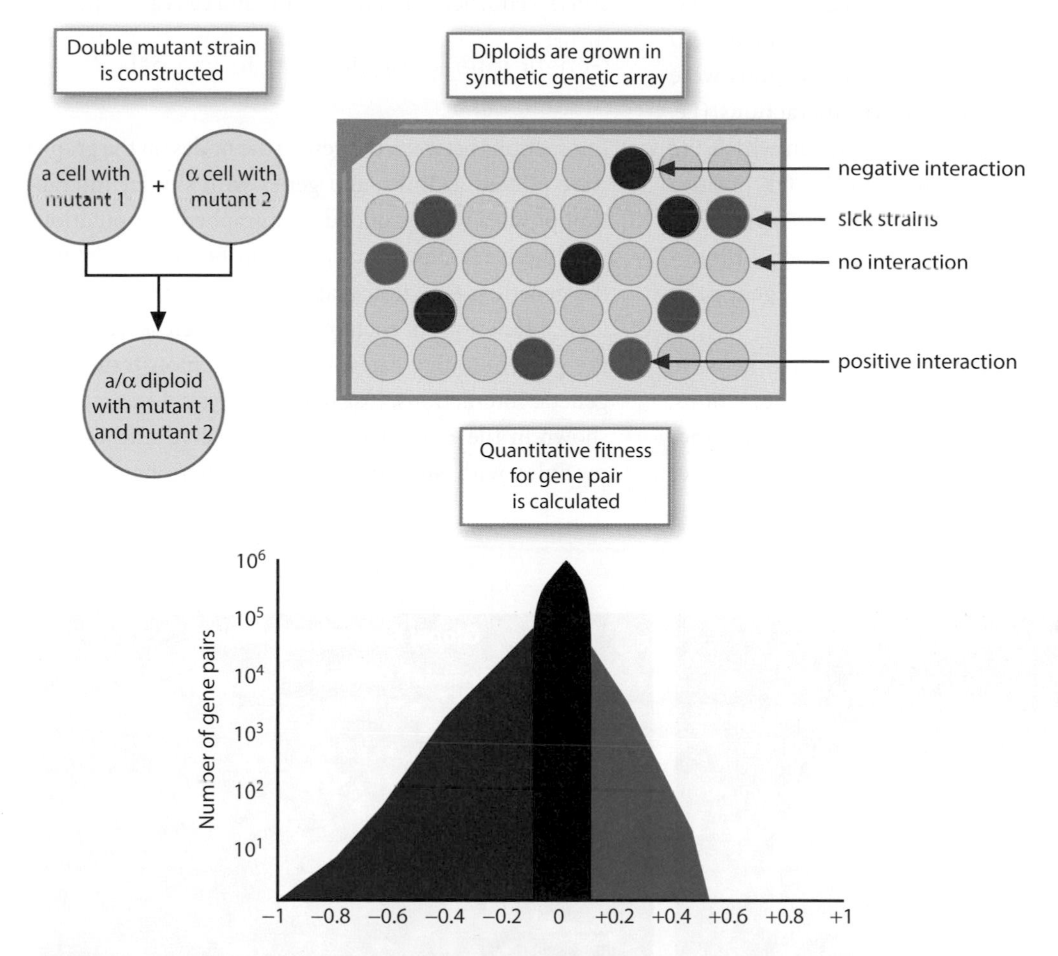

Figure 12.16 The synthetic genetic array to identify genetic interactions. Yeast haploid cells with opposite mating types, each carrying a mutation in a defined gene, were mated to produce a diploid cell with both mutations. The resultant cells were grown individually in the wells of a microtiter dish, and the amount of growth was measured and normalized for overall colony growth. The relative growths of each individual mutant and of the double mutant were calculated, and the genetic interaction score was calculated as the log ratio of double mutant growth to the product of the growth of each single mutant. Thus, if the double mutant grew as well as expected from the growth of the two single mutants, the genetic interaction score is 0. Positive scores indicate that the double mutant grew better than expected (a suppressor, marked in green), while negative scores indicate that the double mutant grew worse than expected (a synthetic enhancer, marked in red). The results depicted in the microtiter dish are slightly exaggerated, since only about 3.5 percent of double mutants showed a genetic interaction; however, as depicted, negative interactions are more frequent overall than positive interactions. The genetic interaction scores are plotted on a log graph. The data are redrawn from the supplemental figures in Costanzo, M. et al. 2010, The genetic landscape of a cell, *Science* 327: 425–31, which should be consulted for more precise information.

basis of the two single mutants were defined as having a **negative interaction**; we referred to these in Chapter 10 as synthetic enhancers or synthetic lethals. Double mutants that grew better than predicted were defined as having a **positive interaction**; in Chapter 10, we referred to these as suppressers.

More than 23 million double mutant pairs have been analysed for their fitness; the analysis has involved 5,416 genes or slightly more than 90 percent of the protein coding genes in yeast. These represented many biological processes: DNA replication and repair, RNA processing, metabolism, chromatin and transcription, and so on. Approximately 900,000 genetic interactions were detected, 550,000 of them negative interactions (i.e. synthetic enhancement or combinations in which the strain grew less well than expected) and 350,000 positive interactions (i.e. suppressors or combinations in which the strain grew better than expected). Thus genetic interactions are common, and about 3.5 percent of gene pairs showed a genetic interaction.

The "average" gene had about one hundred negative interactions and about sixty-five positive interactions, but there was a wide range in the number of interactions per gene, and a hub structure was apparent. Genes recognized as hubs had about 3.5 times as many interactions as the median number; similarly, essential genes had about five more interactions than non-essential genes. A large majority of the genes with the fewest interactions were also non-essential. The essential gene hubs provide the overall structure for the functional relationships within a cell. Because these genes are essential and their many interactions constrain how much variability might be acceptable, these hubs and the interactions among them are likely to be the ones most evolutionarily conserved when other organisms and cells are explored.

Genetic interactions indicate functional relationships

Interactions between the genes (the nodes on the graph) were plotted as links, and genes with similar interaction profiles were grouped and graphed computationally; these are shown on the left in Figure 12.17. Once the genetic interaction graphs were compiled, the biological functions of each of the genes were added using the gene ontology categories. This demonstrated that genetic interactions, just like the physical interactions shown in the Y2H interactome, clustered into distinct modules with similar or related functions; some of

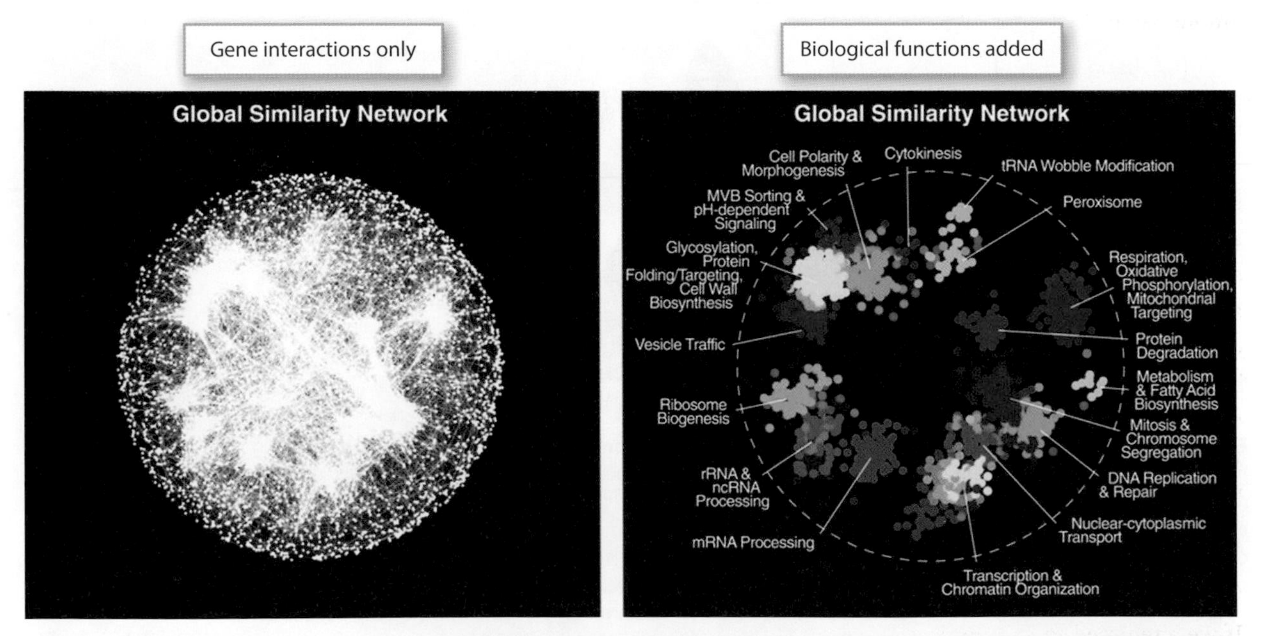

Figure 12.17 Genetic interactions cluster into functional modules. Costanzo et al. 2016, examining more than 23 million pairs of yeast mutants for genetic interactions. The interacting pairs were plotted on the basis of the similarity of their interactions, without prior regard to biological function. Distinct clusters or modules of interacting genes were found, as seen on the left. The functions of genes in each module were inferred from the Gene Ontology (GO) database and superimposed on the genetic interactions. The results, with the biological functions added, are shown on the right. Note that related biological functions clustered near each other. Source: Figure 1 from Costanzo et al. 2016, A global genetic interaction network maps a wiring diagram of cellular function, *Science* 353: aaf1420 reproduced with permission.

their data are shown in the graph on the right in Figure 12.17. An analysis of the clusters reveals that genes belonging to the similar biological processes also share many genetic interactions. Because the results appear to confirm our intuition about how genes affecting the same function should behave, it is important to stress how these clusters were found. Only the genetic interactions and not the functions were used to create a cluster. For example, the genes encoding proteins involved in DNA repair and replication have many genetic interactions among themselves and share many interactions with other genes; thus genes cluster together and, once the biological functions are added, this was recognized as the DNA replication–repair cluster. In addition, the DNA replication–repair cluster lies close to (that is, shares interactions with) clusters involved in mitosis and chromosome segregation. Again, this supports our intuition and the information we had previously about the processes of DNA replication–repair and mitosis. On the other hand, there are few (if any) genetic interactions between these genes and those involved in process we expect to be independent, such as glycosylation or cell wall biosynthesis. This clustering fits and greatly expands the genetic interactions seen with the secretory pathway, namely that most genetic interactions occur among genes that affect the same function or closely related functions.

Some genes did not fit into a particular cluster, or shared genetic interactions with more than a single cluster. Among these genes were ones encoding proteins such as HS90, a negative regulator of the RAS pathway, an E3 ubiquitin ligase, and some others. From the analysis of the phenotypes of single mutants, all these genes are known to be highly pleiotropic and affect more than one process. (Pleiotropy is discussed in Section 6.2.) Thus, although it may not have been a particular focus on the study, the genetic interaction clusters also provided a means to detect genes with pleiotropic effects on the basis of their genetic interaction profile.

Connecting genome-wide results to previous conclusions

If we attempted to summarize the data fully and capture all their subtleties, we would certainly miss some point that may turn out to be significant. In addition, for investigators whose primary interest is a particular gene or a particular biological process, there may not be much interest in knowing, for example, that DNA repair genes don't show many genetic interactions with cell wall biosynthesis genes. Thus we will draw out a few observations and show how they appear to be connected to topics that we considered in Chapters 10 and 11 when we discussed interactions with specific genes, or interactions governing a particular process. To what extent are the conclusions we drew from results with these specific cases supported (or contradicted) when the more extensive wiring diagram of the yeast cell is considered?

Most significantly, we note that genetic interactions, and particularly negative interactions, correspond very closely to the functional relationships and even to the subcellular locations in which the proteins are found. As we noted with examples from the protein interaction networks, these functional relationships may not have been recognized previously. Our premise for using genetic interaction analysis at the beginning of Chapter 10 was that this approach could identify functionally related genes, and this is clearly true. The genetic interaction network is highly organized and strongly predictive for functional relatedness.

This is particularly evident with the negative interactions or synthetic enhancement, the less studied of the principal types of genetic interactions we considered. The essential genes and the non-essential genes (that is, those that, on their own, showed a mutant phenotype and those that did not) were a bit different in this regard, but consistent with our earlier examples. Synthetic enhancement among the essential genes—that is, those mutations that made a mutant phenotype more severe—were preferentially among genes whose products function in the same complex. In effect, the mutation in the original gene impaired the function of the complex, while the second (negative interaction gene) knocked its function out altogether. This is similar to what was observed with *tub1* and *tub2* mutations, whose proteins are both components of tubulin, as described in the case study for Chapter 10. Synthetic enhancement among the non-essential genes—that is, among those that had only a mutant phenotype in combination—tended to be seen as enhancement among genes in the same biological pathway or process but not in same complex. This is similar to what was observed with synthetic enhancement between *lin-8* and *lin-9* and among other genes in those pathways that affect vulval development in *C. elegans*, described in Section 10.3. Genetic interactions among genes that affect the yeast secretory pathway were also consistent with these genome-wide results; the majority of interactions occurred between genes involved at the same step in the pathway (i.e. between some of the same protein complexes) or between genes affecting the same pathway but different steps.

The positive interactions (i.e. the suppressors) did not correlate as strongly as the negative interactions genes involved in the same complex, or even in the same biological process, although a significant relation to function was still observed; this was particularly true for combinations of a non-essential gene (the suppressor in our terminology) with an essential gene (the original mutation). Often positive interactions, especially those between essential genes, occurred between genes in different processes. There is no obvious explanation for this, although we noted in Chapter 10 that suppressors tended to be a diverse group, and even presented a flow chart for sorting them out. The phenomenon may indicate an unknown regulatory relationship between the genes or the processes, but this remains to be examined.

Genetic landscapes

The genetic landscape of the yeast cell is an extraordinarily detailed and nuanced picture of biology; in fact, even a casual reading of these papers is guaranteed to make a well-informed molecular or cellular biologist go online to search for information about some unfamiliar processes. A parallel but less extensive set of experiments has been done with the fission yeast *S. pombe*, which is evolutionarily diverged from *S. cerevisiae*, so it will be possible to test some hypotheses about the evolution of genetic interaction networks.

Literature Link. Frost et al. 2010, *Cell* 149: 1339–52

It is difficult, if not impossible, to imagine that such a comprehensive analysis will be undertaken for any other organism. The virtues of yeast for genetic analysis have been thoroughly on display throughout this book; most significantly, no other eukaryotic organism has a complete set of deletions for every gene, and this forms the basis for the project. The lessons about gene interactions learned from yeast have yet to be applied fully to gene interactions in other organisms, but they also provide us with an intellectual framework for doing so. Most fundamentally, the work of Costanzo et al. has demonstrated on a broad, genome-wide scale that the inferences about genetic interactions drawn from specific genes and specific biological processes have substantial applicability to other genes and processes.

What have we learned so far about connecting genotypes with phenotypes in the yeast cell? We summarize the results from yeast in Figure 12.18. While these experiments focus on a single phenotype—fitness or cell division and growth—the phenotype is broad and encompasses many biological processes and other specific phenotypes. In fact, as demonstrated by Costanzo et al., clustering genes on the basis of their interactions, using fitness as a quantitative phenotype, reveals functional relatedness about many other phenotypes that were not directly being examined. So perhaps the results here are general; they are certainly the most extensive that we have so far. Slightly less than 10 percent of the genes, about 600 of them, have not been tested because mutations are not represented in the current collections and they have proved refractory to analysis. About 19 percent of yeast genes, slightly more than 1,100 genes, have a detectable effect on the phenotype in single gene mutants in homozygous diploids. As we noted in

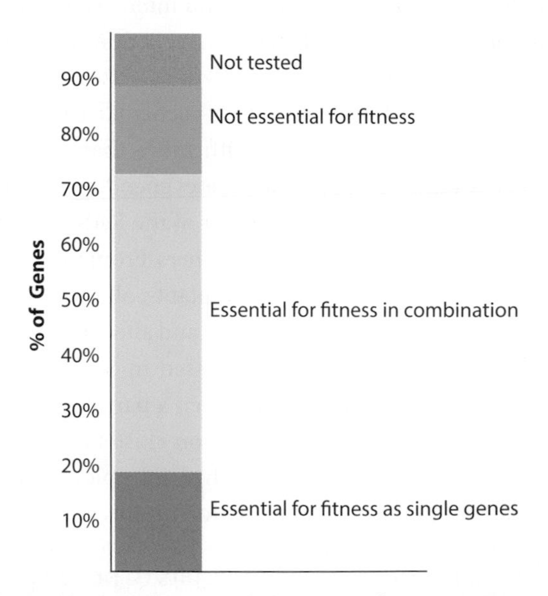

Figure 12.18 A summary of phenotypes from yeast mutants. As determined in many different screens with the yeast knockout collection, about 19 percent of yeast strains are essential in homozygous diploids. However, more than 50 percent of genes are essential for fitness in combination with another gene, although in many cases neither type of gene was essential on its own. About 15 percent of the genes (1,000 in total) are not essential for fitness under laboratory conditions, and about 10 percent could not be tested.

Chapters 6 and 9, these are the genes that geneticists have studied in detail for decades and that make up most of the examples for genetics students. From the genetic landscape analysis, another 3,300 genes, or more than 50 percent of yeast genes, have a mutant phenotype in combination with other genes but do not have a phenotype in single mutants. About 10,000 genetic interactions, or slightly more than 1 percent of the total interactions observed, were synthetic lethal interactions between genes with no mutant phenotypes. If these numbers are typical of other organisms such as humans, then most of our genes only have a mutant phenotype (for instance, result in a disease syndrome) when combined with variation in another gene, which also may not result in a mutant phenotype on its own.

But that leaves about 1,000 genes in yeast that do not have mutant phenotypes on their own or in pairwise combinations so far. These genes are sparsely connected in the genetic interaction networks, so perhaps we can think of them as "genetic hermits." They are not essential on their own, at least using the types of alleles that have been studied in yeast. Perhaps a dominant allele or a different type of genetic variant would result in a phenotype, either alone or in combination with another gene. Moreover, as has been noted many times, only one phenotype—fitness under laboratory conditions—has been measured. Perhaps under different conditions, such as in the natural environment rather than in the lab, some of these genes have essential functions, either alone or in combination. In support of this perspective, six genes that were loosely connected, but not to each other, are needed for growth under high atmospheric pressure or low temperatures. Different growth conditions or phenotypic assays could reveal functions for some of these genetic hermits. Furthermore, the genetic interactions have been limited so far to pairs of genes, so it is possible that some have functions in triplets or high numbers of combinations. Of course, it may also be the case that the genes have no function in present *S. cerevisiae* strains but have functions in natural isolates or in ancestors of the current strain. Genomic analysis of additional natural isolates, which is underway, might reveal whether the concept of the metagenome, common in bacterial genetics, is also relevant to yeast genetics. In our analogy with assembling a desk, perhaps the box simply included some leftover parts.

Literature Link. Goldman and Landweber 2016, *PloS Genetics* 12: e1006181

As impressive and informative as these results are, even yeast has its limits. It is, after all, a single-celled organism. To consider a genetic landscape for worms, flies, or for *Arabidopsis* (let alone a mammal), one would need to consider all of the many cell types. While one might dream of compiling a genetic interaction map for an *Arabidopsis* root cell, that may not be very helpful, in a genetic landscape, for understanding flowering. In addition, even for yeast, we don't know how the genetic interactions change as the environment changes. We know that they must change, since environmental factors such as temperature change alter gene expression patterns and the fitness (i.e. a phenotype) of the cell. But those types of changes have not been probed yet. We think we have the assembling instructions, or at least a significant portion of them, but we still can't quite construct the complete phenotype.

12.7 Gene expression, biological noise, and phenotypes

Furthermore, in Section 6.4 we introduced another contribution from the genotype to the phenotype that might need to be considered. Our discussion of gene activity and the pathway diagrams in Chapters 10 and 11 depicted genes as being ON or OFF. We know that this is not an accurate picture of gene expression, since genes can be expressed (i.e. transcribed and translated) at many different levels, once they are considered ON. To add to the complexity of connecting gene expression to phenotypes, a gene might be considered to be OFF in terms of gene activity but still being transcribed. Like others before us, we have routinely depicted genes as binary switches in contributing to a phenotype, but a more biologically accurate picture would show them more like dimmer switches, with many different levels of gene activity between fully OFF and fully ON. How do these small differences in gene expression levels contribute to changes in the phenotype?

There are definitely individual genes and biological processes in which small changes in gene expression levels result in noticeable changes in the phenotype. In Section 6.4 we cited examples from the chromosomal signal that governs *Drosophila* sex determination, trisomic syndromes, and some human diseases and syndromes in which twofold or smaller differences in gene dosage and gene expression levels affect phenotypes in significant ways. We also noted that there are many counterexamples in which even large changes in gene expression levels seem not to make much difference to the phenotypes of cells or organisms. In these counterexamples, the variation in gene expression levels may be thought of as biological noise, which we defined in Section 6.4 as the naturally occurring fluctuation in gene expression levels. While we can now assay for and detect very small changes in transcript levels even in individual cells, the effects of these small differences on recognizable phenotypes is still uncertain. Not much is known about how small changes in gene expression levels can lead to noticeable effects on the phenotype, at least once we try to draw general conclusions from the individual genes and processes that have been studied.

Literature Link. Newman et al. 2006, *Nature* 441: 840–6; Cao et al. 2017, *Science* 357: 661–7; Karaiskos et al. 2017, *Science* 358: 194–9

Perhaps that lack of a general conclusion is itself the best way to think about the phenotypic effects of small changes in gene expression levels. We know that some genes and processes are very sensitive, and we know that other genes and processes are relatively insensitive to such differences. The effects of such differences may very well depend on the individual gene and on its role in the broader biological process. We can now detect small changes in transcript levels, and we need to be alert to the possibility that these small changes will affect phenotypes; however, we should not assume that all of them will have more general effects on the phenotype. To put it simply, even if this is somewhat of a cliché, some of these differences in gene expression levels are likely to be signals for phenotype differences while others will be the noise. For any given gene or process, the signal and the noise may be hard to tell apart.

The origins of biological noise

While most investigators recognize the existence of biological noise, we may not think much about where it comes from. Broadly speaking, random fluctuations in transcript levels arise because every step in transcription, from pre-initiation to mRNA transport loading onto the ribosome, has some amount of imprecision. These steps are the outcome of millennia of evolution, so they must occur precisely enough for organisms to survive and reproduce successfully. On the other hand, the steps are carried out by the activities of macromolecules, all of which have some inherent error rate. We can summarize some of the known or likely origins of biological noise in Figure 12.19.

Literature Link. Munsky et al. 2012, *Science* 336: 183–7

One source of biological noise is the fact that transcript accumulation does not occur linearly at a continuous and constant rate in either bacteria or eukaryotes. Instead, transcript accumulation happens in bursts or pulses. These bursts could be the result of assembly and disassembly of RNA polymerase on promoter regions, changes between different RNA polymerases during transcription, pauses that occur during elongation, or other sources. We can return to our highway analogy to illustrate the effect of having pulses. Cars along the highway do not move at a constant rate because there are stops, such as a traffic light or a sudden braking by another car. Once the stop is removed, the cars move together in a pulse or a pack and accumulate in certain places. Thus, if one were to count the cars as they pass a certain point, the count would not increase at a uniform rate, as would happen if the cars were evenly distributed along the highway.

Another likely source of fluctuations in the levels of transcript accumulation is the fact that mRNAs have both different lengths and different half-lives from each other. Furthermore, the half-lives of mRNAs are generally shorter than the half-lives of proteins; but, for any given mRNA and its polypeptide product, these may be related in different ways. Our experiments average mRNA accumulation over time, but this will be different for different transcripts. So time-averaging their accumulation creates apparent fluctuations in transcript levels. Related to both these sources of biological noise in

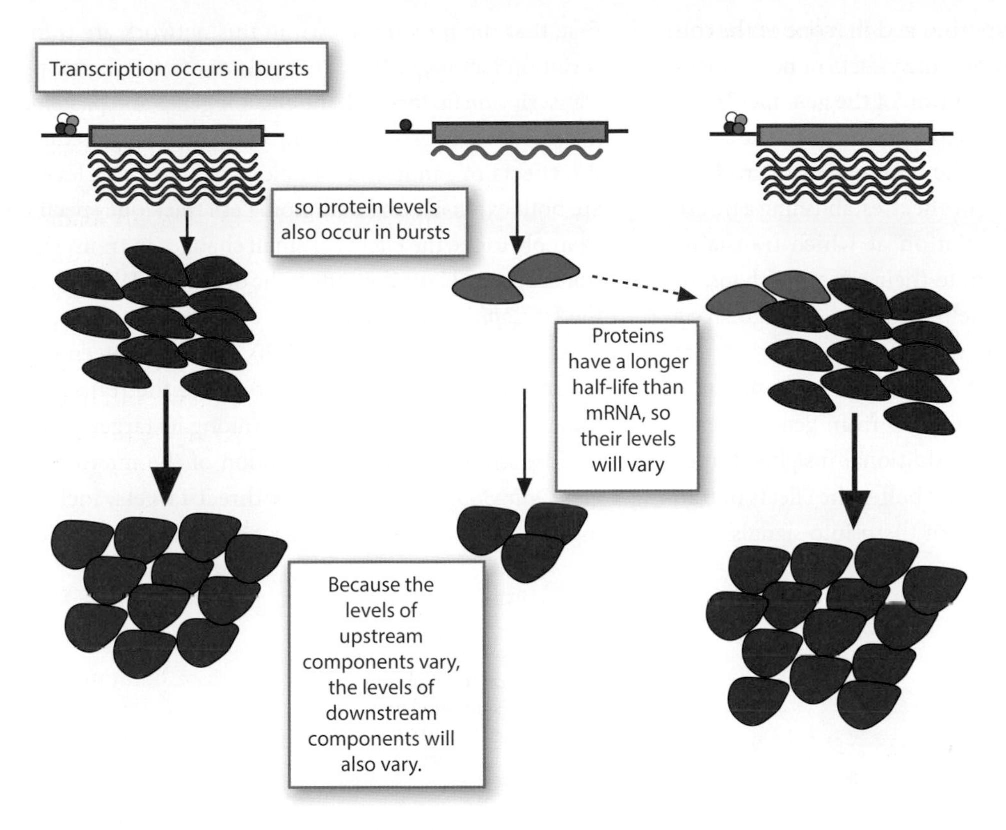

Figure 12.19 The origins of biological noise. Three different origins of biological noise, or small fluctuations in gene expression, are shown; each of them is related to the others. First, transcription is known to occur in bursts, as shown across the top of the figure. Bursts in transcription result in fluctuations in protein levels, as summarized in the second line. Second, because the half-lives of mRNA and proteins are not the same, proteins made in a previous burst cycle can persist into a later burst cycle. This is represented by the different colors of the protein, and results in fluctuations in protein levels. Third, bursts in RNA and protein expression at early steps in a pathway result in bursts in expression at later steps in the pathways, by signal propagation at the lowest line of the figure.

transcription is the fact that all the steps are connected to one another and often dependent on the conclusion of a previous step. Thus, small changes in any part of the transcription pathway or network could be propagated to other parts of the pathway or network, which themselves have variable rates.

There is also at least one more aspect of the small changes that occur in gene expression levels on phenotypes. We tend to use "gene expression levels" to refer to transcription levels, which are easier to study. We can use very sensitive single-cell transcript assays to quantify RNA levels. For those genes and processes in which a non-coding RNA is the active macromolecule, the RNA level reflects the level of gene expression. For those genes in which a polypeptide is the active macromolecule, we may not be able to assume that the RNA level reflects the level of gene expression since translation rates (as well as polypeptide length and half-life) appear also to vary among mRNAs, as discussed in Section 6.4. In whatever way it is analysed, gene expression involves a series of noisy processes, with fluctuations arising from many different sources.

The phenotypic consequences of small changes in gene expression

From carefully controlled quantitative expression studies in some model organisms (both bacteria and eukaryotes) and in humans, it is clear that small fluctuations in transcript levels are prevalent, some of which are probably biological noise. From experiments and modeling, we have some ideas about how these small fluctuations arise. But what is not so clear is how much these small fluctuations might affect the phenotype, although there are some hypotheses and examples.

In Section 12.4 we hypothesized that one of the roles of microRNAs is to dampen the effects of noise arising from the pervasive transcription of the genome. That is, some or many of the transcripts produced by the cell are not translated because of the presence of microRNAs. The microRNAs could be, in effect, establishing a threshold of transcript accumulation at which translation begins to occur. To overstate their effects perhaps, the level of microRNAs may help the cell to distinguish signals from the background noise.

As shown in Figure 12.7, a careful examination of the transcription factor networks from genome annotation projects offers a few additional insights into the processes by which cells either buffer the effects of transcriptional fluctuation or turn them into signals. Note, first, that the proteins shown in this network are transcription factors, so this is a regulatory network among transcription factors with the effector genes and proteins not included. Cross-regulation, which might help reduce the effects of random fluctuations in transcript levels, are both extensive and common. Let's track one specific example to see the effects of small changes in transcription levels and how they affect the organism. We will use the *Drosophila* gene *caudal*.

As highlighted in Figure 12.20, at least nineteen different transcription factors up-regulate the transcriptional of the gene *caudal* (*cad*). Among its target genes, Caudal stimulates the transcription of the microRNA gene *mir-305*. *mir-305* has six direct targets, including *caudal* itself, as well as *Kr* and *D*, both of which

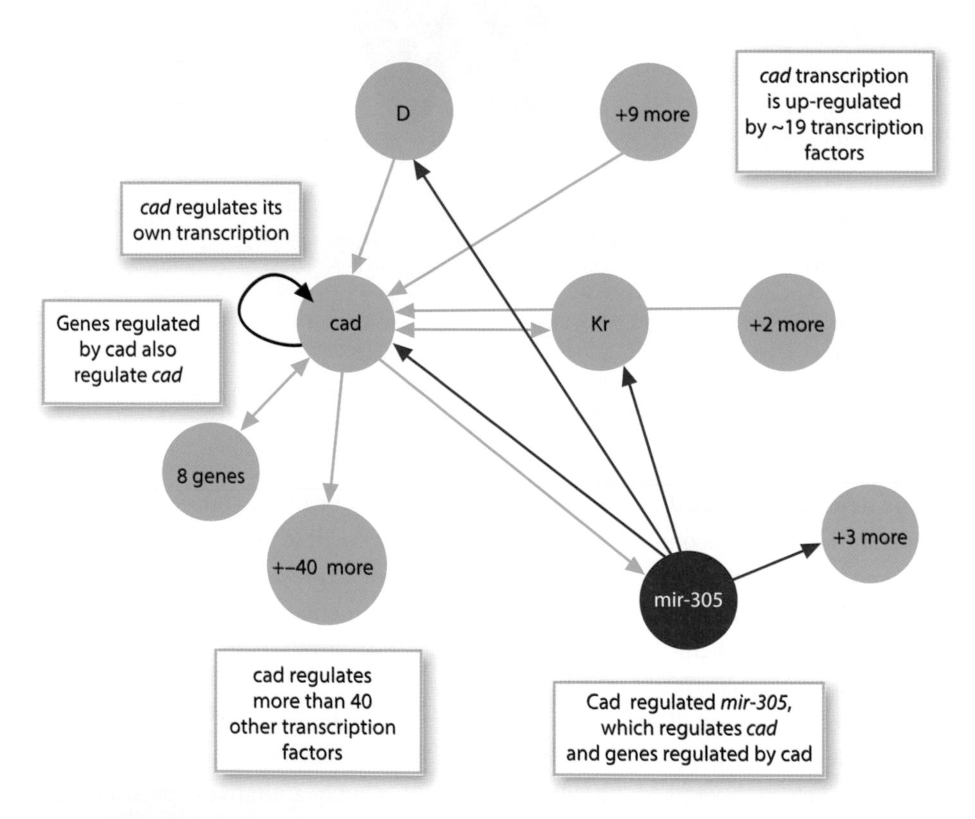

Figure 12.20 The regulatory network of the *caudal* (*cad*) gene in *Drosophila*. *caudal* (*cad*) encodes one of the transcription factors shown in Figure 12.7. Its transcriptional regulation is complex, but many of the features seen with *caudal* are also observed with other genes. First, *caudal* expression is up-regulated by approximately nineteen other transcription factors. Caudal itself regulates the transcription of at least forty additional genes that encode transcription factors. Some of these genes, including *Kr* and at least eight other genes, encode transcription factors that also regulate *caudal* expression, creating a feedback loop that is expected to increase *caudal* expression. In addition, *caudal* regulates the expression of the microRNA gene *mir-305*; *mir-305* down-regulates *caudal* as well as two other genes (*Kr* and *D*) that up-regulate *caudal*, creating additional feedback loops that are expected to decrease *caudal* expression. Furthermore, *caudal* regulates its own transcription, in an autoregulation loop.

up-regulate *caudal* expression. Once this microRNA gene is expressed, it will down-regulate *caudal* mRNA accumulation and will down-regulate two more genes that stimulate *caudal* transcription. Thus *caudal* expression levels are reduced through this feedback loop with the microRNA.

But let's examine the transcriptional regulation of *caudal* expression a bit more. Caudal regulates the expression of the transcription factor gene *Kr*, as well as nearly forty other transcriptional factors. Kr, and some of these other transcription factors, then up-regulate *caudal* transcription, which in turn up-regulates the expression of many other genes. The diagram does not provide information about the time and tissue of *caudal* transcription, since all interactions are reduced to single links, but we can see that *caudal* is being extensively both up-regulated and down-regulated; its transcription level arises from the balance of numerous molecules, both microRNAs and transcription factors; small changes in *caudal* expression could be either propagated or dampened through these series of loops. *caudal* is not unusual in this regard, and many other transcription factors shown in Figure 12.7 have similar regulatory loops that balance their level of expression.

Yet another important source of *caudal* transcription levels is also found in the diagram in Figure 12.20. Note the arrow originating with Caudal that then loops back to *caudal* itself. In other words, the Caudal transcription factor binds to the cis regulatory module of the *caudal* gene and stimulates *caudal* transcription. This is an example of **autoregulation**, in which a gene regulates its own expression.

Autoregulation is not confined to *caudal*, and examples are found throughout the network. In fact autoregulation is exceptionally common in every transcription network examined, both in bacteria and in eukaryotes.

Literature Link. Alon 2007, *Nature Reviews Genetics* 8: 450–61

Recall that the links emanating from a node are referred to as its degrees. For transcription factors, microRNA, and other networks for which the links are directional to a target, we can calculate both an **in-degree** and an **out-degree**. That is, we can find the number of links coming into a node or in which the node is the target of regulation; we can also find the number of links going out from the node, in which the node targets something else. For *caudal*, the in-degree is 19, while the out-degree is about 40. With the help of mathematical models we will not discuss, we can use the average in-degree and the average out-degree to calculate the expected number of times a gene will regulate its own expression; that is, we can calculate how often the same link is expected to be both an in-degree and an out-degree for a node. Depending on the network and the precise modeling method used, the observed number of such examples of autoregulation is at least ten times greater than predicted. Many genes regulate their own transcription directly.

Autoregulation can take one of two forms, depending on the activity of the transcription factor protein. In some cases, the transcription factor negatively regulates or silences further transcription of the gene. Thus, in these cases, the transcription factor acts as a pulse generator; the gene is transcribed, the protein is made, and this then shuts off further transcription, so that expression of the gene is limited to a particular time and location. More commonly, the transcription factor acts as an amplifier. That is, the gene is transcribed, possibly at low levels, the protein is made, and the transcription factor stimulates transcription of the gene to a higher level. Either of these could be affected by small fluctuations in gene expression.

This is what happens with *caudal* itself, in which the amplification is developmentally important, as shown in Figure 12.21.

Literature Link. Schultz and Tautz 1995, *Development* 121: 1023–8; modENCODE.org; Flybase.org

caudal is transcribed at low levels in the mother and the untranslated mRNA is sequestered in the ovum. After fertilization, the sequestered *caudal* mRNA in the zygote is translated. The newly synthesized Caudal protein binds to the cis regulatory module of the *caudal* gene, and this results in high levels of *caudal* transcription in the developing embryo, which then turns on the transcription of the target genes of Caudal. Thus, most (but not all) the transcription factors that up-regulate *caudal* transcription, the in-degree arrows in Figure 12.20, act in the tissues of the mother; the autoregulation of *caudal* occurs in the early embryo; and the regulation of other genes by Cad, the out-degree arrows to other genes, occurs in the developing embryo.

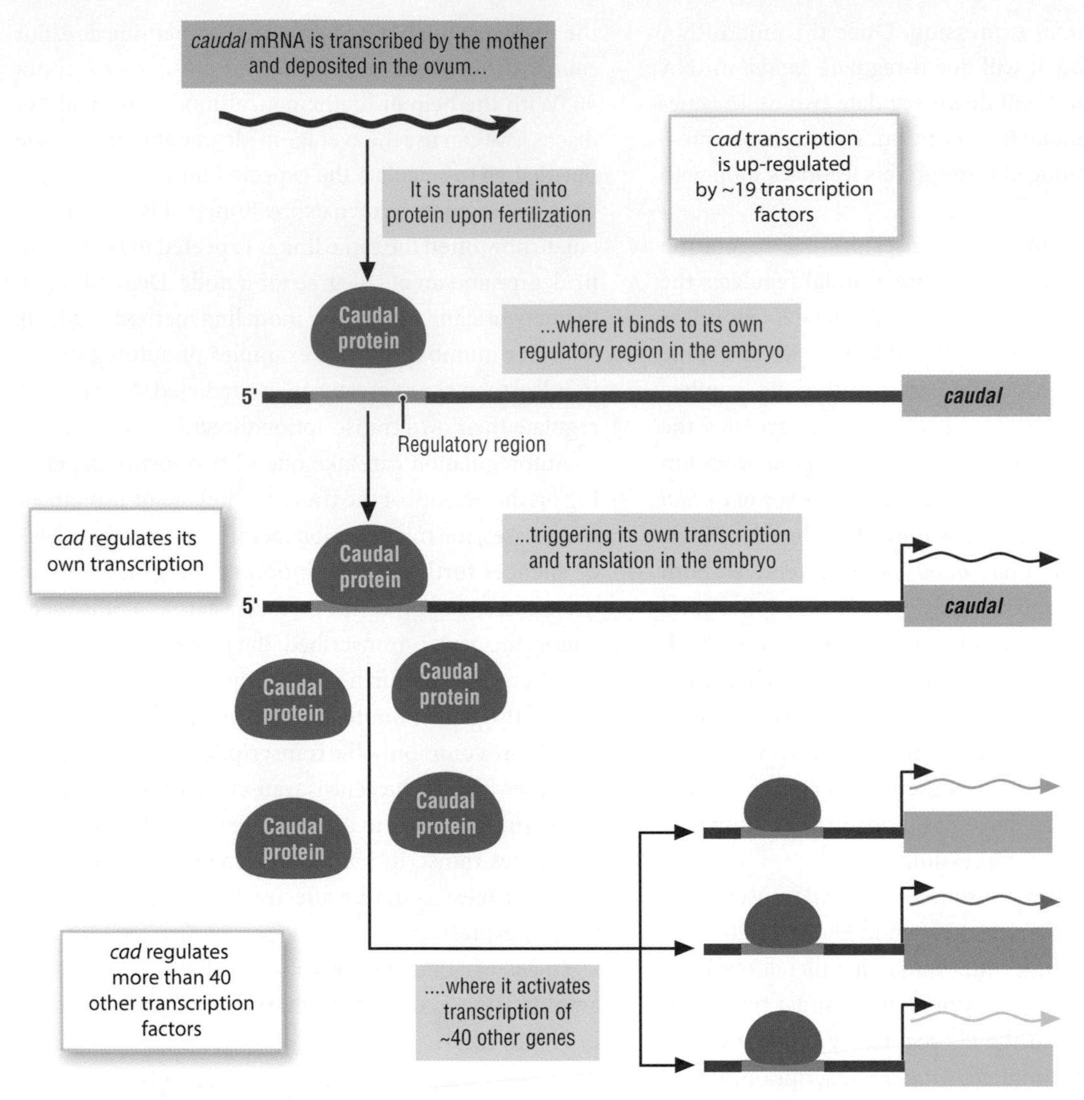

Figure 12.21 ***caudal*** **regulation at different times.** The diagram of *caudal* regulation in Figure 12.20 does not show when or where these regulatory steps occur, but these data are crucial for *caudal* function. The transcription of *caudal* occurs initially in ovarian tissues in the mother, and *caudal* mRNA is sequestered in the ovum. Thus, many of the nineteen genes regulating *caudal* transcription are expressed in the mother. Once the ovum is fertilized, *caudal* mRNA is translated and the Cad protein binds to the gene's own regulatory region. The autoregulatory loop from Figure 12.20 occurs in the very early embryo. *caudal* transcription is now stimulated to a much higher level, when it up-regulates about forty other genes in the developing embryo. Modified from Meneely et al., *Genetics: Genes, genomes and evolution* (Oxford University Press, 2017), Figure 14.23.

Let's return to the question about the biological significance of changes in the levels of gene expression. For a gene like *caudal*, these small changes are crucial to its biological function and to understanding its mutant phenotype. But what our naïve question overlooks is that these changes in *caudal* expression do not occur at the same time; they do not occur even in the same cells (the mother's tissues, the early embryo, and later stages embryos). *Caudal* expression is both up-regulated and down-regulated, but both the timing and the balance of these regulatory mechanisms are necessary for the gene to carry out its function. The network diagram in Figures 12.7 and 12.20 give us critical information about *caudal*, but only the analysis of the individual gene and its mutants can give us the full story about how the phenotype arises from the activity of this gene.

12.8 Pathways, networks, and phenotypes: A summary

At the beginning of the chapter we asked this question: Can we really expect to be able predict the complete phenotype from the genotype – that is, by knowing the genome sequence of an organism? In attempting to answer that question, it is helpful to consider both how far we have progressed and some of what we know remains to be discovered. Let's return to the analogy between the genome and the blueprint.

The first part of the answer requires that we have an inventory of all of the parts. With genome sequencing and annotation methods, with sophisticated methods to assay the expression levels of genes even in a single cell, and with insights gained from all of the genomes annotated previously, it seems reasonable to think that we are very close to being able to accomplish such an inventory. We should not underestimate the amount of effort and the insights that have gone into these annotation or inventory methods, and we know that they can still be improved. Nonetheless, with every new genome from some unusual plant or animals, it seems like we are getting very close to having full inventories of all of the component parts, simply from the genome information. We always learn something new, but what is learned is usually about some process or phenotype in a particular organism rather than something more fundamental.

The second and more demanding part of the answer requires that we have an understanding of how these parts interact with each other. We have to think about how the parts contribute to the whole, of whether organisms are more than the sum of their parts. (We use "sum" in a non-mathematical sense, since the parts can clearly interact in non-additive ways.) We have made a great deal of progress on this in a relatively short amount of time. We have physical interaction networks for some of the most important macromolecules, such as proteins and RNA. Except for yeast, none of these networks shows all the physical interactions for all these macromolecules, and we certainly have not tested all the cell and tissue types. But we have learned many general principles about these physical interactions and many, if not most, will be applicable to the physical interactions that have not been tested yet.

Even more impressive is that we have an extensive and nearly complete set of genetic interactions for an organism: the yeast *Saccharomyces cerevisiae*. This allows us to see how closely the physical interactions and genetic interactions parallel each other in producing a phenotype. They also allow us to test whether our thinking and interpretation about genetic interactions from particular genes and biological processes holds up when applied on a broader scale of genes and phenotypes, even to the fitness of an entire organism. It is encouraging to recognize that we have been on the right track.

The genetic interaction networks perhaps even provide a more advanced template for thinking about how the component parts encoded in the genome are organized with respect to each other. For yeast, as summarized in Figure 12.18, about 19 percent of the genes are essential in homozygous diploids; this number will be different in other eukaryotes, but our genome-wide mutant screens (discussed in Chapter 9) do not suggest that the percentage will be dramatically different. These screens include most of the genes that are highly conserved during evolution, so many of them can be recognized from sequence annotation and we can have an idea about how they contribute to the phenotype. Of the genes that are not essential when evaluated as single mutants, some are members of gene families. In yeast, about 8.5 percent of non-essential genes have a recognized paralog and are probably functionally redundant; this number may be higher in other eukaryotes with more extensive gene families. Because of their sequence similarity to one another, these might also be recognized during sequence annotation, and we can make some prediction as to how they contribute to the phenotype. However, we learned from yeast that more than half of the non-essential genes are needed only in combination with another gene; we can use the existence of an interaction to annotate functions for many of the genes, but we cannot yet predict the occurrence of these interactions. Nonetheless, these form the largest set of genes, so unless we can accurately predict their interactions and their contributions *in combination*, we will not yet be able to predict the phenotype from the genotype. Even in yeast, there are still about 1,000 genes (15 percent of the total) whose functions (if any) cannot be determined by current methods. So we have some more to learn before we can claim that we can predict a phenotype from the genotype.

Because our discussion has focused on genes and the phenotype, we have left out one of the most important factors—for some phenotypes, the single most important factor—in predicting phenotypes. We have not considered environmental impacts, or the even more subtle interactions between the genotype and the environment. For model organisms raised in laboratory incubators or greenhouses, these environmental effects can be generally ignored. For organisms in nature, including humans, they cannot be. Interactions between genotypes and environments will probably always be a frontier in our understanding; that is, after all, the essence of what Darwin recognized about biological systems.

While environmental effects and interactions between genotypes and environments may never be easy to predict when thinking about phenotypes, it is worth our while to reflect on the genetic interactions that have been studied just a bit more, particularly as they relate to humans. The yeast genetic landscape was based on testing 23 million pairwise interactions, an impressive number that probably will not be matched in any other organism or experiment. But in every case two genes had genetic variation (and known genetic variation at those defined loci) and the remaining 6,000 or so genes were isogenic and had no additional genetic variation. This does not describe our genomes. In yeast, it was more often the combination of genetic variants, and not the individual genetic variants, that resulted in a recognizable phenotypic difference. This is almost certainly true for other organisms, including humans. Not only the individual genes that affect our phenotype are the ones that we study most thoroughly, but the combination of genetic variants as well.

It seems clear that we carry in our genomes many genetic variants that do not by themselves result in a significant phenotypic difference. However, each of these variants might be capable of interacting with other variants in the genome that will produce a phenotypic difference, probably variants in scores of other genes; for yeast, a gene interacted with about 100 other genes, and it is hard to imagine that a multicellular organism with a long life span, such as a human organism, will have any fewer than that. (Our genome also has about four times as many protein-coding genes as yeast, so the number of interactions seems likely to be many more than 100 per gene.) In addition, unlike in yeast, these other genetic variants in our genomes are usually not recognized, and they would not be limited to pairwise combinations.

So we have learned a substantial amount about genotypes and phenotypes, enough that "personalized medicine," which uses our individual genome to predict our phenotypes (such as a drug response), is promising in some cases. But we should perhaps be a little more skeptical or cautious about being able to predict most aspects of our phenotype purely from our well-annotated genotype.

But maybe that is not so bad. Think once more about our genome as a blueprint. Every building has some type of blueprint—the Taj Mahal, the Sistine Chapel, and every other grand edifice. Some of us might be able to look closely at the blueprints for those buildings and recognize some of the possibilities. But, with our genomes as with ourselves, almost none of us has the imagination, insight, and genius to examine those blueprints, no matter how closely, and envision the majesty and glory of the final product.

Chapter Capsule

A well-annotated genome sequence provides an inventory of the component parts of a cell and of an organism. In order to predict phenotypes by using these parts, the interactions among them, both direct and indirect, have to be known. Direct interactions, such as transcription factors with their binding sites, microRNAs with their target mRNAs, and proteins with other proteins, have been studied on a genome-wide scale. Indirect interactions are best studied using mutants, alone and in combination with each other. Pairwise mutant combinations have been analysed for nearly all the genes of *Saccharomyces cerevisiae*. Such interaction maps provide some of the assembly instructions for constructing how the component parts interact to produce a phenotype. The results have been informative and encouraging, but still do not capture the subtleties that arise from levels of gene expression or types of genetic variation that may contribute to the phenotype.

Additional Reading

Bar-Even, A. et al. 2006. Noise in protein expression scales with natural protein abundance. ***Nature Genetics*** 38 (6): 636–43.

Chandler, C. H., S. Chari, and I. Dworkin. 2013. Does your gene need a background check? ***Trends in Genetics*** 29: 358–66

Eldar, A. and M. B. Elowitz. 2010. Functional roles for noise in genetic circuits. ***Nature*** 467 (7312): 167–73.

Hartman, J. L. t. et al. 2001. Principles for the buffering of genetic variation. ***Science*** 291 (5506): 1001–4.

Hartwell, L. 2004. GENETICS: Robust Interactions. ***Science*** 303 (5659): 774–5.

Lehner, B. 2013. Genotype to phenotype: Lessons from model organisms for human genetics. ***Nature Reviews Genetics*** 14 (3): 168–78.

Lenski, R. E. et al. 2006 Balancing robustness and evolvability. ***PLoS Biology*** 4: e428.

Mitra, K. et al. 2013. Integrative approaches for finding modular structure in biological networks. ***Nature Reviews Genetics*** 14: 719–32.

Munsky, B. et al. 2012. Using gene expression noise to understand gene regulation. ***Science*** 336 (6078): 183–7.

Newman, J. R. et al. 2006. Single-cell proteomic analysis of ***S. cerevisiae*** reveals the architecture of biological noise. ***Nature*** 441 (7095): 840–6.

Van Leeuwen, J et al. 2016 Exploring genetic suppression interactions on a global scale. ***Science*** 354: 599.

Case Study 12.1

A systems analysis of TGF-β signaling in *C. elegans*

The genetic pathway that leads to dauer larva formation in *C. elegans* was developed in the case study for Chapter 11. As a brief summary, the dauer larva is an alternative to the third larval stage that forms under starvation or other stress conditions. Many genes involved in dauer formation have been described. Mutations in some genes result in constitutive dauer formation even when food is abundant; such genes are known as Daf-C (dauer-constitutive) genes. Mutations in other genes result in no dauer formation even when the population is starved; such genes are known as Daf-D (dauer-defective) genes.

The pathway that leads to dauer formation has two branches, as described in the Chapter 11 Case Study. We will focus on the lower branch, for which the genes encode proteins in a TGF-β pathway; Figure CS.12.1 shows the correspondence between this branch of a part of the dauer pathway and a canonical TGF-β pathway. TGF-β pathways are important in many processes in animal biology, and the proteins are evolutionarily conserved from worms to humans.

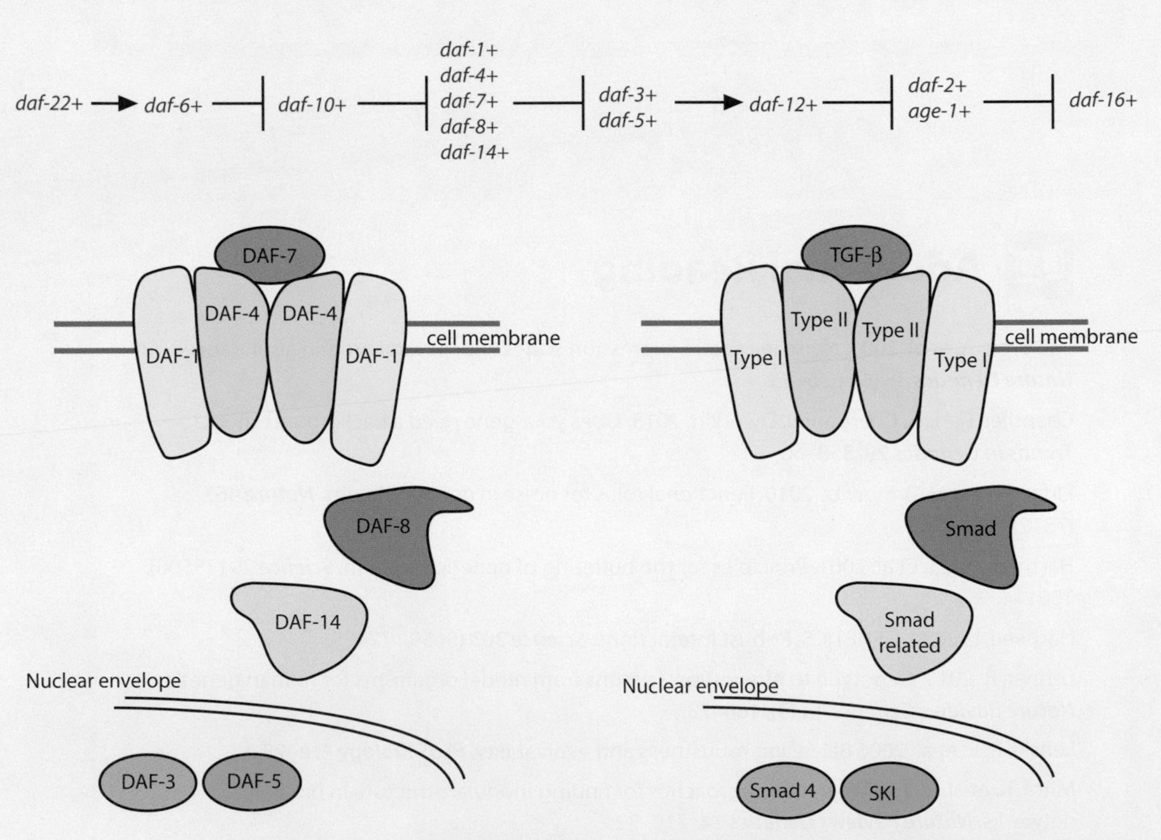

Figure CS12.1 Dauer formation and the TGF-β pathway in *C. elegans*. The epistatic pathway for dauer formation was worked out using gene interactions, as described in the case study for Chapter 11. One branch of that pathway is shown at the top of the figure; the genes in red are needed to prevent dauer formation and the genes in blue are needed to form dauers. The portion of the pathway involving the genes from *daf-1* to *daf-3* and *daf-5* are the same as a TGF-β pathway in mammals. The correspondence between the genes in worms and the genes in a canonical TGF-β pathway are shown at the bottom. At the time when this work was initiated, the *daf-5* gene in worms had not been cloned.

Thinking through the Experiment

1. Do a quick literature search to find some examples of biological processes in other animals that involve a TGF-β pathway.
2. Review the signaling process that occurs during a TGF-β pathway.

Because the TGF-β pathway has been studied in many different animals by a variety of genetic, biochemical, and molecular techniques, dauer formation is a good example of how network approaches can be integrated into our understanding of a fundamental biological process. To begin the analysis, a protein interaction network was constructed by using six of the known proteins from the dauer pathway in *C. elegans* as baits in yeast two-hybrid screens. The prey proteins were found from among a cDNA library with a comprehensive collection of worm genes. The six bait proteins with their number of interactions are shown in Figure CS12.2. Two other proteins could not be used as baits; *daf-8* auto-activates when used as a bait, so no interactions could be detected, whereas *daf-5* had not been molecularly identified when the study began.

Thinking through the Experiment

3. Outline how the Y2H experiment was done, defining terms such as "bait" and "prey" proteins.
4. What does it mean to say that "*daf-8* auto-activates when used as a bait"? Would this affect its ability to be recovered as a prey?
5. What information could this physical interaction network provide that the well-defined genetic pathway could not provide?
6. Before reading ahead, summarize some of the Y2H interactions that you would expect to find, and some that you would expect not to find.
7. *daf-7* encodes the ligand for the pathway. Rather surprisingly, no interactions were found for it. Thinking carefully about the types of interactions identified by Y2H, explain this result.

Interactions were found for four of the six bait proteins, but no interactions were found for either *daf-7* or *daf-12*. The lack of interactions with these two proteins probably illustrates two of the known limitations of Y2H discussed in Chapter 10. DAF-7 is the ligand that binds to a receptor encoded by a complex of DAF-1 and DAF-4; since Y2H finds only binary interactions, a tri-molecular complex among DAF-7, DAF-1, and DAF-4 would not be found. DAF-12 encodes an apparent nuclear hormone receptor. In other organisms, this protein is post-translationally modified by phosphorylation. Y2H misses interactions that require phosphorylation, which probably explains the failure to find interactions with DAF-12.

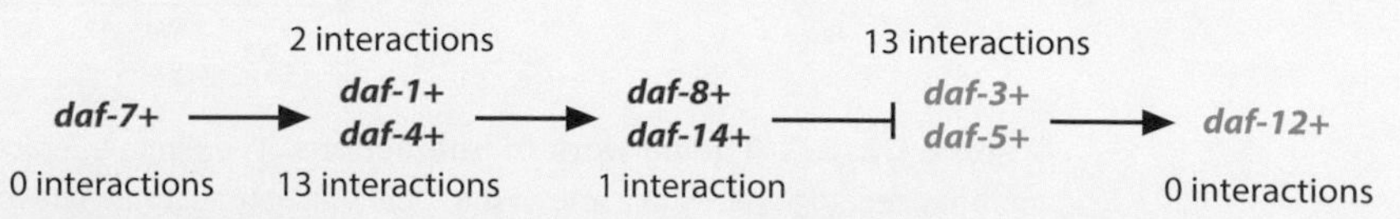

Figure CS12.2 Interactions in the pathway using Y2H. The *daf* genes used as baits in a yeast two-hybrid assay and the number of interacting prey proteins are shown for each. *daf-8* could not be used as a bait.

The network identified known members of the TGF-β Pathway

With the other four bait proteins, twenty-eight interacting proteins were identified. Nineteen of these interactors were then used as baits in a further screen, which recovered twenty-seven additional interacting proteins. In total, seventy-one interactions were found involving fifty-nine proteins, as diagrammed in Figure CS12.3.

In order to confirm that these protein interactions detected by Y2H are physiologically relevant as well, epitope-tagged versions of some of the bait and prey proteins were co-expressed in mammalian cells, and co-IP experiments were done. Among the Y2H interactions tested, 89 percent were also found through the co-affinity precipitation assay by using the epitope tags. This provides further evidence that most of these interactions are capable of occurring in the organism.

Thinking through the Experiment

8. Describe how these co-IP experiments were done. Did these experiments show that these proteins interact in vivo? Why—or why not?

Interactions between proteins in the TGF-β pathway in other organisms had already been described before this set of experiments was done. One of the first tests of the approach was to ask whether protein interactions previously detected by other means in other organisms are also found by the Y2H assays in worms. At least two examples of known interactions were also found by Y2H experiments in worms. For example, in mammalian cells, a protein known as filamen is found in a protein complex with SMAD2. The apparent equivalent of

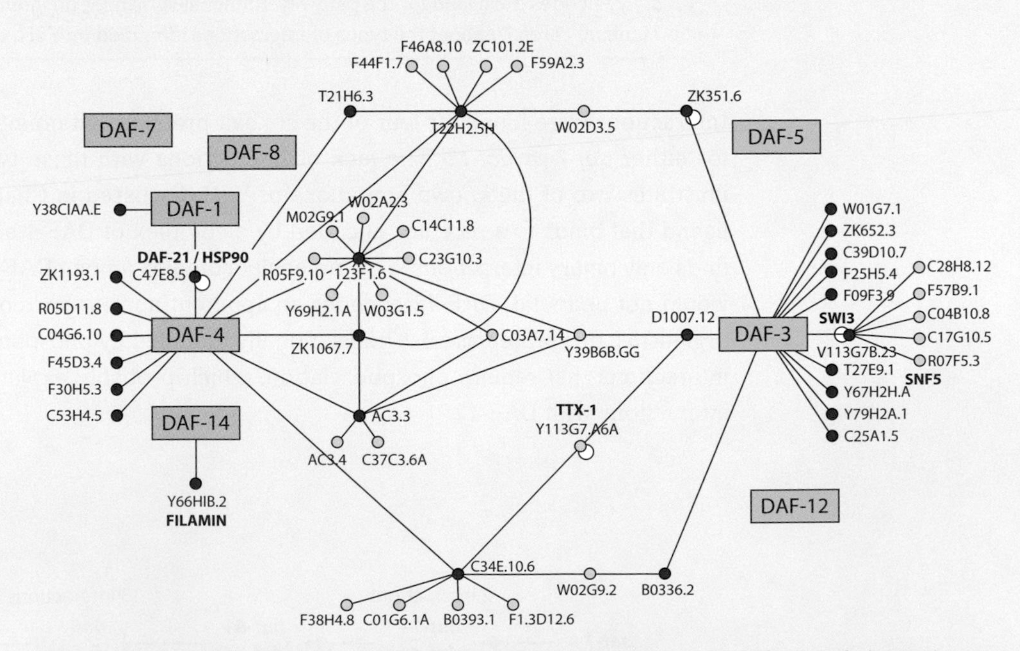

Figure CS12.3 The network of interactions. The direct interactions with these six bait proteins are summarized, together with some of the additional interactions. The interaction with DAF-14 and filament and the interaction of DAF-3 with W01G7.1 are discussed further in the text. Source: Figure 1 from Tewari et al. 2004, Systematic interactome mapping and genetic perturbation analysis of a *C. elegans* TGF-β signaling network, *Molecular Cell* 13: 469–82, used with permission.

the SMAD2 protein in the worm is DAF-14; as predicted from the results in mammals, one of its interacting proteins in worms is a nematode orthologue of filamen that had not been previously recognized, as seen in Figure CS12.3. Also in mammalian cells, the heat-shock protein HSP90 interacts with many proteins including the TGF-β receptor; in the Y2H assay, the nematode HSP90 protein DAF-21 interacts with the co-receptor DAF-1 and DAF-4. The *daf-21* gene had been tentatively placed in the other branch of the dauer formation pathway on the basis of the behavior of an unusual dominant allele. However, the network shows that the gene belongs in both branches of the pathway. Other proteins in the network also are consistent with what is known or inferred from the TGF-β pathway in other cells, including the observation of chromatin remodeling proteins SWI-3 and SNF-5 as part of the module with the DAF-3 protein.

Functional interactions were found by genetic assays

Many of the genes detected in this interaction network did not have known functions. Each of the genes for which no mutations were known were tested for effects on dauer formation by RNAi. If the gene is involved in dauer formation, a knockdown of its function by RNAi is expected to produce a Daf-C or Daf-D phenotype. The approach is depicted on the left half in Figure CS12.4. Wild-type worms were placed onto bacteria expressing the double-stranded RNA that corresponds to one of the genes from the network. The offspring were tested for their ability to form dauers. Three of the uncharacterized genes were found to produce Daf-C phenotypes, which identifies them as genes whose roles in dauer formation were previously unknown. Thus the network approach found new genes that had been missed by other approaches. All three of these genes have mammalian orthologues, but

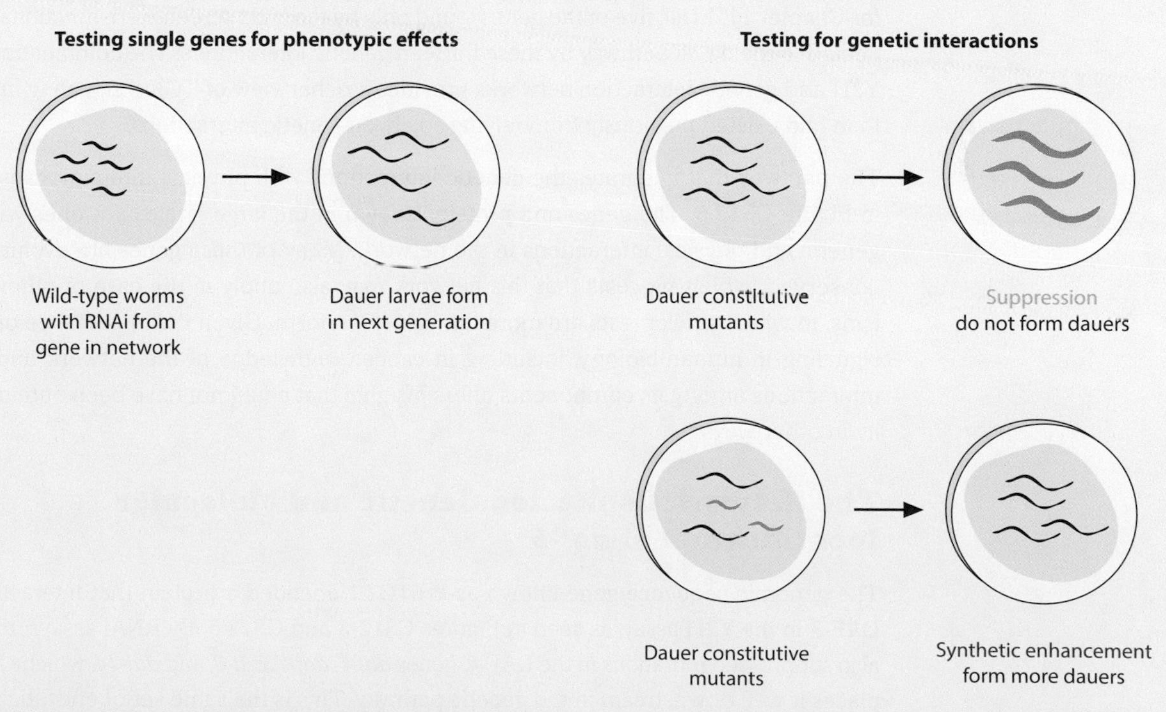

Figure CS12.4 Genetic effects of these genes. The genes whose functions were not known were tested using RNAi for mutant phenotypes. Each newly arisen gene was tested for its effect on wild-type worms and for its effect on one or more of the known dauer-constitutive genes.

none was thought to play any role in TGF-β signaling; an effect of these genes on TGF-β signaling remains to be tested on other organisms.

Comparing genetic interactions and physical interactions

A crucial test of the network approach is to test for genetic interactions among genes whose relationship was found by physical interactions. This test was done by performing RNAi assays on worms that also had a mutation in one of the known dauer genes. The assay is diagrammed on the right half in Figure CS12.4. Known dauer-constitutive mutant worms were placed on RNAi plates, as before, and suppression or synthetic enhancement of their phenotype was noted. Because the genetic pathway has been thoroughly analysed using mutations and the mutant phenotypes for many of the genes were clearly defined and quantified, as described previously, both suppression and synthetic enhancement could be tested. The results were a beautiful confirmation and extension of the previous research.

Thirteen different genetic interactions were found involving the known and unknown genes in the DAF pathway. For example, all three of the newly identified genes with Daf-C phenotypes from RNAi also showed suppression of *daf-12* in genetic assays, and thus were placed upstream of *daf-12*.

Thinking through the Experiment

9. Using the logic outlined in Chapters 10 and 11, explain why these three genes were placed upstream of *daf-12* in the pathway.

Synthetic enhancement was also found, again by the methods described for the case study for Chapter 11. In all, five of the genes found only by the Y2H screen were functionally connected to the dauer pathway by these thirteen genetic interactions. The combination of the Y2H and genetic interaction networks provides a richer view of TGF-β signaling in worms than had existed previously from relying solely on genetic interactions.

The pathway that integrates the genetic interactions with physical interactions is shown in Figure CS12.5. The genes and proteins shown in the large circles are ones with both genetic and physical interactions in the network. Many of these genes are evolutionarily conserved, which suggests that this network may also apply in the case of other organisms, in which similar tests are more difficult to perform. Given the importance of TGF-β signaling in human biology, including in cancer, knowledge of the network and of the interactions among its components offers insights that could not have been obtained easily in other ways.

The Network Connected Genetic and Molecular Identification of *daf-5*

The genomic sequence gene known as W01G7.1 encoded a protein that interacted with DAF-3 in the Y2H assay, as seen in Figures CS12-3 and CS12.5. By RNAi assays, this gene also suppressed mutations in the DAF-C genes *daf-1*, *daf-7*, *daf-8*, and *daf-14*, whicha fact that places it well downstream in the genetic pathway. This is the same set of epistatic interactions shown by a.mutation in the dauer-defective gene *daf-5*. At the time of these experiments, *daf-5* had not yet been cloned. W01G7.1 and *daf-5* map to the same location in the

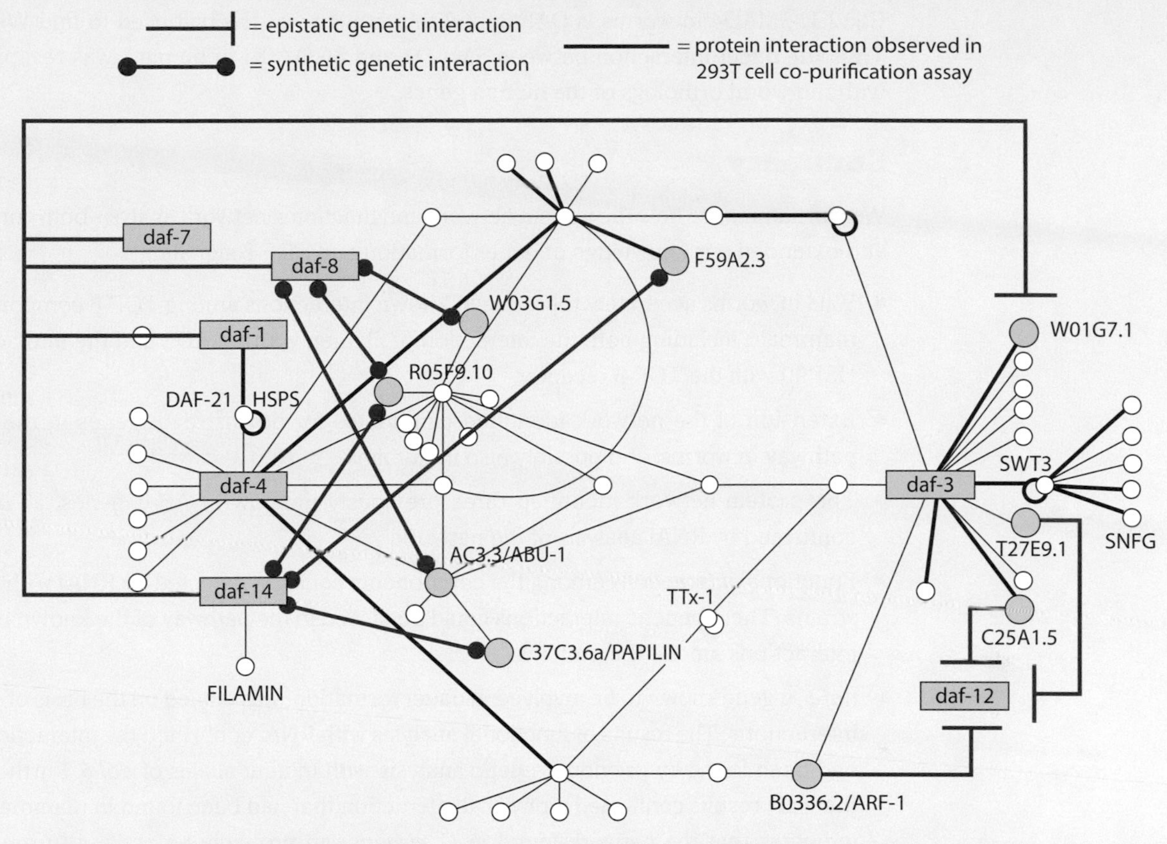

Figure CS12.5 An integrated network of protein and genetic interactions. The protein interactions confirmed by Co-IP are shown in black lines, while those detected only by Y2H are shown in gray lines and genetic interactions are shown in red. Source: Figure 4 from Tewari et al. 2004, Systematic interactome mapping and genetic perturbation analysis of a *C. elegans* TGF-β signaling network, *Molecular Cell* 13: 469–82, used with permission.

genome, so it was natural to test whether W01G7.1 (the gene detected from the sequence of the genome) corresponds to *daf-5* (the gene recognized for its mutant phenotype).

Thinking through the Experiment

10. On the basis of information in Chapter 5, how would you prove that the genomic sequence W01G7.1 corresponds to the gene *daf-5*?

Confirmation of this correspondence was done by two of the methods described in Chapter 5. First, the DNA from W01G7.1 was sequenced from a *daf-5* mutant allele and compared to the wild-type W01G7.1 DNA sequence. The mutant had a base change in the region of W01G7.1 by comparison to the wild type, and this change would have resulted in a missense allele of the gene. Second, the mutant phenotype of *daf-5* was rescued when the wild-type W01G71.1 sequence was injected into *daf-5* mutant worms, which shows that the wild-type W01G7.1 sequence could complement or rescue the *daf-5* mutant phenotype.

The sequence of the wild-type gene indicates that the DAF-5 protein is a worm ortholog of the mammalian oncogene SNO/SKI. The SNO/SKI proteins in mammals are known to associate physically with the SMAD4 protein in the TGF-β pathway, as shown in Figure

CS12.1. SMAD4 in worms is DAF-3, and this protein was the bait used to find W01G7.1. Thus the direct interaction between SMAD4 and SNO/SKI in humans was recapitulated with the worm orthologs of the human genes.

Summary

We can summarize how the protein network and functional network analysis both confirmed and extended our knowledge of dauer formation and TGF-β signaling.

- Y2H in worms accurately reproduced known interactions among TGF-β components in mammals, including both the interaction of filamen with SMAD2 and the interaction of HSP90 with the TGF-β receptor.
- Extension of the network identified candidates for downstream genes in the TGF-β pathway in worms, and possibly also in humans.
- The protein network identified three previously unknown DAF-C genes, all of them confirmed by RNAi analysis of the network.
- Functional interactions among the components could be detected by RNAi with mutant strains. These genetic interactions could be placed in the pathway of the known physical interactions among the proteins.
- *daf-5*, a gene known to be involved in dauer formation, was cloned on the basis of its Y2H interactions. The results of functional analysis with RNAi confirmed the interactions that had been found by previous genetic analysis with mutant alleles of *daf-5*. Furthermore, the Y2H results confirmed a physical interaction that had been found in mammals. This indicates that the network found in *C. elegans* can probably be applied to the TGF-β network in mammals.

Thinking through the Experiment

11. The TGF-β pathway and all its components were known to be highly conserved during animal evolution. The interactions among the proteins were also conserved between humans and worms. Would you predict that a similar conservation of interactions would be found for other signaling pathways? Why—or why not?

Reference

Tewari, M. et al. 2004. Systematic interactome mapping and genetic perturbation analysis of a *C. elegans* TGF-β signaling network. *Molecular Cell* 13: 469–82.

GLOSSARY

allele one of the many alternative forms of a gene. Used very broadly to include both the phenotypic variation, such as round or wrinkled peas, and the molecular variation, such as a sequence alteration in a gene. See also **polymorphism**.

allelic series a collection of different mutant alleles for a gene that have different amounts of residual gene activity, and thus slightly different mutant phenotypes.

alternative splicing the processing of an RNA transcript to yield different mRNA sequences. In higher organisms, it is thought that most transcripts are alternatively spliced to yield related but different mRNA transcripts.

amorphic allele also known as a **null allele**. A mutant allele that eliminates the function of a gene. Most amorphic mutations are recessive.

annotation the process of identifying the functional components of a sequenced genome.

antimorphic mutation see dominant negative mutation.

antisense technology one of several different methods that knock out or reduce expression from a gene using a nucleic acid sequence complementary to the mRNA from the gene.

apoptosis the genetically programmed death of particular cells in multicellular animals.

ascus the sac structure that holds the haploid spores of yeast and other fungi.

attack vulnerability an inherent property of networks that are comprised of hubs. Any process that eliminates a hub renders the entire network unstable or non-functional.

autoregulation A process by which a gene regulates its own expression, for example, by encoding a transcription factor protein that binds to the gene's own cis regulatory module.

auxotroph a mutant that cannot synthesize some essential nutrient for growth. Auxotrophic mutants can only grow if the nutrient is provided in the growth medium. The opposite of auxotroph is **prototroph**.

bait plasmid in a yeast two-hybrid experiment, the plasmid encoding the protein whose interactions are being determined. The proteins being queried are encoded on the prey plasmids.

balanced heterozygote a heterozygous genotype in which a recessive mutation of interest is maintained with marker mutations on the other homologous chromosome. For example, *aB/Ab* is a balanced heterozygote.

balancer chromosome a genetically altered chromosome that cannot recombine with its normal homologous chromosome. Any recessive mutations on the normal homologous chromosome will be inherited together if they are maintained as a heterozygote with a balancer chromosome.

balancing the procedure to maintain a genetic strain or stock as a stable heterozygote. See **balanced heterozygote**.

biological noise the naturally occurring fluctuations in gene expression levels, which may or may not affect a phenotype.

boundary mutations a description of the mutant phenotypes for genes that recognizes that many mutations exhibit their phenotypes at definable stages in the life cycle, such as molts or instars.

bypass suppressor a mutation that overcomes the phenotypic effect of another mutation by affecting a separate genetic pathway than the original mutation.

Cas9 the RNA- directed endonuclease from *Streptococcus pyogenes* commonly used in gene editing by CRISPR.

cell-autonomous a gene or mutation that exerts its effect within a cell.

cell-autonomous effect the effect of a gene whose phenotype corresponds to the genotype of the cell, regardless of the genotype or phenotype of adjacent cells. Most genes whose activity is cell autonomous encode proteins that work intracellularly.

cell-non-autonomous a gene or mutation that exerts its effect outside of the cell in which it is expressed.

chimera an organism with cells of two or more different genotypes, produced by physically aggregating the cells. Compare with **mosaic**.

chromatin the complex of DNA with associated proteins and RNA molecules that comprise the structure of a eukaryotic chromosome.

chromatin immunoprecipitation (ChIP) a procedure in which an antibody against a transcription factor protein is used to identify the binding sites for that protein in the genome.

chromatin remodeling one of several different biochemical processes that alter the three-dimensional structure of a chromosome thereby affecting function.

cis regulatory module a collection of enhancer and silencer sequences that affect where and when a gene is transcribed. See **regulatory region**.

co-immunoprecipitation (co-IP) a biochemical method to detect protein–protein interactions by using an antibody against one protein to also precipitate other proteins in a macromolecular complex.

cold-sensitive mutation a mutation that has a mutant phenotype at a slightly reduced growth temperature but a wild-type phenotype at a higher-than-normal growth temperature. This is the reverse of a **temperature-sensitive** mutation.

Complementation see **complementation test**.

complementation test a genetic test to determine if two recessive mutations are alleles of the same gene. Mutations that are alleles of the same gene fail to complement.

compound heterozygotes individuals with two different recessive alleles for the same gene. Although the individual might be considered a homozygote, the underlying molecular lesions are different on the two homologous chromosomes. See **heteroallelic.**

conditional mutation a mutation that exhibits a mutant phenotype only under certain environmental conditions, such as a slightly elevated growth temperature or particular nutritional conditions. **Temperature-sensitive** mutations are an example of conditional mutations.

counter-selection a genetic screen that allows selection for both a particular genotype and against the same genotype.

Cre a recombination enzyme from the bacteriophage P1 that catalyzes site-specific crossovers at the nucleotide sequences known as *loxP* sites. The Cre–*lox* recombination system is widely used to insert sequence or produce recombination at specific sites.

CRISPR a collection of sequences in a bacterial genome separated by highly similar palindromic repeats. "CRISPR" is an acronym for "clustered interspersed short palindromic repeat." While the term specifically refers to the array of repeats, it is often used more generally to refer to the process of genome editing using this array of repeats and the associated enzyme Cas9.

CRISPR RNA (crRNA) the transcript made from the CRISPR array that, in *S. pyogenes,* directs the Cas9 protein to its target sequence.

CRISPR-associated (*Cas*) genes genes located adjacent to the CRISPR array, and involved with its function to degrade foreign DNA sequences, such as from viruses. Cas9 is CRISPR associated protein 9 in *S. pyogenes*.

crossover suppressor a chromosome rearrangement such as an inversion that cannot recombine with the normal homologous chromosome. Since no recombinant gametes are produced, the chromosome is a crossover suppressor.

crossover suppressors are often used as **balancer chromosomes.**

dauer larva a stage in the life cycle of *C. elegans* that occurs under conditions of starvation, environmental stress, or crowding. A worm can enter the dauer larva stage as an alternative to the third larval stage and remain in the dauer stage indefinitely.

degree in a network diagram, the number of connections or links present for a particular node.

degree distribution the histogram displaying the degrees for all of the nodes in a network.

deletion also known as a **deficiency.** A chromosome rearrangement in which a contiguous piece of the chromosome has been removed and the remaining pieces fused together.

diagnostic landmark an easily recognized event or structure in a cell or tissue that can be used to stage events. The term is frequently used in conjunction with the analysis of temperature sensitive mutants.

dominant negative mutation also known as **antimorphic mutation.** A dominant mutation that eliminates or reduces the function of a gene, usually by interfering with the function of the normal product of the same gene.

double-stranded break (DSB) a cut affecting both strands of a DNA molecule. A DSB is the initiating event for many types of recombination.

duplication a chromosome rearrangement in which a contiguous piece of a chromosome has become duplicated or repeated. Duplications can be found adjacent to the normal site, in which case they are usually referred to as tandem duplications, or they may be found at another site.

ectopic in an unusual location or at an unusual time.

effective mutant phase the stage in the life cycle at which a mutant phenotype can be observed.

embryonic stem (ES) cells cells isolated from the inner cell mass of a mammalian embryo at the blastula stage which can be cultured in vitro.

emergent property a characteristic of a biological system that arises from the interactions of its components as well as from the characteristics of those components themselves.

endogenous from within the organism.

enhance a general term meaning to cause to increase. Two distinct uses are common in genetics. First, "enhance" can be used with an increase in transcription; see enhancer sequence. Second, "enhance" can be used with mutant phenotypes to describe a mutation that results in a more severely mutant organism; see synthetic enhancer.

enhancer see "enhancer sequence" if referring to transcription, or "synthetic enhancer" if referring to mutant phenotypes.

enhancer sequence a sequence in the regulatory region of a gene that confers the specificity of transcription for that gene by activating transcription. Enhancer sequences are the binding sites of transcription factor proteins, and a gene may have many or few enhancer sequences.

epigenetic change a heritable change in the phenotype of an organism without an underlying change in the DNA sequence. Epigenetic changes usually arise from stable modifications in chromatin. Because there is no change in the DNA sequence, epigenetic changes are often erased in the F1 generation.

epistasis a genetic interaction between two genes in which the mutant phenotype of one gene masks the genotype and phenotype of **allele** the other gene.

epitope tagging a process attaching a short amino acid sequence for which antibodies are available (the epitope) to another protein to allow for its purification or analysis.

essential gene a gene whose function is required for the normal growth, development, and fertility of an organism. Often recognized because a mutation in the gene is lethal or sterile.

euchromatin a cytologically detected region of the chromosome that is less densely packed and thus stains more lightly with DNA dyes. Most protein-coding regions are found in euchromatin. Chromatin that stains more intensely is referred to as **heterochromatin**.

execution point the term used in the analysis of the yeast cell cycle for when the cell carries out a particular cellular event.

exogenous from outside the organism.

exome sequencing the collection of all of the exons in a genome in a genome is called the exome. Because the exons comprise only about 1–2 percent of the human genome, some DNA sequencing strategies choose to sequence all of the exons, or the exome, rather than the total DNA.

expression-based cloning any of several methods that rely on some aspect of the expression of a gene, such as its transcription, to identify the underlying DNA sequence of the gene.

extragenic suppressor A mutation in a second gene that overcomes the phenotypic effects of a mutation in another gene. Often referred to simply as a suppressor.

F1 screen also known as a **non-complementation screen**. A mutant screen that uses a pre-existing allele in a gene of interest to find additional alleles of the same gene. The screen looks for mutations in the F1 generation that fail to complement the pre-existing allele.

floxed the nickname for a gene that is flanked by *loxP* sites. It is a specific target for deletion or alteration by the Cre recombinase.

forward genetic analysis see **genetic analysis**.

gain-of-function mutation a mutation that over-expresses or produces a novel function for a gene. Many gain-of-function mutations are dominant.

gap gene a category of the mutant phenotypes observed for genes affecting segmentation in the Drosophila embryo, in which multiple adjacent segments are affected.

gene activity a broad term to refer to the effects of a gene as revealed by its mutant phenotype.

gene cloning any of several methods that identify the underlying DNA sequence of a gene, often a gene only recognized by its mutant phenotype.

gene disruption a change that interrupts the coding region of a gene, often by intentionally inserting or deleting sequences.

gene homology genes related in DNA sequence because of evolutionary descent from a common ancestor.

gene replacement a process of replacing the copy of a gene within the chromosome of a cell or organism with a copy that has been manipulated in vitro.

gene-specific suppressor a suppressor mutation that works on one specific gene, but usually affects several different alleles of that gene.

genetic analysis the process of understanding the biological function of a gene using its mutant phenotype. Also called mutant analysis, or occasionally forward genetic analysis to distinguish from reverse genetic analysis.

genetic landscape one of the terms used to describe the genes, their functions, and their interactions with each other that give rise to the functions of a cell.

genetic selection a process often employed during mutant screens in which only certain genotypes will grow and reproduce.

genome editing any of several methods used to directly alter the DNA sequence of a gene or a chromosome in order to understand its functions. CRISPR-Cas9 is a commonly used genome editing method.

genome-wide genetic analysis a screen in which genome sequence information is used to design specific molecular tools for the analysis of all or nearly all of the identified genes.

genomic imprinting An effect in which the phenotype of an individual depends upon which parent contributed each allele, or differences in the expression of the alleles of a gene depending on whether the allele has been inherited from the mother or the father. Imprinting is also termed a **parent-of-origin effect**. Genomic imprinting is usually interpreted as an example of an epigenetic change.

graph theory a field of mathematics that describes a process using the connections between two points, referred to as nodes. In biology, graph theory is used to describe interactions between two macromolecules.

haplo-insufficient gene a gene whose function is dose dependent, such that eliminating the function of one allele results in a mutant phenotype. A haplo-insufficient gene is recognized by dominant loss-of-function alleles.

hetero-allelic term describing an individual who is heterozygous for two different mutant alleles of a gene, such as *a1/a2*. In human genetics, such individuals are often referred to as **compound heterozygotes**.

heterochromatin a cytologically detected region of the chromosome that is more densely packed and thus stains more intensely with DNA dyes. **Constitutive heterochromatin** stains more intensely at all stages of the cell cycle and in all cells, whereas **facultative heterochromatin** stains more intensely at only certain times in the cell cycle or in only some cells. Chromatin that stains less intensely is referred to as **euchromatin**.

high copy suppressor a gene whose high copy number or over-expression of the wild-type gene suppresses a mutation in another gene. High-copy suppressors are functionally equivalent to over-producing dominant mutations because it is the wild-type function of the gene that produces the suppression.

histone one of the highly conserved, small, positively charged proteins that make up the main protein component of eukaryotic chromosomes.

histone code a hypothesis that many of the functional elements of a genome, such as transcription start sites, origins of replication, and so on, depend on the particular constellation of histone modifications in the chromatin at that location.

holocentric chromosomes in which the spindle microtubules attach along the length rather than at one localized centromere. Chromosomes in *C. elegans* and other nematodes are holocentric, which has implications for methods for transformation and for producing mosaics.

homology-directed repair a DNA repair process within a cell in which a sequence that is highly similar to the target site is used as a template to make the repair.

hub a node in a network with a very large number of links.

hybrid dysgenesis the syndrome of genetic effects arising from the movement of P-transposable elements in *D. melanogaster*.

hypermorphic allele also known as a **hypermorph**. A mutant allele that overproduces the normal function of the gene. Most hypermorphic mutations are dominant.

hypersensitive sites the regions of isolated chromatin that are reproducibly digested by low concentrations of a nuclease such as DNase I or micrococcal nuclease.

Hypersensitive sites are an indication that the chromatin at that site is open or loosely packed so that the DNA is exposed. Also known as DNase hypersensitive sites and abbreviated **HSSs**.

hypomorphic allele also known as a **hypomorph**. A mutant allele that reduces but does not completely eliminate the function of a gene. Most hypomorphic mutations are recessive. Compare with **null allele**.

imaginal disk a pocket of undifferentiated cells in *Drosophila* larvae and pupae that will give rise to structures such as the legs, wings, and genitalia in the adult. Different imaginal disks are the progenitors of different adult structures.

in-degree the number of connections or links that have that node as a target. The in-degree is the number of nodes (genes or macromolecules in molecular biology) that have that gene or macromolecule as one of their targets for interaction; the out-degree is the number of nodes that gene or macromolecule targets for its interactions.

indel a contraction of "insertion-deletion," indicating that the genes or genomes of two individuals differ by an insertion or deletion of the sequence at a particular site. Indel is used rather than attempting to distinguish whether the difference is an insertion into one genome or a deletion from the other genome. Other common polymorphisms are single-nucleotide polymorphisms (SNPs) and copy-number variations (CNVs).

informational suppressor a suppressor mutation that affects a specific type of molecular lesion, such as a nonsense mutation, regardless of the function of the gene being suppressed. Informational suppressors affect many different genes, but will only affect certain alleles of the genes they suppress.

interactional suppressor a suppressor mutation that affects only specific alleles of a specific gene, typically because the suppressor gene and the target gene encode interacting gene products.

interactome the catalog of protein–protein interactions that occur in a tissue or an organism.

intragenic complementation a situation in which two mutations are known to lie in the same gene, but give a wild-type phenotype when heterozygous with each other, and so complement one another.

intragenic suppressor a suppressor mutation located as a separate molecular lesion in the same gene as the original mutation.

knock-in, knock-in mutation a procedure in which the wild-type gene on the chromosome is replaced by a version that has been engineered in vitro with a specific mutation. The procedure is widely used in mice.

knockout mutation a mutant allele in which the wild-type allele has been severely disrupted or deleted. A knockout mutation is a null allele.

landmark event an easily recognized event or structure in a cell or tissue that can be used to stage events.

lateral inhibition an inter-cellular process by which one cell or tissue prevents a neighboring cell or tissue from adopting the same developmental fate.

link in networks, the interaction that occurs between two or more **nodes**. A network graph is a diagram of nodes with the links connecting them. A link is also called an edge.

long non-coding RNAs (long ncRNAs) a diverse category of transcripts that are longer than 200 bp in length, transcribed by RNA polyermase II, and capped and processed similarly to mRNA transcripts but which have no sustained open reading frame that could encode for a protein. The functions of most long ncRNAs are not known, although they are widely believed to have roles in the regulation of gene expression.

loss-of-function mutation a mutation that eliminates or reduces the function of a gene. Most loss-of-function mutations are recessive, and most recessive mutations arise from a loss or reduction in the function of a gene. See **null allele** and **hypomorphic allele**.

loxP the 34 base-pair sequence that is the target for recombination by the Cre recombinase. The Cre–*lox* recombination system is widely used to insert sequence or produce recombination at specific sites.

mapping a general term for identifying the location of a gene on a chromosome and in the genome.

microRNA (miRNA) an RNA product of approximately 22 nucleotides that regulates the expression of another gene by making a double-stranded hybrid with a complementary sequence on the target gene's mRNA. The double-stranded RNA hybrid either targets the mRNA for degradation or blocks its translation.

maternal-effect mutation a mutation that affects a process within the mother to exert a phenotypic effect on the offspring.

missense mutation a mutation in which one amino acid in the protein sequence is replaced by another amino acid.

model organisms organisms that are widely studied, often for their suitability for research rather than their intrinsic economic or biological importance. Lessons learned from the easily studied model organisms are then applied to related, but less well-studied organisms.

molecular bar code a nucleotide sequence not found in the genome that is used to specifically tag and identify a gene in genome-wide screens for mutations.

morpholino a synthetic molecule that is used to block translation or RNA processing by forming a double-stranded hybrid with an mRNA, as a type of antisense technology. Morpholinos have a base sequence corresponding to the complementary strand of approximately 25–40 nucleotides in the mRNA but with a morpholine ring rather than ribose, which prevents them from being degraded by cellular nucleases. Morpholino™ is a trade name held by Gene Tools, but is often used as a general term for such molecules.

mosaic an organism with cells of two different genotypes produced by genetic methods such as somatic recombination or chromosome fragment loss. Compare with **chimera**.

mutagen an agent that induces mutations.

mutant analysis see genetic analysis.

mutant screen any of many experimental strategies designed to find mutations that affect a specific biological process.

mutant selection or genetic selection a procedure in mutant screens that allow only particular genotypes to grow, development, and reproduce.

mutation a change in the genotype that results in a heritable change in the phenotype.

mutator a phenotype in which the organism has a high frequency of mutations. Often, mutator strains have very active transposable elements. Mutations that affect DNA repair also result in a mutator phenotype.

negative interaction an interaction between two or more genes that results in a more severely mutant phenotype than produced by another of the genes alone. Similar to synthetic enhancement, but usually used when the pairwise interactions of many genes are being studied.

neomorphic allele a mutant allele that produces a novel phenotype, often by expressing the gene product ectopically, i.e. in a tissue or at a time that is different from wild-type.

Network a general term to describe a set of interactions among different components. The individual components are called nodes, while the interactions are called links.

node in networks, the individual components, such as a gene or protein, that interact with other components. A network graph is a diagram of nodes with **links** connecting them. A node is also called a vertex.

non-allelic non-complementation a situation in which two recessive mutations that are not alleles of the same gene give a mutant phenotype when both are heterozygous.

non-coding RNA (ncRNA) a general term for an RNA molecule that does not encode a polypeptide as its functional product. Distinctions are often made between **microRNAs** (sometimes abbreviated to miRNAs), for which the mature product is 22 nucleotides long, and **long ncRNAs** for which the mature product is more than 200 bases long. The roles of ncRNAs are diverse.

non-homologous end joining the process by double-stranded break is repaired by attached two DNA sequences not related in sequence.

nonsense mutation a mutation in which one codon in the mRNA is changed to become a stop codon, thus terminating the translation of the polypeptide.

null allele synonymous with **amorphic allele**. A mutant allele that eliminates the function of a gene. Most null mutations are recessive.

off-target effects genes or sequences that are unintentionally and usually unknowingly affected, in addition to the intended target gene or sequence.

outcrossing a series of matings in which a mutagenized strain is crossed to a non-mutagenized strain so that regions of the mutagenized genome are replaced by the same region from an non-mutagenized genome. Outcrossing is done to removing extraneous mutations from a genetic strain.

out-degree the number of nodes that act as targets for the activity of a particular node. The in-degree is the number of nodes (genes or macromolecules in molecular biology) that have that gene or macromolecule as one of their targets for interaction; the out-degree is the number of nodes that gene or macromolecule targets for its interactions.

overproducer a type of mutant allele that makes too much of a gene product. Also known as a hypermorphic mutation or a hypermorph.

P element a transposable element found in *D. melanogaster*, and widely used as vector for transformation.

pair-rule gene a category of the mutant phenotypes observed for genes affecting segmentation in the

Drosophila embryo, in which alternating pairs of segments were affected.

palindromic a sequence that reads the same in either direction.

parent-of-origin effect see genomic **imprinting**.

path links connecting any set of nodes in a graph.

penetrance the percentage of genotypic mutant individuals that exhibit a mutant phenotype. Penetrance ranges from zero to 100 percent.

permissive temperature the temperature at which a temperature-sensitive or cold-sensitive mutation exhibits a wild-type phenotype.

pfam abbreviation for "protein family," a large and centrally curated database of protein families based on sequence alignments.

phage or bacteriophage viruses that infect bacteria.

phenocopy a morphological mimic of a genetic phenotype that is produced by environmental means rather than a mutation.

phenocritical phase the stage in the life cycle when the first effects of a mutant are observed. In yeast, this is sometimes referred to as the **diagnostic landmark**.

pleiotropic see **pleiotropy**.

pleiotropy multiple different phenotypic effects that occur together as a result of a single mutation.

polymorphism genetic variation at a locus. The terms "allele," "mutation," and "polymorphism" have similar meanings but different connotations, with polymorphism being the most inclusive term. Polymorphisms use naturally occurring variations in the DNA sequence and may not have a detectable phenotypic variation. "Mutation" usually implies a genetic change that has been experimentally induced, that occurs rarely, or that substantially alters the phenotype. "Allele" often implies a detectable phenotype difference, which may arise from natural or experimentally induced variation.

polytene chromosome the large chromosome in certain insect tissues, particularly the salivary glands in *D. melanogaster*. The chromosomes in these cells undergo many rounds of DNA replication without strand separation, so that a polytene chromosome is large enough to have easily visible bands and other structural features. These are cytological landmarks for positioning genes on the *Drosophila* chromosomes.

pooling a process of analyzing groups of cells or organisms together rather than individually. If an event is rare, most of the pools will not have any affected cells so screening can occur expeditiously. If a pool of cells shows an effect, small groups from the pool or individual cells in the pool can be tested for the effect.

position effect variegation a mutant phenotype that arises when a gene is moved to a different location on the chromosome and becomes inactivated in certain cells. The gene itself is not altered, but the surrounding chromatin affects its expression. Since the effect varies between cells (for example, in the eye of Drosophila), the mutant phenotype is only seen in some cells and not others.

positional cloning a cloning procedure based on the map location of the gene.

positive interaction an interaction between two or more genes that results in a more less severely mutant phenotype than produced by another of the genes alone. Similar to intergenic suppression but usually used when the pairwise interactions of many genes are being studied.

post-transcriptional gene silencing (PTGS) similar or identical to RNAi, the name of the process first identified in plants.

private variants also known as **individual variants** or **mutations**. Private variants or mutations are those found at very low frequency (much less than 1 percent) in a population and typically within a single family of related individuals.

Protospacer the sequences between the palindromic repeats in a CRISPR array, not including the PAM sequence.

protospacer-adjacent motif (PAM) the di-or tri-nucleotide found at the end of the spacer that marks the sequence as a target for degradation or inactivation.

prototroph a strain that can grow in the absence of an added nutrient because it has the wild-type function to synthesize that nutrient. The opposite of prototroph is **auxotroph**.

reductionist approach analysis that attempts to understand a complicated process from the analysis of its individual component parts.

regulatory region the entire DNA sequence necessary for the gene to be transcribed in the proper tissue at the proper time and in the proper amount. The regulatory region includes all of the **enhancer** and **silencer sequences**. For most genes, the regulatory region is found upstream of the core promoter, but many exceptions are known in which the regulatory region includes sequences downstream of the gene or internal to the coding sequence, such as intronic sequences. See **cis regulatory module**.

restrictive temperature the temperature at which a temperature-sensitive or cold-sensitive mutation exhibits a mutant phenotype.

reverse genetic analysis see **reverse genetics**.

reverse genetics the process of beginning with a cloned gene and finding a mutant phenotype for it. The distinction between "reverse genetics," which begins with a cloned gene, and "genetic analysis," which begins with a mutant phenotype, is blurred, and the term is falling out of use.

reverse phenotyping the process of using the DNA sequence to infer the correct phenotype, such as distinguishing among clinically similar syndromes.

RNA interference (RNAi) a widely used technique to reduce the expression of a gene using double-stranded RNA corresponding to part of the coding region of the gene. RNAi produces the equivalent of a mutant phenotype without altering the DNA sequence for the gene.

RNA-seq a method to analyze transcription pattern by sequencing transcripts. The RNA is usually converted to cDNA and the cDNA is sequenced rather than the RNA itself.

saturation a situation in which all of the genes that could be identified by a particular mutant screen have been found.

segment polarity gene a category of the mutant phenotypes observed for genes affecting segmentation in the Drosophila embryo, in which structures within a segment or the orientation of the segment was affected.

segmental aneuploid a chromosome constitution that has extra or missing copies of portions of a chromosome as a result of a **duplication** or **deletion**. It is only that segment that is aneuploid rather than the entire chromosome.

selectable marker a gene, often one whose function is unrelated to the gene of interest, that can be used in a genetic selection so that only certain genotypes can grow. For example, drug resistance genes are used as selectable markers for plasmids in bacteria.

selection see **genetic selection**.

sex reversal a genetic mutation that causes the mutant to differentiate into a sexual phenotype that is different from its chromosomal phenotype. For example, the sex reversal mutation *tra* in *Drosophila* results in flies with two X chromosomes becoming males rather than females.

shift down in the analysis of temperature-sensitive mutants, the process of moving the cultures of growing organisms from the (higher) restrictive temperature to the (lower) permissive temperature.

shift up in the analysis of temperature-sensitive mutants, the process of moving the cultures of growing organisms from the (lower) permissive temperature to the (higher) restrictive temperature.

short-interfering RNA (siRNA) the functional 22–25-nucleotide double-stranded RNA molecules involved in RNAi.

shuttle vector a plasmid that has genetic markers to allow it to be grown in both *E. coli* and yeast.

silencer see **silence sequence**.

silencer sequence a sequence in the regulatory region of the gene that represses transcription. Silencer sequences are the functional opposite of **enhancer sequences**.

single guide RNA (sgRNA) a RNA molecule that includes both the crRNA and the tracrRNA needed by Cas9 for genome editing.

site-specific recombination a type of homologous recombination in which the recombination enzyme recognizes a specific and often short sequence for making the double-stranded break. Events carried out by Cre-loxP are examples of site-specific recombination.

spliceosome the complex of RNA and protein molecules responsible for RNA splicing.

stage of execution see **temperature-sensitive period**.

suppress see **suppression**.

suppression a genetic interaction between mutant alleles of two genes in which the mutation in the second gene, known as the suppressor mutation, overcomes or reverses the mutant phenotype of the first gene.

suppressor see **suppressor mutation**.

suppressor mutation a mutation in a second gene that overcomes the phenotypic effects of another mutation. As a result of a suppressor mutation, the phenotype of the double mutant more closely resembles wild-type than does the phenotype of the single mutation. Compare with **synthetic enhancer mutation**.

synteny a general term for genes that are located on the same chromosome. Genes may be syntenic but sufficiently far apart on the chromosome that they segregate independently and thus are not linked in their inheritance. Synteny is important in comparing the genomes of closely related species because many linkage relationships have been preserved in evolution. A group of genes located together in two different species comprise a **syntenic block**.

synthetic enhancement a genetic interaction between mutant alleles of two genes in which the mutation in the second gene, known as the synthetic enhancer mutation, intensifies the mutant phenotype of the first gene. One common form is synthetic lethality, in which an individual which is mutant for either one of the two genes is viable but a mutant in both genes is inviable.

synthetic enhancer mutation a mutation in a second gene that makes the phenotype of another mutation more severely mutant. In particular, each mutation alone has a wild-type phenotype, but the double mutant has a mutant phenotype. Compare with **suppressor mutation**.

synthetic genetic array (SGA) a method used in yeast for the high-throughput analysis of gene interactions.

synthetic lethality see **synthetic enhancement**.

systems biology a broadly used term that views an organism or a biological process in terms of the interactions of its components.

TALEN a genome editing technique that relies on a fusion protein between transcription activator-like effector (TALE) domains from bacteria and the FokI nuclease. TALE+N for nuclease.

t-crit see **temperature sensitive period**.

temperature-sensitive mutation a mutation that has a mutant phenotype at a slightly elevated temperature but a wild-type phenotype at a lower growth temperature. The temperature that exhibits the wild-type phenotype is known as the **permissive temperature**. The temperature that exhibits the mutant phenotype is known as the **restrictive temperature**.

temperature-sensitive period (TSP) the time during which the activity of a gene is needed, as determined by temperature-shift experiments. The temperature-sensitive period is sometimes referred to as the stage of execution or the t-crit for the activity of a gene.

temperature shift an experiment in which a mutant with a temperature-sensitive mutation is shifted from one growth temperature to another to determine the time when the activity of the gene is needed. Temperature shifts from the permissive to the restrictive temperature are called shift-up experiments, whereas temperature shifts from the restrictive to the permissive temperature are called shift-down experiments.

terminal phenotype the mutant phenotype observed when a lethal mutation causes developmental arrest. Also known as the **effective lethal phase**.

tetrad the four haploid products of a single meiotic division. In yeast and other fungi, these four haploid spores are found together in the **ascus**.

trans-activating CRISRR RNA (tracrRNA) the specific RNA molecule required for the activity of Cas9 from S. pyogenes. Also known as the tracRNA, but the additional "r" was included for pronunciation as "tracer."

transcription factor a protein that binds to enhancer or silencer sequences to regulate transcriptional initiation of a gene.

transformation rescue introducing a wild-type copy of the gene to restore the normal gene function to a mutant.

yeast one-hybrid (Y1H) assay a genetic method using transcription in yeast to identify all of the proteins that bind to the regulatory region of a gene.

yeast two-hybrid (Y2H) assay a genetic method using transcription in yeast to detect protein–protein interactions by co-expressing both genes on plasmids in the yeast cell.

INDEX

Tables are indicated by an italic t following the page number.

D

G

H

T

U

V

W

X

Y

Z